Recent Advancements in Sustainable Solar Photovoltaic Power Technology

Recent Advancements in Sustainable Solar Photovoltaic Power Technology

Editors

Prince Winston David
Praveen Kumar B

MDPI

Basel • Beijing • Wuhan • Barcelona • Belgrade • Novi Sad • Cluj • Manchester

Editors
Prince Winston David
Department of Electrical and Electronics Engineering
Kamaraj College of Engineering and Technology
Madurai
India

Praveen Kumar B
Department of Electrical and Electronics Engineering
Vardhaman College of Engineering
Hyderabad
India

Editorial Office
MDPI
St. Alban-Anlage 66
4052 Basel, Switzerland

This is a reprint of articles from the Special Issue published online in the open access journal *Sustainability* (ISSN 2071-1050) (available at: www.mdpi.com/journal/sustainability/special_issues/N145O2I1F2).

For citation purposes, cite each article independently as indicated on the article page online and as indicated below:

Lastname, A.A.; Lastname, B.B. Article Title. *Journal Name* **Year**, *Volume Number*, Page Range.

ISBN 978-3-0365-8955-8 (Hbk)
ISBN 978-3-0365-8954-1 (PDF)
doi.org/10.3390/books978-3-0365-8954-1

Contents

About the Editors

Prince Winston David

Dr. D. Prince Winston received his B.E. degree in Electrical and Electronics Engineering from R.V.S CET, Dindugal, in 2006 and his M.E. degree in Power Electronics and Drives from MEPCO Schlenk Engg College in 2008. He received his Ph.D. degree from Anna University in 2013. Presently, he is working as a Dean Research and Professor in the Department of Electrical and Electronics Engineering at Kamaraj College of Engineering and Technology, Madurai, India. He is recognized as being within the top 2% of Scientists Worldwide Listed by Elsevier BV, Stanford University. He is also recognized as within the top Scientists Worldwide Listed by the AD Scientific Index. He has 13 years of teaching experience. He is the Research Supervisor at Anna University Chennai. Eight Ph.D. scholars under his guidance have completed their research work; currently, eleven research scholars are pursuing their Ph.D. He has published more than 60 papers in International Journals listed in SCOPUS/SCI/SCIE. He has also published two sole author papers in IEEE Transactions.

He has received an award from IEEE Madras Section during the years 2019 and 2020 for publishing papers in IEEE Transaction. He received a young scientist fellowship from TNSCST (Government of Tamilnadu) during the year 2018-19. Currently, he has an ongoing research project funded by AICTE under the Research Promotion Scheme. He has filed seven patents. He had completed eight consultancy works for various industries. He is a reviewer/editorial board member in various SCI journals. He has published many book chapters/books. He has developed various products for industries/institutions. He is serving as a DC member for many research scholars. His current research interests include solar PV, solar still, energy conservation in electric motor drives, power converters, power quality, and electric vehicles.

Praveen Kumar B

Dr. B. Praveen Kumar (Senior Member, IEEE) received his B.E. degree in electrical and electronic engineering and M.E. and Ph.D. degrees in power systems engineering from Anna University, Chennai, India, in 2014, 2016, and 2019, respectively. He is currently working as an Associate Professor within the Department of EEE, Vardhaman College of Engineering, Hyderabad, India. He is a Senior Member in IEEE and mentor in the IEEE Power and Energy Society. Also, he is the IEEE Vardhaman Student Branch Photonics Society counselor.

Dr. Praveen has published 70+ scientific articles in reputed international journals with a cumulative impact factor of more than 100, and he also has 10 patents to his name. His current research interests include solar photovoltaics, solar stills, and renewable energy systems, AI, IoT, and EV. Most of his research work is on PV and its allied applications in a real-time environment. He has worked on external fault identification in solar PV using thermal image processing, partial shading classification using a low-cost web camera, and power enhancement in an aged solar PV array system with the help of advanced technologies like ANN, SVM, and other classifiers in MATLAB software. Presently, he is working on reconfiguration techniques for partial shaded PV systems and on MPPT and parameter estimation of the system.

Preface

Energy plays a decisive role in realizing the Sustainable Development Goals (SDGs), such as employment growth, industry, innovation, and infrastructure, sustainable cities and communities, clean energy, and climate change. This book, titled "Recent Advancements in Sustainable Solar Photovoltaic Power Technology", presents significant works in the field of solar photovoltaic systems and critical issues in solar power generation technology, as well as the latest achievements and current problems in solar power generation technology and proposing corresponding solutions to key technical challenges in the utilization of solar power, pointing out future directions for achieving the SDGs through solar power generation technology.

This reprint of articles from the Special Issue provides a variety of comprehensive reviews on solar power generation technologies like maximum power point tracking (MPPT) techniques, power electronic converter topologies, and parameter estimation of PV cells. Along with reviews, this book also outlines the latest developments in the field of MPPTs, converters, cooling systems, rooftop, as well as grid-connected PV systems. This reprint will serve as a reference and update for academics, researchers, and practicing engineers to inspire new research and developments that pave the way for next-generation PV systems for residential and small commercial applications.

This reprint includes three extensive reviews and twelve research articles in the field of solar photovoltaic power technology. We hope that you enjoy reading.

Prince Winston David and Praveen Kumar B
Editors

sustainability

MDPI

Review

A Comprehensive Review of Recent Maximum Power Point Tracking Techniques for Photovoltaic Systems under Partial Shading

Muhammed Y. Worku [1,*], Mohamed A. Hassan [1], Luqman S. Maraaba [2], Md Shafiullah [1], Mohamed R. Elkadeem [1], Md Ismail Hossain [1] and Mohamed A. Abido [1,3,4]

1 Interdisciplinary Research Center for Renewable Energy and Power Systems (IRC-REPS), Research Institute, King Fahd University of Petroleum & Minerals, Dhahran 31261, Saudi Arabia; mhassan@kfupm.edu.sa (M.A.H.); shafiullah@kfupm.edu.sa (M.S.); mohamed.elkadeem@kfupm.edu.sa (M.R.E.); mdismail.hossain@kfupm.edu.sa (M.I.H.); mabido@kfupm.edu.sa (M.A.A.)
2 Department of Electrical Engineering, Arab American University, Jenin P.O. Box 240, Palestine; luqman.maraaba@aaup.edu
3 Electrical Engineering Department, King Fahd University of Petroleum & Minerals, Dhahran 31261, Saudi Arabia
4 King Abdullah City for Atomic and Renewable Energy (K. A. CARE), Energy Research & Innovation Center (ERIC) at KFUPM, Dhahran 31261, Saudi Arabia
* Correspondence: muhammedw@kfupm.edu.sa; Tel.: +966-559713973; Fax: +966-138603535

Abstract: To operate photovoltaic (PV) systems efficiently, the maximum available power should always be extracted. However, due to rapidly varying environmental conditions such as irradiation, temperature, and shading, determining the maximum available power is a time-varying problem. To extract the maximum available power and track the optimal power point under these varying environmental conditions, maximum power point tracking (MPPT) techniques are proposed. The application of MPPT for extracting maximum power plays a crucial role in developing efficient PV systems. These MPPT techniques face several issues and limitations, particularly during partial shading conditions caused by non-uniform environmental conditions. Researchers have been focusing more on mitigating the partial shading condition in PV systems for the last few years due to the need to improve power output and efficiency. This paper provides an overview of MPPTs proposed in the literature for uniform and non-uniform environmental conditions broadly categorized as MPPT-based and circuit-based methods. The MPPT-based methods are classified as conventional, soft computing, and hybrid techniques. A critical analysis of each approach regarding tracking speed, algorithm complexity, and dynamic tracking under partial shading is discussed. The literature shows hybrid strategies provide fast-tracking speed and are efficient with a tracking efficiency of around 99% compared to conventional methods; however, their design and practical implementation are complex. This comprehensive review of MPPT methods aims to provide power utilities and researchers with a reference and guideline to select the best MPPT method for normal operation and partially shaded PV systems based on their effectiveness and economic feasibility.

Keywords: photovoltaic; MPPT; partial shading; global peak; MPPT classification

Citation: Worku, M.Y.; Hassan, M.A.; Maraaba, L.S.; Shafiullah, M.; Elkadeem, M.R.; Hossain, M.I.; Abido, M.A. A Comprehensive Review of Recent Maximum Power Point Tracking Techniques for Photovoltaic Systems under Partial Shading. *Sustainability* **2023**, *15*, 11132. https://doi.org/10.3390/su151411132

Academic Editor: Yagang Zhang

Received: 7 June 2023
Revised: 8 July 2023
Accepted: 10 July 2023
Published: 17 July 2023

1. Introduction

Recently, power generation from renewable sources such as solar and wind is receiving more attention as their operation is pollution free to reduce the environmental impact of fossil fuels [1,2]. Photovoltaic (PV) is the fastest-growing renewable system, and it directly converts solar energy to electrical energy. The power generated from the PV source can also be utilized for chemical energy transformation, such as hydrogen fuel cells [3–5]. The power generated from a PV system varies according to the temperature and irradiation received at any instant [6–8]. To generate the maximum available power from the PV system under varying irradiation and temperature, maximum power point tracking (MPPT) methods are integrated [9,10].

On the other hand, the optimal tilt and orientation of the PV panels can improve the solar yield, as reported by a study conducted in the United Arab Emirates (UAE) [11]. However, the optimal tilt and orientation are region-dependent and vary considerably. The Kingdom of Saudi Arabia in the Sun Belt region experiences high irradiance levels between 4.479 kWh/m^2 and 7.004 kWh/m^2, depending on the geographical location [12]. This available abundance of solar power is being utilized and integrated into the grid by the kingdom. The 300 MW Sakaka PV power plant is the first renewable-based power source covering an area of six square kilometers in Saudi Arabia [13]. Power generation based on PV is growing fast, and different developing countries are generating and integrating this power into their respective grids [14,15].

In the case of uniform irradiance, one maximum power point appears in the PV array characteristics curve that the conventional MPPT techniques can track. However, due to shadows and clouds, PV arrays receive non-uniform irradiation, creating multiple maximum points in the PV array curve. Many modern MPPT techniques are proposed to handle the numerous maximum points since most conventional MPPT methods fail under such circumstances. One of the most crucial factors in choosing a proper MPPT method mainly lies within three specifications. The first factor is performance, which is the tracking speed and accuracy. The second factor is the complexity of the control system, voltage and current sensors, parameter tuning or perturbation, and partial shading detections. The third factor is the cost of the entire MPPT system.

Several MPPT techniques for PV systems have been proposed in the last decade, and the methods developed so far can be broadly classified into MPPT-based and circuit-based methods. The MPPT-based method is classified as conventional, soft computing, and hybrid techniques. Some of the techniques classified under the traditional approach include fractional short circuit (FSC) [16,17], fractional open circuit (FOC) [18], perturb and observe (P&O) [19–24], incremental conductance (IC) [25–32], hill climbing (HC) [33–35], curve fitting (CF) [36], constant voltage (CV) [37,38], and ripple correlation control (RCC) [39]. FSC and FOC methods are less accurate and perform better only in low-power applications. Although the popular MPPT techniques such as P&O, HC and IC can track the maximum power under uniform irradiation, they fail to operate under partial shading properly and have slow tracking speed, poor convergence, and high steady-state oscillations. Hence, conventional methods should work with other methods to track the maximum power under partial shading conditions [40–43].

Since PV systems have nonlinear characteristics, soft computing methods have been proposed by researchers to handle non-linearity and are considered the prime choice for nonlinear optimization. Numerous soft computing techniques for MPPT application are proposed, including fuzzy logic control (FLC) [44–48], artificial neural network (ANN) [49–52], genetic algorithm (GA) [53–56], particle swarm optimization (PSO) [57–60], nonlinear control [61], chaotic approach (CA) [62], differential evolution (DE) [63–65], simulated annealing (SA) [66], grey wolf optimization (GWO) [42,67–69], cuckoo search [70–72], bat search algorithm, bee colony search algorithm [73,74], ant colony optimization [75–78], firefly algorithm [79], and random search methods [80]. PSO is the most popular and widely used optimization technique to track the maximum power in PV systems. Although FLC and ANN effectively track the maximum power, they require large memory and data for training and implementation. They also need detailed knowledge of the system while implementing the algorithm.

To enhance the performance of MPPT methods under partial shading, researchers combined conventional techniques with soft computing techniques to form hybrid strategies. Some of the developed hybrid techniques include incremental conductance combined with firefly algorithm (IC-FFA) [81], perturb and observe with artificial neural network (P&O-ANN) [82], perturb and observe with fireworks (P&O- FWA) [83], perturb and observe with grey wolf (P&O-GWO) [84], perturb and observe with genetic algorithm (P&O-GA) [85], perturb and observe with bat search algorithm (P&O-Bat) [86] and perturb and observe with particle swarm optimization (P&O-PSO) [87,88]. Two or more intelligent algorithms such as particle swarm optimization and simulated annealing (PSO-SA) [89], particle swarm optimization and fish swarm [90,91], differential evolution with Jaya algorithm (DE-Jaya) [92], and differential evolution with whale optimization (DE-WO) [93] can also be combined to form hybrid methods.

This paper presents a detailed, organized, and up-to-date review of the different maximum power point tracking (MPPT) algorithms for photovoltaic (PV) systems. The advantages and disadvantages of each method are presented to assist power utilities and power engineers in choosing the proper MPPT method while designing a PV generation system under partial shading conditions. Hybrid MPPT methods based on the combination of soft computing and conventional methods are more efficient than the other methods. However, reducing the complexity of practical implementation is a challenge and a research direction that should be addressed. Moreover, MPPT methods based on optimization face challenges with respect to periodic tuning, accuracy, stability, and the number of send parameters.

The rest of the paper is organized as follows; Section 2 describes the PV configuration, Section 3 presents the details of PV under partial shading, Section 4 describes the dynamic tracking and classification of MPPT methods under partial shading, Section 5 is the discussion, and Section 6 concludes the paper.

2. PV Configuration

The five-parameter electric circuit model of a PV cell is shown in Figure 1 [94–96]. It consists of a light-dependent current source, a p-n junction diode, and two resistances, one in series and the other in parallel.

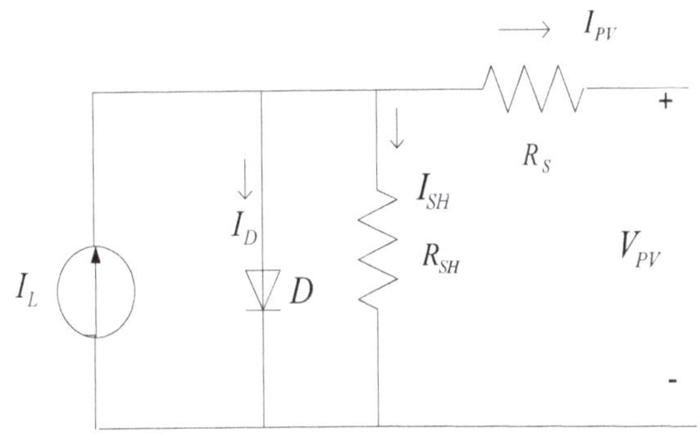

Figure 1. Five-parameter equivalent electric circuit model of a PV cell.

Using simple Kirchhoff's current law:

$$I_{PV} = I_L - I_D - I_{SH} \tag{1}$$

where I_D and I_{SH} depict the diode and shunt branch currents, respectively, and are given by:

$$I_D = I_0\{\exp[\frac{V_{PV} + I_{PV}R_S}{a}] - 1\} \tag{2}$$

$$I_{SH} = \frac{V_{PV} + I_{PV}R_S}{R_{SH}} \tag{3}$$

Putting these expressions of I_D and I_{SH} into Equation (1) gives the complete I-V characteristics of a PV panel:

$$I_D = I_L - I_0\{\exp[\frac{V_{PV} + I_{PV}R_S}{a}] - 1\} - \frac{V_{PV} + I_{PV}R_S}{R_{SH}} \tag{4}$$

where I_{PV} and V_{PV} represent the current and voltage generated from the PV panel. I_L is the light-generated current, I_0 is the diode saturation current, R_S and R_{SH} are the series and parallel resistance, respectively, and factor a is the diode-modified ideality factor, which is given by:

$$a = \frac{NsnKT}{q} \tag{5}$$

where N_s is the number of cells in the PV panel, n is the ideality factor (it has a value between one and two for the real diode), K is Boltzmann's constant, T is the cell temperature, and q is the electronic charge.

A standard PV cell generates a relatively low voltage (around 0.6 V); hence, PV cells are connected in series and parallel to raise the appropriate voltage level for the required application. PV modules are built using the PV cells' series and parallel connections; a PV array consists of PV modules connected in series or parallel [97]. Equation (4) can be modified to represent the I-V relationship of the series and parallel array and written as:

$$I_{PV} = N_{pp} * I_L - N_{pp} * I_0\{\exp[\frac{V_{PV} + I_{PV}R_S * N}{N_{ss} * a}] - 1\} - (\frac{V_{PV} + I_{PV}R_S * N}{R_{SH} * N}) \tag{6}$$

$$N = \frac{N_{ss}}{N_{pp}} \tag{7}$$

where N_{SS} and N_{PP} represent the number of panels connected in series and parallel, respectively. Figure 2 depicts the current–voltage characteristics of PV panels for different irradiation levels from 200 W/m^2 to 1000 W/m^2 and a constant temperature of 25 °C.

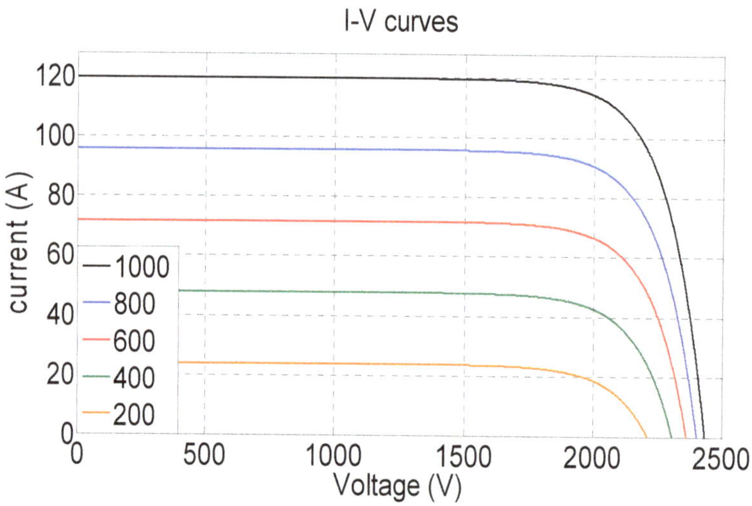

Figure 2. Current–voltage characteristics of a PV panel under different irradiation levels.

3. Maximum Power Point Tracking (MPPT) Methods for Partial Shading

PV arrays consisting of several panels are the most basic units of any PV system. Based on the I–V curve depicted in Figure 3, the PV operating point can vary from zero to the open circuit voltage. The operating point varies with the load variation and does not always stay at the maximum power point. A unique maximum power point (MPP) operating point exists in the I–V and P–V curves for every irradiation and temperature. This point keeps shifting when any atmospheric change occurs [98,99]. Thus, maximum power point tracking (MPPT) controllers are designed to keep tracking MPP, and they form an integral part of PV systems. Figure 3 also depicts the maximum operating point, P_{mp}, V_{mp}, and I_{mp}, for a PV panel. In a uniform insolation case, the total maximum output power of a PV array is equal to the sum of the maximum power values of all individual modules.

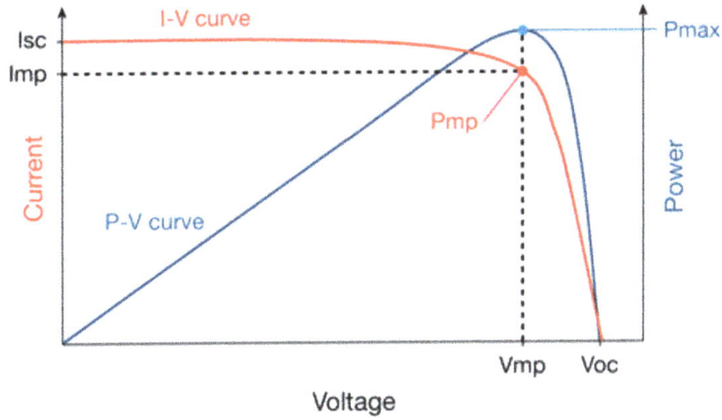

Figure 3. Maximum power point characteristics for I–V and P–V curves.

A significant impact on the operation of PV modules is shading caused by cloud cover, trees, or buildings. When one or more of the modules in a solar panel comes under the effect of shading, the module voltage drops causing it to work as a load rather than as a generator, and this causes a hot spot problem [100–102]. Each PV module is equipped with a bypass diode to overcome the hot spot formation. However, adding the bypass diode creates multiple peak points in the P–V curve.

Among the multiple peaks, one is the global maximum power point (GMPP) and the others are local maxima power points (LMPPs). Multiple maximum points can confuse traditional MPPT schemes as they can easily track and settle at a local maximum, which reduces the available power output from the PV array. A reliable technique is required to track the GMPP appropriately [103–105]. Conventional MPPT techniques cannot identify the GMPP under partial shading conditions (PSC) and usually track local peaks reducing the generated power from the PV system.

Figure 4 shows a PV module where PV modules 3 and 4 are shaded in Figure 4b due to environmental conditions. The bypass diodes provide an alternate path for the current flow, creating multiple peaks. The P–V curve shown in Figure 5 depicts the numerous maxima during PSC. Modern soft computing MPPT techniques are developed to track the GMPP as the conventional methods fail to differentiate between the GMPP and the LMPPs [106–108]. The MPPT methods for partial shading mitigation techniques will be explained in detail in the next section.

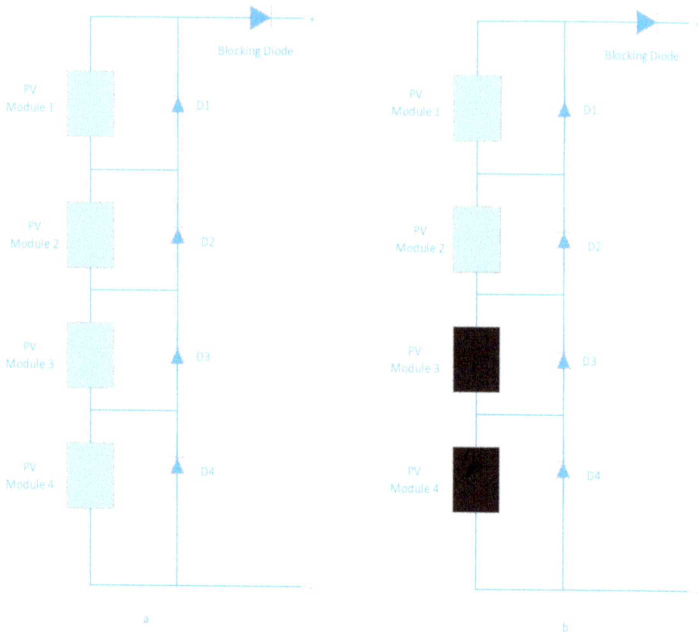

Figure 4. PV modules: (**a**) no shading and (**b**) modules 3 and 4 are shaded.

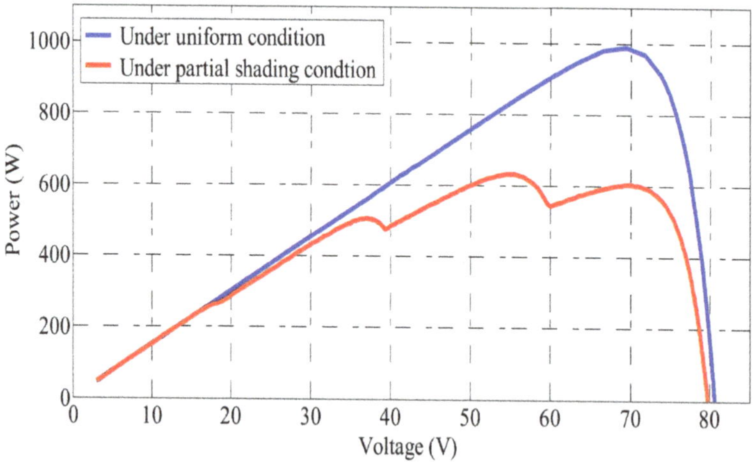

Figure 5. PV array output power characteristics under normal and partial shading.

4. Dynamic Tracking under Partial Shading

Partial shading reduces the overall efficiency of the PV system. To generate the maximum available power from the PV system under this non-uniform environmental condition, partial shading mitigating techniques are essential for the PV system working in grid-connected or standalone modes. Conventional MPPT methods have fixed step sizes and usually become trapped in local peaks, and they fail to dynamically track the MPP under partial shading conditions.

Partial shading mitigation techniques are broadly classified into two major groups. The first group includes all the MPPT-based methods, which have been further classified into modified conventional, soft computing, and hybrid techniques, and the second group comprises circuit-based topologies. The classification of MPPT techniques under partial shading is provided in Figure 6. The conventional MPPT methods under uniform irradiation, such as P&O, IC, and HC, are thoroughly discussed in the literature and will not be covered in this paper. This section provides MPPT techniques under partial shading conditions.

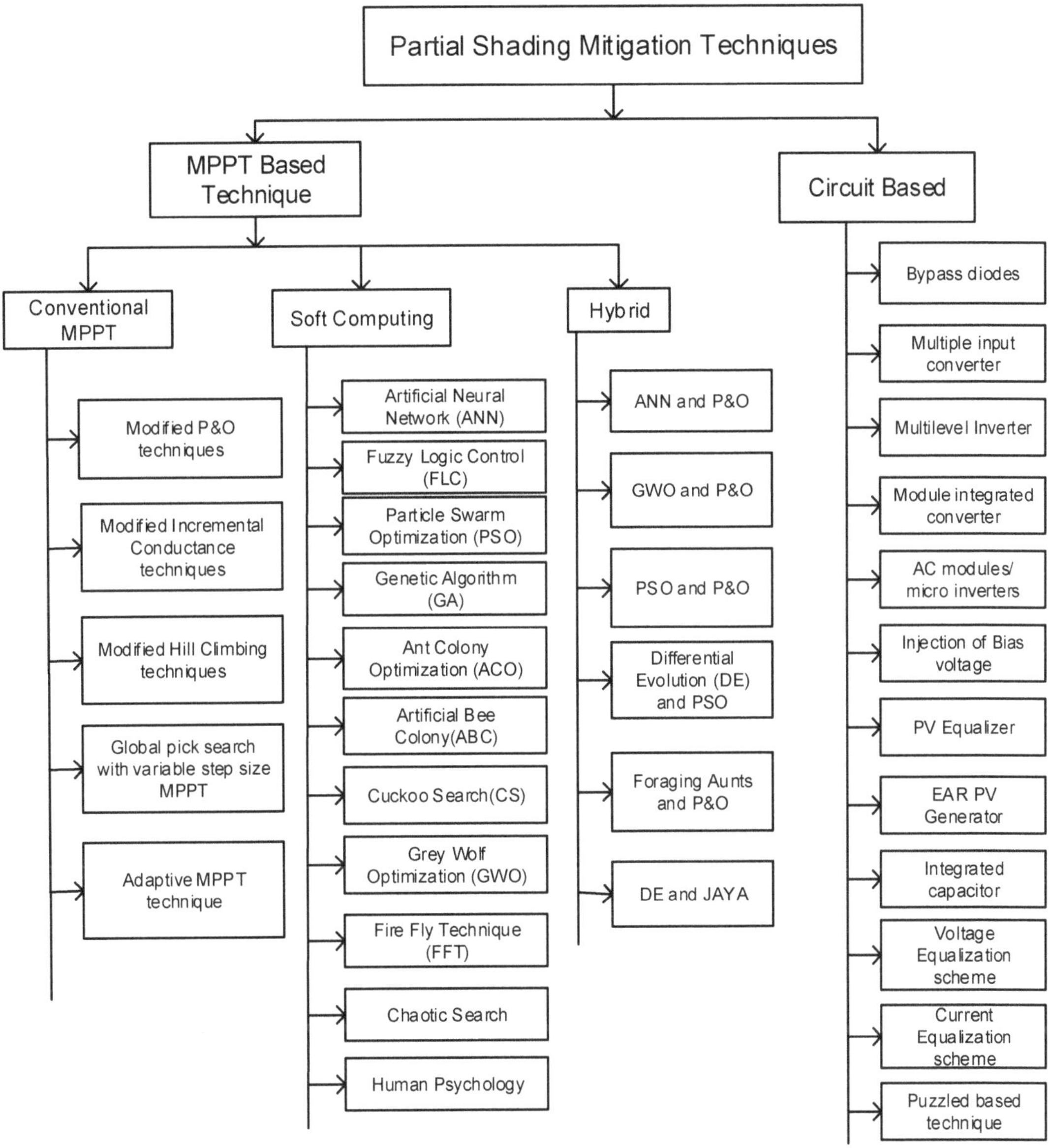

Figure 6. Classification of different MPPT schemes for shading mitigation.

4.1. MPPT-Based Techniques

4.1.1. Conventional MPPT

1. Modified Perturb and Observe (P&O) technique

The conventional P&O MPPT method has limitations during partial shading; hence, overcoming this limitation is required to track the global peak. Figure 7 depicts the flow chart of a modified P&O proposed by [109], where two routines are used. The first routine is the main program and sets a reference voltage close to the open circuit voltage. The main routine scans almost 80% of the P–V curve not to miss the potential global peak. The second is the global peak-tracking routine, which is called into action after executing the main program. Although the proposed method efficiently tracks the international peak, the tracking speed is compromised since the algorithm scans almost the entire P–V curve. Another modified P&O, by comparing two instantaneous power values presented in Equation (8), is proposed in [110].

$$\frac{P_m(t) - P_{ref}(t)}{P_m(t-1)} < \varepsilon \tag{8}$$

where

$P_m(t)$ is the instantaneous measured power and
$P_{ref}(t)$ is the instantaneous reference maximum power.

The algorithm efficiently tracks the global peak; however, new coefficients are introduced that complicate the overall MPPT process. The authors in [111] proposed another modified P&O MPPT method by periodically changing the PV array voltage from maximum to minimum. A microcontroller is used to store the operating voltage and current. The P&O is used to maintain the operation of the PV system after identifying the region of the global peak.

2. Modified Incremental Conductance (IC)

The conventional incremental conductance fails to track and recognize the true MPP as the method is based on derivative characteristics. In both global and local peaks, the derivatives dP/dV or dP/dI are zero; hence, the IC method should be modified to identify the true MPP. A two-stage IC method similar to the modified P&O is proposed in [112], wherein in the first stage, the value of the maximum voltage and current are used to force the PV system to operate close to the global peak. Equation (9) describes the first stage as:

$$R_{MP} = k\frac{V_{MP}}{I_{MP}} \tag{9}$$

where

k is the correction factor, and
V_{MP} and I_{MP} are approximately 80% of V_{OC} and 90% of I_{SC}, respectively.
V_{OC} and I_{SC} are the open circuit voltage and short circuit current, respectively.

The second stage moves the operating point toward the global peak. A linear function to track the global peak is presented in [113] and is expressed as:

$$V* = \frac{V_{grid}}{I_{out}} I(k) \tag{10}$$

where

V_{grid} is the output grid voltage and
I_{out} is the output grid current.

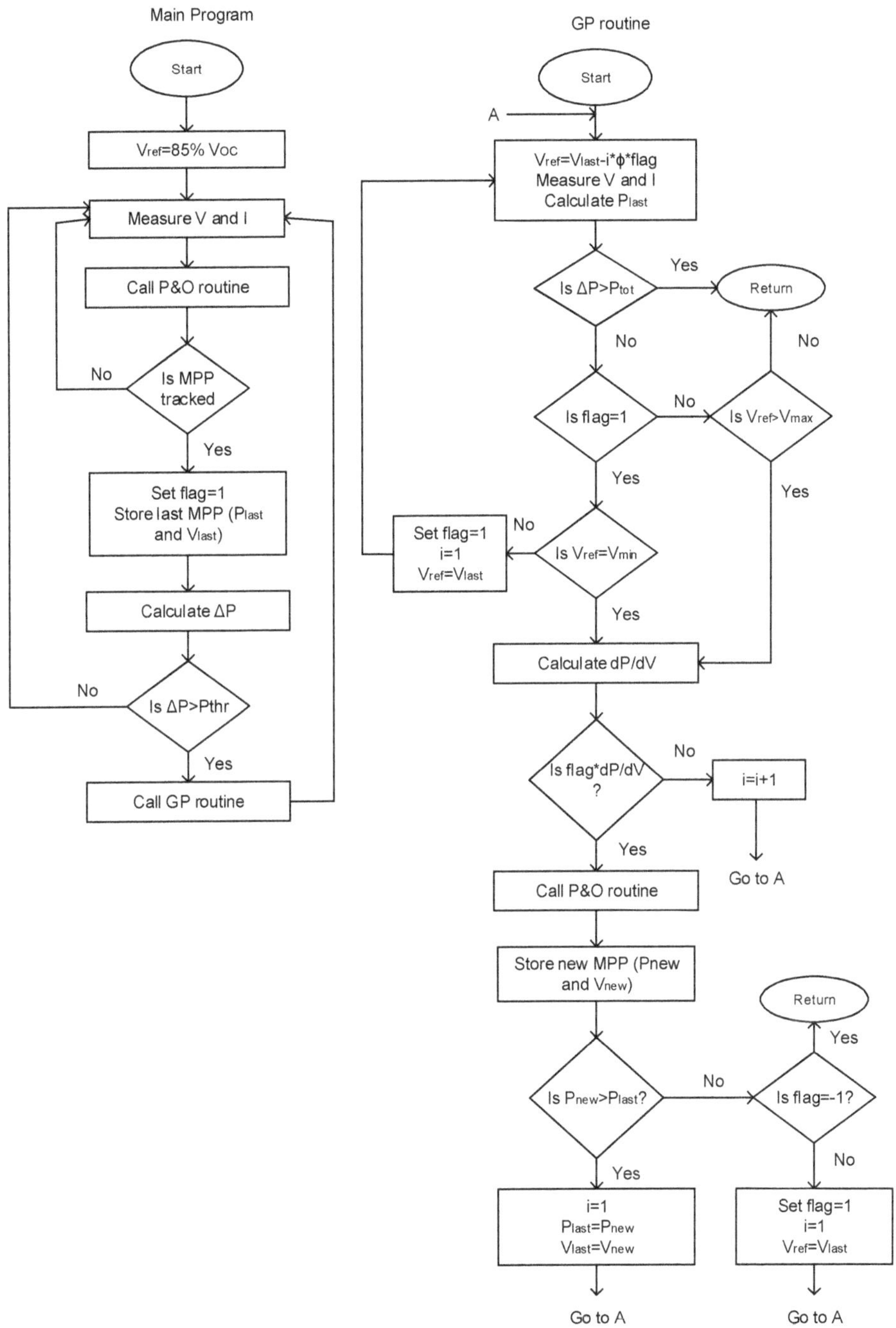

Figure 7. Modified P&O.

Equations (11) and (12) are used to detect the occurrence of partial shading and activation of the linear function.

$$V(k) - V(k-1) < V_{thr} \tag{11}$$

$$\frac{I(k) - I(k-1)}{I(k-1)} < I_{thr} \tag{12}$$

Although the proposed technique efficiently tracks the global peak, it can be applied only for grid-connected PV systems.

3. Modified Hill Climbing (HC)

Like the other conventional methods, the hill climbing method also fails to track the global peak. Several authors proposed a modified HC method to track the maximum available power under partial shading. A modified HC method based on sweeping the duty cycle is presented in [114]. Equation (13) is used to determine the initial value of the duty cycle as:

$$D = 1 - \sqrt{\frac{R_{MP}}{R_{Load}}} \tag{13}$$

where R_{Load} is estimated using the rating of the PV array.

Similar to the modified IC, this method must also scan over 80% of the P–V curve. A multiple-input boost converter for micro-inverters based on modified HC is discussed in [115].

4.1.2. Soft Computing MPPT Techniques

1. Artificial Neural Network (ANN)

Artificial neural networks (ANNs) provide a mechanism to use environmental conditions such as irradiation, temperature, and shading to predict the PV system operation point corresponding to the MPP [116–118]. The input parameters for the ANN are usually PV voltage (V_{PV}), PV current (I_{PV}), irradiation, and temperature. After processing the input variables, the ANN provides an output signal: the optimal voltage V_{MPP}, optimal current $I_{MPP,}$ and duty cycle [119,120]. The ANN is trained based on the experimental measurements and simulation results and mostly uses a back-propagation (BP) training algorithm [121–123]. Figure 8 presents the ANN structure.

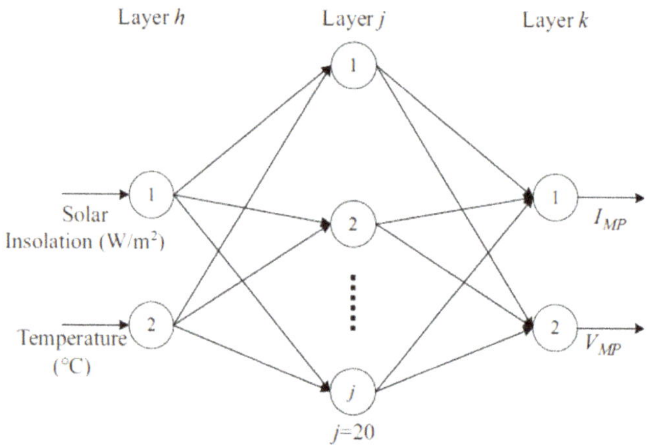

Figure 8. ANN structure [124].

2. Fuzzy Logic Control (FLC)

Similar to ANNs, the FLC does not need the internal parameter and mathematical model of the system. However, prior knowledge of the relationship between input and output is required. Figure 9 presents the control structure of the FLC. Besides input and output, the typical control structure has four control blocks: fuzzification, rules inferences, rule table, and defuzzification. The rule inference performs the calculation based on the rule table [125,126]. Error E and change in error ΔE are the usual input signals for the FLC-based MPPT. Equations (14) and (15) present the error and change error input signal calculation.

$$E(k) = \frac{P(k) - P(k-1)}{V(k) - V(k-1)} \; or \; E(k) = \frac{P(k) - P(k-1)}{I(k) - I(k-1)} \tag{14}$$

$$\Delta E = E(k) - E(k-1) \tag{15}$$

After being converted to a linguistic variable, the error and change in error will be used as input variables to the FLC. The FLC provides output signals in the form of a change in voltage (ΔV), change in current (ΔI), or change in duty cycle (ΔD).

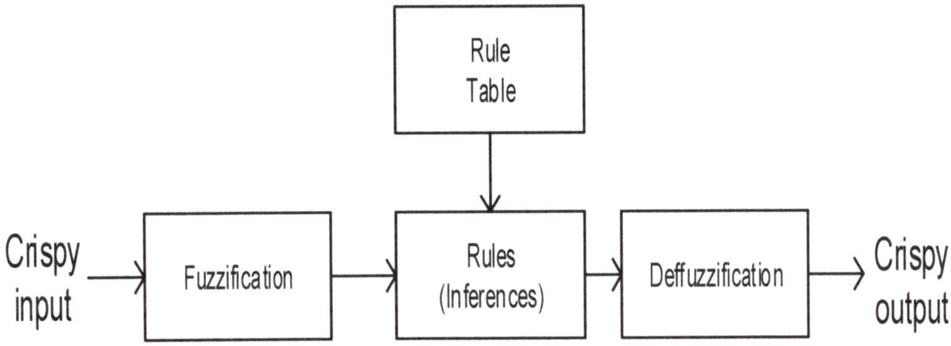

Figure 9. FLC structure.

3. Particle Swarm Optimization (PSO)

PSO is a population-based search method modeled after the behavior of bird flocks [127]. PSO has been popular to optimize and solve nonlinear problems in the last decade. It works by assigning random initial values to the particles in the boundary limits. The particles represent the duty cycle of the DC–DC converter and are optimization solutions. Particles move around the search space, and its best movement in the initial phase is called P_{best}. The overall best movement in the subsequent iteration is called G_{best}. Each particle is represented in the search space by its velocity (V_i) and position (X_i), and these parameters are updated in each iteration until the best solution is found [128,129]. The particles' velocity and position are updated using Equation (16):

$$\begin{aligned} V_i(k+1) &= WV_i(k) + C_1 rand_1 * (Pbest_i(k) - X_i(k)) + C_2 rand_2 * (Gbest_i(k) - X_i(k)) \\ X_i(k+1) &= X_i(k) + V_i(k+1) \end{aligned} \tag{16}$$

where

$V_i(k+1)$ is the particle velocity at k + 1th iteration,
W is the inertia weight,
$V_i(k)$ is the particle velocity at the kth iteration,
C_1 is the acceleration component associated with G_{best},
$X_i(k+1)$ is the particle position at $(k+1)$th iteration,
$X_i(k)$ is the particle position at the kth iteration,

C_2 is the acceleration component associated with P_{best},
$rand_1$ and $rand_2$ are random numbers from zero to one,
G_{best} is the best position of all particles, and P_{best} is the best position of the particle.

The objective function of the PSO optimization is to find the global voltage and power in the P–V characteristics curve. It is started by initializing parameters such as the swarm size (N), maximum iteration, and the voltage and power variable dimension that must be optimized. The global voltage with respect to Equation (16) is given as:

$$\begin{aligned} X_i(k) &= V_g = [V_{g1}, V_{g2}, V_{g3}, V_{g4} \ldots . V_{gj}] \\ j &= 1,2,3, \ldots ..N \end{aligned} \quad (17)$$

The best value of voltage and power that the PSO has found so far will be stored in $P_{best,}$ and the process continues until G_{best}, the best solution, is found.

The disadvantage of PSO is that since the initial position of the search agent is randomly provided depending on the boundary limit, there is a delay in the convergence. This can sometimes trap the algorithm to settle to local MPP during partial shading conditions [130,131].

4. Grey Wolf Optimization (GWO)

The GWO imitates the hunting techniques of grey wolves using a meta-heuristic optimization approach [132–135]. Four parameters, alpha (α), beta (β), delta (δ), and omega (ω), are used to represent the attaching techniques of the wolves. The fittest solution in the optimization is assumed to be α and followed by β. The third and fourth fit solutions are δ and ω, respectively. Figure 10 presents the flow chart of the GWO algorithm. Equation (18) presents the model of the hunting mechanism of grey wolves.

$$\begin{aligned} \underset{E}{\rightarrow} &= | \underset{C.}{\rightarrow}\underset{X_P}{\rightarrow}(t) - \underset{X_P}{\rightarrow}(t) \\ \underset{X}{\rightarrow}(t+1) &= \underset{X}{\rightarrow}(t) - \underset{F}{\rightarrow} . \underset{E}{\rightarrow} \end{aligned} \quad (18)$$

where

E, F, and C are the coefficient vectors,
X_p is the position vector of the hunting prey,
X is the position vector for the Grey wolf, and
t is the current iteration.

The vectors C and F are calculated as follows:

$$\begin{aligned} \underset{F}{\rightarrow} &= 2 \underset{a}{\rightarrow} . \underset{r_1}{\rightarrow} - \underset{a}{\rightarrow} \\ \underset{C}{\rightarrow} &= 2. \underset{r_2}{\rightarrow} \end{aligned} \quad (19)$$

The GWO fitness function is calculated as follows:

$$\begin{aligned} d_i(k+1) &= d_i(k) - F.E \\ P(d_i^k) &> P(d_i^{k-1}) \end{aligned} \quad (20)$$

where

d is the duty cycle,
k represents the iteration count,
i is the number of the current individual grey wolves, and P is the power.

The major advantages of the GWO technique are higher tracking efficiency and elimination of transient and steady-state oscillations [136,137].

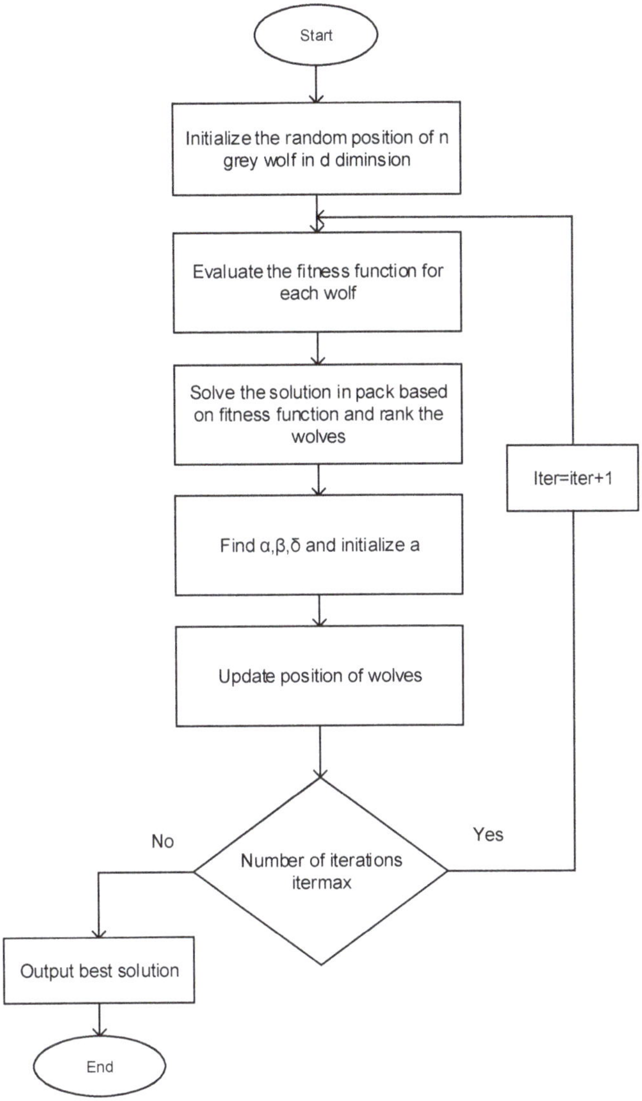

Figure 10. Flowchart of Grey Wolf Optimization.

5. Firefly Algorithm (FA)

The Firefly algorithm is developed from the characteristic relationship between line intensity and fireflies [138]. Different authors proposed a FA that can track the global peak of PV systems under partial shading. Two variables, namely γ, the light absorption coefficient, and α, the random coefficient, are used to randomize the first position of the firefly. A modified version of FA called simplified firefly algorithm (SFA) is proposed in [139,140], where the initial position of the firefly is selected between zero and one. The optimization equation of SFA is represented as:

$$X_i^{t+1} = X_i^t + \beta(X_j - X_i) \quad (21)$$

where

X_i represents the position of the less bright firefly,
X_j represents the position of the brighter firefly, and
β is the firefly attractiveness factor.

The objective function of SFA is to generate the maximum available photovoltaic output power, and the firefly position represents the duty cycle d.

6. Ant Colony Optimization (ACO)

ACO is an optimization technique based on the food-searching behavior of ants. ACO is an efficient and robust MPPT tracking the global peak of PV systems during partial shading conditions [141]. Different researchers evaluated the technique under varying irradiance and different shading patterns. The algorithm has a fast tracking speed of around one-tenth of the conventional MPPT methods for partial shading conditions [142,143].

7. Artificial Bee Colony (ABC)

Similar to ACO, ABC algorithms are an optimization technique based on the food-searching behavior of bees. The advantage of this algorithm is that it uses few parameters, and the convergence criteria are not dependent on the initial condition of the system [74]. The disadvantage of this method is that it is complex for practical implementation, and the tracking speed is slow compared to other MPPT methods used for partial shading conditions. The algorithm sometimes settles at the local peak rather than tracking the global peak [144]. The algorithm classifies the artificial bees into three categories: employed bees, onlooker bees, and the last scouts. Figure 11 depicts the flow chart of ABC, where the algorithm has four phases. The first phase initializes the algorithm by setting the different parameters. The second phase activates the employed bees searching for food, and the third phase activates the onlooker bees waiting in the hive to decide. The fourth phase is the scouting phase, where the bees search for random food sources. All three groups communicate and coordinate to obtain the optimal solution quickly. In the algorithm, the food source is the maximum power, and the duty cycle of the DC–DC converter is the food position. For implementing ABC in MPPT for PV system, the duty cycle for the DC–DC converter is calculated as follows:

$$\begin{aligned} d &= d_{\min} + rand[0,1](d_{\max} - d_{\min}) \\ newd &= d + \varphi_e(d - d_p) \end{aligned} \tag{22}$$

where

d is the current duty cycle,
d_{min} is the minimum value of the duty cycle,
d_{max} is the maximum value of the duty cycle),
ϕ_e is a constant between $[-1, 1]$, and
d_p is the previous duty cycle.

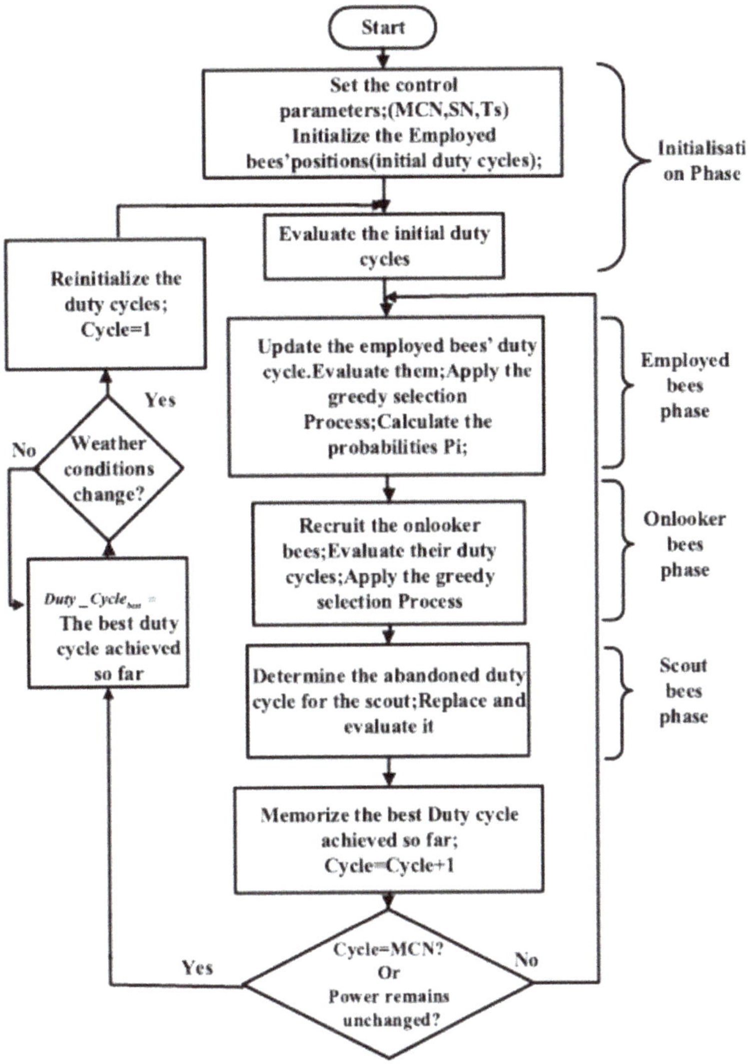

Figure 11. Flow chart of artificial bee colony optimization [145].

8. Cuckoo Search (CS)

The CS method is another optimization technique based on the levy flight mechanism of cuckoo birds [146,147]. The levy flight mechanism algorithm represents the cuckoos' search for a nest. The algorithm is a modified form of PSO with robust performance, high convergence speed, and efficiency. CS needs less tuning variables compared to PSO [148].

9. Jaya Algorithm (JA)

R.V. Rao introduced JA, whichis based on animal activities [149]. The algorithm is based on the distinct feature of animals or humans from the population. Naturally, humans or animals try to mimic the elite members of society and want to distance themselves from the lazy group. Figure 12 presents the flow chart of the JA. The candidate solution moves towards the best solution and tries to move away from the worst solution. The algorithm's simplicity and fast convergence make it the primary choice by different researchers to solve various engineering problems [150,151].

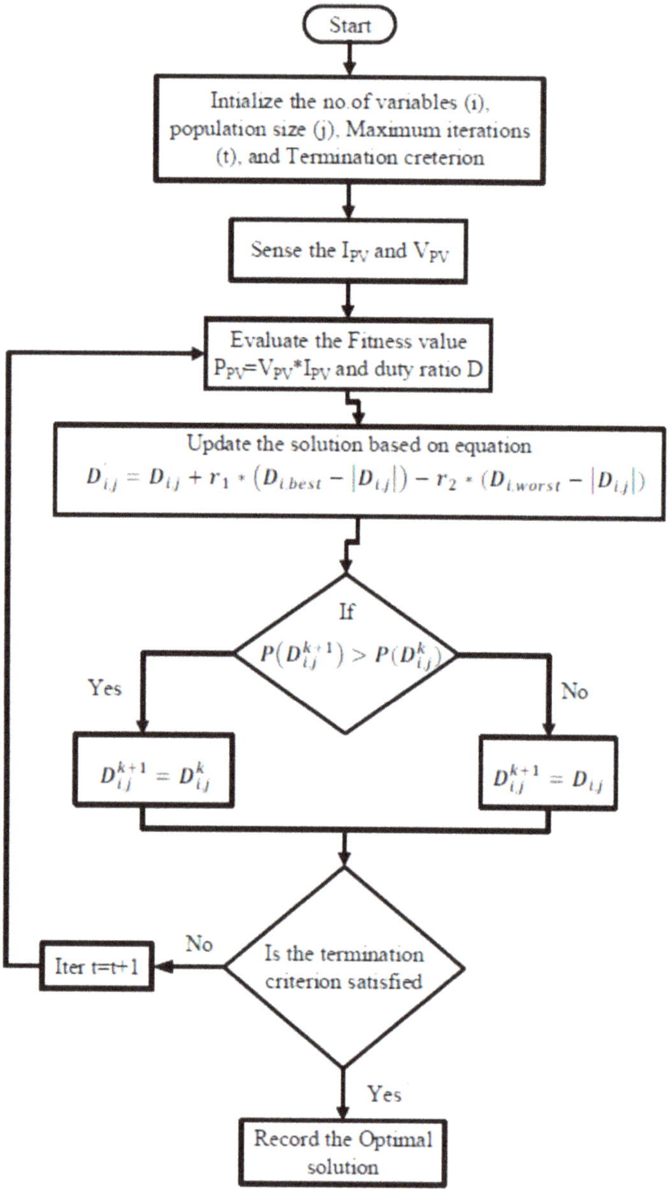

Figure 12. Flow chart of Jaya algorithm [152].

4.1.3. Hybrid MPPT Techniques

1. Hybrid GWO and P&O MPPT Algorithm

The authors in [84] combined P&O and GWO to enhance the performance of the MPPT control of the PV system under partial shading. The method works in two phases. In the first phase, GWO is implemented, and P&O is activated in the second phase to enhance the tracking speed. The computational burden and search space have been reduced by hybridizing the two techniques. This hybrid algorithm of GWO and P&O has several advantages, including fast-tracking speed, high efficiency, and high-tracking capability. Figure 13 presents the flow chart of the hybrid method where GWO is executed in the first phase and P&O in the second phase.

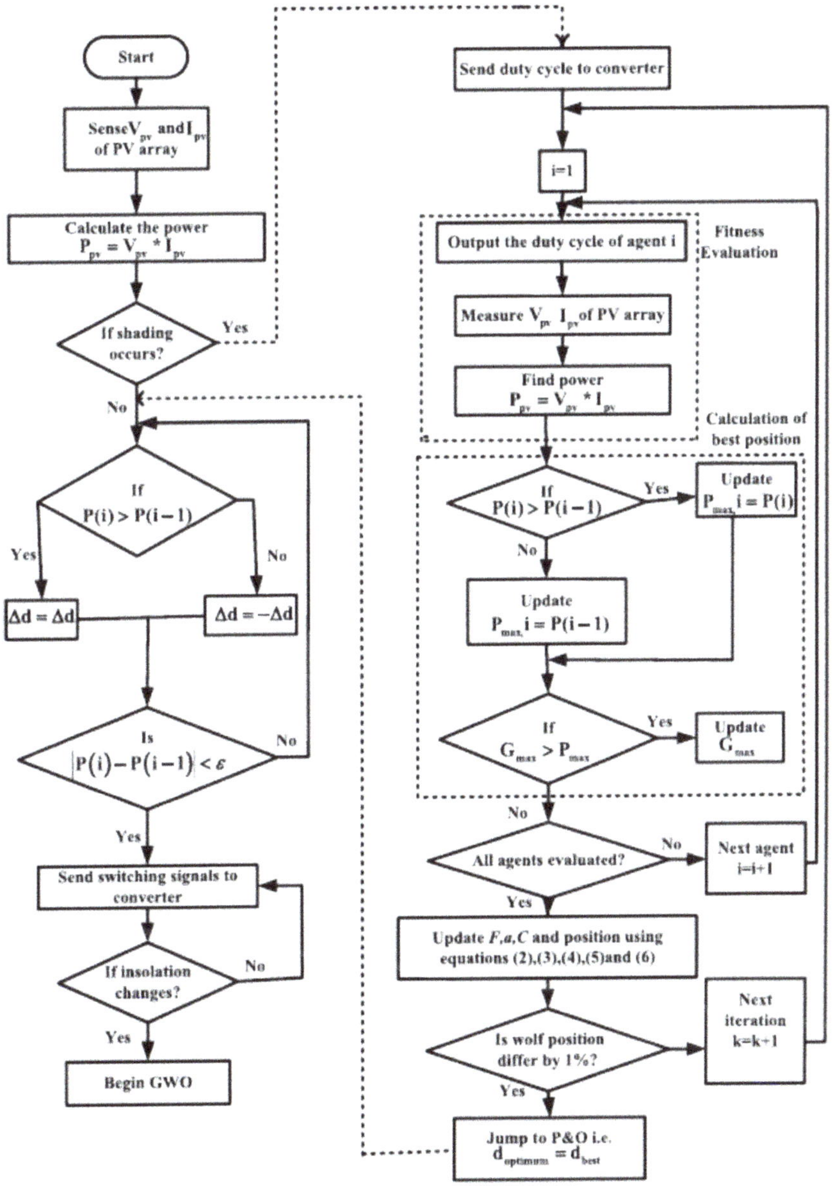

Figure 13. Flowchart of hybrid GWO and P&O algorithm [153].

2. P&O combined with PSO

Another hybrid MPPT technique for partial shading mitigation based on the combination of P&O and PSO is proposed in [154,155]. PSO is used in the initial phase to track the global peak, and then P&O is executed in the final phase. Compared to the conventional PSO, the advantage of this method is that it can track the global peak in a shorter time and has a faster convergence time with better dynamic performance. The hybrid approach has been tested in [154] with different shading scenarios, and to reduce the ripple current, the boost converter is modified to have an interleaved topology.

3. Differential Evolution and PSO (DEPSO)

The PSO, combined with differential evolution (DE), creates an algorithm efficient in tracking the global peak during partial shading conditions [156]. The advantage of the algorithm is that it is system independent and has fast tracking speed. Equation (23) is used to initialize the algorithm using the power fluctuation due to changes in irradiation.

$$\left|\frac{J(X_{q+1}) - J(X_q)}{(X_q)}\right| > \Delta P \tag{23}$$

where $J\ (X_q)$ is the output power of the PV panel. The algorithm efficiently differentiates the local and global peaks using the power mismatch.

4.2. Circuit-Based Approach

Power converters interface the generated power from the PV system to the grid or local loads. These power converters control the power flow and can enable MPPT controllers under partial shading at different levels of the PV system such as the PV cell, module or array [157]. Besides the power converters, changes in the PV system architecture and converter topology improve the performance of the PV system under partial shading. Some of the techniques the researchers implemented under this category are distributed MPPT, monitoring the bypass diode voltage, differential power processing, and power electronics equalizer, as described below.

4.2.1. Monitoring Bypass Diode Voltages

Monitoring the voltage of the bypass diode is effective in detecting the occurrence of partial shading. Under normal operation, the bypass diode is inactive, however during partial shading it will become active, and a voltage drop will appear across it [158]. Conventional (classical) MPPT techniques such as P&O, IC, and HC are implemented during normal irradiation conditions. Global search to track the global peak can be activated once a voltage drop is sensed in the bypass diode. The method works for module-integrated converters where the PV modules are directly connected to the power converters such as DC–DC converters or DC–AC inverters. The advantage of this method is that, unlike other MPPT methods where periodic scanning of the P-V curve is necessary, this technique is activated by sensing the bypass diode voltage. The drawback of this method is that it works for PV systems where the string voltage is accessible for measurement.

4.2.2. Distributed MPPT

Depending on the connection of the DC–DC converter and the DC–AC inverter used to integrate the PV system-generated power into the grid, there are two types of architectures: central and distributed. In central inverter architecture, one highly rated DC–DC converter and one DC–AC inverter are used where the DC–DC converter performs the MPPT, and the inverter is used for grid integration. This type of architecture does not generate the maximum available power in partial shading. On the other hand, distributed architecture alleviates this problem by providing the MPPT converter for each module. This type of arrangement provides greater flexibility, and the power generated from the PV system is better than from central-based architecture. The distributed-based partial shading mitigation technique connects the DC–DC converter and MPPT for each PV system cell, module, or array [159,160]. Each system unit works and tracks the MPP independently as a distributed sub-unit. The distributed and centralized MPPT architecture is provided in Figures 14 and 15, respectively. To reduce the cost and complexity of the MPPT for the distributed architecture, conventional MPPT methods such as P&O and IC are used. The advantage of this method is that the system reliability increases as each unit has its controller, and the failure of one of the sub-units does not affect the entire system.

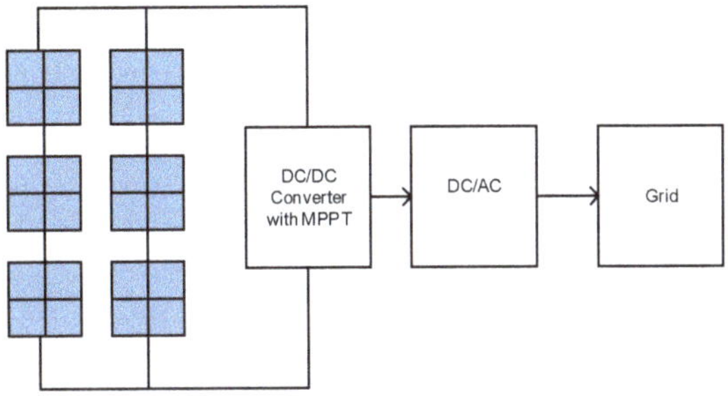

Figure 14. Centralized MPPT architecture.

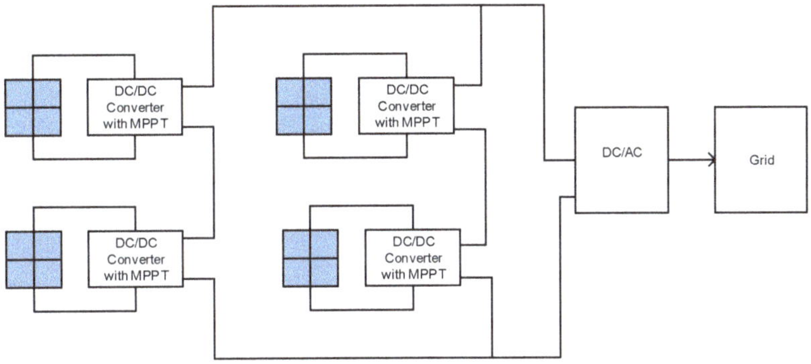

Figure 15. Distributed MPPT architecture.

4.2.3. Differential Power Processing

This method works by placing DC–DC converters between adjacent PV modules [161]. Figure 16 depicts the differential power processing method where the adjacent converters provide the current difference that appears between the current at the MPP of the two modules. The converter will be active when there is a difference in the power generated between the two adjacent modules. Conventional MPPT methods such as P&O and IC are employed for the modules. Compared to the distributed MPPT, this method, where a dedicated converter is connected for each module, minimizes the conversion losses and cost. The differential power processing method also has a better overall conversion efficiency and performs well during the partial shading condition by overcoming the challenges associated with the mismatch MPP current. Moreover, this method tracks the global peak with less power loss as the converter only processes the difference [162].

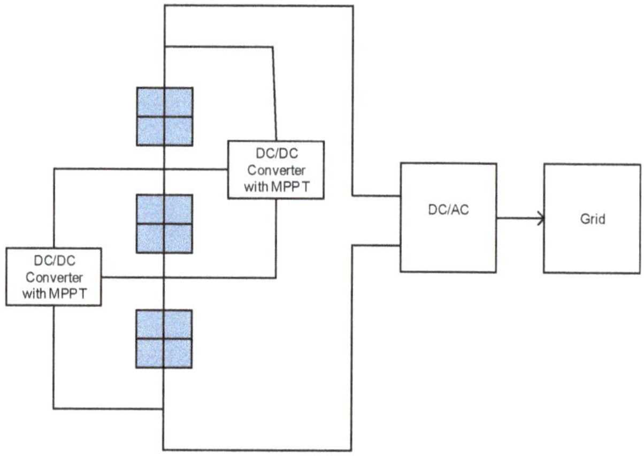

Figure 16. Differential MPPT architecture.

4.2.4. Power Electronics Equalizer

This method works using the power independent principle where series connected cells are operated with different voltages and currents [163]. The power electronics equalizer method works by transferring power from the non-shaded modules to the shaded modules so that all modules work at their respective MPP and exhibit an equal power level across the system. The power electronics equalizer method has a better performance and power harvesting capability as compared to the bypass diode method. The disadvantage of this method is that an extra circuit has to be connected to recover the power from the shaded modules, which increases the complexity of the topology. Energy storage elements such as inductors and capacitors are used to store the power of the non-shaded cells. They will be connected in parallel to distribute the stored power to the cells to have equal power across each cell [164].

5. Discussion

Different MPPT methods have their own merits and demerits. To compare the MPPT methods, different performance evaluation criteria such as tracking speed, dynamic tracking under partial shading, cost and complexity of the method, and differentiation between global and local maxima are used. The partial shading condition affects the power generated from the PV system and hence global maximum power point tracking (GMPPT) are required to increase the efficiency and harvesting capacity. Different GMPP tracking algorithms are discussed in this paper. It can be observed from the discussion that GMPP tracking algorithms based on hybridization of soft computing technique with conventional technique have better performance in terms of tracking speed, high tracking accuracy, and high convergence speed and are effective under partial shading. The hybrid method compensates for the disadvantage of one algorithm with the other; however, the complexity of implementing the technique practically increases. Some algorithms, such as ABC and ACO, have similar performance, and choosing a suitable algorithm depends on the intended application by comparing the evaluation criteria.

Table 1 compares MPPT-based partial shading mitigation techniques concerning efficiency, tracking speed, and level of complexity. Table 2 presents the comparison of circuit-based MPPT mitigation techniques. Table 2 shows that the power electronics equalizer-based MPPT technique has fast tracking speed. However, the method depends on system parameters and is complex. Table 3 provides a comparison of conventional and soft computing methods.

Table 1. MPPT-based methods comparison (soft computing and hybrid methods).

Method	Complexity	Tracking Speed (s)	Efficiency (%)	Converter	Application
ANN	High	-	-	Boost	Island mode
FLC	High	0.3	-	Boost	Island mode
PSO	Medium	1.5	-	Buck-boost	Island mode
ABC	Low	0.2	-	Boost	Island mode
ACO	Low	1.1	≈100	Boost	Island mode
Jaya	Low	1.18	99.99	Boost	Island mode
FA	Low to medium	1.3	98.8	Boost	Island mode
GWO	High	-	99.92	Boost	Island mode
CS	Medium	0.3	-	Boost	Island mode
DE-PSO	Medium	0.44	-	Boost	Island mode
PSO-P&O	Medium	0.9	-	Boost	Island mode
GWO-P&O	Medium	0.015	100	Boost	Island mode

Table 2. Circuit-based MPPT methods comparison.

Method	Complexity	Tracking Speed	Steady-State Oscillation	Dependency on System Parameter
Bypass diode	High	Slow	No	No
Distributed MPPT	Moderate	Variable	Sometimes	No
Differential Power Processing	Moderate	Variable	Sometimes	No
Power Electronics Equalizer	High	Fast	No	Yes

Table 3. Comparison of conventional and soft computing methods.

Method	Advantages	Disadvantages
Conventional Methods		
P&O	• Simple in construction • Easy to implement • Less sensor requirement	• Oscillations around MPP • Increased perturbation rate
IC	• Simple and highly reliable • Highly efficient for uniform irradiation	• Poor convergence • Frequency oscillations around MPP
HC	• Easy to implement • Efficient for slow changes in irradiation	• Slow response during high irradiation • Efficient only for low-power applications
Soft Computing Methods		
Fuzzy Logic Control	• Robust • Effective in error detection • Effective if combined with other conventional methods	• Poor convergence during the dynamic change in irradiation • Rules cannot be changed
Artificial Neural Network	• Accurate • Effective	• Needs large memory • Prior training is required • High computational time
Artificial Bee Colony	• Needs few parameters • Independent convergence criteria with system parameters	• Slow tracking speed • Complex
Particle Swarm Optimization	• Reliable • Simple and effective for handling non-linearity • Effective in tracking global peak • Wide search space usage	• Difficult initializing particle parameters. • Large computation burden for a large population
Cuckoo Search	• Robust • Fast tracking speed • Faster convergence • Less parameters	• Time-consuming calculation • Solution and convergence speed deteriorates
Ant Colony Optimization	• Low cost • Fast convergence speed	• Complex
Grey Wolf Optimization	• Robust • Fast tracking speed • No steady state and transient oscillations	• High cost • High computational time • Needs a large search space

6. Conclusions

This paper presented comprehensive MPPT techniques capable of tracking the global peak during partial shading conditions. The partial shading mitigation technique has been classified as MPPT-based and circuit-based methods. The MPPT-based method is further categorized as modified conventional MPPT, soft computing, and hybrid methods. The modified conventional methods are based on modifying the operation of traditional MPPT methods, such as P&O and IC, so that they can track the global peak efficiently. The soft computing methods are based on optimization and are fast and efficient compared to the modified conventional techniques. Researchers have received PSO well among the optimization methods because of its robustness, simplicity, and easy implementation. FLC and ANN need a lot of training data, and their practical implementation is also complex. The hybrid methods combine soft computing with conventional techniques and are receiving more attention. Circuit-based partial shading mitigation techniques are also discussed. The advantages and disadvantages of different optimization techniques are also discussed to help readers choose a suitable MPPT under partial shading conditions. From the various methods discussed to mitigate partial shading conditions, it is very challenging to pick the best one. The current developed robust methods face high computational time and are complex for practical implementation. Cost of implementation, accuracy, number of sensors required, response time, and efficiency are some of the limitations associated with the currently available MPPT methods and should be addressed in future research. This comprehensive review of the MPPT methods is expected to provide utilities and researchers with a beneficial tool as a reference and guideline to select the best GMPPT method for partially shaded PV systems based on their effectiveness.

Author Contributions: M.Y.W., M.R.E. and M.A.H. contributed to identifying and classifying the MPPTs, manuscript writing, and conclusions; M.A.H., M.S., M.R.E. and L.S.M. contributed to reviewing the MPPTs, manuscript writing, and revisions; M.S., M.R.E., L.S.M. and M.I.H.; contributed to reviewing and classifying MPPTs, manuscript writing, and editing; and M.Y.W., M.I.H. and M.A.A. participated in revising the manuscript, editing, and conclusions. All authors have read and agreed to the published version of the manuscript.

Funding: The authors acknowledge the support provided by the Interdisciplinary Research Center for Renewable Energy and Power Systems (IRC-REPS), Research Institute, King Fahd University of Petroleum and Minerals, through project #INRE2323.

Institutional Review Board Statement: Not applicable.

Informed Consent Statement: Not applicable.

Data Availability Statement: Not applicable.

Conflicts of Interest: The authors declare no conflict of interest.

References

1. Asghar, R.; Ullah, Z.; Azeem, B.; Aslam, S.; Hashmi, M.H.; Rasool, E.; Shaker, B.; Anwar, M.J.; Mustafa, K. Wind Energy Potential in Pakistan: A Feasibility Study in Sindh Province. *Energies* **2022**, *15*, 8333. [CrossRef]
2. Asghar, M.; Anwar, J.; Wadood, H.; Saleem, H.; Rasul, N.; Ullah, Z. Promising Features of Wind Energy: A Glance Overview. In Proceedings of the 2023 4th International Conference on Computing, Mathematics and Engineering Technologies (iCoMET), Sukkur, Pakistan, 17–18 March 2023; pp. 1–6. [CrossRef]
3. Gupta, A.; Likozar, B.; Jana, R.; Chanu, W.C.; Singh, M.K. A review of hydrogen production processes by photocatalytic water splitting—From atomistic catalysis design to optimal reactor engineering. *Int. J. Hydrog. Energy* **2022**, *47*, 33282–33307. [CrossRef]
4. Ješić, D.; Jurković, D.L.; Pohar, A.; Suhadolnik, L.; Likozar, B. Engineering photocatalytic and photoelectrocatalytic CO_2 reduction reactions: Mechanisms, intrinsic kinetics, mass transfer resistances, reactors and multi-scale modelling simulations. *Chem. Eng. J.* **2021**, *407*, 126799. [CrossRef]
5. Kržmanc, M.M.; Daneu, N.; Čontala, A.; Santra, S.; Kamal, K.M.; Likozar, B.; Spreitzer, M. $SrTiO_3/Bi_4Ti_3O_{12}$ Nanoheterostructural Platelets Synthesized by Topotactic Epitaxy as Effective Noble-Metal-Free Photocatalysts for pH-Neutral Hydrogen Evolution. *ACS Appl. Mater. Interfaces* **2021**, *13*, 370–381. [CrossRef]
6. Ahmed, M.; Abdelrahem, M.; Kennel, R. Highly Efficient and Robust Grid Connected Photovoltaic System Based Model Predictive Control with Kalman Filtering Capability. *Sustainability* **2020**, *12*, 4542. [CrossRef]

7. Bubalo, M.; Bašić, M.; Vukadinović, D.; Grgić, I. Hybrid Wind-Solar Power System with a Battery-Assisted Quasi-Z-Source Inverter: Optimal Power Generation by Deploying Minimum Sensors. *Energies* **2023**, *16*, 1488. [CrossRef]
8. Ludin, N.A.; Affandi, N.A.A.; Purvis-Roberts, K.; Ahmad, A.; Ibrahim, M.A.; Sopian, K.; Jusoh, S. Environmental Impact and Levelised Cost of Energy Analysis of Solar Photovoltaic Systems in Selected Asia Pacific Region: A Cradle-to-Grave Approach. *Sustainability* **2021**, *13*, 396. [CrossRef]
9. Uoya, M.; Koizumi, H. A Calculation Method of Photovoltaic Array's Operating Point for MPPT Evaluation Based on One-Dimensional Newton–Raphson Method. *IEEE Trans. Ind. Appl.* **2015**, *51*, 567–575. [CrossRef]
10. El-Khozondar, H.J.; El-Khozondar, R.J.; Matter, K.; Suntio, T. A review study of photovoltaic array maximum power tracking algorithms. *Renew. Wind Water Sol.* **2016**, *3*, 1. [CrossRef]
11. Jafarkazemi, F.; Saadabadi, S.A. Optimum tilt angle and orientation of solar surfaces in Abu Dhabi, UAE. *Renew. Energy* **2013**, *56*, 44–49. [CrossRef]
12. Almasoud, A.H.; Gandayh, H.M. Future of solar energy in Saudi Arabia. *J. King Saud. Univ. Eng. Sci.* **2014**, *27*, 153–157. [CrossRef]
13. SAKAKA PV IPP. Available online: https://acwapower.com/en/projects/sakaka-pv-ipp/ (accessed on 14 March 2023).
14. Liu, L.; Wang, Z.; Zhang, H.; Xue, Y. Solar energy development in China—A review. *Renew. Sustain. Energy Rev.* **2010**, *14*, 301–311. [CrossRef]
15. Ahmed, S.U.; Affan, M.; Raza, M.I.; Hashmi, M.H. Inspecting Mega Solar Plants through Computer Vision and Drone Technologies. In Proceedings of the 2022 International Conference on Frontiers of Information Technology (FIT), Islamabad, Pakistan, 11–12 December 2022; pp. 18–23. [CrossRef]
16. Sher, H.A.; Murtaza, A.F.; Noman, A.; Addoweesh, K.E.; Chiaberge, M. An intelligent control strategy of fractional short circuit current maximum power point tracking technique for photovoltaic applications. *J. Renew. Sustain. Energy* **2015**, *7*, 013114. [CrossRef]
17. Sher, H.A.; Murtaza, A.F.; Noman, A.; Addoweesh, K.E.; Al-Haddad, K.; Chiaberge, M. A New Sensorless Hybrid MPPT Algorithm Based on Fractional Short-Circuit Current Measurement and P&O MPPT. *IEEE Trans. Sustain. Energy* **2015**, *6*, 1426–1434.
18. Baimel, D.; Tapuchi, S.; Levron, Y.; Belikov, J. Improved Fractional Open Circuit Voltage MPPT Methods for PV Systems. *Electronics* **2019**, *8*, 321. [CrossRef]
19. Ahmed, J.; Salam, Z. An improved perturb and observe (P&O) maximum power point tracking (MPPT) algorithm for higher efficiency. *Appl. Energy* **2015**, *150*, 97–108. [CrossRef]
20. Bollipo, R.B.; Mikkili, S.; Bonthagorla, P.K. Hybrid, optimization, intelligent and classical PV MPPT techniques: A Review. *CSEE J. Power Energy Syst.* **2021**, *7*, 9–33.
21. Kumar, N.; Hussain, I.; Singh, B.; Panigrahi, K. Framework of maximum power extraction from solar PV panel using self predictive perturb and observe algorithm. *IEEE Trans. Sustain. Energy.* **2018**, *9*, 895–903. [CrossRef]
22. Jiang, J.-A.; Su, Y.-L.; Kuo, K.-C.; Wang, C.-H.; Liao, M.-S.; Wang, J.-C.; Huang, C.-K.; Chou, C.-Y.; Lee, C.-H.; Shieh, J.-C. On a hybrid MPPT control scheme to improve energy harvesting performance of traditional two-stage inverters used in photovoltaic systems. *Renew. Sustain. Energy Rev.* **2017**, *69*, 1113–1128. [CrossRef]
23. Mousa, H.H.; Youssef, A.-R.; Mohamed, E.E. State of the art perturb and observe MPPT algorithms based wind energy conversion systems: A technology review. *Int. J. Electr. Power Energy Syst.* **2020**, *126*, 106598. [CrossRef]
24. Alik, R.; Jusoh, A.; Sutikno, T. A Review on Perturb and Observe Maximum Power Point Tracking in Photovoltaic System. *TELKOMNIKA (Telecommun. Comput. Electron. Control.)* **2015**, *13*, 745–751. [CrossRef]
25. Al-Dhaifallah, M.; Nassef, A.M.; Rezk, H.; Nisar, K.S. Optimal parameter design of fractional order control based INC-MPPT for PV system. *Sol. Energy* **2018**, *159*, 650–664. [CrossRef]
26. Zakzouk, N.E.; Williams, B.W.; Helal, A.A.; Elsaharty, M.A.; Abdelsalam, A.K. Improved performance low-cost incremental conductance PV MPPT technique. *IET Renew. Power Gener.* **2016**, *10*, 561–574. [CrossRef]
27. Tozlu, Ö.F.; Çalik, H. A Review and Classification of Most Used MPPT Algorithms for Photovoltaic Systems. *Hittite J. Sci. Eng.* **2021**, *8*, 207–220. [CrossRef]
28. Huynh, D.C.; Dunnigan, M.W. Development and Comparison of an Improved Incremental Conductance Algorithm for Tracking the MPP of a Solar PV Panel. *IEEE Trans. Sustain. Energy* **2016**, *7*, 1421–1429. [CrossRef]
29. Elgendy, M.A.; Atkinson, D.J.; Zahawi, B. Experimental investigation of the incremental conductance maximum power point tracking algorithm at high perturbation rates. *IET Renew. Power Gener.* **2016**, *10*, 133–139. [CrossRef]
30. Worku, M.Y.; Abido, M.A. Real-Time Implementation of Grid-Connected PV System with Decoupled P-Q Controllers. In Proceedings of the 22nd Mediterranean Conference on Control & Automation (MED'14), Palermo, Italy, 16–19 June 2014.
31. Kumar, N.; Hussain, I.; Singh, B.; Panigrahi, B.K. Self-Adaptive Incremental Conductance Algorithm for Swift and Ripple-Free Maximum Power Harvesting from PV Array. *IEEE Trans. Ind. Inform.* **2018**, *14*, 2031–2041. [CrossRef]
32. Kandemir, E.; Cetin, N.S.; Borekci, S.A. A comprehensive overview of maximum power extraction methods for PV systems. *Renew. Sustain. Energy Rev.* **2017**, *78*, 93–112. [CrossRef]
33. Jately, V.; Azzopardi, B.; Joshi, J.; Sharma, A.; Arora, S. Experimental Analysis of hill-climbing MPPT algorithms under low irradiance levels. *Renew. Sustain. Energy Rev.* **2021**, *150*, 111467. [CrossRef]
34. Verma, D.; Nema, S.; Shandilya, A.M.; Dash, S.K. Maximum power point tracking (MPPT) techniques: Recapitulation in solar photovoltaic systems. *Renew. Sustain. Energy Rev.* **2016**, *54*, 1018–1034. [CrossRef]

35. Ramli, M.A.M.; Twaha, S.; Ishaque, K.; Al-Turki, Y.A. A review on maximum power point tracking for photovoltaic systems with and without shading conditions. *Renew. Sustain. Energy Rev.* **2017**, *67*, 144–159. [CrossRef]
36. Farayola, A.M.; Hasan, A.N.; Ali, A. Curve Fitting Polynomial Technique Compared to ANFIS Technique for Maximum Power Point Tracking. In Proceedings of the 2017 8th International Renewable Energy Congress (IREC), Amman, Jordan, 21–23 March 2017; pp. 1–6.
37. Lasheen, M.; Rahman, A.K.A.; Abdel-Salam, M.; Ookawara, S. Performance Enhancement of Constant Voltage Based MPPT for Photovoltaic Applications Using Genetic Algorithm. *Energy Procedia* **2016**, *100*, 217–222. [CrossRef]
38. Leedy, A.W.; Guo, L.P.; Aganah, K.A. A constant voltage MPPT method for a solar powered boost converter with DC motor load. In Proceedings of the 2012 IEEE Southeastcon, Orlando, FL, USA, 15–18 March 2012; pp. 1–6.
39. Kimball, J.W.; Krein, P.T. Digital ripple correlation control for photovoltaic applications. In Proceedings of the 2007 IEEE Power Electronics Specialists Conference, Orlando, FL, USA, 17–21 June 2007.
40. Li, G.; Jin, Y.; Akram, M.W.; Chen, X.; Ji, J.I. Application of bio-inspired algorithms in maximum power point tracking for PV systems under partial shading conditions—A review. *Renew. Sustain. Energy Rev.* **2018**, *81*, 840–873. [CrossRef]
41. Balamurugan, M.; Sahoo, S.K.; Sukchai, S. Application of soft computing methods for grid connected PV system: A technological and status review. *Renew. Sustain. Energy Rev.* **2017**, *75*, 1493–1508. [CrossRef]
42. Mohanty, S.; Subudhi, B.; Ray, P.K. A New MPPT Design Using Grey Wolf Optimization Technique for Photovoltaic System Under Partial Shading Conditions. *IEEE Trans. Sustain. Energy* **2015**, *7*, 181–188. [CrossRef]
43. Shams, I.; Mekhilef, S.; Tey, K.S. Improved Social Ski Driver-Based MPPT for Partial Shading Conditions Hybridized with Constant Voltage Method for Fast Response to Load Variations. *IEEE Trans. Sustain. Energy* **2021**, *12*, 2255–2267. [CrossRef]
44. Li, X.; Wen, H.; Hu, Y.; Jiang, L. A novel beta parameter based fuzzy-logic controller for photovoltaic MPPT application. *Renew. Energy* **2019**, *130*, 416–427. [CrossRef]
45. Tang, S.; Sun, Y.; Chen, Y.; Zhao, Y.; Yang, Y.; Szeto, W. An Enhanced MPPT Method Combining Fractional-Order and Fuzzy Logic Control. *IEEE J. Photovolt.* **2017**, *7*, 640–650. [CrossRef]
46. Craciunescu, D.; Fara, L. Investigation of the Partial Shading Effect of Photovoltaic Panels and Optimization of Their Performance Based on High-Efficiency FLC Algorithm. *Energies* **2023**, *16*, 1169. [CrossRef]
47. Aly, M.; Rezk, H. An improved fuzzy logic control-based MPPT method to enhance the performance of PEM fuel cell system. *Neural Comput. Appl.* **2022**, *34*, 4555–4566. [CrossRef]
48. Ram, J.P.; Babu, T.S.; Rajasekar, N. A comprehensive review on solar PV maximum power point tracking techniques. *Renew. Sustain. Energy Rev.* **2017**, *67*, 826–847. [CrossRef]
49. Elobaid, L.M.; Abdelsalam, A.K.; Zakzouk, E.E. Artificial neural network-based photovoltaic maximum power point tracking techniques: A survey. *IET Renew. Power Gener.* **2015**, *9*, 1043–1063. [CrossRef]
50. Srinivasan, S.; Tiwari, R.; Krishnamoorthy, M.; Lalitha, M.P.; Raj, K.K. Neural network based MPPT control with reconfigured quadratic boost converter for fuel cell application. *Int. J. Hydrogen Energy* **2020**, *46*, 6709–6719. [CrossRef]
51. Messalti, S.; Harrag, A.; Loukriz, A. A new variable step size neural networks MPPT controller: Review, simulation and hardware implementation. *Renew. Sustain. Energy Rev.* **2017**, *68*, 221–233. [CrossRef]
52. Karami, N.; Moubayed, N.; Outbib, R. General review and classification of different MPPT Techniques. *Renew. Sustain. Energy Rev.* **2017**, *68*, 1–18. [CrossRef]
53. Zhanga, R.; Ong, S.K.; Nee, A.Y.C. A simulation-based genetic algorithm approach for re manufacturing process planning and scheduling. *Appl. Soft Comput.* **2015**, *37*, 521–532. [CrossRef]
54. Zhang, B.; Li, X.; Wang, S. A novel case adaptation method based on an improved integrated genetic algorithm for power grid wind disaster emergencies. *Expert Syst. Appl.* **2015**, *42*, 7812–7824. [CrossRef]
55. Dou, R.; Zong, C.; Nan, G. Multi-stage interactive genetic algorithm for collaborative product customization. *Knowl. Based Syst.* **2016**, *92*, 43–54. [CrossRef]
56. Daraban, S.; Petreus, D.; Morel, C. Anovel MPPT algorithm based on a modified genetic algorithm specialized on tracking the global maximum power point in photovoltaic systems affected by partial shading. *Energy* **2014**, *74*, 374–388. [CrossRef]
57. Koad, R.B.; Zobaa, A.F.; El-Shahat, A. A Novel MPPT Algorithm Based on Particle Swarm Optimization for Photovoltaic Systems. *IEEE Trans. Sustain. Energy* **2016**, *8*, 468–476. [CrossRef]
58. Sundareswaran, K.; Palani, S. Application of a combined particle swarm optimization and perturb and observe method for MPPT in PV systems under partial shading conditions. *Renew. Energy* **2015**, *75*, 308–317. [CrossRef]
59. Luta, D.N.; Raji, A.K. Fuzzy Rule-Based and Particle Swarm Optimisation MPPT Techniques for a Fuel Cell Stack. *Energies* **2019**, *12*, 936. [CrossRef]
60. Li, H.; Yang, D.; Su, W.; Lü, J.; Yu, X. An Overall Distribution Particle Swarm Optimization MPPT Algorithm for Photovoltaic System Under Partial Shading. *IEEE Trans. Ind. Electron.* **2018**, *66*, 265–275. [CrossRef]
61. Chen, M.; Ma, S.; Wu, J.; Huang, L. Analysis of MPPT Failure and Development of an Augmented Nonlinear Controller fo MPPT of Photovoltaic Systems under Partial Shading Conditions. *Appl. Sci.* **2017**, *7*, 95. [CrossRef]
62. Zhou, L.; Chen, Y.; Liu, Q.; Wu, J. Maximum power point tracking (MPPT) control of a photovoltaic system based on dual carrier chaotic search. *J. Control Theory Appl.* **2012**, *10*, 244–250. [CrossRef]
63. Ramli Makbul, A.M.; Kashif, I.; Faizan, J.; Al-Turki Yusuf, A.; Zainal, S. A modified differential evolution based maximum power point tracker for photovoltaic system under partial shading condition. *Energy Build.* **2015**, *103*, 175–184. [CrossRef]

64. Naim, T.M.F.; Md, A.S.; Zainal, S.; Sazli, S.M. Evolutionary based maximum power point tracking technique using differential evolution algorithm. *Energy Build.* **2013**, *67*, 245–252.
65. Ramasamy, S.; Dash, S.S.; Selvan, T. An Intelligent Differential Evolution Based Maximum Power Point Tracking (MPPT) Technique for Partially Shaded Photo Voltaic (PV) Array. *Int. J. Adv. Soft Comput. Its Appl.* **2014**, *6*, 1–16.
66. Azam, M.A.; Abdullah-Al-Nahid, S.; Kabir, M.A.; Chowdhury, S.M.H. Microcontroller based maximum power tracking of PV using stimulated annealing algorithm. In Proceedings of the International Conference on Informatics, Electronics & Vision (ICIEV), Dhaka, Bangladesh, 18–19 May 2012; pp. 298–303.
67. Li, Q.; Chen, H.; Huang, H.; Zhao, X.; Cai, Z.; Tong, C.; Liu, W.; Tian, X. An Enhanced Grey Wolf Optimization Based Feature Selection Wrapped Kernel Extreme Learning Machine for Medical Diagnosis. *Comput. Math. Methods Med.* **2017**, *2017*, 9512741. [CrossRef]
68. Diab, A.A.Z.; Rezk, H. Optimal Sizing and Placement of Capacitors in Radial Distribution Systems Based on Grey Wolf, Dragonfly and Moth–Flame Optimization Algorithms. *Iran. J. Sci. Technol. Trans. Electr. Eng.* **2018**, *43*, 77–96. [CrossRef]
69. Cherukuri, S.K.; Rayapudi, S.R. Enhanced Grey Wolf Optimizer based MPPT Algorithm of PV system under Partial Shaded Condition. *Int. J. Renew. Energy Dev.* **2017**, *6*, 203–212. [CrossRef]
70. Rezk, H.; Fathy, A.; Abdelaziz, A.Y. A comparison of different global MPPT techniques based on meta-heuristic algorithms for photovoltaic system subjected to partial shading conditions. *Renew. Sustain. Energy Rev.* **2017**, *74*, 377–386. [CrossRef]
71. Jubaer, A.; Zainal, S. A Maximum Power Point Tracking (MPPT) for PV system using Cuckoo Search with partial shading capability. *Appl. Energy* **2014**, *119*, 118–130.
72. Alshareef, M.J. An Effective Falcon Optimization Algorithm Based MPPT Under Partial Shaded Photovoltaic Systems. *IEEE Access* **2022**, *10*, 131345–131360. [CrossRef]
73. Abousoufyane, B.; Aissa, C.; Kamel, K.; Santiago, S.; Ait, S.O. Artificial bee colony based algorithm for maximum power point tracking (MPPT) for PV systems operating under partial shaded conditions. *Appl. Soft Comput.* **2015**, *32*, 38–48.
74. González-Castaño, C.; Restrepo, C.; Kouro, S.; Rodriguez, J. MPPT Algorithm Based on Artificial Bee Colony for PV System. *IEEE Access* **2021**, *9*, 43121–43133. [CrossRef]
75. Belhachat, F.; Larbes, C. A review of global maximum power point tracking techniques of photovoltaic system under partial shading conditions. *Renew. Sustain. Energy Rev.* **2018**, *92*, 513–553. [CrossRef]
76. Titri, S.; Larbes, C.; Toumi, K.Y.; Benatchba, K. A new MPPT controller based on the Ant colony optimization algorithm for Photovoltaic systems under partial shading conditions. *Appl. Soft Comput.* **2017**, *58*, 465–479. [CrossRef]
77. Besheer, A.H. Ant Colony System Based PI Maximum Power Point Tracking for Stand Alone Photovoltaic System. In Proceedings of the IEEE International Conference on Industrial Technology, ICIT-2012, Athens, Greece, 19–21 March 2012.
78. Priyadarshi, N.; Ramachandaramurthy, V.K.; Padmanaban, S.; Azam, F. An Ant Colony Optimized MPPT for Standalone Hybrid PV-Wind Power System with Single Cuk Converter. *Energies* **2019**, *12*, 167. [CrossRef]
79. Sundareswaran, K.; Peddapati, S.; Palani, S. MPPT of PV Systems Under Partial Shaded Conditions through a Colony of Flashing Fireflies. *IEEE Trans. Energy Convers.* **2014**, *29*, 463–472. [CrossRef]
80. Sundareswaran, K.; Peddapati, S.; Palani, S. Application of random search method for maximum power point tracking in partially shaded photovoltaic systems. *IET Renew. Power Gener.* **2014**, *8*, 670–678. [CrossRef]
81. Shi, J.Y.; Ling, L.T.; Xue, F.; Qin, Z.J.; Li, Y.J.; Lai, Z.X.; Yang, T. Combining incremental conductance and firefly algorithm for tracking the global MPP of PV arrays. *J. Renew. Sustain. Energy* **2017**, *9*, 023501. [CrossRef]
82. El-Helw, H.M.; Magdy, A.; Marei, M.I. A Hybrid Maximum Power Point Tracking Technique for Partially Shaded Photovoltaic Arrays. *IEEE Access* **2017**, *5*, 11900–11908. [CrossRef]
83. Chakkarapani, M.; Raman, G.P.; Raman, G.R.; Ganesan, S.I.; Chilakapati, N. Fireworks Enriched P&O Algorithm for GMPPT and Detection of Partial Shading in PV Systems. *IEEE Trans. Power Electron.* **2017**, *32*, 4432–4443. [CrossRef]
84. Mohanty, S.; Subudhi, B.; Ray, P.K. A Grey Wolf-Assisted Perturb & Observe MPPT Algorithm for a Photovoltaic Power System. *IEEE Trans. Energy Convers.* **2017**, *32*, 340–347. [CrossRef]
85. Abdelghani, H.; Sabir, M. Variable step size modified P&OMPPT algorithm using GA-based hybrid offline/online PID controller. *Renew. Sustain. Energy Rev.* **2015**, *491*, 1247–1260.
86. Karagoz, M.K.; Demirel, H. A novel MPPT method for PV arrays based on modified Bat algorithm with partial shading capability. *Int. J. Comput. Sci. Netw. Secur.* **2017**, *17*, 61–66.
87. Hanafiah, S.; Ayad, A.; Hehn, A.; Kennel, R. A hybrid MPPT for quasi-Z-source inverters in PV applications under partial shading condition. In Proceedings of the 11th IEEE International Conference on Compatibility, Power Electronics and Power Engineering, Cadiz, Spain, 4–6 April 2017.
88. Yang, Z.; Duan, Q.; Zhong, J.; Mao, M.; Xun, Z. Analysis of improved PSO and perturb & observe global MPPT algorithm for PV array under partial shading condition. In Proceedings of the 29th Chinese Control and Decision Conference (CCDC), Chongqing, China, 28–30 May 2017.
89. Guan, T.; Zhuo, F. An improved SA-PSO global maximum power point tracking method of photovoltaic system under partial shading conditions. In Proceedings of the IEEE Conference on Environment and Electrical Engineering, Milan, Italy, 6–9 June 2017.
90. Duan, Q.; Mao, M.; Duan, P.; Hu, B. An intelligent algorithm for maximum power point tracking in photovoltaic system under partial shading conditions. *Trans. Inst. Meas. Control.* **2016**, *39*, 244–256. [CrossRef]

91. Mao, M.; Duan, Q.; Duan, P.; Hu, B. Comprehensive improvement of artificial fish swarm algorithm for global MPPT in PV system under partial shading conditions. *Trans. Inst. Meas. Control.* **2017**, *40*, 2178–2199. [CrossRef]
92. Kumar, N.; Hussain, I.; Singh, B.; Panigrahi, B.K. Rapid MPPT for Uniformly and Partial Shaded PV System by Using JayaDE Algorithm in Highly Fluctuating Atmospheric Conditions. *IEEE Trans. Ind. Inform.* **2017**, *13*, 2406–2416. [CrossRef]
93. Kumar, N.; Hussain, I.; Singh, B.; Panigrahi, B.K. MPPT in Dynamic Condition of Partially Shaded PV System by Using WODE Technique. *IEEE Trans. Sustain. Energy* **2017**, *8*, 1204–1214. [CrossRef]
94. Rajasekar, N.; Neeraja, K.K.; Venugopalan, R. Bacterial Foraging Algorithm based solar PV parameter estimation. *Sol. Energy* **2013**, *97*, 255–265. [CrossRef]
95. Worku, M.Y.; Abido, M.A. Grid Connected PV System Using ANFIS Based MPPT Controller in Real Time. In Proceedings of the International Conference on Renewable Energies and Power Quality (ICREPQ'16), Madrid, Spain, 4–6 May 2016.
96. Alam, D.F.; Yousri, D.A.; Eteiba, M.B. Flower Pollination Algorithm based solar PV parameter estimation. *Energy Convers. Manag.* **2015**, *101*, 410–422. [CrossRef]
97. Gao, L.; Dougal, R.A.; Liu, S.; Iotova, A.P. Parallel-Connected Solar PV System to Address Partial and Rapidly Fluctuating Shadow Conditions. *IEEE Trans. Ind. Electron.* **2009**, *56*, 1548–1556. [CrossRef]
98. Abido, M.A.; Sheraz, M.; Worku, M.Y. An Efficient ANFIS-Based PI Controller for Maximum Power Point Tracking of PV Systems. *Arab. J. Sci. Eng.* **2015**, *40*, 2641–2651. [CrossRef]
99. Kamarzaman, N.A.; WeiTan, C. A comprehensive review of maximum power point tracking algorithms for photovoltaic systems. *Renew. Sustain. Energy Rev.* **2014**, *37*, 585–598. [CrossRef]
100. Ishaque, K.; Salam, Z.; Syafaruddin Tahei, H. Modeling and simulation of pho- tovoltaic (PV) system during partial shading based on a two-diode model. *Simul. Model Pract. Theory* **2011**, *19*, 1613–1626. [CrossRef]
101. Kadri, R.; Andrei, H.; Gaubert, J.P.; Ivanovici, T.; Champenois, G.; Andrei, P. Modeling of the photovoltaic cell circuit parameters for optimum connection model and real-time emulator with partial shadow conditions. *Energy* **2012**, *42*, 57–67. [CrossRef]
102. Sánchez Reinoso, C.R.; Milone, D.H.; Buitrago, R.H. Simulation of photovoltaic centrals with dynamic shading. *Appl. Energy* **2013**, *103*, 278–289. [CrossRef]
103. Rajalakshmi, M.; Chandramohan, S.; Kannadasan, R.; Alsharif, M.H.; Kim, M.-K.; Nebhen, J. Design and Validation of BAT Algorithm-Based Photovoltaic System Using Simplified High Gain Quasi Boost Inverter. *Energies* **2021**, *14*, 1086. [CrossRef]
104. Alturki, F.A.; Al-Shamma'a, A.; Farh, H.M.H. Simulations and dSPACE Real-Time Implementation of Photovoltaic Global Maximum Power Extraction under Partial Shading. *Sustainability* **2020**, *12*, 3652. [CrossRef]
105. Ali, A.; Almutairi, K.; Padmanaban, S.; Tirth, V.; Algarni, S.; Irshad, K.; Islam, S.; Zahir, M.H.; Shafiullah, M.; Malik, M.Z. Investigation of MPPT Techniques Under Uniform and Non-Uniform Solar Irradiation Condition–A Retrospection. *IEEE Access* **2020**, *8*, 127368–127392. [CrossRef]
106. Almutairi, A.; Abo-Khalil, A.; Sayed, K.; Albagami, N. MPPT for a PV Grid-Connected System to Improve Efficiency under Partial Shading Conditions. *Sustainability* **2020**, *12*, 10310. [CrossRef]
107. Islam, H.; Mekhilef, S.; Shah, N.; Soon, T.; Wahyudie, A.; Ahmed, M. IImproved Proportional-Integral Coordinated MPPT Controller with Fast Tracking Speed for Grid-Tied PV Systems under Partially Shaded Conditions. *Sustainability* **2021**, *13*, 830. [CrossRef]
108. Pandiyan, P.; Saravanan, S.; Prabaharan, N.; Tiwari, R.; Chinnadurai, T.; Babu, N.; Hossain, E. Implementation of Different MPPT Techniques in Solar PV Tree under Partial Shading Conditions. *Sustainability* **2021**, *13*, 7208. [CrossRef]
109. Patel, H.; Agarwal, V. Maximum power point tracking scheme for PV systemsoperating under partially shaded conditions, industrial electronics. *IEEE Trans.* **2008**, *55*, 1689–1698.
110. Carannante, G.; Fraddanno, C.; Pagano, M.; Piegari, L. Experimental performanceof MPPT algorithm for photovoltaic sources subject to in homogeneous insolation. *IEEE Trans. Ind. Electron.* **2009**, *56*, 7. [CrossRef]
111. Koutroulis, E.; Blaabjerg, F. A New Technique for Tracking the Global Maximum Power Point of PV Arrays Operating Under Partial-Shading Conditions. *IEEE J. Photovoltaics* **2012**, *2*, 184–190. [CrossRef]
112. Kobayashi, K.; Takano, I.; Sawada, Y. A study of a two stage maximum power point tracking control of a photovoltaic system under partially shaded insolation conditions. *Sol. Energy Mater. Sol. Cells* **2006**, *90*, 2975–2988. [CrossRef]
113. Young-Hyok, J.; Doo-Yong, J.; Jun-Gu, K.; Jae-Hyung, K.; Tae-Won, L.; Chung-Yuen, W. A Real maximum power point tracking method for mismatching compensation in PV array under partially shaded conditions, power electronics. *IEEE Trans.* **2011**, *26*, 1001–1009.
114. Lie, M.; Yaojie, S.; Yandan, L.; Zhifeng, B.; Liqin, T.; Jieqiong, S. A high performance MPPT control method. In Proceedings of the 2011 International Conference on Materials for Renewable Energy & Environment (ICMREE), Shanghai, China, 20–22 May 2011; pp. 195–199.
115. Dhople, S.V.; Ehlmann, J.L.; Davoudi, A.; Chapman, P.L. Multiple-input boost converter to minimize power losses due to partial shading in photovoltaic modules. In Proceedings of the 2010 IEEE Energy Conversion Congress and Exposition (ECCE), Atlanta, GA, USA, 12–16 September 2010; pp. 2633–2636.
116. Kiran, S.R.; Basha, C.H.H.; Singh, V.P.; Dhanamjayulu, C.; Prusty, B.R.; Khan, B. Reduced Simulative Performance Analysis of Variable Step Size ANN Based MPPT Techniques for Partially Shaded Solar PV Systems. *IEEE Access* **2022**, *10*, 48875–48889. [CrossRef]

117. Roy, R.B.; Rokonuzzaman, M.; Amin, N.; Mishu, M.K.; Alahakoon, S.; Rahman, S.; Mithulananthan, N.; Rahman, K.S.; Shakeri, M.; Pasupuleti, J. A Comparative Performance Analysis of ANN Algorithms for MPPT Energy Harvesting in Solar PV System. *IEEE Access* **2021**, *9*, 102137–102152. [CrossRef]
118. Allahabadi, S.; Iman-Eini, H.; Farhangi, S. Fast Artificial Neural Network Based Method for Estimation of the Global Maximum Power Point in Photovoltaic Systems. *IEEE Trans. Ind. Electron.* **2022**, *69*, 5879–5888. [CrossRef]
119. Sheraz, M.; Abido, M.A. An Efficient MPPT Controller Using Differential Evolution and Neural Network. In Proceedings of the 2012 IEEE International Conference on Power and Energy (PECon), Kota Kinabalu, Malaysia, 2–5 December 2012; pp. 378–383.
120. Padmanaban, S.; Dhanamjayulu, C.; Khan, B. Artificial Neural Network and Newton Raphson (ANN-NR) Algorithm Based Selective Harmonic Elimination in Cascaded Multilevel Inverter for PV Applications. *IEEE Access* **2021**, *9*, 75058–75070. [CrossRef]
121. Alabedin, A.M.; El-Saadany, E.F.; Salama, M.M. Maximum power point tracking for Photovoltaic systems using fuzzy logic and artificial neural networks. In Proceedings of the 2011 IEEE Power and Energy Society General Meeting, Detroit, MI, USA, 24–18 July 2011; pp. 1–9.
122. Jinbang, X.; Anwen, S.; Cheng, Y.; Wenpei, R.; Xuan, Y. ANN based on IncCond algorithm for MPP tracker. In Proceedings of the 2011 Sixth International Conference on Bio-Inspired Computing: Theories and Applications (BIC-TA), Penang, Malaysia, 27–29 September 2011; pp. 129–134.
123. Islam, M.A.; Kabir, M.A. Neural network based maximum power point tracking of photovoltaic arrays. In Proceedings of the TENCON 2011–2011 IEEE Region 10 Conference, Bali, Indonesia, 21–24 November 2011; pp. 79–82.
124. Ishaque, K.; Salam, Z. A review of maximum power point tracking techniques of PV system for uniform insolation and partial shading condition. *Renew. Sustain. Energy Rev.* **2013**, *19*, 475–488. [CrossRef]
125. Bendib, B.; Belmili, H.; Krim, F. A survey of the most used MPPT methods: Conventional and advanced algorithms applied for photovoltaic systems. *Renew. Sustain. Energy Rev.* **2015**, *45*, 637–648. [CrossRef]
126. Ali, M.N.; Mahmoud, K.; Lehtonen, M.; Darwish, M.M.F. An Efficient Fuzzy-Logic Based Variable-Step Incremental Conductance MPPT Method for Grid-Connected PV Systems. *IEEE Access* **2021**, *9*, 26420–26430. [CrossRef]
127. Miyatake, M.; Veerachary, M. Maximum Power Point Tracking of Multiple Photovoltaic Arrays: A PSO Approach. *IEEE Trans. Aerosp. Electron. Syst.* **2011**, *47*, 367–380. [CrossRef]
128. Ibrahim, A.W.; Shafik, M.B.; Ding, M.; Sarhan, M.A.; Fang, Z.; Alareqi, A.G.; Almoqri, T.; Al-Rassas, A.M. PV maximum power-point tracking using modified particle swarm optimization under partial shading conditions. *Chin. J. Electr. Eng.* **2020**, *6*, 106–121. [CrossRef]
129. Liu, Y.-H.; Huang, S.-C.; Huang, J.-W.; Liang, W.-C. A Particle Swarm Optimization-Based Maximum Power Point Tracking Algorithm for PV Systems Operating Under Partially Shaded Conditions. *IEEE Trans. Energy Convers.* **2012**, *27*, 1027–1035. [CrossRef]
130. Ballaji, A.; Dash, R.; Subburaj, V.; Kalvakurthi, J.R.; Swain, D.; Swain, S.C. Design & Development of MPPT Using PSO with Predefined Search Space Based on Fuzzy Fokker Planck Solution. *IEEE Access* **2022**, *10*, 80764–80783.
131. Kashif, I.; Zainal, S. An improved Particle Swarm Optimization (PSO)–based MPPT for PV with reduced steady-state oscillation. *IEEE Trans. Power Electron.* **2013**, *27*, 3627–3638.
132. Millah, I.S.; Chang, P.C.; Teshome, D.F.; Subroto, R.K.; Lian, K.L.; Lin, J.-F. An Enhanced Grey Wolf Optimization Algorithm for Photovoltaic Maximum Power Point Tracking Control under Partial Shading Conditions. *IEEE Open J. Ind. Electron. Soc.* **2022**, *3*, 392–408. [CrossRef]
133. Rajkumar, M.V.; Mahakumar, M.; Manojkumar, M.; Hemaraj, M.; Kumaravel, E. A New DC-DC Converter topology with grey wolf MPPT algorithm for photovoltaic system. *Int. J. Emerg. Technol. Eng. Res.* **2017**, *5*, 54–59.
134. Rezaei, H.; Bozorg-Haddad, O.; Chu, X. Grey Wolf Optimization (GWO) Algorithm. In *Advanced Optimization by Nature-Inspired Algorithms*; Springer: Berlin/Heidelberg, Germany, 2018; pp. 81–91.
135. Qais, M.H.; Hasanien, H.M.; Alghuwainem, S. A Grey Wolf Optimizer for Optimum Parameters of Multiple PI Controllers of a Grid-Connected PMSG Driven by Variable Speed Wind Turbine. *IEEE Access* **2018**, *6*, 44120–44128. [CrossRef]
136. Guo, K.; Cui, L.; Mao, M.; Zhou, L.; Zhang, Q. An Improved Gray Wolf Optimizer MPPT Algorithm for PV System with BFBIC Converter Under Partial Shading. *IEEE Access* **2020**, *8*, 103476–103490. [CrossRef]
137. Kumar, M.; Panda, K.P.; Rosas-Caro, J.C.; Valderrabano-Gonzalez, A.; Panda, G. Comprehensive Review of Conventional and Emerging Maximum Power Point Tracking Algorithms for Uniformly and Partially Shaded Solar Photovoltaic Systems. *IEEE Access* **2023**, *11*, 31778–31812. [CrossRef]
138. OmPrakash, V.; Deepti, A.; Tejna, P. Opposition and dimensional based modified firefly algorithm. *Exp. Syst. Appl.* **2016**, *44*, 168–176.
139. Huang, Y.P.; Huang, M.Y.; Ye, C.E. A Fusion Firefly Algorithm with Simplified Propagation for Photovoltaic MPPT Under Partial Shading Conditions. *IEEE Trans. Sustain. Energy* **2020**, *11*, 2641–2652. [CrossRef]
140. Safarudin, Y.M.; Priyadi, A.; Purnomo, M.H.; Pujiantara, M. Maximum power point tracking algsorithm for photovoltaic system under partial shaded condition by means updating β firefly technique. In Proceedings of the 6th International Conference on Information Technology and Electrical Engineering (ICITEE 2014), Yogyakarta, Indonesia, 7–8 October 2014; pp. 1–4. [CrossRef]
141. Yun-Chia, L.; Smith, A.E. An Ant Colony Optimization Algorithm for the Redundancy Allocation Problem (RAP). *IEEE Trans. Reliab.* **2004**, *53*, 417–423.

142. Lian, L.; Maskell, D.L.; Patra, J.C. A novel ant colony optimization-based maximum power point tracking for photovoltaic systems under partially shaded conditions. *Energy Build.* **2013**, *58*, 227–236.
143. Katche, M.L.; Makokha, A.B.; Zachary, S.O.; Adaramola, M.S. A Comprehensive Review of Maximum Power Point Tracking (MPPT) Techniques Used in Solar PV Systems. *Energies* **2023**, *16*, 2206. [CrossRef]
144. Sundareswaran, K.; Vigneshkumar, V.; Sankar, P.; Simon, S.P.; Nayak, P.S.R.; Palani, S. Development of an Improved P&O Algorithm Assisted through a Colony of Foraging Ants for MPPT in PV System. *IEEE Trans. Ind. Inform.* **2016**, *12*, 187–200.
145. Ghasemi, M.A.; Foroushani, H.M.; Parniani, M. Partial Shading Detection and Smooth Maximum Power Point Tracking of PV Arrays under PSC. *IEEE Trans. Power Electron.* **2016**, *31*, 6281–6292. [CrossRef]
146. Ramin, R. Cuckoo Optimization Algorithm. *Appl. Soft Comput.* **2011**, *11*, 5508–5518.
147. Liu, L.; Meng, X.; Liu, C. A review of maximum power point tracking methods of PV power system at uniform and partial shading. *Renew. Sustain. Energy Rev.* **2016**, *53*, 1500–1507. [CrossRef]
148. Nugraha, D.A.; Lian, K.L.; Suwarno. A Novel MPPT Method Based on Cuckoo Search Algorithm and Golden Section Search Algorithm for Partially Shaded PV System. *Can. J. Electr. Comput. Eng.* **2019**, *42*, 173–182. [CrossRef]
149. Rao, R.V.; Saroj, A. A self-adaptive multi-population based Jaya algorithm for engineering optimization. *Swarm Evol. Comput.* **2017**, *37*, 1–26. [CrossRef]
150. Warid, W.; Hizam, H.; Mariun, N.; Abdul-Wahab, N.I. Optimal Power Flow Using the Jaya Algorithm. *Energies* **2016**, *9*, 678. [CrossRef]
151. Bhukya, L.; Annamraju, A.; Nandiraju, S. A novel maximum power point tracking technique based on Rao-1 algorithm for solar PV system under partial shading conditions. *Int. Trans. Electr. Energy Syst.* **2021**, *31*, e13028. [CrossRef]
152. Bhukya, L.; Kedika, N.R.; Salkuti, S.R. Enhanced Maximum Power Point Techniques for Solar Photovoltaic System under Uniform Insolation and Partial Shading Conditions: A Review. *Algorithms* **2022**, *15*, 365. [CrossRef]
153. Mohapatra, A.; Nayak, B.; Das, P.; Mohanty, K.B. A review on MPPT techniques of PV system under partial shading condition. *Renew. Sustain. Energy Rev.* **2017**, *80*, 854–867. [CrossRef]
154. Lian, K.L.; Jhang, J.H.; Tian, I.S. A Maximum Power Point Tracking Method Based on Perturb-and-Observe Combined with Particle Swarm Optimization. *IEEE J. Photovolt.* **2014**, *4*, 626–633. [CrossRef]
155. Manickam, C.; Raman, G.R.; Raman, G.P.; Ganesan, S.I.; Nagamani, C. A Hybrid Algorithm for Tracking of GMPP Based on P&O and PSO with Reduced Power Oscillation in String Inverters. *IEEE Trans. Ind. Electron.* **2016**, *63*, 6097–6106.
156. Seyedmahmoudian, M.; Rahmani, R.; Mekhilef, S.; Maung Than, O.A.; Stojcevski, A.; Soon, T.K.; Ghandhari, A.S. Simulation and hardware implementation of new maximum power point tracking technique for partially shaded PV system using hybrid DEPSO method. *IEEE Trans. Sustain. Energy* **2015**, *6*, 850–862. [CrossRef]
157. Lyden, S.; Haque, M.E. Maximum Power Point Tracking techniques for photovoltaic systems: A comprehensive review and comparative analysis. *Renew. Sustain. Energy Rev.* **2015**, *52*, 1504–1518. [CrossRef]
158. Dhople, S.V.; Bell, R.; Ehlmann, J.; Davoudi, A.; Chapman, P.L.; Domínguez-García, A.D. A global maximum power point tracking method for PV module integrated converters. In Proceedings of the IEEE Energy Conversion Congress and Exposition (ECCE), Raleigh, NC, USA, 15–20 September 2012; pp. 4762–4767.
159. Elasser, A.; Agamy, M.; Sabate, J.; Steigerwald, R.; Fisher, R.; Harfman-Todorovic, M. A comparative study of central and distributed MPPT architectures for megawatt utility and large scale commercial photovoltaic plants. In Proceedings of the IEEE Industrial Electronics Society Conference (IECON), Glendale, AZ, USA, 7–10 November 2010; pp. 2753–2758.
160. Poshtkouhi, S.; Palaniappan, V.; Fard, M.; Trescases, O. A General Approach for Quantifying the Benefit of Distributed Power Electronics for Fine Grained MPPT in Photovoltaic Applications Using 3-D Modeling. *IEEE Trans. Power Electron.* **2012**, *27*, 4656–4666. [CrossRef]
161. Shenoy, P.S.; Kim, K.A.; Johnson, B.B.; Krein, P.T. Differential power processing for increased energy production and reliability of photovoltaic systems. *IEEE Trans. Power Electron.* **2013**, *28*, 2968–2979. [CrossRef]
162. Olalla, C.; Deline, C.; Clement, D.; Levron, Y.; Rodriguez, M.; Maksimovic, D. Performance of Power-Limited Differential Power Processing Architectures in Mismatched PV Systems. *IEEE Trans. Power Electron.* **2015**, *30*, 618–631. [CrossRef]
163. Villa, L.F.L.; Tien-Phu, H.; Crebier, J.C.; Raison, B. A Power Electronics Equalizer Application for Partially Shaded Photovoltaic Modules. *IEEE Trans. Ind. Electron.* **2013**, *60*, 1179–1190. [CrossRef]
164. Salam, Z.R.; Amli, M.Z. A simple circuit to improve the power yield of PV array during partial shading. In Proceedings of the IEEE Energy Conversion Congress and Exposition (ECCE), Raleigh, NC, USA, 15–20 September 2012; pp. 1622–1626.

sustainability

Article

Innovative Methodologies for Higher Global MPP of Photovoltaic Arrays under PSCs: Experimental Validation

Belqasem Aljafari [1], Rupendra Kumar Pachauri [2], Sudhakar Babu Thanikanti [3,*] and Bamidele Victor Ayodele [4,*]

[1] Department of Electrical Engineering, College of Engineering, Najran University, Najran 11001, Saudi Arabia; bhaljafari@nu.edu.sa
[2] Electrical and Electronics Engineering Department, School of Engineering, University of Petroleum and Energy Studies, Dehradun 248007, India; rpachauri@ddn.upes.ac.in
[3] Department of Electrical and Electronics Engineering, Chaitanya Bharathi Institute of Technology, Hyderabad 500075, India
[4] Department of Chemical Engineering, Universiti Teknologi PETRONAS, Seri Iskandar 32610, Malaysia
* Correspondence: sudhakarbabu@ieee.org (S.B.T.); bamidele.ayodele@utp.edu.my (B.V.A.)

Abstract: Partial shading conditions (PSCs) are responsible for the root causes of photovoltaic (PV) system performance deprivation such as hotspots (damaged PV cells), mismatch power losses and multiple power maxima. Recently, PV array reconfiguration strategies have proven to be beneficial in improving PV system performance and achieving improved shade dispersion properties. This research analyzes the improved Su-Do-Ku (I-SDK) PV array configuration in order to counteract the shading effect. This approach implements a 6 × 6 size PV array configuration and performance evaluation under different realistic shading scenarios. The performance of the I-SDK configuration is assessed and compared to that of the total-cross-tied (TCT) and Su-Do-Ku (SDK) arrangements. The performance indices such as power loss (PL), power at global maximum power point (GMPP), fill-factor (FF), performance ratio (PR), power enhancement (PE) and execution ratio (ER) are analyzed to show comprehensive comparison. An experimental analysis confirms the MATLAB/Simulink findings, demonstrating that the I-SDK configuration outperforms both the TCT and SDK array setups. The GMPP values of 143.5 W, 141.7 W, 138.1 W and 129.3 W also show the superiority of I-SDK during four shading instances compared to conventional SP, TCT, SDK and SM arrangements. Moreover, under similar PSCs, higher %FF (74.61%, 76.10%, 77.1%, 75.92%) and lower PL (36.7 W, 38.5 W, 42.1 W, 50.9 W) support the adoptability of I-SDK for experimental validation/commercial viability.

Keywords: power loss; photovoltaic system; shading scenarios; fill factor; improved Su-Do-Ku; mismatch loss

Citation: Aljafari, B.; Pachauri, R.K.; Thanikanti, S.B.; Ayodele, B.V. Innovative Methodologies for Higher Global MPP of Photovoltaic Arrays under PSCs: Experimental Validation. *Sustainability* **2023**, *15*, 11852. https://doi.org/10.3390/su151511852

Academic Editors: Domenico Mazzeo and Cristina Ventura

Received: 9 May 2023
Revised: 20 June 2023
Accepted: 27 July 2023
Published: 1 August 2023

1. Introduction

As a climate-sustainable solution, the world needs urgent and rapid incorporation of renewable energy (RE) into the global energy scenario. RE sources are commonly known as clean energy solutions, gaining the deep attention of energy users in commercial and domestic applications. The PV energy conversion method has received immense attention from researchers in recent days [1]. Concerns about the worldwide energy problems and the threat of climate change posed by existing energy sources have spurred the research for alternative energy sources. The most popular renewable energy source is solar PV, which is nonpolluting and requires no maintenance [2].

During the first half of 2020, thirteen countries pledged the largest amount of new RE ever, nearly 50 GW, to be installed in the year span of 2021–2024. Global demand in 2021 is expected to be 25% higher than in 2020 [3,4]. The researchers performed different studies to establish an effective and stable conversion of solar PV energy. Module mismatching and PSCs are two key contributors to PV system losses. Under these conditions, the PV system

(P-V and I-V characteristics) becomes more complex and has many power peaks. The amount of energy made by a PV array is greatly reduced when it is in the shade [5–7]. To boost the efficiency of PV systems, a number of modules are coupled in specified combinations. An evaluation of the most recent research papers is conducted in this work in order to determine the research gap. The many varieties of PV array configurations' performance, reliability, precision, resilience, efficiency and operation are all investigated. There are several things that can cause PSCs, such as passing clouds, high-rise buildings, telecommunications towers, adjacent trees, dead leaves, and so on.

The PV power output is reduced, allowing shaded modules to waste a significant percentage of the power produced by unshaded modules, resulting in hotspots that might eventually destroy the PV cell/module [8,9]. When designing an array, several PV modules are arranged in series and parallel to satisfy the load power demand. Parallel and series configurations of PV modules are used to meet a PV array's power requirements. Traditional and game puzzle-based reconfiguration approaches such as series, series-parallel (SP), bridge-link (BL), total cross-tied (TCT), honey-comb (HC), Latin square (LS), magic square (MS) and Su-Do-Ku (SDK) puzzles have been adopted to design PV array systems.

Because of the way these algorithms are designed, the operative point is the earliest peak in the PV characteristic, which may or may not be a global power peak. Several advancements in the literature have been recorded to promote these algorithms in order to find the global power peak under PSCs. This is accomplished by adding more stages to the algorithm, making it more complex and restricting monitoring speed. Recently, an improved SDK-puzzle-based algorithm for MPP monitoring was discovered to yield good results.

1.1. Literature Review

The paper is a survey of the most recent research publications with the goal of identifying important research gaps. [10–31]. We examined several PV array topologies for consistency, stability and easy implementation.

The authors in [10] observed the effects of PV faults on the power grid under various faulty conditions using the MATLAB/Simulink-based study. A comprehensive analysis is investigated in terms of power at GMPP with selected SP, BL, TCT and reconfigured methodology (RM) configurations as 3.99 kW, 4.01 kW, 4.6 kW and 5.02 kW. In [11], an auto-reconfiguration approach is performed through switching-based PV array electrical connections from SP to TCT under PSCs. For shading pattern-1, the power at GMPP of conventional setups (SP and TCT) is 25.22 W and 26.19 W, respectively. In addition, PL values are observed for SP and TCT configurations under similar shadowing conditions as 3.75 W and 2.78 W. In [12], a detailed study is carried out with various PV array interconnections, including SP, HC and TCT. In the adopted PV array interconnections, power at GMPP of existing setups (SP, HC and TCT) is found as 4.39 kW, 4.44 kW and 4.72 kW, respectively. It is observed that the TCT configuration has a higher rating compared to conventional configurations. The authors of [13] presented a comprehensive study to mitigate shading effects, a novel LS-puzzle-based configuration compared to the TCT configuration. Locations of GMPP for shading pattern-1 exist as 1976 W and 2279 W. During all shading cases, Latin square—TCT (LS-TCT) has the best performance compared to TCT configuration. The Futoshiki puzzle (FP) is investigated in [14] and obtained results show that the power produced by the FP configuration is highest and minimized ML under various shading instances. Power generated as 40.07 W, 53.93 W and 53.93 W in TCT, EAR and Futoshiki configurations for a short wide (SW) shading scenario. For a long narrow (LN) shading condition, the generated power at GMPP is 59.41 W, 60.56 W and 60.56 W. In [15], the authors developed a novel method to reconfigure the PV module's interconnection and compared it to the TCT configuration. The behavior of the P-V characteristic generated from the reconfigured PV array is found to be smoother than that of TCT-configured PV arrays under-considered PSCs. The half-reconfigured PV array (HRPVA) and full-reconfigured PV array (FRPVA) have increased the coherence between the obtained GMPP. With a performance ratio of 0.93, the generated power increased by 14.75% in HRPVA.

Furthermore, with a unity performance ratio, FRPVA enhanced generated power by 23.3%. The proposed MS configuration in [16] is tested using a 3 × 3 PV array coupled to the TCT pattern under SN, LW, LN and SW shadowed conditions. TCT and MS configurations are investigated under SN shading scenarios, with GMPP locations of 1.8576 kW and 2.3734 kW being observed, respectively. For the shading pattern subjected to LN, the GMPP locations are assessed as 2.4278 kW and 2.8849 kW, respectively. The authors of [17] investigated the PV array arrangements, e.g., SP, TCT, BL, HC and hybrid series parallel-total cross tied (SP–TCT) and BL–TCT, and game-puzzle-based configurations such as nonsymmetrical-1, 2 (NS-1 and NS-2) are reported. The shading pattern-3 subjected to TCT array configuration has power at GMPP as 5260 W, 4545 W, 4332 W, 4182 W for four distinguished shading cases, respectively. In a similar shading scenario, the NS-2 configuration has power levels at GMPP locations of 5260 W, 5013 W, 4811 W and 4182 W, which are higher than conventional methods.

In [18], the performance of PV array designs of 4 × 4 sizes, i.e., MATLAB/Simulink is used to investigate TCT, hybrid SP-TCT, BL-TCT, BL-HC and MS, RSP-TCT, RBL-TCT and RBL-HC. The power at GMPP is observed as 2279 W, 1976 W, 2279 W, 1976 W, 2197 W, 1976 W, 2233 W, 1976 W and 2255 W for the respective topologies. The authors of [19] analyzed TCT configuration for performance improvement under PSCs. The power at GMPP under shading case-4 is observed as 377.2 W, 468.9 W and 468.9 W for TCT, Optimal TCT and Novel TCT configurations, respectively. Performance indices such as FF are observed at 40.46%, 60.26% and 60.28%, respectively. The authors of [20] discussed the result and test conducted on a 4 × 5 size PV array configuration with distinguishing shading scenarios. The power at GMPP for shading scenarios is observed as 797.24 W, 823.75 W, 819.22 W, 850.25 W, 851.57 W for SP, BL, HC, TCT and Novel PV array topologies. Furthermore, the calculated PL are found as 1.88%, 4.13%, 2.32%, 2.59% and 1.30%, respectively. The authors of [21] proposed a shade dispersion scheme (SDS) arrangement compared to other conventional PV module interconnections such as SP, BL and TCT under realistic shading patterns. The power at GMPP of PV array for SP, BL, TCT and SDS configurations is 1644.36 W, 1689.84 W, 1721.55 W and 1746.33 W, respectively. Also, results in PL for the same are 301.40 W, 256.12 W, 224.41 W and 199.63 W. The SDS configuration has a higher performance among all the PV array configurations. In [22], the recently developed cross diagonal view (CDV) configuration is proposed and compared to the conventional SP and TCT configuration for a 9 × 9 size PV array module. The power at GMPP is observed for SP, SDK, TCT and TCT-CDV as 6307.5 W, 6307.5 W, 4861.2 W and 7492.5 W. In [23], new physical PV array setups are introduced to reduce the shading impact and generated power at GMPP for SP, TCT and modified TCT (M-TCT) configurations under the considered shading conditions 174.60 W, 185.33 W, 215.74 W and 523.80 W, respectively.

Interconnection methods are proposed in [24] to reduce PL during PSCs. The TCT and proposed configurations have increased the power output by 7.8% and 6.9% compared to the conventional SP arrangement. The GMPP locations are found as 4065.3 W, 4419.3 W and 4393.1 W for SP, TCT and proposed topologies, respectively. The authors of [25] calculated the comparative result in terms of maximum power for parallel and series configurations under distinguished shading effects as 544.2 W, 492.45 W, 381.1 W and 446.5 W. For minimizing the shadowing effect, the parallel configuration is convenient. In [26], SDK arrangement was compared to TCT and validated experimentally on a 5 × 5 size array. Power of 266.4 W and 280.2 W was generated for the proposed Su-Do-Ku and TCT configuration subjected to two types of PSCs. People who came up with this idea say it makes P-V curves more even and eliminates the MPPT algorithm as well as the financial risk. In [27], the authors proposed a column-index-based topology for PV reconfiguration scheme. The generated power at GMPP for PV array topologies was such as proposed, DS, TCT and SP are 5338 W, 5101 W, 5066 W and 4815 W under LN-based shading pattern. The proposed technique was tested on a 9 × 9 size PV array and the results showed that it increased the global maximum power when compared to DS, TCT and SP array configurations.

In [28], the I-SDK arrangement enhanced the global power peaks (28.6%, 22.1%, 22.8%, 17.2%, 6.2% and 5.2%) compared to conventional (SP, BL, HC, TCT) and puzzle-based (SDK and optimal SDK) PV array setups under a realistic shading scenario. The authors of [29] designed 6 × 4 size PV array setups for performance investigation under shading scenarios. The SP, TCT, BL, HC, BL-HC, BL-TCT and SP-TCT arrangements were adopted for performance investigation and the generated power at GMPP was observed as 2177 W, 2394 W, 235.2 W, 2235 W, 2187 W, 2389 W, 2306 W and 2298 W, respectively. Moreover, TCT and BL-TCT configurations have shown a better response in terms of the highest power at GMPP, minimum PL and improved FF. The authors of [30] reviewed the performance of the proposed SRBL-TCT configuration compared to SP, TCT, BL, HC and BL-TCT results obtained under the realistic moving clouds phenomenon. At each of the five shading instants, the obtained GMPP location of the SRBL-TCT configuration was found to be superior to the SP, TCT, BL, HC and BL-TCT configurations. The obtained power at GMPP for the BL-TCT PV array configuration subjected to all five shading scenarios was 39.57 W, 23.89 W, 38.85 W, 38.05 W, 26.5 W. In the SRBL-TCT PV array configuration, the power at GMPP for all five considered shading scenarios as 44.31 W, 38.63 W, 42.44 W, 41.04 W and 40.93 W, respectively. The skyscraper methodology was simulated along with its application to a 9 × 9 size PV array and was matched to the TCT, DS and SDK arrangement using MATLAB/Simulink in [31]. Under PSCs, the PL for skyscraper, TCT, DS and SDK was 0.305%, 0.35%, 0.325% and 0.316%.

1.2. Novelty of Work

The I-SDK PV array enhances PV system performance in terms of a higher GMPP, FF and minimal PL. Furthermore, the I-SDK PV array arrangement is a viable solution for a number of reasons:

- The suggested I-SDK configuration improves performance compared to standard setups because it more evenly disperses the shadow impact throughout the PV array.
- The utility and functionality of the proposed I-SDK configuration are tested by a comprehensive experimental study under distinctive shading scenarios and validated the MATLAB/Simulink results, i.e., GMPP locations, PL, ER and FF.

2. PV Modelling and Array System

2.1. PV Cell Modeling

The PV module's mathematical analysis was used for MATLAB/Simulink modeling of the PV cell and electrical equivalent circuit of a single diode model, as shown in Figure 1.

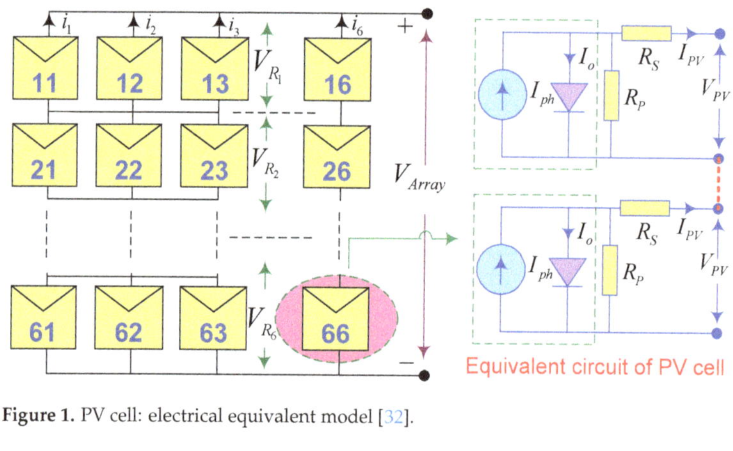

Figure 1. PV cell: electrical equivalent model [32].

The output current of a solar PV cell is given in Equation (1) [19] as

$$I_{PV} = N_{pp}\left\{I_{PV_n} - I_O\left[\exp\left(\frac{V_{PV} + I_{PV}R_S}{V_t N_{SS}}\right) - 1\right]\right\} - \left(\frac{V_{PV} + I_{PV}R_S}{R_P}\right) \quad (1)$$

where 'I_{PV}'—PV cell current, 'R_p'—parallel resistance, 'S'—series resistance, 'I_O'—reverse saturation current, 'V_t'—thermal voltage, 'N_{SS}' and 'N_{pp}' are the number of cells connected in systematic order (series and parallel).

2.2. PV Array Configurations: Conventional

(a) Series-parallel configuration

In a PV array, a finite number of PV modules are arranged in parallel strings to increase the voltage and current to meet the required load power demand. The electrical arrangements are used to show the 6 × 6 size SP configuration in Figure 2 as

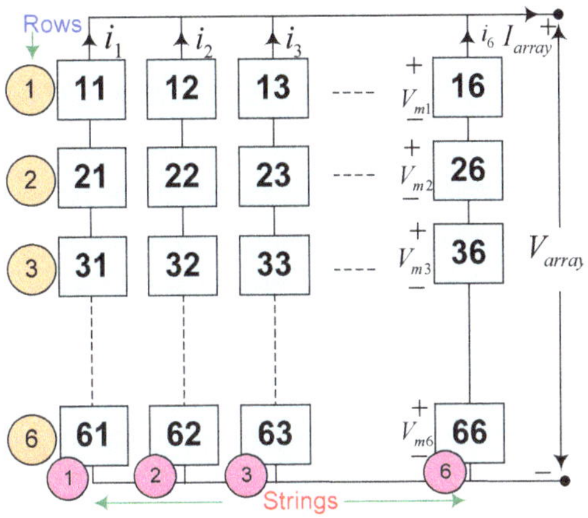

Figure 2. SP arrangement of PV array.

(b) Total-cross-tied configuration

The TCT electrical arrangement is an extended modification of the SP configuration by means of fixing ties across the parallel strings. This cross-tied-based modification is responsible for enhancing the current through each parallel string and constant voltage during PSCs. Mathematically; voltage analysis is expressed for the PV array in Equation (2).

$$\mathrm{V_{array}} = \sum\nolimits_{\mathrm{k}=1}^{6} \mathrm{V_{mk}} \quad (2)$$

where $\mathrm{V_{mk}}$ refers to the voltage (maximum) at the $\mathrm{k^{th}}$ row. Each string of the PV modules is linked in parallel; therefore, the total current drawn by the PV array is the sum of the individual currents drawn by each module in the array. Furthermore, a mathematical approach is applied to each node using Kirchhoff's current law. As a consequence, the array current (I_{array}) can be expressed in Equation (3) as

$$I_{array} = \sum\nolimits_{q=1}^{6}\left(I_{kq} - I_{(k+1)q}\right) = 0,\ p = 1,2,3,\ldots 9 \quad (3)$$

where k and q are the number of rows and columns in the considered 6 × 6 size PV array. Figure 3 depicts the TCT array's electrical configuration of PV modules.

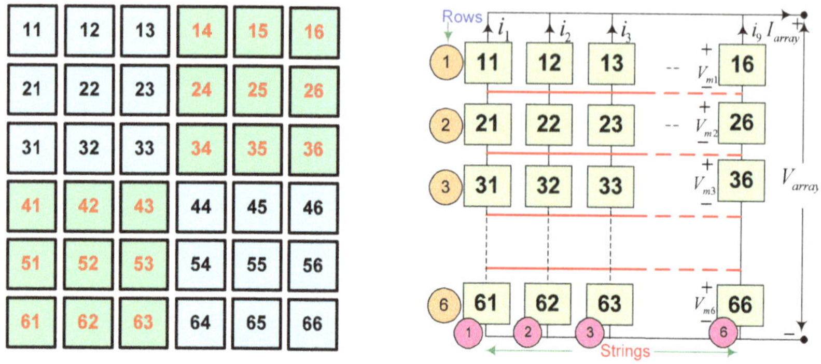

(**a**) (**b**)

Figure 3. PV module arrangements in TCT configuration (6 × 6 size). (**a**) Nomenclature, (**b**) placement of PV module.

2.3. Game-Theory-Based PV Array Configurations

(a) SDK and I-SDK configurations

As the shadow is diffused over the array, the game-theory-based SDK layout guarantees that mismatch losses are minimized. The wiring connections are completed once the panels have been properly arranged and they stay unmodified. This decreases computing difficulties while also preventing the overuse of sensors and switches [33].

The SDK change in columns 2–6 is known as the I-SDK puzzle. By repositioning the PV modules without disturbing the electrical arrangement, the suggested I-SDK layout is implemented in the TCT PV array. In this game puzzle, higher dispersion is found based on the optimal placement of all the integer numbers in an array. This integer placement modification is responsible for higher shade dispersion. The SDK and I-SDK puzzles and approach are depicted in Figure 4a–d as

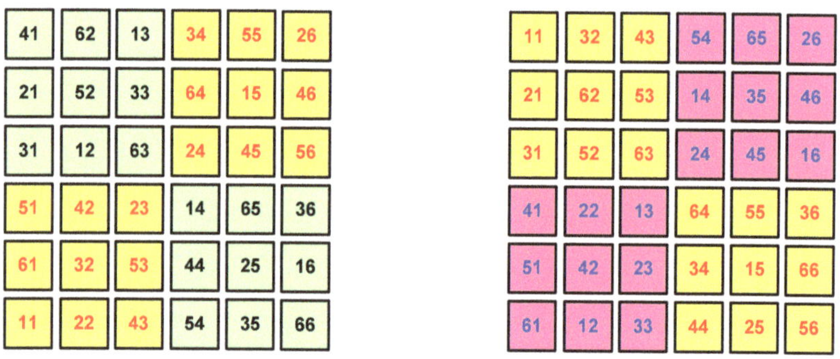

(**a**) (**b**)

Figure 4. *Cont.*

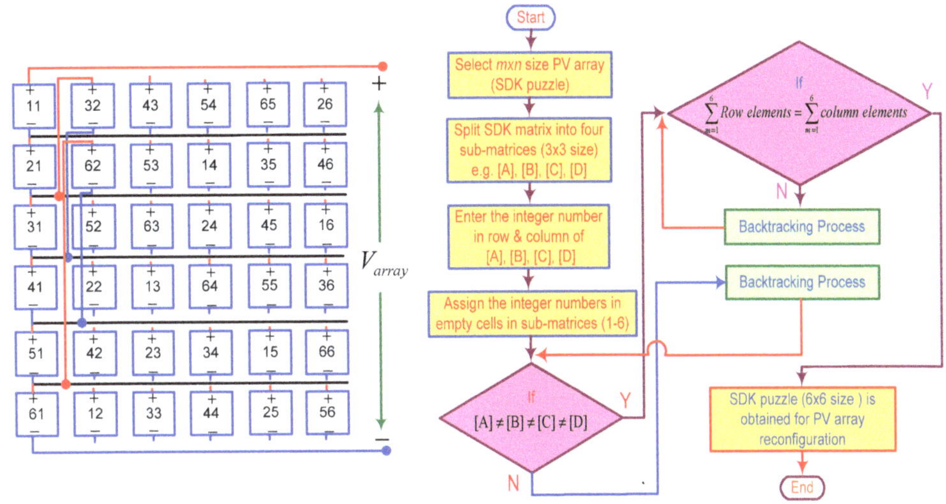

(**c**) (**d**)

Figure 4. (**a**) Number placement for I-SDK; (**b**) number placement for SDK; (**c**) electrical connections for SDK (**d**) methodology to achieve SDK game theory.

(b) Symmetric matrix-based configurations

Cyclic arrangement of integer numbers from 1 to 6 is carried out to establish the SM game-theory-based arrangement. The summation of the considered integer numbers in each row and column is found to be equal as per SM development guidelines. In addition to that, either of the diagonal elements keeps repeating within it. Figure 5a depicts all of the assets of the 6 × 6 size SM as

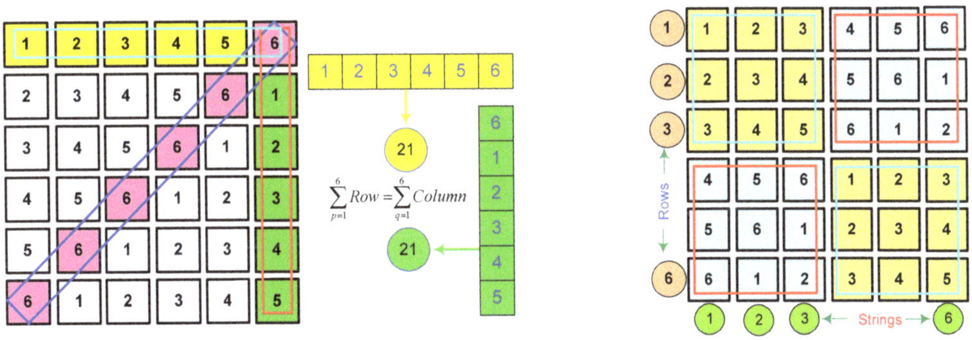

(**a**) (**b**)

Figure 5. (**a**) Row, column and single diagonal property; (**b**) repeated submatrix elements.

The summation of all the items in each particular row/column, according to SM characteristics, is 21. Furthermore, in Figure 5b, there is a repetition of 3 × 3 size square submatrices.

To represent the row-column summation rules, mathematical assumptions are made. In this context, the SM size is considered in the order of $p \times q$. Moreover, the nth element can be placed corresponding to the pth and q^{th} row-column, respectively. So, the location of the PV module (n_{pq}) in an array can be written in a generalized way and expressed in Equation (4) as

$$n_{pq}, where \begin{cases} p = no.\ of\ row & (p = 1,2,\ldots.6) \\ q = no.\ of\ column & (q = 1,2,\ldots.6) \end{cases} \quad (4)$$

In Figure 5a,b, we see the mathematical equations for four distinct cases of row-wise summing and that are accomplished using Equation (5) as follows:

$$\sum_{p=1}^{6} n_{pq}(Summation\ for\ p^{th} row) = \sum_{q=1}^{6} n_{pq}(Summation\ for\ q^{th} column) \quad (5)$$

The above Equations (4) and (5) are involved in the guidelines to achieve the 6 × 6 size SM setup, and the scientific method to establish the SM is shown in Figure 6 as

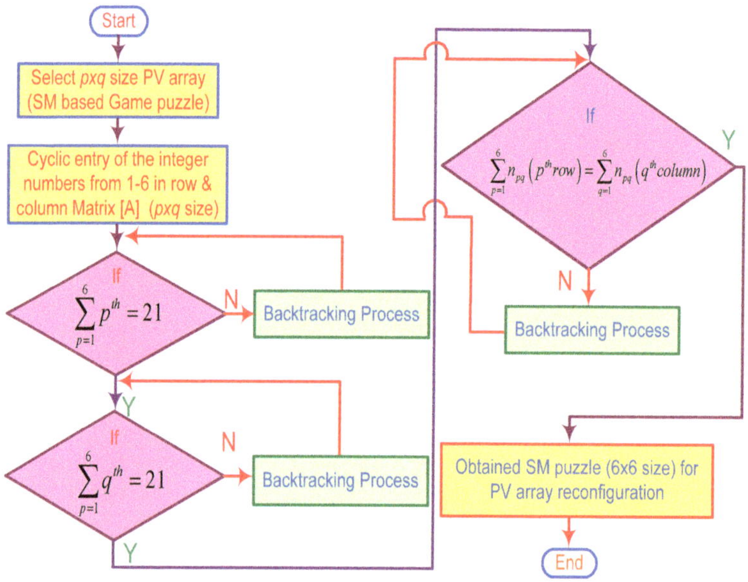

Figure 6. Methodology to achieve SM game theory.

In SM, there are six rows and columns in a 6 × 6 PV array, respectively. As per the nomenclature shown in Figure 7a, the first digit of each individual PV module depicts the row count, while the second digit depicts the column count. It is an easier nomenclature-based methodology to understand the electrical arrangements of PV modules in an array. In Figure 7b, the PV module locations are migrated using the recommended SM structure but the electrical contacts of the PV panels within PSCs remain unchanged.

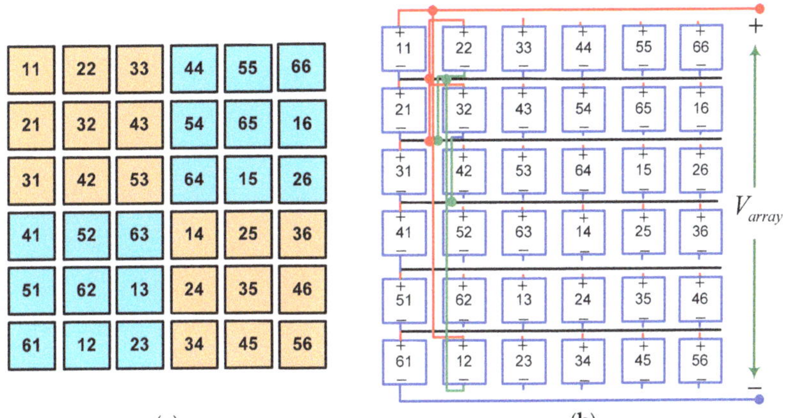

Figure 7. (**a**) Nomenclature of PV modules; (**b**) PV module arrangements for SM configuration.

2.4. Experimental Setup

MATLAB/Simulink modeling and experimentation utilize a commercially available 5 W PV module, as shown in Table 1.

Table 1. PV module specifications (Manf. Universal solar: 5 W, poly-crystalline).

Parameters	Values
Maximum power (P_m)	5 W
Maximum current (I_m)	0.52 A
Maximum voltage (V_m)	9.62 V
Open circuit voltage (V_{OC})	11.25 V
Short circuit current (I_{SC})	0.55 A

For performance validation, an experimental setup is established with 6 × 6 size PV array configurations with the assistance of an embedded-based data acquisition system (DAS) for logging real-time electrical parameters. The performance characterization in terms of I-V and P-V curves is performed for extensive performance analysis in different realistic shading scenarios. The above experimental setup comprises mainly four sections such as (a) PV system (6 × 6 size), arranged in SP, SDK and I-SDK configurations; (b) variable resistive load (Rheostat: 800 Ω, 6 A); (c) self-developed DAS. The developed experimental setup is shown in Figure 8.

Using the AT-mega microcontroller (8051), a self-designed data logger is utilized, which includes analog voltage and current sensors for electrical data investigation. Realistic electrical performance parameters in terms of voltage and current are stored in a micro-SD card. These recorded data are used for I-V and P-V characterization and rigorous investigation of performance parameters. Figure 9 also depicts the working of the constructed data logger. The code has processes for measuring electrical characteristics that have been fine-tuned based on an assessment and data gathering that use a simple algorithm.

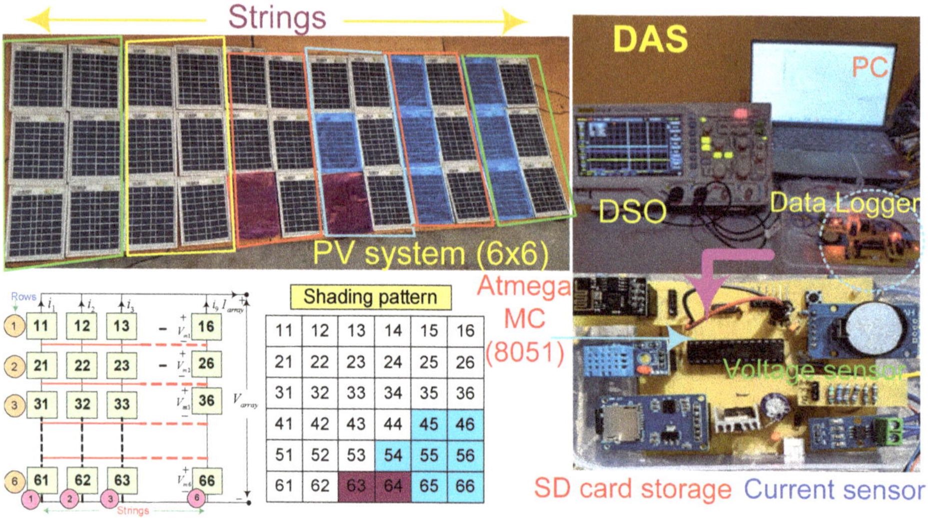

Figure 8. Developed experimental setup for comprehensive study during PSCs.

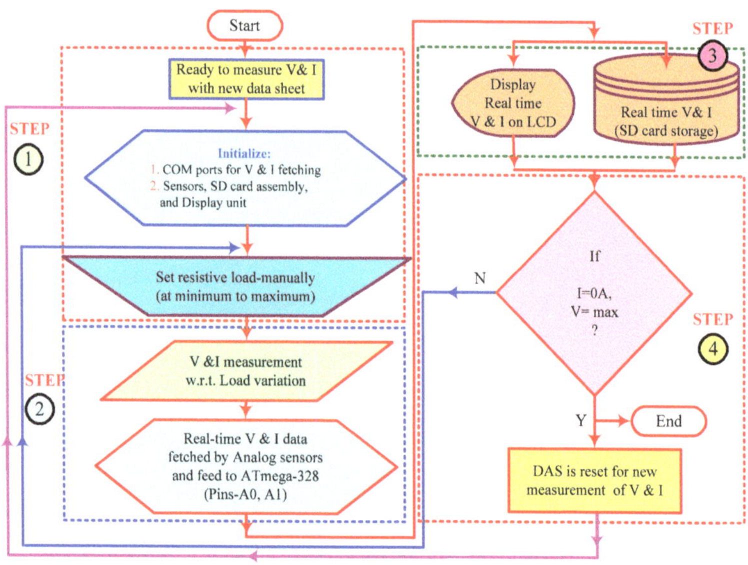

Figure 9. Flow chart for data logger operation.

3. Performance Parameters and Shading Scenarios

This unique MPP is due to P-V curves and PSCs in PV modules in an array. The MPP tracking device function refuses to give the maximum power to the load because of redundant GMPP and LMPP. In the performance assessment under standard test conditions, maximum power and voltage were found to be 180.2 W and 58.09 V, respectively. Figure 10 shows the performance characteristics under uniform irradiances as a result of the P-V and I-V curves.

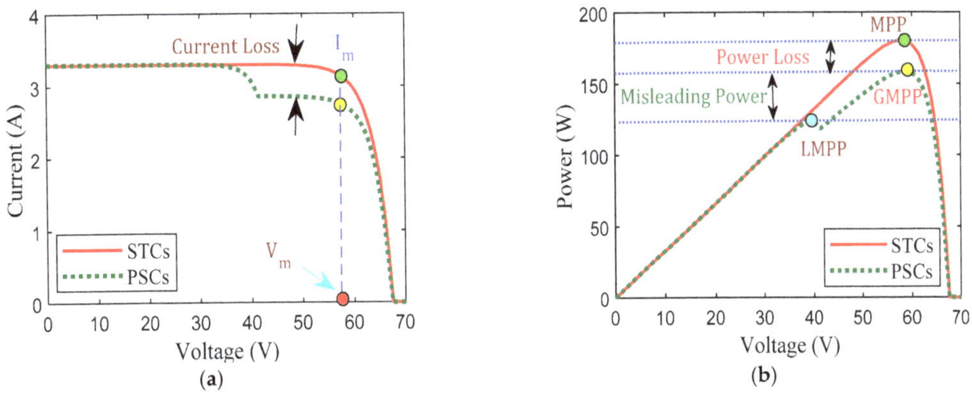

Figure 10. (**a**) I-V (**b**) P-V curves under STC and nonuniform irradiation levels.

3.1. Power and Voltage at GMPP

During the PSCs, multiple power points, i.e., GMPP and LMPP, were observed on the P-V curve. The higher power value is known as "power at GMPP (P_{GMPP})". Furthermore, at which value of voltage, the higher power at GMPP is also called the "voltage at GMPP (V_{GMPP})".

3.2. Power Mismatch Loss

The power ML is represented as ΔP_L and it is calculated based on the maximum power generated during the ideal irradiance scenario and PSCs [34]. The theoretical assessment of power ML is expressed in Equation (6) [17].

$$\%\Delta P_L = \frac{P_{uniform\ irradiance} - P_{PSC}}{P_{uniform\ irradiance}} \times 100 \tag{6}$$

3.3. Power Loss

Theoretically, the evaluation of PL is measured by the difference in maximal power produced between perfect and nonuniform irradiation. The calculated PL is expressed in Equation (7) [17] as

$$PL = P_{max}\ at\ Ideal\ Irradiance - P_{GMPP}\ at\ PSCs \tag{7}$$

3.4. Fill Factor

The FF is defined in Equation (8) [17] as the ratio of the generated GMPP at PSCs to the maximum rated capacity of the PV facility

$$FF = \frac{V_{mpp} \times I_{mpp}}{V_{oc} \times I_{sc}} \tag{8}$$

3.5. Performance Ratio

This is a quality factor that indicates how much solar energy is efficiently used. It compares the actual and theoretical power production of a PV installation. The closer the PR is to 100%, the more efficient the PV plant. This nondimensional factor is defined in Equation (9) [18] as

$$PR = \frac{Y_{\text{final}}}{Y_{\text{ref}}} \tag{9}$$

where Y_{final} uses the following formula shown in Equation (10) to express the relationship between the total energy produced by an array (EA) and the absolute power produced by a PV panel P_{DC} as

$$Y_{\text{final}} = \frac{EA}{P_{DC}} \tag{10}$$

Whereas the "Y_{ref}" calculated by dividing the amount of insolation (HA) received by a panel over a given time period by the amount of irradiance received over the same time period at STC as shown in Equation (11) as

$$Y_{ref} = \frac{HA}{G} \tag{11}$$

3.6. Execution Ratio

This is the ratio between the power at GMPP achieved at PSCs (P_{GMPP}) and power achieved at STCs (P_m). The mathematical representation of ER is given in Equation (12) [18] as

$$\%ER = \frac{P_m \ at \ PSCs}{P_{STC}} \times 100 \tag{12}$$

3.7. Power Enhancement

The power enhancement of game-puzzle-based configurations is evaluated with respect to the existing TCT scheme and expressed in Equation (13) [18] as

$$\%PE = \frac{P_{GMPP}(SM/SDK/I\text{-}SDK) - P_{GMPP}(TCT)}{P_{GMPP}(SM/SDK/I\text{-}SDK)} \times 100 \tag{13}$$

3.8. Shading Patterns Analysis

The obtained P-V and I-V curves during PSCs are described using MATLAB/Simulink modeling and experimentation studies.

(a) Shadowing pattern-I

Based on the different methodologies and placement of integer numbers, these are responsible for developing the game puzzle with shade dispersion capability as shown in Figure 11a–e. With the consideration of nonuniform shade profiles from a minimum to maximum irradiance range such as 200 W/m^2, 400 W/m^2, 600 W/m^2, 800 W/m^2 and 1000 W/m^2, an inclusive study was carried out with conventional (SP, TCT) and game theory (SDK, I-SDK and SM) based configurations. The considered shading pattern showed a highly nonuniform nature in irradiance. Furthermore, the game-theory-based reconfigurable methodologies are beneficial in terms of performance improvement due to the higher shade dispersion factor.

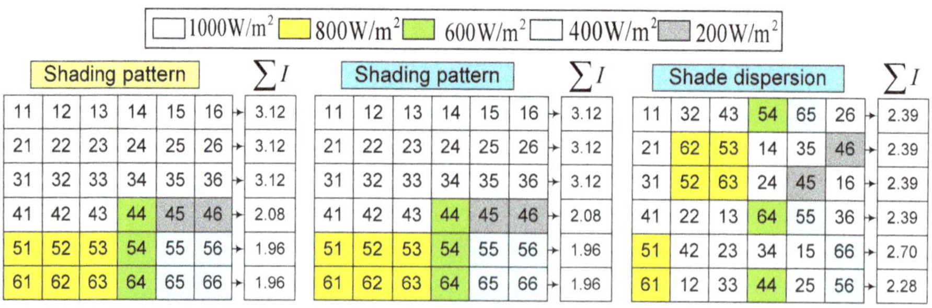

(**a**) SP (**b**) TCT (**c**) SDK

Figure 11. *Cont.*

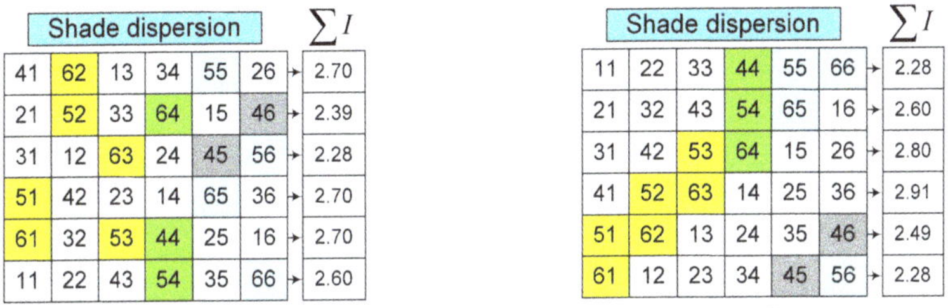

(**d**) I-SDK (**e**) SM

Figure 11. (**a**–**e**) Shade dispersion profiles for pattern-I.

To understand the feasibility of considered reconfiguration methodologies, a theoretical valuation of row-wise current was performed. The theoretical valuation of the produced row-wise current for the conventional SP configuration is expressed in Equations (14)–(16). Table 2 depicts the theoretical current assessment of other game-theory-based PV array systems.

$$I_{r1} = I_{r2} = I_{r3} = \left.\begin{array}{r}\left(\frac{1000}{1000}\right)I_m + \left(\frac{1000}{1000}\right)I_m + \left(\frac{1000}{1000}\right)I_m + \left(\frac{1000}{1000}\right)I_m + \left(\frac{1000}{1000}\right)I_m + \\ \left(\frac{1000}{1000}\right)I_m = 6I_m\end{array}\right\} \quad (14)$$

$$I_{r4} = \left.\begin{array}{r}\left(\frac{1000}{1000}\right)I_m + \left(\frac{1000}{1000}\right)I_m + \left(\frac{1000}{1000}\right)I_m + \left(\frac{600}{1000}\right)I_m + \left(\frac{200}{1000}\right)I_m + \\ \left(\frac{200}{1000}\right)I_m = 4I_m\end{array}\right\} \quad (15)$$

$$I_{r5} = I_{r6} = \left.\begin{array}{r}\left(\frac{800}{1000}\right)I_m + \left(\frac{800}{1000}\right)I_m + \left(\frac{800}{1000}\right)I_m + \left(\frac{600}{1000}\right)I_m + \left(\frac{400}{1000}\right)I_m + \\ \left(\frac{400}{1000}\right)I_m = 3.8I_m\end{array}\right\} \quad (16)$$

Table 2. Theoretical assessment: PV performance under shading pattern-I.

Row	1	2	3	4	5	6
			SP			
V (V)	V_m	$2V_m$	$3V_m$	$4V_m$	$5V_m$	$6V_m$
I (A)	$6I_m$	$6I_m$	$6I_m$	$4I_m$	$3.8I_m$	$3.8I_m$
P (W)	$6I_mV_m$	$12I_mV_m$	$18I_mV_m$	$16I_mV_m$	$19I_mV_m$	$22.8I_mV_m$
			TCT			
V (V)	V_m	$2V_m$	$3V_m$	$4V_m$	$5V_m$	$6V_m$
I (A)	$6I_m$	$6I_m$	$6I_m$	$4I_m$	$3.8I_m$	$3.8I_m$
P (W)	$6I_mV_m$	$12I_mV_m$	$18I_mV_m$	$16I_mV_m$	$19I_mV_m$	$22.8I_mV_m$
			SDK			
V (V)	V_m	$2V_m$	$3V_m$	$4V_m$	$5V_m$	$6V_m$
I (A)	$4.59I_m$	$4.59I_m$	$4.59I_m$	$4.59I_m$	$5.19I_m$	$4.38I_m$
P (W)	$4.59I_mV_m$	$9.18I_mV_m$	$13.77I_mV_m$	$18.36I_mV_m$	$25.95I_mV_m$	$26.28I_mV_m$
			I-SDK			
V (V)	V_m	$2V_m$	$3V_m$	$4V_m$	$5V_m$	$6V_m$
I (A)	$5.19I_m$	$4.59I_m$	$4.38I_m$	$5.19I_m$	$5.19I_m$	$5.01I_m$
P (W)	$5.19I_mV_m$	$9.18I_mV_m$	$13.14I_mV_m$	$20.76I_mV_m$	$25.95I_mV_m$	$30.06I_mV_m$
			SM			
V (V)	V_m	$2V_m$	$3V_m$	$4V_m$	$5V_m$	$6V_m$
I (A)	$4.38I_m$	$5.01I_m$	$5.38I_m$	$5.59I_m$	$4.78I_m$	$4.38I_m$
P (W)	$4.38I_mV_m$	$10.01I_mV_m$	$16.14I_mV_m$	$22.36I_mV_m$	$23.91I_mV_m$	$26.28I_mV_m$

(b) Shadowing pattern-II

Based on the different methodologies and placement of integer numbers, these are responsible for developing the game puzzle with shade dispersion capability as shown in Figure 12a–e. With the consideration of nonuniform shade profiles from minimum to maximum irradiance ranges such as 300 W/m^2, 600 W/m^{2,} 700 W/m^2 and 900 W/m^2, a comprehensive study was carried out with existing and game-theory-based arrangements.

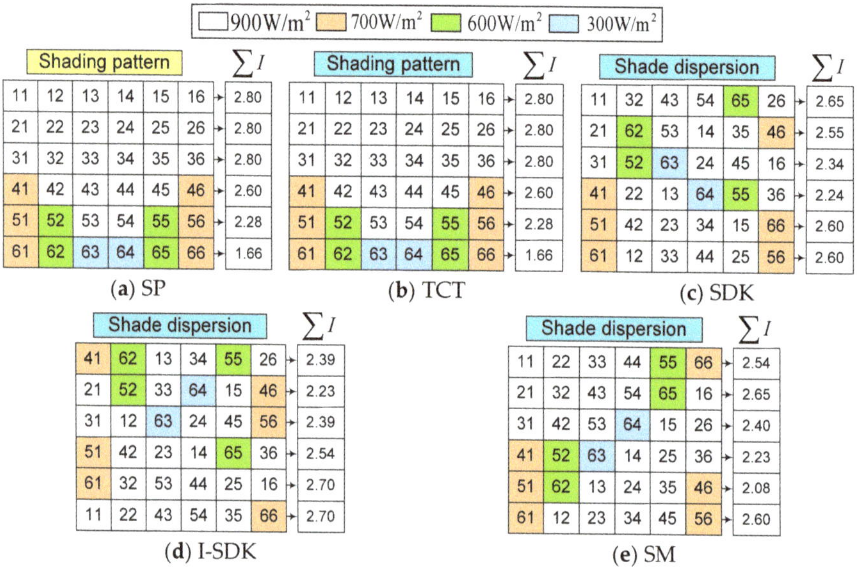

Figure 12. (**a**–**e**) Shade dispersion profiles for pattern-II.

To understand the feasibility of the considered reconfiguration methodologies, a theoretical assessment of row-wise current was performed. The theoretical row-wise current generated by the usual SP design is expressed in Equations (17)–(20). Table 3 depicts the theoretical current assessment of other game-theory-based PV array systems.

$$I_{r1} = I_{r2} = I_{r3} = \left.\begin{matrix}\left(\frac{900}{1000}\right)I_m + \left(\frac{900}{1000}\right)I_m + \left(\frac{900}{1000}\right)I_m + \left(\frac{900}{1000}\right)I_m + \left(\frac{900}{1000}\right)I_m + \\ \left(\frac{900}{1000}\right)I_m = 5.4I_m\end{matrix}\right\} \quad (17)$$

$$I_{r4} = \left.\begin{matrix}\left(\frac{700}{1000}\right)I_m + \left(\frac{900}{1000}\right)I_m + \left(\frac{900}{1000}\right)I_m + \left(\frac{900}{1000}\right)I_m + \left(\frac{900}{1000}\right)I_m + \\ \left(\frac{700}{1000}\right)I_m = 5I_m\end{matrix}\right\} \quad (18)$$

$$I_{r5} = \left.\begin{matrix}\left(\frac{700}{1000}\right)I_m + \left(\frac{600}{1000}\right)I_m + \left(\frac{900}{1000}\right)I_m + \left(\frac{900}{1000}\right)I_m + \left(\frac{600}{1000}\right)I_m + \\ \left(\frac{700}{1000}\right)I_m = 4.38I_m\end{matrix}\right\} \quad (19)$$

$$I_{r6} = \left.\begin{matrix}\left(\frac{700}{1000}\right)I_m + \left(\frac{600}{1000}\right)I_m + \left(\frac{300}{1000}\right)I_m + \left(\frac{300}{1000}\right)I_m + \left(\frac{600}{1000}\right)I_m + \\ \left(\frac{700}{1000}\right)I_m = 3.19I_m\end{matrix}\right\} \quad (20)$$

Table 3. Theoretical assessment: PV performance under shading pattern-II.

Row	1	2	3	4	5	6
			SP			
V (V)	V_m	$2V_m$	$3V_m$	$4V_m$	$5V_m$	$6V_m$
I (A)	$5.38I_m$	$5.38I_m$	$5.38I_m$	$5I_m$	$4.38I_m$	$3.19I_m$
P (W)	$5.38I_mV_m$	$10.76I_mV_m$	$16.14I_mV_m$	$20I_mV_m$	$21.9I_mV_m$	$19.14I_mV_m$

Table 3. *Cont.*

Row	1	2	3	4	5	6
			TCT			
V (V)	V_m	$2V_m$	$3V_m$	$4V_m$	$5V_m$	$6V_m$
I (A)	$5.38I_m$	$5.38I_m$	$5.38I_m$	$5I_m$	$4.38I_m$	$3.19I_m$
P (W)	$5.38I_mV_m$	$10.76I_mV_m$	$16.14I_mV_m$	$20I_mV_m$	$21.9I_mV_m$	$19.14I_mV_m$
			SDK			
V (V)	V_m	$2V_m$	$3V_m$	$4V_m$	$5V_m$	$6V_m$
I (A)	$5.09I_m$	$4.90I_m$	$4.5I_m$	$4.30I_m$	$5.0I_m$	$5.0I_m$
P (W)	$5.09I_mV_m$	$9.80I_mV_m$	$13.5I_mV_m$	$17.2I_mV_m$	$25I_mV_m$	$30I_mV_m$
			I-SDK			
V (V)	V_m	$2V_m$	$3V_m$	$4V_m$	$5V_m$	$6V_m$
I (A)	$4.59I_m$	$4.28I_m$	$4.59I_m$	$4.88I_m$	$5.19I_m$	$5.19I_m$
P (W)	$4.59I_mV_m$	$8.56I_mV_m$	$13.77I_mV_m$	$19.52I_mV_m$	$25.95I_mV_m$	$31.14I_mV_m$
			SM			
V (V)	V_m	$2V_m$	$3V_m$	$4V_m$	$5V_m$	$6V_m$
I (A)	$4.88I_m$	$5.09I_m$	$4.61I_m$	$4.28I_m$	$4.0I_m$	$5.0I_m$
P (W)	$4.88I_mV_m$	$10.18I_mV_m$	$13.8I_mV_m$	$17.12I_mV_m$	$20.0I_mV_m$	$30.0I_mV_m$

(c) Shadowing pattern-III

Based on the different methodologies and placement of integer numbers, these are responsible for developing the game puzzle with shade dispersion capability as shown in Figure 13a–e. With the consideration of nonuniform shade profiles from minimum to maximum irradiance ranges such as 400 W/m^2, 800 W/m^2 and 900 W/m^2, a comprehensive study was carried out with conventional and game-theory-based configurations. Furthermore, the game-theory-based reconfigurable methodologies are beneficial in terms of higher GMPP under shading profiles.

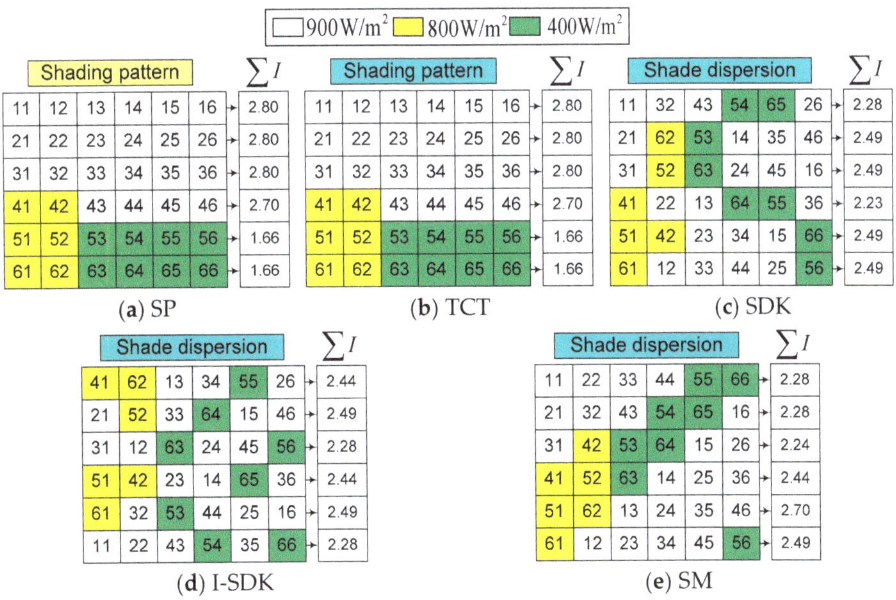

Figure 13. (**a**–**e**) Shade dispersion profiles for the pattern-III.

The theoretical row-wise current generated by the usual SP design is expressed in Equations (21)–(23). Table 4 depicts the theoretical current assessment of other game-theory-based PV array systems.

$$\left.\begin{array}{r} I_{r1} = I_{r2} = I_{r3} = \left(\frac{900}{1000}\right) I_m + \left(\frac{900}{1000}\right) I_m + \left(\frac{900}{1000}\right) I_m + \left(\frac{900}{1000}\right) I_m + \left(\frac{900}{1000}\right) I_m + \\ \left(\frac{900}{1000}\right) I_m = 5.4 I_m \end{array}\right\} \quad (21)$$

$$\left.\begin{array}{r} I_{r4} = \left(\frac{800}{1000}\right) I_m + \left(\frac{800}{1000}\right) I_m + \left(\frac{900}{1000}\right) I_m + \left(\frac{900}{1000}\right) I_m + \left(\frac{900}{1000}\right) I_m + \\ \left(\frac{900}{1000}\right) I_m = 5.2 I_m \end{array}\right\} \quad (22)$$

$$\left.\begin{array}{r} I_{r5} = I_{r6} = \left(\frac{800}{1000}\right) I_m + \left(\frac{800}{1000}\right) I_m + \left(\frac{400}{1000}\right) I_m + \left(\frac{400}{1000}\right) I_m + \left(\frac{400}{1000}\right) I_m + \\ \left(\frac{400}{1000}\right) I_m = 3.2 I_m \end{array}\right\} \quad (23)$$

Table 4. Theoretical assessment: PV performance under shading pattern-III.

Row	1	2	3	4	5	6
			SP			
V (V)	V_m	$2V_m$	$3V_m$	$4V_m$	$5V_m$	$6V_m$
I (A)	$5.38I_m$	$5.38I_m$	$5.38I_m$	$5.19I_m$	$3.19I_m$	$3.19I_m$
P (W)	$5.38I_mV_m$	$10.76I_mV_m$	$16.14I_mV_m$	$20.76I_mV_m$	$15.95I_mV_m$	$19.14I_mV_m$
			TCT			
V (V)	V_m	$2V_m$	$3V_m$	$4V_m$	$5V_m$	$6V_m$
I (A)	$5.38I_m$	$5.38I_m$	$5.38I_m$	$5.19I_m$	$3.19I_m$	$3.19I_m$
P (W)	$5.38I_mV_m$	$10.76I_mV_m$	$16.14I_mV_m$	$20.76I_mV_m$	$15.95I_mV_m$	$19.14I_mV_m$
			SDK			
V (V)	V_m	$2V_m$	$3V_m$	$4V_m$	$5V_m$	$6V_m$
I (A)	$4.38I_m$	$4.78I_m$	$4.78I_m$	$4.28I_m$	$4.78I_m$	$4.78I_m$
P (W)	$4.38I_mV_m$	$9.56I_mV_m$	$14.34I_mV_m$	$17.12I_mV_m$	$23.9I_mV_m$	$28.68I_mV_m$
			I-SDK			
V (V)	V_m	$2V_m$	$3V_m$	$4V_m$	$5V_m$	$6V_m$
I (A)	$4.69I_m$	$4.78I_m$	$4.38I_m$	$4.69I_m$	$4.78I_m$	$4.38I_m$
P (W)	$4.69I_mV_m$	$9.56I_mV_m$	$13.14I_mV_m$	$18.76I_mV_m$	$23.9I_mV_m$	$26.28I_mV_m$
			SM			
V (V)	V_m	$2V_m$	$3V_m$	$4V_m$	$5V_m$	$6V_m$
I (A)	$4.38I_m$	$4.38I_m$	$4.30I_m$	$4.69I_m$	$5.19I_m$	$4.78I_m$
P (W)	$4.38I_mV_m$	$8.76I_mV_m$	$12.90I_mV_m$	$18.76I_mV_m$	$25.95I_mV_m$	$28.68I_mV_m$

(d) Shadowing pattern-IV

Based on the different methodologies and placement of integer numbers, these are responsible for developing the game puzzle with shade dispersion capability as shown in Figure 14a–e. With the consideration of nonuniform shade profiles from minimum to maximum irradiance ranges such as 300 W/m^2, 535 W/m^2 and 840 W/m^2, a comprehensive study was carried out with conventional and game-theory-based configurations. The shading pattern appears to be building a corner shading on the PV array system.

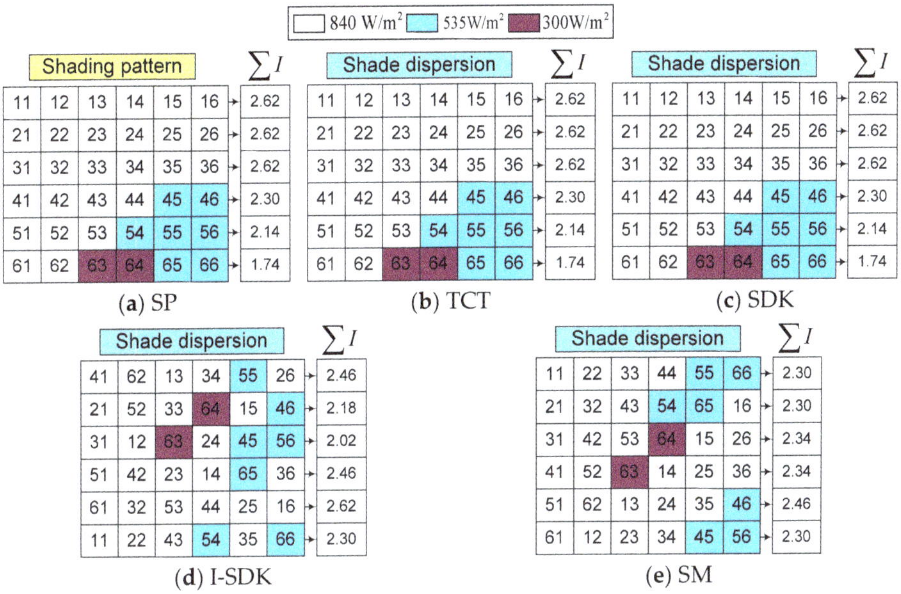

Figure 14. (**a**–**e**) Shade profiles based on the reconfigured PV array.

Equations (24)–(26) give a theoretical evaluation of the generated row-wise current for the typical SP configuration. Table 5 depicts the theoretical current assessment of other game-theory-based PV array systems.

$$I_{r1} = I_{r2} = I_{r3} = I_{r4} = \left(\frac{840}{1000}\right)I_m + \left(\frac{840}{1000}\right)I_m + \left(\frac{840}{1000}\right)I_m + \left(\frac{840}{1000}\right)I_m + \left(\frac{840}{1000}\right)I_m + \left(\frac{840}{1000}\right)I_m = 5.04I_m \tag{24}$$

$$I_{r5} = \left(\frac{840}{1000}\right)I_m + \left(\frac{535}{1000}\right)I_m + \left(\frac{535}{1000}\right)I_m + \left(\frac{535}{1000}\right)I_m + \left(\frac{535}{1000}\right)I_m + \left(\frac{535}{1000}\right)I_m = 3.51I_m \tag{25}$$

$$I_{r6} = \left(\frac{200}{1000}\right)I_m + \left(\frac{200}{1000}\right)I_m + \left(\frac{200}{1000}\right)I_m + \left(\frac{200}{1000}\right)I_m + \left(\frac{535}{1000}\right)I_m + \left(\frac{535}{1000}\right)I_m = 1.87I_m \tag{26}$$

Table 5. Theoretical assessment: PV performance under shading pattern-IV.

Row	1	2	3	4	5	6
			SP			
V (V)	V_m	$2V_m$	$3V_m$	$4V_m$	$5V_m$	$6V_m$
I (A)	$5.03I_m$	$5.03I_m$	$5.03I_m$	$4.42I_m$	$4.11I_m$	$3.34I_m$
P (W)	$5.03I_mV_m$	$10.06I_mV_m$	$15.09I_mV_m$	$17.68I_mV_m$	$20.55I_mV_m$	$20.04I_mV_m$
			TCT			
V (V)	V_m	$2V_m$	$3V_m$	$4V_m$	$5V_m$	$6V_m$
I (A)	$5.03I_m$	$5.03I_m$	$5.03I_m$	$4.42I_m$	$4.11I_m$	$3.34I_m$
P (W)	$5.03I_mV_m$	$10.06I_mV_m$	$15.09I_mV_m$	$17.68I_mV_m$	$20.55I_mV_m$	$20.04I_mV_m$
			SDK			
V (V)	V_m	$2V_m$	$3V_m$	$4V_m$	$5V_m$	$6V_m$
I (A)	$4.42I_m$	$4.73I_m$	$4.19I_m$	$4.19I_m$	$4.73I_m$	$4.73I_m$
P (W)	$4.42I_mV_m$	$9.46I_mV_m$	$12.57I_mV_m$	$16.76I_mV_m$	$23.65I_mV_m$	$28.38I_mV_m$

Table 5. *Cont.*

Row	1	2	3	4	5	6
			I-SDK			
V (V)	V_m	$2V_m$	$3V_m$	$4V_m$	$5V_m$	$6V_m$
I (A)	$4.73I_m$	$4.19I_m$	$3.68I_m$	$4.73I_m$	$5.03I_m$	$4.42I_m$
P (W)	$4.73I_mV_m$	$8.38I_mV_m$	$11.04I_mV_m$	$18.92I_mV_m$	$25.15I_mV_m$	$26.52I_mV_m$
			SM			
V (V)	V_m	$2V_m$	$3V_m$	$4V_m$	$5V_m$	$6V_m$
I (A)	$4.42I_m$	$4.42I_m$	$4.5I_m$	$4.5I_m$	$4.73I_m$	$4.42I_m$
P (W)	$4.42I_mV_m$	$8.84I_mV_m$	$13.5I_mV_m$	$18I_mV_m$	$23.65I_mV_m$	$26.52I_mV_m$

4. Results and Discussion

The suggested PV array configurations were used to estimate performance in all four shading circumstances. The preferred PV array setups were used to estimate performance under all four distinct shading scenarios. An investigation into the PV system's performance was carried out using MATLAB/Simulink, while experimentation was used to verify the results.

4.1. MATLAB/Simulink Study: P-V and I-V Curves under Shading Case I–IV

A thorough examination of the achieved performance of SP, TCT, SDK, I-SDK and SM arrangements was considered. The behavior of the characterized P-V and I-V curves for PV array topologies under shading scenarios I-IV is depicted in Figures 15 and 16.

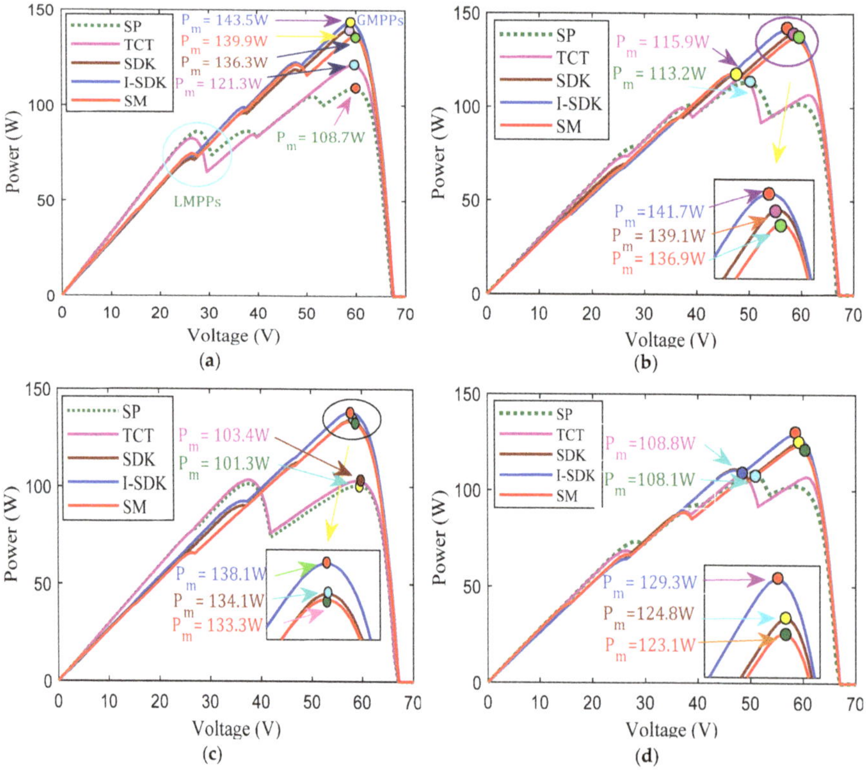

Figure 15. (**a**–**d**) P-V curves of SP, TCT, SDK, I-SDK and SM models under shading scenarios: I–IV.

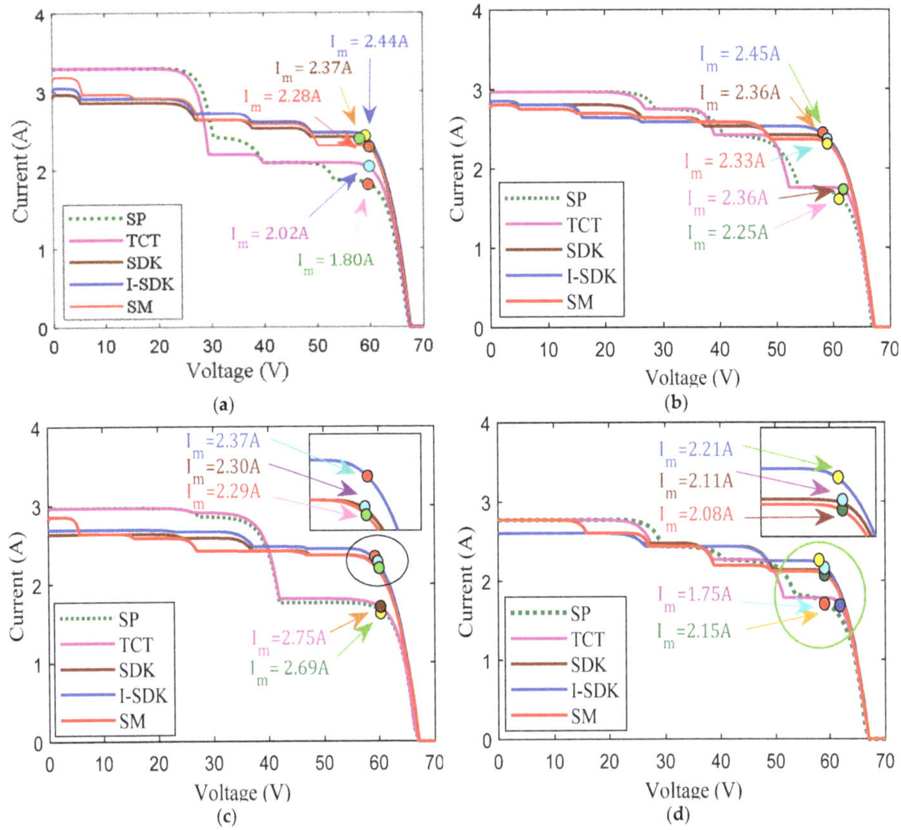

Figure 16. (**a**–**d**) I-V curves for SP, TCT, SDK, I-SDK and SM models under shading scenarios: I–IV.

Shade losses are greater in the SP and TCT configurations due to a lack of coherence between the maximum power of the module and the GMPP of the PV array. The SP and TCT configurations, GMPPs are 108.7 W and 121.3 W for shading case-I at nonuniform irradiation levels: 1000 W/m^2 -200 W/m^2, respectively. Moreover, for other reconfigured PV arrays such as SDK, I-SDK and SM, the GMPPs are investigated as 139.9 W, 143.5 W and 136.3 W.

During the shade case-II, the SP and TCT electrical arrangements had inferior power performance at GMPPs of 113.2 W and 115.9 W, respectively. SDK, I-SDK and SM configurations feature different GMPP sites for equivalent climatic conditions, including 139.1 W, 141.7 W and 136.9 W, respectively, in terms of uniform irradiance levels (1000 W/m^2 -200 W/m^2).

P-V curves with several maximum points show that GMPP power is present. Under the shading case-III, the TCT, SDK and SM models performed better than the SP configuration in terms of GMPP, with values of 103.4 W, 134.1 W and 133.3 W, respectively. Because of its shade dispersion characteristics, the I-SDK design had the maximum power among the examined PV array topologies at GMPP of 138.1 W.

The SP setup had a low power output at GMPP of 108.1 W when using shading case-IV. For identical environmental needs, TCT, SDK, I-SDK and SM setups with multiple GMPP locations such as 108.8 W, 124.8 W, 129.3 W and 123.1 W are available.

In Case-I, the I-SDK configuration had smoother I-V characteristics than the SP, TCT, SDK and SM versions. When compared to other configurations, the I_{SC} for the I-SDK configuration was judged to be 2.90 A after the examination. The values for I_{SC} for all

PV array configurations were SP (3.29 A), TCT (3.29 A), SDK (2.87 A) and SM (2.96 A) arrangements.

For shading case II, the I-V characteristic of the I-SDK setup exhibited smooth behavior, achieving I_{SC} and V_{OC} of 2,80 A and 66.65 V, respectively. For the SP, TCT, SDK and SM models, the I_{SC} values were 2.96 A, 2.96 A, 2.80 A and 2.77 A, respectively. When compared to other settings, the shading effects of SP and TCT settings enhanced volatility.

During shading case-III, the nature of the I-V curve for I-SDK was confirmed to be smoother and reported as 2.63 A when compared to SP, TCT, SDK and SM arrangements. The I_{SC} was 2.97 A (SP), 2.97 A (TCT), 2.63 A (SDK) and 2.85 A for other setups (SM).

The nature of the I-V curves was investigated in shading case-IV. In this scenario, the I-SDK setup exhibited smooth behavior to achieve I_{SC} and V_{OC} in the 2.60 A and 65.5 V ranges. In addition, the I_{SC} values for the 2.77 A, 2.77 A, 2.77 A and 2.77 A configurations were obtained.

The MATLAB and simulation study was analyzed from a critical perspective. The efficiency of solar photovoltaic systems decreased dramatically when realistic shading patterns were used. Voltage and power at GMPP, PL and FF can be calculated using P-V and I-V curves. Table 6 provides the quantitative findings of the MATLAB/Simulink study for shading profiles I–IV.

Table 6. MATLAB/Simulink quantitative study of shading scenarios I–IV.

Performance Parameters	**Case-I**					**Case-II**				
	SP	**TCT**	**SDK**	**I-SDK**	**SM**	**SP**	**TCT**	**SDK**	**I-SDK**	**SM**
P_{GMPP} (W)	108.7	121.3	139.9	143.5	136.3	113.2	115.9	139.1	141.7	136.9
V_{GMPP} (V)	60.35	59.82	58.82	58.81	59.73	50.15	49	58.71	57.81	58.73
I_m (A)	1.80	2.027	2.378	2.44	2.28	2.25	2.36	2.36	2.45	2.33
V_{OC} (V)	65.9	66.3	66.5	66.5	66.5	66.2	66.2	66.55	66.55	66.55
I_{SC} (A)	3.299	3.299	2.87	2.90	2.96	2.969	2.969	2.80	2.80	2.77
PL (W)	71.5	58.3	41	36.7	43.9	67	64.9	41.1	38.5	43.3
P_{loss} (%)	39.6	32.35	22.75	20.36	24.36	37.18	36.01	22.80	21.36	24.02
FF (%)	50.13	55.45	73.30	74.61	69.24	57.76	58.97	74.64	76.10	74.21
PR (%)	60.32	67.31	77.63	79.63	75.63	62.81	64.31	77.19	78.63	75.97
PE (%) w.r.t TCT	-	11.59	28.70	32.62	25.39	-	2.38	22.87	25.17	20.93
Best topology			I-SDK					I-SDK		
Performance Parameters	Case-III					Case-IV				
	SP	TCT	SDK	I-SDK	SM	SP	TCT	SDK	I-SDK	SM
P_{GMPP} (W)	101.3	103.4	134.1	138.1	133.3	108.1	108.8	124.8	129.3	123.1
V_{GMPP} (V)	37.6	37.6	58.18	58.17	58.10	50.55	61.0	59.08	58.43	59.04
I_m (A)	2.69	2.75	2.30	2.37	2.29	2.15	1.75	2.11	2.21	2.08
V_{OC} (V)	65.5	65.5	66.1	66.1	66.1	65.4	65.5	65.5	65.5	65.5
I_{SC} (A)	2.97	2097	2.63	2.63	2.85	2.77	2.77	2.77	2.60	2.77
PL (W)	78.9	76.8	46.1	42.1	46.9	72.1	71.4	55.4	50.9	57.1
P_{loss} (%)	43.78	42.61	25.58	23.36	26.02	40.01	39.62	30.74	28.24	31.68
FF (%)	52	57.4	77.1	79.4	70.7	59.6	59.96	68.78	75.92	67.8
PR (%)	56.21	57.38	74.41	76.63	73.97	59.98	60.37	69.25	71.75	68.31
PE (%) w.r.t TCT	-	2.07	32.37	36.32	31.58	-	0.64	15.44	19.61	13.87
Best topology			I-SDK					I-SDK		

4.2. Power and Voltage at GMPP

GMPP's power evaluation results are given in Figure 17. The I-SDK arrangement delivered the greatest power at GMPP of 143.5 W, 141.7 W, 128.3 W and 129.3 W.

Power distribution to the load side involves many factors, including voltage at GMPP. The voltage at GMPP has distinct values in the MATLAB/Simulation research for SP, TCT, SDK, I-SDK and SM setups under shading case-I (60.35 V, 59.82 V, 8.82 V, 58.81 V and 59.73 V), case-II (50.15 V, 49 V, 58.71 V, 57.81 V and 58.73 V), case-III (37.6 V, 37.6 V, 58.18 V, 58.17 V and 58.10 V) and case-IV (50.55 V, 61 V, 59.08 V, 58.43 V and 59.04 V).

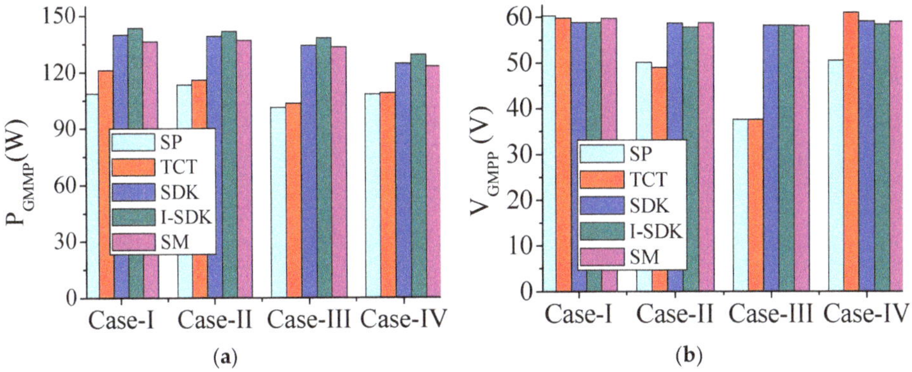

Figure 17. (**a**) Power; (**b**) voltage at GMPP.

4.3. PL and FF Analysis

Power losses due to shade on PV systems such as SP, TCT, SDK, SM and I-SDK configurations were observed in the MATLAB/Simulink study. In order to do this, the I-SDK configuration had a minimum PL of 20.36%, 21.36%, 23.36% and 28.24% under shading cases I–IV, respectively.

Deviation in the FF due to different shadowing scenarios are given in Figure 18 when comparing SP, TCT, SDK, SM and I-SDK layouts. Shade instances I-IV in the MATLAB/Simulink study, according to I-SDK, demonstrated the greatest gains in shading efficiency in terms of FF at 74.61%, 76.10%, 79.4% and 75.72%, respectively.

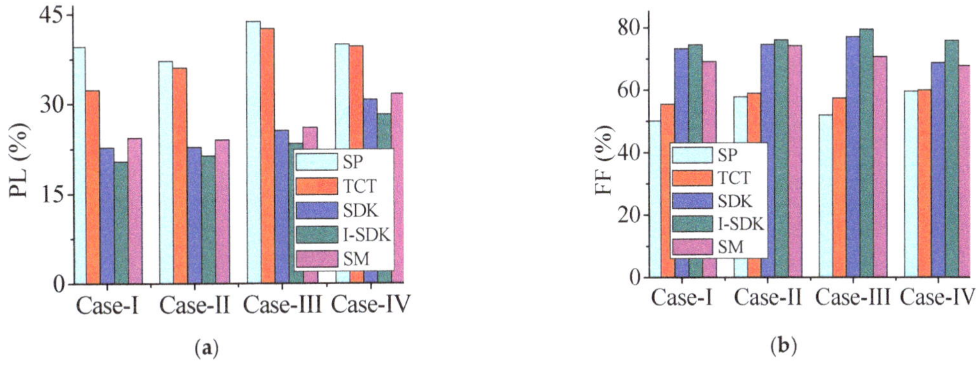

Figure 18. (**a**) PL (**b**) FF analysis for case I-IV.

4.4. PR and PE Analysis

The PR and PE analysis was performed using MATLAB/Simulink, as illustrated in Figure 19. When compared to SP, TCT, SDK and SM arrangements, I-SDK had the highest PR of 79.63%, 78.63%, 76.63% and 71.75% in shading patterns I-IV.

In addition to this, the PE was investigated under shading scenarios I-IV, though MATLAB/Simulink analysis was observed to be highest as 32.62%, 25.17%, 36.32% and 19.61% compared to TCT (11.59%, 2.38%, 2.07% and 0.64%), SDK (28.17%, 22.87%, 32.37%, 15.44%) and SM (25.39%, 20.93%, 31.58%, 13.87%) configurations.

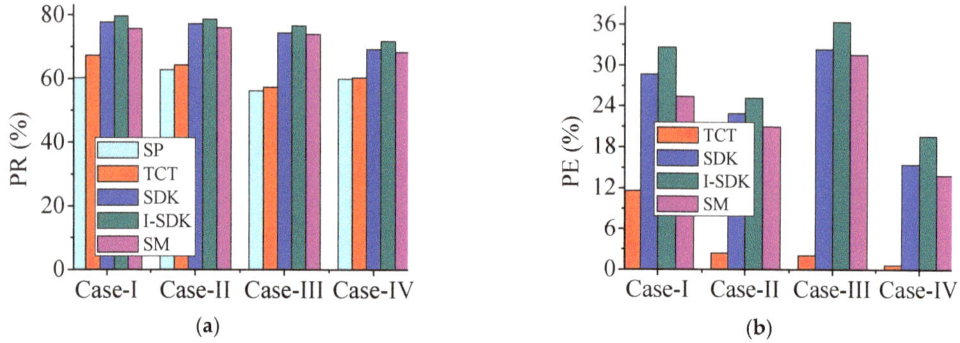

Figure 19. (**a**) PR (**b**) PE analysis for case I-IV.

4.5. Experimental Study: P-V and I-V Curves under Shading Case-IV

This research investigated the effect of shading on the electrical performance of typical SP, TCT, SDK, SM and I-SDK puzzle-based designs. The P-V curve for each of the four PV array installations is given in Figure 20a,b for shading scenarios I–IV.

The presence of numerous power maximum points on P-V curves defined the position of the GMPP. In terms of shading case-IV, the GMPP for the SP, TCT, SDK and SM models were 104.2 W, 105 W, 122.1 W and 120.3 W, respectively. The I-SDK-based setup outperformed the other PV topologies in terms of power at GMPP (127.9 W) due to its shade dispersion capabilities.

I-V characteristics of the I-SDK configuration were smoother than those of the SP, TCT, SDK and SM puzzle-based setups. Under the scrutiny of shading scenario-IV, the ISC values were discovered to be 2.61 A for the I-SDK configuration. Furthermore, the I_{SC} values for all PV arrays based on SP, TCT, SDK and SM configurations were 2.8 A, 2.8 A, 2.62 A and 2.62.A, respectively. In addition, the quantitative observation during experimentation is reported in Table 7.

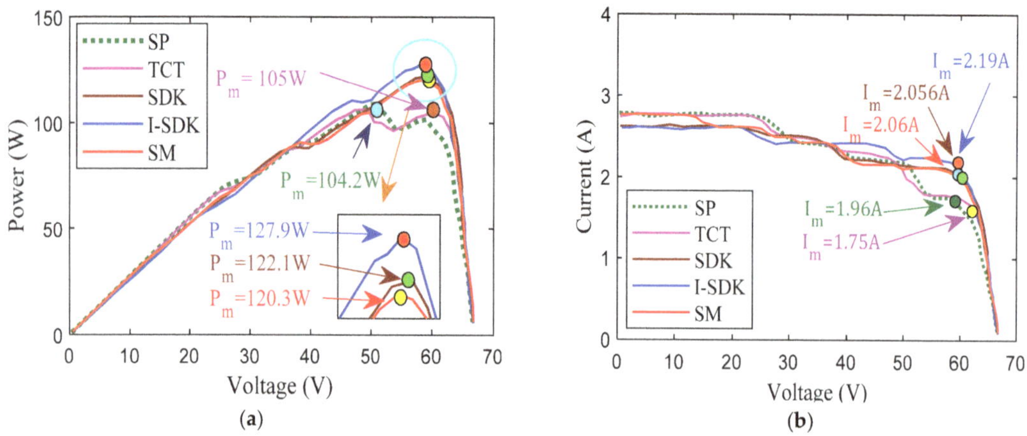

Figure 20. (**a**,**b**) I–V curves for SP, TCT, SDK, I-SDK and SM models under shading scenarios-IV.

Table 7. Quantitative analysis with experimental shading study- IV.

Performance Parameters	Case-IV				
	SP	TCT	SDK	I-SDK	SM
P_{GMPP} (W)	104.2	105	122.1	127.9	120.3
V_{GMPP} (V)	52.90	59.97	59.36	58.34	58
I_m (A)	1.96	1.75	2.05	2.19	2.06
V_{OC} (V)	66.1	66.2	66.3	66.3	66.1
I_{SC} (A)	2.8	2.8	2.62	2.61	2.62
PL (W)	76	75.2	58.1	52.3	59.9
P_{loss} (%)	42.17	41.73	32.24	29.02	33.24
FF (%)	56.29	56.64	70.29	73.91	69.46
PR (%)	57.82	58.26	67.75	70.97	66.75
PE (%) w.r.t TCT	-	0.767	17.17	22.74	15.45
Best topology			I-SDK		

A transient analysis of electrical performance characteristics was monitored during experimental activities to verify the results. Under shading pattern-I, the maximum current, voltage and power were settled from ideal/rated power (180 W) to 104.2 W (SP), 105 W (TCT) and 122.1 W (SDK), 127.9 W (I-SDK) and 120.3 W (SM), respectively, and are shown in Figure 21 as

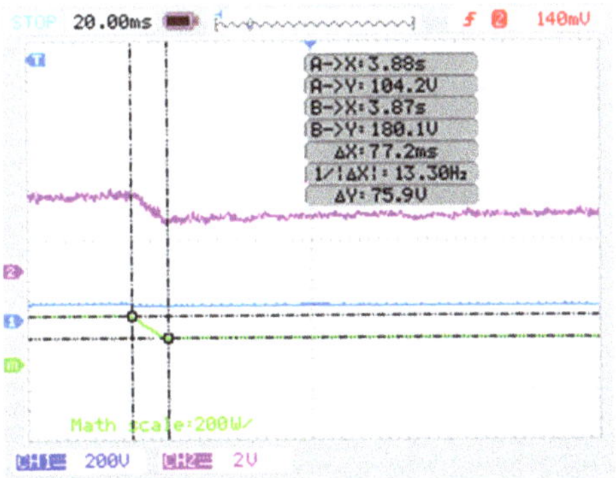

(a)

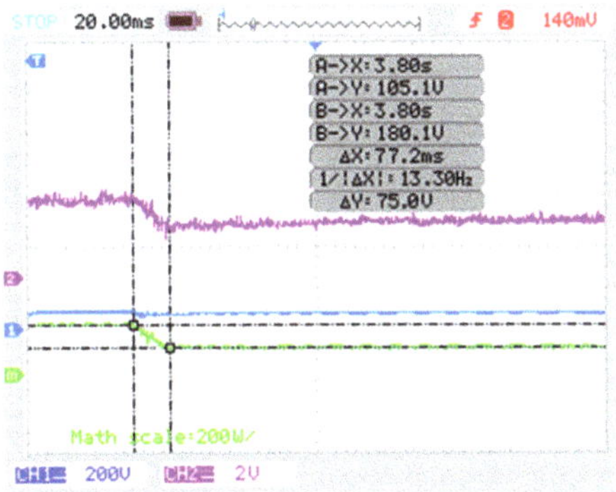

(b)

Figure 21. *Cont.*

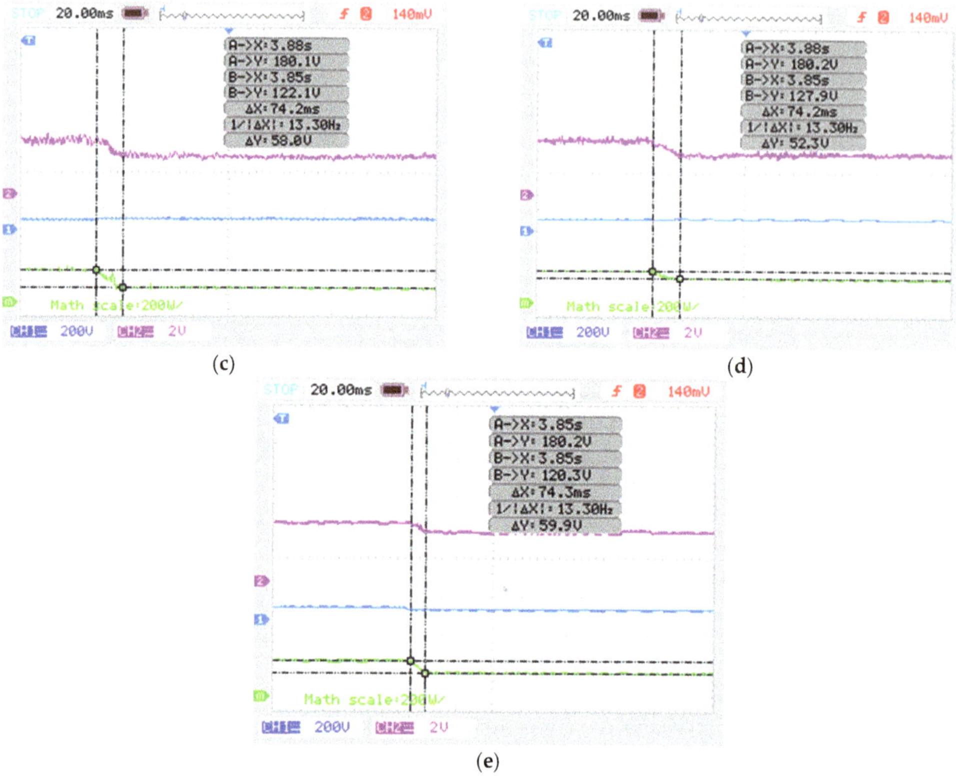

Figure 21. (**a**–**e**) Steady-state analysis of PV array configurations under shading case-IV.

4.6. Power and Voltage at GMPP

In addition, when compared to traditional SP, TCT, SDK and SM configurations, I-SDK setups produced more power at GMPP of 127.9 W.

Under shading case-IV, the SP, TCT, SDK, I-SDK and SM setups had different voltages at GMPP (52.90 V, 59.97 V, 59.36 V, 58.34 V and 58 V). Figure 22 is a bar chart depicting the power and voltage at GMPP.

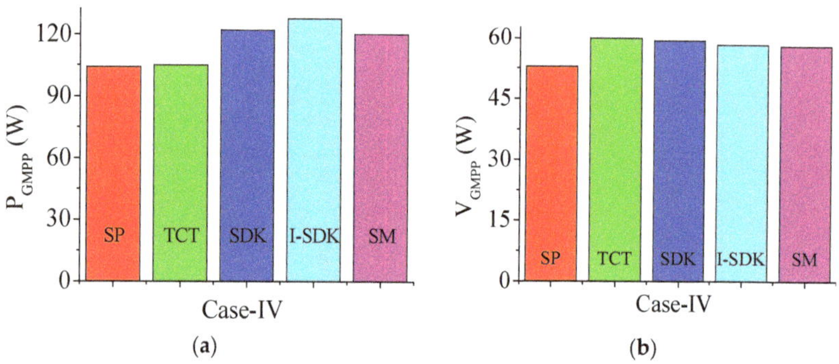

Figure 22. (**a**) Power (**b**) voltage at GMPP.

4.7. PL and FF Analysis

According to a recent experimental investigation, the I-SDK configuration had the lowest PL of 29.02% for performance evaluation during shading case-IV. I-SDK had lower PL values than the SP, TCT, SDK and SM setups according to the bar chart analysis in Figure 23.

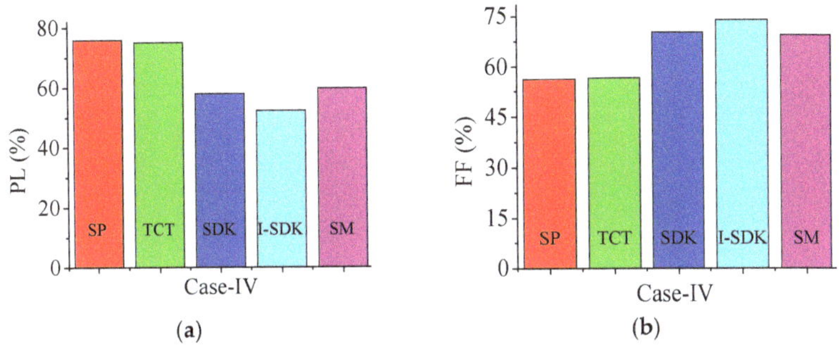

Figure 23. (**a**) PL (**b**) FF analysis case IV.

Experimental research was carried out under comparable lighting conditions as in Case-IV. As a result, the I-SDK setup had a higher FF of 73.91%, validating the MATLAB/Simulink study results.

4.8. PR and PE Analysis

The experimental study was conducted for PR assessment. When compared to SP, TCT, SDK and SM arrangements, I-SDK had the highest PR of 70.97% in shading patterns IV.

The experimental value of PE was detected and certified as 22.74% during the shading case-IV experimental investigation. For experimental research, PR and PE analysis are depicted as a bar chart in Figure 24.

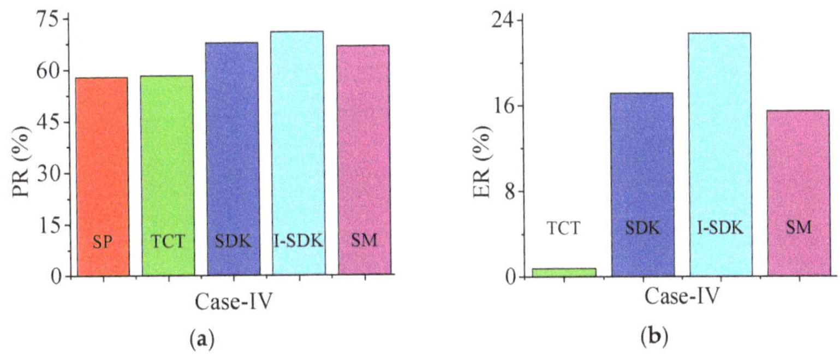

Figure 24. (**a**) PR (**b**) PE analysis for case IV.

4.9. Comparison of Simulation and Experimental Results under Shading Case-IV

The key performance parameters were investigated during the MATLAB/Simulink study and validated through an experimental study under shading case-IV. Table 8 was explored to show the difference between the key parameters during both the studies under shading case-IV as

Table 8. Parameters for Simulink and experimental studies under shading case-IV.

Parameters	Simulink Study	Experimental Study
GMPP (W)	129.3	127.9
FF (%)	75.92	73.91
PR (%)	71.75	70.97
PL (W)	50.9	52.3

5. Conclusions

An I-SDK shade dispersion configuration approach was proposed in this study to increase a PV array's power generation under PSCs. A MATLAB/Simulink model was used to assess the performance of several 6×6 size PV array configurations, such as TCT, SDK, I-SDK and SM, for various parameters such as power at GMPP, FF, PL and PE.

- In shading scenario-I, the minimized PL for I-SDK configuration was quite a bit less as 36.7 W compared to SP (71.5 W), TCT (58.3 W), SDK (41 W) and SM (43.9) configurations. Furthermore, the maximum FF was found to be 74.61% for the I-SDK configuration, which is more than the SP (50.13%), TCT (55.45%), SDK (73.30%) and SM (69.24%) configurations, respectively. The power at GMPP of the I-SDK configuration was found to be highest compared to conventional configurations, at 143.5 W
- In shading scenario-II, the PL in the I-SDK configuration was observed to be smaller, at 38.5 W, compared to SP (67 W), TCT (64.9 W), SDK (41.1 W) and SM (43.3 W) configurations. In the I-SDK configuration, the FF also had a maximum value of 76.10%, which is more than other existing PV array configurations.

When compared to the SP, TCT, SDK and SM configurations under different shading patterns, the improved I-SDK configuration enhanced the global maximum power, uniformly dispersed the shading influence and minimized the PL. In addition, an experimental inquiry was conducted to analyze the data and determine the practicality of the proposal. Using machine learning and artificial intelligence techniques, new puzzle-solving algorithms can be developed for dealing with realistic shading situations on a more generic scale. More research and development into reconfigurable PV array technology is needed to get us closer to commercial viability in the future using metaheuristics techniques to enhance the shade dispersion factor.

Author Contributions: Conceptualization, B.A.; Methodology, B.A.; Software, S.B.T.; Formal analysis, R.K.P.; Resources, B.V.A.; Data curation, R.K.P. and S.B.T.; Writing—original draft, B.A.; Writing—review & editing, R.K.P. and S.B.T.; Visualization, B.V.A.; Project administration, B.V.A. All authors have read and agreed to the published version of the manuscript.

Funding: Authors would like to acknowledge the support of the Deputy for Research and Innovation-Ministry of Education, Kingdom of Saudi Arabia for this research through a grant (NU/IFC/2/SERC/-/25) under the Institutional Funding Committee at Najran University, Kingdom of Saudi Arabia.

Institutional Review Board Statement: Not applicable.

Informed Consent Statement: Not applicable.

Data Availability Statement: No new data were created or analyzed in this study. Data sharing is not applicable to this article.

Conflicts of Interest: The authors declare no conflict of interest.

Copyright License Disclaimer: MATLAB software (R2023a) license is registered with the UPES, Dehradun and Simulink study is carried out in the manuscript.

References

1. Bai, J.; Cao, Y.; Hao, Y.; Zhang, Z.; Liu, S.; Cao, F. Characteristic output of PV systems under partial shading or mismatch conditions. *Sol. Energy* **2015**, *112*, 41–54. [CrossRef]
2. Report Is Available IEA, Renewables 2020, IEA, Paris. Available online: https://www.iea.org/reports/renewables-2020 (accessed on 28 March 2022).
3. Fu, X.; Zhou, Y. Collaborative Optimization of PV Greenhouses and Clean Energy Systems in Rural Areas. *IEEE Trans. Sustain. Energy* **2023**, *14*, 642–656. [CrossRef]
4. Chow, T.T. Review on photovoltaic/thermal hybrid solar technology. *Appl. Energy* **2010**, *87*, 365–379. [CrossRef]
5. Fu, X. Statistical machine learning model for capacitor planning considering uncertainties in photovoltaic power. *Prot. Control Mod. Power Syst.* **2022**, *7*, 51–63. [CrossRef]
6. Ibrahim, A.; Othman, M.Y.; Ruslan, M.H.; Mat, S.; Sopian, K. Recent advances in flat plate photovoltaic/thermal (PV/T) solar collectors. *Renew. Sustain. Energy Rev.* **2011**, *15*, 352–365. [CrossRef]
7. Raju, V.B.; Chengaiah, C. A Comprehensive Study on Re-arrangement of Modules Based TCT Configurations of Partial Shaded PV Array with Shade Dispersion Method. *Trends Renew. Energy* **2020**, *6*, 37–60. [CrossRef]
8. Nasiruddin, I.; Khatoon, S.; Jalil, M.F.; Bansal, R.C. Shade diffusion of partial shaded PV array by using odd-even structure. *Sol. Energy* **2019**, *181*, 519–529. [CrossRef]
9. Pachauri, R.K.; Mahela, O.M.P.; Member, S. Impact of Partial Shading on Various PV Array Configurations and Different Modeling Approaches: A Comprehensive Review. *IEEE Access* **2020**, *8*, 181375–181403. [CrossRef]
10. Ul-Haq, A.; Alammari, R.; Iqbal, A.; Jalal, M.; Gul, S. Computation of Power Extraction from Photovoltaic Arrays under Various Fault Conditions. *IEEE Access* **2020**, *8*, 47619–47639. [CrossRef]
11. Pachauri, R.; Singh, R.; Gehlot, A.; Samakaria, R.; Choudhury, S. Experimental analysis to extract maximum power from PV array reconfiguration under partial shading conditions. *Eng. Sci. Technol. Int. J.* **2019**, *22*, 109–130. [CrossRef]
12. Gul, S.; Haq, A.U.; Jalal, M.; Anjum, A.; Khalil, I.U. A unified approach for analysis of faults in different configurations of PV arrays and its impact on power grid. *Energies* **2019**, *13*, 156. [CrossRef]
13. Pachauri, R.; Yadav, A.S.; Chauhan, Y.K.; Sharma, A.; Kumar, V. Shade dispersion-based photovoltaic array configurations for performance enhancement under partial shading conditions. *Int. Trans. Electr. Energy Syst.* **2018**, *28*, e2556. [CrossRef]
14. Sahu, H.S.; Nayak, S.K.; Mishra, S. Maximizing the Power Generation of a Partially Shaded PV Array. *IEEE J. Emerg. Sel. Top. Power Electron.* **2016**, *4*, 626–637. [CrossRef]
15. Pareek, S.; Dahiya, R. Enhanced power generation of partial shaded photovoltaic fields by forecasting the interconnection of modules. *Energy* **2016**, *95*, 561–572. [CrossRef]
16. Samikannu, S.M.; Namani, R.; Subramaniam, S.K. Power enhancement of partially shaded PV arrays through shade dispersion using magic square configuration. *J. Renew. Sustain. Energy* **2016**, *8*. [CrossRef]
17. Yadav, A.S.; Pachauri, R.K.; Chauhan, Y.K. Comprehensive investigation of PV arrays with puzzle shade dispersion for improved performance. *Sol. Energy* **2016**, *129*, 256–285. [CrossRef]
18. Yadav, A.S.; Pachauri, R.K.; Chauhan, Y.K.; Choudhury, S.; Singh, R. Performance enhancement of partially shaded PV array using novel shade dispersion effect on magic-square puzzle configuration. *Sol. Energy* **2017**, *144*, 780–797. [CrossRef]
19. Vijayalekshmy, S.; Bindu, G.R.; Iyer, S.R. A novel Zig-Zag scheme for power enhancement of partially shaded solar arrays. *Sol. Energy* **2016**, *135*, 92–102. [CrossRef]
20. Bana, S.; Saini, R.P. Experimental investigation on power output of different photovoltaic array configurations under uniform and partial shading scenarios. *Energy* **2017**, *127*, 438–453. [CrossRef]
21. Satpathy, P.R.; Sharma, R.; Jena, S. A shade dispersion interconnection scheme for partially shaded modules in a solar PV array network. *Energy* **2017**, *139*, 350–365. [CrossRef]
22. Bosco, M.J.; Mabel, M.C. A novel cross diagonal view configuration of a PV system under partial shading condition. *Sol. Energy* **2017**, *158*, 760–773. [CrossRef]
23. Djilali, N.; Djilali, N. PV array power output maximization under partial shading using new shifted PV array arrangements. *Appl. Energy* **2017**, *187*, 326–337.
24. Pareek, S.; Chaturvedi, N.; Dahiya, R. Optimal interconnections to address partial shading losses in solar photovoltaic arrays. *Sol. Energy* **2017**, *155*, 537–551. [CrossRef]
25. Vengatesh, R.P.; Rajan, S.E. Analysis of PV module connected in different configurations under uniform and non-uniform solar-radiations. *Int. J. Green Energy* **2016**, *13*, 1507–1516. [CrossRef]
26. Malathy, S.; Ramaprabha, R. Reconfiguration strategies to extract maximum power from photovoltaic array under partially shaded conditions. *Renew. Sustain. Energy Rev.* **2018**, *81*, 2922–2934. [CrossRef]
27. Pillai, D.S.; Ram, J.P.; Nihanth, M.S.S.; Rajasekar, N. A simple, sensorless and fixed reconfiguration scheme for maximum power enhancement in PV systems. *Energy Convers. Manag.* **2018**, *172*, 402–417. [CrossRef]
28. Krishna, G.S.; Moger, T. Improved SuDoKu reconfiguration technique for total-cross-tied PV array to enhance maximum power under partial shading conditions. *Renew. Sustain. Energy Rev.* **2019**, *109*, 333–348. [CrossRef]
29. Premkumar, M.; Subramaniam, U.; Babu, T.S.; Elavarasan, R.M.; Mihet-Popa, L. Evaluation of mathematical model to characterize the performance of conventional and hybrid PV array topologies under static and dynamic shading patterns. *Energies* **2020**, *13*, 3216. [CrossRef]

30. Sagar, G.; Pathak, D.; Gaur, P.; Jain, V. A Su-Do-Ku puzzle based shade dispersion for maximum power enhancement of partially shaded hybrid bridge-link-total-cross-tied PV array. *Sol. Energy* **2020**, *204*, 161–180. [CrossRef]
31. Siva, M.; Nihanth, S.; Ram, J.P.; Pillai, D.S.; Ghias, A.M.Y.M. Enhanced power production in PV arrays using a new skyscraper puzzle based one-time recon fi guration procedure under partial shade conditions (PSCs). *Sol. Energy* **2019**, *194*, 209–224.
32. Thanikanti, S.B.; Kumar, P.; Devakirubakaran, S.; Aljafari, B.; Colak, I. A dynamic mismatch loss mitigation algorithm with dual input dual output converter for solar PV systems. *Sol. Energy Mater. Sol. Cells* **2023**, *251*, 112163. [CrossRef]
33. Rani, B.I.; Ilango, G.S.; Nagamani, C. Enhanced Power Generation From PV Array Under Partial Shading Conditions by Shade Dispersion Using Su Do Ku Configuration. *IEEE Trans. Sustain. Energy* **2013**, *4*, 594–601. [CrossRef]
34. Aljafari, B.; Balachandran, P.K.; Samithas, D.; Thanikanti, S.B. Solar photovoltaic converter controller using opposition-based reinforcement learning with butterfly optimization algorithm under partial shading conditions. *Environ. Sci. Pollut. Res.* **2023**, *30*, 72617–72640. [CrossRef] [PubMed]

sustainability

MDPI

Article

A Hybrid Particle Swarm Optimization with Butterfly Optimization Algorithm Based Maximum Power Point Tracking for Photovoltaic Array under Partial Shading Conditions

Yonggang Wang [1], Shengnan Dai [1,*], Pinchi Liu [1,2] and Xinyu Zhao [1]

1 College of Information and Electronic Engineering, Shenyang Agricultural University, Shenyang 110866, China; wygvern@syau.edu.cn (Y.W.); lpc816@stu.syau.edu.cn (P.L.); zhaoxinyu@stu.syau.edu.cn (X.Z.)

2 Substation Operation and Inspection Office, State Grid Fushun Power Supply Company, Fushun 113008, China

* Correspondence: dlan58@stu.syau.edu.cn; Tel.: +86-156-4034-6096

Abstract: The key objective of this paper is to develop a photovoltaic (PV) maximum power point tracking (MPPT) algorithm based on particle swarm optimization–butterfly optimization algorithm (PSO-BOA) that is adapted for partial shading conditions (PSCs). Generally, conventional MPPT techniques are often unable to accurately locate the global maximum power point (GMPP) generated by partial shading in PV systems. As a result, a significant decrease in power output occurs. The traditional particle swarm optimization (PSO) algorithm traps the local maxima point easily, while the butterfly optimization algorithm (BOA) has slow convergence speed and large oscillations during its use in research. To address the limitations of the aforementioned PSO and BOA algorithms, the MPPT strategy of PV systems combining PSO-BOA is presented, which can ameliorate the efficiency and accuracy in PSCs. In this paper, the control parameter of sensory modality in the BOA can be acquired based on logistic mapping, and the self-adaptive adjustment of the inertial weight of the PSO algorithm is designed. According to the simulation findings, the suggested method is more suitable than PSO and BOA with respect to intricate shading-induced variations in irradiance and changes in external temperatures. The average tracking time is less than 0.5 s, and the tracking accuracy is not less than 99.94%. Especially under sudden variations in irradiance and temperature conditions, the tracking time of the PSO-BOA algorithm is only 49.70% of that of the PSO algorithm and 55.63% of that of the BOA. Therefore, the MPPT method presented has the ability to improve the oscillations and result in less convergence speed, which in turn accurately tracks the GMPP.

Keywords: maximum power point tracking; photovoltaic generation; butterfly optimization algorithm; particle swarm optimization; partial shading conditions

check for updates

Citation: Wang, Y.; Dai, S.; Liu, P.; Zhao, X. A Hybrid Particle Swarm Optimization with Butterfly Optimization Algorithm Based Maximum Power Point Tracking for Photovoltaic Array under Partial Shading Conditions. *Sustainability* **2023**, *15*, 12402. https://doi.org/10.3390/su151612402

Academic Editors: Prince Winston David and Praveen Kumar B

Received: 27 June 2023
Revised: 25 July 2023
Accepted: 8 August 2023
Published: 15 August 2023

1. Introduction

In the past few years, the use of non-renewable energy sources has resulted in serious environmental pollution. Hence, finding renewable energy sources has become an imminent task. Among all the alternative energy resources, sun-powered energy as an abundant and clean source of power has been extensively applied in photovoltaic (PV) power generation [1]. Furthermore, the reduction in the manufacturing costs of PV modules and the improvement in equipment efficiency in PV systems have led to an increase in applications [2]. However, under normal conditions, the conversion of light-to-electricity efficiency of PV cells is barely around 11–28%, which restricts the development of PV systems [3]. To increase the output power, it should remain steady at the Maximum Power Point (MPP) for the PV systems. Hence, maximum power point tracking (MPPT) techniques are the crucial concerns of solar PV systems.

Classic MPPT techniques, including hill climbing (HC) [4–6], open circuit voltage (OCV) [7], incremental conductance (INC) [8], perturb and observe (P&O) [9–11], and constant voltage (CV) [12,13], are widely used due to their low complexity and cost-effectiveness. These algorithms are effective in uniform irradiation conditions and can accurately track the MPP [14]. However, they may experience oscillations when searching around the MPP, resulting in slow convergence and power loss. Moreover, classic MPPT algorithms have a limited capacity to respond quickly to changes in shading conditions, preventing them from effectively tracking the MPP. As a matter of fact, the output power for solar PV systems can be influenced by environmental and weather factors such as the shadows of trees, buildings around the PV power station, moving clouds, and temperature. This situation is defined as partial shading conditions (PSCs), where each PV panel may simultaneously encounter differing solar irradiance and temperatures. Under different PSCs, PV systems exhibit nonlinear power–voltage (P-V) characteristics with multiple peaks of power [15]. These peaks correspond to local maximum power points (LMPPs), with a sole global maximum power point (GMPP) also present. The conventional MPPT algorithm has many problems in PSCs. The problems include failing to jump out of the LMPP, low optimization efficiency, and inaccuracy. In recent years, many scholars have effectively solved the problem of tracking GMPP through the utilization of metaheuristic optimization methodologies, with examples such as the ant colony optimization algorithm (ACO) [16], grey wolf algorithm (WOA) [17,18], firefly algorithm (FA) [19,20], artificial bee colony algorithm (ABC) [21,22], etc.

The particle swarm optimization (PSO) algorithm [23] is extensively used in the field of MPPT. The PSO algorithm has the advantages of low memory requirement and comparatively fast convergence speed. In the reference [24], a novel approach was proposed where each PV module was treated as a particle, and the MPP was considered as the moving element. Compared with the P&O method, this method improved the efficiency by over 12% in the transient state. The modified PSO (MPSO) method was proposed for a multilevel inverter-based PV system in the reference [25]. This MPSO method introduced cognitive components and worst-experience social components to enhance the speed of searching for the MPP. A combination of the modified PSO and P&O methods in the reference [26] was applied. This method used the adaptive sensitivity parameter to detect the GMPP and tracked GMPP faster and more accurately. An MPV-PSO algorithm based on modified particle velocity of PV systems under PSCs was discussed in the reference [27], which achieves a balance between adaptive and deterministic features. Moreover, it could solve problems like particles getting trapped in LMPPs. A logarithmic particle swarm optimization (LPSO) method in PV systems was proposed in the reference [28], which updates particle velocity solely based on the direction of the GMPP. It should be noted that the PSO algorithm requires multiple iterations to converge. Furthermore, the main drawback of the PSO algorithm is that it frequently tends to adhere to the first local peak rather than effectively tracking the dynamic movement of the global peak, particularly when shading conditions vary over time.

In 2018, the Butterfly Optimization Algorithm (BOA) [29] was presented by the authors Arora and Singh. The BOA has been widely recognized for its strong search capabilities and effectiveness in converging toward the global maximum point with a high degree of accuracy. It has been extensively discussed by many scholars. For instance, the work in [30] is presented to validate the proposed chaotic algorithm on single mode, multimodal, and engineering design problems. The work in [31] has been proposed to optimize the analysis of annual cost, energy consumption, energy efficiency, and pollutant reduction. Despite numerous applications of the BOA in various areas, such as microgrid optimization scheduling, parameter adjustment, and other domains, its application in MPPT is still relatively limited. In the reference [32], the BOA method was applied to PV systems with the aim of mitigating the negative impact of shading and improving the tracking speed. A modified version of the BOA was proposed in the reference [33], which used a single dynamic variable as the tuning parameter, resulting in reduced algorithm complexity. In the reference [34], a method was presented to solve power point fluctuations between GMPP and LMPPS, leveraging an opposition-based reinforcement learning methodology in conjunction with the BOA. Based on the above research, it can be found that there exist some deficiencies in the BOA, such as long convergence time, as well as large oscillations during optimization.

Motivated by the research mentioned above, this paper introduces an innovative PSO-BOA algorithm for solving the slow convergence issues in the BOA while incorporating the advantages and strong robustness of the PSO algorithm. Compared to the traditional PSO and BOA methods, the proposed PSO-BOA algorithm effectively combines the benefits of both approaches while overcoming their respective shortcomings, such as the low convergence accuracy of PSO and the slow convergence and large oscillation of the BOA. Under PSCs, the PSO-BOA algorithm demonstrates remarkable accuracy in tracking the GMPP, exhibiting superior tracking speed, efficiency, and reduced oscillation. With the goal of enhancing the optimization performance of the PSO-BOA algorithm, this article introduces two modifications: a control parameter of sensory modality based on logistic mapping and the self-adaptive adjustment of the inertial weight. The simulation results suggest that the proposed algorithm effectively addresses the shortcomings of existing MPPT algorithms and offers a promising alternative for practical application in various renewable energy systems.

The subsequent sections of this article are structured as follows: Section 2 provides an overview of the multi-peak output characteristics under PSCs. Section 3 introduces the PSO-BOA algorithm utilized to control the PV system. Section 4 presents an analysis of the simulation results from MPPT techniques based on PSO-BOA, BOA, and PSO, respectively. Finally, Section 5 provides a summary of the key findings discussed in this paper.

2. Characteristics of Photovoltaic Array under Partial Shading Conditions

2.1. Mathematical Model of Photovoltaic Cells

A PV cell is a semi-conductor material that absorbs energy from sunlight and allows its electrons to jump to higher energy states. The liberated electrons subsequently undergo free movement along connected conductive wires, resulting in the generation of an electric current. This phenomenon of PV conversion is called the PV effect [35]. It harnesses the PV effect to directly transform solar energy into electrical energy. Figure 1 represents the single-diode model of a solar cell.

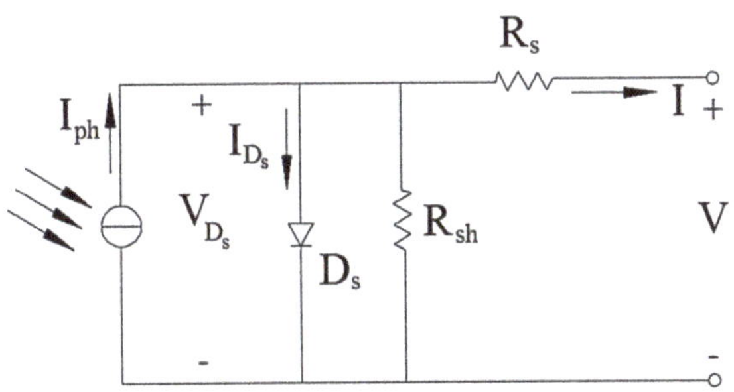

Figure 1. Single-diode model of a PV cell.

Applying Kirchhoff's current law [36], the output current is represented as follows:

$$I = I_{ph} - I_{D_s} - I_{sh} = I_{ph} - I_{D_s} - \frac{V + IR_s}{R_{sh}} \tag{1}$$

where I represents the current through series resistor R_s, I_{sh} is the current through shunt resistor R_{sh}, I_{ph} is the photo-generated current, V_{D_s} is the voltage across D_s, and I_{D_s} is the current through the diode D_s. The output voltage of the solar cell is represented by V.

The following Shockley equation [37] can be used to express the electric current:

$$I_{D_s} = I_0\left(e^{\frac{qVD_s}{\eta KT}} - 1\right) = I_0\left(e^{\frac{q(V+R_S I)}{\eta KT}} - 1\right) \tag{2}$$

where η is the ideality factor of the diode; K is the Boltzmann constant, $K = 1.38 \times 10^{-23}\ J/K$; T is the ambient temperature; q is the electronic charge constant, $q = 1.6 \times 10^{-19}$ C; R_s is the equivalent series resistance; V is the output voltage of the PV array; and I is the output current of the PV array.

By substituting Equation (2) into Equation (1), the I-V characteristic model is derived and represented as follows:

$$I = I_{ph} - I_0\left(e^{\frac{q(V+R_S I)}{\eta KT}} - 1\right) - \frac{V + IR_S}{R_{sh}} \tag{3}$$

A PV module commonly consists of numerous solar cells arranged in a series or in parallel to effectuate increased power, voltage, and current output levels. This unique design also allows PV modules to adapt to various system requirements and environmental conditions. The series configuration enables the module to generate higher voltages, making it suitable for applications that demand higher voltage levels. In contrast, the parallel arrangement ensures that sufficient current is provided, which is ideal for situations where higher current output is essential. The output characteristic is influenced by both its internal parameters and external factors, such as temperature and light intensity [38]. The equivalent circuit is depicted in Figure 2.

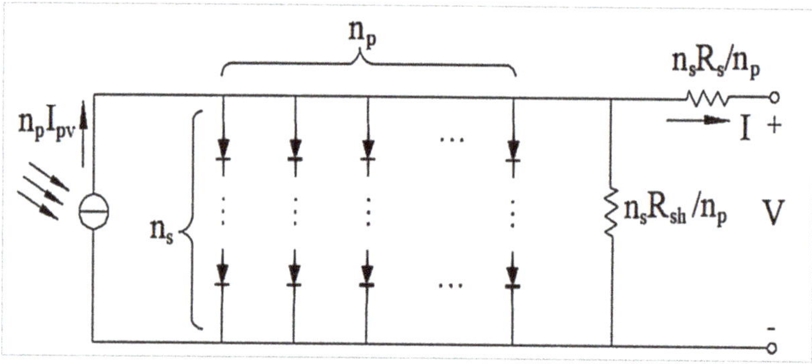

Figure 2. Equivalent circuit diagram of PV cells.

The current-voltage characteristics of a PV module model [39] are represented by Equation (4):

$$I = n_p I_{ph} - n_p I_{sc}\left(exp\left[\frac{q\left(V + I\frac{n_s}{n_p}\right)}{n_s \eta KT}\right] - 1\right) - \frac{\frac{n_p V}{n_s} + IR_s}{R_{sh}} \tag{4}$$

where n_p is the number of lateral PV panels, n_s is the number of vertical PV panels, I_{sc} is the saturation current of the diode.

The value of I_{ph} is dependent on the intensity of the light source and temperature.

$$I_{ph} = I_{ph\_STC} + K_i\left(T - T_{ref}\right)\frac{G}{G_{STC}} \tag{5}$$

where I_{ph_STC} is the short-circuit current under standard temperature and irradiance intensity; K_i is the temperature coefficient of current change, $K_i = 0.003$; T_{ref} is the standard temperature, T_{ref} = 25 °C; and G is the current irradiance intensity, while G_{STC} is the standard irradiance intensity, G_{STC} = 1000 W/m^2.

As stated in Equation (4), the I-V characteristic of PV modules undergoes significant alterations by external factors, such as solar irradiance and ambient temperature. The I-V and P-V curves presented in Figure 3a illustrate how different temperatures, while maintaining the same irradiance level, can influence the performance of the system. With increasing temperature, the I-V curve shifts toward a lower voltage, whereas a decrease in temperature causes the I-V curve to shift toward a higher voltage. Additionally, as indicated by the P-V curve, the power exhibits a negative correlation with the increase in temperature. Figure 3b shows the I-V curve and P-V curve at constant temperatures with varying irradiance. It is evident that a rise in solar irradiance leads to a corresponding increase in current and power. Observing Figure 3a,b, it is apparent that the MPP varies with changes in irradiance and module temperature. Hence, it is imperative to consider the impact of temperature and irradiance on PV solar systems.

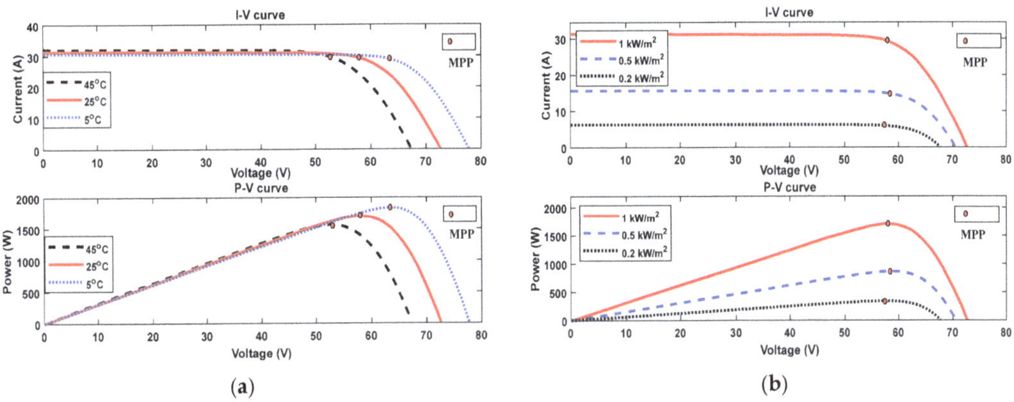

Figure 3. I-V and P-V curves of PV modules. (**a**) Diverse temperatures and (**b**) different irradiances.

2.2. Output Characteristics of Photovoltaic Array under Partial Shading Conditions

Under PSCs, when the incident irradiance on the PV panels decreases, the shaded areas experience heating, leading to the hot spot effect, which can potentially cause damage to the entire panel. To address these issues, bypass diodes are usually connected in parallel with PV cells to prevent temperature rise caused by the hot spot effect. Furthermore, the multi-peak characteristics arising from partial shading are associated with the PV cells connected in series within the array. This paper presents a simulation analysis of the output characteristics of a PV array considering five PV cells under PSCs. The PV array structure is illustrated in Figure 4.

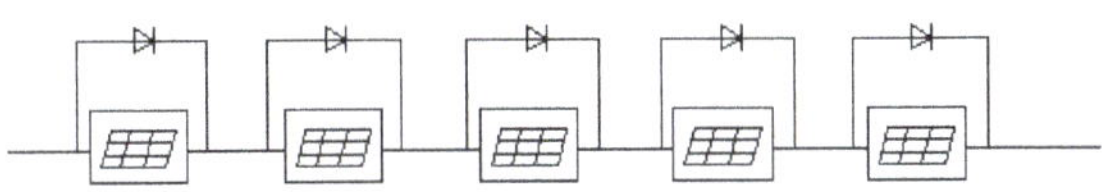

Figure 4. Structure schematic diagram of PV arrays.

Each PV module configures the parameters in the simulation model described in Table 1. The simulation tests are carried out under standard irradiance conditions of 1000 W/m^2 and a standard ambient temperature of 25 °C. These tests are conducted under three different conditions, as detailed in Table 2: example 1 remains unshaded, while examples 2 and 3 are subjected to PSCs. Figure 5 illustrates a simulated diagram of the photovoltaic components tracking the MPP theory, while Figure 6 presents the P-V characteristic curve. The pseudo-code for obtaining the theoretical MPP is outlined in

Algorithm 1, where the theoretical MPP and the corresponding P-V curve are derived using the same method as described in this paper.

Algorithm 1. Pseudo-code of getting theory MPP

Setting the irradiance and temperature received by five solar panels.
Initialize the parameter of $P_{max} = 0$, $I_{max} = 0$, $V_{max} = 0$.
Input *power current voltage* of five solar panels
If $P_{max} < power$
    $P_{max} = power$
    $I_{max} = current$
    $V_{max} = voltage$
End if
Return P_{max}, I_{max}, V_{max}.

Table 1. Simulation model parameters for each PV module.

Parameter	Value
Short-Circuit Current: I_{sc}	7.84 A
Open circuit voltage: U_{oc}	36.3 V
The voltage of MPP: U_m	29 V
The current of MPP: I_m	7.35 A

Table 2. PV panels subjected to different irradiances.

Example	S1 (W/m^2)	S2 (W/m^2)	S3 (W/m^2)	S4 (W/m^2)	S5 (W/m^2)
1	1000	1000	1000	1000	1000
2	1000	1000	900	800	600
3	800	800	600	600	400

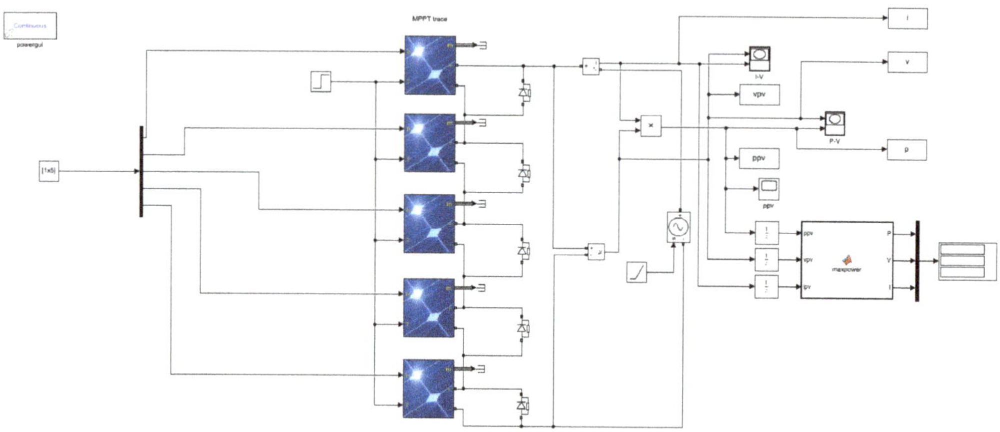

Figure 5. Photovoltaic components for tracking MPP theory.

It can be observed that under non-standard irradiance intensity, the PV array demonstrates the occurrence of multiple peaks in its power output characteristic curve. The number of peaks and the power vary with the degree of shadowing. Therefore, accurately tracking the GMPP is crucial under PSCs.

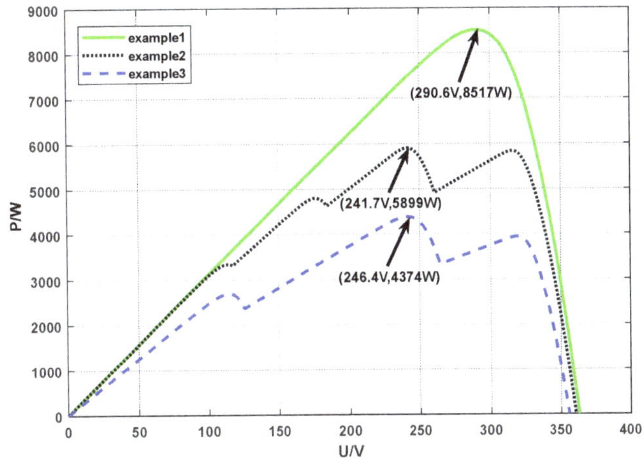

Figure 6. P-V characteristic curve of PV array output.

3. The Proposed Algorithm of Particle Swarm Optimization–Butterfly Optimization Algorithm

3.1. The Particle Swarm Optimization (PSO)

PSO [23] is a swarm intelligence optimization algorithm that mimics the search for food by a moving flock of birds in a multidimensional search space. PSO is widely adopted in solving optimization problems of nonlinear systems, owing to its advantages of rapid convergence and straightforward implementation. The velocity and position update formulas in the PSO algorithm are, respectively, expressed in Equations (6) and (7):

$$v_i^{t+1} = \omega \cdot v_i^t + c_1 \cdot rand_1 \left(p_{best} - x_i^t\right) + c_2 \cdot rand_2 \left(g_{best} - x_i^t\right) \tag{6}$$

$$x_i^{t+1} = x_i^t + v_i^{t+1} \tag{7}$$

where v_i^t and v_i^{t+1} are the velocities of the i-th particle at the (t) and (t + 1) iterations; p_{best} and g_{best} represent the local and global optimal positions of particles. Generally, c_1 and c_2 are the acceleration factors. In the reference [23], the stochastic factors ($rand_1$ and $rand_2$) were multiplied by a factor of 2 to achieve a mean value of 1. This adjustment was made to ensure that the particles would "overfly" the target about half of the time. Therefore, the c_1 and c_2 are equal to 2 in this paper; $rand_1$ and $rand_2$ are the random numbers that range from 0 to 1; ω represents the inertia weight.

3.2. The Butterfly Optimization Algorithm (BOA)

The BOA [29] mimics the habits of butterflies searching for food and seeking mates in their natural habitat. Setting the BOA apart from other optimization algorithms is that each butterfly in the algorithm is equipped with its own unique odor, leading to the generation of distinct odor intensities between individuals. By releasing a higher level of odor intensity, a butterfly can attract and be perceived by neighboring butterflies. The intensity of an individual's odor is perceived by other butterflies, which is denoted by Equation (8):

$$f(x) = cI^a \tag{8}$$

where $f(x)$ represents the perceived magnitude of fragrance; c is the sensory modality; I corresponds to the stimulus intensity; and a is the power exponent that relates to the degree of fragrance absorption and is limited to [0, 1].

In theory, the sensory modality coefficient c can be assigned a value within the range $[0, \infty]$. However, in the iterative process, the specific value of c is determined by the particular optimization problem. During the optimal search phase of the algorithm, the sensory modality c is expressed using the following formulation:

$$c_{t+1} = c_t + \left[\frac{0.025}{c_t \cdot T_{\max}} \right] \tag{9}$$

where $T_{\max}$ represents the upper bound of the number of iterations; generally, the initial value of parameter c_t is 0.01 [40].

Butterflies are capable of finding food and mating partners through both global and local search strategies in nature. The BOA utilizes a switching probability, denoted as "p", that governs the shift from a wide-ranging global exploration to a concentrated local exploration. Based on a comparison of the switching probability 'p' with a random number, the BOA decides whether to execute a local search or a global search. The position updating formula is demonstrated by Equation (10), as follows:

$$x_i^{t+1} = \begin{cases} x_i^t + (r^2 \cdot g^* - x_i^t) f_i \; p < rand \\ x_i^t + (r^2 \cdot x_a^t - x_b^t) f \; p \geq rand \end{cases} \tag{10}$$

where g^* is the current best-performing one in all the solutions that have been generated in the current stage; x_a^t and x_b^t represent the spatial positions of the a-th and b-th butterflies in the t-th iteration, and when $a = b$, the butterfly performs a local random search; r is a number that is generated randomly, $0 < r < 1$; and f_i is the fragrance produced by the i-th butterfly.

3.3. *The Particle Swarm Optimization–Butterfly Optimization Algorithm*

In the process of searching, it is simple for the PSO algorithm to fall into the local optimal solution. The main drawback of the BOA is its extended convergence time and significant oscillations during the process. To address the limitations of the PSO and BOA algorithms, the PSO-BOA algorithm incorporates the BOA search mechanism into the PSO algorithm. Specifically, the algorithm selects the search method based on comparing the generated random number with the predetermined switching probability 'p'. To enhance the algorithm's ability to identify the global optimal value, the PSO-BOA algorithm integrates two strategies. On the one hand, the strategy randomizes the spatial position of individual particles during the local search process, which explores a diverse search space and optimizes the quality of individual particles. On the other hand, the strategy utilizes both local and global optimal particles to update the position and speed of particles in the global search process, enabling the algorithm to exploit the current best solutions and refine the search trajectory toward the global optimal value. Simultaneously, the PSO-BOA algorithm adjusts the sensory modality c and the inertia weight ω in the iteration process, which accelerates the convergence rate and improves the local search performance.

3.3.1. Global Search

The updating criterion of the position for the global search stage in the PSO-BOA algorithm can be represented by Equation (11) as follows:

$$x_i^{t+1} = x_i^t + \left(\omega \cdot v_i^t + c_1 \cdot rand_1 \cdot \left(p_{best} - x_i^t \right) + c_2 \cdot rand_2 \cdot \left(g_{best} - x_i^t \right) \right) f_{\max} \tag{11}$$

where v_i^t is the velocity of the i-th particle at the t-th iteration; p_{best} and g_{best} represent the local and global optimal positions of particles; and $f_{\max}$ represents the current optimal scent intensity value. Generally, $c_1 = c_2 = 2$; $rand_1$ and $rand_2$ generate a random number that falls between 0 and 1. ω represents the inertia weight.

3.3.2. Local Search

The updating of the position formula for the local search stage used in the PSO-BOA algorithm can be represented by Equation (12) as follows:

$$x_i^{t+1} = x_i^t + \left(r^2 \cdot \left(x_a^t - x_i^t\right) - \omega \cdot \left(x_b^t - x_i^t\right)\right) f_i \tag{12}$$

where x_a^t and x_b^t are the spatial positions of the a-th and b-th butterflies in the t-th iteration; the parameter ω represents inertia weight; r generates a random number that falls between 0 and 1.

3.3.3. Parameter Control Strategy

Chaos theory has numerous applications in intelligent optimization algorithms. Logistic mapping [41] is one of the classic chaotic mapping methods in chaos theory, and its representation is shown in Equation (13):

$$z_{l+1} = \mu\, z_l (1 - z_l) \tag{13}$$

where μ is the chaotic parameter, and the value falls in [0, 4], l can be defined as the iteration count of the chaotic map.

The Lyapunov index [42] is a measure for distinguishing chaotic characteristics. A larger value of the Lyapunov exponent indicates a higher degree of chaos and stronger chaotic characteristics. The Lyapunov exponent is calculated by Equation (14):

$$\lambda = \lim_{n \to \infty} \frac{1}{n_h} \sum_{i=0}^{n_h - 1} \ln\left|f'(z_i)\right| \tag{14}$$

where λ is the Lyapunov exponent; n_h is the number of iterations of the map function; and $f'(\cdot)$ is the first derivative of the chaotic map function.

Produce a logistic diagram and a Lyapunov exponent curve of the logistic map where parameter μ is within the interval (0, 4], as illustrated in Figure 7.

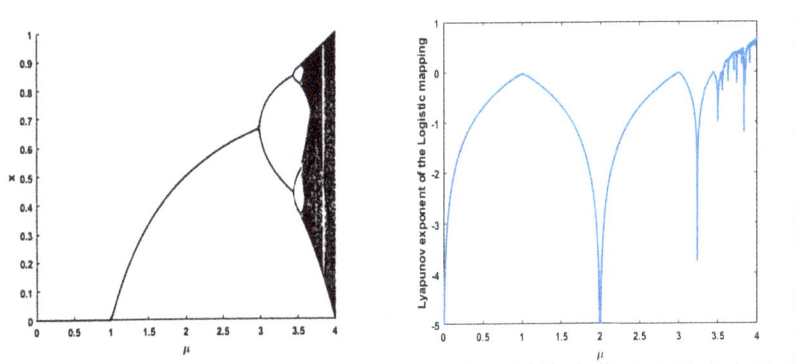

(a) (b)

Figure 7. Logistic mapping. (**a**) Logistic mapping bifurcation diagram and (**b**) Lyapunov exponent curve.

As illustrated in Figure 7, the bifurcation of the logistic map occurs at μ= 3.55, and with an increase in the parameter value, the range of the map gradually expands to (0, 1). When μ = 4, the logistic map exhibits chaotic behavior, leading to a sequence within the range (0, 1). The maximum Lyapunov exponent of the logistic map is calculated to be 0.6839. Consequently, the parameter μ is set to 4.

According to the logistic mapping expression, the sensory modality c in the PSO-BOA algorithm can be represented by Equation (12) as follows:

$$c(t) = 4 \cdot c_0(t-1)(1 - c_0(t-1)) \tag{15}$$

The coefficient of inertia weight directly affects the particle flight speed of the PSO algorithm. A dynamic tuning strategy is utilized to alter the local and global search capabilities of the algorithm, as depicted in Equation (16):

$$\omega = \omega_1 - (\omega_1 - \omega_2)\left(\frac{t}{T_m}\right)^2 \tag{16}$$

where ω_1 represents the initial inertia weight; ω_2 represents the inertia weight at the maximum number of iterations; t represents the current number of iterations; and T_m represents the maximum number of iterations. In this paper, the initial value of the inertia weight is set to 0.9, and the inertia weight value of the last iteration is set to 0.2. As the iteration progresses, the inertia weight decreases from 0.9 to 0.2. A larger inertia weight in the initial stage of the iteration can maintain the strong global search ability of the algorithm, while a smaller inertia weight in the later stage of the iteration is conducive to accurate local search and facilitates algorithm convergence.

3.4. Particle Swarm Optimization–Butterfly Optimization Algorithm for Maximum Power Point Tracking of Photovoltaic Arrays System

The input variables are the current I and voltage V of PV arrays, and the duty cycle D is the particle of individuality. Simultaneously, the duty cycle D is the output variable, which is controlled by regulating the *MOSFET* switching behavior to achieve the desired on and off states. The flowchart of the PSO-BOA algorithm proposed in this paper is presented in Figure 8. The pseudo-code of the PSO-BOA algorithm in this paper is presented in Algorithm 2 [43].

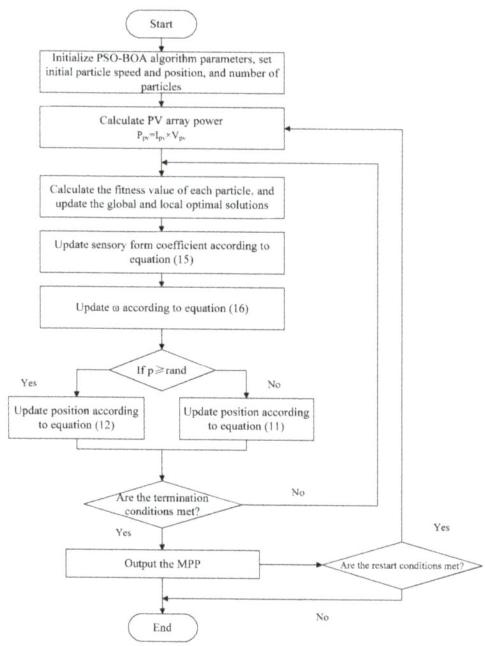

Figure 8. Flowchart of MPPT for PV arrays system based on PSO-BOA algorithm.

Algorithm 2. Pseudo-code of PSO-BOA algorithm

1.Generate the starting population of the particles Xi (i = 1, 2, ..., n) randomly
2.Initialize the acceleration factors $c_1, c_2,$ power exponent a and switch probability p
3.Calculate the value (f_i) of each particles
4.Calculate PV array power $P_{pv} = I_{pv} \times U_{pv}$
5.**While** t = 1: the upper limit of iterations
6. **For** each search of the particles
7. Update the fragrance f according to Equation (8)
8. **End for**
9. Find the best f (f_{max})
10. **For** each search of the particles
11. Set a random number "*rand*" in [0, 1]
12. **If** *rand* < p then
13. Move to the best position according to Equation (12)
14. **Else**
15. Move to adjust positions according to Equation (11)
16. **End if**
17. **End for**
18. Update the velocity v according to Equation (6)
19. Calculate the new fitness f value of each particles
20. **If** $f_{new} < f_{max}$
21. Update the position of best f according to Equation (7)
22. **End if**
23. Update the value of sensory modality c according to Equation (15)
24. Update the value of inertia weight ω according to Equation (16)
25. $t = t + 1$
26.**End while**
27.**If** reach the restart condition as Equation (17)
28.Return to line 5
29. **End if**
30.Output the MPP

When the termination condition is reached, the value of maximum power for the PV array is output; otherwise, the search continues. Additionally, during dynamic local shading, the output of the P-V characteristics also changes. To determine whether the algorithm needs to be restarted by detecting the degree of power change, this article defines the restart condition as shown in Equation (17).

$$\Delta P = \frac{P_m^k - P_m}{P_m} \tag{17}$$

where P_m^k is the effective output power value after local shading changes; P_m is the maximum power output value before local shading changes.

4. Simulation Results

A comparative analysis is performed for the PSO-BOA, PSO, and BOA algorithms proposed in this study in order to assess their effectiveness for MPPT under four different irradiance conditions, namely, standard irradiance conditions, local shading conditions, abrupt alterations for irradiance conditions, and sudden variations for irradiance and temperature conditions. When analyzing scenarios with uniform irradiance, we compared the PSO-BOA algorithm with the P&O, PSO, and BOA algorithms. In this case, the P&O algorithm is configured with a perturbation step size of 0.005. The population size is set to 10, and maximum iteration count is set to 15 for all three algorithms. The basic parameters of these algorithms are presented in Table 3. This study utilizes a boost circuit for the PV array, which was controlled by MPPT, as depicted in Figure 9. In this study, the system uses a PV array comprising five PV panels in series. The parameters of the boost circuit are designed as follows: Cpv = 500 µF, L = 0.85 mH, C = 200 µF, R = 30 Ω, and the MOSFET frequency is set at 0.1 MHz.

Table 3. Parameters of three algorithms.

Algorithm	Related Parameters
PSO-BOA	$a = 0.4, c_0 = 0.35, \omega_1 = 0.9, \omega_2 = 0.2, c_1 = 2, c_2 = 2, p = 0.8$
PSO	$\omega = 0.4, c_1 = 2, c_2 = 2$
BOA	$a = 0.4, c_0 = 1, p = 0.8$

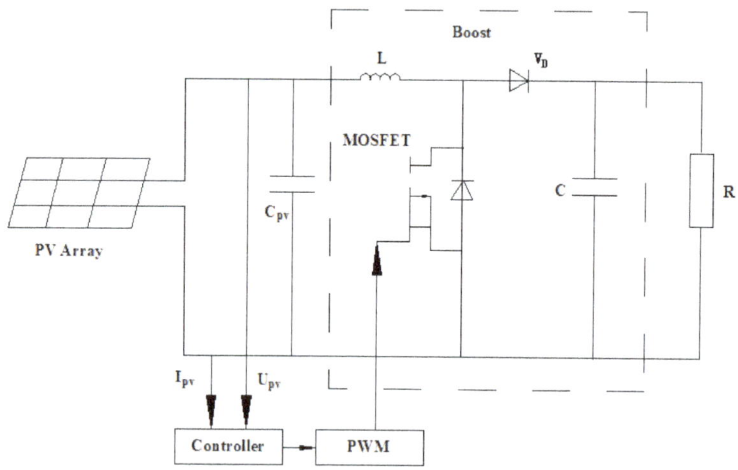

Figure 9. Structure of the MPPT system.

4.1. Optimization Results under Uniform Irradiance

Under a standard light intensity of 1000 W/m^2 and a standard ambient temperature of 25 °C, it is observed that the output power demonstrated a single peak characteristic. The P-V characteristic of the output is depicted in Figure 10. Specifically, the GMPP of the PV arrays is observed to be 8517 W.

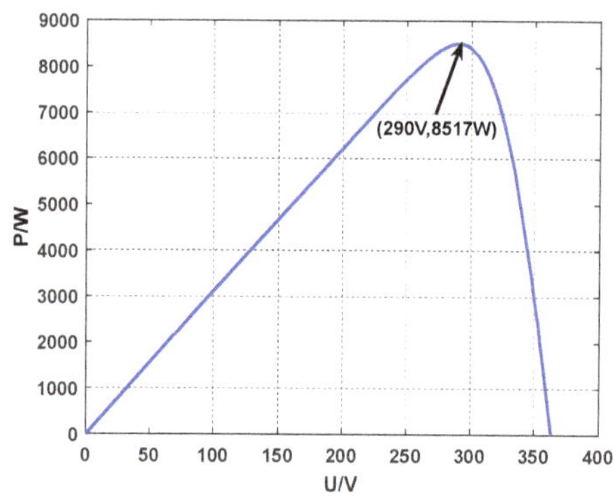

Figure 10. P-V characteristic of PV array under uniform irradiance.

The GMPP is searched using the aforementioned four algorithms in this article. Figure 11 presents the simulation results for these four algorithms, with a simulation time of 2 s.

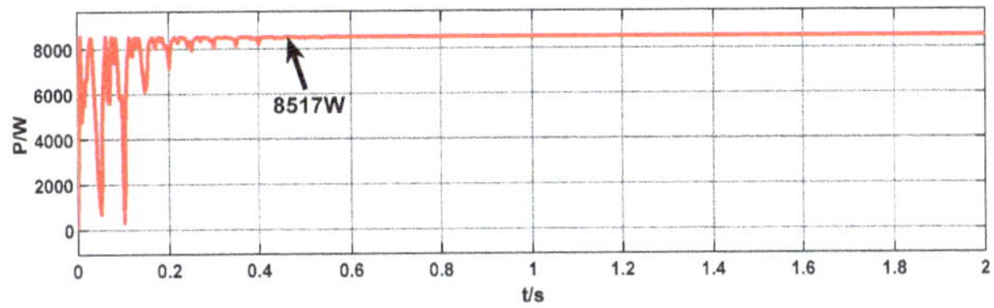

(**a**)

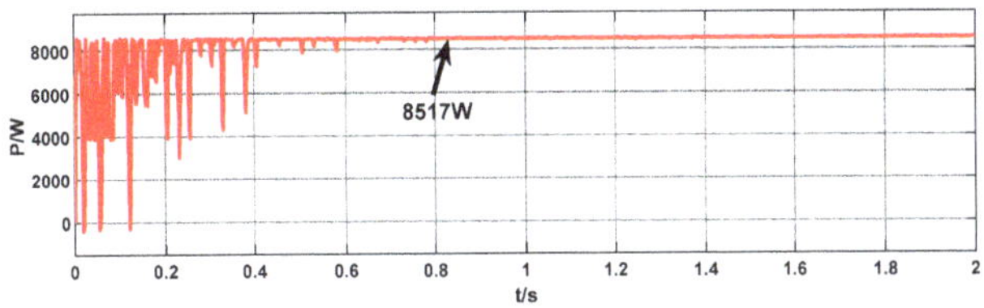

(**b**)

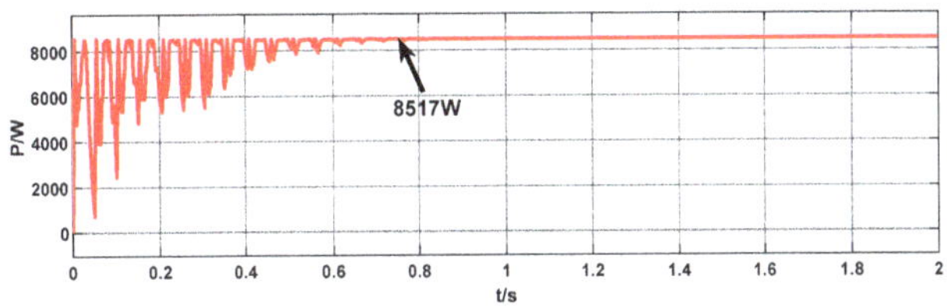

(**c**)

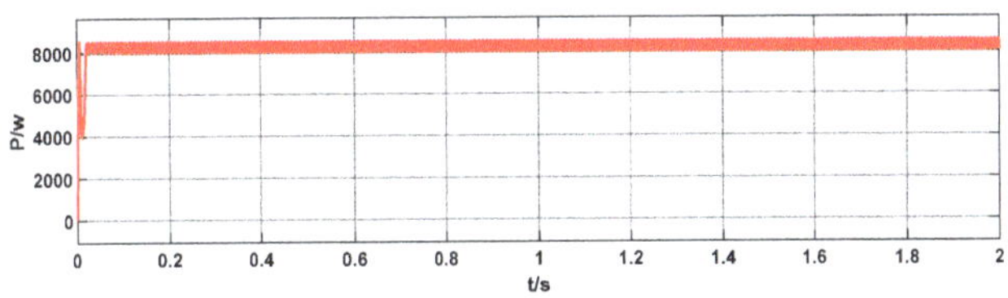

(**d**)

Figure 11. Power outputs of three algorithms under no shading condition at the MPP. (**a**) PSO-BOA algorithm; (**b**) PSO algorithm; (**c**) BOA; and (**d**) P&O algorithm.

Figure 11 demonstrates that all four algorithms (PSO-BOA, PSO, P&O, and BOA) are capable of tracking GMPP under uniform irradiance. In this situation, the P&O algorithm can track the MPP relatively quickly, but it suffers from significant oscillations and fails to converge to the MPP. Therefore, this paper does not provide further comparisons for the other complex conditions. When reaching the stable state, the power tracked by the other three algorithms is 8517 W, which is the theoretical maximum power. However, the convergence rate of the three algorithms varies. The PSO algorithm converges rapidly but tends to exhibit fluctuations around the maximum power point for an extended period of time, whereas the BOA has the slowest convergence rate. The PSO-BOA algorithm requires the least amount of time and significantly improves the convergence speed.

4.2. Optimization Results during Static Shading

In the setting of standard ambient temperature conditions at 25 °C, each of the five PV panels is subjected to varying light intensities: 800 W/m^2, 800 W/m^2, 600 W/m^2, 600 W/m^2, and 400 W/m^2. In this situation, the output power of the PV array exhibits multi-peak characteristics, with the GMPP measuring at 4374 W, as depicted in Figure 12.

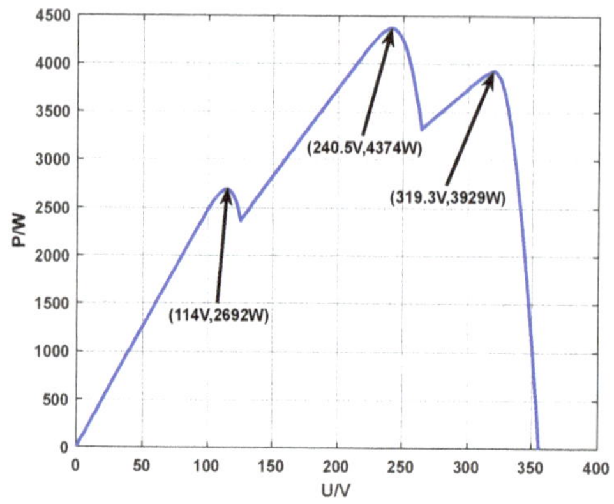

Figure 12. P-V characteristics of array output under static shading.

The GMPP at this time is determined using the three algorithms mentioned earlier. The simulation curves of these algorithms with a simulation time of 2 s are depicted in Figure 13.

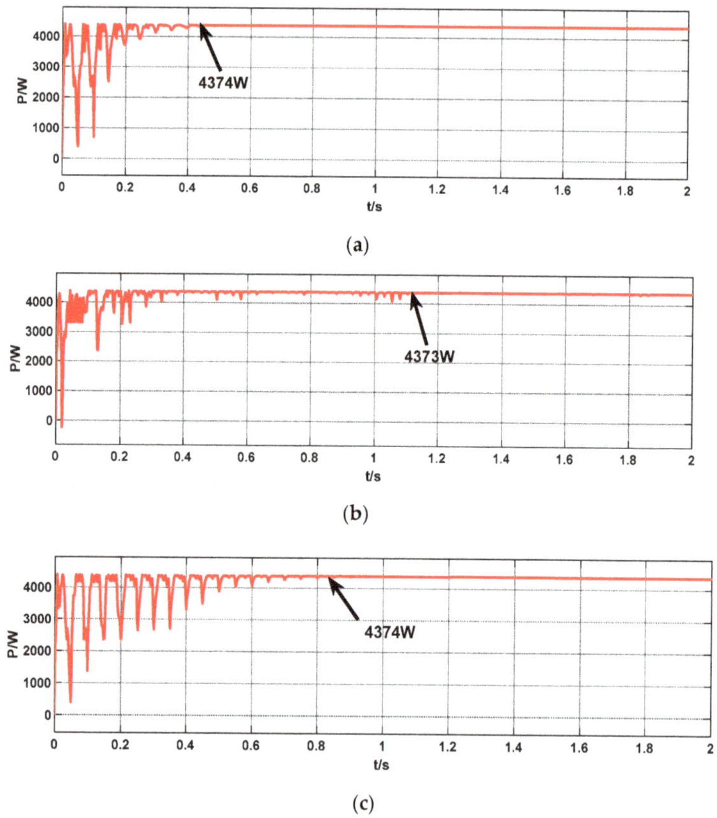

Figure 13. Power outputs of three algorithms under static shading. (**a**) PSO-BOA algorithm; (**b**) PSO algorithm; and (**c**) BOA algorithm.

Figure 13 shows that both the PSO-BOA and BOA algorithms have the capability to track the theoretical GMPP accurately. However, the PSO algorithm tracks a slightly lower GMPP of 4373 W, with a deviation value of 1 W, and exhibits small oscillations even reaching the steady state (after 0.4 s). In contrast, the BOA has slower convergence and larger power oscillations. Under static shading conditions, the PSO-BOA algorithm displays significant improvement in convergence speed and reduction in power oscillation.

4.3. Optimization Results under Abrupt Alterations for Irradiance Conditions

To test the response of a PV array to rapid changes in light intensity, this paper conducts a series of tests involving exposing the array to different light intensities at specific time intervals. Specifically, the array is devised to varying light intensities of 1000 W/m^2, 1000 W/m^2, 800 W/m^2, 800 W/m^2, and 400 W/m^2 from 0 to 0.8 s, and is then designed by 800 W/m^2, 800 W/m^2, 600 W/m^2, 400 W/m^2, and 400 W/m^2 from 0.8 to 2 s. These simulations are carried out under the environmental temperature of 25 °C, and the resulting P-V characteristics are depicted in Figure 14. During the two stages, the corresponding GMPP values of the array are 4606 W and 3337 W. Further evaluation of the system's performance is conducted by comparing the dynamic shading simulations for three algorithms with a simulation time of 2 s. The comparison is depicted in Figure 15.

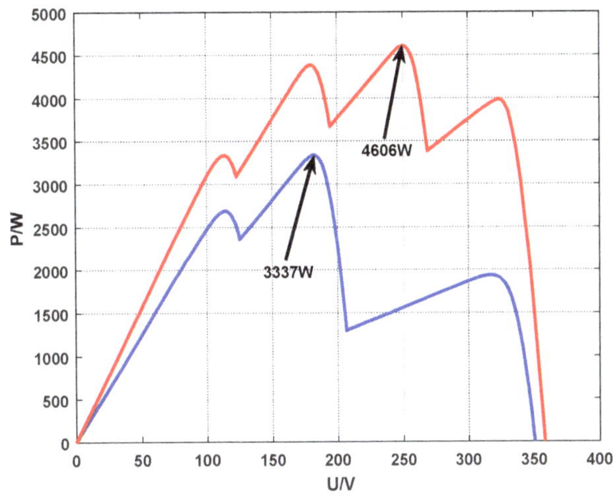

Figure 14. P-V characteristics of PV array output under abrupt alterations for irradiance conditions.

Figure 15 shows that the PSO-BOA algorithm accurately tracks the theoretical GMPP under varying irradiance conditions. The BOA also displays good performance in this regard, albeit with a slight tracking error. However, the PSO algorithm exhibits a significant deviation from the theoretical GMPP and is susceptible to local optima, thus resulting in low convergence accuracy. Furthermore, in terms of convergence time, the BOA requires around 0.7 s to converge, with more oscillations during varying irradiance conditions. In contrast, the PSO algorithm has relatively faster convergence, requiring about 0.4 s. Meanwhile, the PSO-BOA algorithm exhibits the fastest convergence time of about 0.3 s, accompanied by less oscillation, thereby demonstrating its superior tracking performance under dynamic local shading conditions. Overall, compared to both the PSO and BOA algorithms, the PSO-BOA algorithm offers improved tracking accuracy and less oscillation.

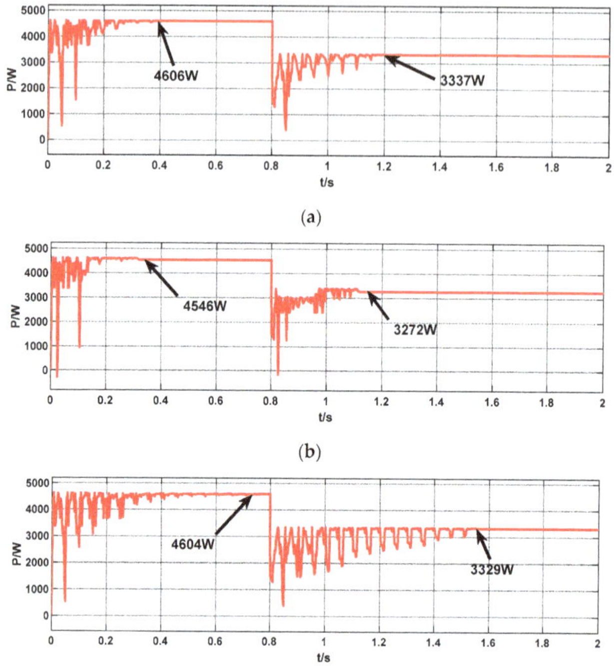

(**a**)

(**b**)

(**c**)

Figure 15. Power outputs of three algorithms under abrupt alterations for irradiance conditions. (**a**) PSO-BOA algorithm; (**b**) PSO algorithm; and (**c**) BOA.

4.4. Optimization Results under Sudden Variations for Irradiance and Temperature Conditions

To evaluate the output power characteristics under harsh environmental conditions considering the influence of temperature on MPP, this paper sets the irradiance intensity of the array as 800 W/m^2, 800 W/m^2, 600 W/m^2, 400 W/m^2, and 400 W/m^2 at 0–0.8 s, while the ambient temperature is maintained at 25 °C. Later, from 0.8 to 2 s, the array experience suddenly changes light intensity to 800 W/m^2, 600 W/m^2, 400 W/m^2, 200 W/m^2, and 200 W/m^2, while the environmental temperature is increased to 30 °C. The resulting P-V characteristics outputs are depicted in Figure 16, where the GMPP for the two stages is 3798 W and 2237 W. Figure 17 shows the dynamic shading simulations for three algorithms under harsh environmental conditions.

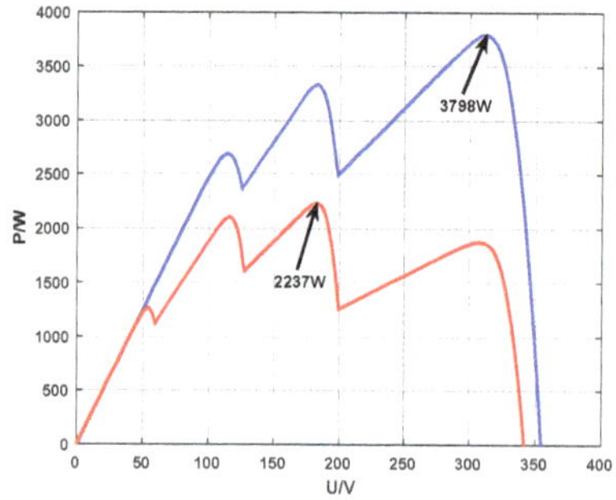

Figure 16. P-V characteristics of PV array outputs under sudden variations for irradiance and temperature conditions.

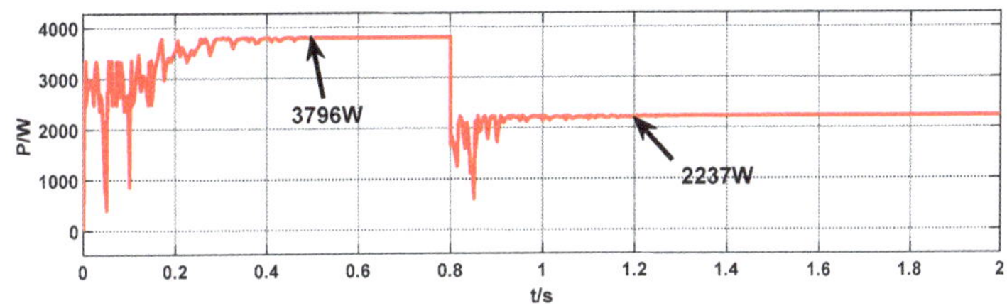

(**a**)

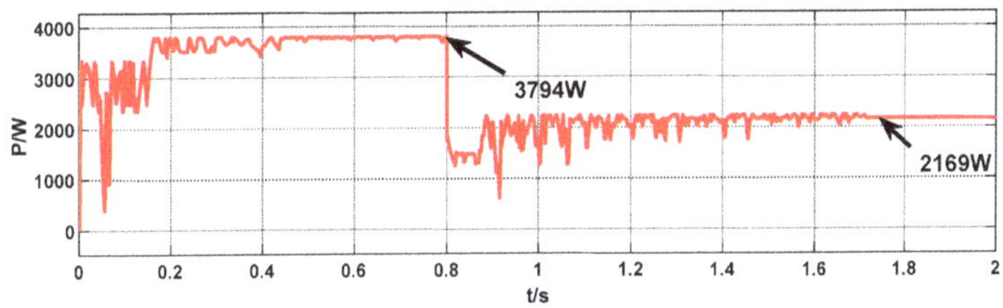

(**b**)

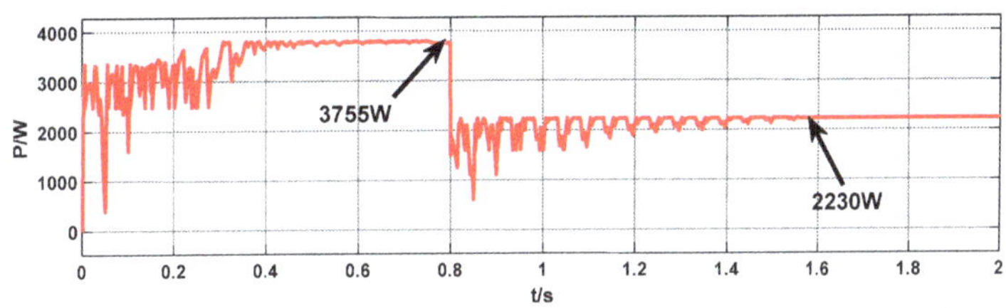

(**c**)

Figure 17. Power outputs of three algorithms under sudden variations for irradiance and temperature conditions. (**a**) PSO-BOA algorithm; (**b**) PSO algorithm; and (**c**) BOA.

According to the findings illustrated in Figure 17, the PSO-BOA algorithm successfully tracks the theoretical GMPP with high precision. In contrast, the PSO algorithm is prone to falling into local optima after abrupt changes in irradiance and temperature, leading to significant deviations from the theoretical GMPP. Although the BOA can track the theoretical GMPP, the error is still greater than that of the PSO-BOA algorithm. As for convergence speed, the PSO-BOA algorithm shows the fastest convergence speed and the least oscillation. Conversely, the PSO algorithm converges extremely slowly after a sudden change in conditions, oscillating around GMPP. The BOA converges slowly with a large power oscillation amplitude. Notably, in harsh environmental conditions, the PSO-BOA algorithm outperforms different algorithms in the context of both convergence speed and power oscillations.

4.5. Statistics of Results and Analysis

To provide a clearer insight into the performance indicators of each algorithm, statistical tables are created summarizing the results obtained using the above-mentioned three algorithms. The performance of each algorithm is assessed originating from various evaluation metrics, including convergence time, optimization value, projected annual loss cost, and tracking efficiency, as shown in Table 4.

Table 4. Performance comparison of algorithms.

Example	Statistics	Units	PSO-BOA	PSO	BOA
Uniform irradiance	Convergence time	s	0.43	0.79	0.75
	Optimization value	W	8517	8517	8517
	Theoretical	W	8517	8517	8517
	Projected annual loss cost	¥	0	0	0
	Tracking efficiency	%	100	100	100
	Oscillation situation	/	small	big	big
Static shading	Convergence time	s	0.42	1.12	0.82
	Optimization value	W	4374	4373	4374
	Theoretical	W	4374	4374	4374
	Projected annual loss cost	¥	0	2.19	0
	Tracking efficiency	%	100	99.98	100
	Oscillation situation	/	small	Less small	big
Abrupt alterations for irradiance conditions	Convergence time	s	0.37/0.35	0.32/0.35	0.74/0.71
	Optimization value	W	4606/3337	4546/3272	4604/3329
	Theoretical	W	4606/3337	4606/3337	4606/3337
	Projected annual loss cost	¥	0/0	131.40/142.35	4.38/17.52
	Tracking efficiency	%	100/100	98.70/98.05	99.96/99.76
	Oscillation situation	/	Less small	small	bigger
Sudden variations in irradiance and temperature conditions	Convergence time	s	0.47/0.37	0.77/0.92	0.76/0.75
	Optimization value	W	3796/2237	3794/2169	3755/2230
	Theoretical	W	3798/2237	3798/2237	3798/2237
	Projected annual loss cost	¥	4.38/0	8.76/148.92	94.17/15.33
	Tracking efficiency	%	99.94/100	99.89/96.96	98.87/99.69
	Oscillation situation	/	small	bigger	bigger

As illustrated in Table 4, the PSO-BOA algorithm outperforms both the BOA and PSO in all four simulations, achieving faster and smoother convergence to GMPP. It is worth noting that the PSO-BOA algorithm shows a faster convergence speed. The convergence time of PSO-BOA is 54.43% relative to PSO and 57.33% relative to the BOA under uniform irradiance, and is reduced to 37.5% relative to PSO and 51.22% relative to the BOA under local shading conditions. Based on the local electricity prices and assuming an average daily sunlight duration of 12 h, this study calculates the annual cost of losses. In scenarios with abrupt lighting condition changes, the PSO algorithm results in the highest annual loss cost, whereas the PSO-BOA algorithm proves effective in mitigating these losses and reducing costs. Meanwhile, the PSO algorithm achieves faster convergence in cases of sudden changes in irradiance and temperature, but is more prone to falling into local optima, resulting in the worst convergence accuracy. The convergence rate of the BOA is sluggish, with many oscillations occurring over the course of convergence. In contrast, the PSO-BOA algorithm is able to maintain stable and higher optimization accuracy than both the BOA and PSO across all four simulations. These results suggest that the PSO-BOA algorithm effectively addresses the issues of large search oscillation in the BOA and low optimization accuracy in PSO while also improving convergence speed.

5. Conclusions

In response to the issues of low tracking accuracy and susceptibility to local optima in classical PSO algorithms, as well as the problems of slow convergence speed and large oscillation in the BOA, this study introduces a novel PSO-BOA algorithm based on the PSO and BOA. The paper simulated four different scenarios, and the simulation results demonstrate that the PSO-BOA algorithm outperforms the PSO and BOA in terms of convergence accuracy, with a tracking accuracy of no less than 99.94%. In contrast, the PSO algorithm is prone to becoming trapped in local optima, resulting in a convergence accuracy of only 96.96% when both irradiation and temperature undergo abrupt changes. The PSO-BOA algorithm also surpasses both the PSO and BOA algorithms in handling oscillations. In terms of convergence time, the PSO-BOA algorithm shows a significant improvement. Particularly, in scenarios of abrupt changes in irradiation and simultaneous changes in temperature and irradiation, the convergence time of PSO-BOA is less than 0.5 s, while the BOA takes approximately double the time compared to the PSO-BOA. Moreover, the convergence time of the PSO algorithm is relatively longer, and it tends to converge quickly but may be trapped in local optima. Therefore, the proposed algorithm exhibits faster convergence speed, higher tracking accuracy, and smaller oscillations compared to both the PSO and BOA algorithms, which can effectively enhance power supply reliability and safety.

Author Contributions: Conceptualization, Y.W. and S.D.; methodology, Y.W.; software, S.D.; validation, Y.W. and S.D.; formal analysis, Y.W. and S.D.; investigation, P.L. and X.Z.; resources, P.L.; data curation, P.L. and X.Z.; writing—original draft preparation, S.D.; writing—review and editing, Y.W.; visualization, P.L. and X.Z.; supervision, Y.W.; project administration, P.L. and X.Z.; funding acquisition, Y.W. All authors have read and agreed to the published version of the manuscript.

Funding: This project was supported by the National Key Research and Development Program "Key Special Project on Intergovernmental Cooperation for National Scientific and Technological Innovation" (2019YFE0197700), the National Natural Science Foundation of China (NSFC) (61673281, 61903264) and Scientific Research Funding Project of Liaoning Province, China (LJKZ0689), and the Natural Science Foundation of Liaoning Province (2019-KF-03-01).

Data Availability Statement: Data are unavailable due to privacy.

Acknowledgments: The authors sincerely thank the National Key Research and Development Program for their financial support.

Conflicts of Interest: The authors declare no conflict of interest.

References

1. Yu, S.L.; Zhang, L.J.; Lu, H.H.; Fernando, T.; Wong, K.P. A DSE-Based Power System Frequency Restoration Strategy for PV-Integrated Power Systems Considering Solar Irradiance Variations. *IEEE Trans. Ind. Inform.* **2017**, *13*, 2511–2518. [CrossRef]
2. Khemili, F.Z.; Bouhali, O.; Lefouili, M.; Chaib, L.; El-Fergany, A.A.; Agwa, A.M. Design of Cascaded Multilevel Inverter and Enhanced MPPT Method for Large-Scale Photovoltaic System Integration. *Sustainability* **2023**, *15*, 9633. [CrossRef]
3. He, Z.; Tian, Y. Design and Analysis of Comprehensive Solar Utilization System Based on Photovoltaic Concentration and Spectral Splitting. *Processes* **2023**, *11*, 1944.
4. Carreon-Bautista, S.; Eladawy, A.; Nader Mohieldin, A.; Sánchez-Sinencio, E. Boost converter with dynamic input impedance matching for energy harvesting with multi-array thermoelectric generators. *IEEE Trans. Ind. Electron.* **2014**, *61*, 5345–5353. [CrossRef]
5. Xu, W.; Mu, C.X.; Jin, J.X. Novel Linear Iteration Maximum Power Point Tracking Algorithm for Photovoltaic Power Generation. *IEEE Trans. Appl. Supercond.* **2014**, *24*, 1–6.
6. Liu, H.D.; Lu, S.D.; Lee, Y.L.; Lin, C.H. A Novel Photovoltaic Module Quick Regulate MPPT Algorithm for Uniform Irradiation and Partial Shading Conditions. *Processes* **2021**, *9*, 2213.
7. Kumar, N.; Singh, B.; Panigrahi, B.K. PNKLMF-Based Neural Network Control and Learning-Based HC MPPT Technique for Multiobjective Grid Integrated Solar PV Based Distributed Generating System. *IEEE Trans. Ind. Inform.* **2019**, *15*, 3732–3742.
8. Pathak, P.K.; Padmanaban, S.; Yadav, A.K.; Alvi, P.A.; Khan, B. Modified incremental conductance MPPT algorithm for SPV-based grid-tied and stand-alone systems. *IET Gener. Transm. Distrib.* **2022**, *16*, 776–791. [CrossRef]
9. Ahmed, J.; Salam, Z. An improved perturb and observe (P&O) maximum power point tracking (MPPT) algorithm for higher efficiency. *Appl. Energy* **2015**, *150*, 97–108.
10. Goud, J.S.; Kalpana, R.; Singh, B.; Kumar, S. A Global Maximum Power Point Tracking Technique of Partially Shaded Photovoltaic Systems for Constant Voltage Applications. *IEEE Trans. Sustain. Energy* **2019**, *10*, 1950–1959. [CrossRef]
11. Lasheen, M.; Rahman, A.K.A.; Abdel-Salam, M.; Ookawara, S. Adaptive reference voltage-based MPPT technique for PV applications. *IET Renew. Power Gener.* **2017**, *11*, 715–722. [CrossRef]
12. Femia, N.; Petrone, G.; Spagnuolo, G.; Vitelli, M. Optimization of perturb and observe maximum power point tracking method. *IEEE Trans. Power Electron.* **2005**, *20*, 963–973. [CrossRef]
13. Bounechba, H.; Bouzid, A.; Nabti, K.; Benalla, H. Comparison of perturb & observe and fuzzy logic in maximum power point tracker for PV systems. *Energy Procedia* **2014**, *50*, 677–684.

14. Padmanaban, S.; Priyadarshi, N.; Bhaskar, M.S.; Holm-Nielsen, J.B.; Hossain, E.; Azam, F. A Hybrid Photovoltaic-Fuel Cell for Grid Integration with Jaya-Based Maximum Power Point Tracking: Experimental Performance Evaluation. *IEEE Access* **2019**, *7*, 82978–82990. [CrossRef]
15. Rezk, H.; Fathy, A. Simulation of global MPPT based on teaching-learning-based optimization technique for partially shaded PV system. *Electr. Eng.* **2017**, *99*, 847–859. [CrossRef]
16. Chao, K.H.; Rizal, M.N. A Hybrid MPPT Controller Based on the Genetic Algorithm and Ant Colony Optimization for Photovoltaic Systems under Partially Shaded Conditions. *Energies* **2021**, *14*, 2902. [CrossRef]
17. Mohanty, S.; Subudhi, B.; Ray, P.K. A New MPPT Design Using Grey Wolf Optimization Technique for Photovoltaic System Under Partial Shading Conditions. *IEEE Trans. Sustain. Energy* **2016**, *7*, 181–188. [CrossRef]
18. Darcy Gnana Jegha, A.; Subathra, M.S.P.; Manoj Kumar, N.; Subramaniam, U.; Padmanaban, S. A High Gain DC-DC Converter with Grey Wolf Optimizer Based MPPT Algorithm for PV Fed BLDC Motor Drive. *Appl. Sci.* **2020**, *10*, 2797. [CrossRef]
19. Watanabe, R.B.; Ando, O.H., Jr.; Leandro, P.G.M.; Salvadori, F.; Beck, M.F.; Pereira, K.; Brandt, M.H.M.; de Oliveira, F.M. Implementation of the Bio-Inspired Metaheuristic Firefly Algorithm (FA) Applied to Maximum Power Point Tracking of Photovoltaic Systems. *Energies* **2022**, *15*, 5338. [CrossRef]
20. Teshome, D.F.; Lee, C.H.; Lin, Y.W.; Lian, K.L. A modified firefly algorithm for photovoltaic maximum power point tracking control under partial shading. *IEEE Trans. Emerg. Sel. Topics Power Electron.* **2017**, *5*, 661–671. [CrossRef]
21. Benyoucef, A.S.; Chouder, A.; Kara, K.; Silvestre, S. Artificial bee colony based algorithm for maximum power point tracking (MPPT) for PV systems operating under partial shaded conditions. *Appl. Soft Comput.* **2015**, *32*, 38–48. [CrossRef]
22. Gong, L.; Hou, G.; Huang, C. A two-stage MPPT controller for PV system based on the improved artificial bee colony and simultaneous heat transfer search algorithm. *ISA Trans.* **2023**, *132*, 428–443. [CrossRef] [PubMed]
23. Kennedy, J.; Eberhart, R. Particle Swarm Optimization. In Proceedings of the IEEE International Conference on Neural Networks, Perth, Australia, 27 November–1 December 1995.
24. Chen, L.R.; Tsai, C.H.; Lin, Y.L.; Lai, Y.S. A biological swarm chasing algorithm for tracking the PV maximum power point. *IEEE Trans. Energy Convers.* **2010**, *25*, 484–493. [CrossRef]
25. Panda, K.; Anand, A.; Bana, P.; Panda, G. Novel PWM Control with Modified PSO-MPPT Algorithm for Reduced Switch MLI Based Standalone PV System. *Int. J. Emerg. Electr. Power Syst.* **2018**, *19*, 20180023. [CrossRef]
26. Mathi, D.K.; Chinthamalla, R. A hybrid global maximum power point tracking method based on butterfly particle swarm optimization and perturb and observe algorithms for a photovoltaic system under partially shaded conditions. *Int. Trans. Electr. Energy Syst.* **2020**, *30*, e12543. [CrossRef]
27. Sen, T.; Pragallapati, N.; Agarwal, V.; Kumar, R. Global maximum power point tracking of PV arrays under partial shading conditions using a modified particle velocity-based PSO technique. *IET Renew. Power Gener.* **2018**, *12*, 555–564. [CrossRef]
28. Makhloufifi, S.; Mekhilef, S. Logarithmic PSO-Based Global/Local Maximum Power Point Tracker for Partially Shaded Photovoltaic Systems. *IEEE J. Emerg. Sel. Top. Power Electron.* **2021**, *10*, 375–386. [CrossRef]
29. Arora, S.; Singh, S. Butterfly optimization algorithm: A novel approach for global optimization. *Soft Comput.* **2019**, *23*, 715–734. [CrossRef]
30. Arora, S.; Singh, S. An improved butterfly optimization algorithm with chaos. *J. Intell. Fuzzy Syst.* **2017**, *32*, 1079–1088. [CrossRef]
31. Yuan, Z.; Wang, W.; Wang, H.; Khodaei, H. Improved Butterfly Optimization Algorithm for CCHP Driven by PEMFC. *Appl. Therm. Eng.* **2019**, *173*, 114766.
32. Aygül, K.; Cikan, M.; Demirdelen, T.; Tumay, M. Butterfly optimization algorithm based maximum power point tracking of photovoltaic systems under partial shading condition. *Energy Sources Part A Recovery Util. Environ. Eff.* **2023**, *45*, 8337–8355. [CrossRef]
33. Shams, I.; Mekhilef, S.; Tey, K.S. Maximum Power Point Tracking Using Modified Butterfly Optimization Algorithm for Partial Shading, Uniform Shading, and Fast Varying Load Conditions. *IEEE Trans. Power Electron.* **2021**, *36*, 5569–5581. [CrossRef]
34. Aljafari, B.; Balachandran, P.K.; Samithas, D. Solar photovoltaic converter controller using opposition-based reinforcement learning with butterfly optimization algorithm under partial shading conditions. *Environ. Sci. Pollut. Res.* **2023**, *30*, 72617–72640. [CrossRef] [PubMed]
35. Tiwari, G.N.; Tiwari, A. *Handbook of Solar Energy—Theory, Analysis and Applications*; Springer: Singapore, 2017.
36. Humada, A.M.; Hojabri, M.; Mekhilef, S.; Hamada, H.M. Solar cell parameters extraction based on single and double-diode models: A review. *Renew. Sustain. Energy Rev.* **2016**, *56*, 494–509. [CrossRef]
37. Sedra, A.; Smith, K. *Microelectronic Circuits*, 6th ed.; Oxford University Press: New York, NY, USA, 2011.
38. Vimalarani, C.; Kamaraj, N.; Chitti Babu, B. Improved method of maximum power point tracking of photovoltaic (PV) array using hybrid intelligent controller. *Optik* **2018**, *168*, 403–415.
39. Fathy, A.; El-arini, M.; Othman, A. A New evolutionary algorithm for the optimal sizing of stand-alone photovoltaic system based on genetic algorithm. *Int. Rev. Electr. Eng.* **2013**, *8*, 1067–1075.
40. Singh, B.; Anand, P. A novel adaptive butterfly optimization algorithm. *Int. J. Comput. Mater. Sci. Eng.* **2019**, *7*, 1850026. [CrossRef]
41. May, R. Simple mathematical models with very complicated dynamics. *Nature* **1976**, *261*, 459–467. [CrossRef]

42. Aguila-Camacho, N.; Duarte-Mermoud, M.A.; Gallegos, J.A. Lyapunov functions for fractional order systems. *Commun. Nonlinear Sci. Numer. Simul.* **2014**, *19*, 2951–2957. [CrossRef]
43. Zhang, M.; Long, D.; Qin, T.; Yang, J. A Chaotic Hybrid Butterfly Optimization Algorithm with Particle Swarm Optimization for High-Dimensional Optimization Problems. *Symmetry* **2020**, *12*, 1800. [CrossRef]

sustainability

MDPI

Article

Horse Herd Optimized Intelligent Controller for Sustainable PV Interface Grid-Connected System: A Qualitative Approach

Anupama Ganguly [1], Pabitra Kumar Biswas [1], Chiranjit Sain [2], Ahmad Taher Azar [3,4,5,*], Ahmed Redha Mahlous [3,4] and Saim Ahmed [4]

1 Department of Electrical and Electronics Engineering, National Institute of Technology Mizoram, Aizawl 796012, India; anupama.babi@gmail.com (A.G.); pabitra.eee@nitmz.ac.in (P.K.B.)

2 Electrical Engineering Department, Ghani Khan Choudhury Institute of Engineering and Technology, Malda 732141, India; chiranjit@gkciet.ac.in

3 College of Computer and Information Sciences, Prince Sultan University, Riyadh 11586, Saudi Arabia; armahlous@psu.edu.sa

4 Automated Systems and Soft Computing Lab (ASSCL), Prince Sultan University, Riyadh 11586, Saudi Arabia; sahmed@psu.edu.sa

5 Faculty of Computers and Artificial Intelligence, Benha University, Benha 13518, Egypt

* Correspondence: aazar@psu.edu.sa or ahmad.azar@fci.bu.edu.eg or ahmad_t_azar@ieee.org

Abstract: The need for energy is always increasing as civilization evolves. Renewable energy sources are crucial for meeting energy demands as conventional fuel resources are slowly running out. Researchers are working to extract the most amount of power possible from renewable resources. Numerous resources are in demand, including solar, wind, biomass, tidal, and geothermal resources. Solar energy outperformed all the aforementioned resources in terms of efficiency, cleanliness, and pollution freeness. Intermittency, however, is the resource's main shortcoming. Maximum power point tracking algorithm (MPPT) integration is required for the system to achieve continuous optimum power by overcoming the feature of intermittency. However, generating electrical energy from solar energy has presented a significant problem in ensuring the output power's quality within a reasonable range. Total harmonic distortion (THD), a phenomenon, may have an impact on the power quality. Depending on the properties of the load, variables like power factor, voltage sag/swell, frequency, and unbalancing may occur. The quality of power and its criterion exhibits a non-linear connection. The article's primary objective is to analyze the PV interface grid-linked system's qualitative and quantitative performance. With respect to varying solar irradiation conditions, partial shading conditions, and solar power quality within the acceptable dimension, a novel intelligent multiple-objective horse herd optimization (HHO)-based adaptive fractional order PID (HHO-AFOPID) controller is used to achieve this goal. Adaptive fractional order PID (AFOPID), conventional FOPID, and PID controllers were used to evaluate the performance of the suggested controller, which was then validated using a commercially available PV panel in MATLAB/Simulink by varying the productivity of non-conventional resources, the inverter's level of uncertainty, and the potential at the grid's end. In order to realize the features of the system, sensitivity examination is also carried out for solar energy's sensitive parameters. The stability analysis of the proposed control topology is also carried out in terms of the integral absolute error (IAE) and integral time absolute error (ITAE). The examination of the sensitivity of variations in solar radiation in kilowatt per square meter per day is based on the total net present cost (TNPC) and levelized cost of energy (LCOE), as optimal dimension and energy cost are both aspects of priority. The suggested control methodology is an approach for the qualitative and quantitative performance analysis of a PV interface grid-oriented system.

Keywords: horse herd optimization; adaptive fractional order PID controller; sensitivity analysis; power quality and quantity analysis; maximum power point tracking

Citation: Ganguly, A.; Biswas, P.K.; Sain, C.; Azar, A.T.; Mahlous, A.R.; Ahmed, S. Horse Herd Optimized Intelligent Controller for Sustainable PV Interface Grid-Connected System: A Qualitative Approach. *Sustainability* **2023**, *15*, 11160. https://doi.org/10.3390/su151411160

Academic Editors: Prince Winston David and Praveen Kumar B

Received: 27 May 2023
Revised: 6 July 2023
Accepted: 7 July 2023
Published: 18 July 2023

1. Introduction

With increasing awareness of how quickly conventional energy sources like coal and petroleum products are depleting, renewable sources of energy are gaining relevance on a global scale today [1]. According to data from the U.S. Energy Information Administration's monthly energy review article as of June 2021, natural gas accounts for 58.8% of all energy usage, followed by electricity at 39.1% and petroleum at 9.6% and 7%. The situation in India indicates a greater reliance on coal and oil. As a result, conventional resources are quickly running out. Due to the availability of non-conventional energy sources in nature, they are prepared to offer services at permissible tweaks. As previously noted, solar energy is the most practical resource for creating electrical energy because it does not generate any greenhouse emissions or other hazards. However, due to the inability to deliver optimum power under load without interruption, it is necessary to implement the maximum power point tracking (MPPT) method. In addition to maximizing the power generated by the PV source, MPPT also helps the PV system last longer [2].

There are numerous ways to obtain the maximum power possible out of a photovoltaic source [3–7]. The ability to follow the actual maximum power point (MPP), speed of convergence, robustness, efficiency, cost, and hardware implementation is the criteria used to classify MPPTs. MPPTs are divided into three categories based on the aforementioned requirements:online, hybrid, and offline methods. Offline techniques are dependent on the solar cell model's parameters. It is also known as a "model-based" approach. Online approaches are known as "model free" methods, which denote that they are independent of the solar model's parameters. The two previously described strategies are combined to create the hybrid method [8]. The most popular online MPPT techniques are perturb and observe (P&O). Incremental conductance (IC) and hill climbing (HC) are two techniques that function effectively when ambient temperatures and solar irradiation do not fluctuate quickly [9–13], but offline approaches are effective when solar irradiation is changing rapidly. Popular techniques include ANN-based MPPT, PI, fuzzy logic controller (FLC), GA- and ACO-optimized MPPT, and others [14]. Hybrid MPPTs are used to retain the maximum amount of PV panel power production in order to circumvent the drawbacks of the aforementioned approaches [15–17]. A grid-connected system's power quality should also be evaluated in addition to power quantity monitoring. The interfacing of nonlinear loads, such as power electronics components, results in harmonic content, which lowers the quality of the power sent to the grid. In order to manage the power quality within a certain range, DVR and D STATCOM are introduced [18]. IEEE Std (1250-2011) [19], states that the maximum voltage deviation is 10% of the base value, the maximum frequency deviation is ± 0.1 Hz, and the maximum voltage/current deviation is 5% of the base value. According to IEEE Std (519-2014) [20], THD shall not exceed 5%, while IEC [60831-1/2] standards specify that the power factor must be more than or equal to 0.9 [20]. When operating a PV system connected to the grid, good power quality must be satisfied while taking into account all of the aforementioned constraints. Due to the fractional order PID controller's quick convergence and response for both linear and nonlinear loads, the theory of factional calculus gained popularity [19,20]. The enhanced version of AFO+PID, known as adaptive fractional order PID controller, is applicable for both linear and non-linear loads and has a high degree of efficiency when operating in perturbed conditions. The system is interfaced with an adaptive FOPID controller to increase gain and robustness. The ideal system's dimension and energy cost must also be taken into consideration in order to ensure the successful functioning of a system, in addition to power quantity and quality study. By adjusting peak demand when carrying a fixed energy demand and vice versa, the grid-interfaced system is optimized to reduce energy costs (COEs), with a reduction in peak load conditions [21]. The implications of various storage capacities on the performance analysis of a hybrid micro-grid system are described in [22] along with a sensitivity analysis. In [23], the performance evaluation of HRES was provided, and it was proven that the pumped storage hydropower plant was the best option in terms of cost savings. By talking about the constraints, a novel, multi-objective HHO-AFOPID control topology is provided to overcome the problems, and the evaluations of sensitive SPV parameters that are dependent on LCOE and TNPC are carried out.

The work's main contributions are noted below:

- The recommended controller works with the property of adaptation, and an intelligent algorithm calibrates the settings. Horse herd optimization uses partial shade and different irradiances. By using the specified (HHO-AFOPID) controller, the system was able to successfully collect 100 kW of solar energy. By considering the adaptive strategy of the AFOPID controller from a hybrid PO-NDPID MPPT controller, dc link voltage and current control logic are implemented, along with an illustration of the quadrature axis that is also implemented.
- To improve the power quality, the combined impacts of voltage deviation, total harmonic distortion (THD), and frequency fluctuation have been researched and managed.
- Undershoot, settling time, ripples, integral absolute error (IAE), and integral time absolute error (ITAE) have all been included in the evaluation of the suggested controller.
- To determine how the PV interface grid-linked system would react to changes in irradiation data, sensitivity analysis based on LCOE and TNPC has also been conducted. To evaluate the system's performance in light of the changes to sensitive parameters, it is crucial to carry this out.

The organization of this article is as follows: Section 2: System Model Simulation; Section 3: Solar Photovoltaic (SPV) Control Implementation Adaptive Fractional Order PID Controller Design; Section 4: Results and Clarifications; Section 5: Sensitivity Analysis; Section 6: Conclusions and Future Directions.

2. System Model Simulation

2.1. Photovoltaic Modeling

The PV system is a nonlinear source since it consists of a parallel-connected current source and a diode, where R_s denotes the metal junction loss.

The PV model is shown in Figure 1. The generation of photovoltaic current results from the action of electron flow, and its magnitude is exactly proportional to the quantity of irradiance (G), with low variations due to ambient temperature (T) and the use of the Shockley equation by [24,25].

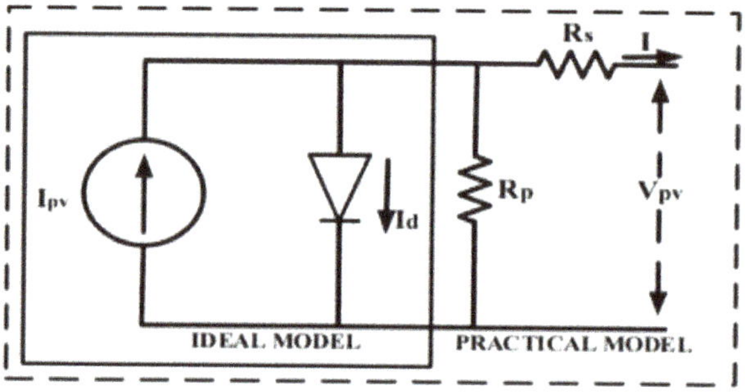

Figure 1. Combined model of photovoltaic generation.

$$\mathrm{I} = I_{pv} - I_d \tag{1}$$

The PV system output can be addressed by:

$$I_{pv} = I_{ph} - \mathrm{I}\left(e^{\frac{qvp}{KT}} - 1\right) \tag{2}$$

The output power of the PV panel is formulated by:

$$P_{pv} = V_{pv} I_{pv} \tag{3}$$

The output of solar energy is a property of the photovoltaic panels' rated capacity, and the PV de-rating factor, a measuring factor, permits the impact of many losses that may cause the PV module's output to be less than what it was originally [26]. The power generated by solar photovoltaics' (SPV) can be visualized as [26] follows:

$$Power_{sp} = P_s D_s \left(\frac{I_a}{I_{aSTC}} \right) [1 + \beta_a (T_a - T_{a,STC})] \tag{4}$$

P_s = estimated capacity of the SPV array (kW);
D_s = solar de-rating factor (%);
I_a = Incident SPV irradiation on the SPV array in present time (kW/m^2);
I_{aSTC} = incident SPV irradiation at the standard test condition (1 kW/m^2);
β_a = coefficient of the power temperature (%);
T_a = temperature of the SPV cell (°C);
$T_{a,STC}$ = SPV cell temperature at the standard test condition (25 °C).

If β_a is overlooked in the PV interface grid system, then the power outcome may be conveyed as follows:

$$Power_{sp} = P_s D_s \left(\frac{I_a}{I_{aSTC}} \right) \tag{5}$$

Partial Shading and the Impact of Bypass Diode

When SPV arrays are connected in series, they are subjected to constant illumination, and P-V I-V curves have a single maximum power point. However, they are not meant to receive homogenous irradiation when they are combined in a series–parallel fashion. Due to this occurrence, some arrays are unable to produce the intended results. Poor efficiency is the end result, and hot spot and non-matching problems are also introduced [27]. The purpose of the bypass diode is to decrease the effects of mismatching-related problems [28]. Figure 2 depicts a solar photovoltaic system with a blocking and bypass diode in a series-shunt configuration.

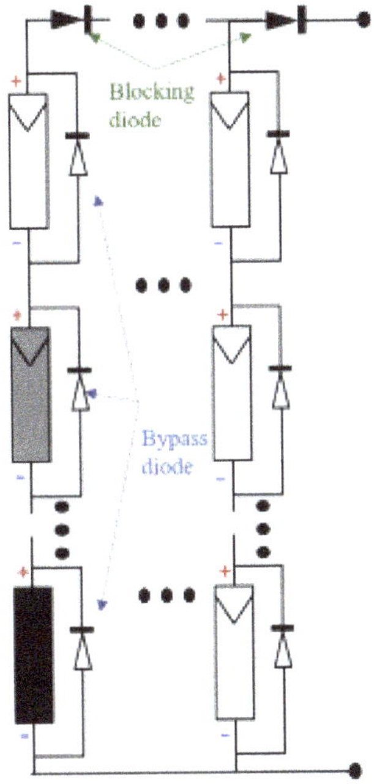

Figure 2. SPV with a bypass diode and blocking diode designed in series-shunt combination.

2.2. Converter

Due to intermittent problems with generating solar photovoltaic power, it is possible that the DC voltage produced is not large enough to support further research. It needs a DC-DC boost converter attached to it in order to increase to the appropriate level. The boost converter is modeled using typical calculations based on the system rating. An inverter with a 2 L voltage source performs the DC-AC conversion. The inverter's output is represented as follows:

$$P_{inverter,output} = \eta_{inverter} \times P_{DC} \tag{6}$$

$\eta_{inverter}$ = efficacy of the inverter;
P_{DC} = power outcome from DC-DC boost converter;
$P_{inverter,output}$ = Power delivered from the inverter.

2.3. LCL Filter

The LCL filter is primarily employed in grid-interfaced solar photovoltaic systems, and its an amalgamation of $L_1 + R_1$, $L_2 + R_2$ and C + R_c, which is applied in order to mitigate the harmonic effect introduced during PWM operations. Equation (7) depicts the numerical expressions:

$$\begin{cases} L_1\frac{di_1}{dt} + R_1 i_1 = v_i - v_c - R_c i_c \\ L_2\frac{di_g}{dt} + R_2 i_g = v_c - v_g - R_c i_c \\ C\frac{dv_c}{dt} = i_c \\ i_1 = i_g + i_c \end{cases} \tag{7}$$

v_g = potential of the grid end;
v_i = voltage of the inverter;
R_1 and R_2 represent the resistors of the inductors L_1 and L_2 consecutively.

Neglecting the resistors from the aforementioned equations, the Laplace transform of the LCL filter is derived using Equation (8).

$$G_{LCL\_}\mathrm{f(s)} = \frac{1}{s^3 L_1 L_2 C + s(L_1 + L_2)} \tag{8}$$

The resonant frequency of the used filter is shown in Equation (9).

$$f_{resonant} = \frac{1}{2\pi}\sqrt{\frac{L_1 + L_2}{L_1 L_2 C}} \tag{9}$$

The current and voltage of the grid end can be related using the transfer equation of the system model, and this is shown in Equation (10).

$$\mathrm{H(s)} = \frac{i_g}{v_i} = \frac{RCs + 1}{L_1 L_2 C s^3 + RC(L_1 + L_2)s^2 + (L_1 + L_2)s} \tag{10}$$

3. Solar Photovoltaic (SPV) Control Implementation Adaptive Fractional Order PID Controller Design

In this section, the three phases of the proposed HHO-AFOPID's execution are shown as a photovoltaic source's interface with the grid system.

$$\mathrm{V} = \frac{2}{3}\left(v_{aN} + \exp\left(\frac{j2\pi}{3}\right) v_{bN} + \exp\left(\frac{j4\pi}{3}\right) v_{cN}\right) \tag{11}$$

v_{aN}, v_{bN}, and v_{cN} are termed as voltages falling between phase and neutral. The grid infrastructure of the grid-interfaced dc-to-ac inverter in a d-q frame may be arranged using the following differential equations [29].

$$\frac{di_a}{dt} = -\frac{R}{L}i_a + \frac{1}{L}v_{aN} - \frac{1}{L}e_{aN} \tag{12}$$

$$\frac{di_b}{dt} = -\frac{R}{L}i_b + \frac{1}{L}v_{bN} - \frac{1}{L}e_{bN} \tag{13}$$

$$\frac{di_c}{dt} = -\frac{R}{L}i_c + \frac{1}{L}v_{cN} - \frac{1}{L}e_{cN} \tag{14}$$

The *dc* end of the inverter can be expressed in the form of a differential equation while disregarding losses that happened in the switches of the inverter. The following diagram illustrates how the inductor's resistance and rate of energy fluctuate [30].

$$C\frac{dv_{dc}}{dt} = i_p - i_{dc} = i_p - \frac{e_d i_d + e_q i_q}{v_{dc}} \tag{15}$$

e_d, and e_q and i_d and i_q shows the potentials and currents of the grid end in a movable d-q frame. C represents the capacitor of the dc bus, i_p exhibits the photovoltaic current, v_{dc} is the potential of the dc link.

The state equation of the SPV system can be depicted by applying abc-dq transformations in Equation (16) as follows.

$$\begin{cases} \dot{x_1} = -a_1x_1 + a_2x_2 - a_3 + a_4u_1 \\ \dot{x_2} = -a_2x_1 - a_1x_2 - a_5 + a_4u_2 \\ \dot{x_3} = a_6 - \frac{a_7x_1 + a_8x_2}{a_9x_3} \end{cases} \tag{16}$$

The notations are provided below.

$x_1 = I_d$, $x_2 = I_q$, $x_3 = v_{dc}$, $u_1 = v_d$, $u_2 = v_q$, $a_1 = \frac{R}{L}$, $a_2 = \omega$, $a_3 = \frac{e_d}{L}$, $a_4 = \frac{1}{L}$, $a_5 = \frac{e_d}{L}$, $a_6 = \frac{i_p}{C}$, $a_7 = e_d$, $a_8 = e_q$, $a_9 = C$.

The non-linear features of solar photovoltaic systems are shown in Equations (17a) and (17b):

$$\dot{x} = f(x) + g_1(x)u_1 + g_2(x)u_2 \tag{17a}$$

$$\dot{x} = \begin{pmatrix} -a_1x_1 & a_2x_2 & a_3 \\ -a_2x_1 & -a_1x_2 & -a_5 \\ a_6 & -\frac{a_7x_1 + a_8x_2}{a_9x_3} & 0 \end{pmatrix} \tag{17b}$$

The suggested controller's main goal is to send the best switching signal possible to the inverter in order to draw the most amount of power possible from the PV source. The maximum power point is followed by the control topology in this article to complete the research component. Table 1 displays the PV parameter, while the attributes of the grid are shown in Table 2.

Table 1. Parameter of the PV system.

Quantity	Value
Rating of PV system	100 kW
Optimum power of one module (W)	215
Potential at maximum power point (V_{MPP}) (V)	39.8
Current at maximum power point (I_{MPP}) (A)	6.4

Table 2. Parameters of the grid's infrastructure.

Parameter	Parametric Value
R (ohm)	0.15
L (mHenry)	1.85
C (mFarad)	4.8
E (Volt)	77.8
f (cps/Hz)	50

The execution of the three-phase inverter approach in the d-q frame can be analyzed using

$$v_d = e_d + \mathrm{R}i_d + \mathrm{L}\frac{di_d}{dt} + \omega \mathrm{L}i_q \quad (18)$$

$$v_q = e_q + \mathrm{R}i_q + \mathrm{L}\frac{di_q}{dt} - \omega \mathrm{L}i_d \quad (19)$$

e_d, and e_q; i_d, and i_q; and v_d, and v_q represent the currents, potentials, and output potentials of a PV-interfaced inverter, respectively. The alternating quantity's frequency is represented by the symbol. The equation for AC and DC power stability is given as follows:

$$e_d i_d + e_q i_q = v_{dc} i_{dc} \quad (20)$$

where v_{dc} and i_{dc} represent the input parameters of the PV-interfaced inverter.

3.1. Design an Adaptive Fractional Order PID Controller

FOPID controllers with adaptive properties are initiated to generate the output power, which is more prominent and showed robustness [31–34].

$$\alpha.D_t^{\alpha} = \begin{cases} \frac{d}{dt}\frac{\alpha}{t^{\alpha}},\ \alpha > 0 \\ 1,\ \alpha = 0 \\ \int_{\alpha}^{t} d\tau^{\alpha}, \alpha < 0 \end{cases} \quad (21)$$

For the requirements of the system, the upper and lower boundaries are measured by α and t when $\alpha \in \mathrm{R}$ with respect to the order of operation. The performance was compared using conventional PID and FOPID and without optimized AFOPID. The suggested controller showed better non-erroneous responses that are applicable to both linear and non-linear loads.

The FOPID is the modified version of a conventional PID controller with two parameters added: fractional integrator order (λ) and fractional derivative order (μ). As a result, it exhibits the outcome's quality [35,36].

$$\mathrm{G(S)} = k_P + \frac{k_I}{S^{\lambda}} + k_D S^{\mu} \quad (22)$$

3.2. Computational Formation of the HHO Optimization Algorithm

Horse herd optimization (HHO) was developed based on the way horses behave in their original habitat. A few common behavioral traits of horses include hierarchy, grazing, imitation, sociability, defense mechanism, and roaming [37,38]. The method is motivated by these six attitudes towards horses of various ages. At each phase, horses are moved in accordance with Equation (23):

$$P_m^{iter,age} = Vel_m^{iter,age} + P_m^{(iter-1),age} \quad (23)$$

Age = α, β, γ, and δ.

For the above equation, we have the following:

- $P_m^{iter,age}$ represents the position of the mth horse;
- The range of each horse is shown by age;
- The current number of iterations is given by iter;
- $Vel_m^{iter,age}$ provides the velocity of the particular horse.

During their life span, the horses show different behaviors. The average life span of a horse is 25–30 years [38]. Where Δ represents horses that age between 0 and 5; γ shows those aged between 5 and 10; β denotes ages between 10 and 15; α denotes those older than 15 years old. Each iteration should have a thorough examination to ascertain the horses' ages. The top 10% of horses in the ordered matrix are picked as horses, with the remaining horses chosen from the remaining 90% of horses in the ordered matrix. In total, 20% of the population after that makes up the group. Moreover, 30% and 40% of the remaining horses belong to the groupings, respectively. The techniques that statistically mimic the six movements of various groups of horses are used to compute the velocity vector.

Considering the following behavioral trends [39], Equations (24)–(27) might be considered the motion vectors of various aged horses throughout every iteration cycle.

$$Vel_m^{iter,\alpha} = Gra_m^{iter,\alpha} + DefenseMec_m^{iter,\alpha} \tag{24}$$

$$Vel_m^{iter,\beta} = Gra_m^{iter,\beta} + H_m^{iter,\beta} + Soc_m^{iter,\beta} + DefenseMec_m^{iter,\beta} \tag{25}$$

$$Vel_m^{iter,\gamma} = Gra_m^{iter,\gamma} + H_m^{iter,\gamma} + Soc_m^{iter,\gamma} + Imi_m^{iter,\gamma} + Roam_m^{iter,\gamma} + DefenseMec_m^{iter,\gamma} \tag{26}$$

$$Vel_m^{iter,\delta} = Gra_m^{iter,\delta} + Imi_m^{iter,\delta} + Roam_m^{iter,\delta} \tag{27}$$

These are the key stages of a horse's social and individual intelligence.

3.2.1. Grazing (Gra)

Horses are roving creatures that consume fodder such as grasses and plants. With only a few hours of respite, they graze in pastures for 16 to 20 h every day. This kind of progressive grazing is known as continuous eating; you may have observed mares graze in pastures while carrying their foals [38]. The HHO method is used to represent each horse's grazing space. Each horse grazes in a specific spot due to coefficient g. Horses graze throughout their entire lives at any age. Grazing is carried out along a line using mathematical Equations (28) and (29).

$$Gra_m^{iter,age} = g_{iter}(low + r * upp)\left(P_m^{iter-1}\right),\ \text{age} = \alpha, \beta, \gamma, \delta \tag{28}$$

$$g_m^{iter,age} = w_g \times g_m^{(iter-1)age} \tag{29}$$

$Gra_m^{iter,age}$ indicates the horse's range of motion and shows how well the associated horse can graze. For each cycle, the grazing variable decreases linearly at w_g.

While "low" and "upp" represent the bottom and higher limits of the grazing space, respectively, variable "r" has an arbitrary value between 0 and 1. It is suggested that "low" and "upp" should be adjusted to 0.95 and 1.05, respectively, for all age groups. In all age ranges, coefficient g's value is set to 1.5.

3.2.2. Hierarchy (H)

Horses are dependent. They follow a leader throughout their life, which is a behavior that people frequently exhibit; based on the principle of domination [38], a mature stallion or a filly is also in charge of overseeing a herd of wild horses. Coefficient h_m in HHO is defined as the propensity of a group of horses to follow the direction of the most skilled and powerful horse. Horses follow the law of hierarchy when they are between the ages of

5 and 15 years according to research.Equations (30) and (31) are useful for establishing this.

$$H_m^{iter,age} = h_m^{iter,age} \left(P_{lbh}^{iter-1} - P_m^{iter-1} \right) \tag{30}$$

$$h_m^{iter,age} = h_m^{iter-1,age} \times w_g \tag{31}$$

$H_m^{iter,age}$ shows the optimal horse's location using the velocity variable. P_{lbh}^{iter-1} identifies where the finest horse is standing.

3.2.3. Sociability (Soc)

Horses may cohabit with various animal species and require social contact. Living in a herd protects wild horses from predators that may pursue them. Pluralism increases their chances of surviving and makes escaping simpler. Due to their social nature and the fact that they are so unique, horses regularly fight with one another.

Some horses seem to prefer being around other animals like sheep and cattle, but they despise being by themselves [38].

The following calculations demonstrate that horses between the ages of 5 and 15 years are most interested in being with a herd.

$$Soc_m^{iter,age} = soc_m^{iter,age} \left[\left(\frac{1}{N} \sum_{j=1}^{N} P_j^{iter-1} \right) - P_m^{iter-1} \right] \text{age} = \beta, \gamma \tag{32}$$

$$soc_m^{iter,age} = soc_m^{iter-1,age} \times \omega_{Soc} \tag{33}$$

The above-mentioned equations reveal the following:

$Soc_m^{iter,age}$ explains the social motion vector that the ith horse presents.

$soc_m^{iter,age}$ demonstrates how the horse is facing the direction of group ith.

iter, which has a parameter of ω_s, reduces the iteration with each cycle.

The total number of horses is expressed by N.

Age is a representation of each horse's age range.

By evaluating these factors, the derivations of coefficients γ and β are carried out.

3.2.4. Imitation (Im)

Horses mimic one another and learn from one another's good and bad habits, such as where the finest feeding area is [38]. Young horses have a tendency to imitate elder ones, and this practice is sustained until the end of their life span, as explained in Equations (33) and (34).

$$Im_m^{iter,age} = im_m^{iter,age} \left[\left(\frac{1}{pN} \sum_{j=1}^{pN} P_j^{iter-1} \right) - P^{iter-1} \right] \tag{34}$$

$$im_m^{iter,age} = im_m^{iter-1,age} \times \omega_{im} \tag{35}$$

The contributions from the above set of equations are listed as follows.

$Im_m^{iter,age}$ expresses the motion vector that shows the ith horse among the best choice of horses at P position.

$im_m^{iter,age}$ represents the inclination of that particular horse in the orientation of the group on the ith cycle.

N shows the best position's horse number. p is the category of the 10% of chosen horses.

ω_{im} Factor denotes the factor of reduction/iteration for i_{iter}.

3.2.5. Defense Mechanism (Defense Mec)

Since they have historically been preyed upon, the horses' behavior reflects this. Horses fight for food and water to keep rivals at bay and to avoid hazardous areas where foes such as wolves may lurk. They also buck when caught. Horses engage in a fight-or-flight reaction to defend themselves [38]. In the HHO approach, the horses' defense mechanism allows them to escape from any other horses that display inappropriate or unfavorable behavior. This characteristic describes their main line of defense. As mentioned earlier, horses must flee from or engage in combat with their adversaries. A young or adult horse has such a defense system in place whenever it is practical. Equations (35) and (36) describe the defensive strategies of horses that do not allow other animals to enter dangerous regions.

$$DefenseMec_m^{iter,age} = defensemec_m^{iter,age}\left[\left(\frac{1}{qN}\sum_{j=1}^{qN} P_j^{iter-1}\right) - P^{iter-1}\right] \quad (36)$$

Age = α, β, γ, δ

$$defensemec_m^{iter,age} = defensemec_m^{iter-1,age} \times \omega_{defensemec} \quad (37)$$

From the above equations, we have the following:

$Defense\ Mec_m^{iter,age}$, based on the average location of a horse in the worst P position, describes the escape vector of the ith horse.

Here, q is equal to 20% of the total number of horses, and qN displays the number of horses in the poorest situations.

$\omega_{defense\ mec}$ indicates the earlier determined reduction factor per cycle for iter.

3.2.6. Roam(Roam)

Horses travel and graze around the countryside in search of nourishment, moving from pasture to pasture. Although they maintain the aforementioned feature, most horses are kept in stables. A horse could rapidly switch where it grazes. Due to their intense curiosity, horses routinely visit other pastures to familiarize themselves with their environment. Through the side walls of their enclosures, the horses can see one another, and a suitable stable satisfies their need for socialization [38].

The program mimics this behavior by using factor r, which is nothing more than a random movement. When horses are young, roaming is almost never observed, and it gradually decreases as they grow older. Equations (37) and (38) depict the variables of roaming.

$$Roam_m^{iter,age} = roam_m^{iter,age}\ \delta \mathrm{P}(\mathrm{iter}-1) \quad (38)$$

Age γ, δ

$$roam_m^{iter,age} = roam_m^{iter-1,age} \times \omega_{roam} \quad (39)$$

$Roam_m^{iter,age}$ is the ith horse's arbitrary velocity vector for local searches and escapes from local minima.

ω_{roam} = factor of reduction in the $roam_m^{iter,age}$/cycle.

Substituting the outcomes from (28) to (39) into Equations (24)–(27), the generic velocity vector may be established.

The velocity of the horses belonging to the age group of 0–5 years is defined as follows, which is defined as δ.

$$Vel_m^{iter,\delta} = \left[\omega_g \times g_m^{iter-1,\delta}\ (\mathrm{low} + \mathrm{r}\ \times\ \mathrm{upp})\left(P_m^{iter-1}\right)\right] + im_m^{iter-1,\delta} \times \omega_{im} \times \left[\left(\frac{1}{pN}\sum_{j=1}^{pN} P_j^{iter-1}\right) \times \left(P^{iter-1}\right)\right] + roam_m^{iter-1,\delta} \times \omega_{roam} \times \delta P^{iter-1} \quad (40)$$

The velocity of the horses belonging to the 5–10-year age group is described as follows, and it is defined as γ:

$$\begin{aligned} Vel_m^{iter,\gamma} = & \left[\omega_g \times g_m^{iter-1,\gamma}(\text{low} + \text{r} \times \text{upp})\left(P_m^{iter-1}\right)\right] + h_m^{iter-1,\gamma} \times \omega_h \times \\ & \left[P_{lbh}^{iter-1,\gamma} - P_m^{iter-1}\right] \\ & + soc_m^{iter-1,\gamma} \times \omega_{soc} \times \left[\frac{1}{N}\sum_{j=1}^{N} P_j^{iter-1,\gamma} - P^{iter-1}\right] im_m^{iter-1,\gamma} \times \omega_{im} \times \\ & \left[\left(\frac{1}{pN}\sum_{j=1}^{pN} P_j^{iter-1}\right) - P^{iter-1}\right] + \\ & roam_m^{iter-1,\gamma} \times \omega_{roam} \times \delta P^{iter-1} \end{aligned} \quad (41)$$

The velocity of the horses belonging to the 5–10 year age group is described as follows, and it is defined as β.

$$\begin{aligned} Vel_m^{iter,\beta} = & \left[\omega_g \times g_m^{iter-1,\beta}(\text{low} + \text{r} \times \text{upp})\left(P_m^{iter-1}\right)\right] + h_m^{iter-1,\beta} \times \omega_h \times \left[P_{lbh}^{iter-1,\beta} - P_m^{iter-1}\right] \\ & + soc_m^{iter-1,\beta} \times \omega_{soc} \times \left[\frac{1}{N}\sum_{j=1}^{N} P_j^{iter-1,\beta} - P^{iter-1}\right] - \\ & defensemec_m^{iter-1,\beta} \times \omega_{defensemec} \times \\ & \left[\left(\frac{1}{qN}\sum_{j=1}^{qN} P_j^{iter-1}\right) - P^{iter-1}\right] \end{aligned} \quad (42)$$

The velocity of the horses belonging to the >15 age group is described below, and it is defined as α.

$$\begin{aligned} Vel_m^{iter,\alpha} = & \left[\omega_g - g_m^{iter-1,\alpha}(\text{low} + \text{r} \times \text{upp})\left(P_m^{iter-1}\right)\right] - defensemec_m^{iter-1,\alpha} \times \\ & \omega_{defensemec} \times \left[\left(\frac{1}{qN}\sum_{j=1}^{qN} P_j^{iter-1}\right) - P^{iter-1}\right] \end{aligned} \quad (43)$$

From the above elaborative narration, the benefits of the horse herd optimization algorithm are summarized below:

- The α horses will act as the role model for the other age group and provide the best reactions. They will serve as a coach when they start their search for optimized reactions and build up an exploited plan of action. This behavior happens when grazing traits and protective measures are required.
- β horses meticulously scan the area for the most likely perfect spots, paying close attention to α.
- γ horses' natural behaviors are all used to construct "The" horses. Despite their forceful and arbitrary movements, they appear to be effective for both the exploratory and exploitative phases.
- Young horses appear to be more excitable and animated, making them better suited for the exploration stage.

The HHO algorithm is established to derive the best parameter value of AFOPID by minimizing the integral absolute error (IAE) and integral time absolute error (ITAE) objective functions.

3.3. Implementation of the Grid-Interfaced Controller

It is necessary to perform d-q analysis to make the essential data from the grid end relatable with the controller so that it can send the switching signal to the inverter to ensure that the PV source can receive the maximum amount of power. The exhaustive control methodology is illustrated in Figure 3.

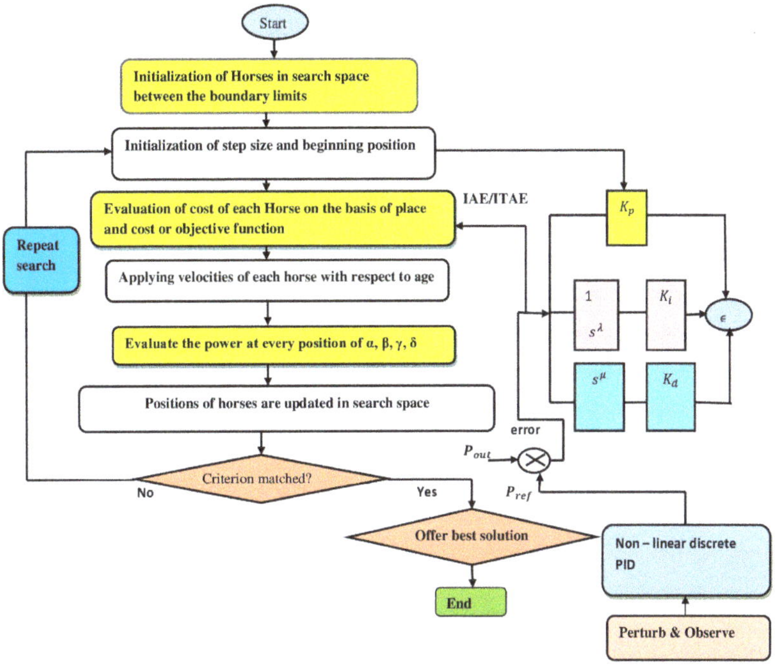

Figure 3. Flowchart of the proposed HHO-AFOPID control algorithm.

For the PV-connected grid's infrastructure, resistance R, and inductance (L), the potential of the system is depicted as v_d and v_q, and i_d, and i_q show the grid's currents. Conventional MPPT algorithms are the best fit for uniform environmental conditions. To overcome the shortcomings, a hybrid MPPT is proposed. The suggested control methodology is initiated by combining perturb and observe (P&O) with non-linear discrete PID (NDPID) to induce reference power (P_{ref}) during varying solar irradiations and ambient temperature constants. The error is calculated by subtracting the reference power from the output power.

$$\text{Error} = \left| P_{out} - P_{ref} \right| \tag{44}$$

The error is termed as the integral absolute error (IAE) and integral time absolute error (ITAE), which can be described as:

$$\text{IAE} = \int_0^t |e(t)|dt \tag{45}$$

$$\text{ITAE} = \int_0^t t|e(t)|dt \tag{46}$$

where t is the simulation time.

By introducing the optimization, the superior value of fitness is achieved and, thereafter, the law of adaptive control can be set. The preliminary amplitude of the controller is obtained using horse herd optimization, which initiates the control logic expansion. The hybrid PO-NDPID MPPT controller is incorporated to induce the voltage of dc link v_{dc}^* under changing irradiation from the PV source. Accordingly, to achieve the power factor of unity, the reference of quadrature axis i_q^* is calculated using thePV inverter. Later, by defining the state vector as $\dot{x} = (x_1, x_2, x_3)^T$, which indicates that the parameters are $(i_q, i_d, v_{dc})^T$, output matrix y can be defined as y = $(y_1, y_2)^T = (i_q, v_{dc})^T$, and input u = $(u_1, u_2)^T = (v_{dc}, v_q)^T$. The photovoltaic inverter's state equation can be written as:

$$\dot{x} = \begin{bmatrix} -\frac{R}{L}x_1 & -\omega x_2 & \frac{e_d}{L} \\ -\frac{R}{L}x_2 & \omega x_1 & -\frac{e_q}{L} \\ \frac{i_p}{C} & \frac{e_d x_1 + e_q x_2}{C x_3} & 0 \end{bmatrix} + \begin{bmatrix} \frac{1}{L} & 0 \\ 0 & \frac{1}{L} \\ 0 & 0 \end{bmatrix} \text{u} \quad (47)$$

$$\text{the trailing error is defined as e} = [e_1, e_2]^T = \left[i_q - i_q^*, v_{dc} - v_{dc}^*\right]^T \quad (48)$$

The control input u is obtained by differentiating error e.

$$\begin{bmatrix} \dot{e} \\ \ddot{e} \end{bmatrix} = \begin{bmatrix} f_{1(x)} \\ f_{2(x)} \end{bmatrix} + \text{B(x)} \begin{bmatrix} u_1 \\ u_2 \end{bmatrix} - \begin{bmatrix} i_q^* \\ v_{dc}^* \end{bmatrix} \quad (49)$$

$$\begin{bmatrix} f_{1(x)} \\ f_{2\,(x)} \end{bmatrix} = \text{Matrix A} \quad (50)$$

Matrix A is further elaborated:

$$f_1(\text{x}) = -\frac{R}{L} i_q + \omega i_d - \frac{e_q}{L} \quad (51)$$

$$f_2(\text{x}) = \frac{i_p}{C} - e_d\left(-\frac{R}{L} i_d - \omega i_q - \frac{e_d}{L}\right) + e_q\left(-\frac{R}{L} i_q + \omega i_d - \frac{e_q}{L}\right)/C v_{dc} - \frac{\left(e_{d i_d + e_q i_q}\right)}{C^{2 v_{dc}^2}} i_p + \frac{\left(e_d i_d + e_q i_q\right)^2}{C^2 v_{dc}^3} \quad (52)$$

where

$$\text{B(x)} = \begin{bmatrix} 0 & 0 \\ -\frac{e_d}{LC v_{dc}} & -\frac{e_q}{LC v_{dc}} \end{bmatrix} \quad (53)$$

The controlling input $[u_1,\ u_2\ ]$ can be achieved for photovoltaic inverters based on numerous HHO-AFOPID objectives.

The following formulation is obtained from (49).

$$\begin{aligned} u_1 \ = & -\frac{LC v_{dc}}{e_d}\left(v_{dc}^* - v_{dc} + v_{dc}^*\right) + \frac{e_q}{LC V_{dc}}\left(k_{p1} + \frac{k_{i1}}{s^{\lambda 1}} + k_{d1} s^{\mu 1}\right) - \frac{i_p}{C} + \\ & \frac{e_d\left(-\frac{R}{L} i_d - \omega i_q - \frac{e_d}{L}\right) + e_q\left(-\frac{R}{L} i_q + \omega i_d - \frac{e_q}{L}\right)}{C v_{dc}} + \left\{\left(e_d i_d + e_q i_q\right)/C v_{dc}^2\right\} i_p - \left(k_{p2} + \frac{k_{i2}}{s^{\lambda 2}} + k_{d2} s^{\mu 2}\right) \end{aligned} \quad (54)$$

$$u_2 = \text{L} i_q^* - \omega \text{L} i_d + \text{R} i_q^* + e_q + \left(v_{dc} - v_{dc}^*\right)\left(i_q - i_q^*\right)\left[\left(k_{p1} + k_{p2}\right) + \left(\frac{k_{i1}}{s^{\lambda 1}} + \frac{k_{i2}}{s^{\lambda 2}}\right) + \left(k_{d1} s^{\mu 1} + k_{d2} s^{\mu 2}\right)\right] \quad (55)$$

To attain the target, two control inputs (u_1, and u_2) are executed. The mentioned control topology is termed as numerous objectives due to its two fold adaptive features combined with the current of the q-axis (i_q) and the potential of the dc-link (v_{dc}). The focus of the proposed HHO-AFOPID is to reduce the objective function or fitness function as follows:

$$\text{Reducing F(x)} = \int_0^{T_{sim}} \left(\left(v_{dc} - v_{dc}^*\right) + \left(i_q - i_q^*\right)\right) dt \quad (56)$$

$$\text{Subjected to} \begin{cases} k_{pi}^{min} \le k_{pi} \le k_{pi}^{max} \\ k_{di}^{min} \le k_{di} \le k_{di}^{max} \\ k_{ii}^{min} \le k_{ii} \le k_{ii}^{max} \\ \mu_{imin} \le \mu_i \le \mu_{imax} \\ \lambda_{imin} \le \lambda_i \le \lambda_{imax} \end{cases} \text{for i} = 1,2 \quad (57)$$

With a view to affirm the stability of the proposed HHO-AFOPID control topology, the single-input single-output (SISO) system is analyzed in terms of error reduction.

The transfer function of the SISO system is written as follows:

$$\dot{e_1} + \left(k_{p1} + k_{p2} + \frac{R}{L}\right)e_1 + \left(\frac{k_{i1}}{s^{\lambda 1}} + \frac{k_{i2}}{s^{\lambda 2}}\right)e_1 + \left(k_{d1}s^{\mu 1} + k_{d2}s^{\mu 2}\right)e_1 \tag{58}$$

$$\ddot{e} + \frac{1}{CR_{dc}}e_2 + k_{p2}e_2 + k_{p1}e_2 + \left(\frac{k_{i2}}{s^{\lambda 2}} - \frac{k_{i1}}{s^{\lambda 1}}\right)e_2 + \left(k_{d2}s^{\mu 2} - k_{d1}s^{\mu 1}\right)e_1 \tag{59}$$

$$G_1(s) = \frac{1}{1 + \frac{s + \frac{R}{L}}{(k_{p1}+k_{p2}) + (k_{d1}s^{\mu 1} + k_{d2}s^{\mu 2}) + \left(\frac{k_{i1}}{s^{\lambda 1}} + \frac{k_{i2}}{s^{\lambda 2}}\right)}} \tag{60a}$$

$$G_2(s) = \frac{1}{1 + \frac{CR_{dc}s^2 + 1}{CR_{dc}\left(k_{p1} + k_{p2} + k_{d2}s^{\mu 2} - k_{d1}s^{\mu 1} + \frac{k_{i2}}{s^{\lambda 2}} + \frac{k_{i1}}{s^{\lambda 1}}\right)}} \tag{60b}$$

The investigation of the stability of the advanced HHO-AFOPID control topology is analyzed in terms of IAE and ITAE. Figure 4 shows a grid-connected PV system where HHO-AFOPID generates the gate signal to operate the inverter in order to achieve the desired goal.

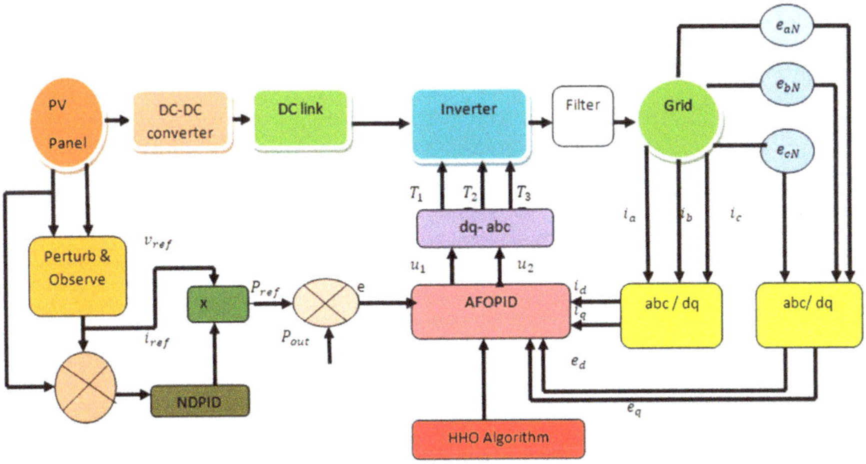

Figure 4. Block diagram of the proposed methodology.

The other parameters that are supplied to the same controller from the grid end are translated into d-q values in order to meet the control operation. The error produced from the differentiation between reference power and output power is delivered at the input of the AFOPID controller. To adjust the controller's settings, k_p, k_i, k_d, μ, and λ, an intelligent optimizing algorithm, horse herd optimization, is introduced on the basis of the cost function or integral absolute error (IAE) and integral time absolute error (ITAE) objective functions.

4. Results and Discussions

4.1. Case Study 1

A small number of experiments are conducted to confirm the present controller's successful operation and resilience. The initial test involves using the PV source to produce 100 kW while maintaining irradiation of 1000 W/m^2. The result in Figure 5a can be used to demonstrate that the proposed HHO-AFOPID controller is superior to conventional controllers in terms of quick convergence, short settling times, and minimal oscillation on MPPT. The next test is carried out to generate the same amount of power from the PV source by changing the irradiation from 1000 W/m^2 to 700 W/m^2. Here, the proposed controller also proves to be the best and shows its robustness in Figure 5b. Here, the controller also outperforms the aforementioned point of the criterion.

Later, a test was conducted by first turning down the PV source's irradiation intensity to 500 W/m^2 and then back up to 850 W/m^2. The suggested HHO-AFOPID controller in Figure 5c appears to be the most effective and reliable under these conditions.

As observed in the results, HHO-AFOPID exhibits the highest performance relative to following the MPP with high efficiency and little power loss under both constant and variable irradiation conditions. The magnitudes of k_{pi}, k_{di}, k_{ii} , μ_i, and λ_i are tabulated in Table 3 by incorporating various controllers with the suggested controller.

Table 3. Optimally tuned parameters of HHO-AFOPID acquired by horse herd optimization.

Algorithms	Voltage or Potential of DC Link (u_1) k_{pi}, k_{di}, k_{ii}	Current of q-Axis (u_2) k_{pi}, k_{di}, k_{ii}
HHO-AFOPID	k_{p1} = 100, k_{p2} = 129 k_{d1} = 105, k_{d2} = 115 k_{i1} = 76, k_{i2} = 178 μ_1 = 1.25, μ_2 = 1.05 λ_1 = 0.98, λ_2 = 1.35	k_{p1} = 177, k_{p2} = 147 k_{d1} = 94, k_{d2} = 121 k_{i1} = 116, k_{i2} = 120 μ_1 = 1.41, μ_2 = 1.23 λ_1 = 0.99, λ_2 = 1.76
AFOPID	k_{p1} = 109, k_{p2} = 139 k_{d1} = 115, k_{d2} = 145 k_{i1} = 96, k_{i2} = 192 μ_1 = 1.18, μ_2 = 1.35 λ_1 = 1.98, λ_2 = 1.5	k_{p1} = 187, k_{p2} = 177 k_{d1} = 104, k_{d2} = 141 k_{i1} = 136, k_{i2} = 129 μ_1 = 1.51, μ_2 = 1.33 λ_1 = 1.02, λ_2 = 1.95
FOPID	k_{p1} = 119 k_{d1} = 143 k_{i1} = 106 μ_1 = 1.68 λ_1 = 2.1	k_{p2} = 190 k_{d2} = 155 k_{i2} = 140 μ_2 = 1.48 λ_2 = 2.2
PID	k_{p1} = 139 k_{d1} = 166 k_{i1} = 126	k_{p2} = 200 k_{d2} = 175 k_{i2} = 159

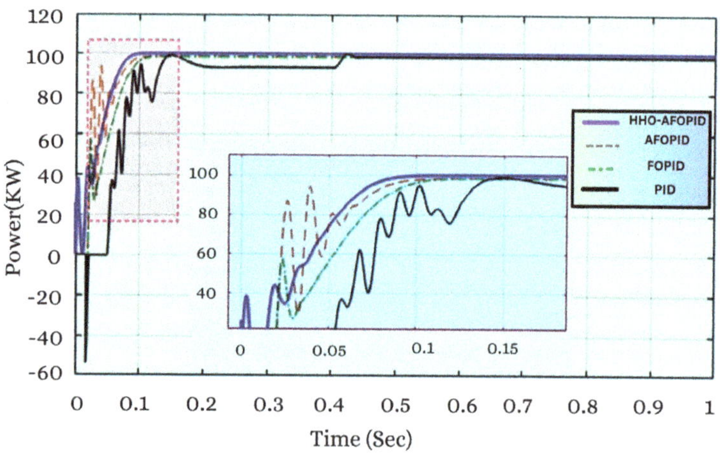

(a)

Figure 5. *Cont.*

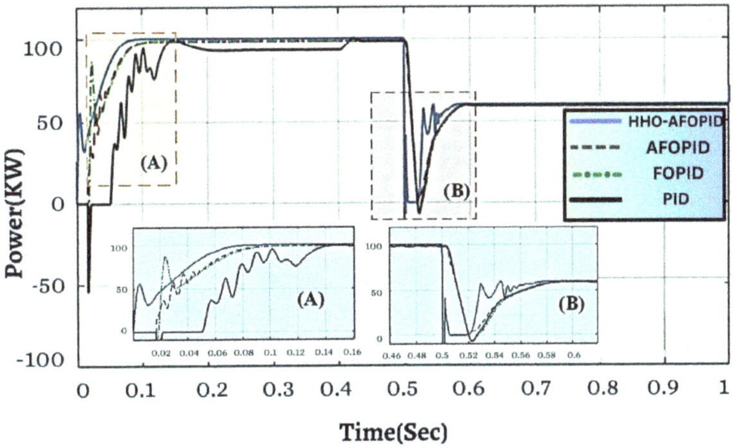

(b)

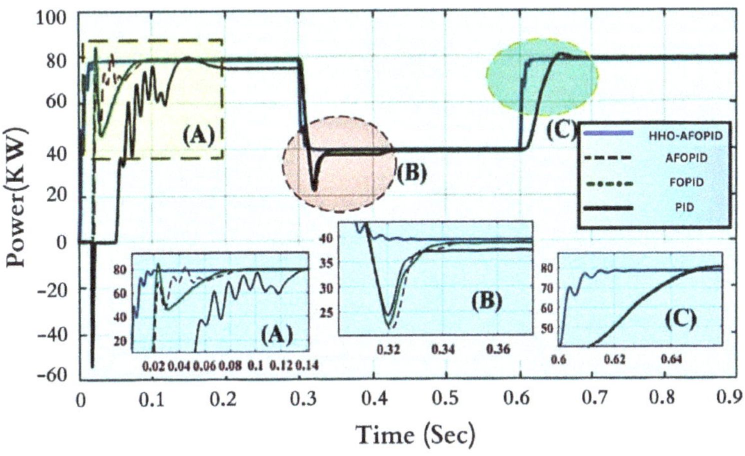

(c)

Figure 5. Experiment of the proposed HHO-AFOPID controller (**a**) under constant irradiation, (**b**) under changing irradiation, and (**c**) under various irradiation levels: 850 W/m^2 and 500 W/m^2.

4.2. Case Study 2

Another criterion applies to vindicate the achievement of the derived controller that is tracking the voltage of the DC link under constant and changing irradiation levels of the PV source. After analysis, it was observed that the HHO-AFOPID again satisfies the robustness. In Figure 6a, it shows that the proposed controller is the best one to meet the aim of the constant irradiation condition, and in Figure 6b, satisfactory performance is observed under varying irradiation conditions.

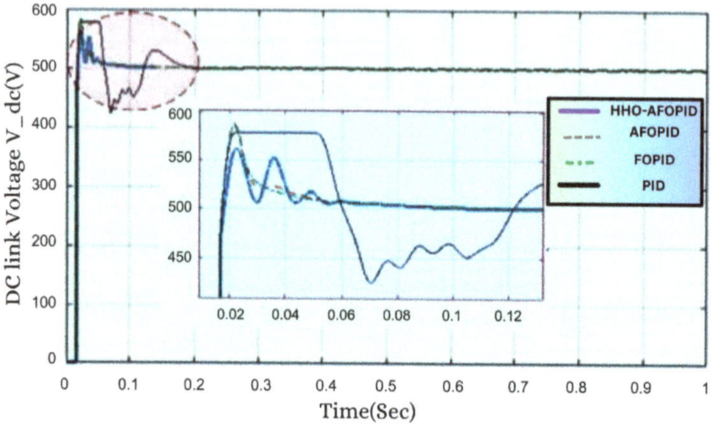

(**a**)

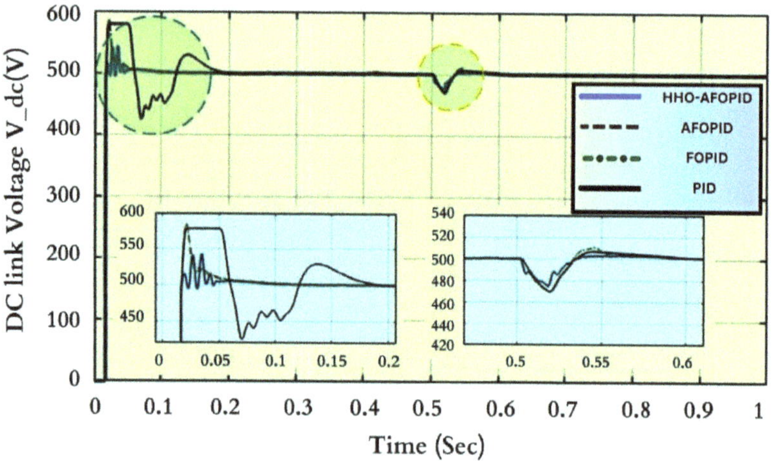

(**b**)

Figure 6. Features of the voltage of the dc link (**a**) under constant irradiation level and (**b**) under changing irradiation levels.

4.3. Case Study 3

Additionally, the system is evaluated within partial shading circumstances. When compared to other controllers, the suggested controller performs better in this instance and extracts 100 kW from the system. In this instance, a 4 × 1 array configuration with a right-skewed half-plane MPP position is chosen. This is carried out to confirm that the proposed HHO-AFOPID MPPT control topology is repeatable and can successfully manage the partial shading condition. In addition to this, another goal is to show how the methodology with skewed global MPP varies in performance on the characteristics graph. The power to validate competing methodologies is displayed in the comparison in Figure 7.

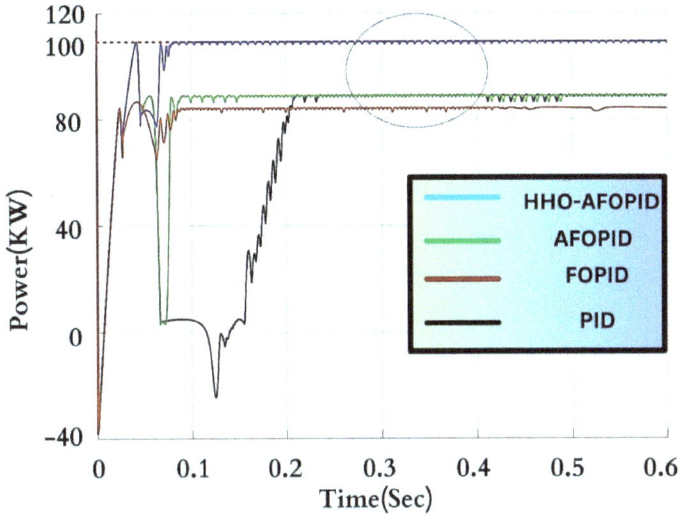

Figure 7. Experiment under the partial shading condition.

The methodical details are revealed in Table 4. Due to the inclusion of the method of the fractional order, it is clear from the analytical observation that HHO-AFOPID exhibits the lowest fitness function and the shortest time of convergence, which highlights its superior performance index and robustness compared to other control methodologies

Table 4. Comparative performance analysis of different controllers.

Control Topology	Objective Function (p.u)			Time of Convergence (h)			Number of Iterations for Convergence		
	High	Low	Mean	High	Low	Mean	High	Low	Mean
HHO-AFOPID	1.02	0.88	0.95	0.62	0.33	0.475	125	100	112.5
AFOPID	1.67	1.28	1.475	0.48	0.25	0.365	114	89	101.5
FOPID	1.88	1.45	1.68	0.26	0.19	0.225	98	59	78.5
PID	2.8	2.25	2.525	0.09	0.01	0.05	39	9	24

4.4. Case Study 4

Power quality was also assessed in terms of the voltage variation, THD, and frequency for verifying the controller's effectiveness. HHO-AFOPID exhibits the best outcome and upholds its reputation in this instance as well.

4.4.1. Voltage Deviation

Power quality is one of the main issues of grid-connected systems. The root mean square (RMS) value of the voltage can be expressed as an equation based on the peak value and sample/cycle (61) [20]:

$$v_i^{rms} = \sqrt{\frac{1}{M}\sum_{k=1}^{i+M-1} v_k^2} \tag{61}$$

M = sample/cycle of the initial;
v_k = k-th specimen of the registered potential waveform;
v_i^{rms} = i-th specimen of the measured r.m.s voltage.

The value of the root mean square voltage lags behind the phase voltage by $(M-1)$ cycles since there are M cycles per second.

4.4.2. THD

The total harmonic distortion (THD) can be defined as the ratio of the root mean square value of harmonics to the elemental signal and is manifested in Equation (62) [20].

$$THD_v = \frac{\sqrt{v_{rms}^2 - v_1^2}}{v_1} \tag{62}$$

4.4.3. Frequency

Equation illustrates how the frequency at the grid's end is confirmed to be roughly 50 Hz using PV panels (63) [20].

$$\Delta f = -(\Delta P_{PV})R \tag{63}$$

R is the frequency droop coefficient and is restricted up to 5%.

In reality, there is no way to completely solve power quality problems, but they can be managed or improved to the required level. In this study, a novel HHO-AFOPID controller is used and successfully simulated while retaining the aforementioned power quality issues in the grid-connected system with an SPV interface under normal and perturbed conditions. System perturbation is created from 0.95 s to 1.9 s. To achieve a balanced system quickly, the unbalanced three-phase voltage and current have been controlled using inverter switching pulses. The proposed controller gains fast triggering. The system's voltage is immediately balanced by the gate pulses that are fired in comparison to the other mentioned controllers, i.e., AFOPID, FOPID, and PID.

Figure 8a–c depict the comparative performance assessment in terms of deviation in voltage, frequency, and total harmonic distortion, respectively. While it is clear from the voltage deviation profile that FOPID and AFOPID have shown less deviation than PID, the suggested HHO-AFOPID shows smooth and little change in %VD. The suggested controller's frequency and THD profile also display no frequency fluctuation and practically no THD, which supports the high-quality power produced by the grid-connected PV system and boosts the system's effectiveness.

Tables 5 and 6 compare the performance of several controllers with the proposed controller under changing irradiation and partial shading conditions, respectively. The recommended controller is put to the test against various controllers in terms of undershoot, settling time, ripples, and stability under varying irradiation conditions as measured by IAE and ITAE. In comparison to the other described controllers, the suggested controller exhibits the least undershoot, settling time, and ripple content. The smallest IAE and ITAE further demonstrate HHO-AFOPID's stability across a variety of irradiation conditions.

Table 5. Comparative performance analysis of different controllers under changing irradiation conditions.

Index of Performance	PID	FOPID	AFOPID	HHO-AFOPID
Undershoot (%)	10.1	5.4	1.4	0.031
Settling time (ms)	58.43	19.5	19.46	19.36
Ripples	1.33	0.134	0.010	0.0057
IAE	0.2567	0.2243	0.1978	0.1657
ITAE	0.1765	0.1533	0.1243	0.1211

Table 6. Comparative performance analysis of different controllers under the partial shading condition.

Index of Performance	PID	FOPID	AFOPID	HHO-AFOPID
Undershoot (%)	10.09	5.45	1.6	0.027
Settling time (ms)	60.09	19.1	18.49	19.29
Ripples	1.4	0.14	0.011	0.0055
IAE	0.2657	0.2133	0.1320	0.0061
ITAE	0.1543	0.1324	0.1109	0.0056

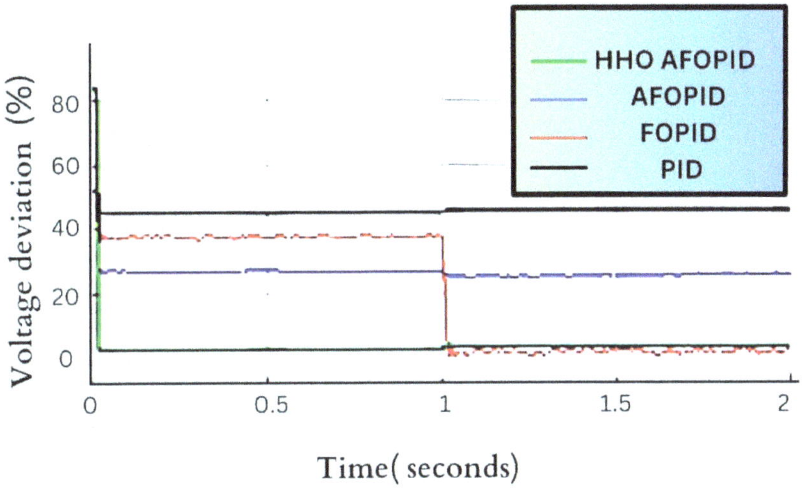

(a)

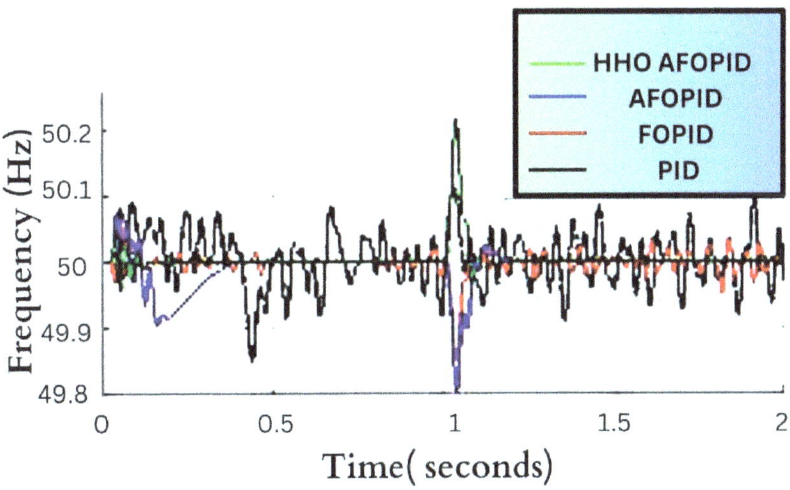

(b)

Figure 8. *Cont.*

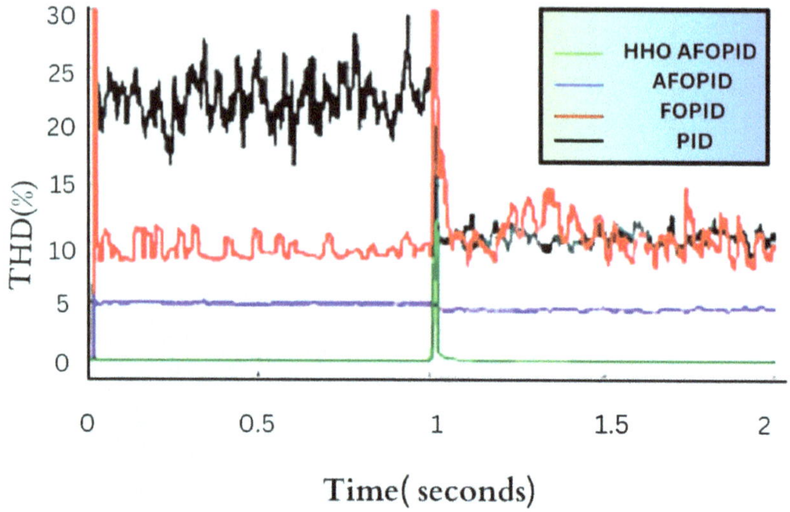

(c)

Figure 8. Power quality analysis on the basis of (**a**) voltage deviations under HHO-AFOPID, (**b**) frequency under HHO-AFOPID, and (**c**) THD analysis under HHO-AFOPID.

In comparison to PID, FOPID, and AFOPID, Table 6 demonstrates that the suggested HHO-AFOPID controller maintains consistency under partial shade conditions in terms of least undershoot, lower settling time, and lower ripple content. When compared to PID, FOPID, and AFOPID, the lowest IAE and ITAE also have the significance of being stable under the given conditions.

In addition to these trials, a literature review was conducted to verify the HHO-AFOPID controller's satisfactory performance analysis, which was based on the case studies listed below. Since earlier times, experts have concentrated on extracting the most energy possible from solar energy. MPPTs are therefore relevant.. Researchers' attention has been drawn to the hybridization of the MPPT algorithm as their study has progressed. Here, numerous other hybrid and non-hybrid controllers are compared to the suggested hybrid HHO-AFOPID MPPT controller in order to provide a comparison of them based on power efficacy and oscillations, and these studies assert that the proposed controller is the best in the aforementioned areas. The suggested controller's performance study with different non-hybrid and hybrid MPPT control topologies is shown in Table 7.

Table 7. Comparative performance analysis of the hybrid HHO-AFOPID with other hybridized and non-hybridized MPPT algorithms.

Algorithms	References	Power Efficacy	Oscillations
P&O	[40]	93–97%	Excessive
P&O-PSO	[41]	94–98%	Excessive
P&O-Fuzzy	[42]	92%	Intermediate
P&O-Fuzzy	[43]	99%	Truncated
PSO-PID	[44]	97%	Intermediate
PSO-NDPID	[44]	99.5%	Excessive Low
GA-PID	[44]	95%	Mitigated
GA-NDPID	[44]	99%	Truncated
ANN	[45]	92–98%	Intermediate
ANN-PO	[46]	99.75%	Excessive Low
HOA	[39]	99.8%	Excessive Low
HHO-AFOPID (Proposed)		99.98%	Almost negligible

5. Sensitivity Analysis

Sensitivity analysis is used to examine how well a system performs in relation to changes in sensitive parameters, such as solar irradiation in the case of a PV-interfaced system. Analyzing the disturbance in sensitive variables is crucial. The variations in sensitive factors and their magnitudes are shown in Table 8.

Table 8. Variation of sensitive factors with the magnitude of SPV.

Parameter	Magnitude (Solar Radiation Intensity)
Sensitive variables	5.01, 7, 9, 10

The amount of solar radiation needed to generate electrical energy is solely reliant on the sun's radiation output. The average solar radiation availability each day may be within the range of 5.01 to 10 $kW/m^2/day$. The optimal dimension of a system with the lowest cost of energy is the most desired criterion; hence, this phenomenon has been analyzed on the basis of factors like the total net present cost (TNPC) and levelized cost of energy (LCOE). Equations (64)–(68) can be used to define the cost function of this study, which seeks to minimize TNPC and LCOE [24].

$$TNPC_{min} = \frac{\text{TAC}}{CRF(i_r, n_p)} \tag{64}$$

TAC is the total annum cost, which includes the capital cost (CC), replacement cost (RC), and maintenance cost (MC) described in Equation (65).

$$\text{TAC} = C_{CC} + C_{RC} + C_{MC} \tag{65}$$

CRF = the factor of capital recovery depending on the basis of the original interest rate (i_r) and project span (n_p) is described in Equation (66).

$$\text{CRF}\ (i_r, n_p) = \frac{i_r(1+i_r)^{n_p}}{(1+i_r)^{n_p} - 1} \tag{66}$$

$$i_r = \frac{i_n - f}{1+f} \tag{67}$$

i_r can be calculated using Equation (67) with i_n, which is the magnitude dependent on ROI, and f symbolizes the inflation rate:

$$LCOE_{min} = \frac{TAC}{ESPA} \tag{68}$$

ESPA = energy served per annum.

The simulation result shows that solar irradiation has are ciprocating effect on TNPC and LCOE, as shown in Figure 9.

The increase in solar irradiation is inversely related to the overall cost parameters according to a thorough analysis of the sensitivity variable of SPV generation or solar radiation with respect to cost parameters. It is possible to deduce cost parameters from a thorough examination of several sensitivity factors in the system, namely net present cost and cost of energy, with an increase in sunray values.

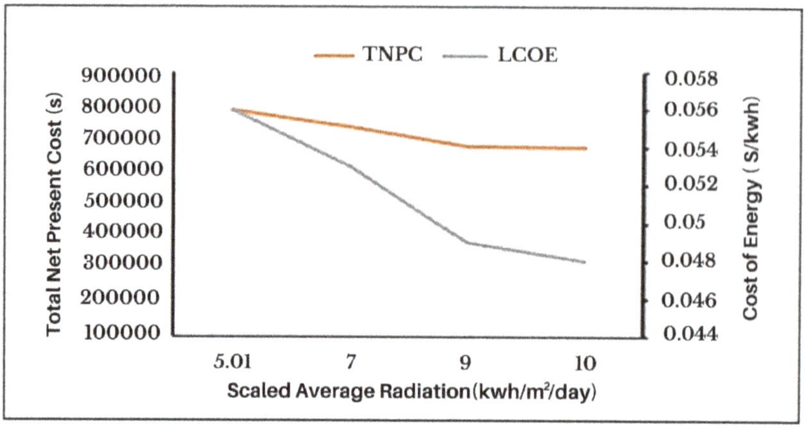

Figure 9. Effect of varying solar irradiation on the total net present cost and levelized cost of energy.

6. Conclusions and Future Directions

The conducted experiments make it abundantly clear that the suggested controller exhibits the greatest results in terms of qualitative and quantitative analysis when the environment is changing. The reference voltage was produced using the traditional P&O algorithm, which then allowed the nonlinear discrete PID (NDPID) controller to calculate the reference power. The forward Euler formula was used to combine the traditional PID controller with discretized integral and derivative portions. Additionally, the difference between reference power and output power is used to calculate inaccuracy. The HHO further optimizes the AFOPID controller to obtain the best settings for the controller to create an adaptive control law. The switching signal for the inverter is provided by this error. In addition, it has been demonstrated that the resulting topology performs better in terms of the lowest fitness value, improved settling time, and least oscillations under varying environmental conditions and partial shading conditions.To examine the stability of the proposed control methodology, the controller is judged in terms of the integral absolute error (IAE) and integral time absolute error (ITAE) under variable solar radiation intensity and partial shading conditions. In both scenarios, the control topology outperforms by achieving the lowest IAE and ITAE. Due to its fractional calculus property and multi-control strategy with respect to handling inverter switching, the suggested controller exhibits greater robustness than existing controllers according to the study.

Power quality is assessed in addition to power quantity on the basis of voltage variation, THD, and frequency. In this area, the proposed controller exhibits a suitable response as well. Low THD, lower frequency fluctuation, and lower voltage deviation all point to the suggested control methodology, which improves the system's performance. The suggested control topology achieves the best result even though the impact of power quality issues cannot be totally eliminated or minimized.

Sensitivity analysis was also carried out on the basis of TNPC and LCOE, as the optimal design and reducing energy cost is of utmost priority in grid-interfaced systems. Here, the experiment shows that an increase in solar irradiation results in a reduction in the per unit cost of energy. This analysis will inspire the nation to use solar energy in various applications for techno-economic sustainability.

Furthermore, many cascaded controllers with a number of newly introduced optimization algorithms like mount gazelle optimization (MGO), honey badger algorithm (HBA), and Ebola algorithm may be incorporated to extract the maximum power from a PV source and other renewable energy resources as well. This concept will be helpful for microgrids as well.

Author Contributions: Conceptualization, A.G., P.K.B., C.S. and A.T.A.; Data curation; Formal analysis, A.G., P.K.B., C.S., A.T.A. and A.R.M.; Funding acquisition, A.R.M.; Investigation, A.T.A., A.R.M. and S.A.; Methodology, A.G., P.K.B., C.S., A.T.A. and A.R.M.; Resources, A.G., P.K.B., C.S., A.T.A., A.R.M. and S.A.; Software, A.G., C.S. and S.A.; Supervision, P.K.B.; Validation, A.T.A., A.R.M. and S.A.; Writing—original draft, A.G., P.K.B., C.S. and A.T.A.; Writing—review & editing, A.G., P.K.B., C.S., A.T.A., A.R.M. and S.A. All authors have read and agreed to the published version of the manuscript.

Funding: This research was funded by Prince Sultan University, Riyadh, Saudi Arabia.

Institutional Review Board Statement: Not applicable.

Informed Consent Statement: Not applicable.

Data Availability Statement: Not applicable.

Acknowledgments: The authors would like to thank Prince Sultan University, Riyadh, Saudi Arabia for funding the Article Processing Charges (APCs) of this publication. Special acknowledgments are given to Automated Systems & Soft Computing Lab (ASSCL), Prince Sultan University, Riyadh, Saudi Arabia.

Conflicts of Interest: The authors declare no conflict of interest.

Abbreviations

Variables	Abbreviations
j	Imaginary quantity
MPPT	Maximum power point tracking
P&O	Perturb and observe
NDPID	Nonlinear discrete proportional integral derivative controller
AFOPID	Adaptive fractional order proportional integral derivative controller
FOPID	Fractional order proportional integral derivative controller
PID	Proportional integral derivative controller
HHO	Horse herd optimization
PV	Photovoltaic
THD	Total harmonic distortion

References

1. Babaei, E.; Mahmoodieh, M.E.S.; Mahery, H.M. Operational modes and output voltage ripple analysis and design considerations of buck-boost dc-dc converters. *IEEE Trans. Ind. Electron.* **2012**, *59*, 381–391. [CrossRef]
2. Baghat, A.B.G.; Helwab, N.H.; Ahmadb, G.E.; El Shenawyb, E.T. Maximum power point tracking controller for PV systems using Neural Networks. *Renew. Energy* **2005**, *30*, 1257–1268.
3. Salas, V.; Olías, E.; Barrado, A.; Lázaro, A. Review of maximum power point tracking algorithms for stand-alone photovoltaic systems. *Sol. Energy Mater. Sol. Cells* **2006**, *90*, 1555–1578. [CrossRef]
4. Esram, T.; Chapman, P.L. Comparison of photovoltaic array maximum power point tracking techniques. *IEEE Trans. Energy Convers.* **2007**, *22*, 439–449. [CrossRef]
5. Subramaniam, U.; Reddy, K.S.; Kaliyaperumal, D.; Sailaja, V.; Bhargavi, P.; Likhith, S. A MIMO–ANFIS-Controlled Solar-Fuel-Cell-Based Switched Capacitor Z-Source Converter for an Off-Board EV Charger. *Energies* **2023**, *16*, 1693. [CrossRef]
6. Premkumar, M.; Subramaniam, U.; Babu, T.S.; Elavarasan, R.M.; Mihet-Popa, L. Evaluation of Mathematical Model to Characterize the Performance of Conventional and Hybrid PV Array Topologies under Static and Dynamic Shading Patterns. *Energies* **2020**, *13*, 3216. [CrossRef]
7. Darcy Gnana Jegha, A.; Subathra, M.S.P.; Manoj Kumar, N.; Subramaniam, U.; Padmanaban, S. A High Gain DC-DC Converter with Grey Wolf Optimizer Based MPPT Algorithm for PV Fed BLDC Motor Drive. *Appl. Sci.* **2020**, *10*, 2797. [CrossRef]
8. Reisi, A.R.; Moradi, M.H.; Jamasb, S. Classification and Comparison of maximum power point tracking techniques for photovoltaic system: A Review. *Renew. Sustain. Energy Rev.* **2013**, *19*, 433–443. [CrossRef]
9. Ammar, H.H.; Azar, A.T.; Shalaby, R.; Mahmoud, M.I. Metaheuristic Optimization of Fractional Order Incremental Conductance (FO-INC) Maximum Power Point Tracking (MPPT). *Complexity* **2019**, *2019*, 7687891. [CrossRef]
10. Ramadan, H.; Youssef, A.R.; Mousa, H.H.H.; Mohamed, E.E.M. An efficient variable-step P&O maximum power point tracking technique for grid-connected wind energy conversion system. *SN Appl. Sci.* **2019**, *1*, 1658. [CrossRef]
11. Delavari, H.; Zolfi, M. Maximum power point tracking in photovoltaic systems using indirect adaptive fuzzy robust controller. *Soft Comput.* **2021**, *25*, 10969–10985. [CrossRef]

12. Kamal, N.A.; Ibrahim, A.M. Conventional, Intelligent, and Fractional-Order Control Method for Maximum Power Point Tracking of a Photovoltaic System: A Review. In *Fractional Order Systems Optimization, Control, Circuit Realizations and Applications, Advances in Nonlinear Dynamics and Chaos (ANDC)*; Elsevier: Amsterdam, The Netherlands, 2018; pp. 603–671.
13. Ibrahim, A.; Obukhov, S.; Aboelsaud, R. Determination of Global Maximum Power Point Tracking of PV under Partial Shading Using Cuckoo Search Algorithm. *Appl. Sol. Energy* **2019**, *55*, 367–375. [CrossRef]
14. Govinda Chowdary, V.; Udhay Sankar, V.; Mathew, D.; Hussaian Basha, C.; Rani, C. Hybrid Fuzzy Logic-Based MPPT for Wind Energy Conversion System. In *Soft Computing for Problem Solving. Advances in Intelligent Systems and Computing*; Das, K., Bansal, J., Deep, K., Nagar, A., Pathipooranam, P., Naidu, R., Eds.; Springer: Singapore, 2020; Volume 1057. [CrossRef]
15. Dadfar, S.; Wakil, K.; Khaksar, M.; Rezvani, A.; Miveh, M.R.; Gandomkar, M. Enhanced control strategies for a hybrid battery/photovoltaic system using FGS-PID in grid connected mode. *Int. J. Hydrogen Energy* **2019**, *44*, 14642–14660. [CrossRef]
16. Meghni, B.; Dib, D.; Azar, A.T.; Saadoun, A. Effective Supervisory Controller to Extend Optimal Energy Management in Hybrid Wind Turbine under Energy and Reliability Constraints. *Int. J. Dyn. Control* **2018**, *6*, 369–383. [CrossRef]
17. Mohapatra, A.; Byamakesh, N.; Priti, D. A review on MPPT techniques of PV system under partial shading condition. *Renew. Sustain. Energy Rev.* **2017**, *80*, 854–867. [CrossRef]
18. Kaushal, J.; Basak, P. Power Quality Control based on voltage sag/swell, unbalancing, frequency, THD and power factor using artificial neural network in PV integrated AC microgrid. *Sustain. Energy Grid Netw.* **2020**, *23*, 100365. [CrossRef]
19. *IEEE Std 1250-2011*; IEEE Guide for Identifying and Improving Voltage Quality in Power Systems—Redline; Revision of IEEE Std 1250-1995—Redline. IEEE: New York, NY, USA, 2011; pp. 1–70. [CrossRef]
20. *IEEE Std 519-2014*; IEEE Recommended Practice and Requirements for Harmonic Control in Electric Power Systems; Revision of IEEE Std 519-1992. IEEE: New York, NY, USA, 2014; pp. 1–29. [CrossRef]
21. Rajbongshi, R.; Borgohain, D.; Mahapatra, S. Optimization of PV-biomass-diesel and grid base hybrid energy systems for rural electrification by using HOMER. *Energy* **2017**, *126*, 461–474. [CrossRef]
22. Sharma, S.; Sood, Y.R. Optimal planning and sensitivity analysis of green microgrid using various types' storage systems. *Wind Energy* **2021**, *45*, 939–952. [CrossRef]
23. Arevalo, P.; Jurado, F. Performance analysis of a PV/HKT/WT/DG hybrid autonomous grid. *Electr. Eng.* **2021**, *103*, 227–244. [CrossRef]
24. Nasir, A.; Rasool, I.; Sibtain, D.; Kamran, R. Adaptive Fractional Order PID Controller Based MPPT for PV Connected Grid System Under Changing Weather Conditions. *J. Electr. Eng. Technol.* **2021**, *16*, 2599–2610. [CrossRef]
25. Meenakshi, S.B.; Manikandan, B.V.; Praveen Kumar, B.; Prince Winston, D. Combination of novel converter topology and improved MPPT algorithm for harnessing maximum power from grid connected solar PV system. *J. Electr. Eng. Technol.* **2019**, *14*, 733–746. [CrossRef]
26. Chandra, S.; Yadav, A.; Khan, M.A.R.; Pushkarna, M.; Bajaj, M.; Sharma, N.K. Influence of Artificial and Natural Cooling on Performance Parameters of a Solar PV System: A Case Study. *IEEE Access* **2021**, *9*, 29449–29457. [CrossRef]
27. Li, K.; Liu, C.; Jiang, S.; Chen, Y. Review on hybrid geothermal and solar power systems. *J. Clean. Prod.* **2020**, *250*, 119481. [CrossRef]
28. Dhimish, M.; Badran, G. Current limiter circuit to avoid photovoltaic mismatch conditions including hot spots and shading. *Renew. Energy* **2020**, *145*, 2201–2216. [CrossRef]
29. Lalili, D.; Mellit, A.; Lourci, N.; Medjahed, B.; Berkouk, E.M. Input output feedback linearization control and variable step size mppt algorithm of a grid connected photovoltaic inverter. *Renew. Energy* **2011**, *36*, 3282–3291. [CrossRef]
30. Lekouaghet, B.; Boukabou, A.A. Control of PV grid connected systems using MPC technique and different inverter configuration models. *Electr. Power Syst. Res.* **2018**, *154*, 287–298. [CrossRef]
31. Podlubny, I. *Fractional Differential Equations*; Academic Press: New York, NY, USA, 1999.
32. Meghni, B.; Dib, D.; Azar, A.T.; Ghoudelbourk, S.; Saadoun, A. Robust Adaptive Supervisory Fractional order Controller For optimal Energy Management in Wind Turbine with Battery Storage. *Stud. Comput. Intell.* **2017**, *688*, 165–202.
33. Gorripotu, T.S.; Samalla, H.; Jagan Mohana Rao, C.; Azar, A.T.; Pelusi, D. TLBO Algorithm Optimized Fractional-Order PID Controller for AGC of Interconnected Power System. In *Soft Computing in Data Analytics. Advances in Intelligent Systems and Computing*; Nayak, J., Abraham, A., Krishna, B., Chandra Sekhar, G., Das, A., Eds.; Springer: Singapore, 2019; Volume 758.
34. Ibraheem, G.A.R.; Azar, A.T.; Ibraheem, I.K.; Humaidi, A.J. A Novel Design of a Neural Network-based Fractional PID Controller for Mobile Robots Using Hybridized Fruit Fly and Particle Swarm Optimization. *Complexity* **2020**, *2020*, 3067024. [CrossRef]
35. Jeba, P.; Immanuel Selvakumar, A. FOPID based MPPT for photovoltaic system. *Energy Sources Part A Recovery Util. Environ. Eff.* **2018**, *40*, 1591–1603. [CrossRef]
36. Mohanty, A.; Viswavandya, M.; Mohanty, S. An optimized FOPID controller for dynamic voltage stability and reactive power management in a stand-alone micro grid. *Int. J. Electr. Power Energy Syst.* **2016**, *78*, 524–536. [CrossRef]
37. Mills, D.S.; McDonnell, S.M. *The Domestic Horse: The Origins, Development and Management of Its Behavior*; Cambridge University Press: Cambridge, UK, 2005.
38. MiarNaeimi, F.; Azizyan, G.; Rashki, M. Horse Herd optimization algorithm: A nature inspired algorithm for high dimensional optimization problems. *Knowl. Based Syst.* **2021**, *213*, 106711. [CrossRef]
39. Sarwar, S.; Hafeez, M.A.; Javed, M.Y.; Asghar, A.B.; Ejsmont, K. A horse herd optimization algorithm (HOA) based MPPT technique under partial and complex partial shading conditions. *Energies* **2022**, *15*, 1880. [CrossRef]

40. Elmelegi, A.; Aly, M.; Ahmed, E.M.; Alharbi, A.G. A simplified phase shift PWM based feed forward distributed MPPT method for grid connected cascaded PV inverters. *Sol. Energy* **2019**, *187*, 1–12. [CrossRef]
41. Lian, K.L.; Jhang, J.H.; Tian, I.S. A maximum power point tracking method based on perturbs and observe combined with Particle Swarm Optimization. *IEEE J. Photovolt.* **2014**, *4*, 626–633. [CrossRef]
42. Zainuri, M.A.A.M.; Radzi, M.A.M.; Soh, A.C.; Rahim, N.A. Development of adaptive perturbs and observe-fuzzy control maximum power point tracking for photovoltaic boost dc-dc converter. *IET Renew. Power Gener.* **2014**, *8*, 183–194. [CrossRef]
43. Bataineh, K.; Eid, N. A hybrid maximum power point tracking method for photovoltaic systems for dynamic weather conditions. *Resources* **2001**, *7*, 68. [CrossRef]
44. Pathak, D.; Sagar, G.; Gaur, P. An application of intelligent nonlinear discrete–PID controller for MPPT of PV System. *Procedia Comput. Sci. ICCIDS* **2019**, *167*, 1574–1583. [CrossRef]
45. Kok, B.C.; Goh, H.H.; Chua, H.G. Optimal Power tracker for standalone photovoltaic system using artificial neural network (ANN) and Particle Swarm Optimization (PSO). In Proceedings of the International Conference on Renewable Energy and Power Quality (ICREPQ'12), Batu Pahat, Malaysia, 28–30 March 2012; pp. 440–445.
46. Ayad, A.I.; Elwarakki, E.; Baghdadi, M. Intelligent Perturb and Observe Based MPPT Approach Using Multilevel DC-DC Converter to Improve PV Production System. *J. Electr. Comput. Eng.* **2021**, *2021*, 6673022. [CrossRef]

sustainability

MDPI

Article

An Intensified Marine Predator Algorithm (MPA) for Designing a Solar-Powered BLDC Motor Used in EV Systems

Rajesh Kanna Govindhan Radhakrishnan [1], Uthayakumar Marimuthu [2,3], Praveen Kumar Balachandran [4], Abdul Majid Mohd Shukry [3] and Tomonobu Senjyu [5,*]

1 Department of EEE, Kalasalingam Academy of Research & Education, Virudhunagar 626125, Tamil Nadu, India
2 Department of Automobile Engineering, Kalasalingam Academy of Research & Education, Virudhunagar 626125, Tamil Nadu, India
3 Faculty of Mechanical Engineering & Technology, University Malaysia Perlis (UniMAP), Arau 02600, Perlis, Malaysia
4 Department of EEE, Vardhaman College of Engineering, Hyderabad 501218, Telangana, India
5 Faculty of Engineering, University of the Ryukyus, Okinawa 903-0213, Japan
* Correspondence: b985542@tec.u-ryukyu.ac.jp

Abstract: Recently, due to rapid growth in electric vehicle motors, used and power electronics have received a lot of concerns. 3ϕ induction motors and DC motors are two of the best and most researched electric vehicle (EV) motors. Developing countries have refined their solution with brushless DC (BLDC) motors for EVs. It is challenging to regulate the 3ϕ BLDC motor's steady state, rising time, settling time, transient, overshoot, and other factors. The system may become unsteady, and the lifetime of the components may be shortened due to a break in control. The marine predator algorithm (MPA) is employed to propose an e-vehicle powered by the maximum power point tracking (MPPT) technique for photovoltaic (PV). The shortcomings of conventional MPPT techniques are addressed by the suggested approach of employing the MPA approach. As an outcome, the modeling would take less iteration to attain the initial stage, boosting the suggested system's total performance. The PID (proportional integral derivative) is used to govern the speed of BLDC motors. The MPPT approach based on the MPA algorithm surpasses the variation in performance. In this research, the modeling of unique MPPT used in PV-based BLDC motor-driven electric vehicles is discussed. Various aspects, which are uneven sunlight, shade, and climate circumstances, play a part in the low performance in practical scenarios, highlighting the nonlinear properties of PV. The MPPT technique discussed in this paper can be used to increase total productivity and reduce the operating costs for e-vehicles based on the PV framework.

Keywords: brushless DC motor (BLDC); maximum peak point tracking (MPPT); photovoltaic (PV) systems; electric vehicle (EV) applications; marine predator algorithm (MPA)

Citation: Radhakrishnan, R.K.G.; Marimuthu, U.; Balachandran, P.K.; Shukry, A.M.M.; Senjyu, T. An Intensified Marine Predator Algorithm (MPA) for Designing a Solar-Powered BLDC Motor Used in EV Systems. *Sustainability* **2022**, *14*, 14120. https://doi.org/10.3390/su142114120

Academic Editors: Nien-Che Yang and Shuhua Fang

Received: 11 September 2022
Accepted: 21 October 2022
Published: 29 October 2022

1. Introduction

The advancement of electric vehicles is driven by the ambition to reduce emissions to increase the consumption of fuels [1]. In this situation of India and China, the shortage of energy is anticipated to happen rapidly because of a reduction in the future availability of fossil fuels and a 76% hike in necessity in the period from 2020 to 2045 [2]. Carbon emissions and waste are decreased by employing renewable energy [3]. Due to this, there is an increase in demand for clean, pollution-free renewable energy that emits only 30 carbon dioxides [4]. Industry and researchers have utilized advanced PV modules for many purposes due to consecutive reductions in the price of PV panels and power electronics components [5]. To maximize a PV array's capacity, the MPPT approach with a DC-DC converter topology is commonly utilized [6]. No carbon emissions are produced. Industry and researchers have utilized the advanced solar PV array for many applications because of continuous reductions of price in power electronics components and PV panels [7]. The MPPT approach with DC-DC converters typically maximizes a PV array's capacity [8]. Various MPPT control techniques have been proposed, with fractional open/short-circuit control methods, incremental conductivity (INC), and perturbation and observation (P&O) being the most often used conventional techniques. These techniques result in a high turnout in a steady-state activity [9]. These algorithms were verified to be not effective as when the weather is bad, conversion ratios are slow, and bigger variances prevent them from obtaining an overall maximum power point (MPP) in settings with partial shading conditions. To deal with these problems, MPPT with a bioinspired optimization algorithm has been proposed.

The artificial immune system (AIS) and the metaheuristic genetic algorithm (GA) were applied to overcome such nonlinear uncertain conditions because of the appropriate particular sensor and the complicated circuitry [10]. However, immune cells have a huge population structure and adaptive machinery, which results in a poor conversion rate and a lengthy conversion time for AIS and GA algorithms [11]. Crossover procedures are used in conjunction with computational convergence time to enhance the mutation. Many MPPT techniques with bioinspired optimization are implemented to challenge such difficulty [12]. FSA is a fish life-inspired methodology designed to reduce grade point average assessment (GMPP) oscillations. Numerous control settings are needed for PSO's random accelerating value choosing, and it may be a significant drawback. The bioinspired optimization techniques presently have more tracking efficiency, a high convergence rate, and low transients [13]. Gray wolf (GW), ant colony (AC), glowworm optimization algorithm, and fish swarm algorithm (FSA) are a few examples. However, due to less bee availability and the weather being unpredictable, the poor conversion rate in ABC approaches [14,15].

Due to a shortfall of contingency and a heavy nest population, the cuckoo search algorithm is a more productive way for nonlinear-based issues, although its rate of melting is moderate. Due to this, several researchers have implemented this bioinspired approach based on photovoltaic system investigation. Considering the difficulty present in MPPT techniques, this paper proposed a novel MPPT control technique for MPA. It does not need hardware data from PV, as it can exactly and rapidly search to find the GMPP. This work object is to enhance the overall performance of PV-powered electric vehicles. In this technology, the BLDC motor is used in PV-powered e-vehicles. The MPA technique has been implemented to increase the complete performance of the system. The MPA technique features a faster convergence rate and a better method to locate GMPP. An observation of MPPT output has been illustrated to determine the effectiveness of the suggested approach in this framework.

For maximum power tracking from solar PV, making use of combined MPPT, different techniques were presented by scientists. In the following segment, the different MPPT approach for EVs driven by BLDC is surveyed, and it was designed for maximum power tracking.

The complete paper's structure is provided below: The literature review is presented in Section 2. the proposed work is explained in Section 3. Sections 4 and 5 give the control techniques which were used in the proposed work, followed by its results in Section 6 and conclusion in Section 7.

2. Literature Review

Himabindu et al. [16] presented the partially solar-powered EV. The EV's energy efficiency is greater than that of fuel-powered vehicles without taking electricity generation, transmission, and efficiency into account. Moreover, for a limited solar-powered EV, the unique prototype of a lightweight EV was elaborated on in this paper. The development of the unique energy-efficient prototype of EV and the possibility of a limited solar-based EV was discussed. Lakshmiprabha et al. [17] presented the BLDC motor with a PV-based electric vehicle approach. The approach for developing the BLDC driven with PV-powered EV, which was a potential solution for the lake of impending, was explained in this work.

The approaches to finding the right parts of this application were explored, and both of them were tested and simulated in a real-world application. The integrated system of the PV-powered EV features the BLDC motor, batteries, battery charger controller, solar module, and a DC-DC boost converter. Ahmad et al. [18] demonstrated that the nature of the autoindustry was changing as a result of worries about oil supply, foreign relations, and fuel prices. There were numerous hybrid technologies available at the time, due to the availability of hydrogen. Among the oldest vehicles using alternative fuel, the vehicle integrated with solar power has several applications in the expanding EV market. The development of the solar-powered telemetry system for high-speed cars helps in improving the understanding of the vehicle's power aspects and the operation implemented in EVs. This work inspected the position and history of electric vehicles and solar energy, in addition to a standard solar vehicle.

García et al. [19] conferred on e-rickshaws driven by a BLDC motor a fuzzy logic controller (FLC)-based technique to develop ideal power management for regenerative braking. The FLC was adapted to control the separate power management for the battery and for the supercapacitor, to supply the output of the e-rickshaw driven by BLDC. E-rickshaw enhanced operating time by the solar-powered approach to boost the operation, and using simulated testing rickshaws was verified, which exposed the examination of the BLDC's performance under several operating conditions. If the need for power increases suddenly in a temporary situation, the supercapacitor manages the complete need for power. The power ratio is divided to enable the battery to be deeply discharged, increasing battery life. Ho et al. [20] explained the integration of electric power systems for the EV. The objective of this work was to introduce the theoretical arrangement to successfully integrate EVs into electric interconnected networks. The advanced structure was split into power market environments and the grid technical operations. Participants in both processes, as well as their actions, were all considered and fully explained. Moreover, various simulations, with the dynamic and analysis of steady-state behavior, were explained to make clear the impacts and benefits originating from the EVs and integration of the grid using the cited methodology. Oubelaid et al. [21] demonstrated the controlling techniques for hybrid electric vehicles. Global optimization techniques and dynamic programming were mainly employed to evaluate the powertrain configuration's prospective fuel efficiency. These control procedures cannot be applied directly until advanced driving conditions could be likely at the time of real-world application; even so, the results obtained with this noncausal method delivered the criteria for analyzing the best possible control technique that is attainable.

Lan et al. [22] conferred the creation of the Japanese government's EV policy. The scope of this work was to inspect the policy for the creation of alternative vehicles to traditional vehicles, the outcomes of government actions, and the requirement of a technological adaptability program supported by the government. The effects of this scheme on the methods of innovation were explored through the use of this viewpoint and technological literature improvement. The complete network with the assistance was investigated, further to the context in which this different policy has been used since the early 1970s. Saha et al. [23] demonstrated for EVs with BLDC Motors that are electric, hybrid, and plug-in hybrid an effective regenerative braking system using battery/ultracapacitor. The ultracapacitor used a suitable inverter switching template for energy regeneration and/or regenerative braking to store the vehicle's kinetic energy. Due to this, no extra power electronics interfaces were needed. Simultaneously, the EV's front and back wheels received braking force from the artificial neural network controller, which is responsible for distributing it. To attain steady torque braking, additionally, the PI controller was used to vary the PWM operating cycle. Li et al. [24] explained that the BLDC motors are controlled by a hybrid sliding-mode system without a position sensor (HSMC). This research gave effective and reliable control techniques for the position-sensorless EV using the BLDC motor. To adopt the BLDC motor sensorless control of the BLDC motor, the back EMF finding technique was initially implemented and enhanced. The corresponding circuits of regulating systems were presented, as well as the creation of energy regeneration and standard driving mathematical models. A technique for the EV HSMC approach was implemented to promise by integrating both the system effectiveness and using the high-order sliding-mode approach; the nonsingle terminal sliding mode has sustained stability.

Gupte et al. [25] have conferred on transmission a selectively aligned surface (PM-BLDC) for the HEV motor. A programmable and adjustable generated voltage constant in an axial-flux PMBLDCM was used to achieve the field weakening. This quality was exclusively suitable in motors for driving vehicles with vast ratios of constant-power speeds, where it was imperative to get rid of gear shifts and shrink the overall motor drive's size. The advantages of this method's high pole count were discussed, and the simulation's impact on the kilovolt-ampere motor drive, acceleration, maximum speed, and efficiency was described over regular driving cycles. The e-vehicles with BLDC motors used in this system are energized using solar PV. The MPA technique is derived concerning enhancing the system's total efficiency. For efficient MPP tracking from a PV array, scientists took help of the MPA technique.

3. Modeling and Description of the Entire System

Figure 1 illustrates the BLDC motor with a PV-powered, battery-operated architectural arrangement for EV applications. From left to right in Figure 1, the system consists of a solar PV array, a DC-DC converter, a battery, a VSI, and a BLDC motor. The PV array produces power and is given as an input to the DC-DC boost converter, which utilizes the MPA technique to operate the MOSFET switch. To get more power output from the PV array, the MPPT algorithm is implemented. The boost converter is used to provide the essential power for battery charging; the PV is serially connected to the battery bank, to operate the BLDC motor through an inverter. The electronics commutation is used to generate the switching pulse required for an inverter connected with a BLDC motor. In the following sections, we will discuss how to create and control the suggested system.

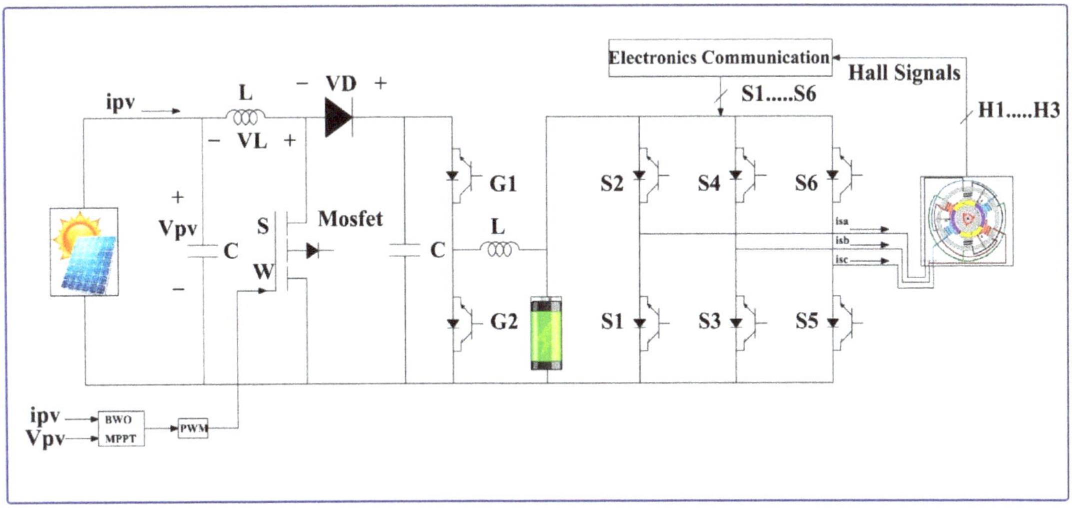

Figure 1. Circuit arrangement for the advanced battery-operated, solar PV-powered, BLDC motor-operated system.

3.1. Proposed System's Design Configuration

In Figure 1, the alignment of the suggested system configuration has been presented. The most essential parts of the system configuration are the SPV array, DC-DC converter, battery bank, and brushless DC. This system is built to function satisfactorily regardless of changes in the amount of solar irradiation.

3.2. Arrangement of Solar PV Array

The number of PV modules connected in parallel and series in a solar array is used to estimate the current, voltage, open-circuit voltage, and short-circuit current. The comparable circuit of PV cells is illustrated in Figure 2. The parallel diode, current source, and series resistors are the components required. To create photovoltaic modules, the PV cells are built simultaneously. The required power is based on a combination of parallel and series supply. U* a and U* p represent the number of parallel and series photovoltaic cells, respectively. The voltage and current output relationship can be expressed as

$$I'_{PV} = N'_P I'_G - N'_P I'_S \left(\exp\left[\frac{q^*}{AKT_C} \left(\frac{V'_{PV}}{N'_S} + \frac{R_S I'_{PV}}{N'_P} \right) \right] - 1 \right) \quad (1)$$

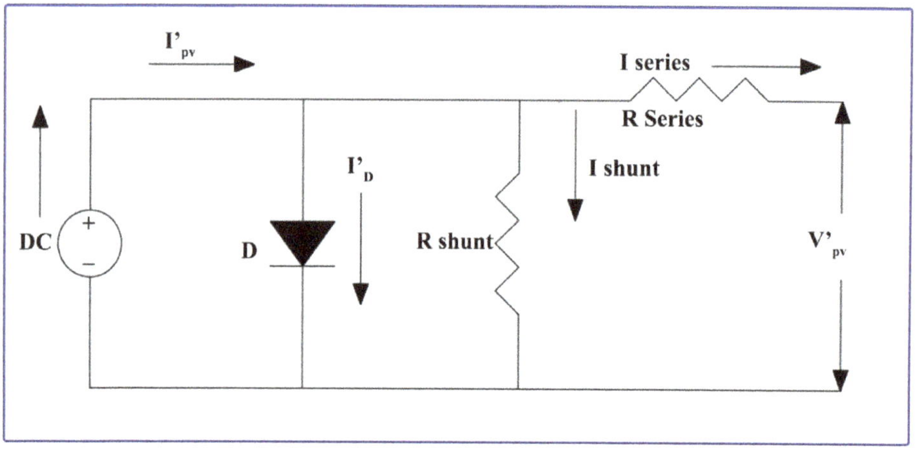

Figure 2. PV cell equivalent circuit.

Photocurrent I'_G is produced by solar irradiation, as shown below:

$$I'_G = I'_{sc} + k_1 \left(T_C - T_{ref} \right) \frac{S}{1000} \quad (2)$$

I'_S is explicit as the PV cell saturation current and temperature variation based on the following relationship:

$$I'_S = I'_{rs} \left[\frac{T_C}{T_{ref}}]^3 \exp[\frac{q' E_G}{AK} \left(\frac{1}{T_{ref}} - \frac{1}{T_C} \right) \right] \quad (3)$$

3.3. PV Characteristics

The PV array's nonlinear characteristics are dependent on temperature and irradiance. Variation in temperature and irradiation causes a change in them. The V-I and P-V characteristics at different irradiation and constant temperature (1000 W/m^2, 800 W/m^2, 500 W/m^2, 350 W/m^2) are depicted in Figure 3, as well as the V-I and P-V characteristics at changing temperature and constant irradiation are depicted in Figure 4.

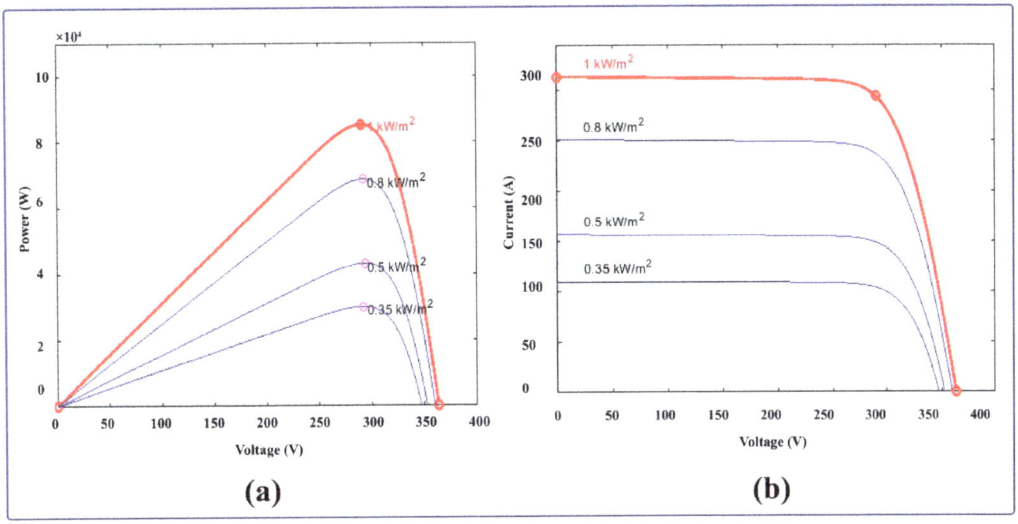

Figure 3. (**a**) P-V curve, during constant temperature (**b**), V-I curve, during constant temperature.

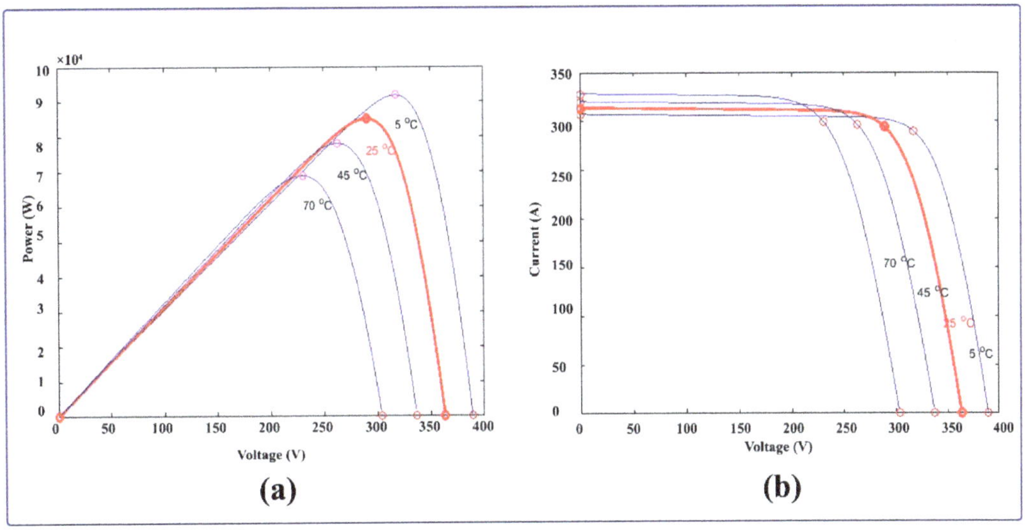

Figure 4. (**a**) Continuous irradiation of the P-V curve, (**b**) continuous irradiation of the V-I curve.

3.4. DC-DC Boost Converter Equivalent Circuit

The equivalent circuit of the DC-DC converter is shown in Figure 5. At the initial stage Switch, sw1 and sw2 are in closed and open positions, respectively, and the inductor current (IL) will be raised from zero. Consequently, switch sw1 and sw2 are in open and closed positions respectively; at that time, the inductor current will supply the load, and the charges will be stored in the capacitor. The voltage in proportion to the duty cycle of the input and output of a DC-DC converter is depicted in this equation.

$$\frac{V'_O}{V'_{IN}} = \frac{1}{1 - d'_{DUTY}} \tag{4}$$

$$\frac{V'_O}{V'_{IN}} = \frac{T'_{RISE}}{T'_{FALL}} + 1 \quad (5)$$

$$d'_{DUTY} = \frac{T'_{RISE}}{T'_{RISE} + T'_{FALL}} \quad (6)$$

where, d'_{DUTY} = duty cycle, T'_{RISE} = switch sw1 is in closed at the moment of raising the inductor current, T'_{FALL}= switch sw1 is open at the moment when the inductor current is falling.

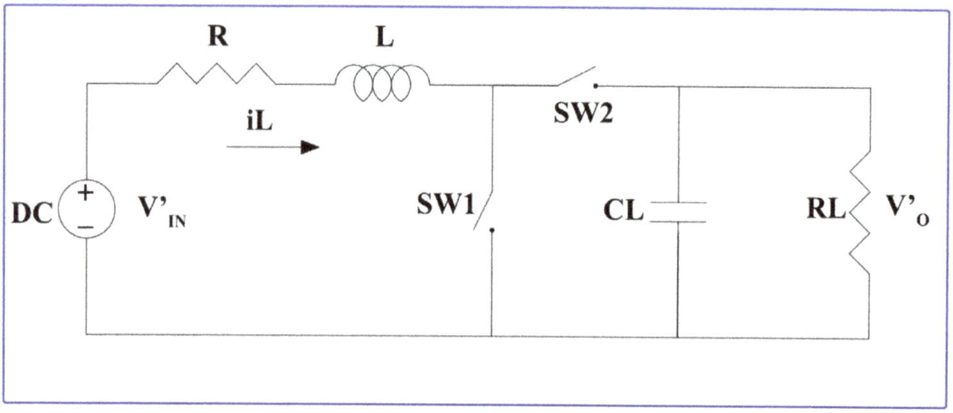

Figure 5. DC-DC boost converter equivalent circuit.

3.5. BESS Equivalent Circuit

The presented battery equivalent circuit contains polarization capacitor Cpl, polarization resistor Rpl, and ohmic internal resistor R, where the battery transient feedback is simulated using Rpl and Cpl in both charging and draining modes. In that, V (h(t)) represents the nonlinear function of the SoC for h(t)'s. The terminal voltage is taken as the calculated output, and the current is considered as a control input. Figure 6 depicts a lithium-ion battery, and MATLAB/Simulink software is used to perform the simulation. Figure 7 shows the input characteristics of the battery energy storage system where E0 is constant voltage (V), K is polarization constant in (Ah^{-1}), A is exponential voltage (V), and B is the exponential capacity (Ah^{-1}).

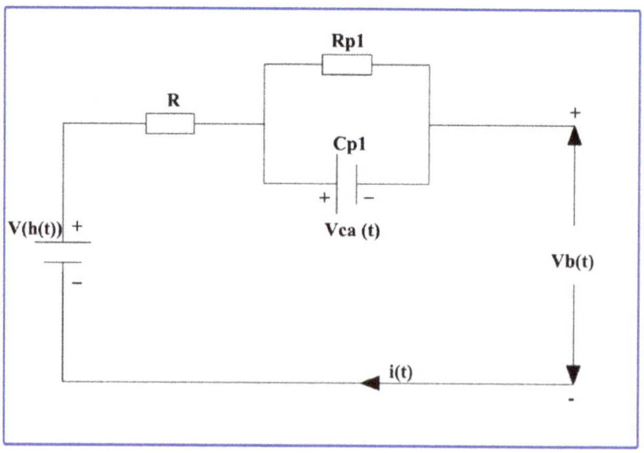

Figure 6. BESS equivalent circuit.

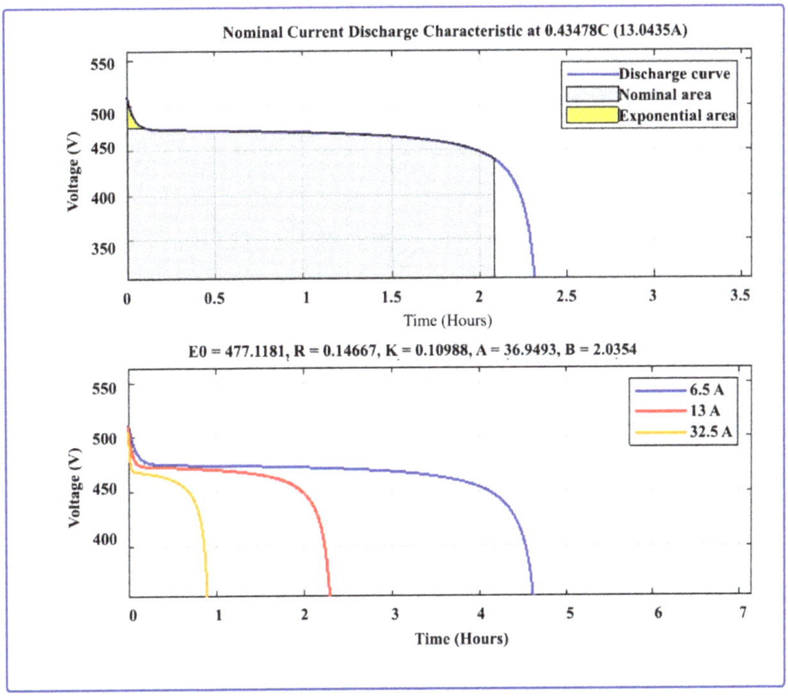

Figure 7. Input characteristics of battery energy storage system.

To build the 440 V/30 Ah and a 100% SoC, interconnected modules are used to actualize the battery. The circuit dynamics is expressed by applying Kirchhoff's law:

$$V_b(t) = V(h(t)) - R * i(t) - V_{ca}(t) \tag{7}$$

$$\frac{dV_{ca\ (t)}}{dt} = -\frac{1}{c_{pi}R_{pi}}V_{ca}(t) + i(t) \tag{8}$$

Here, $V_b(t)$ = terminal voltage, $i(t)$ = terminal current, and $V_{ca}(t)$ = voltage across RC, which cannot be directly computed.

3.6. Modeling and Motor Choice

In recent years, many electric motors have been used in electric vehicles. In the e-rickshaw, the DC motor's dynamic properties are better; the main disadvantage of the DC motor is that it needs more maintenance due to the brush and commutator. Induction motors are therefore a better option, because they are often suitable for such circumstances, but the induction motor needs huge control. Thus, in automotive applications, the induction motor is not usually employed. After that, the researchers take an alternative and find a trustworthy and effective motor. The BLDC motors are easy to regulate, require less maintenance, and have a high roughness. It has high torque, fast dynamic responses, a low operating voltage range, and a good performance ratio.

The BLDC motor consists of a permanent magnet stator and three-phase windings in the rotor. The currents generated in the rotor can be neglected, and there is no need to model damper windings if the stainless-steel retaining sleeves and magnet have high resistance. The analogous circuit of a BLDC motor is depicted in Figure 8, where R is a stator resistance, L is self-inductance and mutual inductance, and e is phase back-EMF voltage of A, B, and C, respectively. The 3 ϕ winding governing equation for the phase variables is

$$\begin{bmatrix} V_a^* \\ V_b^* \\ V_c^* \end{bmatrix} = R^* \begin{bmatrix} 1 & 0 & 0 \\ 0 & 1 & 0 \\ 0 & 0 & 1 \end{bmatrix} \begin{bmatrix} i_a^* \\ i_b^* \\ i_c^* \end{bmatrix} + \begin{bmatrix} l-m & 0 & 0 \\ 0 & l-m & 0 \\ 0 & 0 & l-m \end{bmatrix} \frac{d}{dt} \begin{bmatrix} i_a^* \\ i_b^* \\ i_c^* \end{bmatrix} + \begin{bmatrix} E_a^* \\ E_b^* \\ E_c^* \end{bmatrix} \tag{9}$$

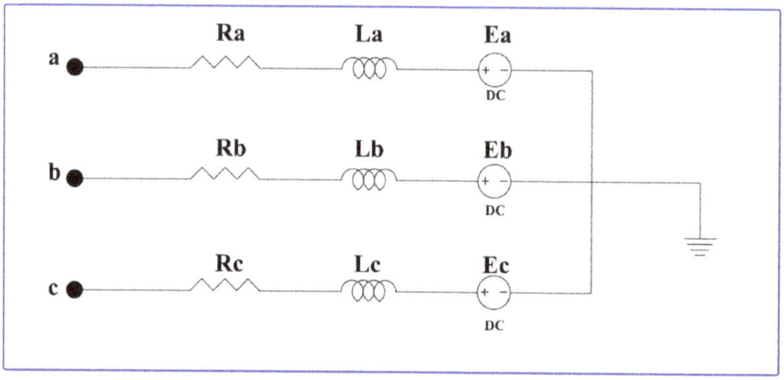

Figure 8. BLDC motor equivalent circuit.

R^* = phase resistance, m = mutual inductance, l = phase inductance.

The mechanical equation is shown below:

$$J^* \cdot \frac{d\omega_r^*}{dt} = T_e^* - T_l^* - f_r^* \omega_r^* \tag{10}$$

Finite element analysis is used to calculate the three back-EMFs, and Fourier series equations are used to display the results. It is a ratio of speeds.

4. Control Method Using MPA Technique

As shown in Figure 1, to carry out the required operation and to get the output from PV, MPPT with a DC-DC converter is needed. In the implementation of MPPT, a control variable (duty cycle) is controlled by the MPPT controller. This generates a control signal in the range [0, 1] which is given in Equations (11) and (12):

$$V_{out} = \frac{V_{in}}{1-d} \tag{11}$$

$$d = \frac{T_{on}}{T_{Switching}} \tag{12}$$

where, V_{out} and V_{in} are boost converter output and input voltages, and d denotes the duty cycle. This article gives a new bioinspired algorithm based on marine predators' social behavior pattern.

4.1. Marine Predator Algorithm

The marine predator algorithm (MPA) is a bioinspired, metaheuristic optimization technique [26] that has been applied to various optimization problems. A few of the applications of MPA are estimating the parameters of solar PV cells [27], MPPT for solar PV systems [28], and many more. In this section, the MPA is applied in MPPT in an optimized way to the optimal expected output for EVs.

The key points of MPA are (i) the Levy motion for a prey environment of low concentration given in Equation (13), (ii) the Brownian motion for a prey environment of high concentration given by Equation (14), and (iii) the very decent memory in recalling their partners and the location of successful hunting shown in Figure 9. These features make the marine predator's technique more advanced compared to other bioinspired techniques. The population can be started by Equation (15). Dmin and Dmax are the lower and upper limits for the variables, and rand is the random number.

$$\text{Lévy}\ (\alpha) = 0.05 \times \frac{x}{|y|^{\frac{1}{\alpha}}} \tag{13}$$

$$f(x;\ \mu,\ \sigma) = \frac{1}{\sqrt{2\pi}} e^{-\frac{x^2}{2}} \tag{14}$$

$$D_0 = D_{min} + \text{rand}\ (D_{max} - D_{min}) \tag{15}$$

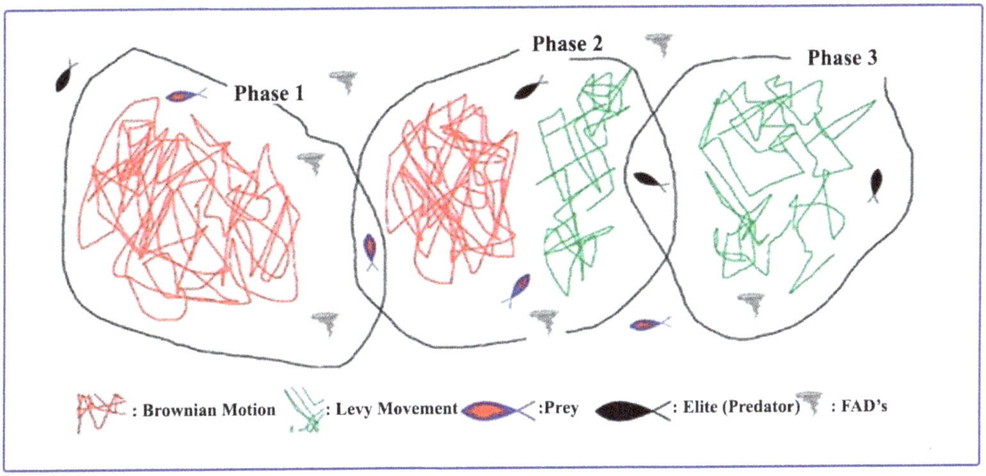

Figure 9. Three phases in marine predator algorithm (MPA) optimization.

An elite matrix is developed by the fittest solutions among the marine predators following the survival of the fittest idea. Naturally, the topmost predators (denoted by de) are brilliant in hunting and is given in Equation (16). The position of the predator gets updated from time to time. The prey matrix is developed in which $d_{i,j}$ gives the jth position of the prey and is given by Equation (17).

$$\text{Elite} = \begin{bmatrix} de_{1,1} & \cdots & de_{1,n} \\ \vdots & \ddots & \vdots \\ de_{m,1} & \cdots & de_{m,n} \end{bmatrix}_{m \times n} \tag{16}$$

$$\text{Prey} = \begin{bmatrix} d_{1,1} & \cdots & d_{1,n} \\ \vdots & \ddots & \vdots \\ d_{m,1} & \cdots & d_{m,n} \end{bmatrix}_{m \times n} \tag{17}$$

Optimization Process of MPA

There are three phases in optimization, as shown in Figure 9. Depending upon the velocity ratio and time, the phases are classified. Phase 1: predator is traveling slower than the prey (increased velocity ratio). Phase 2: predator and prey are at the almost same pace (unity velocity ratio). Phase 3: predator is traveling faster than the prey (decreased velocity ratio).

The prey is traveling faster than the predator. This is called as exploration phase, and it happens only in the initial or starting iterations of the algorithm and is given by Equations (18) and (19). Here, R is a rand [0, 1]. Set in a high exploration phase, this phase happens for the first three of the iterations. Prey is accountable for the exploration, and it is given by the Equations (20)–(26). That CF is a step-size controlling parameter for a predator:

$$\overrightarrow{\text{Stepsize}_a} = \overrightarrow{R_B} \times \left(\overrightarrow{\text{Elite}_a} - \overrightarrow{R_B} \times \overrightarrow{\text{Prey}_a} \right);\ a = i \ldots n \tag{18}$$

$$\overrightarrow{\text{Prey}_a} = \overrightarrow{\text{Prey}_a} + P\,\vec{R} \times \overrightarrow{\text{Stepsize}_a} \tag{19}$$

For the predator population:

$$\overrightarrow{\text{Stepsize}_a} = \overrightarrow{R_L} \times \left(\overrightarrow{\text{Elite}_a} - \overrightarrow{R_L} \times \overrightarrow{\text{Prey}_a} \right); a = i \ldots \frac{n}{2} \tag{20}$$

$$\overrightarrow{\text{Prey}_a} = \overrightarrow{\text{Prey}_a} + P\,\vec{R} \times \overrightarrow{\text{Stepsize}_a} \tag{21}$$

For the prey population:

$$\overrightarrow{\text{Stepsize}_a} = \overrightarrow{R_B} \times \left(\overrightarrow{R_B} \times \overrightarrow{\text{Elite}_a} - \overrightarrow{\text{Prey}_a} \right); a = \frac{n}{2} \ldots n \tag{22}$$

$$\overrightarrow{\text{Prey}_a} = \overrightarrow{\text{Elite}_a} + P\,\overrightarrow{CF} \times \overrightarrow{\text{Stepsize}_a} \tag{23}$$

$$\overrightarrow{\text{Stepsize}_a} = \overrightarrow{R_L} \times \left(\overrightarrow{R_L} \times \overrightarrow{\text{Elite}_a} - \overrightarrow{\text{Prey}_a} \right);\ a = 1 \ldots n \tag{24}$$

$$\overrightarrow{\text{Prey}_a} = \overrightarrow{\text{Elite}_a} + P\,\overrightarrow{CF} \times \overrightarrow{\text{Stepsize}_a} \tag{25}$$

To avoid the eddy formation or fish aggregating devices (FADs), which may change the marine predators' behavior, marine predators may take a long jump, as is given in Equation (26):

$$\overrightarrow{\text{Prey}_a} = \begin{cases} \overrightarrow{\text{Prey}_a} + CF\left(\overrightarrow{D_{min}} + \vec{R} \times \left(\overrightarrow{D_{max}} - \overrightarrow{D_{min}} \right) \times \vec{U} \right) & \text{if } r \leq FAD_1 \\ \overrightarrow{\text{Prey}_a} + (FAD_s \times (1 - r) + r)\left(\overrightarrow{\text{Prey}_{r1}} - \overrightarrow{\text{Prey}_{r2}} \right) & \text{if } r \leq FAD_s \end{cases} \tag{26}$$

4.2. *Implementation of MPA for MPPT during PSCs*

In the search space between Dmin and Dmax [0 to 1], the particles should be initialized for the implementation of MPPT using the MPA optimization technique with a population size of 4, since the work uses four panels in the array. Due to irradiance change, the power will change at that time the code will automatically restart or reinitialize and is given by the conditional Equation (27).

$$\text{if } \frac{\left| P_{PV_{new}} - P_{PV_{old}} \right|}{P_{PV_{old}}} \geq P_{PV}(\%) \tag{27}$$

The flow chart of the proposed bioinspired MPA technique-based MPPT is given in Figure 10, which will validate Equation (27).

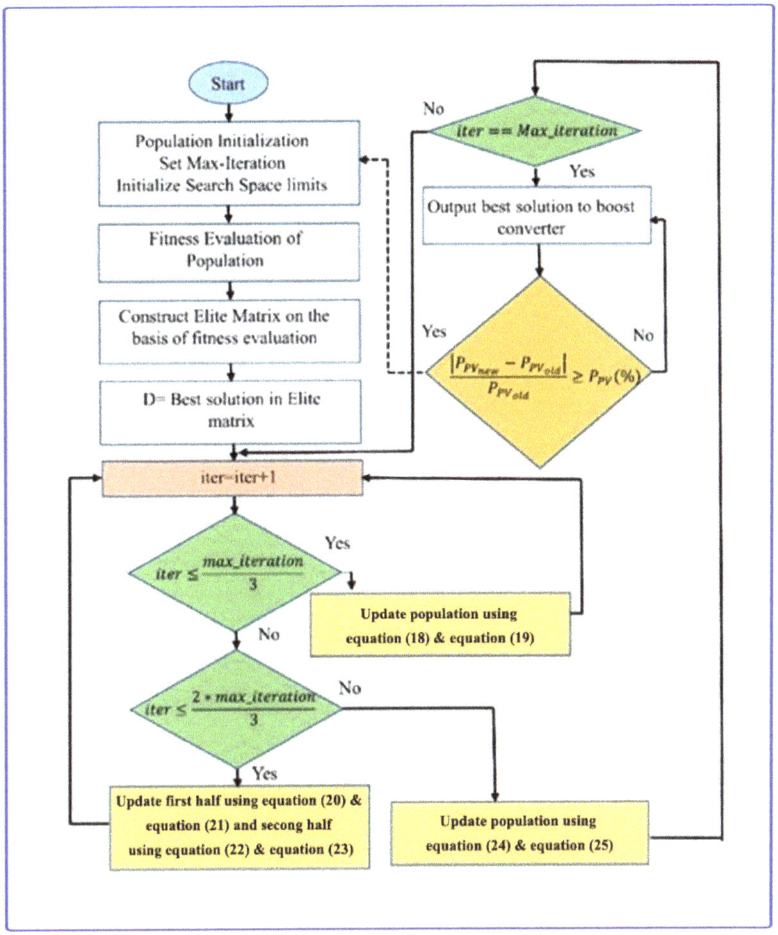

Figure 10. Flowchart for marine predator algorithm-based MPPT.

To provide maximum power for the EV, the above-mentioned process using MPA will track and extract the maximum power from the solar PV.

5. BLDC Motor Control Based on PID

PID consists of a set of conditions that could be applied to give a closed-loop control system precise regulation. In a closed-loop control process, the controlling device receives continuous real-time measurements of the process being controlled to ensure it reaches the desired range. The computed value, also known as the "process variable", is made by the controlling device to resemble the indented value, also known as the "set point". To complete the desired work effectively, the PID control algorithm is adapted. The most important of these is proportionate control, which measures the error value and creates proportionate changes to lower the error in the control variable. Proportional control is mostly used in many control systems. The PID controller continuously evaluates the difference between the process variable and the set point and makes the necessary corrections. Derivative control monitors the process variable's rate of change and modifies the output variable to take unexpected changes into account.

Each of the three control functions is directed by a user-defined parameter. These characteristics may differ from one control system to another, and as a result, they must be modified for the best control precision. Finding the values of these parameters is known as PID tuning. Although many people think of PID tuning as "black magic", it is always known as a precise mathematical process.

There are numerous ways to accomplish PID tuning, and any of the methods can be used to tune any system. While some PID controlling methods require more devices than others, they typically produce more accurate output with low effort. The fundamental objective of the PID controller is to execute algorithm-based tuning constants. The control engineer delivers the current plant process value and the operator's intended operating value (set point). In most situations, the controller will conduct to bring the process value as close to the set point as is practical. To perform a simple process control loop, PID algorithms will be implemented by the control engineer.

PID Controlling

The main goal of the PID controller is to maintain the constant output level so that there is no difference (error) between the process variable (Pv) and the set point (SP). The valve may control the flow of gas to a heater, the water level of the tank, the temperature of a chiller, the flow through a pipe, the pressure of the pipe, or any other method for process control and shown in Figure 11.

$$obj = k_p^* e_t^* dqO + k_i^* \int_0^1 e_t^* dqOdt + k_d^* \frac{de_t^*,\, dqO}{dt} \tag{28}$$

$$Min(X) = Min(ITAE^*) \tag{29}$$

where, X = total controller error, $ITAE^*$= absolute integral-time error.

$$ITAE^* = \int_0^\infty T|e_t^* dpO|dt \tag{30}$$

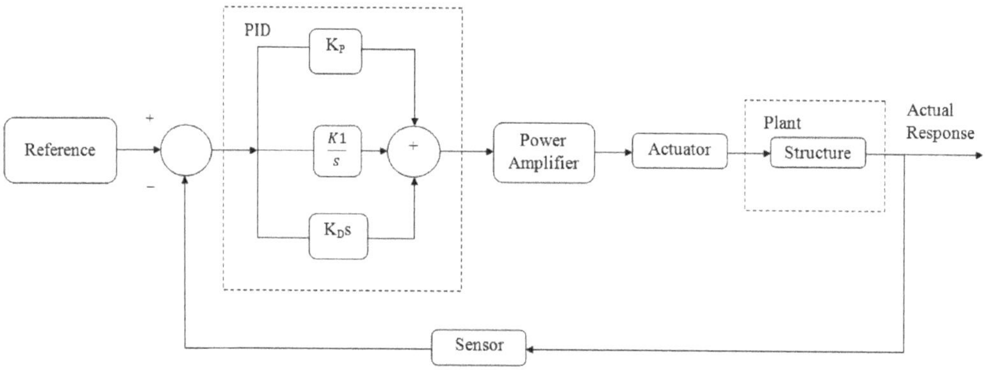

Figure 11. PID controller block diagram.

$e_t^* dpO$ = error signal between a reference voltage and load.

The MPA-based control method is implemented in the proposed work. It is explained in Section 4 and provides a detailed explanation of the PID approach utilized for BLDC motors with electronic commutation. It is the most common strategy for controlling the 3ϕ AC motor speed and torque by utilizing the current control method. When the BLDC motors are operated at both high and low speeds, at that time there is no precise speed control and ripple in torque. For applications such as washing machines, PID is very important. The VSI switching signals were produced by the motor's "electronic commutation". The Hall effect sensor is attached to the stator and is used to find the rotor position angle. These Hall effect signals are changed into six switching pulses, which are used to control the switches of the voltage source inverter.

6. Results and Discussion

The solar PV panels used to generate the power of 500 W, in addition to power connecting the DC-DC converter, are included in the proposed work; it was made by using MATLAB/Simulink software. To charge a vehicle battery, the power extract from PV is used. In this configuration, we are powering a BLDC motor rating of 3000 rpm, 256 V, 1 kW, and a BLDC motor with a series batteries rating of 220 V, 90.4348 Ah. The current and voltage extract from the solar array is given as MPPT input, and the switch in the DC-DC converter is activated by PWM signal, enabling the implementation of the MPPT from the solar array. The proposed MPA-based MPPT control algorithm is correlated to the GWO and WOA MPPT control bioinspired algorithms to assess how well it performs.

The findings of this study are examined in four modes, which are described in the following section:

Mode 1: Constant Motor Speed and Constant Irradiance

The MATLAB Simulink software is used to simulate the suggested work under 25 °C constant temperature and a constant irradiation of 1000 W/m^2, as shown in Figure 12. Meanwhile the settling time is 0.06 s and 0.035 s for power through the GWO and WOA, respectively. This produces a high fluctuating signal. The output power through the MPA is settled in 0.02 s. In the first case, the PV panel temperature is 25 °C constant, and the PV irradiance is 1000 W/m^2. In affixing, the BLDC motor speed is set at 3000 rpm constant, and the battery voltage and PV power outputs are analyzed in Figure 13. The PV current, PV voltage, and PV power are illustrated in Figure 13; the solar PV power reaches 62 kW and settles in 0.02 s, and the PV current and voltage are obtained at 185 A and 340 V, respectively. Figure 13 demonstrates the outputs of the battery, which are the current, SOC, and output voltage of the battery. The battery current and voltage are obtained at 480 A and 360 V, respectively, and battery SOC is reached at 100% in discharging mode.

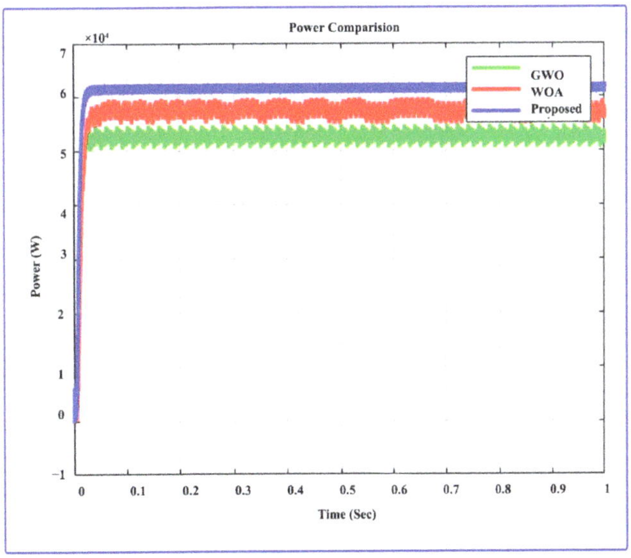

Figure 12. PV output power while the solar irradiation and motor speed are both constant.

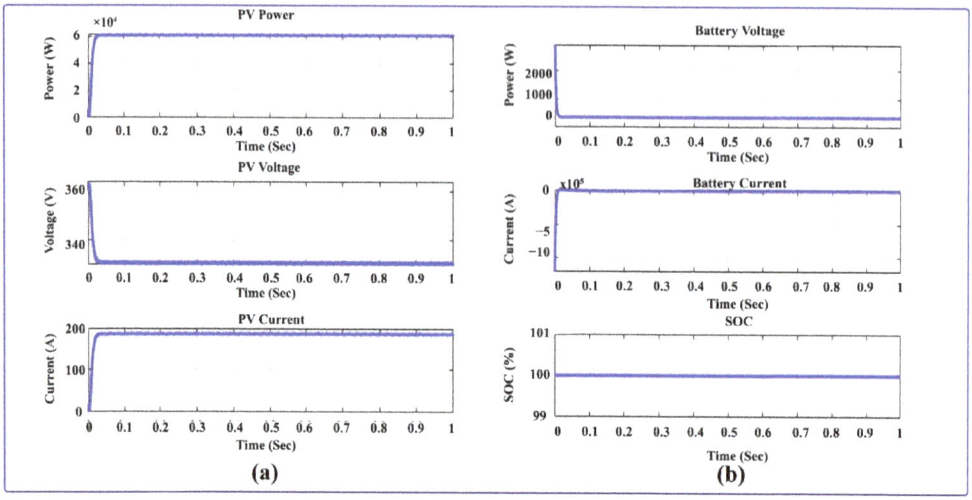

Figure 13. (**a**) PV outputs at constant speed and irradiation (**b**) battery outputs at constant speed and irradiation.

7. BLDC Motor Outputs

The BLDC motor output is illustrated in Figures 14 and 15. Here, Figure 14 shows the speed and the BLDC motor comparison. At first, 3000 rpm is set as a reference speed for one second. Figure 14a illustrates the BLDC motor speed, which is set to a constant speed of 3000 rpm, and Figure 14b demonstrates the speed and reference speed comparison. Figure 15 shows the stator current and torque of the BLDC motor. Here, a stator current of 3 ϕ is used to demonstrate how the BLDC motor's torque is reached at 1.2 Nm. Figure 15a at the beginning increases at 100 Nm, and at 0.02 s it decreases to 1.2 Nm. Figure 16 illustrates the Hall signal and back EMF of the BLDC motor.

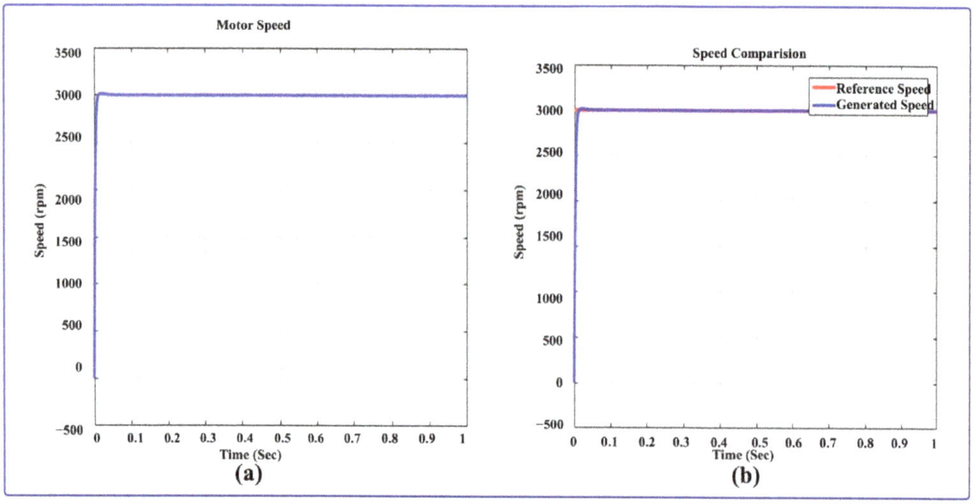

Figure 14. (**a**) Speed of the BLDC motor (**b**) comparison of motor speed and reference speed.

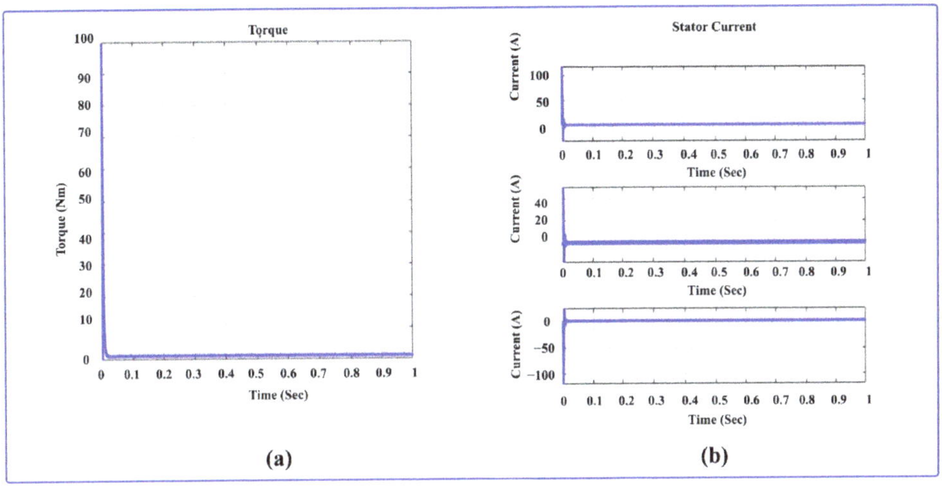

Figure 15. (**a**) BLDC motor torque at constant speed (**b**) BLDC motor stator current at constant speed.

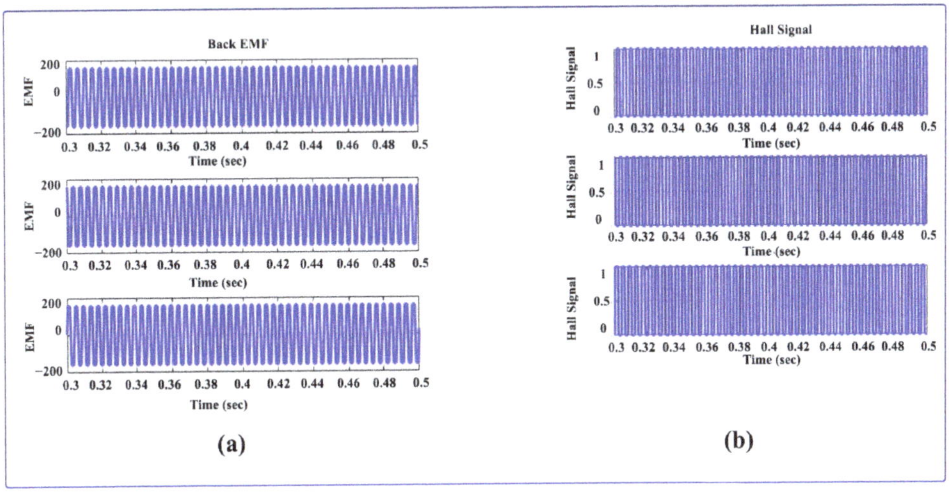

Figure 16. (**a**) BLDC motor back EMF at constant speed (**b**) BLDC motor Hall signal at constant speed.

Mode 2: Constant motor speed and varying irradiation

The MATLAB Simulink software is used to simulate the suggested work for 0–0.5 s in 1000 W/m^2 irradiation, 0.5–1 s in 500 W/m^2, with 3000 rpm motor speed and 25 °C constant temperature. In Figure 17, the power flow across the MPA is settled in 0.02 s. The power through the GWO and WOA is settled in 0.06 and 0.35 s, respectively. This produces a highly fluctuated signal. The suggested power from solar PV reaches 63 kW in 0–0.5 s at an irradiance of 1000 W/m^2, as shown in the graph. Then, the power is changed to 40 kW after 0.5–1 s at an irradiance of 500 W/m^2. The suggested approach is related to the traditional approach, such as the GWO and WOA MPPT algorithm, and proves the advancement of the suggested method. The battery and PV output is depicted in Figure 18. Figure 18a illustrates the PV power, voltage, and current attained at 62 kW, 340 V, and 185 A, respectively, that were generated in solar PV at 0–0.5 s time range. The irradiance is 1000 W/m^2, and then suddenly PV power, PV current, and PV voltage raise to 61 kW, 175 A, and 325 V, respectively, and irradiance to 500 W/m^2.

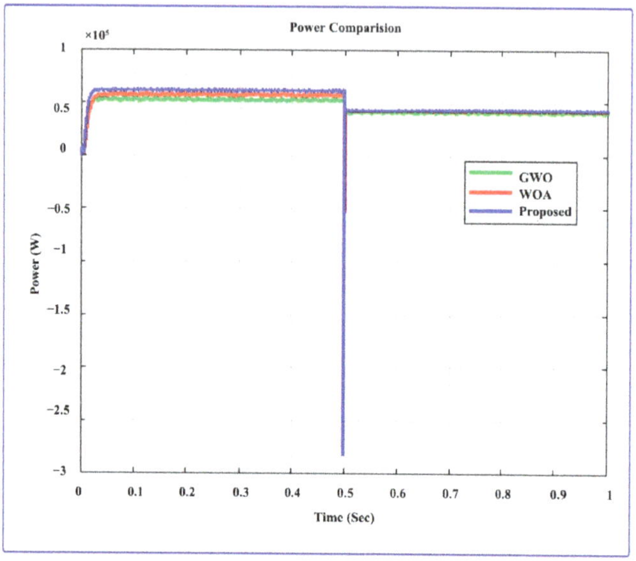

Figure 17. PV panel output power with variable irradiance.

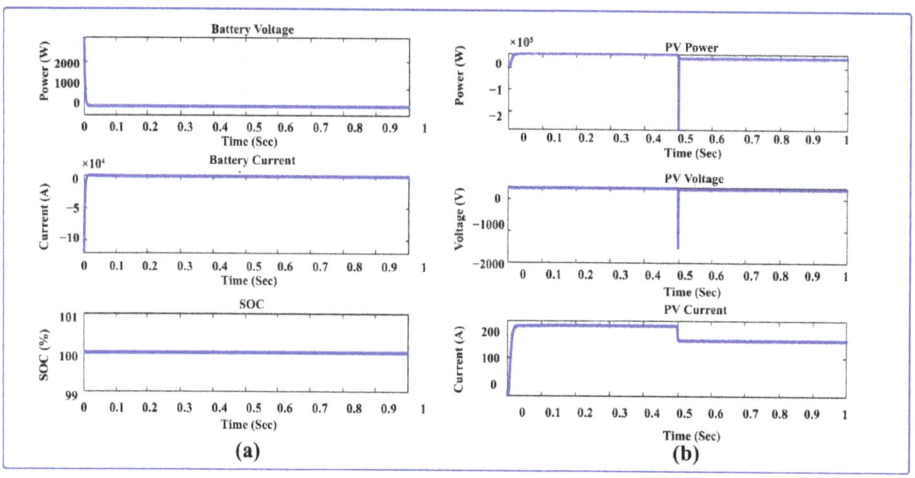

Figure 18. (**a**) Battery outputs at variable irradiance (**b**) PV outputs at variable irradiance.

The BLDC motor speed comparison at various irradiances is shown in Figure 19. In Figure 19a, the BLDC motor speed in regard to setting the reference speed of 3000 rpm for 0–0.5 s at PV irradiance is 1000 W/m^2. After 0.5 s at the same motor speed, PV irradiance is set to 500 W/m^2. The speed of the actual speed-to-reference speed comparisons is depicted in Figure 19b. The torque of the BLDC motor and stator current is shown in Figure 20; Figure 20a shows the high torque starting stage of the motor after 0.01 s, which dropped immediately to settle at 0.01 s on 1.2 Nm. The comparison of speed during the constant speed and variable irradiance of the motor is shown in Figure 20b. The Hall signal and back EMF of all phases are shown in Figure 21a,b.

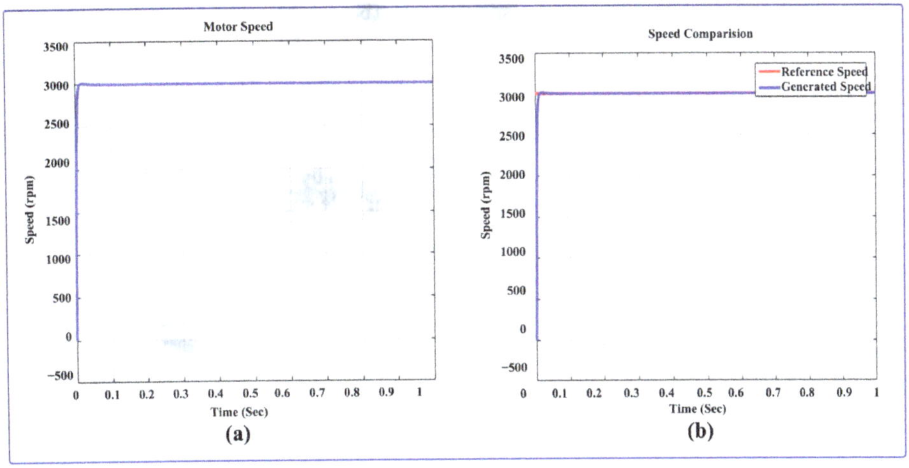

Figure 19. (**a**) Speed of the BLDC motor (**b**) comparison of motor speed and reference speed at varying irradiation.

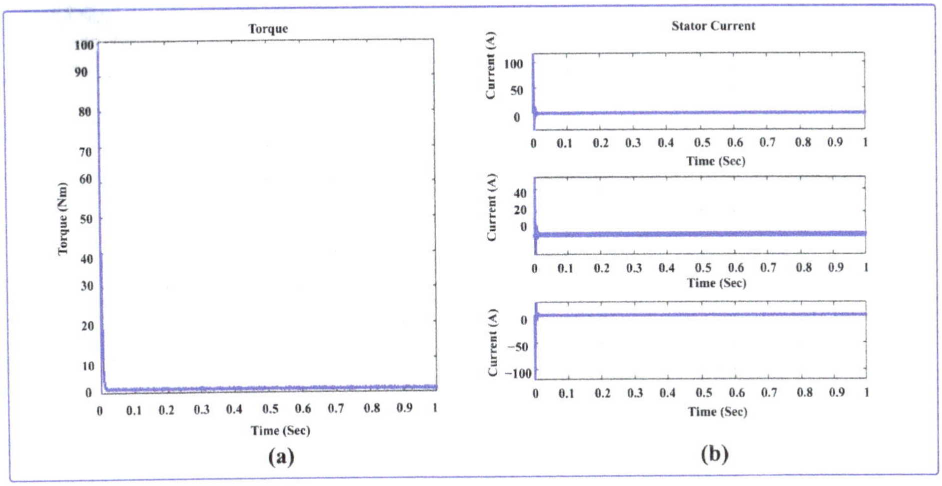

Figure 20. (**a**) BLDC motor torque with variable irradiation and constant speed (**b**) BLDC motor stator current with variable irradiation and constant speed.

Mode 3: Variable motor speed and constant irradiation

The input for this mode is variable motor speed and constant irradiance; 1000 W/m^2 for constant irradiance, 3000 W/m^2 for 0–0.2 s, 1000 W/m^2 for 0.2–0.4 s, 1500 W/m^2 for 0.6–0.8 s, 2500 W/m^2 for 0.6–0.8 s, and 2000 W/m^2 for 0.8–1 s for BLDC motor speed, respectively. Figure 22 illustrates the PV power at the variable motor speed and constant irradiance comparison. The suggested approach is more admirable than the GWO and WOA. According to the variation in the motor speed, the power will be varied. When the motor speed is 3000 rpm, the suggested technique power will reach 62 kW. When the motor speed is decreased to 1000 rpm, the suggested technique power is increased from 62 KW to reach a power of 81 kW. Accordingly, for motor speeds of 1500 rpm, 2000 rpm, and 2500 rpm, the power of the suggested approach will be 78 kW, 70 kW, and 68 kW, respectively. Figure 23 depicts the PV output and battery output at 1000 W/m^2 of constant irradiance, and variable speeds of 3000 between 0 and 0.2 s, 1000 from 0.4–0.6 s, 2500 from 0.6–0.8 s, and 2000 from 0.8–1 s are used. The PV current, power, and voltage are shown in Figure 23a. In that PV power for the regular interval of (0.2, 0.4, 0.6, 0.8), high changes may occur for the regular interval. In the meantime, the voltage of PV is drained to 325 V, and the current of PV is increased to 220 A. The battery SOC, battery current, and voltage of the battery are depicted in Figure 24. The changes in battery current and voltage, as well as battery charging and discharging, are caused by the variable motor-speed input. The BLDC motor Hall signal and back EMF at variable speed and constant irradiance are depicted in Figure 25.

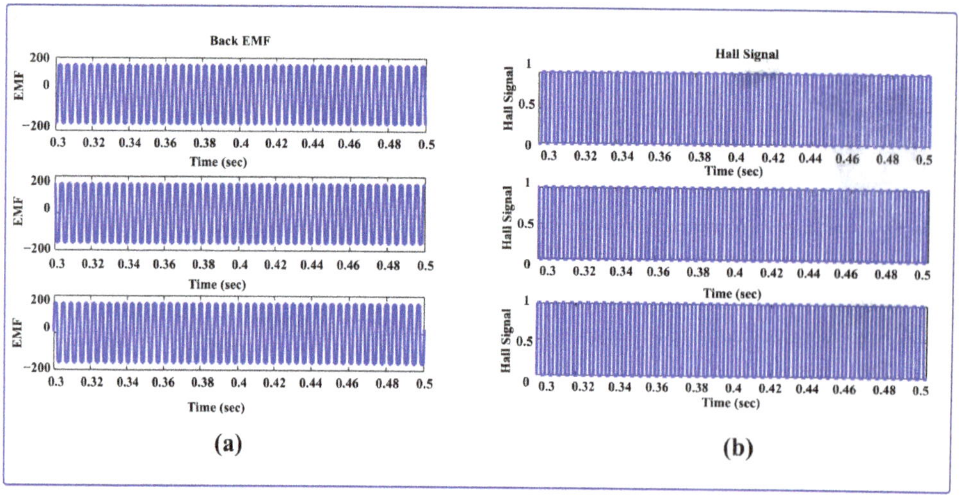

Figure 21. (**a**) BLDC motor back EMF with variable irradiation and constant speed (**b**) BLDC motor Hall signal with variable irradiation and constant speed.

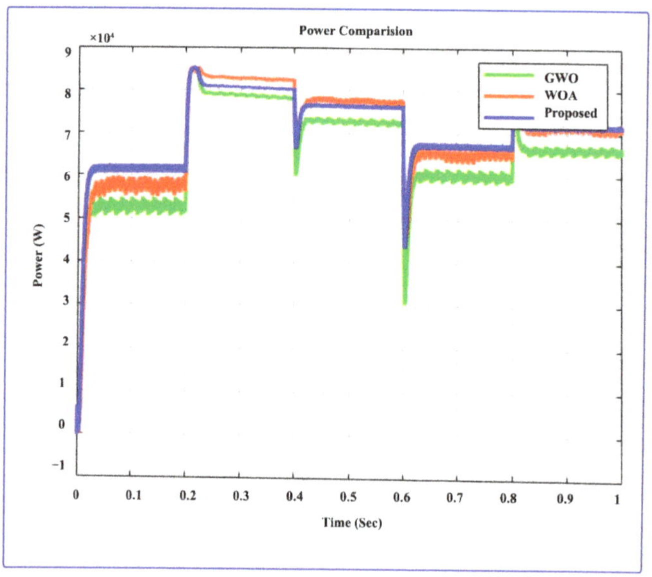

Figure 22. PV power comparison with variable motor speed and constant irradiation.

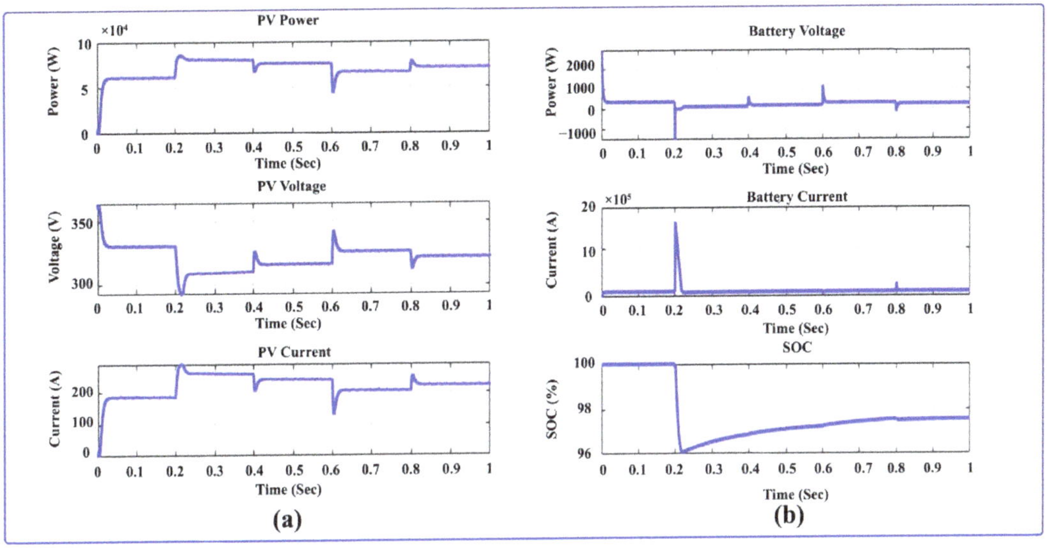

Figure 23. (**a**) PV outputs with varying speeds and constant irradiance (**b**) battery outputs with varying speeds and constant irradiance.

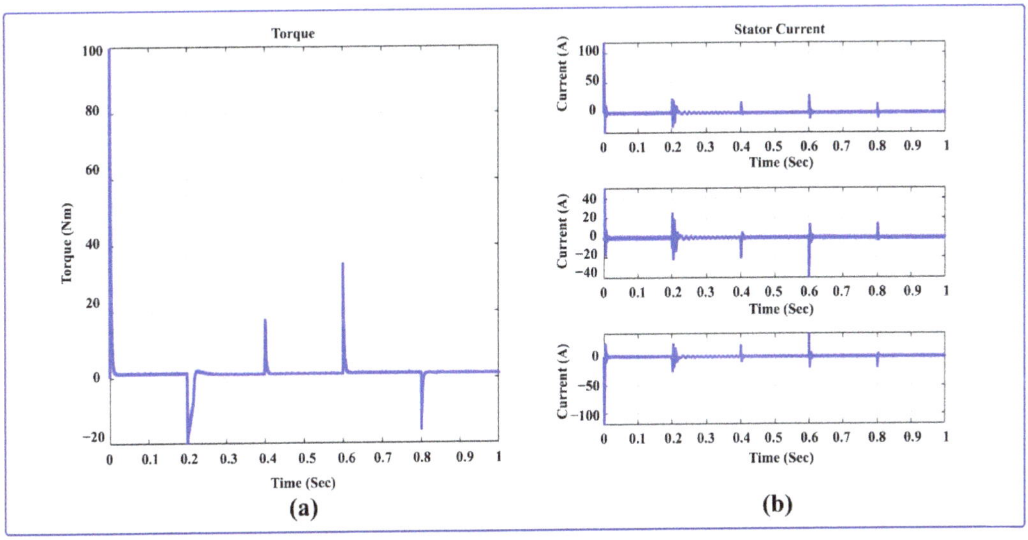

Figure 24. (**a**) BLDC motor torque at varying speeds and constant irradiation (**b**) stator current at varying speeds and constant irradiation..

Mode 4: Variable Motor Speed and Variable Irradiance

Figure 26 illustrates the comparison of power output at variable motor speed and variable irradiance. In this mode, variable irradiance is 1000 W/m^2 for 0–0.5 s before shifting to 500 W/m^2 for 0.5–1 s. Additionally, the speed is fixed at 3000 rpm for 0–0.2 s, 1000 rpm for 0.4–0.6 s, 1500 rpm for 0.6–0.8 s 2500 rpm, and 2000 rpm for 0.8–1 s. The advanced approach is related to the traditional approach to prove its advantages. The PV and battery output are depicted in Figure 27. The output of the solar PV at variable speed and irradiance is illustrated in Figure 27a. Power from PV varies concerning varying irradiance and variable speed. Between 0 and 0.5 s, the variable irradiance is changed to 1000 W/m^2 and 500 W/m^2 between 0.5–1 s. Furthermore, the BLDC motor speed is modified, from 0 to 0.2 s, 3000 rpm, 1000 to 0.4 s, 1500 to 0.6 s, 2500 to 0.8 s, and 2000 to 1 s. All outcomes from PV are changed in favor of the variable speed and variable irradiance. Figure 27b shows the outputs of the battery current, battery voltage, and battery SoC. The BLDC motor speed and a comparison of its speed under different irradiation conditions are illustrated in Figure 28. The motor's speed relative to the reference speed for 1000, 1500, 2000, 2500, and 3000 rpm is demonstrated in Figure 28a. The comparison speed to a reference speed is illustrated in Figure 28b. The reference speed is denoted as a straight red line, and the BLDC motor's actual speed is mentioned as a blue line. The variable speed is given to 0, 0.2, 0.4, 0.6, and 0.8 s time intervals at 3000, 1000, 1500, 2500, and 2000 rpm. Figure 29 depicts the BLDC motor stator current and torque. Figure 29a illustrates the torque. Due to modifications made to the speed of the BLDC motor and irradiance of the solar PV panel, the peak and dip on the torque are now visible. Figure 29b illustrates all three-phase stator currents. Solar energy is used by the boost converter to power the 3000 rpm, 48 V, 1 kW BLDC motor and for battery charging. The operation of a variable speed BLDC motor is shown in Figure 28. We started the motor for 0.2 s and set the reference speed of 3000 rpm. After that, for 0.4 s it is shifted to 1000 rpm, then for 0.6 s it is changed to 1500 rpm, and at the end, it is changed to 2500 rpm. The speed of the motor reaches the designated reference speed in less than 0.01 s. Figure 29a shows the variation in the torque, first supplied at 100 Nm for 0.01 s. Due to speed changes, there are a few spikes in the torque. Figure 29b illustrates the motor current, the result of which is a spike representing a change in speed. Figure 30 illustrates the BLDC motor Hall signal and BLDC motor back EMF; it is varied with variation in speed of 1000, 1500, 2000, 2500, and 3000 rpm. According to the above results, the proposed approach is superior to the existing approach.

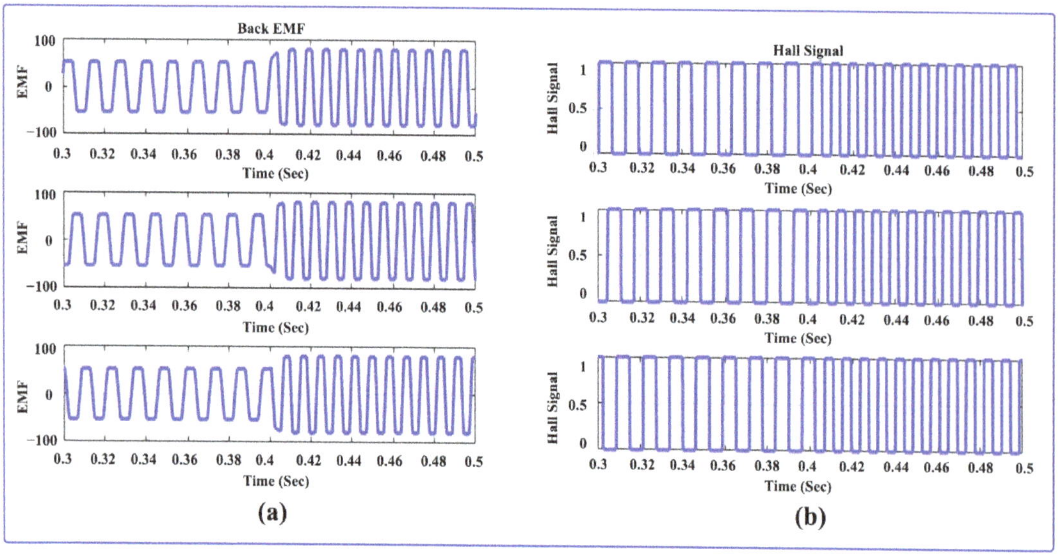

Figure 25. (**a**) Back EMF of BLDC motor (**b**) Hall signal of BLDC motor operating at variable speed and constant irradiation.

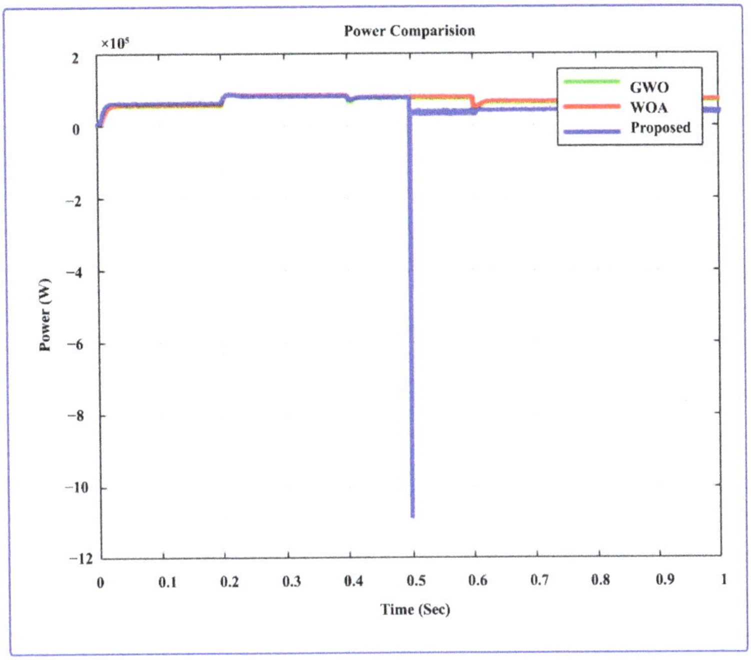

Figure 26. Comparison of output power with varying irradiation and varying motor speed.

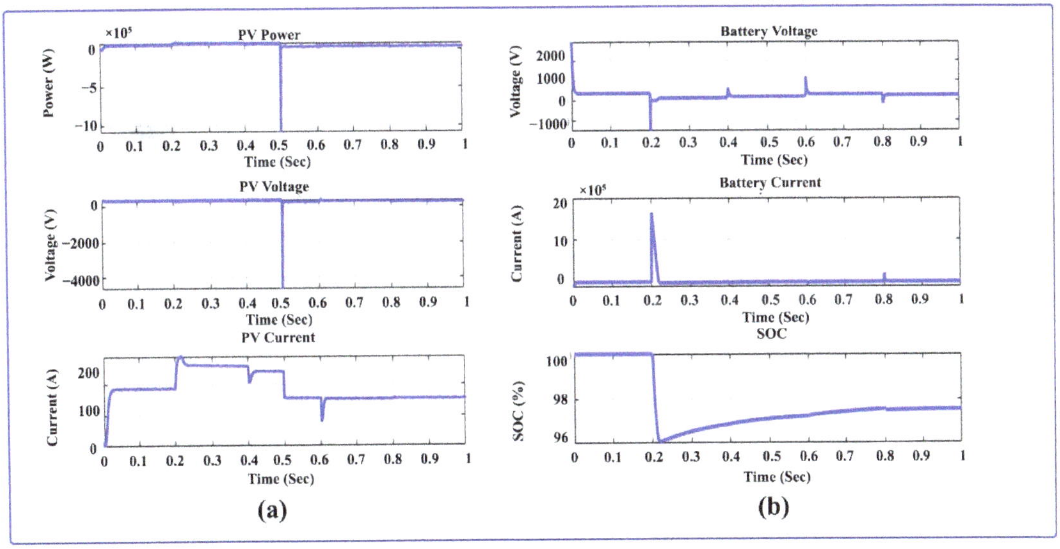

Figure 27. (**a**) PV outputs with varying irradiance and varying speed (**b**) battery outputs with varying irradiance and varying speed.

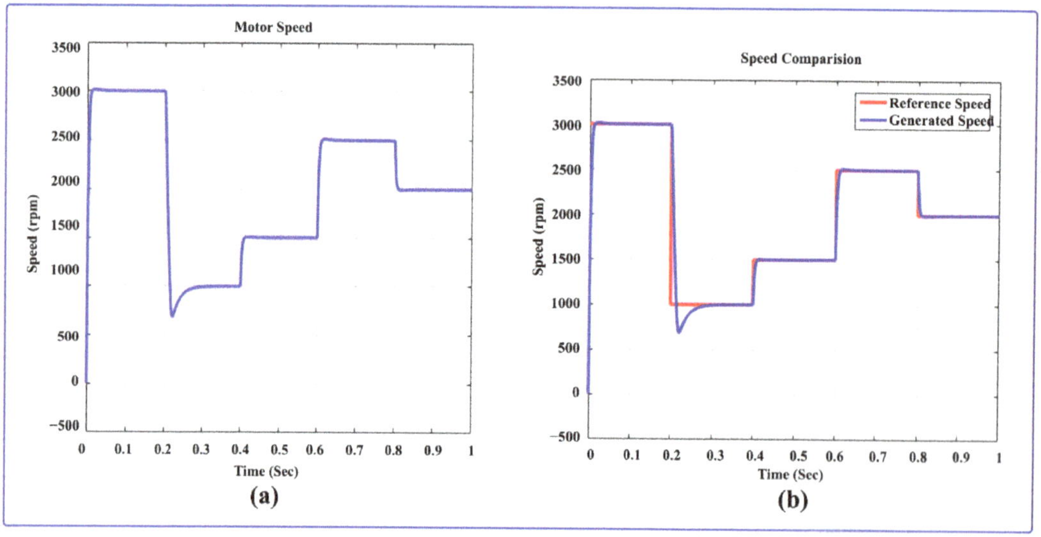

Figure 28. (**a**) BLDC motor speed (**b**) comparison of motor speed and reference speed with variable irradiance and variable speed.

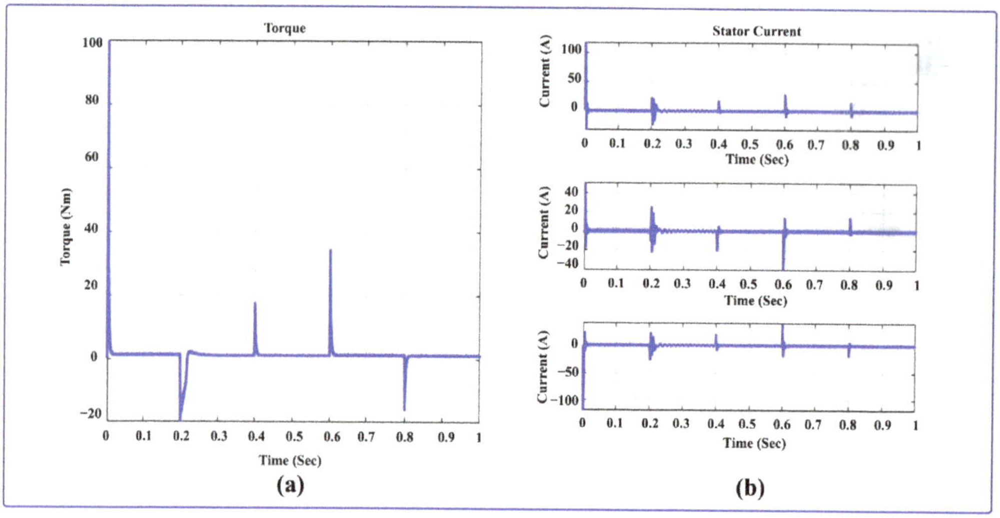

Figure 29. (**a**) BLDC motor torque with varying speed and irradiation (**b**) stator current with varying speed and irradiation.

Tables 1 and 2 present the qualitative and quantitative comparative analysis of the existing [29] and proposed optimization-based MPPT techniques, respectively. During quantitative analysis, the tracking time and efficiency measures are compared with the conventional P&O, FLC, and AFLC mechanisms. Then, the overall performance of the MPPT controlling algorithms is validated and compared during qualitative analysis based on the parameters of tracking speed, complexity, tracking efficiency, reliability, MPP oscillations, and tracking accuracy. The MPA-MPPT controlling technique provides highly improved results compared to the standard MPPT techniques.

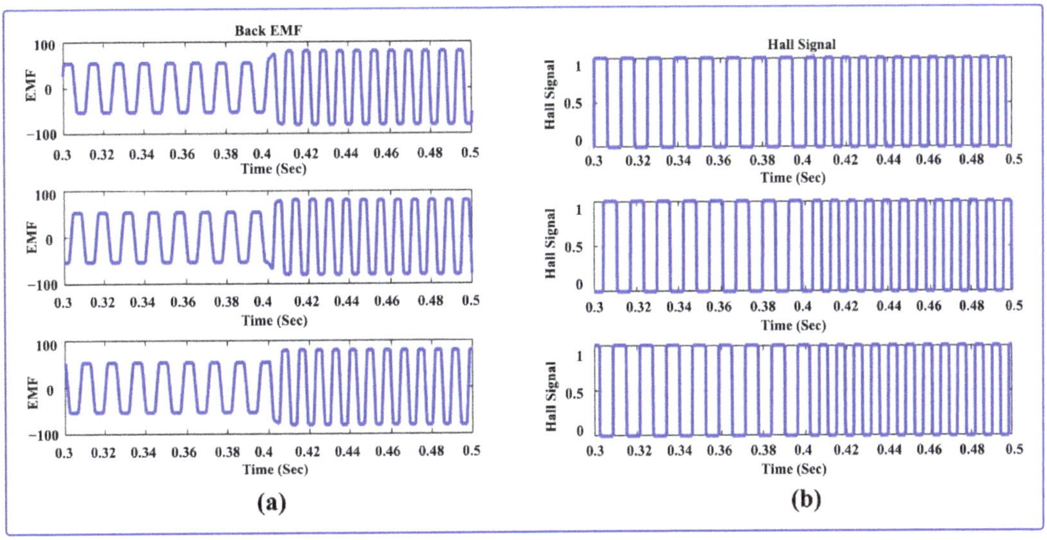

Figure 30. (**a**) Back EMF for BLDC motor with varying speed and irradiation (**b**) Hall signal for BLDC motor with varying speed and irradiation.

Table 1. Quantitative analysis.

Methods	Tracking Time	Efficiency
P&O	0.05	99.94
FLC	0.05	99.96
AFLC	0.038	99.97
Proposed MPA	0.025	99.98

Table 2. Qualitative analysis.

Criteria	P&O	FLC	ACO-FLC	Fuzzy-PSO	GWO-FLC	Proposed MPA
Tracking Speed	Slow	Moderate	Moderate	Moderate	Fast	Very Fast
Complexity	Less	Less	Moderate	Moderate	Less	Very Less
Tracking Efficiency	Less	Less	Medium	Medium	High	Very High
Reliability	Low	Low	Low	High	High	Very High
MPP Oscillations	High	High	Moderate	High	Less	Very Less
Tracking accuracy	Medium	Medium	Medium	Medium	Accurate	High Accurate

Table 3 presents the comparative analysis of existing [30] and proposed optimization-based MPPT controlling techniques based on the parameters of convergence time, settling time, and efficiency. Then, its corresponding graphical illustrations are presented in Figures 31 and 32, respectively. The estimated analysis proves that the time of the proposed MPA technique is greatly increased with high efficiency, which is highly superior to the other MPPT controlling techniques.

Table 3. Comparative analysis.

Methods	Convergence Time (s)	Settling Time (s)	Efficiency (%)
SRA	0.1811	0.2402	99.98
GHO	0.3112	0.5621	99.89
GWO	0.4421	0.6514	99.88
PSOGS	0.3522	0.6112	99.91
CS	0.3801	0.7701	99.84
PSO	0.4501	0.7102	99.86
Proposed MPA	**0.1532**	**0.2187**	**99.99**

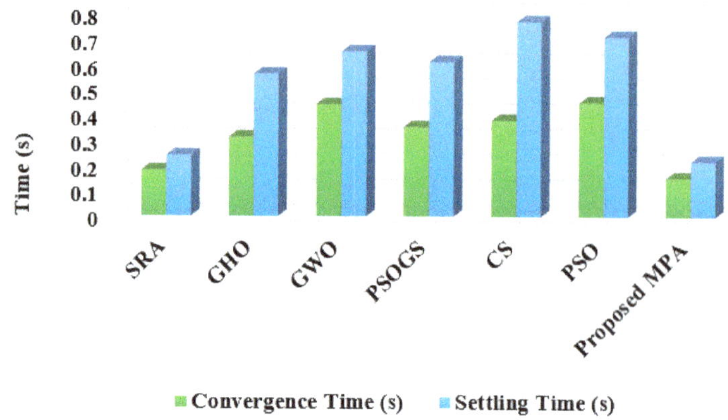

Figure 31. Time analysis.

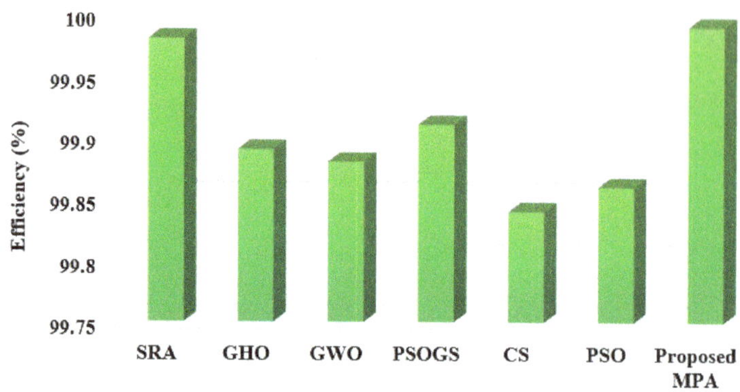

Figure 32. Efficiency analysis.

8. Conclusions

This research proposes an advanced metaheuristic MPA optimization approach used in a "small electric vehicle system", which operates on a "solar-powered BLDC motor" system. An MPA optimization approach is implemented to retain "Maximum Power Point" tracking from partial shadow condition as well as constant irradiation in the PV cell in this case. The traditional "MPPT algorithms", especially "WOA" and "GWO", depending on the results of MPPT, were examined, and the proposed algorithm was compared to them. To charge the battery, PV energy is used, and it is also used to supply power to the BLDC motor. The proposed system is created by using MATLAB software. Through the use of torque and change in speed accelerating and decelerating, in addition to the PID controller, the BLDC motor's initial, dynamic, and steady-state behaviors were evaluated. According to the simulation result, the MPA optimization technique improves the performance of the motor and charges the battery well. Consequently, due to continuous solar charging throughout the daytime, the battery is used to operate the BLDC motor for more distance than any electric vehicle.

Author Contributions: Data curation: R.K.G.R. and P.K.B.; Writing original draft: R.K.G.R.; Supervision: T.S., P.K.B. and U.M.; Project administration: P.K.B., U.M. and T.S.; Conceptualization: U.M. and A.M.M.S.; Methodology: P.K.B. and R.K.G.R.; Validation: A.M.M.S. and U.M.; Visualization: U.M. and A.M.M.S.; Resources: T.S. and P.K.B.; Review & Editing: T.S., P.K.B. and R.K.G.R.; Funding acquisition: T.S. All authors have read and agreed to the published version of the manuscript.

Funding: This research received no external funding.

Institutional Review Board Statement: Not applicable.

Informed Consent Statement: Not applicable.

Data Availability Statement: Not applicable.

Conflicts of Interest: The authors declare no conflict of interest.

References

1. Vertgewall, C.M.; Trageser, M.; Kurth, M.; Ulbig, A. Modeling Probabilistic Driving and Charging Profiles of Commercial Electric Vehicles. *Electr. Power Syst. Res.* **2022**, *212*, 108538. [CrossRef]
2. Sheela, A.; Logeswaran, T.; Revathi, S.; Rajalakshmi, K. Distributed MPPT Configuration for Improving Solar Energy Production. In Proceedings of the 2022 3rd International Conference for Emerging Technology (INCET), Belgaum, India, 27–29 May 2022; pp. 1–5.
3. Kee, S.H.; Chiongson, J.B.V.; Saludes, J.P.; Vigneswari, S.; Ramakrishna, S.; Bhubalan, K. Bioconversion of agro-industry sourced biowaste into biomaterials via microbial factories–A viable domain of circular economy. *Environ. Pollut.* **2021**, *271*, 116311. [CrossRef] [PubMed]
4. Sethi, V.; Sun, X.; Nalianda, D.; Rolt, A.; Holborn, P.; Wijesinghe, C.; Xisto, C.; Jonsson, I.; Gronstedt, T.; Ingram, J.; et al. Enabling Cryogenic Hydrogen-Based CO_2-Free Air Transport: Meeting the demands of zero carbon aviation. *IEEE Electrif. Mag.* **2022**, *10*, 69–81. [CrossRef]
5. Zengin, S. A hybrid current modulated DAB DC/DC converter for connecting PV modules to DC grid considering partial shading. *Comput. Electr. Eng.* **2022**, *101*, 108109. [CrossRef]
6. Yoon, Y.; Gi, H.; Lee, J.; Cho, M.; Im, C.; Lee, Y.; Bae, C.; Kim, S.J.; Lee, Y. A Continuously-Scalable-Conversion-Ratio Step-Up/Down SC Energy-Harvesting Interface With MPPT Enabled by Real-Time Power Monitoring With Frequency-Mapped Capacitor DAC. *IEEE Trans. Circuits Syst. I Regul. Pap.* **2022**, *69*, 1820–1831. [CrossRef]
7. Bansal, N.; Jaiswal, S.P.; Singh, G. Prolonged degradation and reliability assessment of installed modules operational for 10 years in 5 MW PV plant in hot semi-arid climate. *Energy Sustain. Dev.* **2022**, *68*, 373–389. [CrossRef]
8. Pandey, A.; Pattnaik, S. Design and Analysis of Extendable Switched-Inductor and Capacitor-Divider Network Based High-Boost DC-DC Converter for Solar PV Application. *IEEE Access* **2022**, *10*, 66992–67007. [CrossRef]
9. Wu, R.; Hao, J.; Zheng, S.; Sun, Q.; Wang, T.; Zhang, D.; Zhang, H.; Wang, Y.; Zhou, X. N dopants triggered new active sites and fast charge transfer in MoS2 nanosheets for full Response-Recovery NO2 detection at room temperature. *Appl. Surf. Sci.* **2021**, *571*, 151162. [CrossRef]
10. Li, Y.; Han, W.; Shao, W.; Zhao, D. Virtual sensing for dynamic industrial process based on localized linear dynamical system models with time-delay optimization. *ISA Trans.* 2022, *in press*. [CrossRef]
11. Kim, M.-H.; Lee, S.-H.; Kim, S.; Park, B.-G. A Fast Weight Transfer Method for Real-Time Online Learning in RRAM-Based Neuromorphic System. *IEEE Access* **2022**, *10*, 37030–37038. [CrossRef]
12. Zhang, W.; Xu, J.; Yu, T.X. Dynamic behaviors of bio-inspired structures: Design, mechanisms, and models. *Eng. Struct.* **2022**, *265*, 114490. [CrossRef]

13. Padmanaban, S.; Priyadarshi, N.; Bhaskar, M.S.; Holm-Nielsen, J.B.; Ramachandaramurthy, V.K.; Hossain, E. A Hybrid ANFIS-ABC Based MPPT Controller for PV System With Anti-Islanding Grid Protection: Experimental Realization. *IEEE Access* **2019**, *7*, 103377–103389. [CrossRef]
14. Lei, D.; Cui, Z.; Li, M. A dynamical artificial bee colony for vehicle routing problem with drones. *Eng. Appl. Artif. Intell.* **2021**, *107*, 104510. [CrossRef]
15. Rajeshkanna, G.; Sasiraja, R.M.; Winston, D.P. Design and development of Truncated Angle Variant (TAV) controller for multi-source-fed BLDC motor drive. *Electr. Eng.* **2020**, *102*, 1931–1946. [CrossRef]
16. Himabindu, N.; Hampannavar, S.; Deepa, B.; Swapna, M. Analysis of microgrid integrated Photovoltaic (PV) Powered Electric Vehicle Charging Stations (EVCS) under different solar irradiation conditions in India: A way towards sustainable development and growth. *Energy Rep.* **2021**, *7*, 8534–8547.
17. Lakshmiprabha, K.; Govindaraju, C. An integrated isolated inverter fed bldc motor for photovoltaic agric pumping systems. *Microprocess. Microsyst.* **2020**, *79*, 103276. [CrossRef]
18. Ahmad, F.; Khalid, M.; Panigrahi, B.K. An enhanced approach to optimally place the solar powered electric vehicle charging station in distribution network. *J. Energy Storage* **2021**, *42*, 103090. [CrossRef]
19. García, M.; Ponce, P.; Soriano, L.A.; Molina, A.; MacCleery, B.; Romero, D. Lifetime Improved in Power Electronics for BLDC Drives using Fuzzy Logic and PSO. *IFAC-PapersOnLine* **2019**, *52*, 2372–2377. [CrossRef]
20. Ho, J.C.; Huang, Y.-H.S. Evaluation of electric vehicle power technologies: Integration of technological performance and market preference. *Clean. Responsible Consum.* **2022**, *5*, 100063. [CrossRef]
21. Oubelaid, A.; Albalawi, F.; Rekioua, T.; Ghoneim, S.S.M.; Taib, N.; Abdelwahab, S.A.M. Intelligent Torque Allocation Based Coordinated Switching Strategy for Comfort Enhancement of Hybrid Electric Vehicles. *IEEE Access* **2022**, *10*, 58097–58115. [CrossRef]
22. Lan, H.; Hao, D.; Hao, W.; He, Y. Development and comparison of the test methods proposed in the Chinese test specifications for fuel cell electric vehicles. *Energy Rep.* **2022**, *8*, 565–579. [CrossRef]
23. Saha, B.; Singh, B. An Improved Flux Observer Based Position Sensorless Single Stage BLDC Motor Drive With Regenerative Braking For Solar Powered LEV. In Proceedings of the 2022 IEEE Transportation Electrification Conference & Expo (ITEC), Anaheim, CA, USA, 15–17 June 2022; pp. 1248–1253.
24. Li, H.; Ning, X.; Li, W. Implementation of a MFAC based position sensorless drive for high speed BLDC motors with nonideal back EMF. *ISA Trans.* **2017**, *67*, 348–355. [CrossRef] [PubMed]
25. Gupte, S. Experimental Analysis and Feasibility Study of 1400 CC Diesel Engine Car Converted into Hybrid Electric Vehicle by Using BLDC Hub Motors. *Energy Procedia* **2014**, *54*, 177–184. [CrossRef]
26. Faramarzi, A.; Heidarinejad, M.; Mirjalili, S.; Gandomi, A.H. Marine Predators Algorithm: A nature-inspired metaheuristic. *Expert Syst. Appl.* **2020**, *152*, 113377. [CrossRef]
27. Soliman, M.A.; Hasanien, H.M.; Alkuhayli, A. Marine Predators Algorithm for Parameters Identification of Triple-Diode Photovoltaic Models. *IEEE Access* **2020**, *8*, 155832–155842. [CrossRef]
28. Vankadara, S.K.; Chatterjee, S.; Balachandran, P.K.; Mihet-Popa, L. Marine Predator Algorithm (MPA)-Based MPPT Technique for Solar PV Systems under Partial Shading Conditions. *Energies* **2022**, *15*, 6172. [CrossRef]
29. Laxman, B.; Annamraju, A.; Srikanth, N.V. A grey wolf optimized fuzzy logic based MPPT for shaded solar photovoltaic systems in microgrids. *Int. J. Hydrog. Energy* **2021**, *46*, 10653–10665. [CrossRef]
30. Zafar, M.H.; Khan, N.M.; Mirza, A.F.; Mansoor, M.; Akhtar, N.; Qadir, M.U.; Khan, N.A.; Moosavi, S.K.R. A novel meta-heuristic optimization algorithm based MPPT control technique for PV systems under complex partial shading condition. *Sustain. Energy Technol. Assess.* **2020**, *47*, 101367.

sustainability

MDPI

Review

A Review of Multilevel Inverter Topologies for Grid-Connected Sustainable Solar Photovoltaic Systems

Shaik Nyamathulla and Dhanamjayulu Chittathuru *

School of Electrical Engineering, Vellore Institute of Technology, Vellore 632014, India; shaik.nyamathulla2020@vitstudent.ac.in
* Correspondence: dhanamjayulu.c@vit.ac.in

Abstract: Solar energy is one of the most suggested sustainable energy sources due to its availability in nature, developments in power electronics, and global environmental concerns. A solar photovoltaic system is one example of a grid-connected application using multilevel inverters (MLIs). In grid-connected PV systems, the inverter's design must be carefully considered to improve efficiency. The switched capacitor (SC) MLI is an appealing inverter over its alternatives for a variety of applications due to its inductor-less or transformer-less operation, enhanced voltage output, improved voltage regulation inside the capacitor itself, low cost, reduced circuit components, small size, and less electromagnetic interference. The reduced component counts are required to enhance efficiency, to increase power density, and to minimize device stress. This review presents a thorough analysis of MLIs and a classification of the existing MLI topologies, along with their merits and demerits. It also provides a detailed survey of reduced switch count multilevel inverter (RSC-MLI) topologies, including their designs, typical features, limitations, and criteria for selection. This paper also covers the survey of SC-MLI topologies with a qualitative assessment to aid in the direction of future research. Finally, this review will help engineers and researchers by providing a detailed look at the total number of power semiconductor switches, DC sources, passive elements, total standing voltage, reliability analysis, applications, challenges, and recommendations.

Keywords: Renewable Energy Sources (RESs); Photovoltaic (PV) systems; Switched Capacitor (SC); Reduced Switch Count Multilevel Inverter (RSC-MLI); Common Ground Switched-Capacitor (CGSC); Reliability

check for updates

Citation: Nyamathulla, S.; Chittathuru, D. A Review of Multilevel Inverter Topologies for Grid-Connected Sustainable Solar Photovoltaic Systems. *Sustainability* **2023**, *15*, 13376. https://doi.org/10.3390/su151813376

Academic Editor: Mohamed A. Mohamed

Received: 26 July 2023
Revised: 22 August 2023
Accepted: 29 August 2023
Published: 6 September 2023

1. Introduction

Recently, there has been a remarkable rise in the use of grid-supplied power. This can be attributed to an increased number of users and the expansion of high-power sectors. Traditional power production has led to a significant surge in global emissions, thereby causing detrimental effects on the environment. Significant progress has been made in integrating renewable energy sources such as solar and wind into the grid. Welcome to the world of photovoltaic (PV) systems, which have become the top choice for harnessing energy thanks to their incredible potential. In fact, worldwide, grid-connected solar PV capacity has soared to over 635 GW, satisfying approximately 2% of the global energy consumption [1].

Solar energy is a rapidly growing field, and one crucial aspect that has gained significant importance is power electronics. Researchers are actively engaged in the pursuit of developing highly efficient power electronic converters to enhance the overall performance of solar energy systems. In applications requiring medium and high power, MLIs are increasingly being employed. This is because MLIs provide several inherent advantages over two-level inverters, as mentioned in Table 1.

Table 1. Comparison between two-level and multilevel inverters (MLIs).

Parameter	Two-Level Inverter	MLI
Function at a fundamental frequency	Fails	Operate
Operate at high voltage and current	Operate	Operate
Fault-tolerant operation	Impossible	Possible
Harmonic content	Low	High
Stress on power electronic switches	More	Less
Switching losses	High	Low
Power quality performance	Low	High
Voltage variation rate	High	Low
Generation of voltage in common mode	Higher	Lower
Generation of variable voltage	Not possible	Possible
Capability of functioning without a transformer	No	Yes
Efficiency	Low	High
Input current distortions	High	Low
Voltage applications	Low	High
Structure	Complicated	Modular
Electromagnetic interference	High	Low

It has been predicted that renewable energy would contribute 29% of worldwide power output in 2020, up from 27% in 2019, that renewable energy generation would increase by more than 8% to 8300 TWh by 2021, and that solar PV and wind would account for two-thirds of the growth in renewable energy. The increase in renewable energy alone in China in 2021 was about half of what was predicted, followed by the United States, the European Union, and India, as shown in Figure 1a. China has continued to be the largest PV market, although there is growth in the United States due to continuous federal and state legislative support. In India, new solar PV capacity additions have recovered quickly from COVID-19-related delays in 2021. According to the IEA's 2021 Renewable Energy Market Update, by 2020, renewable energy was the only type of energy whose consumption increased despite the pandemic. To increase worldwide renewable power in 2021 and 2022, the renewable energy sector has looked at new additions. In addition, 270 GW went online in 2021 and 280 GW went online in 2022, continuing the remarkable level of renewable energy additions that are anticipated. This expansion has exceeded the yearly capacity rise record set in 2017–2019 by more than 50%, indicating that renewables have been responsible for 90% of the increase in global capacity in 2021 and 2022, as shown in Figure 1b.

Flexible alternating current transmission systems (FACTSs), customized power devices (CPDs), variable-speed drives (VSDs), active front-end converters (AFCs), and renewable energy sources for power generation are just a few of the many uses for MLIs [2–5]. MLIs can be classified as classical if they use the most common topologies, such as the diode-clamped multilevel inverter (DCMLI), flying capacitor multilevel inverter (FCMLI), and cascaded H-bridge (CHB) multilevel inverter, mentioned in Figure 1c. There has been a lot of interest in these topologies, but their practical implementation is highly impacted by the application, the system that is designed, and costs. The fundamental disadvantage of a DCMLI is its asymmetrical loss distribution. This, in turn, results in an irregular distribution of junction temperature, which, in turn, results in constraints on the inverter's power, current, and switching frequency at maximum junction temperature [6,7].

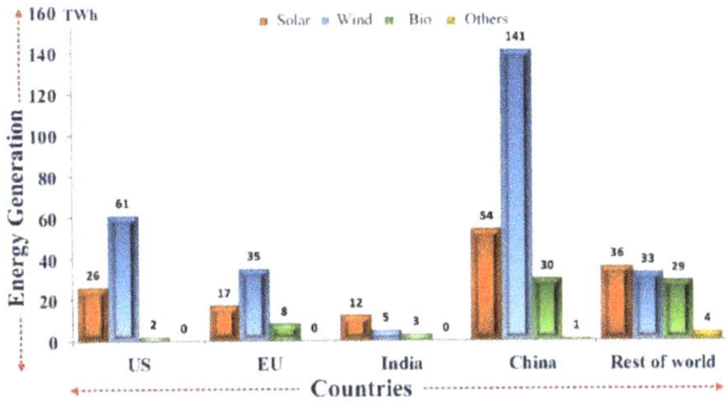

(a)

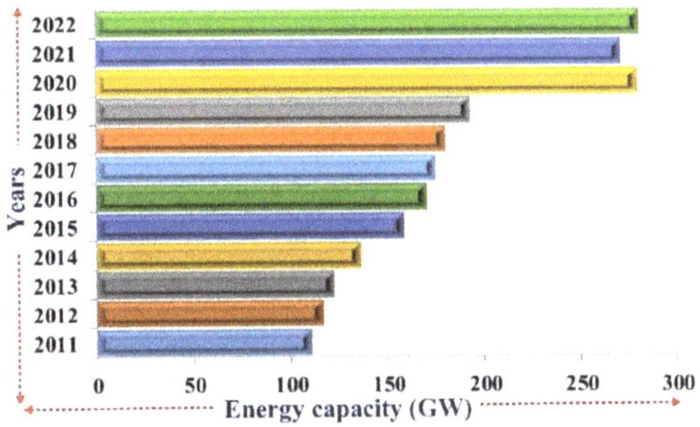

(b)

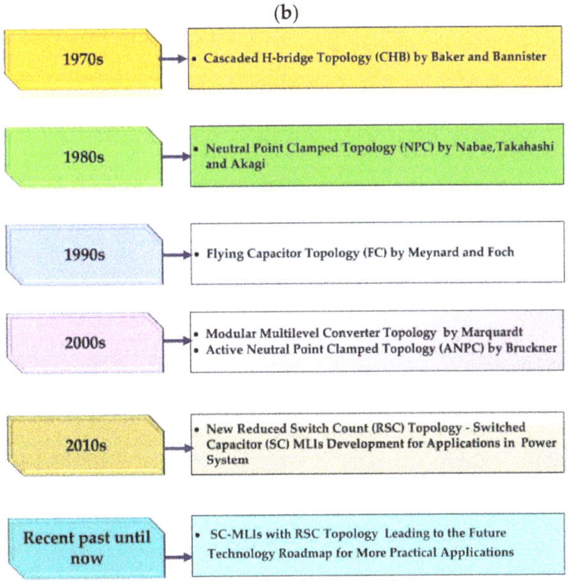

(c)

Figure 1. (**a**) Worldwide renewable power generation in 2020–2021; (**b**) net renewable capacity additions, by renewable energy market update 2021—IEA; (**c**) an outlook on the development of various MLI topologies.

Traditional MLIs, on the other hand, need a larger number of components to achieve the same number of output levels, have issues with capacitor voltage balance, and cannot increase their voltages [8]. Different reduced device count MLIs have been presented to address traditional MLIs' high component count [9–14]; however, these MLIs have lacked a voltage-boosting function. To eliminate capacitor voltage imbalance in typical MLIs, complicated control algorithms or multi-output boost converters have been implemented. These methods increase an inverter's weight, complexity, and expense. SC-MLIs minimize the drawbacks of traditional and reduced device count MLIs [15–22].

Researchers have continued to investigate and to develop additional topologies by implementing little or major modifications to conventional MLIs. As a result, MLIs with a lower device count have been developed, and this subset of MLIs has been dubbed RSC-MLIs, which have recently been the subject of many reviews. Newly designed RSC-MLIs for integrating renewable energy sources and driving applications are addressed in [23–25]. The incorporation of switched-capacitor (SC)-based circuits is one of the most widely used methods for designing better MLIs, and rapid progress has been made in the area of SC-MLI development since 2010. Pure SC-based switching circuits use a series–parallel switching conversion technique to take the available fixed DC-link voltage and produce a multilevel voltage using a reduced number of capacitors, power switches, and/or diodes. SC-MLIs are a valuable and interesting solution for many new applications due to the voltage step-up feature they offer, which includes self-voltage balancing for the involved capacitors, which is the result of a single-stage switching operation that eliminates the need for inductors and transformers [26–35]. The following is a list of the primary factors and propensities for SC-MLIs:

1. Maximizing the number of output voltage levels while minimizing the number of semiconductor devices needed.
2. Increasing the overall output voltage gain with single or multiple DC-source configurations.
3. Reducing or controlling the current stress/loss profile of switches with soft charging or pulse-width modulation (PWM)-based techniques for better power density and to improve efficiency.

A wide range of new issues, design requirements, and real-world constraints of conventional MLIs have been highlighted in recent articles [36–45]. Different circuit designs are used to build MLIs using the SC concept. These include single, multiple, mid-point-clamped, and common-grounded (CG)-based DC-source structures [46–55]. Hybrid topological designs that integrate well-known integrated methods such as flying capacitor (FC) and switched boost (SB) technologies into the SC framework significantly raise their performance [56–82].

The significant contributions of this review include:

(1) A complete literature overview and a rigorous analysis of about 200 recently published papers regarding the development, classification, and use of MLI topologies;
(2) A thorough analysis of MLIs and a classification of the existing MLI topologies, along with their merits and demerits.
(3) A detailed survey of reduced switch count multilevel inverter (RSC-MLI) topologies, including their designs, typical features, limitations, and criteria for selection.
(4) A critical analysis and classification of the existing SC-MLI topologies with a qualitative assessment of the merits and downsides of SC-MLIs with well-known applications, and a future roadmap is explored.
(5) An effective summary of multilevel inverters, highlighting the necessity for new or modified multilevel inverters for grid-connected sustainable solar PV systems.
(6) Finally, this review study should help engineers and researchers by providing a detailed look at the total number of power semiconductor switches, DC sources, passive elements, total standing voltage, reliability analysis, applications, challenges, and recommendations.

This review paper includes the following: Section 2 describes grid-connected solar PV systems and MLI background including MLI applications; different types of energy sources integrated with MLI-based systems, motivational factors for generating an RSC-MLI, and assessment parameters are discussed in great length. Section 3 presents the detailed MLI categorization and description of the structure and working principle of features for each reported RSC-MLI topology, and a variety of SC-MLIs with single or multiple DC-source, mid-point-clamped, and CGSC configurations are examined and then compared on several criteria, including the total number of power semiconductor switches, DC sources, passive elements, total standing voltage, efficiencies relative to the number of levels, and the structural motivations behind each concept. Section 4 provides a reliability assessment study to estimate the lifespan of an MLI. Section 5 provides the present challenges and recommendations. Finally, Section 6 concludes the article with some final thoughts.

2. Grid-Connected Solar Photovoltaic System

Massive worldwide energy demand has led to significant usage of fossil fuels, which has affected the environment by increasing greenhouse gas emissions. So, renewable energy resources have gained popularity and growth through producing clean electricity [83–87].

PV cells are used in solar-based technologies to transform the sun's energy into usable power. Figure 2 describes the operation of photovoltaic cells, converters, inverters, and energy control units that make up a system for converting solar energy. Nevertheless, efforts are being made to better understand how to incorporate renewable energy sources into the electricity grid. There has been an increased focus on power converters and their controls because of the importance of their work in transforming electricity and controlling the output power. DC–DC converters are typically used in the initial stage of integrating renewable energy sources into a DC grid. Due to the output voltage variations of renewable energy sources such as wind and solar PVs, this stage must operate at peak efficiency. Hence, it is imperative that the DC–DC converters in the front-end stages exhibit responsive behavior towards such fluctuations in order to operate at their optimal efficiency [88–97]. In small-scale industrial or utility applications, these inverters are frequently employed because of their elevated voltage stress, poor efficiency, elevated operating temperature, and increased pressure capabilities. Multiple inverters are commonly utilized in large-scale, high-power, grid-connected renewable energy systems due to their advantageous characteristics [98–102].

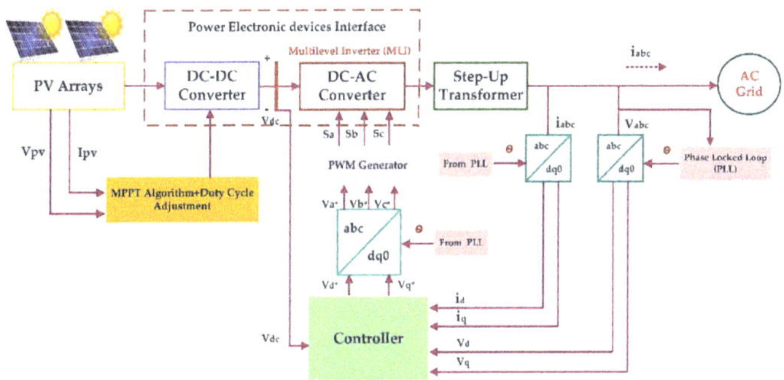

Figure 2. Grid-connected multilevel inverter for solar PV application [103].

An MLI is selected for medium- and high-power applications based on its capability to generate voltage waveforms of superior quality while functioning at a low switching frequency [104–108]. Figure 3 indicates how multilevel inverters have a wide variety of uses in the emerging field of renewable energy, and Figure 4 exhibits the MLI-based system integration of various renewable energy sources being employed and discussed [109–113].

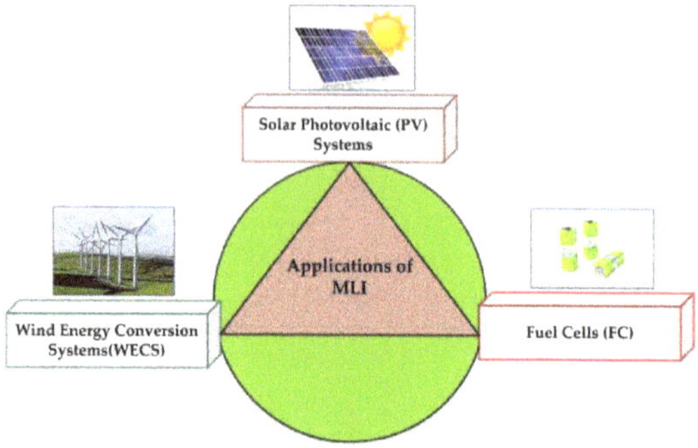

Figure 3. Multilevel inverters have a wide variety of uses in the emerging field of renewable energy.

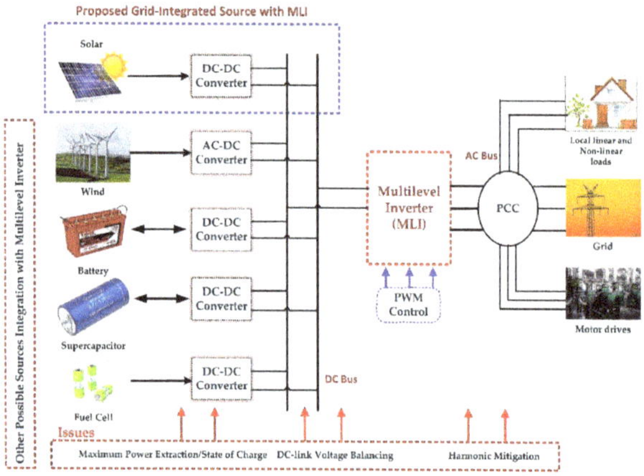

Figure 4. MLI-based system integration of various renewable energy sources [103].

A. Reduced Switch Count Multilevel Inverters (RSC-MLIs): Background

In order to overcome the size and complexity limitations of conventional MLIs, RSC-MLIs have been developed. However, their structural and operational features are affected by changes in their topological arrangement.

(i) Factors Contributing to Motivation

Researchers often consider the following qualities while building a novel RSC-MLI architecture, as shown in Figure 5. The most salient features are enumerated below.

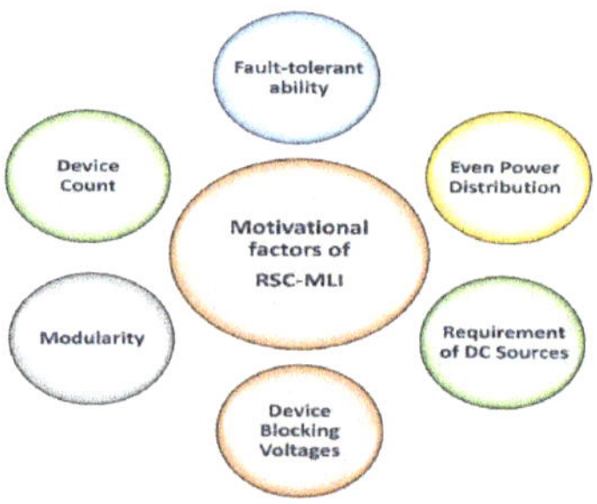

Figure 5. Factors contributing to the motivation for an RSC-MLI.

(ii) Classification

In any physical design, switches and DC-link voltages can be connected in any topology. Often, there is no required architecture, but other times there are ladders, staircases, columns, U-shaped structures, and cascade structures. As can be seen in Figure 6, the resulting RSC-MLIs can be categorized in accordance with their topological and functional properties, as discussed in [114–116].

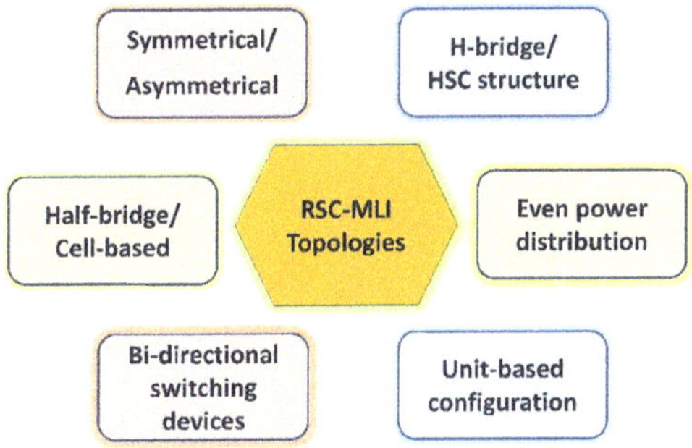

Figure 6. Classification of RSC-MLI topologies.

(iii) The Evaluation Criteria for an MLI

In contrast, this review study takes into consideration broad criteria for rating the proposed topologies:

Several MLI evaluation parameters are context dependent, as shown in Figures 7 and 8, and some of the key features of an MLI that contribute to the different capabilities of the power system are discussed [117–121].

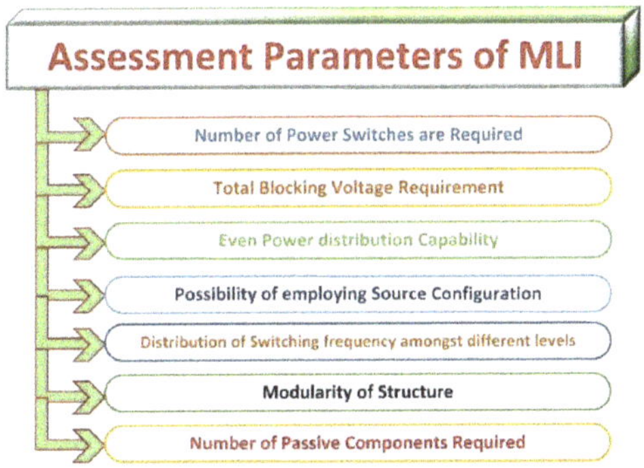

Figure 7. MLI evaluation criteria [103].

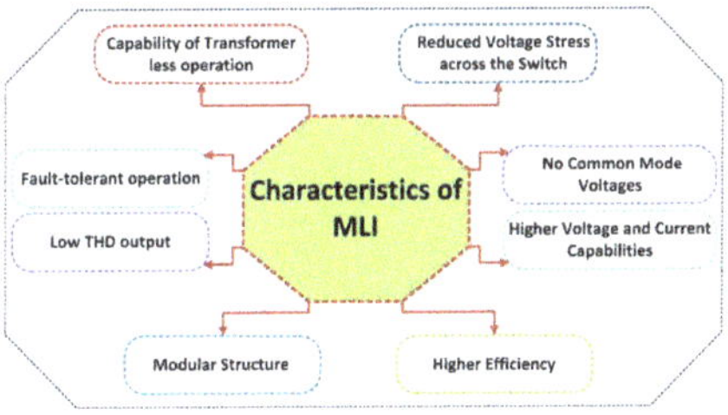

Figure 8. Characteristics of MLIs.

An interconnected multilevel converter system can use renewable energy sources, including solar PV, wind energy, and fuel cells. Their operation, effectiveness, improved power quality, and applications are mostly determined by the control scheme used in the MLI-PWM. Multiple MLI topologies have been suggested in recent years [122–125]. Based on the number of DC sources in their topology, MLIs have been classified as shown in Figure 9, and based on a categorization of the reduced switch in their topology, MLIs have been classified as shown in Figure 10. The NPC-MLI or DC-MLI, FC-MLI, and CHB-MLI are the most prevalent industrial topologies [126–130].

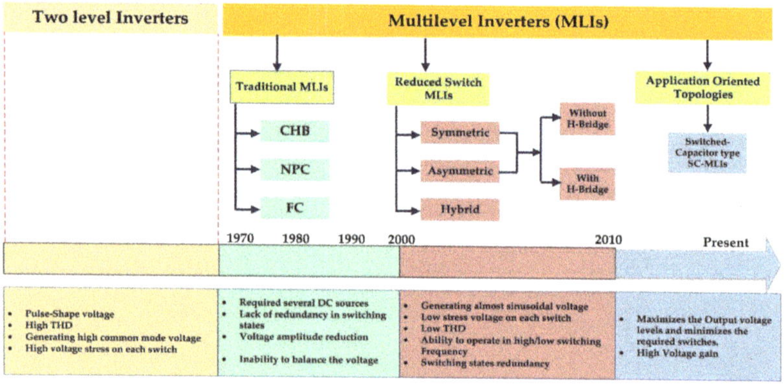

Figure 9. Simplified classification of multilevel inverters.

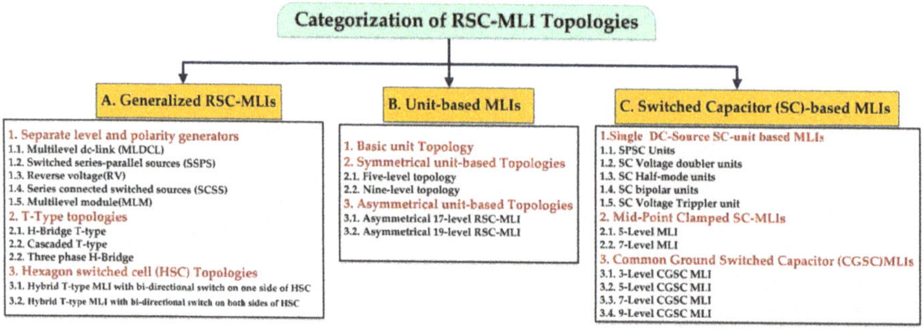

Figure 10. Categorization of reduced switch count MLI topologies.

- Cascaded H-bridge multilevel inverter (CHB-MLI): This topology, which was initially patented by Baker and Bannister [1,39], was thought to be a good alternative to previously reported topologies since it required fewer power components. The series connection of H-bridges with independent DC sources makes up the topology known as the CHB-MLI. Many series-connected H-bridge constructions combine to produce the multilevel stepped waveform. A generic H-bridge cell can be cascaded to create the CHB-MLI that theoretically has an infinite number of layers. Due to its modular design, it effectively corrects the voltage imbalance that can be seen in NPC and FC settings. A CHB typically consists of power conversion cells connected in series on the AC side and individually powered by an isolated DC source from a battery, ultra-capacitor, or fuel cell on the DC side.
- Diode-clamped multilevel inverter (DC-MLI): Nabae, Takashi, and Akagi proposed the diode-clamped multilevel inverter (DC-MLI), also known as the NPC-MLI, in 1981 [73]. The widespread adoption of these inverters can be attributed to their tremendous competency in high-power and medium-voltage operations as well as their relatively high efficiency.

- Flying capacitor multilevel inverter (FC-MLI): Meynard and Foch [25,39] suggested the FC-MLI topology in 1992 to address the issue of static and dynamic sharing of the voltage between semiconductor switches as implemented in the NPC-MLI architecture.

3. Reduced Switch Count Multilevel Inverters (RSC-MLIs) Topologies

A. Generalized RSC-MLI Topologies

Generalized RSC-MLI topologies can be further divided into subcategories depending on the similarity of their structures and the switching devices used. The categorization is as follows:

(1) Separate Level and Polarity Generator Topologies

Each phase-voltage level has its own independent polarity and level generators, which results in an unusually high number of levels. Popular combinations include the MLDCL, SSPS, RV, SCSS, and MLM topologies, according to [2]. Figure 11 depicts the per-phase architecture of these three-source arrangements. An isolated DC supply is used in each basic unit, which uses bidirectional switches to generate levels in the MLM topology. It is important to note that MLDCL and SSPS topologies need identical device-blocking voltages. Adding a new basic unit to an RV, SCSS, or MLM topology increases voltage stress. As a result, the total DC-link voltage is equal to the blocking voltage of each device. Except for the SCSS and MLM topologies, all of these topologies offer symmetric and asymmetric configurations.

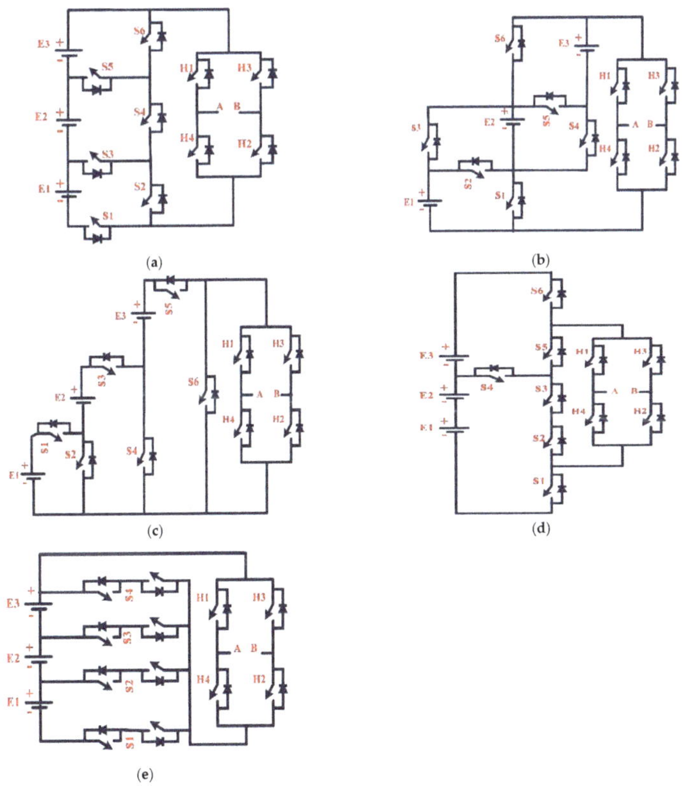

Figure 11. Separate level and polarity generator topologies: (**a**) MLDCL; (**b**) SSPS; (**c**) SCSS; (**d**) RV; (**e**) MLM.

It is acquired from the polarity generator in the SSPS topology since the level generator cannot make it. Only two devices in the level generator are in conduction to obtain any positive or negative voltage level in the SSPS topology depicted in Figure 11b. Because the output voltage is raised by charging all of the DC-link capacitors in series or parallel operation, the output voltage is also increased. This SSPS function is ideal for charging batteries and storing energy. For grid-connected PV systems, series or parallel operation maximizes the use of DC sources. Adding an H-bridge to an SSPS topology RSC-MLI reduces switch counts and further losses. An asymmetrical improvement architecture with a voltage gain is presented by the SSPS topology with minimal modifications. To expand the RV architecture to higher levels, just duplicate the encircling intermediate stage of the level generator illustrated in Figure 11d. In the topologies presented in Figure 11, level generators can only use additive DC source combinations.

(2) T-Type Structure Topologies

A T-type topological structure interconnects numerous DC-link nodes by a phase-leg of full-/half-bridge structures on the burden side. These designs use unidirectional and bidirectional switches. Figure 12 depicts typical T-type topology combinations through complete bridge, cascaded, and half-bridge structures. The T-type with complete bridge construction is the most common, owing to its simplicity and reduced switch count [131,132].

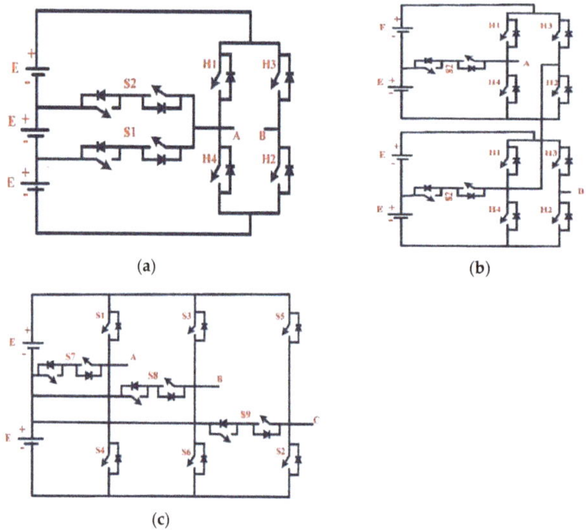

Figure 12. Topologies with T-type structures: (**a**) H-bridge T-type; (**b**) cascaded T-type; (**c**) T-type three-phase half-bridge.

(*i*) T-type MLI: This type of architecture is based on H-bridges and was suggested by [2,133]. The H-bridge is built from the unidirectional switches, and the bidirectional switches connect the H-midpoint bridges to the DC-link voltages. Figure 12a depicts the T-type per-phase configuration with three DC voltage sources. With no redundancies, this topology is merely symmetrical. To increase the topology's flexibility, the number of DC sources can be increased with bidirectional switches or cascade many T-type modules, which allows for uneven voltage ratios and switching redundancy. T-type modules should have the same DC-link voltage. A T-type MLI with two five-level T-type modules cascading is shown in Figure 12b.

(*ii*) Half-leg T-type MLI: Separately, the phase leg is linked to the DC link by bidirectional switches in half-bridge T-type topologies. A three-level topology is shown in Figure 12c as an example of this structure. To obtain larger levels, just increase the DC sources using bidirectional switches, which may provide even and odd phases of voltage. In Figure 12c, two devices per leg are in conduction at any one moment, and the voltage rating of bidirectional switches is smaller than the devices in a phase-leg. Due to minor conduction losses and a reduced total blocking voltage, this design is preferred over a DCMLI and ANPC in terms of efficiency. Many PV and grid-connected applications have used this design. When an open-circuit switch malfunction occurs, this inverter can be reconfigured to withstand the problem [2,132].

(3) HSC Structure Topologies

A hexagon switch cell (HSC) structure is constructed using unidirectional switches, and the DC link is connected to this structure using bidirectional switches.

(*i*) Topology-I: A mix of T-type with HSC RSC-MLI topology with two stiff DC sources, i.e., ES and ER, on each side of the HSC. Figure 13a illustrates that more bidirectional switches or several modules cascaded together may expand this topology's capabilities further. According to Figure 13a, this topology is similar to the five-level T-type MLI when short-circuiting and open-circuiting the unidirectional switches H5 and H2, as shown. Since there are now unidirectional switches, we can work in asymmetrical configurations with the H-bridge to HSC. Asymmetrical behavior occurs only when ES is less than or equal to the number of elements in the configuration. DC-link capacitors "n" in the asymmetrical design of this architecture can provide (4n + 1) levels of phase voltage. Other voltage levels can be operated by using an asymmetrical arrangement with suitable voltage ratios.

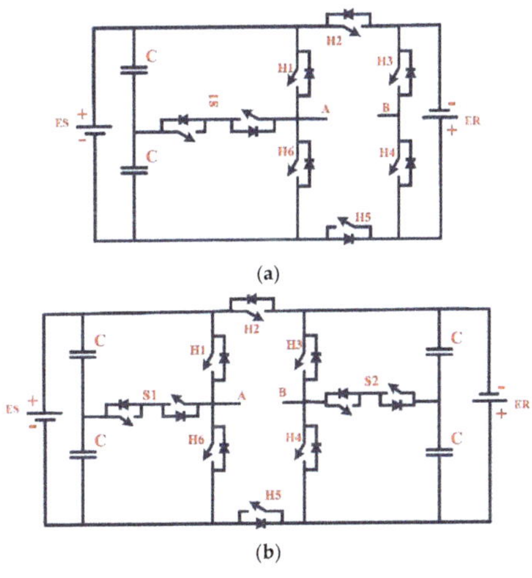

Figure 13. Hybrid T-type topologies and extended HSC structures: (**a**) Hybrid T-type MLI with a bidirectional switch on one side of the HSC; (**b**) hybrid T-type MLI with a bidirectional switch on both sides of the HSC.

(*ii*) Topology-II: Figure 13b shows a configuration that is identical to Topology-I but uses bidirectional switches to connect both sides of the HSC and DC connections. This topology preserves many of the same properties and functions as Topology-I through a pair of unidirectional devices [2].

B. Unit-Based RSC-MLIs

A few authors have developed feasible topologies with drastically reduced switch numbers compared to conventional MLIs to reduce the topological size, price, and complexity. There are a limited number of output voltage values that these topologies can provide. These setups function as RSC-MLIs with fixed topologies and output voltage values.

(1) Basic Unit RSC-MLI Topology

Separate polarity and level generators are used in this H-bridge topology [108,110]. A five-level unipolar voltage can be obtained by using an RSC-MLI basic unit. The basic unit in Figure 14 is made up of three-cell and one-cell structures. The three-cell structure has three voltage sources coupled by five unidirectional switches.

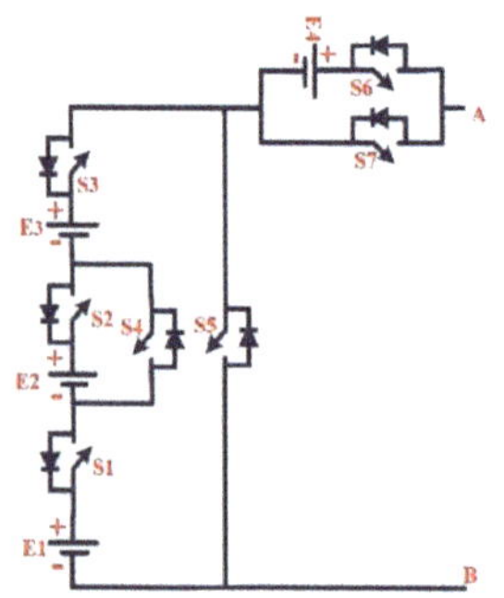

Figure 14. Basic unit RSC-MLI topology.

(2) Symmetrical Unit-Based Topology

There are two unit-based topologies that function with a reduced device sum for a specified number of output voltage levels [2]. These setups are detailed below.

(i) Five-level configuration: To produce nine-level inverters, just cascade two units as indicated in Figure 15a. In each cycle, the cascaded units exchange switching pulses. Consequently, the units perform uniformly.

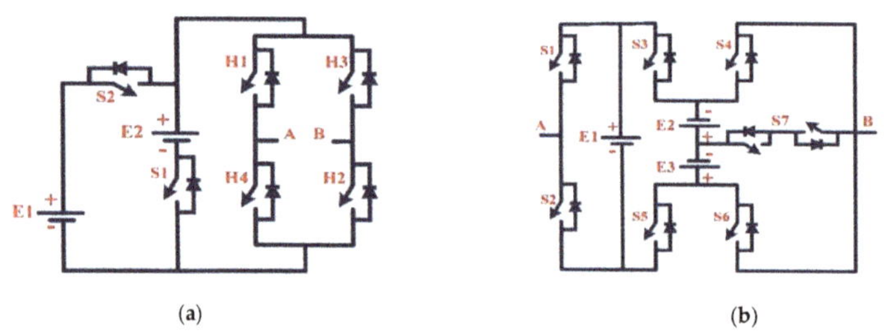

Figure 15. Symmetrical unit-based topologies: (**a**) Five-level; (**b**) nine-level.

(ii) Nine-level configuration: To compare, Topology-I of the hybrid T-type design seems to be comparable, with two DC-link capacitors of identical voltage, as illustrated in Figure 15b.

(3) Asymmetrical Unit-Based Topologies

Figure 16 shows a 17-level MLI circuit with an asymmetrical design. It has ten unidirectional switches with anti-parallel diodes and three asymmetrical DC voltage sources in a 1:2:5 ratio to produce the expected 17-level output voltage. It is described in detail in the aforementioned [134].

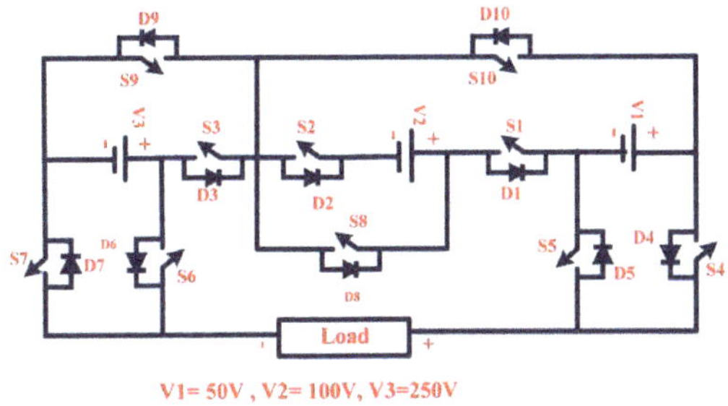

Figure 16. An asymmetrical 17-level RSC MLI.

To produce the desired output voltages of 19 levels in a 1:3:5 ratio [135], an asymmetrical design of the MLI circuit is shown in Figure 17. The design employs ten unidirectional switches with anti-parallel diodes and three asymmetric DC voltage sources. Tables 2 and 3 provide information on how 17-level and 19-level MLIs are built, how they compare with each other, and summarize the various recent topologies. Similarly, Figure 18 compares the parameters of the 17-level MLI topologies that have recently been developed, while Figure 19 compares the parameters of the 19-level MLI topologies that have recently been developed [134–137].

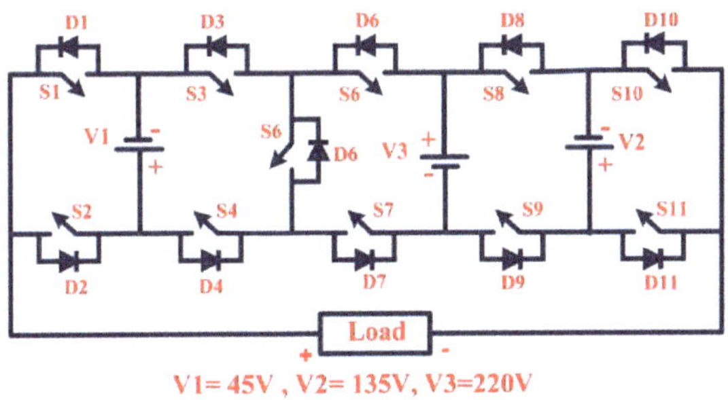

Figure 17. An asymmetrical 19-level RSC MLI.

Table 2. Parametric comparisons of the recently developed 17-level MLIs.

Topologies	N_{SWT}	N_{DCS}	N_L	N_{CAP}	N_{DK}	TSV_{PU}	Transformer-Less Interfacing Capability	Leakage Current Limiting Capability	FCC/L	CF/L	
										$\alpha = 1.5$	$\alpha = 0.5$
[3]	14	4	17	4	14	11	No	No	3.05	4.02	3.38
[138]	10	2	17	4	20	-	No	No	3.88	-	-
[139]	20	2	17	4	20	-	No	No	3.88	-	-
[140]	24	2	17	4	24	-	No	No	4.48	-	-
[5]	16	4	17	4	14	11	No	No	3.17	4.14	3.5
[141]	10	4	17	0	10	36	No	No	1.41	9.18	4.94
[142]	20	8	17	0	20	36	No	No	4	7.17	5.05
[143]	10	2	17	0	10	40	No	No	1.35	9.88	5.18
[134]	10	3	17	0	10	12	No	No	1.94	3.92	3.66
17-Level SC-MLI [144]	13	0	17	4	13	5.6	Yes	Yes	2.11	2.02	1.69

Table 3. Parametric comparisons of the recently developed 19-level MLIs.

Topologies	N_L	N_{SWT}	N_{DK}	N_{DCS}	FCC/L	TSV_{PU}	Efficiency (%)	CF/L	
								$\alpha = 0.5$	$\alpha = 1.5$
[145]	19	12	11	1	1.26	7.55	87.72	1.46	1.85
[146]	19	10	19	2	1.63	8.88	-	1.86	2.33
[147]	19	22	22	8	2.73	8.22	93.49	2.95	3.38
[148]	19	11	19	5	1.84	8	-	2.05	2.47
[149]	19	10	10	2	1.15	6.88	-	1.33	1.7
[150]	19	11	11	4	1.36	5.77	97.38	1.52	1.82
[136]	19	13	13	3	1.52	6.66	93.67	1.70	2.05
[135]	19	11	11	3	1.31	4.66	97.38	1.43	1.68

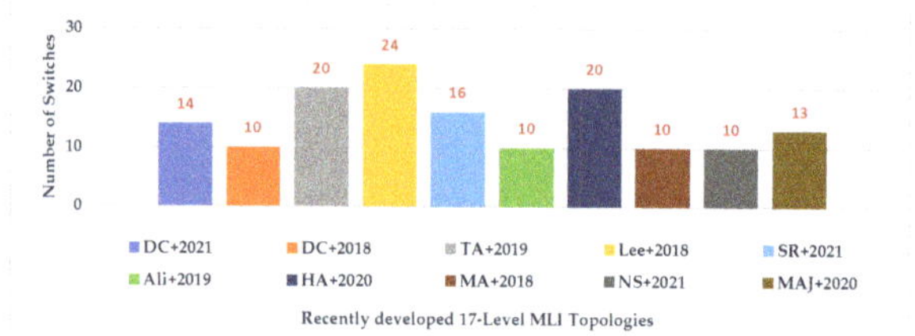

(**a**) Number of switches

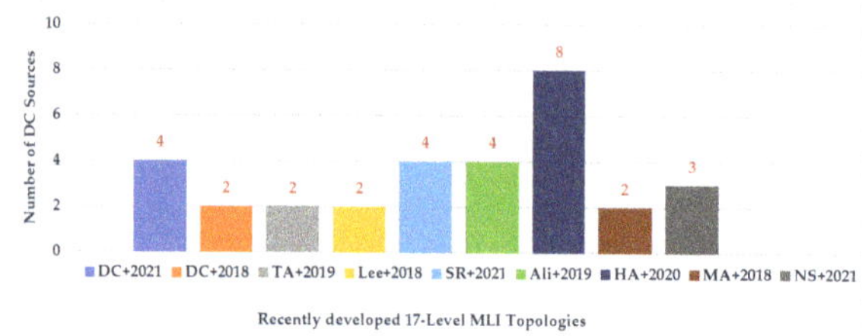

(**b**) Number of DC sources

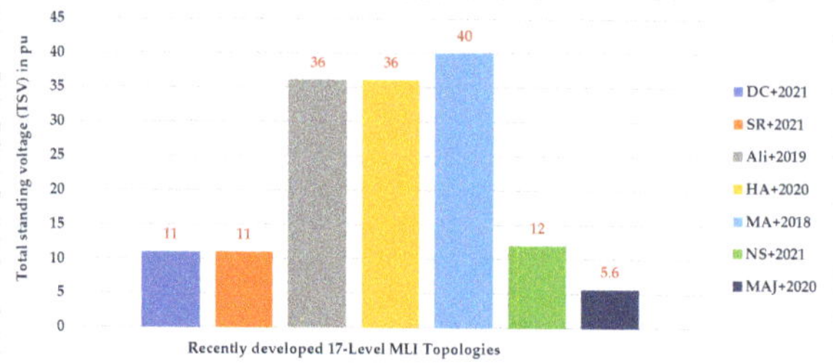

(**c**) Total standing voltage

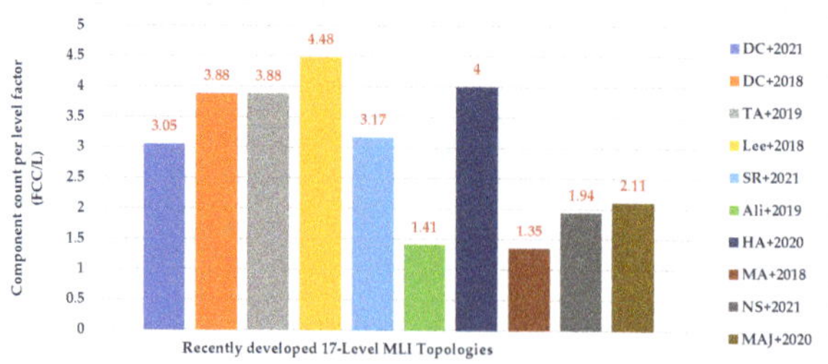

(**d**) Factor of component count per level

Figure 18. *Cont.*

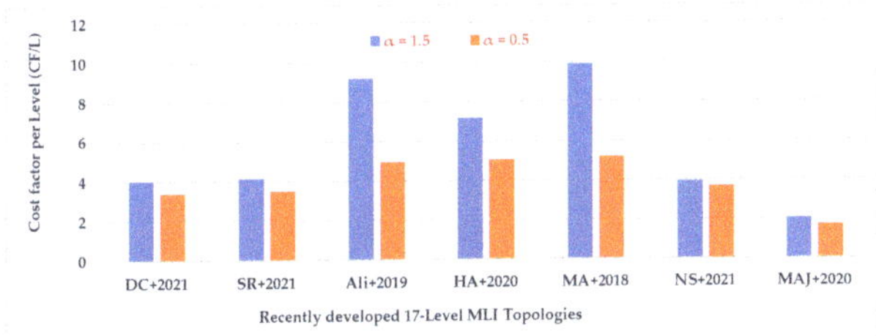

(**e**) Cost factor per level

Figure 18. Parametric comparisons of recently developed 17-level MLI topologies [3,5,134,138–144].

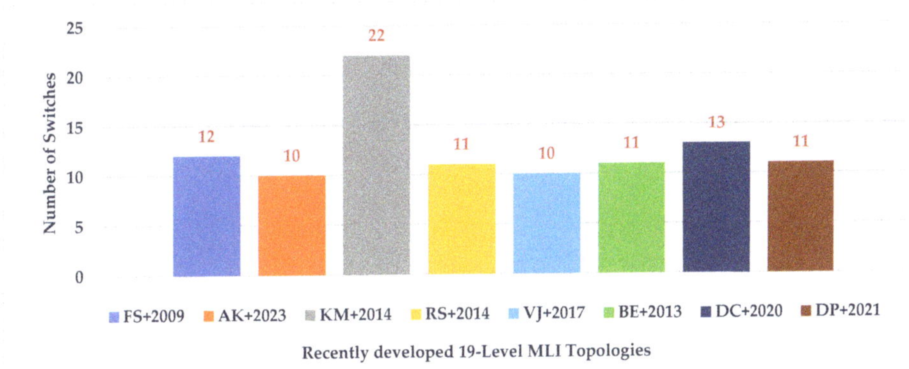

(**a**) Number of switches

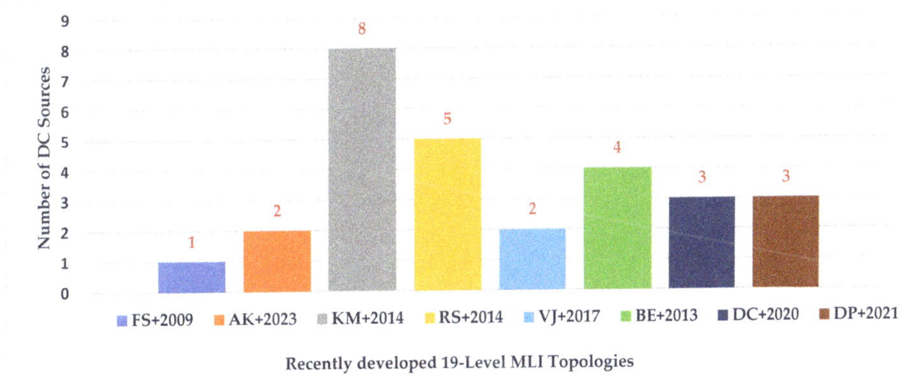

(**b**) Number of DC sources

Figure 19. *Cont.*

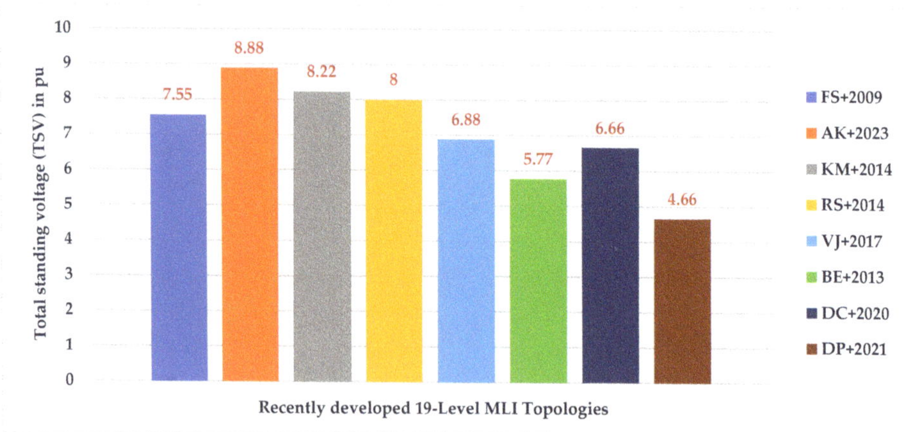

(**c**) Total standing voltage

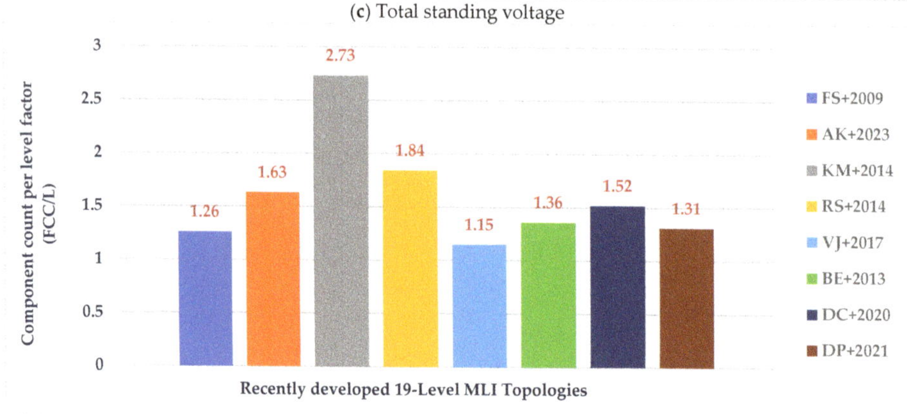

(**d**) Factor of component count per level

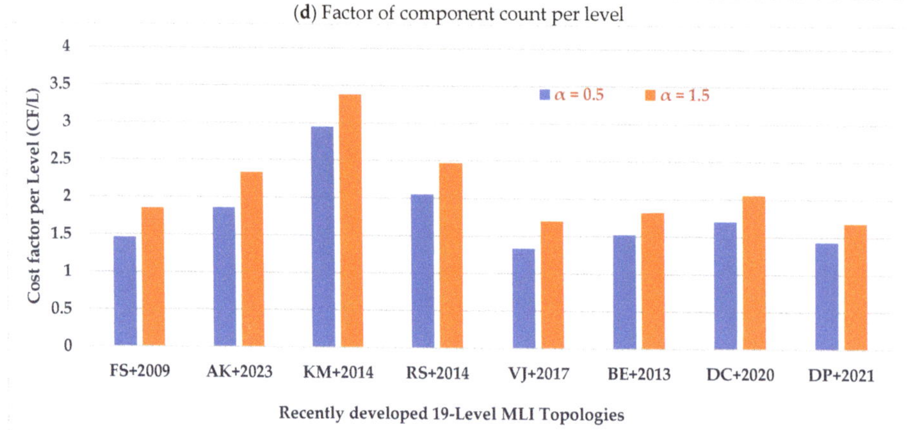

(**e**) Cost factor per level

Figure 19. *Cont.*

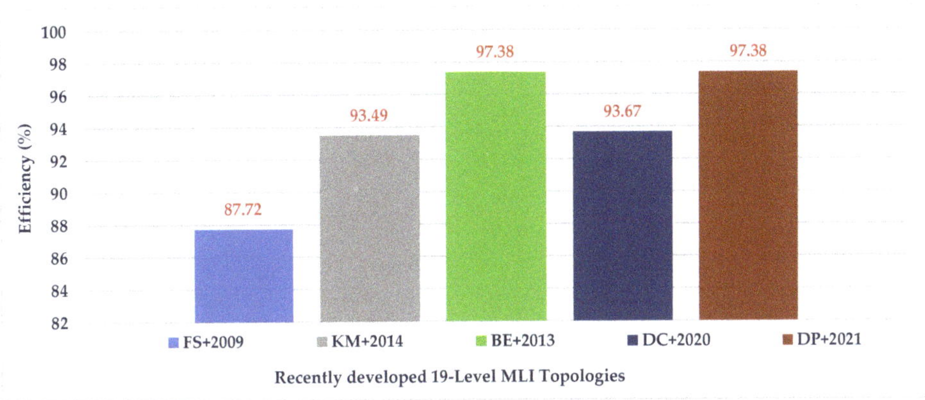

(f) Efficiency

Figure 19. Parametric comparisons of recently developed 19-Level MLI topologies [135,136,145–150].

a. Power loss efficiency calculations:

Power losses are the main constraints in inverters, such as conduction losses (P_{Cond}) and switching losses (P_{Swi}) [134–136]. The net amount of conduction losses can be obtained by considering losses in the IGBT switch (P_{CIGBT}) and anti-parallel diode (P_{CDI}) in the current conduction state and is represented as follows:

$$P_{Cond}(t) = P_{CIGBT}(t) + P_{CDI}(t) \tag{1}$$

$$P_{Cond}(t) = \left\{\left[V_{IGBT} + R_{DI} i_n^{\beta}(t)\right] + \left[V_{DI} + R_{DI} i_n(t)\right]\right\} i_n(t) \tag{2}$$

where V_{IGBT}, V_{DI}, and i_n are the *IGBT* threshold voltages and peak current, respectively. If the diodes (N_{DI}) and switches (N_{IGBT}) are conducted at the same intervals (t), R_{IGBT} and R_{DI} are the IGBT and diode on-state resistance, respectively, β is the IGBT constant. The average power loss is:

$$P_{Cond} = \frac{1}{2\pi}\int_0^{2\pi}\{N_{IGBT}(t)P_{CIGBT}(t) + N_{DI}(t)P_{CDI}(t)\}dt \tag{3}$$

The energy losses such as energy turn-on (E_{non}) and turn-off (E_{noff}) for IGBT turn-on and -off states during power consumption are:

$$E_{noff} = \frac{1}{6} V_{IGBTj} I t_{off} \tag{4}$$

$$E_{non} = \frac{1}{6} V_{IGBTj} I' t_{on} \tag{5}$$

The j is the loss in *IGBT* and t_{noff} and t_{non} are the turn-on and -off, E_{noff} and E_{non} I and I' of the *IGBT* switches, respectively.

$$P_{Swi} = f\left\{\sum\nolimits_{j=1}^{N_{IGBT}}\left[\sum\nolimits_{j=1}^{N_{nonj}} E_{nonji} + \sum\nolimits_{j=1}^{N_{noffj}} E_{noffji}\right]\right\} \tag{6}$$

The N_{nonj} and $Nnoffj$ are IGBT turn-on and -off jth time intervals with fundamental (f) in one complete cycle.

$$P_{Tloss} = P_{cond} + P_{swi} \tag{7}$$

To calculate the inverter efficiency by Equation (8):

$$\%\eta = \frac{P_{outn}}{P_{inn}} = \frac{P_{outn}}{P_{outn} + P_{Tloss}} \times 100 \quad (8)$$

where both P_{outn} and P_{inn} are abbreviations used to denote output and input power, respectively. The output power can be calculated using Equation (9) as follows:

$$P_{outn} = V_{rms} \times I_{rms} \quad (9)$$

b. Switch stress total standing voltage (TSV) calculations:

To produce the largest blocking voltage via each switch, the multilevel inverter is crucial, and the TSV is the most essential factor in switch selection. There is a pairing between the maximum voltage across the switches and the TSV values. A voltage-blocking stress has been applied across the switch. Differences in voltage stress exist between unidirectional and bidirectional switches.

According to [134], it is possible to calculate the TSV per unit (TSV_{PU}) as:

$$TSV_{PU} = \frac{V_{TSV}}{V_{omax}} \quad (10)$$

c. Cost function (CF) parameter calculations:

The cost function (CF) can be used to make educated guesses about the financial viability of various MLI design alternatives, which is useful for highlighting budgetary constraints and showcasing design tradeoffs. The following equation provides a means through which the cost factor can be determined:

$$CF = (N_{SWT} + N_{DD} + N_{CAP} + N_{DCS} + N_{DK} + \alpha TSV_{PU}) \quad (11)$$

where N_{SWT} indicates the number of switches, N_{DD} indicates the number of diodes, N_{CAP} indicates the number of capacitors, N_{DCS} indicates the number of DC sources, N_{DK} indicates the number of driver circuits, and TSV_{PV} indicates the total standing voltage, if TSV_{PU} is multiplied by the "α" weighting factor. In order to calculate the cost function (CF) can be used Equation (12) can be used as follows:

$$CF = (N_{SWT} + N_{DK} + N_{DCS} + \alpha TSV_{PU}) \quad (12)$$

For the best cost factor computation, 0.5 and 1.5 will be explored. The component count per level factor (FCC/L) can be calculated by using Equation (13):

$$\mathrm{FCC/L} = \frac{(N_{SWT} + N_{DCS} + N_{CAP} + N_{DD} + N_{DK})}{Levels} \quad (13)$$

C. *Switched capacitor (SC) unit-based topologies*

Basically, a DC source, diodes, capacitors, and switches make up the building blocks of an "SC unit". SPSC units, SC voltage doubler units, SC half-mode units, SC bipolar units, and SC voltage triple units are all subsets of basic SC units. The SC-MLIs can be categorized as single and multiple DC-source SC-MLIs, mid-point-clamped SC-MLIs, common ground switched-capacitor (CGSC)-based MLIs, and hybrid SC-MLIs [151–161].

1. Single DC-source SC-Unit-based MLIs

 a. SPSC Units

There are two main types of SPSC units utilized in SC-MLIs, and they are depicted in Figure 20; Figure 20a depicts the minimal component count for Type-I of this device, which consists of just two switches, one capacitor, and one power diode [162–166]. The output voltage can be set to one of two discrete positive values, V_{DC} or $2V_{DC}$, depending on the value of the input DC source. SPSC Unit-II, represented in Figure 20b, employs the same capacitor charging and discharging principle, although with an extra capacitor and a power switch in place of the diode and four-quadrant switch. The SPSC Unit-I is different from the SPSC Unit-II in that it can only send power in one direction. Furthermore, unlike the SPSC Unit-I, the SPSC Unit-II uses charged capacitor voltages to create both discrete voltage levels, which eliminates the possibility of a DC offset during the formation of the output voltage level in SC-MLIs. In this case, in addition to the paralleled conventional power switches in SPSC Unit-II, four-quadrant power switches with a back-to-back connection of two standard MOSFETs can be employed [167–171].

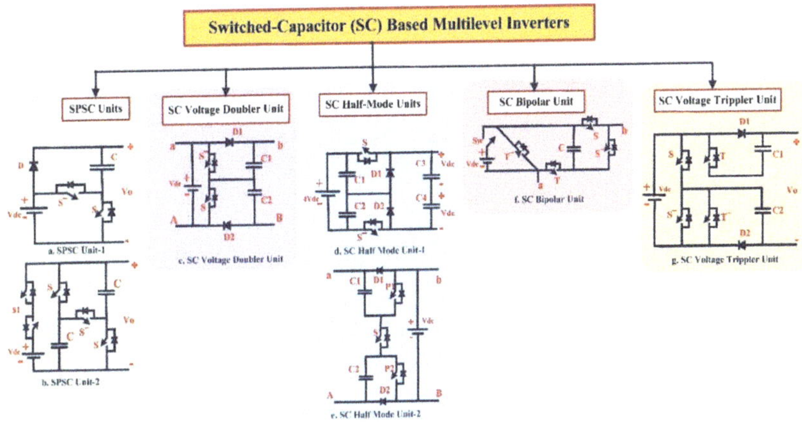

Figure 20. Categorization of different SC-based basic units.

b. SC Voltage Doubler Unit

A voltage doubler SC unit, as shown in Figure 20c, is a two-port converter that utilizes a single DC source, two capacitors, two complementary power switches, and two power diodes. Each capacitor has its own charging channel, which includes a diode and a power switch. Being a two-port SC-based basic unit, it can provide five different DC-link voltages, including 0 V_{DC}, $\pm V_{DC}$, and $\pm 2V_{DC}$. This fundamental SC unit's adaptability to operation comes at the expense of a lack of bidirectional power flow capacity [172–176].

c. SC Half-Mode Units

When using an SC half-mode device, the DC-link capacitors can be charged to a voltage that is only a small multiple of the DC-input source's voltage. Several other SC half-mode units have been introduced recently, as illustrated in Figure 20d. This is connected to the charging activities of the capacitors in the SC half-mode Unit-I; two fixed values of discrete DC-link voltages are required at its output, and this requires four DC-link capacitors, two complementary switches, and two diodes. Using a capacitive charging channel consisting of two diodes and a single power switch, the capacitors in this setup are charged to half the main DC-link voltage, earning this configuration the designation SC half-mode Unit-II, as shown in Figure 20e.

d. SC Bipolar Unit

The SC bipolar unit, shown in Figure 20f, has the ability to generate bipolar output voltage levels, such as $0V_{DC}$, $\pm V_{DC}$. In this device, just one DC-link capacitor is required to be charged in parallel to the input DC-source voltage, whereas five power switches are required for the entire operation.

e. SC Voltage Tripler Units

Basic units based on SC technology give SC-MLI topologies with voltage increases that are three times the normal value. Figure 20g shows the functioning concept of the most common type of SC voltage tripler unit. It is comprised of two DC-link capacitors, two power diodes, and four power switches [177–181]. As can be observed, both capacitors in this device are charged by the DC input. Table 4 provides a comparative study of different single DC source SC-MLIs [151], Table 5 describes the comparative study of different SC-MLIs with two asymmetric DC sources, and Table 6 provides a comparative study of cross-connected asymmetrical 15-level SC-MLIs [182–186].

Table 4. Comparative study of different single DC source SC-MLIs.

Type of SC-MLI	No. of Levels/THD	Overall Voltage Gain/Caps Voltage	TSV (pu)/MVS	Reported Rated Efficiency	No. of Components		
					S	D	C
FB based [153]	7/19.5%	3/V_{DC} (3)	6/3V_{DC}	85%@1kHz/5W	10	0	3
HB based [169]	7/7%	3/V_{DC} (2)	5.66/3V_{DC}	92.2%@50Hz/500W	9	4	2
HB/NPP based [172]	7/16.2%	1.5/0.5V_{DC} (2)	6/V_{DC}	95.5%@50Hz/250W	10	0	2
HB based [173]	7/11.2%	4/V_{DC} (2), 2V_{DC} (2)	5.5/4V_{DC}	96.5%@50Hz/270W	12	0	4
HB based [162]	7/13.7%	1.5/V_{DC} (2)	5.33/V_{DC}	96.6%@50Hz/600W	10	0	2
FB based [151]	7/2.8%	3/V_{DC} (2)	6/3V_{DC}	92.1%@50Hz/150W	9	1	2
FB based [174]	9/3.1%	2/V_{DC} (2)	5.75/2V_{DC}	94.2%@1kHz/200W	9	2	2
FB based [175]	9/13.8%	4/V_{DC} (1), 2V_{DC} (1)	5.25/4V_{DC}	NA%@50Hz/NA	9	2	2
HB/NPP based [176]	9/12.5%	2/0.5V_{DC} (2)	5.5/2V_{DC}	NA%@50Hz/400W	11	0	2
HB/NPP based [36]	9/8.8%	2/V_{DC} (1), 0.5V_{DC} (2)	5.5/2V_{DC}	97.4%@50Hz/1kW	12	0	3
HB/NPP based [37]	9/10.2%	2/V_{DC} (1), 0.5V_{DC} (2)	5.5/2V_{DC}	98%@50Hz/1kW	10	1	3
HB based [165]	11/6.8%	5/V_{DC} (2), 3V_{DC} (2)	5/6V_{DC}	95.5%@50Hz/220W	9	4	4
HB based [166]	13/11%	6/V_{DC} (2), 3V_{DC} (2)	5.5/6V_{DC}	95.5%@50Hz/500W	10	4	4
HB/NPP based [177]	13/5.3%	3/V_{DC} (2), 0.5V_{DC} (2)	6/3V_{DC}	NA%@50Hz/1kW	12	4	3
HB based [178]	13/7.2%	6/V_{DC} (2), 3V_{DC} (1)	6/3V_{DC}	NA%@50Hz/NA	13	2	3
HB based [38]	13/7.7%	6/V_{DC} (1), 2V_{DC} (2)	5/3V_{DC}	94%@50Hz/1kW	15	0	3
HB/NPP based [160]	17/NA	8/V_{DC} (2), 2V_{DC} (2), 4V_{DC} (2)	4.25/8V_{DC}	95.5%@50Hz/1kW	10	4	6
HB based [179]	17/3.9%	8/V_{DC} (1), 2V_{DC} (2), 4V_{DC} (2)	4.25/8V_{DC}	94.5%@50Hz/80W	10	5	5
HB based [180]	21/4.8%	10/V_{DC} (2), 2V_{DC} (4), 4V_{DC} (2)	5/2V_{DC}	NA%@50Hz/NA	20	8	8
HB based [181]	21/4.5%	10/V_{DC} (2), 2V_{DC} (4), 4V_{DC} (2)	5/2V_{DC}	NA%@50Hz/NA	20	12	10

Table 5. Comparative study of different SC-MLIs with two asymmetric DC sources.

Type of SC-MLI	No. of Levels/THD	Overall Voltage Gain/CapsVoltage	TSV (pu)/MVS	Asymmetric Amplitude of DC-Sources	No. of Components		
					S	D	C
SCC-Mode-I [43]	11/NA	1.25/V_{DC}	4.2/4V_{DC}	V_{DC} and 3V_{DC}	11	0	1
SCC-Mode-II [43]	11/9.3%	1.67/2V_{DC}	4.4/2V_{DC}	V_{DC} and 2V_{DC}	11	0	1
Figure 1 [49]	15/4.7%	1.75/V_{DC}	8.5/7V_{DC}	V_{DC} and 3V_{DC}	10	1	1
Figure 1 [47]	17/4.8%	1.6/2V_{DC}, 1.5V_{DC} (1)	4.5/8V_{DC}	2V_{DC} and 3V_{DC}	12	1	3
Figure 1 [44]	19/8.9%	1.8/4V_{DC} (2)	4.89/9V_{DC}	V_{DC} and 4V_{DC}	13	0	2
Figure 1 [47]	21/4.3%	2/V_{DC} (2), 4V_{DC} (2)	5/10V_{DC}	V_{DC} and 4V_{DC}	14	0	4
Figure 1 [52]	25/1.8%	2/V_{DC}, 4V_{DC}	10/10V_{DC}	V_{DC} and 5V_{DC}	12	2	2
Figure 3 [44]	29/NA	2/V_{DC} (2), 2.5V_{DC}	4.64/14V_{DC}	V_{DC} and 2.5V_{DC}	18	0	3
Figure 12 [182]	31/NA	4/V_{DC} (2), 4V_{DC} (2)	5.5/15V_{DC}	V_{DC} and 4V_{DC}	14	2	4
Figure 13f SCC [183]	31/NA	3/V_{DC} (2), 4V_{DC} (2)	5.6/15V_{DC}	V_{DC} and 4V_{DC}	16	0	4
Figure 13g SCC [184]	49/NA	4/V_{DC} (3), 4V_{DC} (3)	7.25/24V_{DC}	V_{DC} and 4V_{DC}	18	2	6
Figure 3 [185]	49/NA	2/V_{DC} (2), 5V_{DC} (2)	6/24V_{DC}	V_{DC} and 5V_{DC}	18	2	4
Figure 4 [44]	49/NA	2/V_{DC} (2), 5V_{DC} (2)	5/24V_{DC}	V_{DC} and 5V_{DC}	20	0	4

Table 6. Parametric comparisons of cross-connected 15-level SC-MLIs.

Topologies	N_{Lev}	N_{Swt}	N_{DK}	N_{DC}	N_C	N_{Dio}	V_{TB}/N_L	Efficiency (%)
[187]	15	10	10	5	0	1.26	-	93.73
[188]	15	14	14	1	4	1.63	4.86	-
[189]	15	10	9	4	0	2.73	4.6	97.5
[190]	15	8	8	3	0	1.84	2	95.2
[191]	15	10	9	3	0	1.15	2.26	-
[192]	15	10	10	5	0	1.36	1.06	90
Figure 1 [186]	15	10	8	2	2	1.31	3.6	96.3

4. Mid-Point-Clamped SC-MLIs

High-frequency variable CMV, which results from the varying numbers of HB legs used in the aforementioned single and multiple DC-source SC-MLIs to invert the SC unit/generalized SCC output voltage polarity, is one of the main issues that prevents their widespread use, for example, in grid-tied PV systems. Since their output voltage is monitored at the neutral point of the DC connection, mid-point-clamped MLIs are a common choice in this scenario. While a multilevel output voltage waveform is generated by a single input DC source, the leakage current problem in grid-tied PV applications is significantly reduced. The identical capacitors used in the DC links of single-phase MLIs may be used in three-phase systems. Table 7 provides a comparative study of different mid-point-clamped SC-MLIs [187–197].

Table 7. Comparative study of different mid-point-clamped SC-MLIs.

Type of SC-MLI	No. of Levels/THD	TSV (pu)/MVS	Reported Rated Efficiency	No. of Components		
				S	D	C
SC [56,159]	4/41.4%	2.66/1.5V_{DC}	97%@1kW	4	2	4
ABNPC [57]	5/NA	5/V_{DC}	98.5%@1.2kW	6	2	3
ABNPC [60]	5/NA	6/0.5V_{DC}	NA@50W	10	0	3
ABNPC [62]	5/NA	6/0.5V_{DC}	97.5%@800W	6	2	4
ABNPC [81]	5/NA	6/1.5V_{DC}	97.1%@1kW	8	2	4
ABNPC [82]	6/20.2%	4.4/3V_{DC}	95.8%@450W	6	4	5
Sym SC [172]	7/12.2%	5/V_{DC}	97%@150W	9	1	3
ABNPC [61]	7/NA	5.33/V_{DC}	96%@50W	9	0	3
ABNPC [66]	7/19.3%	5.3/2V_{DC}	96.7%@250W	9	0	3
Dual T-type [70]	7/NA	7.33/2V_{DC}	98%@100W	10	0	4
ABNPC [71]	7/NA	6.66/V_{DC}	97%@100W	10	0	4
ABNPC [74]	7/NA	4.66/V_{DC}	97.8%@400W	8	2	4
Dual T-type [61]	9/NA	10/V_{DC}	96%@50W	12	0	3
ABNPC [72]	9/NA	5/0.5V_{DC}	97%@500W	10	4	4
ABNPC [73]	9/NA	10/2V_{DC}	98%@400W	11	4	3
ABNPC [75]	9/4.1%	5/V_{DC}	97.1%@400W	10	2	2
Asym SC [56]	15/5.5%	5/2V_{DC}	97%@150W	12	2	4
9-Level SC-MLI [198]	9/1.07%	8.5 V_{DC}	98.03%@583.91W	10	4	4
15-Level MC-MLI [199]	15/5.66%	7 V_{DC}	94.1%@113.75W	9	4	3
MM-STC [200]	9/9.28%	4 V_{DC}	98.65%@321.35W	8	8	0
17-Level SC-MLI [201]	17/NA	5/2V_{DC}	96.5%@434.7W	16	10	4

a. Five-Level mid-point-clamped SC-based inverter

The five-level mid-point clamped-based MLI approach, as shown in Figure 21a, involves adding two capacitors, C3 and C4, and a four-quadrant power switch "*p*" to a standard 3L NPC-based inverter to provide five distinct levels of output voltage.

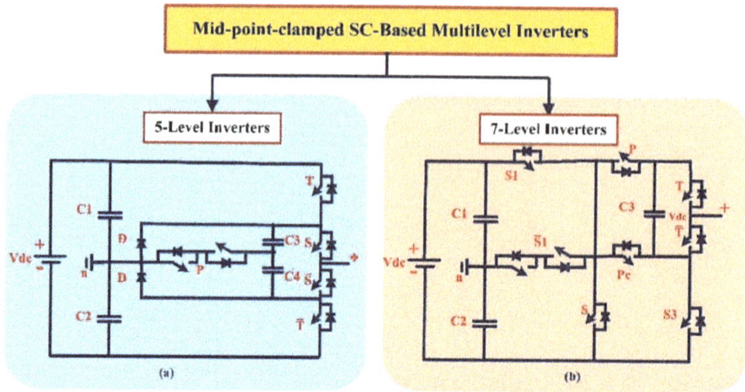

Figure 21. Mid-point-clamped SC-based inverters: (**a**) 5-Level inverter, (**b**) 7-Level inverter.

b. Seven-Level mid-point-clamped SC-based inverter

The seven-level mid-point-clamped inverter proposed in [79] and depicted in Figure 21b is another example of this topology, although one that employs nine rather than eight switches. Figure 22 shows the comparison of the efficiency of SC-MLI with multi-source MLIs [198], and Figure 23 shows the measured efficiency of the 19-level SC-MLI at different frequency ranges [144,198–201].

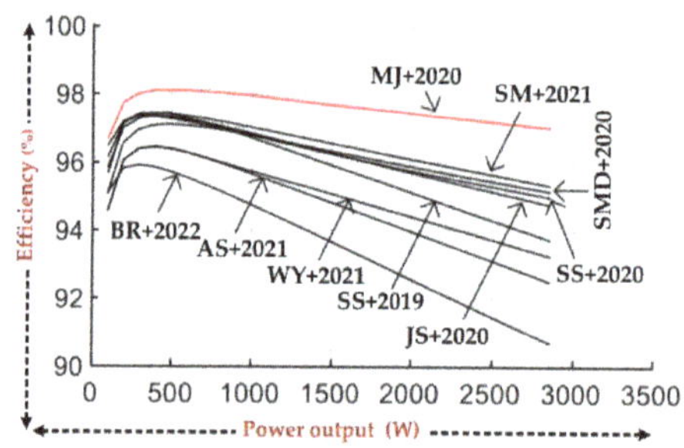

Figure 22. Comparison of the efficiency of SC-MLI with multi-source MLIs [61,68,72–74,151,172,198].

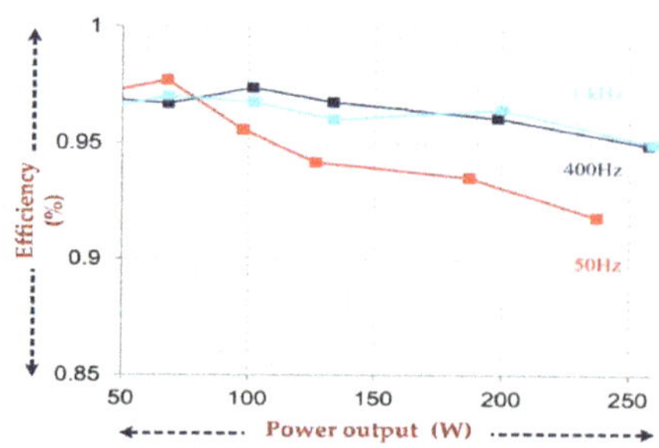

Figure 23. Measured efficiency of the 19-level SC-MLI at different frequency ranges.

5. Common Ground Switched-Capacitor (CGSC)-Based MLIs

Mid-point clamped MLIs can minimize leakage current in transformer-less grid-tied PV systems. HF-CMV in midpoint-clamped MLIs is from DC-link capacitors. To eliminate CMV, CGSC-based MLIs have been proposed, where the input DC source's ground and the grid's neutral point are directly coupled. No leakage current can flow through the system because the parasitic capacitance between the negative terminal of the input DC source, such as PV panels, and the ground sees a grounded potential instead of a fluctuating HF-CMV. This converter is also called a three-port single-DC source inverter or a transformer-less inverter with double grounding. Table 8 provides a comparative study of different CGSC-based MLIs, and Table 9 provides comparative study of different hybrid MLIs [194–196].

Table 8. Comparative study of different CGSC-based MLIs.

Type of SC-MLI	No. of Levels/THD	Overall Voltage Gain/Caps Voltage	TSV (pu)/MVS	Reported Rated Efficiency	No. of Components		
					S	D	C
[86]	5/NA	2/V_{DC} (1), 2V_{DC} (1)	4.5/2V_{DC}	98.1%@600W	6	2	2
[89]	5/NA	2/V_{DC} (1), 2V_{DC} (1)	6/2V_{DC}	96%@40W	7	2	2
[90]	5/35.4%	2/V_{DC} (2)	5/2V_{DC}	98%@600W	8	1	2
[92]	5/NA	2/V_{DC} (1), 2V_{DC} (1)	5/2V_{DC}	98%@600W	11	0	4
[93]	5/NA	2/V_{DC} (1), 2V_{DC} (1)	6.5/2V_{DC}	97.5%@600W	8	2	3
[94]	5/NA	2/V_{DC} (2)	5.5/2V_{DC}	98.3%@600W	8	0	2
[99]	5/36.4%	2/V_{DC} (1), 2V_{DC} (1)	6.5/2V_{DC}	97.5%@330W	8	1	2
[102]	5/NA	2/V_{DC} (2)	6/2V_{DC}	96.7%@1kW	9	0	2
[102]	5/NA	1/0.5V_{DC} (2)	6/V_{DC}	97%@500W	6	1	2
[94]	7/NA	3/V_{DC} (3)	6/3V_{DC}	98.3%@600W	11	0	2
[95]	7/NA	3/V_{DC} (2), 2V_{DC} (1), 3V_{DC} (1)	6/3V_{DC}	98%@800W	6	4	4
[97]	7/NA	3/V_{DC} (1), 2V_{DC} (2)	5.33/3V_{DC}	NA@1kW	8	4	3
[98]	9/NA	4/V_{DC} (4)	6/4V_{DC}	NA@275W	17	4	4
[102]	9/NA	4/V_{DC} (4)	6/4V_{DC}	NA@1kW	17	0	4

Table 9. Comparative study of different hybrid MLIs.

Type of SCMLI	No. of Levels/THD	Overall Voltage Gain/Caps Voltage	TSV (pu)/MVS	Reported Rated Efficiency	No. of Components		
					S	D	C
CGSC based [105]	5/NA	1/V_{DC} (1), 0.5V_{DC} (1)	4/V_{DC}	95.8%@1.2kW	6	1	2
ABNPC based [106]	7/NA	1/0.5V_{DC} (2), V_{DC} (1), 0.5V_{DC} (1)	5/V_{DC}	98%@2.2kW	8	2	4
ABNPC based [107]	7/NA	1/0.5V_{DC} (2), V_{DC} (1), 0.5V_{DC} (1)	5.5/V_{DC}	NA	10	0	4
ABNPC based [112]	7/NA	0.5/0.5V_{DC} (2), 0.33V_{DC} (2)	6/0.5V_{DC}	NA	11	0	4
HB based [108]	9/NA	2/V_{DC} (2), 0.5V_{DC} (1)	6/2V_{DC}	96.4%@500W	8	2	3
HBSC based [109]	9/13.5%	2/V_{DC} (1), 0.5V_{DC} (1)	6/V_{DC}	97.3%@330W	11	0	2
HBSC based [110]	9/NA	2/V_{DC} (1), 0.5V_{DC} (1)	5/2V_{DC}	96.5%@330W	8	1	2
HBSC based [111]	9/NA	2/V_{DC} (1), 0.5V_{DC} (1)	5.5/2V_{DC}	96.6%@600W	8	1	2
HBSC based [114]	9/9.4%	2/V_{DC} (1), 0.5V_{DC} (1)	5.5/V_{DC}	96.5%@800W	10	0	2
CGSC based [116]	9/NA	2/V_{DC} (2), 0.5V_{DC} (1)	5/2V_{DC}	97.5%@1.2kW	9	1	3
CGSC based [117]	5/NA	2/V_{DC} (1), 2V_{DC} (1)	6/2V_{DC}	97.5%@700W	7	2	2

a. Three-level CGSC-based inverter

The inverters are based on CGSC and have been recently practiced. Taking into consideration Figure 24a, [202] proposed a three-level CGSC-based inverter with five power switches and a virtual DC-link capacitor, where the capacitor, C, was charged to the input DC source, Vdc, during the positive and zero levels of the output voltage and discharged during the generation of the negative output voltage. Siwakoti-H inverters were proposed in [203] that reduced the number of switching devices for this type of three-level CGSC-based inverter, as illustrated in Figure 24b,c, where one additional diode was employed in Type-I of this converter and two RB-IGBTs were used in its Type-II variation. In [204], a three-level, four-switch CGSC-based inverter is presented; the virtual DC-link capacitor is charged indirectly to Vdc through a diode-aided CPC cell.

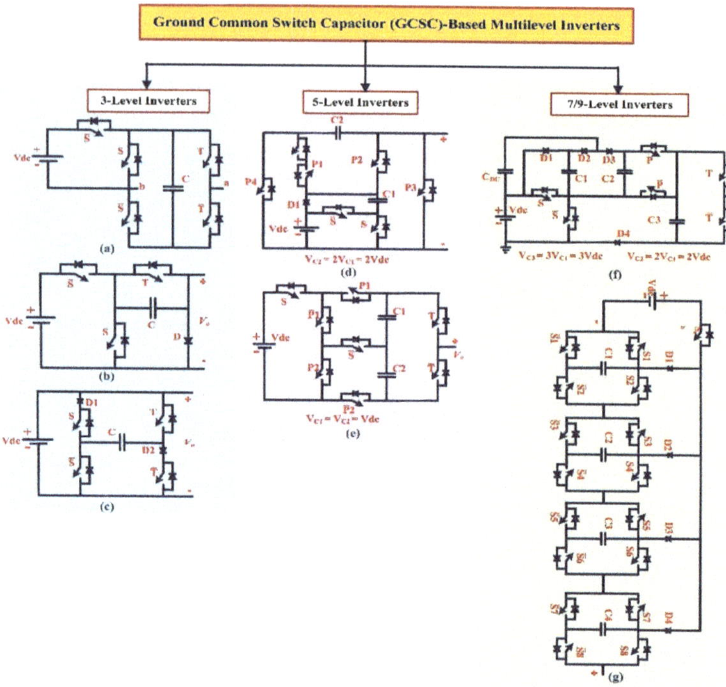

Figure 24. CGSC-based inverters: (**a–c**) Three-level inverters; (**d**,**e**) five-level inverters; (**f**) seven-level inverter; (**g**) nine-level inverter.

b. Five-level CGSC-based inverter

There are two distinct CGSC-based inverter topologies, as can be seen in Figure 24d,e, one employing six power switches and the other employing eight power switches, both of which are capable of producing five levels of double voltage conversion gain output. A five-level CGSC-based inverter, recently presented by Ardashir et al. [197], is comprised of six power switches. The negative output voltage levels are generated using a similar virtual DC-link SC approach, but the total gain of the voltage conversion is unity.

c. Seven-level CGSC-based inverter

The authors of [134] described a further inverter design based on a seven-level CGSC by considering Figure 24f, with fewer switches. By charging capacitors C1, C2, and C3 to Vdc, 2Vdc, and 3Vdc, respectively, the whole seven-level inverter output voltage range can be made with a voltage gain of three. This construction is based on the virtual DC-link principle, just like the aforementioned five-level CGSC-based inverters. Only six power switches are needed, but an extra four power diodes are essential. Since C2 and the input DC source are connected in series, C3 can be charged from either. This charging action is feasible both at zero and at the highest positive output voltage level, +3 Vdc, just as in the five-level CGSC-based inverters given in [86–91]. Additionally, in positive and negative half-cycles of the output voltage, the direction of the load current charges and discharges the DC-link capacitor Cdc. As a result, a higher Cdc and C3 capacitance may be required to reduce the voltage ripple caused by these lengthy discharging cycles.

d. Nine-Level CGSC-based inverter

A nine-level quadruple-voltage-gain CGSC-based inverter is shown in Figure 24g. The consistent MVSs across all the FB-cell switches and the lowered balanced voltage value of the related capacitors make this a compelling design, despite the high number of switching devices it employs. Because bigger values of the capacitance are required in the case of increased power injection requirements, the converter's lengthy discharging cycle for the capacitors may be a major drawback.

The scope of use for SC-MLIs can be expanded as shown in Figure 25, and SC-MLIs are appealing for grid-tied PV-based low-power applications because of their single-stage voltage step-up capability, despite their pulsing input current [86–88]. Other developed applications of SC-MLIs with restricted output power performance include motor drives, electric vehicles, energy storage systems, and balancing in battery strings.

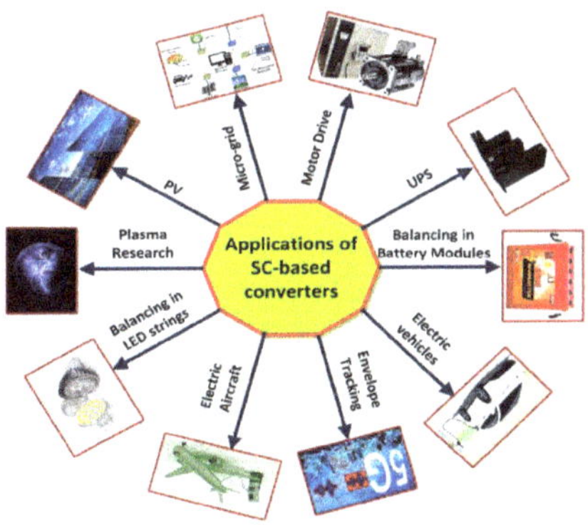

Figure 25. Various applications of SC-based multilevel converters/inverters.

As can be seen in Figure 26, we take into consideration ten characteristics to present a comprehensive qualitative overview of the circuit properties of various SC-MLIs.

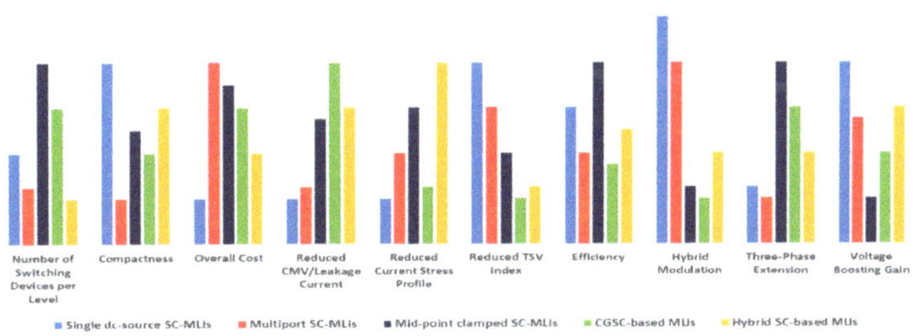

Figure 26. Qualitative comparison analysis of different SC-MLIs in various characteristics.

6. Modulation Techniques

Modulation techniques typically involve a carrier signal and a modulator waveform with different waveform parameters. By adjusting the characteristics of a carrier signal using a reference signal, modulation can be used to control the switching time of the switches in the MLIs. Harmonic reduction and switching losses, both of which can be controlled by modulation methods, are two of the things that affect the overall efficiency of a multilevel inverter. The modulation index plays an important role in all control systems. The THD fluctuates with a modulation ratio (either too much or too little). There are a variety of methods in the literature that may be used based on the switching frequency, whether fundamental or high frequency. Figure 27 shows several MLI modulation control techniques.

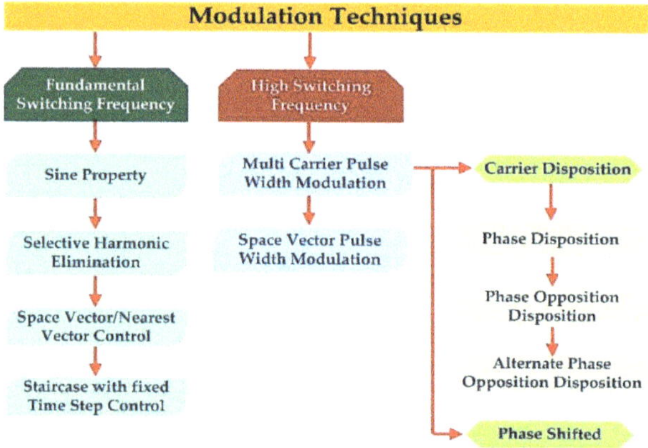

Figure 27. Multilevel inverter modulation control techniques.

Table 10 provides a comprehensive study of conventional MLI topologies; Table 11 offers the merits and demerits of new multilevel inverter topologies; Table 12 provides the applications of MLI topologies; and Table 13 provides information on a comprehensive examination study of traditional and new multilevel inverter topologies.

Table 10. The advantages of switched-capacitor multilevel inverters (SC-MLIs) over conventional multilevel inverters.

Advantages	Switched-Capacitor Multilevel Inverters (SC-MLIs) [36,38,68,90,91,99,151,198,201]	Conventional Multilevel Inverters (DC-MLI, FC-MLI & CHB-MLI) [1,2,11,31,39,106,205]
Reduced component count	Fewer power electronic components required	More components needed
Cost	Fewer components lead to lower costs and increased reliability	A higher component count might lead to increased cost and complexity
Higher efficiency	Fewer components and simplified control can contribute to higher efficiency	Higher component count and more complex control might lead to lower efficiency
Compact design	Modular and compact designs are possible	The size might be larger
Modular structure	Scalable and adaptable design by adding or removing capacitor modules	Limited scalability
Simplified control	Simple control strategy due to switched capacitors	Control complexity may be higher
Voltage balancing	Inherent voltage balancing	Requires active balancing
Improved reliability	A simpler structure and fewer components can result in improved overall reliability	Complex structures and more components might lead to increased failure points

Table 10. *Cont.*

Advantages	Switched-Capacitor Multilevel Inverters (SC-MLIs) [36,38,68,90,91,99,151,198,201]	Conventional Multilevel Inverters (DC-MLI, FC-MLI & CHB-MLI) [1,2,11,31,39,106,205]
Reduced EMI	Potentially lower EMI	EMI considerations may be higher
Low-power applications	Well-suited for applications with lower power requirements	Suitable for a range of power levels, but complexity might be overkill for lower power applications
Improved waveform quality	Inherent voltage balancing leads to improved output waveform quality	Output waveform might require additional filtering to achieve desired quality.

Table 11. Benefits and restrictions of new multilevel inverter topologies.

MLI Topology	Benefits	Restrictions
RVDC-C [205]	• A more modular framework • Asymmetric or symmetric source configurations are possible • Cells may share electricity equally • Can be used at the basic switching frequency	• Independent DC sources are required
Developed H-bridge [31]	• To produce greater levels of output, it only requires a small number of switches	• Why equality of load sharing across sources is impractical. • An asymmetrical arrangement is required • Independent DC sources are required
SCU [151]	• High modularity, equal load sharing across sources • Works with both symmetric and asymmetric sources	• Control complex • Capacitor performance problem • DC sources must be isolated
DCC [1]	• There is the option of using an asymmetric source arrangement • This device has a basic construction and a high degree of modularity	• Cell-to-cell power transfer is impossible • Various voltage switches are required • Requires independent DC.
CPCC [205]	• It may be used with either an asymmetrical or symmetrical source arrangement • It is feasible to distribute electricity equally across cells in a symmetrical design	• No need for fundamental frequency shifting. • Requires various voltage switches.
CIC [205]	• Minimal switching and conductor losses • Modular construction	• It requires both isolated and non-isolated DC supply, and it cannot be used with asymmetric source configurations
Hybridtopology [31,43,51]	• Low-, medium-, and high-voltage applications are all well-served by this component • Switching frequency at the fundamental level can be applied	• A lot of work is required to manage this • Switches with various voltage ratings are required • This does not apply to a trinary-source setup • Isolated DC power supply are required
HERC-C [205]	• Simple and very modular structure • Power distribution across modules may be done on an equal basis • Switching and conduction losses have been • minimized significantly	• It is not possible to use a trinary source configuration
Cascaded half- bridge [11,33]	• Modular and simplified design • All that is needed is a single, isolated DC supply • Switch rating stays constant as the number of levels increases • Totally eliminates common-mode leakage current in solar PV	• Balancing capacitors is a delicate process that needs further attention • High losses in switching • It is not feasible to share authority equally • It is feasible to set up a source in an asymmetrical fashion.

Table 11. *Cont.*

MLI Topology	Benefits	Restrictions
SSSC [151]	• Switches with the same power rating are needed • A reduction in switch count without an increase in switch rating • Equality of power can be achieved	• The structure is complicated • Capacitor dynamics prevent it from operating at its basic switching frequency
SADC [151]	• Simple Architecture and it can function at the highest voltage-rated switch	• No two sources can share the burden equally • A variety of switches are required • A DC source of uneven magnitudes is needed for this application
CCS [11,33]	• Suitable for low-, medium-, and high-voltage applications • Simple construction	• Sources needed to be mandatorily asymmetric
Staircase cascaded [33]	• Structural flexibility • Low conduction losses • Low-, medium-, and high-voltage applications • Non-isolated DC input levels	• Requires switches of different voltage rating • Equal distribution of power among sources cannot be attained • Asymmetric source configuration is not possible
Reduced switch type [205]	• Asymmetric arrangement is also feasible • High efficiency • Simple basic module layout for many levels	• DC power sources must be kept separate • Various switches have different blockings • Voltage ratings
Cascade unit based [11]	• Has a modular design • Can be swapped at the fundamental frequency • Asymmetric source setup is another option.	• It is impossible to have an equal distribution of power • Switches with various voltage ratings are required

Table 12. Applications of MLI topologies.

MLI Type	Applications
NPC [72,79]	• Drives • RESs • Power conversion
FC [205]	• RES • Drives
ANPC [57]	• Solar PV systems • Filters
CHB [11,33]	• Power systems • RES • Motor Drives
HCHB [31,43,51]	• Drives • RES
MLDCL [12]	• Permanent magnet motor drives for below 100 KW • Solar PV and fuel cell incorporation
SSPS [151]	• RES • Drives • Traction

Table 12. *Cont.*

MLI Type	Applications
T-type [68]	• Drives • RES • Railways (traction)
N-type [1]	• RES • Medium voltage level industries
CCHB [1,11,33]	• Drives • Power conversion • RES
RV [2,205]	• Power conversion • High voltage DC transmission systems
MLM [205]	• RES
CCS [11,33]	• Solar PV systems
PUC [205]	• Drives • RES
CBSC [151,205]	• RES
M-type [1]	• High voltage DC transmission systems • Wind energy conversion systems

Table 13. Comprehensive examination study of traditional and new multilevel inverter topologies (n_{level} = number of levels in phase voltage).

Topology	Unidirectional Switches (N_{sw})	Bidirectional Switches	DC Sources (N_{DC})	Capacitors (N_{cap})	H-Bridge	Highest Switch Rating	Total StandingVoltage Requirement (p.u.)
CHB-MLI [11,31]	$2 * (n_{level} - 1)$	0	$\frac{(n_{level} - 1)}{2}$	0	-	VDC	$2 * (n_{level} - 1) * VDC$
NPC-MLI [72,73]	$2 * (n_{level} - 1)$	0	1	$(n_{level} - 1)$	No	VDC	$2 * (n_{level} - 1) * VDC$
FC-MLI [205]	$2 * (n_{level} - 1)$	0	1	$\frac{n_{level}}{2} * (nlevel - 1)$	No	$2 * VDC$	$2 * (n_{level} - 1) * VDC$
RVDC-MLI [1,2]	$3 * (n_{level} - 1)$	0	$\frac{(n_{level} - 1)}{2}$	0	No	$2 * VDC$	$\frac{11}{4} * (n_{level} - 1) * VDC$
H-bridge MLI [31,45,51]	$2 * (\log_2 (n_{level} + 1))$	0	$(\log_2 (nlevel + 1) - 1)$	0	No	$\frac{n\ level - 1}{2} * VDC$	$2 * (n_{level} - 1) * VDC$
SCU-MLI [151,205]	$3 * (\log_3 (\frac{n_{level} + 1}{2})) + 4$	0	$\text{Log}_3 (\frac{n_{level} + 1}{2})$	$\log 3 * (\frac{n\ level + 1}{2})$	Yes	$\frac{n\ level - 1}{2} * VDC$	$\frac{11}{4} * (n_{level} - 1) * VDC$
DCC-MLI [1,2,133]	$\frac{5 * (n_{level} * 21)}{6}$	0	$\frac{(n_{level} - 1)}{3}$	0	Yes	$3 * VDC$	$\frac{7 * n_{level} - 9}{2} * VDC$
CPCC-MLI [11,33,205]	$\frac{2 * (n_{level} - 1)}{3}$	$\frac{(n_{level} - 1)}{3}$	$\frac{(n_{level} - 1)}{2}$	0	No	$\frac{3 * (n_{level} - 7)}{N_{sw}}$	$\frac{10}{3} * (n_{level} - 1) * VDC$
CIC-MLI [11]	$\frac{2 * (n_{level} + 8)}{3}$	0	$\frac{(n_{level} - 1)}{2}$	0	No	$2 * VDC$	$(3 * n_{level} - 7) * VDC$
Hybrid MLI [31,43,51]	$\frac{3 * (n_{level} - 1))}{4}$	$\frac{(n\ level - 1)}{8}$	$\frac{(n_{level} - 1)}{4}$	$\frac{(n_{level} - 1)}{4}$	No	$2 * VDC$	$\frac{13}{8} * (n_{level} - 1) * VDC$
HERC-MLI [11,33,205]	$\frac{(3 * n_{level} - 1)}{2}$	1	$\frac{(n_{level} - 1)}{2}$	0	No	$2 * VDC$	$\frac{15}{4} * (n_{level} - 1) * VDC$

Table 13. *Cont.*

Topology	Unidirectional Switches (N_{sw})	Bidirectional Switches	DC Sources (N_{DC})	Capacitors (N_{cap})	H-Bridge	Highest Switch Rating	Total StandingVoltage Requirement (p.u.)
CHB-MLI [11]	$(n_{level} + 1)$	0	1	$\frac{(n_{level}+1)}{2}$	No	$2 * VDC$	$(n_{level} + 3) * VDC$
SSSC-MLI [151]	$(5 * n_{level} - 1)$	0	1	$(n_{level} - 1)$	No	VDC	$(5 * n_{level} - 1) * VDC$
SADC-MLI [151,205]	$\frac{(n_{level}-1)}{3}$	$\frac{(n_{level}-1)}{12}$	$\frac{(n_{level}-1)}{4}$	0	No	$10 * VDC$	$\frac{9}{4} * (n_{level} - 1) * VDC$
CCS-MLI [33,205]	6	$\frac{(n_{level}-5)}{2}$	2	$\frac{(n_{level}-9)}{2} + 2$	No	$2 * VDC$	$\frac{1}{4} * (n_{level} - 1) * VDC$
RSC-MLI [205]	6	$\frac{(n_{level}-1)}{2}$	$\frac{(n_{level}-1)}{2}$	0	Yes	$2 * VDC$	$\frac{7}{2} * (n_{level} - 1) * VDC$

(' $*$ ' indicates the multiplication operation).

7. Reliability Assessment

Reliability assessment is the process of estimating a device's lifespan and chance of failure. Reliability is vital to a system's seamless operation. Manufacturing companies work with reliability analyses to build durable, high-performing, and low-maintenance goods. This idea of "reliability" includes various aspects for assessing a device's reliability. Figure 28 shows reliability categories and how to calculate system reliability [103,144,206,207].

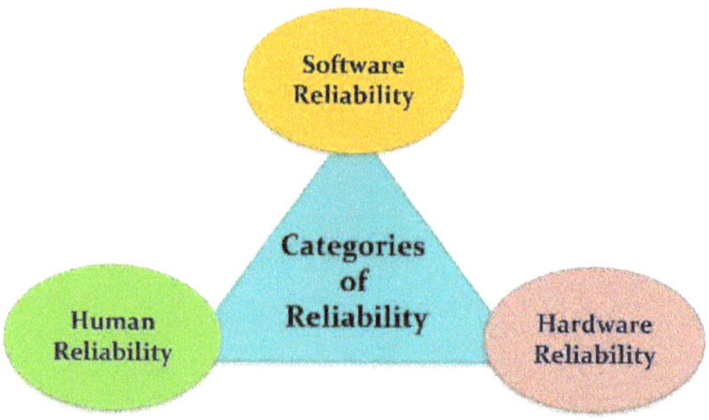

Figure 28. Reliability classifications.

Lifespan estimation is crucial, and a device's or part's lifespan can be estimated by calculating the mean time to failure. A high MTTF suggests reliability. The MTTF can be calculated using MIL-HDBK 217E. These standard handbooks will help calculate a device's failure rate (FR) and mean time between failures (MTTF). Reliability depends on several aspects. Figure 29 shows a system's reliability influence factors.

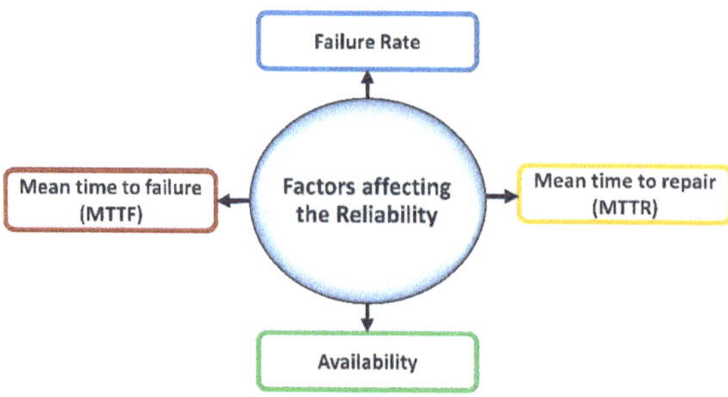

Figure 29. Reliability influence factors.

(a) Reliability

"Reliability" can be defined as the ability of an object to perform its intended function within specified conditions and time frames. This attribute is commonly assessed by quantifying the probability or frequency of failures.

(b) Failure

The system fails when it stops doing the requested task. Thus, the time it takes something to function without breaking down is frequently unpredictable. Failure can be quick or delayed. A sudden failure is called cataleptic failure.

(c) Failure Rate (FR)

The "failure rate" is a crucial aspect in the assessment of system reliability. The chance of failure at a specific moment can be determined by utilizing the "failure rate" function.

(d) Mean Time to Failure (MTTF)

The MTTF measures how long an item or system lasts, on average, before breaking down. This malfunction has rendered the device useless. The MTTF is often provided among components with hourly or thousand-hour service life requirements.

(e) Mean Time to Repair (MTTR)

The MTTR is the typical amount of time needed to repair broken equipment, and its value is directly proportional to the quantity of care it receives [103].

(f) Availability and Average Availability

Availability is the probability that a system will be functional at a particular moment.

The FR and MTTF are the most crucial metrics for this reliability analysis. As the FR is time invariant, it can be used to describe D(t) [103]. The FR is a statistical measure of the frequency with which a failure happens within a certain time frame. Combining the above failure rates, the exponential distribution is utilized to obtain the probability distribution function. The proportion of attempts that fail is also represented by "λ" as follows.

$$P\,(t, \lambda) = \lambda\, e^{-\lambda t} \quad (14)$$

The reliability function can be obtained from Equation (15):

$$D\,(t, \lambda) = e^{-\lambda t} \quad (15)$$

The failure in time (FIT) is a metric for estimating the "failure rate" which is defined as the average number of failures per time interval:

$$1\ \text{FIT} = 10^{-9}\ \text{failure/hour} \quad (16)$$

$$MTTF = \int_{0}^{+\infty} D(t)dt. \quad (17)$$

$$MTTF = \frac{1}{\lambda} \quad (18)$$

Using MIL-HDBK-217E specifications, Table 14 calculates FR [103]. Based on device counts, power electronic circuits can determine their MTTFT [103]. The MTTFT decreases as device numbers increase. The MTTFT increases with the component count. The inverter topologies are evaluated by the number of components needed. The reliability features (FR and MTTF) are calculated using the approximation technique [103] and summarized in Tables 14 and 15, as well as graphically represented in Figure 30.

Table 14. Failure rates of each component by using the approximation method [103].

SI. No.	Components	Failure Rate (Failures/Hour)
1.	Switches	250×10^{-9}
2.	Diodes	100×10^{-9}
3.	Capacitors	300×10^{-9}

Table 15. Expected mean failure time for three standard inverters.

Components	NPC [103]	FC [103]	CHB [103]
IGBTs	4800 (12)	4800 (12)	1200 (12)
Capacitors	600 (2)	1500 (5)	1200 (3)
Diodes	1800 (18)	1200 (12)	1200 (12)
Total FITs	7200	7500	3600
Failure rate (failure/10^6 h)	7.2	7.5	3.6
MTTF (hours)	138,888	133,333	277,777

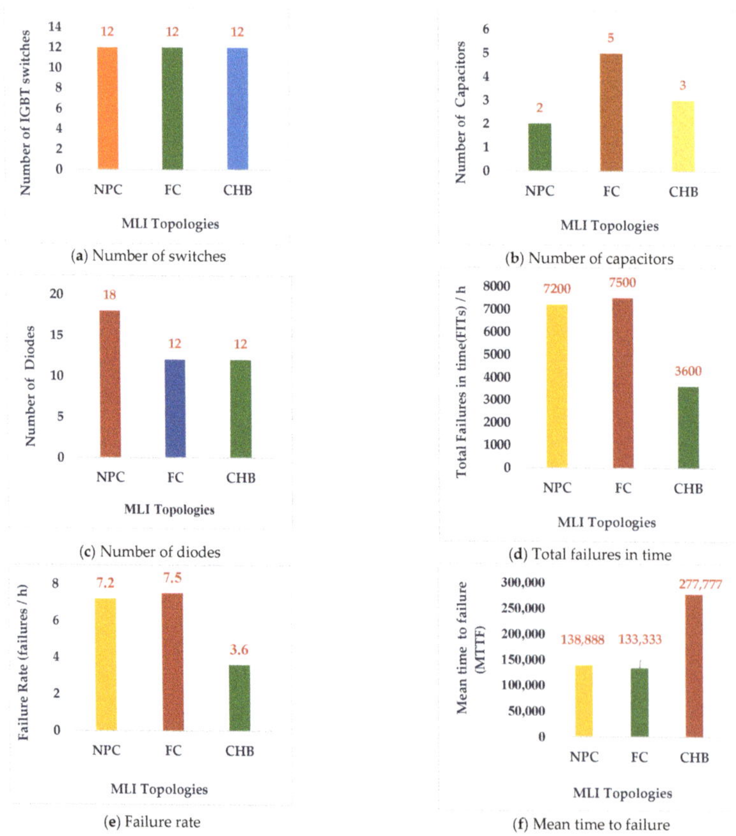

(**a**) Number of switches (**b**) Number of capacitors (**c**) Number of diodes (**d**) Total failures in time (**e**) Failure rate (**f**) Mean time to failure

Figure 30. List of comparisons among three basic MLI topologies [103].

Calculation of the overall *MTTF* for power electronic circuits involves estimation of the cumulative failure rate of the constituent circuit parts. In order to obtain the total failure rate, denoted as λ_{Total}, it is necessary to multiply the number of components such as switches, diodes, and capacitors by their respective FR values, as specified in Equation (19):

$$\lambda_{Total} = (\lambda_S \times N_{SWT}) + (\lambda_D \times N_{DIO}) + (\lambda_C \times N_{CAP}) \tag{19}$$

The total *MTTF* of the circuit can be calculated by Equation (20):

$$MTTF_{Total} = \frac{1}{\lambda_{Total}} \tag{20}$$

The *MTTF* of power electronic circuits can be determined by considering the number of device counts. When there are a large number of device counts, the related total mean time to failure ($MTTF_{Total}$) is reduced. A higher $MTTF_{Total}$ is observed when the number of components is lower. In this study, the main aim is to evaluate the average duration of inverter topologies by considering the number of components needed for every individual topology.

8. Challenges and Recommendations

The utilization of renewable energy systems in power grids has been enhanced due to advancements in power electronics devices and related technologies. However, challenges remain pertaining to electricity quality, grid reliability, and security. In order to ensure the quality of grid power, a multitude of standards and guidelines have been established for grid-connected RES. Based on the reviewed literature, it is understood that additional research is required in the following areas:

Challenges:

- The evaluation of the performance of these novel topologies in grid-integrated applications is imperative, as the majority of them have not yet been examined in the context of grid-connected Renewable Energy Sources (RESs).
- MLI control and modulation systems should be more robust, flexible, and fault tolerant.
- In recent times, researchers and industries have begun to develop hybrid topologies in order to successfully address power quality challenges and to meet demanding grid standards in a cost-efficient manner.
- More research is needed on quantitative approaches for solving MLI nonlinear systems.
- New voltage balancing techniques must be employed in MLIs to minimize capacitor size and to increase inverter power density.
- Resonant converters with single DC source MLIs are suggested.
- It is imperative for smart grid systems to include the integration of microgrid load interactions with MLIs as an essential component.
- However, because of the lower TSV, new RSC-MLI topologies need to be created to boost their appropriateness for both solar PV and wind energy integration.
- Renewable energy sources are increasingly evolving towards a future smart grid as they are integrated into networks utilizing appropriate MLIs, and for MLI topology creation and control, this poses considerable hurdles. There have been many breakthroughs in this sector.

Recommendations:

- The roadmap in Figure 31 shows SC-MLIs' future progress. In addition to exploring new topologies with higher voltage conversion gain, future SC-MLIs can consider factors such as fewer switching devices, reduced MVS and TSV index across switches, improved performance during high pulsating inrush current, and lower cost.
- A modern MLI performance analysis for many practical applications cannot measure all failure prediction parameters and limits the ageing information of PV inverters. Hence, Figures 32 and 33 are the proposed and future road maps for the reliability study of PV inverters.

- In grid-connected solar PV systems, safe and reliable operation of the multilevel inverter depends on the use of suitable safety mechanisms and control strategies, which are listed in Table 16.

Table 16. Grid-fault challenges and recommendations for multilevel inverters in grid-connected solar PV systems.

Grid-Fault Conditions (Challenges)	Multilevel Inverter Response (Recommendations)
Overvoltage [34,163]	• The output voltage magnitude can be controlled by reducing the modulation index • Enable voltage regulation control
Under voltage [34,163]	• The output voltage magnitude can be controlled by reducing the modulation index • Reactive power injection is required to maintain grid voltage
Frequency deviation [34,163,194]	• The output frequency of the inverter can be adjusted by activating the frequency control loop • Frequency adaptive control algorithms are activated
Voltage sag [163,194]	• Voltage support methods, such as reactive power injection, can be used • Inverter may maintain grid voltage by drawing power from DC-link capacitors • Dynamic voltage restorer support during sag periods
Voltage swell [163,194]	• Inverter reduces the output voltage to mitigate excessive power generation during the swell
Grid disconnect [163,194]	• Inverter switch.hes to island mode (if applicable) and operates as a stand-alone system or shuts down safely • Reconnection to the grid after stabilization
Short circuit [34,163,194]	• In order to eliminate faults and restore inverter functionality, fast disconnect and reconnect procedures are required • Short-circuit protection algorithms activated
Grid outage [34,163,194]	• Inverter disconnects from the grid to ensure islanding protection • May switch to an internal control mode to provide power to local loads
Harmonic distortion [34,163,194]	• Activate harmonic filtering control to mitigate harmonics in inverter output • Implement active and passive filtering strategies to mitigate harmonics

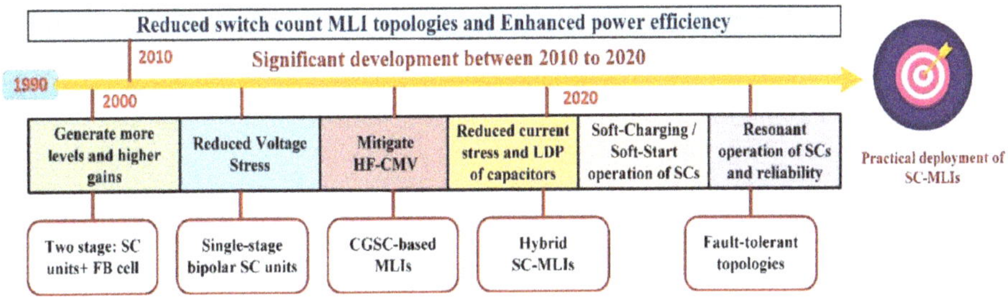

Figure 31. Perception evaluation and future development of SC-MLIs.

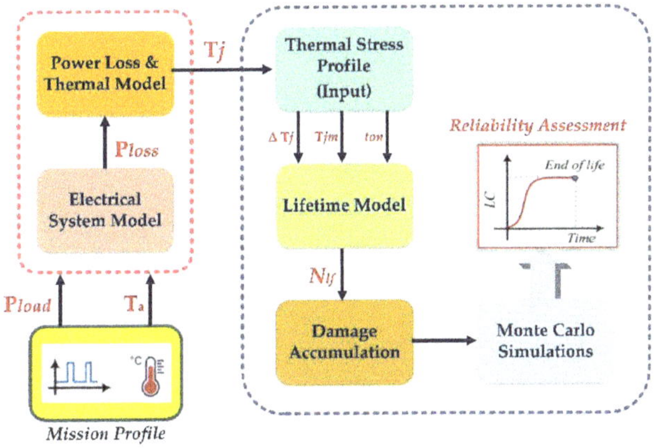

Figure 32. Monte Carlo-based reliability study of PV inverters [103].

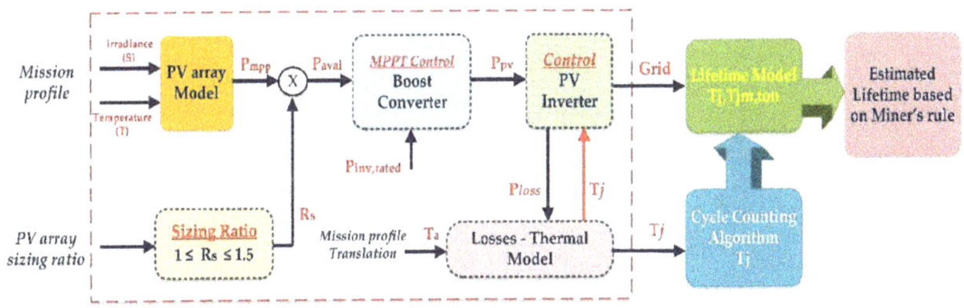

Figure 33. PV system mission profile translation diagram by PV array size ratio Rs consideration [103].

9. Conclusions

This review provides an efficient summary of multilevel inverters to emphasize the necessity for new or modified multilevel inverters for grid-connected sustainable solar PV systems. Firstly, this review presented a detailed survey of reduced switch count multilevel inverter (RSC-MLI) topologies, including their designs, typical features, limitations, assessment parameters, and selection for particular applications. Secondly, this review presented a comprehensive analysis of MLIs and a classification of the existing MLI topologies, along with their merits and demerits. Thirdly, this review also included a survey of SC-MLI topologies with a qualitative assessment to aid in the direction of future research due to their variety of applications such as inductor-less or transformer-less operation, enhanced voltage output, improved voltage regulation, low cost, reduced circuit components, size, and less electromagnetic interference. Lastly, this review serves as a valuable resource for engineers and researchers because it provides a detailed look at parametric comparisons of the total number of power semiconductor switches, DC sources, passive elements, total standing voltage, reliability assessment, applications, challenges, and recommendations.

Author Contributions: Conceptualization and methodology, S.N.; validation, D.C.; formal analysis, S.N.; investigation, D.C.; resources, D.C.; data curation, S.N.; writing—original draft preparation, S.N.; writing—review, S.N.; editing, D.C.; visualization, D.C.; supervision, D.C.; project administration, D.C. All authors have read and agreed to the published version of the manuscript.

Funding: This research received no external funding.

Data Availability Statement: This review has no information related to it.

Acknowledgments: The authors further thank the Vellore Institute of Technology in Vellore, India, for their assistance and also the authors wish to thank the particular copyright holders for allowing approval to use the pictures, graphics, tables, and figures in this work.

Conflicts of Interest: The authors declare no conflict of interest.

Abbreviations

MLIs	Multilevel inverters
RES	Renewable energy sources
NPC	Neutral point clamped
SC-MLI	Switched-capacitor multilevel inverters
DC-MLI	Diode-clamped multilevel inverters
DCC	Developed cascaded cell based
PUC	Packed U-cell
N_{SWT}	Number of switches
N_{DCS}	Number of DC sources
N_L	Number of levels
N_{DIO}	Number of diodes
N_{CAP}	Number of capacitors
N_{DK}	Number of driver circuits
TSV_{PU}	Total standing voltage per unit
CF/L	Cost function
FCC/L	Component count factor per level
CHB-MLI	Cascaded H-bridge multilevel inverter
PV	Photovoltaic
SDCS	Separate DC source
MPPT	Maximum power point tracking
THD	Total harmonic distortion
EMI	Electromagnetic interference
MMC	Modular multilevel converter
FC-MLI	Flying-capacitor multilevel inverter
CSD	Cascaded switched diode
CPCC	Cascaded predictive current control
CCHB	Cross-connected half-bridges
CCS	Cross-connected source based
MCSI	Multilevel current-source inverter
ASD	Adjustable speed drives
AFC	Active front-end converters
CPD	Custom power devices
ANPC	Active neutral point clamped
ABNPC	Active boost neutral point clamped
PWM	Pulse width modulation
CGSC	Common-grounded switched-capacitor
FACTS	Flexible alternating current transmission systems

MLDCL	Multilevel DC-link
SSPS	Switched series-parallel sources
SPSC	Series–parallel switched-capacitor
SSSC	Single-source switched-capacitor
SADC	Symmetric–asymmetric DC sources based
RV	Reverse voltage
RVDC-C	Reduced variety of DC voltage sources based cascaded
HERC-C	Highly efficient and reliable configuration based cascaded
SCSS	vSeries-connected switched sources
SCU	Switched-capacitor Unit
MLM	Multilevel module
HBSC	Half-bridge switched-capacitor
SCC	Switched-capacitor converters

References

1. Choudhury, S.; Bajaj, M.; Dash, T.; Kamel, S.; Jurado, F. Multilevel Inverter: A Survey on Classical and Advanced Topologies, Control Schemes, Applications to Power System and Future Prospects. *Energies* **2021**, *14*, 5773. [CrossRef]
2. Vemuganti, H.P.; Sreenivasarao, D.; Ganjikunta, S.K.; Suryawanshi, H.M.; Abu-Rub, H. A Survey on Reduced Switch Count Multilevel Inverters. *IEEE Open J. Ind. Electron. Soc.* **2021**, *2*, 80–111. [CrossRef]
3. Dhanamjayulu, C.; Prasad, D.; Padmanaban, S.; Maroti, P.K.; Holm-Nielsen, J.B.; Blaabjerg, F. Design and Implementation of Seventeen Level Inverter With Reduced Components. *IEEE Access* **2021**, *9*, 16746–16760. [CrossRef]
4. Dhanamjayulu, C.; Padmanaban, S.; Holm-Nielsen, J.B.; Blaabjerg, F. Design and Implementation of a Single-Phase 15-Level Inverter with Reduced Components for Solar PV Applications. *IEEE Access* **2021**, *9*, 581–594. [CrossRef]
5. Khasim, S.R.; Dhanamjayulu, C.; Padmanaban, S.; Holm-Nielsen, J.B.; Mitolo, M. A Novel Asymmetrical 21-Level Inverter for Solar PV Energy System with Reduced Switch Count. *IEEE Access* **2021**, *9*, 11761–11775. [CrossRef]
6. Dhanamjayulu, C.; Rudravaram, V.; Sanjeevikumar, P. Design and implementation of a novel 35-level inverter topology with reduced switch count. *Electr. Power Syst. Res.* **2022**, *212*, 108641. [CrossRef]
7. Khasim, S.R.; Dhanamjayulu, C. Design and Implementation of Asymmetrical Multilevel Inverter with Reduced Components and Low Voltage Stress. *IEEE Access* **2022**, *10*, 3495–3511. [CrossRef]
8. Roy, T.; Sadhu, P.K.; Dasgupta, A. Cross-switched multilevel inverter using novel switched capacitor converters. *IEEE Trans. Ind. Electron.* **2019**, *66*, 8521–8532. [CrossRef]
9. Gupta, K.K.; Ranjan, A.; Bhatnagar, P.; Sahu, L.K.; Jain, S. Multilevel inverter topologies with reduced device count: A review. *IEEE Trans. Power Electron.* **2016**, *31*, 135–151. [CrossRef]
10. Samanbakhsh, R.; Taheri, A. Reduction of power electronic components in multilevel converters using new switched capacitor-diode structure. *IEEE Trans. Ind. Electron.* **2016**, *63*, 7204–7214. [CrossRef]
11. Barzegarkhoo, R.; Zamiri, E.; Vosoughi, N.; Kojabadi, H.M.; Chang, L. Cascaded multilevel inverter using series connection of novel capacitor based units with minimum switch count. *IET Power Electron.* **2016**, *9*, 2060–2075. [CrossRef]
12. Tayyab, M.; Sarwar, A.; Tariq, M.; Chakrabortty, R.K.; Ryan, M.J. Hardware-in-the-loop implementation of projectile target search algorithm for selective harmonic elimination in a 3-phase multilevel converter. *IEEE Access* **2021**, *9*, 30626–30635. [CrossRef]
13. Iqbal, A.; Siddique, M.D.; Reddy, B.P.; Maroti, P.K. Quadruple boost multilevel inverter (QB-MLI) topology with reduced switch count. *IEEE Trans. Power Electron.* **2021**, *36*, 7372–7377. [CrossRef]
14. Reddy, B.P.; Siddique, M.D.; Iqbal, A.; Mekhilef, S.; Rahman, S.; Maroti, P.K. 7L-SCBI topology with minimal semiconductor device count. *IET Power Electron.* **2020**, *13*, 3199–3203. [CrossRef]
15. Barzegarkhoo, R.; Kojabadi, H.M.; Zamiry, E.; Vosoughi, N.; Chang, L. Generalized structure for a single phase switched-capacitor multilevel inverter using a new multiple DC link producer with reduced number of switches. *IEEE Trans. Power Electron.* **2016**, *31*, 5604–5617. [CrossRef]
16. Siddique, M.D.; Karim, M.F.; Mekhilef, S.; Rawa, M.; Seyedmahmoudian, M.; Horan, B.; Stojcevski, A.; Ahmed, M. Single-Phase Boost Switched-Capacitor-Based Multilevel Inverter Topology With Reduced Switching Devices. *IEEE J. Emerg. Sel. Top. Power Electron.* **2022**, *10*, 4336–4346. [CrossRef]
17. Siddique, M.D.; Mekhilef, S.; Shah, N.M.; Sarwar, A.; Iqbal, A.; Memon, M.A. A New Multilevel Inverter Topology With Reduce Switch Count. *IEEE Access* **2019**, *7*, 58584–58594. [CrossRef]
18. Siddique, M.D.; Iqbal, A.; Memon, M.A.; Mekhilef, S. A New Configurable Topology for Multilevel Inverter With Reduced Switching Components. *IEEE Access* **2020**, *8*, 188726–188741. [CrossRef]
19. Siddique, M.D.; Mekhilef, S.; Rawa, M.; Wahyudie, A.; Chokaev, B.; Salamov, I. Extended Multilevel Inverter Topology With Reduced Switch Count and Voltage Stress. *IEEE Access* **2020**, *8*, 201835–201846. [CrossRef]
20. Panda, K.P.; Bana, P.R.; Panda, G. FPA Optimized Selective Harmonic Elimination in Symmetric–Asymmetric Reduced Switch Cascaded Multilevel Inverter. *IEEE Trans. Ind. Appl.* **2020**, *56*, 2862–2870. [CrossRef]
21. Ali, J.S.M.; Almakhles, D.J.; Ibrahim, S.A.A.; Alyami, S.; Selvam, S.; Bhaskar, M.S. A Generalized Multilevel Inverter Topology With Reduction of Total Standing Voltage. *IEEE Access* **2020**, *8*, 168941–168950. [CrossRef]

22. Siddique, M.D.; Mekhilef, S.; Shah, N.M.; Memon, M.A. Optimal Design of a New Cascaded Multilevel Inverter Topology With Reduced Switch Count. *IEEE Access* **2019**, *7*, 24498–24510. [CrossRef]
23. Koohi-Fayegh, S.; Rosen, M.A. A review of energy storage types, applications and recent developments. *J. Energy Storage* **2020**, *27*, 101047. [CrossRef]
24. Vijeh, M.; Rezanejad, M.; Samadaei, E.; Bertilsson, K. A general review of multilevel inverters based on main submodules: Structural point of view. *IEEE Trans. Power Electron.* **2019**, *34*, 9479–9502. [CrossRef]
25. Bana, P.R.; Panda, K.P.; Naayagi, R.; Siano, P.; Panda, G. Recently developed reduced switch multilevel inverter for renewable energy integration and drives application: Topologies, comprehensive analysis and comparative evaluation. *IEEE Access* **2019**, *7*, 54888–54909. [CrossRef]
26. Kumari, M.; Siddique, M.D.; Sarwar, A.; Tariq, M.; Mekhilef, S.; Iqbal, A. Recent trends and review on switched-capacitor-based single stage boost multilevel inverter. *Int. Trans. Elect. Energy Syst.* **2021**, *31*, e12730. [CrossRef]
27. Khan, M.N.H.; Forouzesh, M.; Siwakoti, Y.P.; Li, L.; Blaabjerg, F. Switched capacitor integrated (2n+1)-level step-up single-phase inverter. *IEEE Trans. Power Electron.* **2020**, *35*, 8248–8260. [CrossRef]
28. Nakagawa, Y.; Koizumi, H. A boost-type nine-level switched capacitor inverter. *IEEE Trans. Power Electron.* **2019**, *34*, 6522–6532. [CrossRef]
29. Barzegarkhoo, R.; Siwakoti, Y.P.; Aguilera, R.P.; Khan, M.N.H.; Lee, S.S.; Blaabjerg, F. A novel dual-mode switched-capacitor five-level inverter with common-ground transformerless concept. *IEEE Trans. Power Electron.* **2021**, *36*, 13740–13753. [CrossRef]
30. Khan, M.N.H.; Siwakoti, Y.P.; Scott, M.J.; Li, L.; Khan, S.A.; Lu, D.D.C.; Barzegarkhoo, R.; Sidorski, F.; Blaabjerg, F.; Hasan, S.U. A common grounded type dual-mode five-level transformerless inverter for photovoltaic applications. *IEEE Trans. Ind. Electron.* **2021**, *68*, 9742–9754. [CrossRef]
31. Ye, Y.; Chen, S.; Zhang, X.; Yi, Y. Half-bridge modular switched capacitor multilevel inverter with hybrid pulsewidth modulation. *IEEE Trans. Power Electron.* **2020**, *35*, 8237–8247. [CrossRef]
32. Saeedian, M.; Adabi, M.E.; Hosseini, S.M.; Adabi, J.; Pouresmaeil, E. A novel step-up single source multilevel inverter: Topology, operating principle, and modulation. *IEEE Trans. Power Electron.* **2019**, *34*, 3269–3282. [CrossRef]
33. Saeedian, M.; Adabi, M.E.; Adabi, J. Step-up switched-capacitor module for cascaded MLI topologies. *IET Power Electron.* **2018**, *11*, 1286–1296. [CrossRef]
34. Jahan, H.K. A new transformerless inverter with leakage current limiting and voltage boosting capabilities for grid-connected PV applications. *IEEE Trans. Ind. Electron.* **2020**, *67*, 10542–10551. [CrossRef]
35. Ye, Y.; Chen, S.; Hua, T.; Lin, M.; Wang, X. Self-balanced multilevel inverter with hybrid double- and half-mode switched capacitor. *IEEE Trans. Ind. Electron.* **2022**, *69*, 5735–5744. [CrossRef]
36. Sarwer, Z.; Siddique, M.D.; Sarwar, A.; Zaid, M.; Iqbal, A.; Mekhilef, S. A switched-capacitor multilevel inverter topology employing a novel variable structure nearest-level modulation. *Int. Trans. Elect. Energy Syst.* **2021**, *31*, e13151. [CrossRef]
37. Anand, V.; Singh, V. A 13-level switched-capacitor multilevel inverter with single dc source. *IEEE J. Emerg. Sel. Topics Power Electron.* **2022**, *10*, 1575–1586. [CrossRef]
38. Roy, T.; Tesfay, M.W.; Nayak, B.; Panigrahi, C.K. A 7-level switched capacitor multilevel inverter with reduced switches and voltage stresses. *IEEE Trans. Circuits Syst. II Exp. Briefs* **2021**, *68*, 3587–3591. [CrossRef]
39. Sharma, P.; Naidu, R.C. Optimization techniques for grid-connected PV with retired EV batteries in centralized charging station with challenges and future possibilities: A review. *Ain Shams Eng. J.* **2023**, *14*, 101985. [CrossRef]
40. Barzegarkhoo, R.; Zamiri, E.; Moradzadeh, M.; Shadabi, H. Symmetric hybridised design for a novel step-up 19-level inverter. *IET Power Electron.* **2017**, *10*, 1377–1391. [CrossRef]
41. Taheri, A.; Rasulkhani, A.; Ren, H. A multilevel inverter using switched-capacitors with reduced components. *IET Power Electron.* **2020**, *13*, 3954–3962. [CrossRef]
42. Rawa, M. Dual input switched-capacitor-based single-phase hybrid boost multilevel inverter topology with reduced number of component. *IET Power Electron.* **2020**, *13*, 881–891. [CrossRef]
43. Iqbal, A.; Siddique, M.D.; Reddy, B.P.; Maroti, P.K.; Alammari, R. A new family of step-up hybrid switched-capacitor integrated multilevel inverter topologies with dual input voltage sources. *IEEE Access* **2021**, *9*, 4398–4410. [CrossRef]
44. Raman, S.R.; Fong, Y.C.; Ye, Y.; Cheng, K.W.E. Family of multiport switched-capacitor multilevel inverters for high-frequency ac power distribution. *IEEE Trans. Power Electron.* **2019**, *34*, 4407–4422. [CrossRef]
45. Raman, S.R.; Cheng, K.W.E.; Ye, Y. Multi-input switched capacitor multilevel inverter for high-frequency AC power distribution. *IEEE Trans. Power Electron.* **2018**, *33*, 5937–5948. [CrossRef]
46. Arif, M.; Siddique, M.; Ahmad, S.; Iqbal, A.; Ashique, R.; Ayob, S. An improved asymmetrical multilevel inverter topology with boosted output voltage and reduced component count. *IET Power Electron.* **2021**, *12*, 2052–2066. [CrossRef]
47. Roy, T.; Tesfay, M.W.; Nayak, B.; Panigrahi, C.K. Step-up switched capacitor multilevel inverter with a cascaded structure in asymmetric DC source configuration. *J. Power Electron.* **2018**, *18*, 1051–1066.
48. Sarebanzadeh, M.; Hosseinzadeh, M.A.; Garcia, C.; Babaei, E.; Hosseinpour, M.; Seifi, A.; Rodriguez, J. A 15-level switched-capacitor multilevel inverter structure with self-balancing capacitor. *IEEE Trans. Circuits Syst. II Exp. Briefs* **2022**, *69*, 1477–1481. [CrossRef]
49. Wang, Y.; Wang, Z.; Liu, W.; Zhang, Y.; Wang, K.; Liang, J. Step-up switched-capacitor multilevel inverter employing multiple inputs with reduced switches. *J. Power Electron.* **2021**, *21*, 986–997. [CrossRef]

50. Roy, T.; Sadhu, P.K. A novel symmetric switched capacitor multilevel inverter using non-isolated power supplies with reduced number of components. *Sadhana* **2020**, *45*, 111. [CrossRef]
51. Cao, L.; Lin, J.; Chen, S.; Ye, Y. Symmetrical cascaded switched capacitor multilevel inverter based on hybrid pulse width modulation. *Energies* **2021**, *14*, 7643. [CrossRef]
52. Murshid, S.; Tayyab, M.; Sarwar, A.; Tariq, M.; Al-Durra, A.; Tomar, A. Self-balanced twenty five level switched capacitor multilevel inverter with reduced switch count and voltage boosting capability. *IEEE Trans. Ind. Appl.* **2022**, *58*, 2183–2194. [CrossRef]
53. Fong, Y.C.; Raman, S.R.; Ye, Y.; Cheng, K.W.E. Generalized topology of a hybrid switched-Capacitor multilevel inverter for high frequency AC power distribution. *IEEE J. Emerg. Sel. Topics Power Electron.* **2020**, *8*, 2886–2897. [CrossRef]
54. Khatoonabad, S.D.; Varesi, K. A novel dual-input switched-capacitor based 27-level boost inverter topology. In Proceedings of the 2020 28th Iranian Conference on Electrical Engineering (ICEE), Tabriz, Iran, 4–6 August 2020; pp. 1–5.
55. Lin, W.; Zeng, J.; Liu, J.; Yan, Z.; Hu, R. Generalized symmetrical step-up multilevel inverter using crisscross capacitor units. *IEEE Trans. Ind. Electron.* **2020**, *67*, 7439–7450. [CrossRef]
56. Ye, Y.; Chen, S.; Sun, R.; Wang, X.; Yi, Y. Three-phase step-up multilevel inverter with self-balanced switched-capacitor. *IEEE Trans. Power Electron.* **2021**, *36*, 7652–7664. [CrossRef]
57. Siwakoti, Y.P. A new six-switch five-level boost-active neutral point clamped (5L-Boost-ANPC) inverter. In Proceedings of the 2018 IEEE Applied Power Electronics Conference and Exposition (APEC), San Antonio, TX, USA, 4–8 March 2018; pp. 2424–2430.
58. Siwakoti, Y.P.; Palanisamy, A.; Mahajan, A.; Liese, S.; Long, T.; Blaabjerg, F. Analysis and design of a novel six-switch five-level active boost neutral point clamped inverter. *IEEE Trans. Ind. Electron.* **2020**, *67*, 10485–10496. [CrossRef]
59. Rathod, A.A.; Subramanian, B. Scrutiny of hybrid renewable energy systems for control, power management, optimization and sizing: Challenges and future possibilities. *Sustainability* **2022**, *14*, 16814. [CrossRef]
60. Lee, S.S.; Lim, C.S.; Siwakoti, Y.P.; Idris, N.R.N.; Alsofyani, I.M.; Lee, K.-B. A new unity-gain 5-level active neutral-point-clamped (UG-5L-ANPC) inverter. In Proceedings of the 2019 IEEE Conference on Energy Conversion (CENCON), Yogyakarta, Indonesia, 16–17 October 2019; pp. 213–217.
61. Lee, S.S.; Lim, C.S.; Lee, K.-B. Novel active-neutral-point-clamped inverters with improved voltage-boosting capability. *IEEE Trans. Power Electron.* **2020**, *35*, 5978–5986. [CrossRef]
62. Dhanamjayulu, C.; Girijaprasanna, T. Experimental Implementation of Cascaded H-Bridge Multilevel Inverter with an Improved Reliability for Solar PV Applications. *Int. Trans. Electr. Energy Syst.* **2023**, *2023*, 8794874. [CrossRef]
63. Khodaparast, A.; Adabi, J.; Rezanejad, M. A step-up switched capacitor multilevel inverter based on 5-level T-type module. *IET Power Electron.* **2019**, *12*, 483–491. [CrossRef]
64. Ye, Y.; Hua, T.; Chen, S.; Wang, X. Neutral-point-clamped five-level inverter with self-balanced switched capacitor. *IEEE Trans. Ind. Electron.* **2022**, *69*, 2202–2215. [CrossRef]
65. Iqbal, A.; Siddique, M.D.; Ali, J.S.M.; Mekhilef, S.; Lam, J. A new eight switch seven level boost active neutral point clamped (8S-7L-BANPC) inverter. *IEEE Access* **2020**, *8*, 203972–203981. [CrossRef]
66. Lee, S.S.; Lim, C.S.; Siwakoti, Y.P.; Lee, K.-B. Hybrid 7-level boost active-neutral-point- clamped (H-7L-BANPC) inverter. *IEEE Trans. Circuits Syst. II: Exp. Briefs* **2020**, *67*, 2044–2048. [CrossRef]
67. Zeng, J.; Lin, W.; Liu, J. Switched-capacitor-based active-neutral point-clamped seven-level inverter with natural balance and boost ability. *IEEE Access* **2019**, *7*, 126889–126896. [CrossRef]
68. Alyami, S.; Ali, J.S.M.; Almakhles, D.; Almutairi, A.; Obeidat, M. Seven level T-type switched capacitor inverter topology for PV applications. *IEEE Access* **2021**, *9*, 85049–85059. [CrossRef]
69. Sathik, M.J.; Sandeep, N.; Blaabjerg, F. High gain active neutral point clamped seven-level self-voltage balancing inverter. *IEEE Trans. Circuits Syst. II Exp. Briefs* **2020**, *67*, 2567–2571. [CrossRef]
70. Lee, S.S.; Lee, K.-B. Dual-T-type seven-level boost active-neutral point-clamped inverter. *IEEE Trans. Power Electron.* **2019**, *34*, 6031–6035. [CrossRef]
71. Lee, S.S.; Bak, Y.; Kim, S.-M.; Joseph, A.; Lee, K.-B. New family of boost switched-capacitor seven-level inverters (BSC7LI). *IEEE Trans. Power Electron.* **2019**, *34*, 10471–10479. [CrossRef]
72. Siddique, M.D.; Iqbal, A.; Sathik, M.J.; Mekhilef, S.; Almakhles, D.J. Design and implementation of a new unity gain nine-level active neutral point clamped multilevel inverter topology. *IET Power Electron.* **2020**, *13*, 3204–3208. [CrossRef]
73. Siddique, M.; Reddy, B.; Iqbal, A.; Mekhilef, S. A new family of active neutral point clamped inverter topology with reduced switch count. *IET Power Electron.* **2021**, *14*, 1433–1443. [CrossRef]
74. Jagabar Sathik, M.; Sandeep, N.; Almakhles, D.; Bhatnagar, K.; Yang, Y.; Blaabjerg, F. Seven-level boosting active neutral point clamped inverter topology using cross-connected switched-capacitor cell. *IET Power Electron.* **2020**, *13*, 1919–1924. [CrossRef]
75. Subramanian, M.S.; Boopathi, C. A switched capacitor based seven level active neutral point clamped (ANPC) inverter topology with reduced switching devices. *J. Elect. Eng. Technol.* **2021**, *16*, 3103–3112. [CrossRef]
76. Sathik, M.J.; Sandeep, N.; Almakhles, D.; Blaabjerg, F. Cross connected compact switched-capacitor multilevel inverter (C3-SCMLI) topology with reduced switch count. *IEEE Trans. Circuits Syst. II Exp. Briefs* **2020**, *67*, 3287–3291.
77. Siddique, M.D.; Mekhilef, S.; Padmanaban, S.; Memon, M.A.; Kumar, C. Single-phase step-up switched-capacitor-based multilevel inverter topology with SHEPWM. *IEEE Trans. Ind. Appl.* **2021**, *57*, 3107–3119. [CrossRef]

78. Sathik, M.J.; Sandeep, N.; Almakhles, D.; Blaabjerg, F. Improved K-type seven-level switched capacitor inverter topology with self-voltage balancing. *J. Elect. Eng. Technol.* **2020**, *48*, 1800–1819.
79. Dhanamjayulu, C.; Sanjeevikumar, P.; Fellow, I.E.T.E.; Prasad, D.; Khasim, S.R.; Blaabjerg, F. A New 29-Level Switched-Diode Multilevel Inverter with Optimal Device Count. *IETE J. Res.* **2023**, 1–16. [CrossRef]
80. Lee, S.S.; Lee, K.-B. Switched-capacitor-based modular T-type inverter. *IEEE Trans. Ind. Electron.* **2021**, *68*, 5725–5732. [CrossRef]
81. Dhara, S.; Hota, A.; Jain, S.; Agarwal, V. A transformerless 1-φ, 5-level half-bridge PV inverter configuration based on switched-capacitor technique. *IEEE Trans. Ind. Appl.* **2021**, *57*, 1619–1628. [CrossRef]
82. Tran, T.-T.; Nguyen, M.-K.; Duong, T.-D.; Lim, Y.-C.; Choi, J.-H. A switched-capacitor-based six-level inverter. *IEEE Trans. Power Electron.* **2022**, *37*, 4804–4816. [CrossRef]
83. Kurdkandi, N.V.; Marangalu, M.G.; Vala, S.S.; Mohseni, P.; Gharehkoushan, A.Z.; Hosseini, S.H.; Sabahi, M. A new transformer-less common grounded three-level grid-tied inverter with voltage boosting capability. *IEEE Trans. Energy Convers.* **2021**, *36*, 1896–1909. [CrossRef]
84. Vosoughi, N.; Hosseini, S.H.; Sabahi, M. Single-phase common grounded transformer-less grid-tied inverter for PV application. *IET Power Electron.* **2020**, *13*, 157–167. [CrossRef]
85. Vosoughi, N.; Hosseini, S.H.; Sabahi, M. A new single-phase transformerless grid-connected inverter with boosting ability and common ground feature. *IEEE Trans. Ind. Electron.* **2020**, *67*, 9313–9325. [CrossRef]
86. Barzegarkhoo, R.; Siwakoti, Y.P.; Long, T.; Blaabjerg, F. Five level grid-tied inverter employing switched-capacitor cell with common grounded feature. In Proceedings of the 2020 IEEE Applied Power Electronics Conference and Exposition (APEC), New Orleans, LA, USA, 15–19 March 2020; pp. 3298–3303.
87. Barzegarkhoo, R.; Siwakoti, Y.P.; Vosoughi, N.; Blaabjerg, F. Six switch step-up common-grounded five-level inverter with switched capacitor cell for transformerless grid-tied PV applications. *IEEE Trans. Ind. Electron.* **2021**, *68*, 1374–1387. [CrossRef]
88. Vosoughi, N.; Hosseini, S.H.; Sabahi, M. A new transformer-less five-level grid-tied inverter for photovoltaic applications. *IEEE Trans. Energy Convers.* **2020**, *35*, 106–118. [CrossRef]
89. Ito, K.; Takahashi, K.; Koizumi, H. A common ground five-level inverter using switched capacitor. In Proceedings of the IECON 2019—45th Annual Conference of the IEEE Industrial Electronics Society, Lisbon, Portugal, 14–17 October 2019; Volume 1, pp. 1485–1489.
90. Barzegarkhoo, R.; Mojallali, H.; Shahalami, S.; Siwakoti, Y. A novel common-ground switched-capacitor five-level inverter with adaptive hysteresis current control for grid-connected applications. *IET Power Electron.* **2021**, *14*, 2084–2098. [CrossRef]
91. Barzegarkhoo, R.; Siwakoti, Y.P.; Blaabjerg, F. A new switched-capacitor five-level inverter suitable for transformerless grid connected applications. *IEEE Trans. Power Electron.* **2020**, *35*, 8140–8153. [CrossRef]
92. Sandeep, N.; Sathik, M.J.; Yaragatti, U.R.; Samy, V.K.; Verma, A.K.; Pota, H.R. Common-ground-type five-level transformerless inverter topology with full DC-bus utilization. *IEEE Trans. Ind. Appl.* **2020**, *56*, 4071–4080.
93. Kumari, S.; Verma, A.K.; Sandeep, N.; Yaragatti, U.R. An eight switch five-level inverter with zero leakage current. *IET Power Electron.* **2021**, *14*, 590–601. [CrossRef]
94. Barzegarkhoo, R.; Lee, S.S.; Khan, S.A.; Siwakoti, Y.P.; Lu, D.D.-C. A novel generalized common-ground switched-capacitor multilevel inverter suitable for transformerless grid-connected applications. *IEEE Trans. Power Electron.* **2021**, *36*, 10293–10306. [CrossRef]
95. Chen, M.; Loh, P.C.; Yang, Y.; Blaabjerg, F. A six-switch seven level triple-boost inverter. *IEEE Trans. Power Electron.* **2021**, *36*, 1225–1230. [CrossRef]
96. Chen, M.; Yang, Y.; Loh, P.C.; Blaabjerg, F. A single-source nine level boost inverter with a low switch count. *IEEE Trans. Ind. Electron.* **2022**, *69*, 2644–2658. [CrossRef]
97. Khasim, S.R.; Dhanamjayulu, C.; Muyeen, S.M. A Single Inductor Multi-Port Power Converter for Electric Vehicle Applications. *IEEE Access* **2023**, *11*, 3367–3385. [CrossRef]
98. Jahan, H.K.; Abapour, M.; Zare, K. Switched-capacitor-based single-source cascaded H-bridge Multilevel inverter featuring boosting ability. *IEEE Trans. Power Electron.* **2019**, *34*, 1113–1124. [CrossRef]
99. Samizadeh, M.; Yang, X.; Karami, B.; Chen, W.; Blaabjerg, F.; Kamranian, M. A new topology of switched-Capacitor multilevel inverter with eliminating leakage current. *IEEE Access* **2020**, *8*, 76951–76965. [CrossRef]
100. Samizadeh, M.; Li, X.; Yang, X.; Karami, B.; Chen, W.; Hassan, A.; Abou Houran, M.; Kamranian, M. A new asymmetric switched-capacitor inverter for photovoltaic systems. In Proceedings of the 2020 IEEE 9th International Power Electronics and Motion Control Conference (IPEMC2020-ECCE Asia), Nanjing, China, 29 November–2 December 2020; pp. 443–447.
101. Khenari, M.; Taghvaei, A.; Adabi, J.; Rezanezhad, M. Multilevel inverter with combined T-type and cross-Connected modules. *IET Power Electron.* **2018**, *11*, 1407–1415. [CrossRef]
102. Kumari, S.; Verma, A.K.; Sandeep, N.; Yaragatti, U.R.; Pota, H.R. A five-level transformer-less inverter with self-voltage balancing and boosting ability. *IEEE Trans. Ind. Appl.* **2021**, *57*, 6237–6245. [CrossRef]
103. Nyamathulla, S.; Chittathuru, D.; Muyeen, S.M. An Overview of Multilevel Inverters Lifetime Assessment for Grid-Connected Solar Photovoltaic Applications. *Electronics* **2023**, *12*, 1944. [CrossRef]
104. Ardashir, J.F.; Gasemi, M.; Peyghami, S.; Rozmeh, B.; Blaabjerg, F. A novel five-level transformer-less inverter topology with common ground for grid-tied PV applications. In Proceedings of the 2021 23rd European Conference on Power Electronics and Applications (EPE'21 ECCE Europe), Ghent, Belgium, 6–10 September 2021; pp. 1–10.

105. Grigoletto, F.B. Five-level transformerless inverter for single-phase solar photovoltaic applications. *IEEE J. Emerg. Sel. Topics Power Electron.* **2020**, *8*, 3411–3422. [CrossRef]
106. Siwakoti, Y.P.; Mahajan, A.; Rogers, D.J.; Blaabjerg, F. A novel seven-level active neutral-point-clamped Converter with reduced active switching devices and DC-link voltage. *IEEE Trans. Power Electron.* **2019**, *34*, 10492–10508. [CrossRef]
107. Mhiesan, H.; Lee, S.S.; Wei, Y.; Mantooth, A. A new family of 7-level boost active neutral point clamped inverter. In Proceedings of the 2019 IEEE 7th Workshop on Wide Bandgap Power Devices and Applications (WiPDA), Raleigh, NC, USA, 29–31 October 2019; pp. 20–24.
108. Lin, W.; Zeng, J.; Hu, J.; Liu, J. Hybrid nine-level boost inverter with simplified control and reduced active devices. *IEEE J. Emerg. Sel. Topics Power Electron.* **2021**, *9*, 2038–2050. [CrossRef]
109. Naik, B.S.; Suresh, Y.; Venkataramanaiah, J. Experimental verification of a hybrid multilevel inverter with voltage-boosting ability. *Int. J. Circuit Theory Appl.* **2020**, *48*, 420–434. [CrossRef]
110. Naik, B.S.; Suresh, Y.; Venkataramanaiah, J.; Panda, A.K. A hybrid nine-level inverter topology with boosting capability and reduced component count. *IEEE Trans. Circuits Syst. II Exp. Briefs* **2021**, *68*, 316–320. [CrossRef]
111. Panda, K.P.; Bana, P.R.; Kiselychnyk, O.; Wang, J.; Panda, G. A single source switched-capacitor-based step-up multilevel inverter with reduced components. *IEEE Trans. Ind. Appl.* **2021**, *57*, 3801–3811. [CrossRef]
112. Wu, X.; Zhao, Y.; Liu, F.; Liu, J.; Wang, J.; Jiang, J. Seven level inverters with switched-capacitors. *IET Power Electron.* **2019**, *19*, 2897–2903. [CrossRef]
113. Silva, G.V.; de Andrade, J.M.; Coelho, R.F.; Lazzarin, T.B. Switched-capacitor differential boost inverter: Design, modeling, and control. *IEEE Trans. Ind. Electron.* **2020**, *67*, 5421–5431. [CrossRef]
114. Naik, B.S.; Suresh, Y.; Aditya, K.; Rao, B.N. A novel nine-level boost inverter with a low component count for electric vehicle applications. *Int. Trans. Elect. Energy Syst.* **2021**, *13*, e13172. [CrossRef]
115. Barzegarkhoo, R.; Farhangi, M.; Lee, S.S.; Aguilera, R.P.; Siwakoti, Y.P.; Pou, J. Nine-level nine-switch common-ground switched-capacitor inverter suitable for high-frequency AC-microgrid applications. *IEEE Trans. Power Electron.* **2022**, *37*, pp. 6132–6143. [CrossRef]
116. Lee, S.S.; Siwakoti, Y.P.; Barzegarkhoo, R.; Blaabjerg, F. A novel common-ground-type nine-level dynamic boost inverter. *IEEE J. Emerg. Sel. Topics Power Electron.* **2021**, *10*, 4435–4442. [CrossRef]
117. Barzegarkhoo, R.; Khan, S.A.; Siwakoti, Y.P.; Aguilera, R.P.; Lee, S.S.; Khan, M.N.H. Implementation and analysis of a novel switched-boost common-ground five-level inverter modulated with model predictive control strategy. *IEEE J. Emerg. Sel. Topics Power Electron.* **2022**, *10*, 731–744. [CrossRef]
118. Bughneda, A.; Salem, M.; Richelli, A.; Ishak, D.; Alatai, S. Review of Multilevel Inverters for PV Energy System Applications. *Energies* **2021**, *14*, 1585. [CrossRef]
119. Gopi, R.R.; Sreejith, S. Converter topologies in photovoltaic applications–A review. *Renew. Sustain. Energy Rev.* **2018**, *94*, 1–14. [CrossRef]
120. Salem, A.; Van Khang, H.; Robbersmyr, K.G.; Norambuena, M.; Rodriguez, J. Voltage source multilevel inverters with reduced device count: Topological review and novel comparative factors. *IEEE Trans. Power Electron.* **2020**, *36*, 2720–2747. [CrossRef]
121. Prasad, D.; Dhanamjayulu, C.; Padmanaban, S.; Holm-Nielsen, J.B.; Blaabjerg, F.; Khasim, S.R. Design and implementation of 31-level asymmetrical inverter with reduced components. *IEEE Access* **2021**, *9*, 22788–22803. [CrossRef]
122. Prasad, D.; Dhanamjayulu, C. Reduced Voltage Stress Asymmetrical Multilevel Inverter With Optimal Components. *IEEE Access* **2022**, *10*, 53546–53559. [CrossRef]
123. Hamidi, M.N.; Ishak, D.; Zainuri, M.A.A.M.; Ooi, C.A. An asymmetrical multilevel inverter with optimum number of components based on new basic structure for photovoltaic renewable energy system. *Sol. Energy* **2020**, *204*, 13–25. [CrossRef]
124. Basu, T.S.; Maiti, S. A hybrid modular multilevel converter for solar power integration. *IEEE Trans. Ind. Appl.* **2019**, *55*, 5166–5177. [CrossRef]
125. Mahfuz-Ur-Rahman, A.; Islam, M.R.; Muttaqi, K.M.; Sutanto, D. Model Predictive Control for a New Magnetic Linked Multilevel Inverter to Integrate Solar Photovoltaic Systems With the Power Grids. *IEEE Trans. Ind. Appl.* **2020**, *56*, 7145–7155. [CrossRef]
126. Prabaharan, N.; Salam, Z.; Cecati, C.; Palanisamy, K. Design and implementation of new multilevel inverter topology for trinary sequence using unipolar pulse width modulation. *IEEE Trans. Ind. Electron.* **2019**, *67*, 3573–3582. [CrossRef]
127. Dhanamjayulu, C.; Arunkumar, G.; Pandian, B.J.; Kumar, C.R.; Kumar, M.P.; Jerin, A.R.A.; Venugopal, P. Real-time implementation of a 31-level asymmetrical cascaded multilevel inverter for dynamic loads. *IEEE Access* **2019**, *7*, 51254–51266. [CrossRef]
128. Kumari, S.; Sandeep, N. A Self-Balancing Switched-Capacitor Based Dual Boosting Nine-Level Inverter. *IEEE J. Emerg. Sel. Top. Power Electron.* **2023**. [CrossRef]
129. Prabaharan, N.; Palanisamy, K. A comprehensive review on reduced switch multilevel inverter topologies, modulation techniques and applications. *Renew. Sustain. Energy Rev.* **2017**, *76*, 1248–1282. [CrossRef]
130. Masoudina, F.; Babaei, E.; Sabahi, M.; Alipour, H. New Cascaded Multilevel Inverter With Reduced Power Electronic Components. *Iran. J. Electr. Electron. Eng.* **2020**, *16*, 107–113.
131. Yousofi-Darmian, S.; Barakati, S.M. A new asymmetric multilevel inverter with reduced number of components. *IEEE J. Emerg. Sel. Top. Power Electron.* **2020**, *8*, 4333–4342. [CrossRef]
132. Meraj, S.T.; Hasan, K.; Masaoud, A. A novel configuration of cross-switched T-type (CT-type) multilevel inverter. *IEEE Trans. Power Electron.* **2019**, *35*, 3688–3696. [CrossRef]

133. Zaid, M.M.; Ro, J.-S. Switch ladder modified H-bridge multilevel inverter with novel pulse width modulation technique. *IEEE Access* **2019**, *7*, 102073–102086. [CrossRef]
134. Nyamathulla, S.; Dhanamjayulu, C. Design of 17-Level Inverter with Reduced Switch Count. In Proceedings of the 2021 Innovations in Power and Advanced Computing Technologies (i-PACT), Kuala Lumpur, Malaysia, 27–29 November 2021; pp. 1–8.
135. Devalraju, P.; Dhanamjayulu, C. A Novel 19-level Asymmetrical Multilevel Inverter for Dynamic Voltage Restorer Applications. In Proceedings of the 2021 Innovations in Power and Advanced Computing Technologies (i-PACT), Kuala Lumpur, Malaysia, 27–29 November 2021.
136. Dhanamjayulu, C.; Kaliannan, P.; Padmanaban, S.; Maroti, P.K.; Holm-Nielsen, J.B. A new three-phase multi-level asymmetrical inverter with optimum hardware components. *IEEE Access* **2020**, *8*, 212515–212528. [CrossRef]
137. Dhanamjayulu, C.; Padmanaban, S.; Ramachandaramurthy, V.K.; Holm-Nielsen, J.B.; Blaabjerg, F. Design and Implementation of Multilevel Inverters for Electric Vehicles. *IEEE Access* **2021**, *9*, 317–338. [CrossRef]
138. Dhanamjayulu, C.; Meikandasivam, S. Implementation and comparison of symmetric and asymmetric multilevel inverters for dynamic loads. *IEEE Access* **2018**, *6*, 738–746. [CrossRef]
139. Tirupathi, A.; Annamalai, K.; Tirumala, V. A new hybrid flying capacitor—Based single-phase nine-level inverter. *Int. Trans. Elect. Energy Syst.* **2019**, *29*, e12139. [CrossRef]
140. Lee, S.S. Single-stage switched-capacitor module (S*CM) topology for cascaded multilevel inverter. *IEEE Trans. Power Electron.* **2018**, *33*, 8204–8207. [CrossRef]
141. Ali, J.S.M.; Alishah, R.S.; Sandeep, N.; Hosseini, S.H.; Babacei, E.; Vijayakumar, K.; Yaragatti, U.R. A new generalized multilevel converter topology based on cascaded connection of basic units. *IEEE J. Emerg. Sel. Topics Power Electron.* **2019**, *7*, 2498–2512.
142. Hamidi, M.N.; Ishak, D.; Zainuri, M.A.A.M.; Ooi, C.A.; Tarmizi, T. Asymmetrical multilevel DC-link inverter for PV energy system with perturb and observe based voltage regulator and capacitor compensator. *J. Mod. Power Syst. Clean Energy* **2020**, *9*, 199–209. [CrossRef]
143. Hannan, M.A.; Hoque, M.M.; Hussain, A.; Yusof, Y.; Ker, P.J. State-of-the-Art and Energy Management System of Lithium-Ion Batteries in Electric Vehicle Applications: Issues and Recommendations. *IEEE Access* **2018**, *6*, 19362–19378. [CrossRef]
144. Majumdar, S.; Jana, K.C.; Pal, P.K.; Sangwongwanich, A.; Blaabjerg, F. Design and Implementation of a Single-Source 17-Level Inverter for a Single-Phase Transformer-Less Grid-Connected Photovoltaic Systems. *IEEE J. Emerg. Sel. Top. Power Electron.* **2022**, *10*, 4469–4485. [CrossRef]
145. Kang, F.S. A modified cascade transformer-based multilevel inverter and its efficient switching function. *Electr. Power Syst. Res.* **2009**, *79*, 1648–1654. [CrossRef]
146. Habib, A.K.M.A.; Hasan, M.K.; Issa, G.F.; Singh, D.; Islam, S.; Ghazal, T.M. Lithium-Ion Battery Management System for Electric Vehicles: Constraints, Challenges, and Recommendations. *Batteries* **2023**, *9*, 152. [CrossRef]
147. Tsang, K.M.; Chan, W.L. Single DC source three-phase multilevel inverter using reduced number of switches. *IET Power Electron.* **2014**, *7*, 775–783. [CrossRef]
148. Alishah, R.S.; Nazarpour, D.; Hosseini, S.H.; Sabahi, M. New hybrid structure for multilevel inverter with fewer number of components for high-voltage levels. *IET Power Electron.* **2014**, *7*, 96–104. [CrossRef]
149. Venkataramanaiah, J.; Suresh, Y.; Panda, A.K. Design and development of a novel 19-level inverter using an effective fundamental switching strategy. *IEEE J. Emerg. Sel. Top. Power Electron.* **2017**, *6*, 1903–1911. [CrossRef]
150. Babaei, E.; Dehqan, A.; Sabahi, M. A new topology for multilevel inverter considering its optimal structures. *Electr. Power Syst. Res.* **2013**, *103*, 145–156. [CrossRef]
151. Barzegarkhoo, R.; Forouzesh, M.; Lee, S.S.; Blaabjerg, F.; Siwakoti, Y.P. Switched-Capacitor Multilevel Inverters: A Comprehensive Review. *IEEE Trans. Power Electron.* **2022**, *37*, 11209–11243. [CrossRef]
152. Babaei, E.; Gowgani, S.S. Hybrid multilevel inverter using switched capacitor units. *IEEE Trans. Ind. Electron.* **2014**, *61*, 4614–4621. [CrossRef]
153. Jena, R.; Dash, R.; Reddy, K.J.; Parida, P.K.; Dhanamjayulu, C.; Swain, S.C.; Muyeen, S.M. Enhancing Efficiency of Grid-Connected Solar Photovoltaic System with Particle Swarm Optimization & Long Short-Term Memory Hybrid Technique. *Sustainability* **2023**, *15*, 8535.
154. Gholami, K.; Azizivahed, A.; Arefi, A. Risk-oriented energy management strategy for electric vehicle fleets in hybrid AC-DC microgrids. *J. Energy Storage* **2022**, *50*, 104258. [CrossRef]
155. Ravi, S.S.; Aziz, M. Utilization of Electric Vehicles for Vehicle-to-Grid Services: Progress and Perspectives. *Energies* **2022**, *15*, 589. [CrossRef]
156. Wang, Y.; Zhou, G.; Li, T.; Wei, X. Comprehensive Evaluation of the Sustainable Development of Battery Electric Vehicles in China. *Sustainability* **2019**, *11*, 5635. [CrossRef]
157. Lipu, M.S.H.; Mamun, A.A.; Ansari, S.; Miah, M.S.; Hasan, K.; Meraj, S.T.; Abdolrasol, M.G.; Rahman, T.; Maruf, M.H.; Sarker, M.R.; et al. Battery Management, Key Technologies, Methods, Issues, and Future Trends of Electric Vehicles: A Pathway toward Achieving Sustainable Development Goals. *Batteries* **2022**, *8*, 119. [CrossRef]
158. Qays, M.O.; Buswig, Y.; Basri, H.; Hossain, M.L.; Abu-Siada, A.; Rahman, M.M.; Muyeen, S.M. An Intelligent Controlling Method for Battery Lifetime Increment Using State of Charge Estimation in PV-Battery Hybrid System. *Appl. Sci.* **2020**, *10*, 8799. [CrossRef]
159. Zhang, Y.; Ma, T.; Yang, H. Grid-connected photovoltaic battery systems: A comprehensive review and perspectives. *Appl. Energy* **2022**, *328*, 120182. [CrossRef]

160. Lipu, M.S.H.; Faisal, M.; Ansari, S.; Hannan, M.A.; Karim, T.F.; Ayob, A.; Hussain, A.; Miah, M.S.; Saad, M.H.M. Review of Electric Vehicle Converter Configurations, Control Schemes and Optimizations: Challenges and Suggestions. *Electronics* **2021**, *10*, 477. [CrossRef]
161. Hasan, M.K.; Mahmud, M.; Habib, A.A.; Motakabber, S.; Islam, S. Review of electric vehicle energy storage and management system: Standards, issues, and challenges. *J. Energy Storage* **2021**, *41*, 102940. [CrossRef]
162. Zhao, J.; Chen, Y.; Zeng, J.; Liu, J. Low-voltage stress seven-level inverter based on symmetrical capacitors. *IEEE J. Emerg. Sel. Topics Power Electron.* **2021**, *10*, 3033–3044. [CrossRef]
163. Bikdeli, E.; Adabi, J.; Rezanejad, M.; Gholamian, S.A. Investigation on fault tolerant capability of a single source switched capacitor multilevel inverter. *IEEE Trans. Ind. Electron.* **2021**, *68*, 7921–7930. [CrossRef]
164. Ye, Y.; Zhang, G.; Wang, X.; Yi, Y.; Cheng, K.W.E. Self-balanced switched-capacitor thirteen-level inverters with reduced capacitors count. *IEEE Trans. Ind. Electron.* **2022**, *69*, 1070–1076. [CrossRef]
165. Venkatasatish, R.; Chittathuru, D. Coyote Optimization Algorithm-Based Energy Management Strategy for Fuel Cell Hybrid Power Systems. *Sustainability* **2023**, *15*, 9638. [CrossRef]
166. Ye, Y.; Chen, S.; Wang, X.; Cheng, K.-W.E. Self-balanced 13-level inverter based on switched capacitor and hybrid PWM algorithm. *IEEE Trans. Ind. Electron.* **2021**, *68*, 4827–4837. [CrossRef]
167. Sandeep, N. A 13-level switched-capacitor-based boosting inverter. *IEEE Trans. Circuits Syst. II Exp. Briefs* **2021**, *68*, 998–1002. [CrossRef]
168. Peng, W.; Ni, Q.; Qiu, X.; Ye, Y. Seven-level inverter with self-balanced switched-capacitor and its cascaded extension. *IEEE Trans. Power Electron.* **2019**, *34*, 11889–11896. [CrossRef]
169. Khodaparast, A.; Hassani, M.J.; Azimi, E.; Adabi, M.E.; Adabi, J.; Pouresmaeil, E. Circuit configuration and modulation of a seven-level switched-capacitor inverter. *IEEE Trans. Power Electron.* **2021**, *36*, 7087–7096. [CrossRef]
170. Sharma, P.; Thangavel, S.; Raju, S.; Prusty, B.R. Parameter estimation of solar PV using Ali Baba and Forty Thieves optimization technique. *Math. Probl. Eng.* **2022**, *2022*, 5013146. [CrossRef]
171. Taghvaie, A.; Adabi, J.; Rezanejad, M. A self-balanced step-up multilevel inverter based on switched-capacitor structure. *IEEE Trans. Power Electron.* **2018**, *33*, 199–209. [CrossRef]
172. Wang, Y.; Yuan, Y.; Li, G.; Ye, Y.; Wang, K.; Liang, J. A T-type switched-capacitor multilevel inverter with low voltage stress and self-balancing. *IEEE Trans. Circuits Syst. I Reg. Papers* **2021**, *68*, 2257–2270. [CrossRef]
173. He, L.; Sun, J.; Lin, Z.; Cheng, B. Capacitor-voltage self-balance seven-level inverter with unequal amplitude carrier-based APODPWM. *IEEE Trans. Power Electron.* **2021**, *36*, 14002–14013. [CrossRef]
174. Liu, J.; Wu, J.; Zeng, J.; Guo, H. A novel nine-level inverter employing one voltage source and reduced components as high-frequency AC power source. *IEEE Trans. Power Electron.* **2017**, *32*, 2939–2947. [CrossRef]
175. Ngo, B.B.; Kim, J.H.; Zare, F. Single-phase multilevel inverter based on switched-capacitor structure. *IET Power Electron.* **2018**, *11*, 1–8. [CrossRef]
176. Ali, J.S.M.; Krishnasamy, V. Compact switched capacitor multilevel inverter (CSCMLI) with self-voltage balancing and boosting ability. *IEEE Trans. Power Electron.* **2019**, *34*, 4009–4013.
177. Zeng, J.; Lin, W.; Cen, D.; Liu, J. Novel K-type multilevel inverter with reduced components and self-balance. *IEEE J. Emerg. Sel. Topics Power Electron.* **2020**, *8*, 4343–4354. [CrossRef]
178. Panda, K.P.; Bana, P.R.; Panda, G. A switched-capacitor self-balanced high-gain multilevel inverter employing a single DC source. *IEEE Trans. Circuits Syst. II Exp. Briefs* **2020**, *67*, 3192–3196. [CrossRef]
179. Deliri, S.; Varesi, K.; Siwakoti, Y.P.; Blaabjerg, F. Generalized diamond type single dc source switched-capacitor-based multilevel inverter with step-up and neutral voltage balancing capability. *IET Power Electron.* **2021**, *14*, 1208–1218. [CrossRef]
180. Talooki, M.F.; Rezanejad, M.; Khosravi, R.; Samadaei, E. A novel high step-up switched-capacitor multilevel inverter with self-voltage balancing. *IEEE Trans. Power Electron.* **2021**, *36*, 4352–4359. [CrossRef]
181. Mohsenzadeh, M.; Rezanejad, M.; Adabi, J. Marxiplier: An innovative Marx-based single-source multilevel inverter with voltage multiplying capability. *IEEE Trans. Ind. Electron.* **2022**, *69*, 357–364. [CrossRef]
182. Alishah, R.S.; Hosseini, S.H.; Babaei, E.; Sabahi, M.; Gharehpetian, G.B. New high step-up multilevel converter topology with self-voltage balancing ability and its optimization analysis. *IEEE Trans. Ind. Electron.* **2017**, *64*, 7060–7070. [CrossRef]
183. Girijaprasanna, T.; Dhanamjayulu, C. A Review on Different State of Battery Charge Estimation Techniques and Management Systems for EV Applications. *Electronics* **2022**, *11*, 1795. [CrossRef]
184. Roy, T.; Sadhu, P.K. A step-up multilevel inverter topology using novel switched capacitor converters with reduced components. *IEEE Trans. Ind. Electron.* **2021**, *68*, 236–247. [CrossRef]
185. Barzegarkhoo, R.; Moradzadeh, M.; Zamiri, E.; Kojabadi, H.M.; Blaabjerg, F. A new boost switched-capacitor multilevel converter with reduced circuit devices. *IEEE Trans. Power Electron.* **2018**, *33*, 6738–6754. [CrossRef]
186. Hussan, M.R.; Sarwar, A.; Siddique, M.D.; Iqbal, A.; Alamri, B. A Cross Connected Asymmetrical Switched-Capacitor Multilevel Inverter. *IEEE Access* **2021**, *9*, 96416–96429. [CrossRef]
187. Mahato, B.; Majumdar, S.; Vatsyayan, S.; Jana, K.C. A new and generalized structure of MLI topology with half-bridge cell with minimum number of power electronic devices. *IETE Tech. Rev.* **2020**, *38*, 267–278. [CrossRef]

188. Hannan, M.; Wali, S.; Ker, P.; Rahman, M.A.; Mansor, M.; Ramachandaramurthy, V.; Muttaqi, K.; Mahlia, T.; Dong, Z. Battery energy-storage system: A review of technologies, optimization objectives, constraints, approaches, and outstanding issues. *J. Energy Storage* **2021**, *42*, 103023. [CrossRef]
189. Saeedian, M.; Pouresmaeil, E.; Samadaei, E.; Rodrigues, E.M.G.; Godina, R.; Marzband, M. An innovative dual-boost nine-level inverter with low voltage rating switches. *Energies* **2019**, *12*, 207. [CrossRef]
190. Taheri, A.; Rasulkhani, A.; Ren, H.-P. An asymmetric switched capacitor multilevel inverter with component reduction. *IEEE Access* **2019**, *7*, 127166–127176. [CrossRef]
191. Narasipuram, R.P.; Mopidevi, S. A technological overview & design considerations for developing electric vehicle charging stations. *J. Energy Storage* **2021**, *43*, 103225.
192. Sarwer, Z.; Siddique, M.D.; Iqbal, A.; Sarwar, A.; Mekhilef, S. An improved asymmetrical multilevel inverter topology with reduced semiconductor device count. *Int. Trans. Electr. Energy Syst.* **2020**, *30*, 1–19. [CrossRef]
193. Rodriguez, J.; Bernet, S.; Steimer, P.K.; Lizama, I.E. A survey on neutral-point-clamped inverters. *IEEE Trans. Ind. Electron.* **2010**, *57*, 2219–2230. [CrossRef]
194. Hemmati, T.; Marangalu, M.G.; Kurdkandi, N.V.; Khoshkbar-Sadigh, A.; Hosseini, S.H.; Jahan, H.K. Topology review of grid-connected multilevel inverters supplied by photovoltaic panels using switched-capacitor based circuits. In Proceedings of the 2021 22nd IEEE International Conference on Industrial Technology (ICIT), Valencia, Spain, 10–12 March 2021; Volume 1, pp. 508–513.
195. Khan, M.N.H.; Forouzesh, M.; Siwakoti, Y.P.; Li, L.; Kerekes, T.; Blaabjerg, F. Transformer less inverter topologies for single-phase photovoltaic systems: A comparative review. *IEEE J. Emerg. Sel. Topics Power Electron.* **2020**, *8*, 805–835. [CrossRef]
196. Mahmoud, G.A.; Orabi, M.; Ibrahim, A.; Kennel, R.; Abdelrahem, M. Common-ground photovoltaic inverters for leakage current mitigation: Comparative review. *Appl. Sci.* **2021**, *11*, 11266.
197. Ardashir, J.F.; Gasemi, M.; Rozmeh, B.; Peyghami, S.; Blaabjerg, F. A six-switch five-level transformer-less inverter without leakage current for grid-tied PV system. In Proceedings of the 2021 IEEE Energy Conversion Congress and Exposition (ECCE), Vancouver, BC, Canada, 10–14 October 2021.
198. Pal, P.K.; Jana, K.C.; Siwakoti, Y.P.; Majumdar, S.; Blaabjerg, F. An Active-Neutral-Point-Clamped Switched-Capacitor Multilevel Inverter With Quasi-Resonant Capacitor Charging. *IEEE Trans. Power Electron.* **2022**, *37*, 14888–14901. [CrossRef]
199. Majumdar, S.; Mahato, B.; Jana, K.C. Implementation of an Optimum Reduced Components Multicell Multilevel Inverter (MC-MLI) for Lower Standing Voltage. *IEEE Trans. Ind. Electron.* **2020**, *67*, 2765–2775. [CrossRef]
200. Paul, S.; Jana, K.C.; Majumdar, S.; Pal, P.K.; Mahato, B. Performance Analysis of a Multimodule Staircase (MM-STC)-Type Multilevel Inverter With Reduced Component Count and Improved Efficiency. *IEEE J. Emerg. Sel. Top. Power Electron.* **2022**, *10*, 6619–6633. [CrossRef]
201. Majumdar, S.; Mahato, B.; Jana, K.C. Analysis and implementation of a generalised switched-capacitor multi-level inverter having the lower total standing voltage. *IET Power Electron.* **2020**, *13*, 4031–4042. [CrossRef]
202. Gu, Y.; Li, W.; Zhao, Y.; Yang, B.; Li, C.; He, X. Transformerless inverter with virtual dc bus concept for cost-effective grid-connected PV power systems. *IEEE Trans. Power Electron.* **2013**, *28*, 793–805. [CrossRef]
203. Siwakoti, Y.P.; Blaabjerg, F. Common-ground-type transformerless inverters for single-phase solar photovoltaic systems. *IEEE Trans. Ind. Electron.* **2018**, *65*, 2100–2111. [CrossRef]
204. Ardashir, J.F.; Sabahi, M.; Hosseini, S.H.; Blaabjerg, F.; Babaei, E.; Gharehpetian, G.B. A single-phase transformerless inverter with charge pump circuit concept for grid-tied PV applications. *IEEE Trans. Ind. Electron.* **2017**, *64*, 5403–5415. [CrossRef]
205. Omer, P.; Kumar, J.; Surjan, B.S. A Review on Reduced Switch Count Multilevel Inverter Topologies. *IEEE Access* **2020**, *8*, 22281–22302. [CrossRef]
206. Anjaneya Vara Prasad, P.; Dhanamjayulu, C. An Overview on Multi-Level Inverter Topologies for Grid-Tied PV System. *Int. Trans. Electr. Energy Syst.* **2023**, *2023*, 9690344. [CrossRef]
207. Alavi, O.; Viki, A.H.; Shamlou, S. A Comparative Reliability Study of Three Fundamental Multilevel Inverters Using Two Different Approaches. *Electronics* **2016**, *5*, 18. [CrossRef]

sustainability

Article

High-Order Sliding-Mode Control Strategy for Improving Robustness of Three-Phase Interleaved Bidirectional Converter

Yifan Jia [1], Dazhi Wang [1,*], Guofeng Sun [2], Yongliang Ni [3], Keling Song [3] and Yanming Li [3]

1 School of Information Science and Engineering, Northeastern University, Shenyang 110819, China; jia1345205145@163.com
2 State Grid Shandong Electric Power Company, Yantai 264000, China; sgf1126758284@163.com
3 China North Vehicle Research Institute, Beijing 100072, China; 15210131012@163.com (Y.N.); skl@vip.sina.com (K.S.); lym811016@163.com (Y.L.)
* Correspondence: prodzw@126.com

Abstract: In response to the era background of "comprehensive electrification" and "dual carbon plan" of electric vehicles, DC/DC converters have a good performance in terms of weight, volume, and efficiency and are widely used in fields such as solar power generation, UPS, communication, computers, and electric vehicles. At present, the DC bus voltage is an important indicator for measuring the safe and stable operation of high-voltage DC power systems in electric vehicles. Therefore, regulating the stability of bus voltage through converters has good economic benefits for the sustainable development of electric vehicles in terms of maintenance costs and effective energy management. In order to solve the problem of bus voltage resonance instability caused by negative impedance characteristics of constant power load in an electric vehicle DC power system, a sliding-mode control design strategy of three-phase interleaved bidirectional converter under constant power load was proposed. Firstly, a GPI observer was designed to estimate the state and concentrated disturbances of the system. Then, the estimated value was introduced into the controller for feedforward compensation, thereby achieving fast-tracking of the output voltage to the reference voltage. Finally, the simulation results show that the controller can effectively maintain the influence of disturbances and better improve tracking characteristics and robustness to disturbances and uncertainties.

Keywords: three-phase interleaved parallel bidirectional converter; electric vehicle; output voltage stability; power quality; high-order sliding-mode control; constant power load; GPI observer

check for updates

Citation: Jia, Y.; Wang, D.; Sun, G.; Ni, Y.; Song, K.; Li, Y. High-Order Sliding-Mode Control Strategy for Improving Robustness of Three-Phase Interleaved Bidirectional Converter. *Sustainability* **2023**, *15*, 9720. https://doi.org/10.3390/su15129720

Academic Editors: Prince Winston David and Praveen Kumar B

Received: 27 May 2023
Revised: 16 June 2023
Accepted: 17 June 2023
Published: 18 June 2023

1. Introduction

With the research boom of renewable DC power sources, research on DC microgrids is gradually expanding. Because of the obvious advantages of DC microgrid technology, its application in the field of vehicle power system is becoming more and more extensive. In the DC microgrid of electric vehicles, converters are often used to realize voltage conversion between the DC bus and load. For DC power systems, maintaining the stability of the DC bus voltage is the foundation for ensuring the stable operation of vehicles. The application of a large number of power conversion devices subject to strict closed-loop control leads to an increase in the proportion of constant power loads in the system, which greatly reduces the stability of the system when the power of such loads fluctuates [1]. The research shows that the constant power load always exhibits negative-impedance characteristics and brings an instability effect to the system. The negative-impedance characteristics of a constant power load can cause significant voltage oscillations in the system when there are significant changes in the CPL (constant power load), thereby reducing power quality and posing safety hazards [2,3].

Therefore, how to keep the DC bus voltage of electric vehicles quickly adjusted and stable is the key problem with the DC microgrid. The DC microgrid of electric vehicles is a system composed of many parts, its instability phenomena are various, and the mechanism of the system instability is also complicated. Since most distributed power generations, energy storage devices, and loads in the DC microgrid need to be connected with the DC bus through converters, and these power electronic converter devices have nonlinear characteristics, the power electronic system composed by them also has nonlinear characteristics [4].

In practical applications, the system model will be affected by various disturbances, such as the uncertainty of inductance and magnetic characteristics, the instability of input voltage, the disturbance of load, etc. At present, many advanced nonlinear control methods have been applied to the converter, such as active disturbance rejection control, adaptive control, sliding-mode control, etc. The sliding-mode control has the advantages of simple operation, high precision, good stability, and robustness in practical applications. Sliding-mode control technology and a DC-DC converter work well together because they are both based on a variable-switching strategy [5]. For the SMC (slide-mode control) method, Reference [6] ensured large signal stability and a fast dynamic response. In order to further improve the transient dynamics of the system, a simple finite-time convergence SMC method is adopted in the converter system [7,8]. However, it is difficult to maintain high accuracy in the event of external disturbances or changes in internal components. Therefore, modern advanced control methods are studied, such as sliding-mode control, adaptive control, optimal control, predictive control, etc. The above control method basically solves the problem of output-voltage instability. However, these methods cannot quickly track and suppress interference. For closed-loop systems, it is difficult to achieve a good voltage output performance under interference.

Considering the perturbations and uncertainties existing in practical applications, it is difficult to measure them with actual sensors, but the designed observer can achieve accurate estimation and compensation of perturbations. In Reference [9], an expanded state observer was designed to realize the estimation of load changes, and a sliding-mode controller was designed to improve the anti-interference performance of the system. Reference [10] designed the unknown input observer, which has low sensitivity to noise and only needs to adjust one parameter, which is easy to implement in the actual system. Reference [11] proposed a sliding-mode control method based on the disturbance observer, which can converge to the neighborhood near the reference voltage in a finite time. However, the above observer can only accurately estimate the slow time-varying perturbations [12]. The perturbations in the actual system are more complex, and there may be higher-order polynomial perturbations. Reference [13] proposed a passive controller based on interconnection and damping allocation, which is robust and easy to implement. However, it can lead to a slow transient response.

In Reference [14], the GPI (generalized proportional integral) observer was designed to achieve an accurate estimation of slow and fast time-varying disturbances, and it was combined with the backstepping method to deal with the unmatched load disturbance. The basic idea of the backstepping method is to decompose a complex system into multiple subsystems, which are recursive from backward to forward through the design of the virtual control law. Interference factors are designed into each subsystem, but in the design process, there may be a high-order derivative of the virtual control function in the controller, which is more complicated to calculate.

In these controllers, in order to maintain the stability of the output voltage, the switch gain is required to be greater than the upper limit of the disturbance. However, in some low-order sliding-mode control laws, an excessive switching gain can lead to significant voltage fluctuations, resulting in unstable output voltage in the practical-implementation literature [15,16]. The interference estimation and compensation technology provide a feasible method to alleviate the chattering phenomenon in Reference [17]. In Reference [18], the nonlinear disturbance observer was used to estimate the uncertain power change, which provides a new way of thinking for dealing with CPL problem. It is difficult for the observer to obtain satisfactory estimation accuracy when dealing with a fast time-varying CPL. There is also no consideration of supply voltage fluctuations. Performance degradation is caused by these factors. Reference [19] proposes a distributed current-sharing control method. The outer loop is the voltage droop control with the purpose of embedding virtual impedance, while the inner loop is the PI control, which can improve the dynamic and steady-state performance of the system. References [20,21] compensated for the virtual impedance coefficient by actively detecting line impedance to achieve current equalization, and they improved the voltage drop through voltage observer feedforward compensation control. Reference [22] proposed an algorithm for compensating for a current imbalance caused by resistance mismatch. By perturbing the duty cycle of one phase and measuring the deviation of other phase duty cycles, the degree of parameter mismatch is estimated, and current balance is achieved through appropriate compensation coefficients.

There are many effective error estimation methods, including the unknown input observer (UIO) [23], disturbance observer (DOB) [24], and extended state observer (ESO) [25]. The disturbance-observer-based control (DOBC) has been proven to effectively reduce unknown external disturbances and system uncertainty. Due to the fact that these observer techniques are model based, a large amount of information needs to be considered when establishing interference observers. However, both DOB and ESO can only estimate constant and stage constant perturbations and cannot estimate polynomial perturbations.

Due to the uncertainty of converter parameters and the influence of concentrated disturbances, there is an increasing amount of research on the precise estimation of disturbances, using the estimated values as feedforward compensation to improve the anti-interference performance of control. Disturbance-observer-based control (DOBC) considers the parameter changes, load changes, and input voltage fluctuations of the filter as external disturbances to the system. The disturbance observer is used to nominal the controlled object and observe these disturbances through the disturbance observer. The observed values are then fed forward to the output of the voltage control loop to counteract the impact of disturbances on the system. Another disturbance estimation technique is the extended state observer (ESO), which treats both internal uncertainties and external disturbances as total disturbances, treats the total disturbance as a new system state, estimates the system state and disturbances through internal calculations, and then designs the controller by combining the estimated values with the improved sliding-mode control method, so that the output voltage of the converter can track the reference signal quickly [26].

In order to suppress load resistance interference and input voltage changes, an SMC method based on GPI observer for three-phase interleaved parallel DC/DC converter is proposed. Estimate the disturbance and state of the system by designing a GPI observer [27]. Then, based on the estimated values obtained from the GPI observer, a composite controller is constructed using SMC technology, which enables the output voltage to asymptotically track the reference voltage [28]. The simulation results show that, compared with the sliding-mode control method based on NDO, this control method can track the reference value faster and improve the steady-state performance of the system. Meanwhile, this method reduces costs in practical systems.

The GPI observer sliding-mode control method based on the second-order sliding-mode algorithm proposed in this article can achieve small switching gain without sacrificing interference suppression by combining interference estimation, ensuring the stability of the output voltage. Compared with other observer control methods, it has strong robustness against disturbances.

The main work content of this paper is divided into three parts:

- The stability of three-phase interleaved parallel DC/DC converter supplying CPL in electric vehicle DC power system is analyzed, and its mathematical model is established.
- The sliding-mode control strategy based on the GPI observer can effectively reduce the switching gain by compensating the lumped disturbance accurately.

- Through the simulation study of MATLAB/Simulink in the interference of input voltage and CPL power and the evaluation of the proposed composite controller, the correctness of the proposed controller is proved.

2. Stability Analysis of DC Power System for Electric Vehicles

The DC power system of electric vehicles is mainly composed of a power generation unit, generator set, energy storage unit, AC/DC load, and power converter connected to each unit module. As shown in Figure 1, in a power generation unit, the energy flows in one direction, the battery is connected to the DC bus by a DC/DC converter, and the generator set provides energy to the bus by an AC/DC converter.

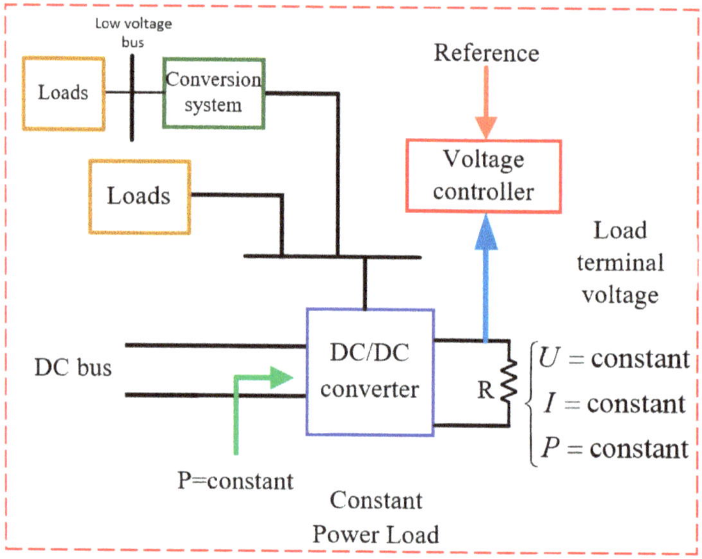

Figure 1. DC/DC converters with constant power characteristics in electrical systems.

All kinds of power electronic devices in the power system of electric vehicles are connected to the on-board high-voltage power supply system of electric vehicles in the form of a cascade, and most of these power electronic devices adopt closed-loop control; when the bus voltage changes, the output power can remain constant. When the input voltage changes, the input current changes in the opposite trend; constant power load has negative impedance characteristics. It is therefore said to have a negative impedance characteristic ($\Delta V/\Delta I < 0$) constant power load.

In a constant power load, P is constant. Thus, as shown in Figure 2, as the voltage at both ends of a constant power load increases/decreases, its current decreases/increases. Because the incremental impedance of CPL is negative ($\Delta V/\Delta I < 0$), in this case, the system will deviate from its stable region, resulting in CPL negative impedance instability [29]. Interaction with other devices may affect the dynamic characteristics and stability of the system.

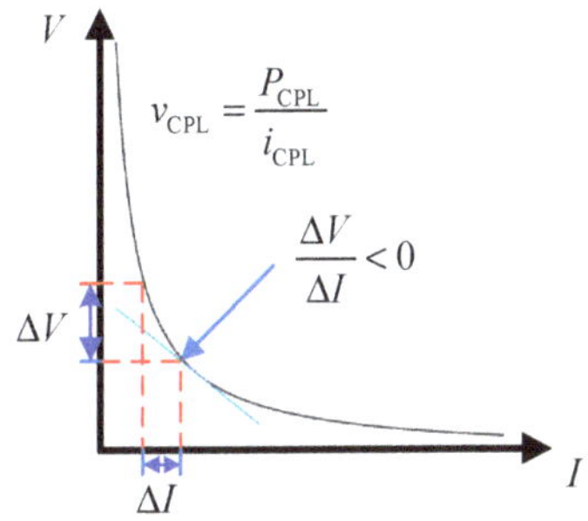

Figure 2. Negative impedance characteristics of constant power load.

As shown in Figure 3a. The midpoint is the initial stable operating point of the system. When the system is subjected to external disturbances causing an increase in the input current of the CPL, as can be seen from Figure 3a, the voltage at both ends of the CPL is at this time, and according to KVL, the voltage at both ends of the filtering inductor is at this time. At this point, the inductance current will further increase, causing the system to move away from the stable operating point. On the contrary, when the input current of the CPL decreases due to external disturbances in the system, the voltage at both ends of the CPL is reduced. From KVL, it can be seen that the voltage at both ends of the filter inductor is reduced, and the inductor current will further decrease, thus keeping the system away from the stable operating point. The obtained volt ampere characteristic curve is shown in Figure 3b for when the load is a pure resistive load, where the point is the initial stable operating point of the system. When external disturbances increase the input current, there is a filter inductance voltage, which can be determined by KVL. At this time, the current will correspondingly decrease, so the system can return to the initial stable operating point; that is, the system is stable. CPL negative impedance was obtained via a small signal analysis [30]. In Reference [30], the equivalent model of CPL was extracted through a small signal analysis and a large signal analysis.

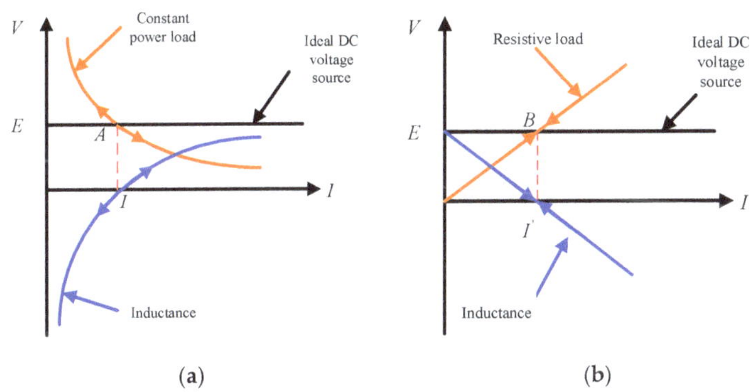

Figure 3. Changes in DC bus voltage during power fluctuations of different types of loads: (**a**) constant power load power fluctuation and (**b**) pure resistance load power fluctuation.

In order to analyze the three-phase interleaved parallel DC/DC converter system supplying CPL, the conjugated model and circuit were extracted for a small signal analysis. Based on the analysis of Reference [3], the destabilizing effect and limit of CPL negative impedance on converter were explained. Since the root of the characteristic equation is on the right-hand side, there is negative impedance instability in the output voltage of the system. In addition, the control performance is severely degraded due to the inevitable voltage fluctuations in the DC supply voltage [31].

According to the research of Reference [3], through the small signal analysis of the system equation of state, the transfer function of the system can be obtained as follows:

$$G(s) = \frac{(1-d)U_{bus} - I_L Ls}{LCs^2 + \frac{L}{R_L}s + (1-d)^2}, \quad (1)$$

From the transfer function, with the increase of CPL power, the negative incremental resistance characteristic of CPL becomes more obvious, and the root of the system characteristic equation begins to move to the right of the complex plane. As shown in Figure 4, once the power consumed by CPL exceeds the power consumed by resistive loads, that is, $P_{CPL} > P_R$, CPL plays a dominant role in the system. The damping coefficient of the corresponding system is less than 0, and the slope of the output characteristic curve is negative. In this case, the DC bus voltage will be in an oscillating state. When the power consumption of CPL is less than that of resistive load, that is, $P_{CPL} < P_R$, the resistive load plays a dominant role in the system, the damping coefficient of the corresponding system is greater than 0, and the slope of the output characteristic curve is positive. Under this condition, the DC bus voltage of the system is in a stable state.

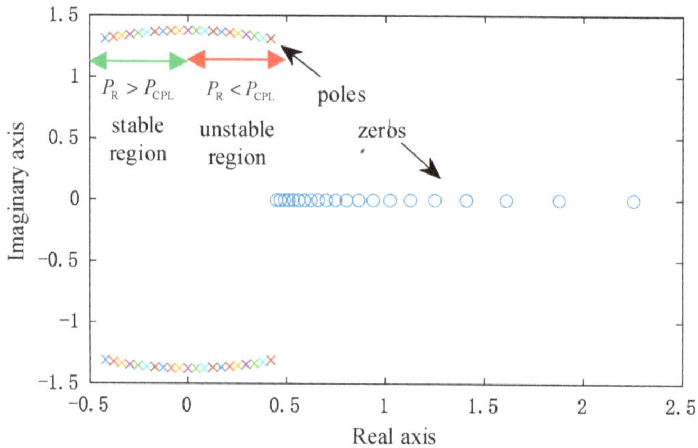

Figure 4. Stable and unstable regions based on small signal theory.

In order to elucidate the impact of CPL power fluctuations on the stability of the DC bus voltage, we first introduced some common CPLs in special vehicles and preliminarily analyzed the dynamic characteristics of negative incremental resistance of constant power loads. Secondly, through a theoretical analysis, we found that the power imbalance between the generating and receiving ends is the fundamental cause of bus voltage fluctuations. Then, based on small signals, the reason for the low-frequency oscillation of the DC bus caused by CPL was obtained. Through a simplified circuit analysis of an ideal voltage source, filter inductor, and CPL in series, it was found that when CPL power fluctuates, it amplifies the power fluctuation, causing the system to move away from the initial operating equilibrium point.

3. Modeling of Three-Phase Interleaved Parallel Bidirectional Converter

The main circuit topology of the three-phase interleaved parallel bidirectional half bridge DC-DC converter is shown in Figure 5, consisting of three bidirectional Bucks–Boosts in parallel. In the same switching cycle, only one switch tube is on the upper and lower bridge arms of the half-bridge switch tube. According to the conduction state of the switch tube, there are two states: Boost and Buck. When the energy storage capacitor releases the stored energy to the load end, the input end of the converter can be approximated as a constant voltage source, and the energy flows from the input end to the load end, where the converter is in a Boost state. When the load side needs to store energy, it operates in Buck mode, and the load-side power flows to the input side to charge the energy storage capacitor. The topology parameters of the three-phase interleaved parallel converter are shown in Table 1.

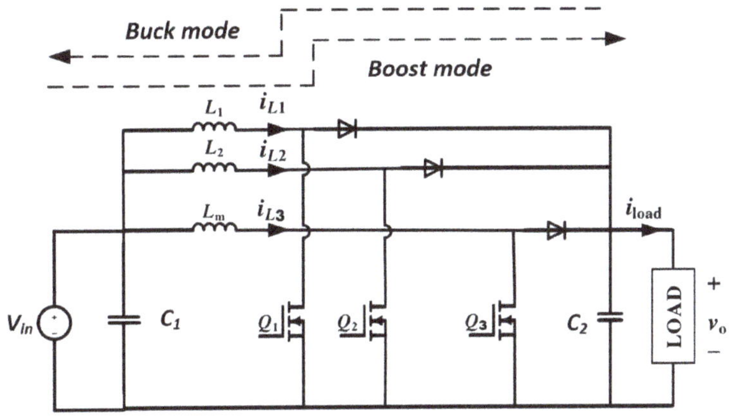

Figure 5. Topology structure of three-phase interleaved parallel bidirectional DC/DC main circuit.

Table 1. Topology diagram parameters of three-phase interleaved parallel converter.

Topology Diagram Parameters	
v_{in}	DC source bus voltage
v_o	DC load bus voltage
i_{L1}, i_{L2}, i_{L3}	Inductance current
R_L	Load resistance
v_{in0}	Nominal value of DC source bus voltage
R_{L0}	Nominal value of load resistance
C_1, C_2	Filter capacitor
L_1, L_2, L_3	Filter inductance
Q_1, Q_2, Q_3	Switch tube components

The advantages of adopting an interleaved parallel structure in bidirectional DC/DC circuits are, on the one hand, under a certain power output, the voltage and current stress of the inductor are reduced, allowing for the selection of smaller inductors, thereby reducing the volume and weight of the converter; and, on the other hand, the difference between the PWM driving waveforms of each phase is 120°, further reducing the input current ripple, reducing the inductance, while also reducing the output voltage ripple and reducing the capacitor voltage and current stress, thus ensuring that the bidirectional DC/DC converter has a higher power density. For the convenience of analysis, if the switching frequency is set to fs and the influence of voltage dead band is ignored, then ws = 2 πfs. Ts = 1/fs. Figure 6 shows the main waveforms of the three-phase interleaved parallel boost converter under different duty ratios, d.

Assuming that the duty cycle, d, of each switch tube is equal and each phase is 120° different in sequence, there are eight switching modes of the converter. Use "1" and "0" to represent the "on" and "off" of the switch tubes, respectively. The switch states of switch tubes Q1, Q2, and Q3 can be represented as corresponding binary numbers: 001 (Mode I), 010 (Mode II), 011 (Mode III), 100 (Mode IV), 101 (Mode V), 110 (Mode VI), 111 (Mode VII), and 000 (Mode VIII). Figure 7 shows the equivalent circuits with 0, 1, 2, and 3 switch tubes on, respectively.

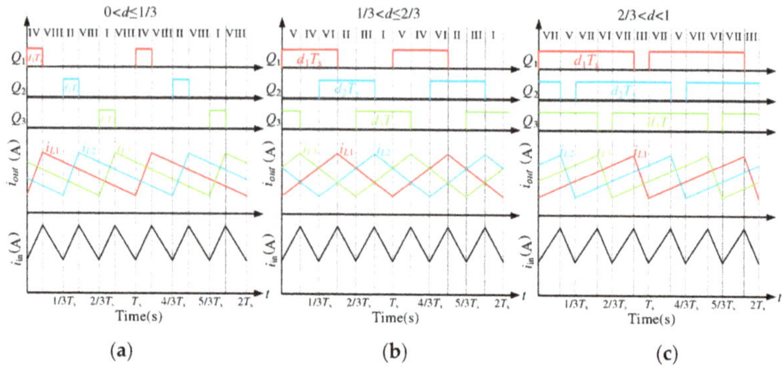

Figure 6. The main waveforms of the three-phase interleaved parallel converter during steady-state operation: (**a**) the main waveform of 0 < d < 1/3, (**b**) the main waveform of 1/3 < d < 2/3, and (**c**) the main waveform of 2/3 < d < 1/3.

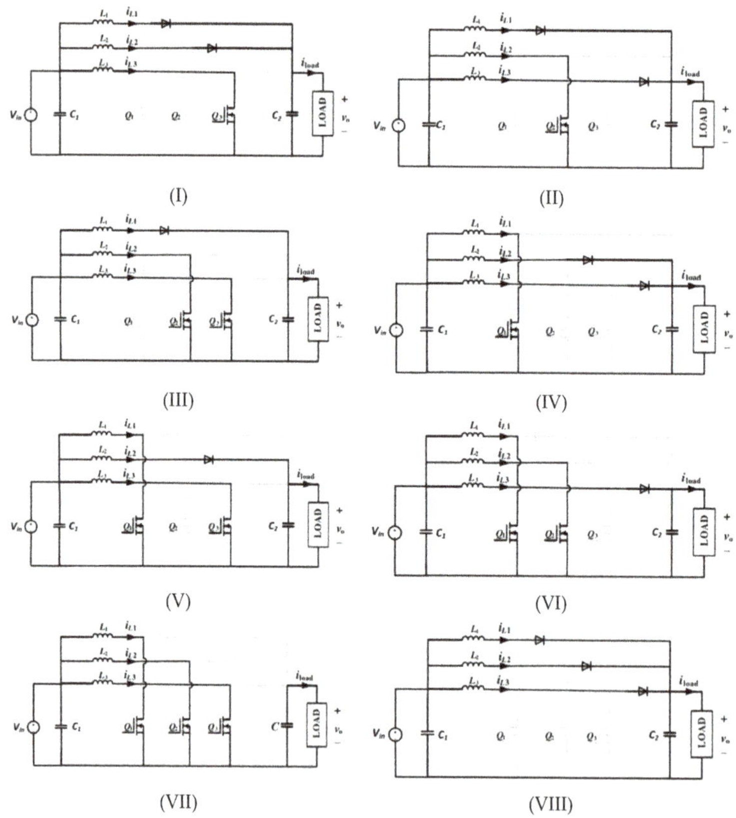

Figure 7. Equivalent topology diagram of three-phase interleaved converter at different working stages: (I) Switch tubes Q1 and Q2 are turned off, and switch tube Q3 is on. (II) Switch tubes Q1 and Q3 are turned off, and switch tube Q2 is on. (III) Switch tube Q1 is off, switch tubes Q2 and Q3 are on. (IV) Switch tube Q1 is on, switch tubes Q2 and Q3 are off. (V) Switch tubes Q1 and Q3 are on, while switch tube Q2 is off. (VI) Switch tubes Q1 and Q2 are on, while switch tube Q3 is off. (VII) The switch tubes Q1, Q2, and Q3 are conducting. (VIII) The switch tubes Q1, Q2, and Q3 are turned off.

For the convenience of description, this article takes one of the situations as an example for analysis, while other situations can be analogized. When the duty cycle is $0 < d < 1/3$, the converter can be divided into six working modes based on the power switch on/off situation. The driving signal and inductance current waveform of the corresponding switch in the system under these six working modes are shown in Figure 6a.

Process 1: (Corresponding Mode IV) The switch Q_1 is in a conductive state, and the current of inductor L_1 continues to increase. The v_{in} end charges the inductor L_1, Q_2 and Q_3 are in the off state, and the current of inductors L_2 and L_3 continues to decrease. Inductors L_2 and L_3 discharge towards the v_o terminal.

Process 2: (Corresponding Mode VIII) Switch tubes Q_1, Q_2, and Q_3 are in the off state, and the current of inductors L_1, L_2, and L_3 continues to decrease. Inductors L_1, L_2, and L_3 discharge towards the v_o terminal.

Process 3: (Corresponding Mode II) The switch tube Q_2 is in a conductive state, and the current of inductor L_2 continues to increase. The v_{in} end charges the inductor L_2, Switch tubes Q_1 and Q_3 are in the off state, and the current of inductors L_1 and L_3 continuously decreases. Inductors L_1 and L_3 discharge towards the v_o terminal.

Process 4: (Corresponding Mode VIII) Switch tubes Q_1, Q_2, and Q_3 are in the off state, and the current of inductors L_1, L_2, and L_3 is continuously decreasing. Inductors L_1, L_2, and L_3 discharge towards the v_o terminal.

Process 5: (Corresponding Mode I) The switch Q_3 is in a conductive state, and the current of inductor L_3 continues to increase. The v_{in} end charges the inductor L_3. The switch tubes Q_1 and Q_2 are in a conductive state, and the current of inductors L_1 and L_2 continues to decrease. Inductors L_1 and L_2 discharge towards the v_o terminal.

Process 6: (Corresponding Mode VIII) Switch tubes Q_1, Q_2, and Q_3 are in the off state, and the current of inductors L_1, L_2, and L_3 continues to decrease. Inductors L_1, L_2, and L_3 discharge towards the v_o terminal.

3.1. Modeling of Three-Phase Interleaved Parallel DC/DC Converter Circuit

Based on the circuit structure and working principle of a three-phase interleaved parallel bidirectional DC-DC converter, this article divides it into three identical Buck–Boost circuits, without considering the parasitic components of capacitors and inductors. The control flowchart of the composite controller is shown in Figure 8.

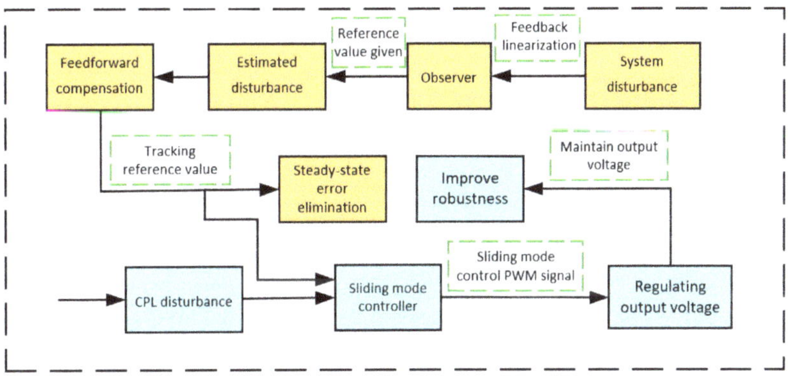

Figure 8. Control flowchart of composite controller.

The equivalent circuit diagram is shown in Figure 9a for when the converter is used as a Boost converter. The equivalent circuit diagram is shown in Figure 9b for when the converter is used as a Buck converter. The topology parameters are shown in Table 2.

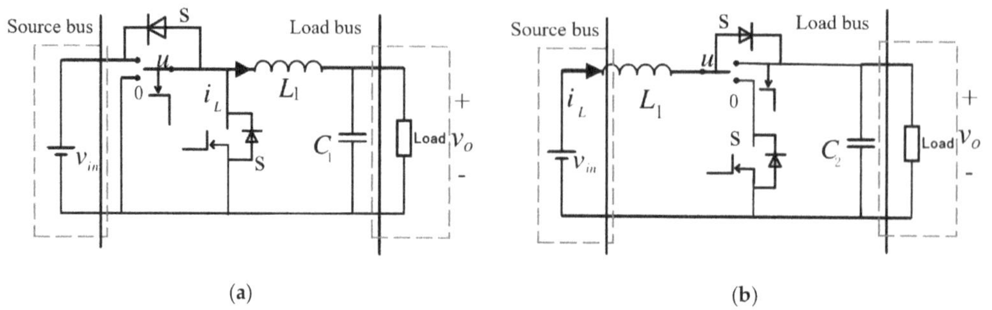

Figure 9. Equivalent variable structure model of Boost converter and Buck converter: (**a**) Boost mode and (**b**) Buck mode.

Table 2. Topology diagram parameters of Buck mode and Boost mode.

Topology Diagram Parameters	
v_{in}	Input voltage
v_o	Output voltage
i_L	Instantaneous inductance current
R_L	load resistance
C_1	Boost circuit capacitance value
C_2	Buck circuit capacitance value
L	Inductance value

The results of variable structure theory analysis can be used to obtain the state equation of the bidirectional DC-DC converter in Buck mode with continuous inductance current as follows:

Firstly, the Buck circuit is modeled and studied, and its equivalent circuit topology is shown in Figure 9. Write the state equation in stages and calculate the average variable.

(1) In $0 \leq t \leq dT_S$, switch the tube S conduction and diode VD cutoff, and, at this time, there is the following equation of state.

$$\begin{cases} L\frac{di_L(t)}{dt} = u_{in}(t) - u_o(t) \\ C\frac{du_o(t)}{dt} = -\frac{u_o(t)}{R_{CPL}} + i_L(t) \end{cases}, \tag{2}$$

(2) In$dT_S \leq t \leq T_S$, switch S off, diode VD conduction, and the inductor L release magnetic field can supply constant power load at the same time to charge the capacitor. The equation of state is as follows.

$$\begin{cases} L\frac{di_L(t)}{dt} = -u_o(t) \\ C\frac{du_o(t)}{dt} = -\frac{u_o(t)}{R_{CPL}} + i_L(t) \end{cases}, \tag{3}$$

By averaging (2) and (3), the following matrix equation can be obtained.

$$\begin{pmatrix} \frac{di_L}{dt} \\ \frac{du_o}{dt} \end{pmatrix} = \begin{pmatrix} 0 & -\frac{1}{L} \\ \frac{1}{C_1} & -\frac{1}{C_1 R} \end{pmatrix} \begin{pmatrix} i_L \\ u_o \end{pmatrix} + \begin{pmatrix} \frac{u_{in}}{L} \\ 0 \end{pmatrix} u, \tag{4}$$

The state space equation in Boost mode with a continuous inductance current is as follows:

$$\begin{pmatrix} \frac{di_L}{dt} \\ \frac{du_o}{dt} \end{pmatrix} = \begin{pmatrix} 0 & -\frac{1}{L} \\ \frac{1}{C_2} & -\frac{1}{C_2 R} \end{pmatrix} \begin{pmatrix} i_L \\ u_o \end{pmatrix} + \begin{pmatrix} \frac{u_o}{L} \\ -\frac{i_L}{C_2} \end{pmatrix} u + \begin{pmatrix} \frac{u_{in}}{L} \\ 0 \end{pmatrix}, \tag{5}$$

The transfer function can be derived through Laplace transform, using the average state space equation:

$$\begin{cases} G_{id}(s) = \frac{v_{in}(1+R_L Cs)}{R_L+Ls+R_L CLs^2} \\ G_{vd}(s) = \frac{R}{1+R_L Cs} \end{cases}, \tag{6}$$

This article first analyzes the Buck pattern.

$$P = \frac{{u_o}^2}{R_L}, \tag{7}$$

where d is the duty cycle of the converter, and T is the switching cycle. A dynamic model of the Buck converter was established using the state-space averaging method. By substituting Equation (7) into (4) and linearizing it, we obtain the following:

$$\begin{cases} \frac{di_L}{dt} = \frac{v_{in}}{L}u - \frac{v_o}{L} \\ \frac{du_o}{dt} = \frac{i_L}{C} - \frac{P}{Cv_o} - \frac{v_o}{R_L C} \end{cases}, \tag{8}$$

The voltage tracking error is defined as $x_1 = e = v_o - v_{ref}$, where v_{ref} is the reference voltage. The dynamic model in Equation (8) can be rewritten as follows:

$$\dot{x}_1 = \frac{i_L}{C} - \frac{v_o}{R_L C} - \dot{v}_{ref} + d_1(t), \tag{9}$$

where $d_1(t) = -\frac{P}{Cv_o}$, and another state variable is defined as $x_2 = \frac{i_L}{C} - \frac{v_o}{R_L C}$, so take the derivative of that and obtain the following:

$$\dot{x}_2 = \frac{u}{LC}v_{in0} - \frac{x_1 + v_{ref}}{LC} - \frac{x_2}{R_L C} + d_2(t), \tag{10}$$

where $d_2(t) = \frac{1}{RC}d_1(t) + \frac{v_{in} - v_{in0}}{LC}$, $d_2(t)$ is a more complex form of time varying, consisting of a constant power load and fluctuations in input voltage. The following equation can be obtained by sorting out Equations (9) and (10):

$$\begin{cases} \dot{x}_1 = x_2 + d_1 \\ \dot{x}_2 = \frac{u}{LC}v_{in0} - \frac{x_1+v_{ref}}{LC} - \frac{x_2}{R_L C} + d_2 \end{cases}, \tag{11}$$

3.2. Observer-Based Sliding-Mode Control (SMC) Design

This section's control objective is to design a generalized proportional integral observer to estimate the time-varying disturbance and update it into the controller in real time, so as to effectively suppress the influence of the disturbance and improve the anti-jamming performance of the whole system.

Sliding-mode control uses the designed control function to make the motion state of the system in "sliding mode", which is a discontinuous switching control, so it is also called sliding-mode variable structure control. The basic idea of the sliding-mode variable structure control theory is to consider a nonlinear system and assume that there is a phase plane, which is called the sliding-mode surface, and a point in the plane is called a balance point. Using this sliding surface as a reference path, through effective design, the state variable of the system, i.e., the controlled trajectory, is attracted to slide along the set trajectory of the reference path and converges to the equilibrium point, regardless of the initial state of the system [32–34].

The sliding-mode control needs to meet the following three basic conditions: existence, accessibility, and stability. Existence refers to the existence of a sliding surface in a system. Reachability refers to the ability of points outside the sliding surface of a system state to move to the sliding surface within a finite time [35–39]. Stability refers to the ultimate stability of the system state under model control.

For the sliding-mode control, the first step is to determine the sliding surface and select the appropriate sliding-surface function, s(x). Under the action of different control functions, the trajectory of the system moves differently. As shown in Figure 10a, by designing appropriate control functions, the system can start from an arbitrary initial point, x_0, in the state space and reach the switching surface (as shown in the $x_0 \rightarrow A$ section) in a finite time. This process is called the approach section. Once the system trajectory reaches the switching surface, it stays on it and continues to move, and this is called the sliding-mode section (as shown in section A→O). The state of the system moving on the switching surface is called the sliding mode. Since the switching surface is designed according to the expected moving target of the system, no matter how the external parameters change, the system trajectory will eventually reach the preset value on the switching surface [40–43].

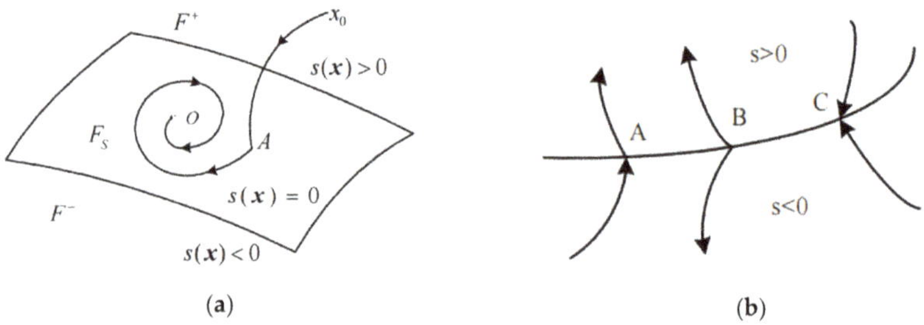

Figure 10. System motion under sliding-mode control: (**a**) the motion trajectory of the system on the sliding surface and (**b**) three types of points on the sliding surface.

In the state space, take s(x) = 0 as the sliding surface, which represents the state, as shown in Figure 10b. The space is divided into two: s(x) > 0 and s(x) < 0. The motion points on the sliding surface can be divided into three categories:

- Usually Point A: After the system moves near the sliding surface, it will pass through this point;
- Starting Point B: After the system motion point reaches the vicinity of the sliding surface, it leaves from both sides of that point;
- Termination Point C: The system moves towards this point from the upper and lower sides of the sliding surface.

In the study of sliding-mode control, the first two types of motion points have little significance for system control and are generally ignored. If a certain area on the sliding-mode surface is all termination points, it means that once the system state moves near that area, it will be attracted to the area, and this area is therefore called the "sliding mode area". Due to the fact that all points on the sliding-mode area are termination points, when the system moves near the sliding surface, there will inevitably be $\lim_{s \to 0} s\dot{s} < 0$ [44–46].

The specific system control is shown in Figure 11. The system block diagram includes four parts: two generalized proportional integral observers, a sliding-mode controller, a pulse width modulator (PWM), and a three-phase interleaved parallel DC/DC converter. The system works as follows: Firstly, two generalized proportional integral observers are constructed based on the feedback values of inductance current and output voltage, and the matched and unmatched disturbances are estimated, respectively. Then, a sliding-mode controller is designed using the estimated values. The controller is compared with the sawblade wave to obtain a PWM wave, and the switching tube of the DC step-down converter is controlled by the PWM wave. The converter can output the desired voltage stably.

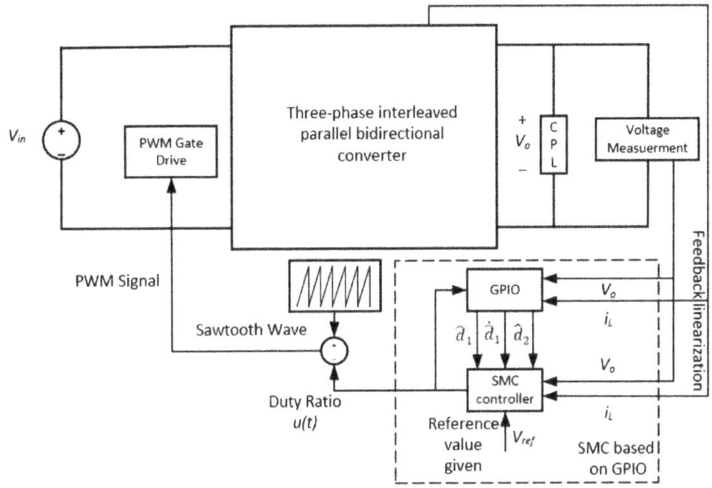

Figure 11. Control block diagram of the whole system.

3.3. GPI Observer Design

The voltage tracking accuracy of the DC-DC converter system will be affected by disturbances, such as input voltage fluctuation, parameter uncertainty, load resistance disturbance, etc. An effective method to eliminate these disturbances is to introduce disturbance estimation to compensate accurately. For the Buck converter, two generalized integral observers are designed to estimate the matched disturbance and the unmatched disturbance, respectively, and the disturbance estimation is introduced into the design of the control law to compensate for the influence of these disturbances and uncertainties. The specific design of the two generalized proportional integral observers is as follows. The GPI observer of CPL perturbation based on the Buck converter can be constructed as follows:

$$\begin{cases} \dot{\hat{x}}_1 = x_2 + \hat{d}_1 + h_{11}(x_1 - \hat{x}_1) \\ \dot{\hat{d}}_1 = \hat{\dot{d}}_1 + h_{12}(x_1 - \hat{x}_1) \\ \cdots \\ \dot{\hat{d}}_1{}^{(n-1)} = \hat{d}_1{}^{(n)} + h_{1n}(x_1 - \hat{x}_1) \\ \dot{\hat{d}}_1{}^{(n)} = h_{1(n+1)}(x_1 - \hat{x}_1) \end{cases}, \tag{12}$$

where $\hat{d}_1{}^{(n)}$ is an estimated value of the nth order derivative of d_1, and h_{1i} (i = 1, 2, ..., n + 1) represents the parameters to be determined.

To estimate the input voltage disturbance, $d_2(t)$, another GPI observer is constructed:

$$\begin{cases} \dot{\hat{x}}_2 = \frac{u}{LC}V_{in0} - \frac{x_1 + V_{ref}}{LC} - \frac{x_2}{R_L C} + z_2 + h_{21}(x_2 - \hat{x}_2) \\ \dot{\hat{d}}_2 = \dot{d}_2 + h_{22}(x_2 - \hat{x}_2) \\ \cdots \\ \dot{\hat{d}}_2{}^{(m-1)} = \hat{d}_2{}^{(m)} + h_{2m}(x_2 - \hat{x}_2) \\ \dot{\hat{d}}_2{}^{(m)} = h_{2(m+1)}(x_2 - \hat{x}_2) \end{cases}, \tag{13}$$

where $\hat{d}_2{}^{(m)}$ is an estimated value of the nth order derivative of d_2, and h_{2j} (j = 1, 2, ..., m + 1) represents the parameters to be determined.

According to Equations (9) and (10), the uncertainties (i = 1, 2) are related to the power of constant power load, so from a practical point of view, their values and derivatives should be bounded.

In a steady state, the power of the constant power load is considered constant. Therefore, the following assumptions can be made:

The uncertain variables, d_i and $\dot{d}_i$, of the system (i = 1, 2) meet the following two conditions [47]:

$$\begin{cases} d_i(t) \in L_\infty, \dot{d}_i(t) \in L_\infty \\ \lim\limits_{t\to\infty} d_i{}^{(n)} = 0 \end{cases}, \tag{14}$$

According to Equation (8), the uncertain term is defined by the following:

$$\begin{cases} \hat{d}_1 = l_1(x_1 - p_1) \\ \dot{p}_1 = x_2 + \hat{d}_1 \end{cases}, \tag{15}$$

where p_1 is the auxiliary state of the observer, and l_1 is a normal number that is expressed as the observer gain. Similarly, the uncertainty term is given by the following:

$$\begin{cases} \hat{d}_2 = l_2(x_2 - p_2) \\ \dot{p}_2 = \frac{u}{LC} v_{in0} - \frac{x_1 + v_{ref}}{LC} - \frac{x_2}{R_L C} + \hat{d}_2 \end{cases}, \tag{16}$$

where p_2 is the observer's auxiliary state, and l_2 is a normal number that is expressed as the observer gain.

Based on Equations (12) and (13), the standard model and observer estimate of the load power can be provided according to the sliding-mode control design of the proposed composite controller. Them we take the switching function of the system as follows:

$$\begin{cases} s = k_1 x_1 + x_2 + \hat{d}_1 \\ \dot{s} = k_1(x_2 + d_1) + \dot{x}_2 + \dot{\hat{d}}_1 \end{cases}, \tag{17}$$

where $k_1 > 0$ is a parameter to be selected.

As a high-order sliding-mode algorithm, the realization of the high-order sliding-mode algorithm usually requires the derivative of sliding-mode variables, while the super-distortion algorithm is a second-order sliding-mode algorithm in nature, so its realization does not require the derivative of sliding-mode variables, thus simplifying the controller structure. Through the design of the control rate, the sliding-mode variable structure rapidly converges within a limited time [30].

The general form of the super-twisting algorithm is as follows:

$$\begin{cases} \frac{da_1}{dt} = -\lambda |a_1|^{\frac{1}{2}} \mathrm{sign}(a_1) + a_2 + \rho_1 \\ \frac{da_2}{dt} = -c\,\mathrm{sign}(a_1) + \rho_2 \end{cases}, \tag{18}$$

In Equation (16), a_1 and a_2 are the state variables; λ and c are the positive constants; and ρ_1 and ρ_2 are the disturbance quantities.

In order to weaken chattering, saturation function is often used to replace the sign function [48]. The form of saturation function is as follows:

$$\mathrm{sat}(s, \delta) = \begin{cases} \frac{s}{\delta} & |s| \le \delta \\ \mathrm{sign}(s) & |s| > \delta \end{cases}, \tag{19}$$

By combining Equations (18) and (19), the form of the super-twisting algorithm becomes the following:

$$\begin{cases} u_{sta} = -\alpha |s|^{\frac{1}{2}} \mathrm{sat}(s) + \omega \\ \frac{d\omega}{dt} = -\beta\,\mathrm{sat}(s) \end{cases}, \tag{20}$$

When the super-twisting algorithm is used to design the sliding-mode control function, let $\omega = a_2 + \rho_1$, and α, β are the parameters to adjust the dynamic velocity and set the steady-state error, respectively. Among them, the parameters α, β affect the convergence rate of the sliding-mode surface. In general, the system can reach the sliding-mode surface faster by taking a larger value and a smaller value.

When the first derivative of the sliding-mode surface is zero, the switching signal, u, is equivalent to a continuous value, u_{eq}.

$$u_{eq} = -\frac{LC}{v_{in0}}\left[k_1\left(x_2 + \hat{d}_1\right) - \frac{x_1 + v_{ref}}{LC} - \frac{x_2}{R_L C} + \hat{d}_2 + \dot{\hat{d}}_1 + \eta \mathrm{sat}(s)\right], \tag{21}$$

where the control parameter, $k_1 > 0$, and the switching gain, $\eta > 0$, are the parameters to be designed. Then, the total switch signal, u, is as follows:

$$u = u_{eq} + u_{sta}, \tag{22}$$

where u_{eq} is used to ensure that the trajectory of the system phase is maintained on the sliding-mode surface, and u_{sta} is used to overcome the disturbance effect and to ensure the robustness of the system. The following proves the existence and accessibility of the switching surface, and the Lyapunov function is used to analyze the switching function, so as to enable the stability of the controller's control voltage and the state curve to quickly converge to the sliding surface.

4. Controller Stability Analysis

Theorem 1. *Consider a DC-DC converter system with both CPL and supply voltage perturbations and combine Equation (14). Under the proposed control law (21), the effect of the time-varying perturbation is removed from the output voltage channel, provided that the switching gain,* $\eta > (k_1 e_1{}^* + e_2{}^* + e_3{}^*)$*, and the observer parameters in selected Equations (12) and (13) are appropriate, such that (25) is the Hurwitz matrix.*

Proof of Theorem 1. For the GPI observer, the estimated error is defined as $e_1 = d_1 - \hat{d}_1$, $e_2 = d_2 - \hat{d}_2$, and $e_3 = \dot{d}_1 - \dot{\hat{d}}_1$. The upper bound of the estimation error is defined as $e_i{}^*$, (i = 1, 2, 3). Then, the estimated error of the observer can be expressed as follows:

$$\begin{cases} e_{11} = x_1 - \hat{x}_1 \\ e_{12} = d_1 - \hat{d}_1 \\ \quad \cdots \\ e_{1n} = d_1{}^{(n-1)} - \hat{d}_1{}^{(n-1)} \\ e_{1(n+1)} = d_1{}^{(n)} - \hat{d}_1{}^{(n)} \end{cases}, \tag{23}$$

$$\begin{cases} e_{21} = x_2 - \hat{x}_2 \\ e_{22} = d_2 - \hat{d}_2 \\ \quad \cdots \\ e_{2m} = d_2{}^{(m-1)} - \hat{d}_1{}^{(m-1)} \\ e_{2(m+1)} = d_2{}^{(m)} - \hat{d}_1{}^{(m)} \end{cases}, \tag{24}$$

where e = $[e_{11}\ e_{12} \ldots e_{1(n+1)}\ e_{21}\ e_{22} \ldots e_{2(m+1)}]^T$ takes the derivative of the estimate error. Then, the observer error can be dynamically expressed as follows:

$$\dot{e} = H_e e + \dot{d}, \tag{25}$$

where $\dot{d} = \left[0\,0 \ldots d_1{}^{(i)}\,0\,0 \ldots d_2{}^{(j)}\right]^T$

$$H_e = \begin{bmatrix} H_{e1} & 0 \\ 0 & H_{e2} \end{bmatrix},$$

$$H_{e1} = \begin{bmatrix} -h_{11} & 1 & 0 & \ldots & 0 \\ -h_{12} & 0 & 1 & \ldots & 0 \\ \ldots & & & & \\ -h_{1n} & 0 & 0 & \ldots & 1 \\ -h_{1(n+1)} & 0 & 0 & 0 & 0 \end{bmatrix},$$

$$H_{e2} = \begin{bmatrix} -h_{21} & 1 & 0 & \ldots & 0 \\ -h_{22} & 0 & 1 & \ldots & 0 \\ \ldots & & & & \\ -h_{2m} & 0 & 0 & \ldots & 1 \\ -h_{2(m+1)} & 0 & 0 & 0 & 0 \end{bmatrix},$$

By selecting parameters correctly in the GPI observer, we can get the Hurwitz stability matrix; that is, the state matrix of the system is the Hurwitz matrix. Then, the error dynamic is asymptotically stable, which means the following:

$$\lim_{t \to \infty} e_i = d_i - \hat{d}_i = 0, (i = 1, 2), \tag{26}$$

Take the Lyapunov function as follows:

$$V = \frac{1}{2}s^2, \tag{27}$$

$$\dot{V} = s\dot{s}, \tag{28}$$

Substituting (17) and (22) into (28) gives the following:

$$\begin{aligned} \dot{V} &= -\eta|s| + (k_1e_1 + e_2 + e_3)s + \tfrac{V_{in0}}{LC}u_{sta}s \\ &\leq -[\eta - (k_1e_1 + e_2 + e_3)]|s| + \tfrac{V_{in0}}{LC}u_{sta}s \\ &\leq -\sqrt{2}[\eta - (k_1e_1{}^* + e_2{}^* + e_3{}^*)]V^{\frac{1}{2}} + \tfrac{V_{in0}}{LC}u_{sta}s \end{aligned} \tag{29}$$

where the coefficient $\alpha > 0, \beta > 0$ of u_{sta} can be seen from Equation (17); when $s < 0$, $u_{sta} > 0$, and when $s > 0$, $u_{sta} < 0$. Then, when the switching gain $\eta > (k_1e_1{}^* + e_2{}^* + e_3{}^*)$ meets the condition, it is satisfied, $\dot{V} < 0$. According to Lyapunov's sliding-mode reachability condition, the system can reach the designed sliding-mode surface in a finite time. The system state will reach the defined sliding surface, $s = 0$, in a finite time.

By integrating Equations (11) and (17), we obtain the following:

$$\dot{x}_1 = -k_1x_1 - \left(d_1 - \hat{d}_1\right), \tag{30}$$

According to Reference [18], if the following system exists,

$$\dot{x} = f(t, x, u), x \in R, u \in R, \tag{31}$$

when the system input reaches stability, if the input signal meets $\lim_{t \to \infty} u = 0$, then the states satisfy $\lim_{t \to \infty} x = 0$. According to Equations (26) and (17), $\lim_{t \to \infty} e_1 = 0, k_1 > 0$. It is easy to arrive at the conclusion that the voltage tracking error converges asymptotically to zero along the sliding surface, thus completing the proof. Similarly, the Boost mode is a similar proof. □

5. Simulation Results and Analysis

In order to verify whether the proposed method has a faster convergence rate and load-resistance performance, for Buck converters with matched and unmatched disturbances, the proposed method was simulated and compared with the control methods in other works from the literature under the condition that the circuit parameters are the same. The results show that the proposed method can effectively suppress the influence of perturbation and ensure the stability of output voltage. It can improve the anti-interference performance of the whole system.

5.1. Simulation Model Construction

The simulations are conducted in Simulink, using GPI observers to estimate the state and lumped disturbances of the system. Then, the estimated values of the observers are introduced into the controller for feedforward compensation, achieving fast and accurate tracking of the output voltage to the reference voltage. Select the input and output voltage, use the inductance current as feedback variables, and perform a high-order sliding-mode control algorithm to limit the calculated amplitude to prevent the switch from being in a continuous on or off state. Build a simulation model of a three-phase interleaved parallel bidirectional DC/DC converter in Simulink and write a sliding-mode control algorithm through the MATLAB function. The circuit uses ideal components, and the simulation diagram is shown in Figure 12.

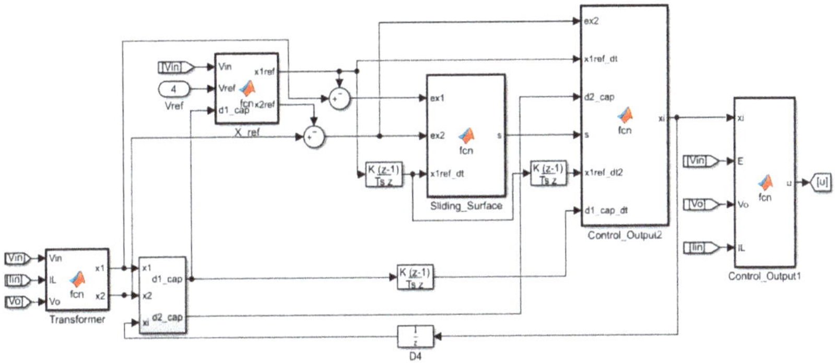

Figure 12. Sliding-mode control simulation model.

5.2. Set System Parameters

The control objective of this simulation experiment is, under the same circuit system, sequentially use two methods to control, observe, and analyze the value of each state variable with time from the figure, compare the transient performance and resistance of the system under the control of the two methods' interference performance. In order to ensure the fairness of the comparison, the two methods should be simulated and compared under the same converter circuit system.

The effectiveness of the proposed control strategy is verified by the following simulation. The parameter values of the simulation experiment are shown in Table 3. To further illustrate the advantages, this paper compares the NDO-Integral SMC scheme with the GPIO-SMC scheme. A MATLAB simulation is used to compare the two methods:

Table 3. Parameters of simulation system.

Descriptions	Parameters	Values
Input voltage	V_{in}	20 V
Reference voltage	V_{ref}	5 V
Inductance	L	1×10^{-4} H
Capacitance	C	1×10^{-3} F
Switching frequency	f_s	20 kHz
The power of CPL	P	6 W
Load resistance	R_L	50 Ω

In Method 2, the switching function of the system can be designed as follows:

$$\begin{cases} s = k_1x_1 + x_2 + d_1 + k_2\int_0^t x_1 dt \\ \dot{s} = k_1(x_2 + d_1) + \dot{x}_2 + \dot{d}_1 + k_2x_1 \end{cases}, \tag{32}$$

Then, the switching signal is as follows:

$$u = -\frac{LC}{v_{in0}}\left[k_1\left(x_2 + \hat{d}_1\right) - \frac{x_1 + v_{ref}}{LC} - \frac{x_2}{R_LC} + \hat{d}_2 + \dot{d}_1 + k_2x_1 + \eta sat(s)\right], \tag{33}$$

Since the observers used for load estimation provide composite controller reference values, it is necessary to first design their parameters to ensure that the observers can accurately estimate the power of the load. The size of gain l_1 and l_2 will affect the accuracy of the observer in tracking load power. In order to investigate the tracking effect of the l_1 size on the observer's load power, the parameters k_1, k_2, and l_2 are first fixed to 200, 1000, and 200, respectively, and the observer's observation effect is observed by setting different values. As shown in Figure 13a, when the CPL power jumps from 5 W to 10 W, the observer can quickly track the fluctuation of load power as the gain l_1 increases. The gain l_1 can be set to 1000.

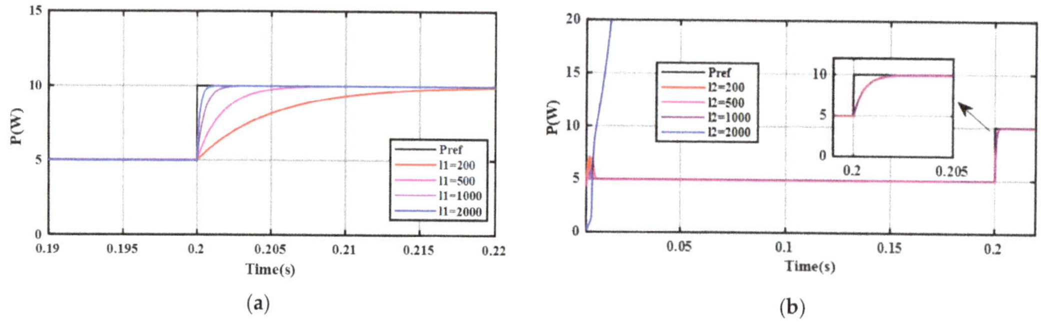

Figure 13. Load estimation performance of l_1 and l_2 under different values: (**a**) load estimation response with different values of l_1 and (**b**) load estimation response with different values of l_2.

The same settings $k_1 = 200$ and $k_2 = 1000$, fix l_1 at 1000, and different values of l_2 will be set to observe the impact of their magnitude on the performance of load power fluctuation estimation. From Figure 13b, we can see that when $l_2 < 2000$, as l_2 increases, the observer's tracking effect on load power fluctuations remains almost unchanged. However, when $l_2 < 2000$, the observer was unable to accurately estimate the fluctuation of load power. To prevent the observer from being unable to track load-power fluctuations, the gain l_2 was set to 200.

Afterwards, it is necessary to adjust the relevant parameters of the composite controller. To obtain the impact of parameter changes on the DC bus voltage, first fix k_1 at 200 and observe the fluctuation of DC bus voltage by setting different values of k_2, as shown in the Figure 14. From Figure 14a, it can be seen that, as k_2 increases, the time it takes for the bus voltage to recover to its steady-state value becomes shorter. However, when k_2 exceeds 1000, the time it takes for the bus voltage to recover to the stable value does not change, so k_2 will be set at 1000.

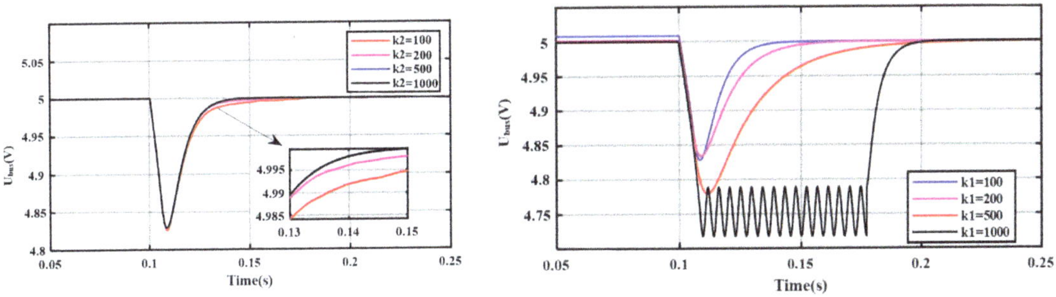

(**a**) (**b**)

Figure 14. Load estimation performance of k_1 and k_2 under different values: (**a**) voltage tracking response with different values of k_2 and (**b**) voltage tracking response with different values of k_1.

After obtaining l_1, l_2, and k_2, analyze the impact of their values on the DC bus voltage by setting different k_1 values. From Figure 14b, it can be seen that as k_1 gradually increases, the fluctuation amplitude of the DC bus voltage increases. When k_1 = 1000, the DC bus voltage even exhibits oscillation, so k_1 can be set at 100.

5.3. Analysis of Simulation Examples

5.3.1. Keep the Constant Power Load and Change the Load Resistance

The load resistance experiences a sudden change during the simulation; that is, the resistance drops from 20 Ω to 10 Ω at 0.04 s and then rises to 20 Ω at 0.08 s.

For three-phase interleaved parallel converter systems with matched and unmatched disturbances, as shown in Figure 15a, both methods can make the output voltage rapidly approach the set reference value, and there is no steady-state error in the output voltage. However, it is clear from Figure 15a that the output voltage response time of Method 1 and Method 2 is about 0.005 s and 0.009 s, respectively, so Method 1 has a faster convergence speed. As shown in Figure 15b, although the overshoot of the inductance current in Method 1 is greater than that in Method 2, the rise time and adjustment time of Method 1 are about 0.002 s and 0.004 s respectively. Both are less than the rise time and adjustment time of Method 2. As shown in the controller output in Figure 15c, the response speed of controller output, u, in Method 1 is higher than that in Method 2, and the convergence time of the controller is shorter.

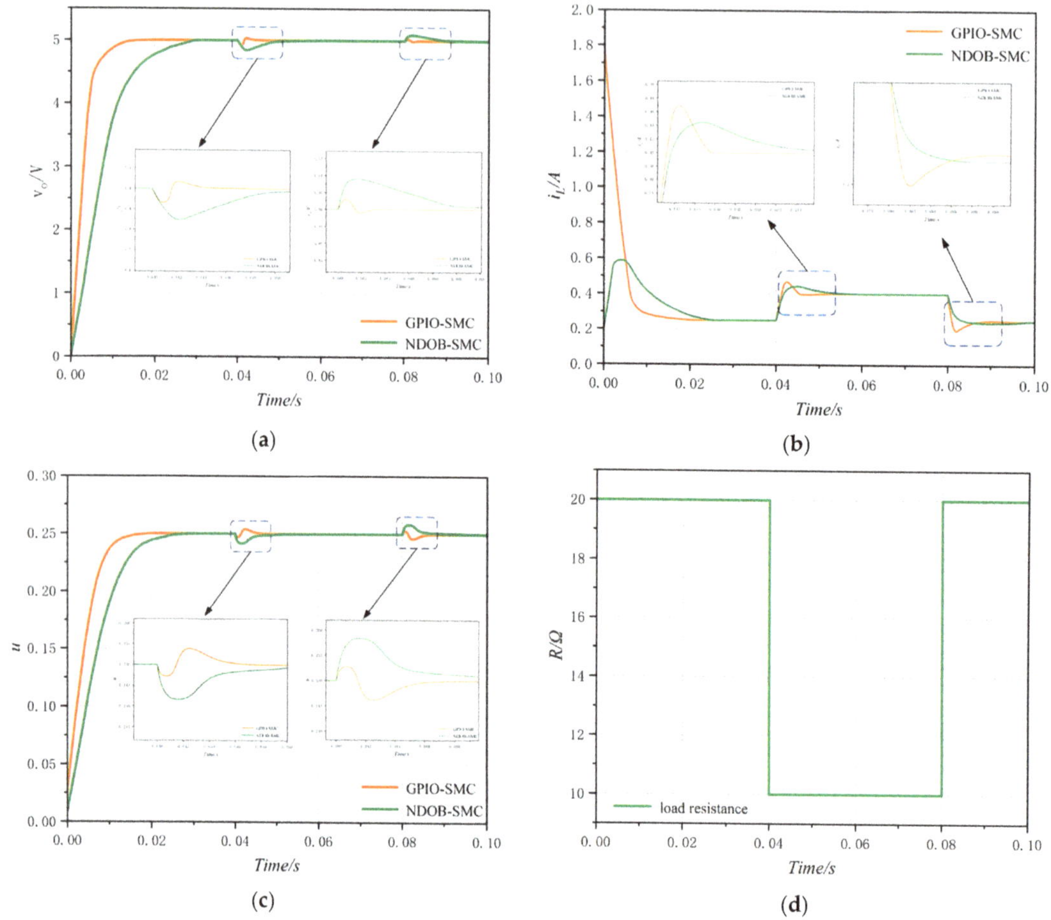

Figure 15. Comparison of output responses under load variation: (**a**) output voltage, (**b**) inductive current, (**c**) control input, and (**d**) load resistance.

Figure 16a,b capture the estimation effect of the GPI observer on the unmatched disturbance and the matched disturbance, respectively. From the comparison results in Table 4, it can be seen that at the moment 0.04 s, the resistance drops from 20 Ω to 10 Ω, and the estimated convergence time of the observer's sum is 0.014 s and 0.006 s, respectively. The resistance increases from 10 Ω to 20 Ω at 0.08 s, and the convergence time of the observer is 0.012 s and 0.007 s, respectively. It can be seen that when the load resistance changes abruptly, the GPI observer can quickly track the value of the disturbance and make an accurate estimation of and match the disturbance, thus further proving that the GPI observer has strong adaptability to the resistance load change in the system; that is, the observer has a fast response speed and good accuracy when estimating the system interference.

Table 4. Comparison of dynamic response of the proposed controller under changes in load resistance.

P = 6 W, V_{ref} = 5 V, R_L = 10~20 Ω		
Method	GPIO	NDOB
Input voltage	20 V	20 V
Setting time	0.04 s~0.08 s	0.04 s~0.08 s
Voltage recovery time	0.005 s	0.009 s
Current recovery time	0.002 s	0.004 s
Voltage overshoot	0.056 V	0.144 V
Current overshoot	0.064 A	0.102 A

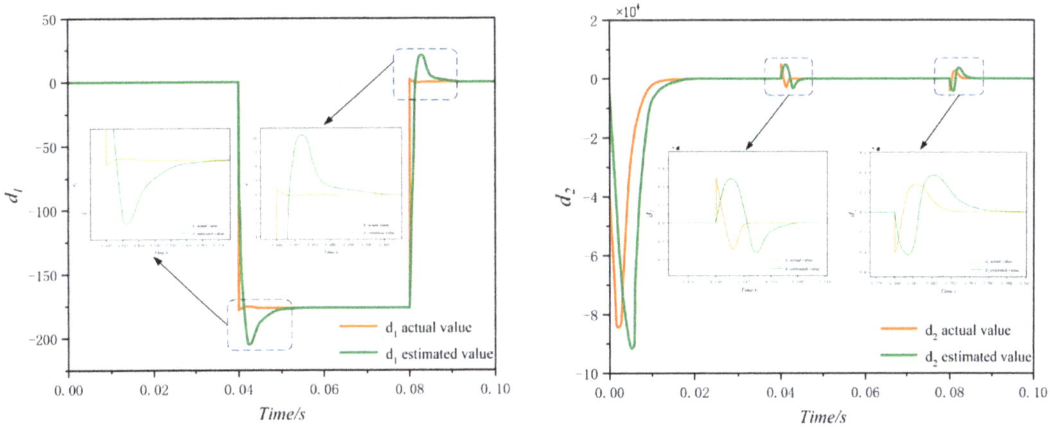

(**a**) (**b**)

Figure 16. Estimate of the observer under load variation: (**a**) estimation of unmatched perturbations, d_1; and (**b**) estimation of matching perturbations, d_2.

5.3.2. Keep the Load Resistance and Change the Constant Power Load

The constant load power was changed during the simulation; that is, CPL rose from 6 W to 8 W at 0.04 s and then dropped to 6 W at 0.08 s.

As can be seen from the output voltage response curve in Figure 17a, Method 1 not only minimizes the voltage drop but also minimizes the recovery time. Although Method 2 can also keep the system stable, it sacrifices control performance to some extent. From the inductor current curve in Figure 17b, it can be seen that the recovery speed of the Method 1 curve is significantly faster than that of Method 2. As shown in the controller output of Figure 17c, the response and convergence speed of the controller output, u, of Method 1 are higher than those of Method 2.

As shown in Figure 18, the convergence time of the observer pairs and estimates are 0.015 s and 0.007 s, respectively. At 0.08 s, the power of the constant power load decreases from 8 W to 6 W, and the convergence time of the observer is 0.014 s and 0.007 s, respectively. The GPI observer can quickly track the value of the disturbance and make accurate estimates of and match the disturbance.

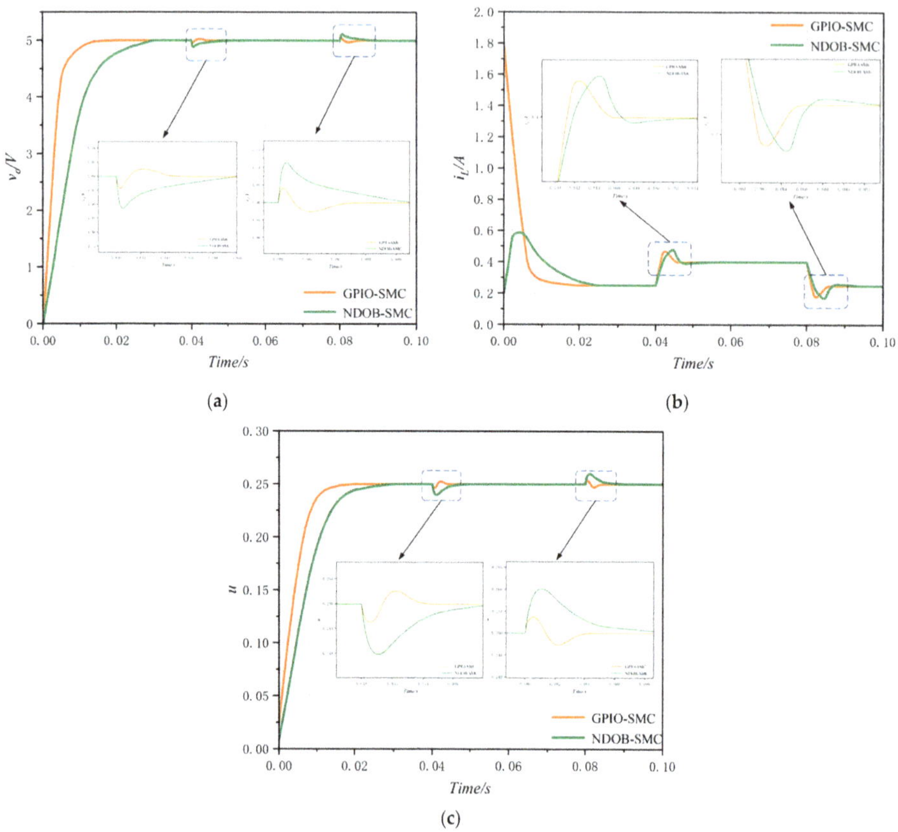

Figure 17. Comparison of output responses under CPL variation: (**a**) output voltage, (**b**) inductive current, and (**c**) control input.

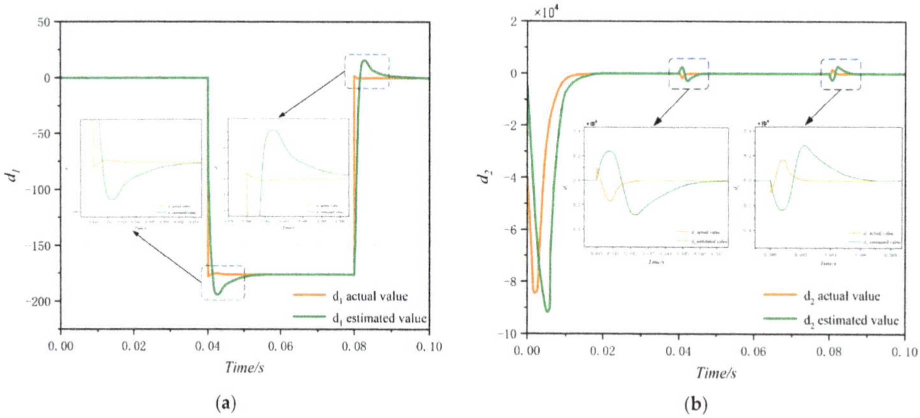

Figure 18. Estimate of the observer under CPL variation: (**a**) estimation of unmatched perturbations, d_1; and (**b**) estimation of matching perturbations, d_2.

In short, from the comparison results in Table 5, it can be seen that the converter under the control of Method 1 has a faster output voltage response, better transient performance, and greater anti-disturbance ability.

Table 5. Comparison of dynamic response of the proposed controller under CPL changes.

$P = 6\sim8$ W, $V_{ref} = 5$ V, $R_L = 20\ \Omega$		
Method	GPIO	NDOB
Input voltage	20 V	20 V
Setting time	0.04 s~0.08 s	0.04 s~0.08 s
Voltage recovery time	0.007 s	0.01 s
Current recovery time	0.005 s	0.009 s
Voltage overshoot	0.025 V	0.108 V
Current overshoot	0.069 A	0.078 A

6. Conclusions

Aiming at the precise power control of three-phase interleaved bidirectional converters in a DC microgrid under the variation of supply voltage and constant power load, a GPI observer sliding-mode control method based on a super-twisting algorithm was proposed. The proposed high-order sliding-mode control strategy can stabilize the output voltage at the expected CPL power value and generate errors between the inductance current and the output voltage of the converter on the sliding-mode surface. By combining the interference estimations, smaller switching gains can be achieved without sacrificing interference suppression, thus ensuring the stability of the output voltage.

Finally, the effectiveness of the control algorithm was verified by comparing the simulation results of the proposed control method with the classical method. The transient recovery index and anti-interference capability of the controller were further improved. The results show that the proposed method can ensure that the output voltage of the system converges to the reference voltage, and this controller further improves the transient recovery index, anti-interference ability, and vibration-reduction performance of the system. The advantage of the controller is that it can quickly improve the rate of convergence of the system state, which is the disadvantage of the traditional observer.

Author Contributions: Y.J. and D.W. described the proposed framework and wrote the whole manuscript; Y.L. implemented the simulation experiments; G.S. and K.S. collected data; Y.J. and Y.N. revised the manuscript. All authors have read and agreed to the published version of the manuscript.

Funding: This research is supported in part by the Natural Science Foundation of China, under Grant 52077027 Study, and in part by the Liaoning Province Science and Technology Major Project No. 2022021000014.

Institutional Review Board Statement: Not applicable.

Informed Consent Statement: Not applicable.

Data Availability Statement: The data presented in this study are available upon request from the corresponding author.

Acknowledgments: The authors thank the chief editor and the reviewers for their valuable comments on how to improve the manuscript.

Conflicts of Interest: The authors declare no conflict of interest.

References

1. Emadi, A.; Khaligh, A.; Rivetta, C.H.; Williamson, G.A. Constant power loads and negative impedance instability in automotive systems: Definition, modeling, stability, and control of power electronic converters and motor drives. *IEEE Trans. Veh. Technol.* **2006**, *55*, 1112–1125. [CrossRef]
2. Wang, Z.; Li, S.; Yang, J.; Li, Q. Current sensorless finite-time control for Buck converters with time-varying disturbances. *Control Eng. Pract.* **2018**, *77*, 127–137. [CrossRef]

3. Yang, J.; Cui, H.; Li, S.; Zolotas, A. Optimized active disturbance rejection control for DC-DC Buck converters with uncertainties using a reduced-order GPI observer. *IEEE Trans. Circuits Syst. I Reg. Pap.* **2018**, *65*, 832–841. [CrossRef]
4. Zolfi, P.; Vahid, S.; EL-Refaie, A. Development of A Family of High Voltage Gain Step-Up Multi-Port DC-DC Converters for Fuel Cell-based Hybrid Vehicular Power Systems. In Proceedings of the 2022 24th European Conference on Power Electronics and Applications (EPE'22 ECCE Europe), Hanover, Germany, 5–9 September 2022; pp. 1–11.
5. Zheng, C.; Dragičević, T.; Zhang, J.; Chen, R.; Blaabjerg, F. Composite Robust Quasi-Sliding Mode Control of DC–DC Buck Converter with Constant Power Loads. *IEEE J. Emerg. Sel. Top. Power Electron.* **2021**, *9*, 1455–1464. [CrossRef]
6. Singh, S.; Gautam, A.R.; Fulwani, D. Constant power loads and their effects in DC distributed power systems: A review. *Renew. Sustain. Energy Rev.* **2017**, *72*, 407–421. [CrossRef]
7. Ding, S.; Zheng, W.X.; Sun, J.; Wang, J. Second-order sliding-mode controller design and its implementation for buck converters. *IEEE Trans. Ind. Inform.* **2018**, *14*, 1990–2000. [CrossRef]
8. Lin, X.; Liu, J.; Liu, F.; Liu, Z.; Gao, Y.; Sun, G. Fractional-Order Sliding Mode Approach of Buck Converters with Mismatched Disturbances. *IEEE Trans. Circuits Syst. I Regul. Pap.* **2021**, *68*, 3890–3900. [CrossRef]
9. Pawar, S.N.; Chile, R.H.; Patre, B.M. Design of generalized extended state observer based control for nonlinear systems with matched and mismatched uncertainties. In Proceedings of the 2017 Indian Control Conference (ICC), Guwahati, India, 4–6 January 2017.
10. Gao, Z.; Liu, X.; Chen, Z.Q. Unknown input observer—Based robust fault estimation for systems corrupted by partially decoupled disturbances. *IEEE Trans. Ind. Electron.* **2016**, *63*, 2537–2547. [CrossRef]
11. Wang, Z.; Li, S.; Wang, J.; Li, Q. Robust control for disturbed Buck converters based on two GPI observers. *Control Eng. Pract.* **2017**, *66*, 13–22. [CrossRef]
12. Wang, Z.; Li, S.; Wang, J.; Li, Q. Generalized proportional integral observer based backstepping control for DC-DC Buck converters with mismatched disturbances. In Proceedings of the IEEE International Conference on Industrial Technology, Taipei, Taiwan, 14–17 March 2016; pp. 1783–1789.
13. Jeung, Y.-C.; Lee, D.-C.; Dragičević, T.; Blaabjerg, F. Design of Passivity-Based Damping Controller for Suppressing Power Oscillations in DC Microgrids. *IEEE Trans. Power Electron.* **2021**, *36*, 4016–4028. [CrossRef]
14. Jianguo, G.; Ningbo, L.; Shengjiang, Y.; Jun, Z. A New Adaptive Sliding Mode Control for A Mismatched Second-order System. In Proceedings of the 2020 Chinese Control and Decision Conference (CCDC), Hefei, China, 22–24 August 2020.
15. Shtessel, Y.; Edwards, C.; Fridman, L.; Levant, A. *Sliding Mode Control and Observation*; Springer: New York, NY, USA, 2013.
16. Chen, W.-H.; Yang, J.; Guo, L.; Li, S. Disturbance-observer-based control and related methods-An overview. *IEEE Trans. Ind. Electron.* **2016**, *63*, 1083–1095. [CrossRef]
17. Xu, X.; Liu, Q.; Zhang, C.; Zeng, Z. Prescribed performance controller design for DC converter system with constant power loads in DC microgrid. *IEEE Trans. Syst. Man Cybern. Syst.* **2018**, *50*, 4339–4348. [CrossRef]
18. Xu, Q.; Zhang, C.; Wen, C.; Wang, P. A novel composite nonlinear controller for stabilization of constant power load in DC microgrid. *IEEE Trans. Smart Grid* **2019**, *10*, 752–761. [CrossRef]
19. Wang, H.; Han, M.; Han, R.; Guerrero, J.M.; Vasquez, J.C. A Decentralized Current-Sharing Controller Endows Fast Transient Response to Parallel DC–DC Converters. *IEEE Trans. Power Electron.* **2018**, *33*, 4362–4372. [CrossRef]
20. Zhang, B. *Research on Equal Current Sharing Parallel Connection Technology of Multiple Converters in Ship DC Microgrid*; Dalian Maritime University: Dalian, China, 2020.
21. Zhang, Q.; Zhuang, X.; Liu, Y.; Guo, H.; Wang, C. A DC Microsource Parallel Current Sharing Control Strategy Based on Line Impedance Compensation. *Grid Technol.* **2021**, *45*, 1912–1921.
22. Foley, R.F.; Kavanagh, R.C.; Egan, M.G. Sensorless Current Estimation and Sharing in Multiphase Buck Converters. *IEEE Trans. Power Electron.* **2012**, *27*, 2936–2946. [CrossRef]
23. Xia, J.; Li, Z.; Gao, X.; Guo, Y.; Zhang, X. Real-Time Sensor Fault Identification and Remediation for Single-Phase Grid-Connected Converters Using Hybrid Observers with Unknown Input Adaptation. *IEEE Trans. Ind. Electron.* **2023**, *70*, 2407–2418. [CrossRef]
24. Hao, W.; Lin, X.; Chen, W.; Ji, Y.; Jing, F.; Gao, Y.; Wang, Z.; Liu, J. Disturbance Observer Based Back-Stepping Control for Buck Converters with Unmatched Disturbances: An RBF Neural Network Approach. In Proceedings of the 2022 China Automation Congress (CAC), Xiamen, China, 25–27 November 2022; pp. 3335–3340. [CrossRef]
25. Wu, B.; Yang, J.; Wang, J.; Li, S. Extended state observer based control for DC-DC buck converters subject to mismatched disturbances. In Proceedings of the 33rd Chinese Control Conference, Nanjing, China, 28–30 July 2014; pp. 8080–8085. [CrossRef]
26. Nan, Y.R.; Yang, C.B.; Lv, W.J. ESO based Approach Law Control of Buck Converter. *Comput. Meas. Control* **2020**, *28*, 84–88.
27. Nan, Y.; Chen, S.; Wang, S. Buck type converter approach law control based on GPI observer. *Comput. Meas. Control* **2021**, *29*, 89–94. [CrossRef]
28. Dehri, K.; Nouri, A.S. A discrete repetitive adaptive sliding mode control for DC-DC buck converter. *Proc. Inst. Mech. Eng. I J Syst. Control. Eng.* **2021**, *235*, 1698–1708. [CrossRef]
29. Boukerdja, M.; Chouder, A.; Hassaine, L.; Bouamama, B.O.; Issa, W. H∞ based control of a DC/DC Buck converter feeding a constant power load in uncertain DC microgrid system. *ISA Trans.* **2020**, *105*, 278–295. [CrossRef] [PubMed]
30. Fulwani, D.; Singh, S. *Mitigation of Negative Impedance Instabilities in DC Distribution Systems*; Springer: Singapore, 2017.
31. Utkin, V.; Poznyak, A.; Orlov, Y.; Polyakov, A. Conventional and high order sliding mode control. *J. Frankl. Inst.* **2020**, *357*, 10244–10261. [CrossRef]

32. Young, K.D.; Utkin, V.I.; Ozguner, U. A control engineer's guide to sliding mode control. *IEEE Trans. Control Syst. Technol.* **1999**, *7*, 328–342. [CrossRef]
33. Utkin, V. Sliding mode control. In *Control Systems, Robotics and Automation–Volume XIII: Nonlinear, Distributed, and Time Delay Systems-II*; EOLSS: Abu Dhabi, United Arab Emirates, 2009; p. 130.
34. Şahın, M.E.; Okumuş, H.İ.; Kahvecı, H. Sliding mode control of PV powered DC/DC Buck-Boost converter with digital signal processor. In Proceedings of the 2015 17th European Conference on Power Electronics and Applications (EPE'15 ECCE-Europe), Geneva, Switzerland, 8–10 September 2015; pp. 1–8.
35. Chen, Z.X.; Wang, J.; Ge, L.S.; Jiang, T.; Liu, Y.-F.; Liu, Y.-F. Double Integral Sliding Mode Control of Paralleled DC-DC Converters. In Proceedings of the Conference on Industrial Electronics and Applications (ICIEA), Auckland, New Zealand, 15–17 June 2015.
36. Wang, Y.; Xia, H.; Cao, Y. Voltage Controller of DC-DC Buck Converter Using Terminal Sliding Mode. In Proceedings of the IECON 2015—41st Annual Conference of the IEEE Industrial Electronics Society, Yokohama, Japan, 9–12 November 2015.
37. Fossas, E.; Ras, A. Second-order Sliding-mode Control of a Buck Converter. In Proceedings of the 41st IEEE Conference on Decision and Control, Las Vegas, NV, USA, 10–13 December 2002.
38. Utkin, V.I.; Poznyak, A.S. Adaptive sliding mode control with application to super twist algorithm: Equivalent control method. *Automatica* **2013**, *49*, 39–47. [CrossRef]
39. Zhao, Y.; Qiao, W.; Ha, D. A Sliding-Mode Duty-Ratio Controller for DC/DC Buck Converters with Constant Power Loads. *IEEE Trans. Ind. Appl.* **2014**, *50*, 1448–1458. [CrossRef]
40. Yu, X.; Man, Z. Fast Terminal Sliding-mode Control Design for Nonlinear Dynamical Systems. *IEEE Trans. Circuits Syst. Part I Fundam. Theory Appl.* **2002**, *49*, 261–264.
41. Rui, L.; Yan, D.; Wu, M.; Chai, Y. High-order Sliding-mode Control for DC-DC Converters. In Proceedings of the 2012 7th International Power Electronics and Motion Control Conference (IPEMC 2012), Harbin, China, 2–5 June 2012.
42. Li, S.; Yang, J.; Wu, B.; Li, Q. Finite-Time disturbance observer based nonsingular terminal sliding mode control for PWM-based DC-DC buck converters with mismatched load disturbances. *IET Power Electron.* **2016**, *9*, 1995–2002.
43. Zaman, H.; Zheng, X.; Khan, S.; Ali, H.; Wu, X. Hysteresis Modulation-based Sliding Mode Current Control of Z-source DC-DC Converter. In Proceedings of the 2016 IEEE 8th International Power Electronics and Motion Control Conference (IPEMC 2016—ECCE Asia), Hefei, China, 22–26 May 2016.
44. Basin, M.V.; Rodriguez-Ramirez, P.C. Sliding-mode filter design for linear systems with unmeasured states. *IEEE Trans. Ind. Electron.* **2010**, *58*, 3616–3622. [CrossRef]
45. Yang, G.; Wang, Y.; Zhao, Y. Observer-Based Non-singular Terminal Sliding Mode Control of Boost converters. In Proceedings of the 2020 39th Chinese Control Conference (CCC), Shenyang, China, 27–29 July 2020; pp. 2096–2100.
46. Wang, Y.; Ying, Z.; Zhang, W. Unified Sliding Mode Control of Boost Converters with Quantitative Dynamic and Static Performances. In Proceedings of the IECON 2020—46th Annual Conference of the IEEE Industrial Electronics Society, Singapore, 18–21 October 2020; pp. 3271–3276.
47. Hu, J.; Wang, Z. Sliding Mode Control for DC-DC Converters with Unknown Constant Power Load in Renewable Energy Systems. In Proceedings of the 2022 7th International Conference on Power and Renewable Energy (ICPRE), Shanghai, China, 23–26 September 2022.
48. Chalanga, A.; Cail Al, S.K.; Fridman, L.; Bandyopadhyay, B.; Moreno, J.A. Implementation of super-twisting Control: Super. The death and higher order Sliding mode observer based printing proaches. *IEEE Trasactions Ind. Electmnics* **2016**, *63*, 3677–3685. [CrossRef]

sustainability

Article

A New Cuk-Based DC-DC Converter with Improved Efficiency and Lower Rated Voltage of Coupling Capacitor

Khaled A. Mahafzah [1,*], Ali Q. Al-Shetwi [2,3,4,*], M. A. Hannan [4], Thanikanti Sudhakar Babu [5,6] and Nnamdi Nwulu [6]

1 Department of Electrical Engineering, Faculty of Engineering, Al-Ahliyya Amman University, Amman 19328, Jordan
2 Electrical Engineering Department, Fahad Bin Sultan University, Tabuk 71454, Saudi Arabia
3 Institute of Power Engineering, Universiti Tenaga Nasional, Kajang 43000, Selangor, Malaysia
4 Department of Electrical and Electronic Engineering, Universiti Tenaga Nasional, Kajang 43000, Selangor, Malaysia; hannan@uniten.edu.my
5 Department of Electrical and Electronics Engineering, Chaitanya Bharathi Institute of Technology, Hyderabad 500075, India; sudhakarbabu@ieee.org
6 Centre for Cyber Physical Food, Energy and Water Systems, University of Johannesburg, Johannesburg 2006, South Africa; nnwulu@uj.ac.za
* Correspondence: k.mahafzah@ammanu.edu.jo (K.A.M.); aalshetwi@fbsu.edu.sa (A.Q.A.-S.)

Abstract: DC-DC converters play a crucial role in recent and advanced applications, enabling efficient power conversion and management for renewable energy systems, electric vehicles, portable devices, and advanced communication systems. In line with this, the objective of this paper is to introduce a new DC-DC configuration based on the Cuk converter named as Mahafzah converter, which utilizes a coupling capacitor with a lower rated voltage. The paper aims to demonstrate the effectiveness of the proposed converter in terms of improved efficiency, reduced size, and reduced semiconductor device currents compared to the conventional Cuk converter. The proposed configuration comprises the same components as the Cuk converter, but in a different arrangement, without any additional elements. The main advantage of the proposed converter is using a coupling capacitor with a much lower rated voltage than the Cuk converter, resulting in a smaller capacitor size, reduced printed circuit board (PCB) size, and manufacturing cost. Additionally, the proposed converter reduces the currents of the semiconductor devices compared to those in the Cuk converter. To demonstrate its effectiveness, the converter is operated under continuous current mode (CCM) with a constant duty cycle and switching frequency. The study provides an in-depth discussion of the various operating modes by making use of equations relating to currents, voltages, duty cycles, and voltage gains. It also provides detailed illustrations of the limits between CCM and discontinuous current mode (DCM). The effectiveness of the proposed converter is demonstrated through a design example with operating parameters of 1 kW, 200 V/−300 V, and 20 kHz. Additionally, a low voltage–low power prototype (12/−18 V, 3.24 W, 20 kHz) is established to verify the operation of the proposed converter. Simulation and experimental verification of the proposed configuration achieved the desired results to improve efficiency and reduce the rate. The results clearly indicate that the efficiency of the proposed converter surpasses that of the conventional Cuk converter under identical operating conditions, reaching approximately 88% at rated load conditions.

Keywords: DC-DC converters; inverted output voltage; electric vehicles; hybrid systems; non-isolated converters; Cuk converter; Mahafzah converter; continuous current mode

check for updates

Citation: Mahafzah, K.A.; Al-Shetwi, A.Q.; Hannan, M.A.; Babu, T.S.; Nwulu, N. A New Cuk-Based DC-DC Converter with Improved Efficiency and Lower Rated Voltage of Coupling Capacitor. *Sustainability* **2023**, *15*, 8515. https://doi.org/10.3390/su15118515

Academic Editor: J. C. Hernandez

Received: 27 April 2023
Revised: 19 May 2023
Accepted: 22 May 2023
Published: 24 May 2023

1. Introduction

DC-DC converters play an essential role in too many different applications, including renewable energy systems [1,2], hybrid or fully electric vehicles (EVs) systems [3,4], micro-grids in power systems [5,6], and voltage regulation applications [7,8]. These converters are mainly divided into two types: First, a linear converter depends on a linear passive device such as a series or shunt resistance to regulate the output voltage. This converter is a very simple converter with low noise in its output voltage. However, using passive elements deteriorates the converter efficiency due to heat generation. Additionally, it is used as a step-down converter only [9–11]. Second, switching converters: these converters are the most common ones. The output of these converters is regulated by using a semi-conductor-controlled switch (at least one switch is used). The presence of controlled switches allows for either step-up or step-down of the output voltage and even enables inversion of the output voltage polarity. Although the use of controlled switches increases complexity and output noise, it improves the overall efficiency of these converters [12–20].

Switching converters can be categorized as either hard-switching or soft-switching resonant converters. The hard-switching converters could be non-isolated or isolated DC-DC converters. The non-isolated converters include Buck, Boost, SEPIC, Buck–Boost, and Cuk converters. These topologies typically consist of a single controlled semiconductor switch, a single diode, one or two inductors, and a low-pass filter [12–20]. In contrast, the non-isolated DC-DC converters employ galvanic isolation equipment such as a transformer-like flyback converter [21] and a forward converter [22]. Hard-switching converters suffer from high switching losses, which limits their ability to achieve a high-efficiency range [23,24]. To address this drawback, soft switching converters have been introduced, significantly reducing switching losses. These converters cover the zero current switching (ZVS) [25] and zero voltage switching (ZCS) [25] converters. More details are illustrated in Figure 1.

Due to the rapid development of renewable energy resources, DC-DC converters with inverted output voltage are commonly used in hybrid solar and wind systems. These converters serve the purpose of providing a constant voltage source when the solar energy or wind speed falls below the desired limits [26]. Additionally, another significant application that requires inverted output voltage is in electric vehicles, which involve two energy storage devices: a power supply with high energy storage and a rechargeable energy storage system that enables two-directional power capability [3]. As a result, converters with inverted output voltage, such as the Buck–Boost converter and the Cuk converter, are highly preferred for these applications [27].

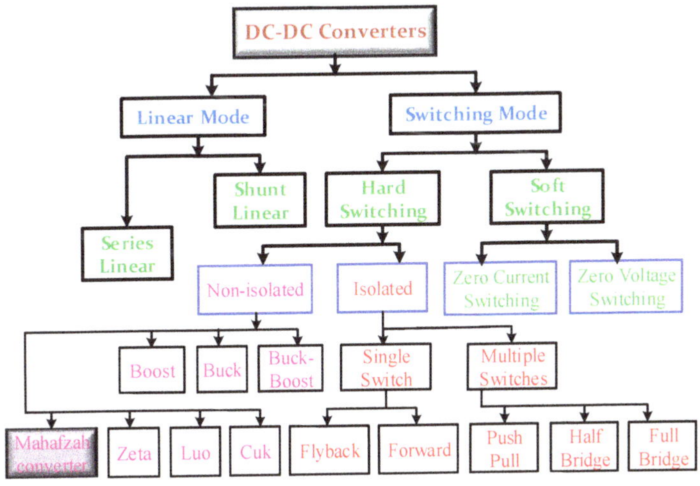

Figure 1. Classification of DC-DC Converters with the location of the Mahafzah converter [28–31].

The Buck–Boost converter can either step-up or step-down the output voltage using a low number of components. Additionally, it offers high efficiency with a low-duty cycle, making it suitable for many applications [13]. However, it cannot achieve high gain without compromising the converter's efficiency. Moreover, the absence of isolation in the converter can lead to instability in certain applications [13]. Furthermore, when the switch is open, the stored energy in the output side inductor (L_2) is transferred back to the supply, which can be undesirable and restrict the converter's usability [16].

The Cuk converter is utilized for both stepping up and stepping down the output voltage. It consists of two inductors that help reduce the ripple in the input/output currents. In addition, this converter has a continuous input/output current. Furthermore, in the Cuk converter, when the switch is closed, the coupling capacitor supplies energy to both the output side inductor L_2 and the load simultaneously. Yet, unlike the Buck–Boost converter, when the switch is opened, the energy stored in L_2 is transferred to the load [16]. Despite these advantages, the Cuk converter does have some drawbacks; for example, the compensation circuit may be added to stabilize the converter, which reduces its response.

According to the authors' best knowledge and after a careful review of the DC-DC converters presented in review papers [28–31], the proposed configuration is not yet presented in the literature. Therefore, this paper proposes a new converter that is designed and verified experimentally and by simulation. The outcomes of this configuration enhance the efficiency and reduce the coupling capacitor voltage rating. Table 1 compares different DC-DC converters with their limitations.

Table 1. Different DC-DC converters topologies with their limitations.

Ref.	Year of Publication	Objective	Limitations
[12]	2011	A buck converter with coupled inductor for ZVS is proposed	Critical design of the coupled inductor
[13]	2020	A different DC-DC converters with average model is presented	It is used for multi-phase applications with coupled inductor
[15]	2018	Design quasi-SEPIC converter with high voltage gain capability	It uses a coupled inductor and the way to improve the magnetic core characteristics
[16]	2019	Proposes a new Cuk converter fed switched reluctance motor	The circuit has additional semiconductor devices and many inductors
[17]	2021	Proposed an interleaved Luo converter	The critical design of the magnetic circuit
[18]	2022	Design a flyback with a ripple free in inductor current	Adding many passive components to the conventional flyback

The main contribution of this paper is the introduction of a new DC-DC converter that offers higher efficiency, a lower rated voltage of coupling capacity, and cost reduction as compared to Cuk converters. Another advantage of the new configuration (Figure 2) is that it utilizes the same components as the well-known Cuk converter but in a different arrangement. Additionally, the proposed converter demonstrates improved efficiency compared to the Cuk converter under similar operating conditions, reaching approximately 88% at rated conditions. Furthermore, the voltage of the coupling capacitor is reduced to (V_m = ±100 V) compared to (V_m = ±500 V) in the Cuck converter. A design example is presented to validate the functionality of the proposed converter, which is suitable for hybrid renewable energy systems and electric vehicle applications. Moreover, a low voltage–low power prototype of 12/−18 V, 3.24 W is established to verify the operation of the proposed converter, showing a close match between measurements and simulations.

The survey above highlighted the two main types of converters: linear converters, which utilize passive elements and have simplicity but lower efficiency, and switching converters, which employ semiconductor-controlled switches for improved efficiency but higher complexity and noise. The survey also discussed the limitations of hard-switching converters and the advantages of soft-switching converters. However, despite the extensive review, the proposed configuration of the Mahafzah converter, which offers higher efficiency, reduced coupling capacitor voltage rating, and cost reduction compared to Cuk converters, has not been presented in the existing literature. This research gap motivates the introduction of the new converter and its experimental and simulation verification, addressing the need for an improved DC-DC converter design in hybrid renewable energy systems and electric vehicle applications.

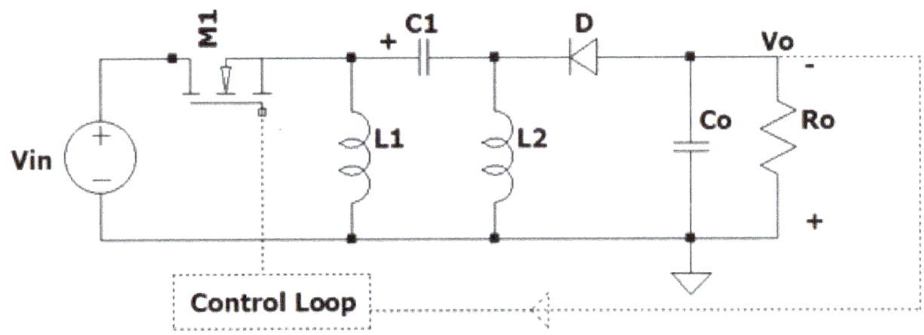

Figure 2. The proposed Mahafzah converter.

The rest of this paper is organized as follows: Section 1 discusses the operating modes, duty cycle, and voltage gain of the proposed converter. A design example and parameters selection is presented in Section 2. Section 3 illustrates the simulation results based on the calculations in the previous section. Section 4 provides experimental results of a low voltage–low power prototype. Finally, the paper is concluded in Section 5.

2. Operating Modes, Duty Cycle, and Voltage Gain of the Proposed Converter

The proposed converter comprises, as shown in Figure 2, one DC power supply, two inductances L_1 and L_2, one coupling capacitance C_1, one controlled switch M_1, one diode D, and also a low pass filter includes C_o in parallel with the load resistance.

2.1. Operating Mode of the Proposed Converter

The proposed converter has two different operating modes as follows:

- When the switch M_1 is ON: the diode D is reversed biased. Figure 3a shows the equivalent circuit and current directions of this mode. The energy is transferred and stored in the coupling capacitor C_1. Meanwhile, both inductors L_1 and L_2 are energized. The current slope in the inductors is given according to the following equations:

$$\frac{dI_{L_1}}{dt} = \frac{V_{in}}{L_1} \tag{1}$$

$$\frac{dI_{L_2}}{dt} = \frac{V_{in} + V_{C_1}}{L_2} \tag{2}$$

From (1) and (2), the switch current is the sum of two inductors' current. This can be written as (3). Instantaneously, the voltage of capacitor C_1 is given by (4).

$$I_{M1} = I_{L1} + I_{L2} \tag{3}$$

$$V_{C1} = V_{in} - V_{L2} \tag{4}$$

- When the switch M_1 is OFF: the diode is forward and conducts the current. Figure 3b shows this mode's equivalent circuit and current directions. All the energy stored in C_1, L_1, and L_2 is transferred to the load. The current slope in the inductors is given according to the following equations:

$$\frac{dI_{L_1}}{dt} = \frac{V_o + V_{C_1}}{L_1} \tag{5}$$

$$\frac{dI_{L_2}}{dt} = \frac{V_{in} + V_{C_1}}{L_2} \tag{6}$$

From (5) and (6), the diode current is the sum of two inductors' current. This can be written as on (7). Instantaneously, the voltage of capacitor C_1 is given by (8).

$$I_D = I_{L1} + I_{L2} \tag{7}$$

$$V_{C1} = V_{L1} - V_o \tag{8}$$

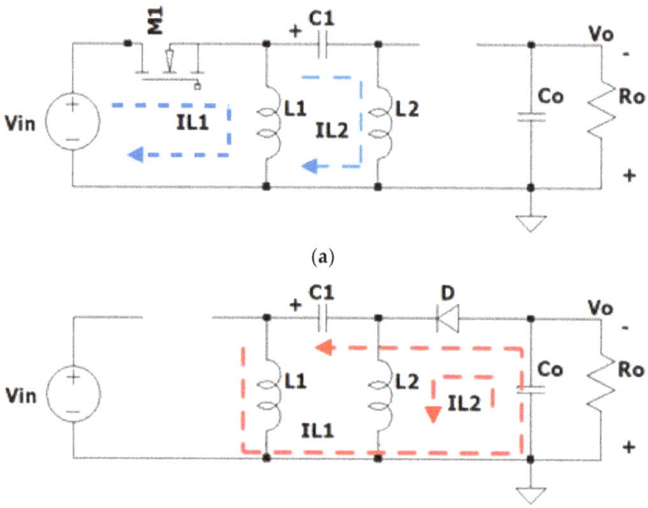

Figure 3. (**a**) Mode 1, when switch M_1 is on, (**b**) Mode 2, when switch M_2 is off.

In the steady state, the average inductors' voltages are zero. Based on (4) and (8) and the capacitance C_1 is large enough, then, the average C_1 voltage has to be equal to the following equation:

$$V_{C1} = V_{L1} - V_{L2} \tag{9}$$

According to (9), it can be seen that the capacitor voltage C_1 depends on the voltage difference between the two inductors. In the proposed converter, the coupling capacitor's average voltage is equal to zero. This means all the energy stored during the turn-on period is dissipated during the turn-off period. More details will be discussed in the Simulation Results Section.

2.2. Duty Cycle and Voltage Gain of the Proposed Converter

The duty cycle of the proposed converter can be calculated based on the same procedure used in other DC-DC converters. In a steady state, the average inductor voltages over one switching cycle (T_s) must equal zero. Then, once the switch M_1 is turned on, the inductor L_1 is energized from the input DC voltage V_{in}. On the other hand, when the switch M_1 is turned off, the energy stored in L_1 is delivered to the load through the coupling capacitor C_1 and the diode D. Based on that, the inductor voltage function is given by:

$$V_{L1}(t) = \begin{cases} V_{in}, & 0 < t < DT_s \\ -V_o & DT_s < t < T_s \end{cases} \tag{10}$$

$$V_{L2}(t) = \begin{cases} V_{in}, & 0 < t < DT_s \\ -V_o, & DT_s < t < T_s \end{cases} \quad (11)$$

Calculating the average voltages of L_1 and L_2 results in (12). By solving (12), the voltage gain can be produced as per (13).

$$\langle V_{L1} \rangle = DT_s V_{in} + (1 - D)T_s V_o = 0 \quad (12)$$

$$V_G = \frac{V_o}{V_{in}} = -\frac{D}{1 - D} \quad (13)$$

where V_G is the voltage gain of the proposed converter.

Equation (12) shows that the voltage gain of the proposed converter is the same as Buck–Boost, Cuk, and SEPIC converters [32]. Nevertheless, the proposed converter has an inverted output voltage, which can be used to step up and down by selecting a proper duty cycle value (see (13)). Additionally, it can be seen that the critical value between the step-up and step-down is D = 50%. Figure 4 plots the voltage gain of the proposed converter and the converter duty cycle.

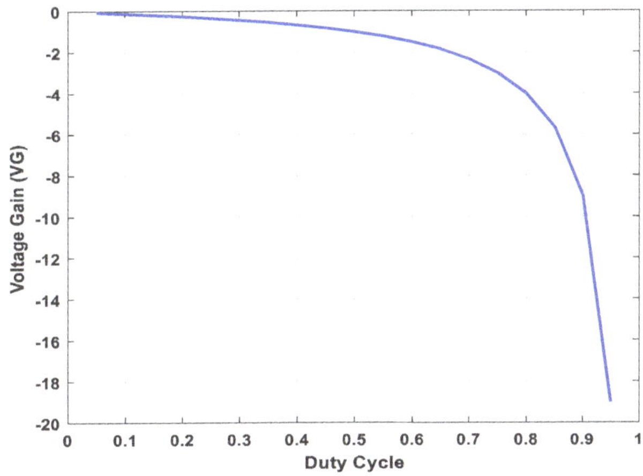

Figure 4. The voltage gain change based on the converter duty cycle.

3. Design Example and Parameters Selection

As seen in Figure 2, the proposed converter comprises a single controlled switch, a single diode, two decoupled inductors, one coupling capacitor, and one low-pass filter (shunt filter capacitor with load resistance). For simplicity, the proposed converter is assumed to work in continuous current mode (CCM). The following discussion confirms a proper selection of the converter parameters to achieve CCM operating mode.

The selected design parameters are based on the data presented in [33]; the DC output voltage (motor voltage) is 48 V. Nevertheless, in this paper, it is changed to 300 V for better indication, and the rated power of the motor is 1 kW. As the input voltage to the circuit in [28] is an AC RMS line voltage, then the average of the rectified voltage is calculated to equal 200 V-DC. In addition, the switching frequency is set to 20 kHz. The ripple current percentage in both inductor currents is set to 20%. In contrast, the ripple in the output voltage should not exceed 10% of the desired output voltage. Table 2 presents the selected parameters for the design along with their corresponding values. These parameters were derived from the application proposed in [33]. The calculated parameters based on the presented equations are included in the fourth and fifth columns.

Table 2. The selected parameters of the design.

These Parameters Are Taken from the Application Proposed in [33]		The Calculated Parameters Based on Presented Equations	
Parameter	**Value**	**Parameter**	**Value**
P_{in}/P_o	1 kW	I_o	3.3 A
V_{in}	200 V	I_{L1}	5 A
V_o	−300 V	I_{L2}	3.3 A
ΔI_{L1}	0.92 A	R_o	90 Ω
ΔI_{L2}	1.3 A	L_1	6.5 mH
ΔV_{C1}	<0.1	L_2	6.5 mH
K_p	0.2	C_1	0.5 µF
K_i	0.001	C_o	5 µF
Duty Cycle (D)	60%		
f_s	20 kHz		

The average inductor L_1, L_2 currents are calculated as:

$$I_{L1} = \frac{P_o}{V_{in}} \tag{14}$$

$$I_{L2} = \frac{P_o}{V_o} \tag{15}$$

The load resistance is set according to (16), and then the selected inductors are given by (17) and (18), (when $D = 0.6$).

$$R_o = \frac{V_o^2}{P_o} \tag{16}$$

$$I_{L1,\max} = \frac{V_{in} D}{f_s \Delta I_{L1}} \tag{17}$$

$$I_{L2,\max} = \frac{V_o (1 - D)}{f_s \Delta I_{L2}} \tag{18}$$

When selecting the coupling capacitor C_1, the ripple in the output voltage should not exceed 20% of V_o, thus it is found as per (19). Finally, the filter capacitor can be calculated by using (20). Thus, C_o has a minimum value calculated as:

$$C_1 = \frac{V_{C1} D}{R_o f_s \Delta V_{C1}} \tag{19}$$

$$C_o = \frac{V_o D}{R_o f_s \Delta V_{C_o}} \tag{20}$$

4. Simulation Results

MATLAB 2020a is used to simulate the proposed converter, which can provide very close results to the real prototype. The simulation parameters are discussed in the previous section, whereas the simulation time is set to 1 s, and the solver is selected to be an ordinary differential equation ode23tb with a maximum step size is 250 µs and continuous simulation type. Moreover, the proposed converter is operating under the CCM mode with a hard-switching technique and constant duty cycle equal to 0.6.

The output voltage is plotted in Figure 5. It is seen that the average output is 297 V DC voltage with a ripple percentage in the voltage of around 6.7%, which is an acceptable value. Moreover, as the load is a pure resistance, the load current has the same voltage pattern but is scaled by (1/90), which gives the average load current 3.3 A. Thus, the current ripple in I_{L1} is given by (21). Based on the design example (21) gives 0.92 A. Otherwise, the ripple in I_{L2} is calculated based on (22). It is also important to point out that the result of (21) is 1.3 A.

$$\Delta I_{L1} = \frac{V_{in} D T_s}{L_1} \tag{21}$$

$$\Delta I_{L2} = \frac{V_o D T_s}{L_2} \tag{22}$$

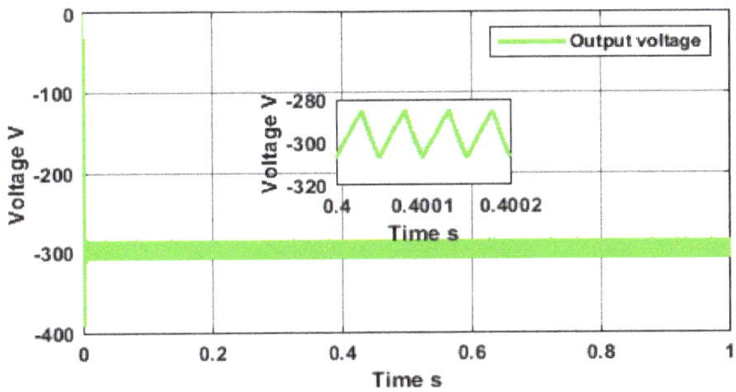

Figure 5. The output voltage of the proposed Mahafzah converter.

The continuous current operation in the proposed converter is clearly seen in Figure 6. As the inductance values of both inductors are the same, the difference in the slopes and their averages are related to the difference in the applied voltage across inductor terminals during the on/off periods. The average inductor current I_{L1} is equal to 5 A, and the average inductor current I_{L2} is equal to 3.3 A, with a ripple current percentage of less than 20% of both currents.

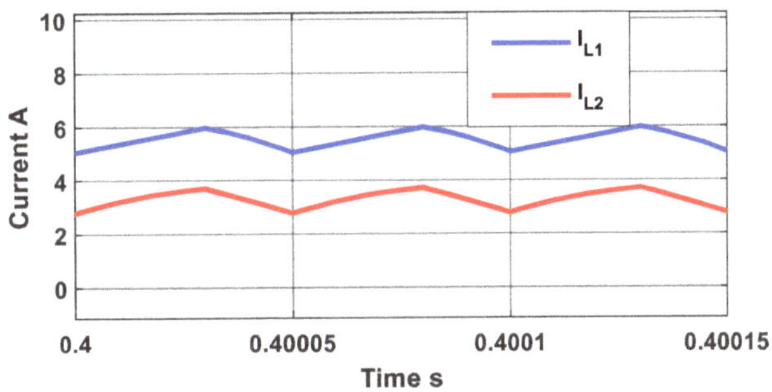

Figure 6. Inductor currents in CCM.

On the other hand, the inductors' voltages are illustrated in Figure 7. During the conduction of switch M_1, the L_1 is clamped to V_{in}. Meanwhile, the L_2 voltage to the difference between V_{L1} and V_{C1}. During the conduction period of the diode, the L_1 has a voltage of V_{L2} + V_{C1} but in the reverse direction, and the L_2 voltage is clamped to the load voltage. In steady-state operation, the average inductor voltages are equal to zero. Figure 8 presents the coupling capacitor voltage and its current. Over one switching cycle, it is noticeable that the capacitor bypasses the energy from the input side to the output side without any remaining voltage across its terminals. This means the average capacitor voltage is zero based on Figure 8a. Additionally, the balance in the capacitor charge is illustrated in Figure 8b. Whereas, the average capacitor current over one switching cycle is zero in steady-state conditions.

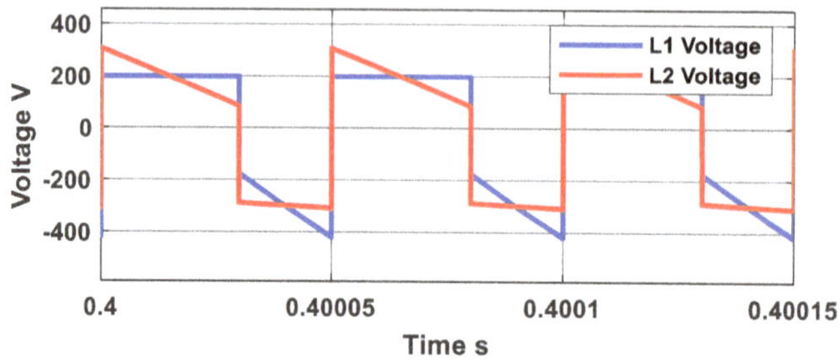

Figure 7. The inductors voltages in CCM.

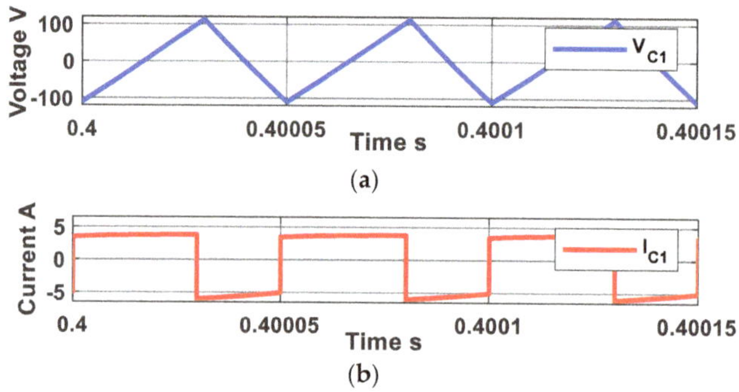

Figure 8. (**a**) The voltage of coupling capacitor C_1 (**b**). The current of coupling capacitor C_1.

The merit of the proposed converter is the existence of an LCL tank connected with the switch M_1, this connection offers a soft switching turn on and turn off. The voltage stress across the switch is illustrated in Figure 9a. The voltage reaches the sum of $V_{in} + V_{C1} + V_o$. The same issue with the output diode. Figure 9b shows the switch and diode currents. The average switch and diode voltages are calculated, respectively, using the following equations:

$$I_{M1} = \frac{V_{in} D^2}{R_o (1 - D)^2} \tag{23}$$

$$I_D = \frac{V_{in} D}{R_o (1 - D)} \tag{24}$$

The proposed converter is compared with the Cuk converter. The Cuk converter is simulated using the same design example discussed above to compare the results. Both converters' output voltages are shown in Figure 10a. It shows that the output voltage in both cases decreased to −300 V, but the proposed converter has more ripple in its voltage than the Cuk converter. Moreover, the proposed converter has an unrecognizable overshoot higher than the Cuk converter, but the proposed converter is faster than the Cuk converter in achieving the steady state period, see Figure 10b.

The coupling capacitor C_1 plays an important role in energy transfer in Cuk, SEPIC, Buck–Boost, and Luo converters, as well as it has a role in the proposed Mahafzah converter. The selected capacitor must be sized so that it has a rated voltage value that is higher than twice the voltage across its terminal. The higher rated voltage results in a higher size capacitor. Furthermore, the large size of this capacitor holds a rather large place on the PCB, thus reducing the cost of circuit manufacturing.

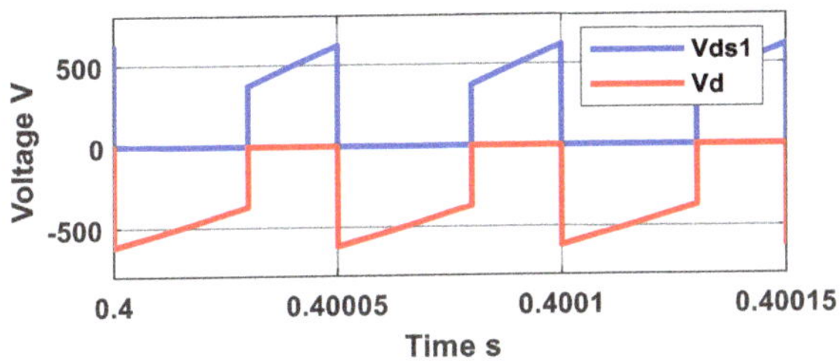

(**a**)

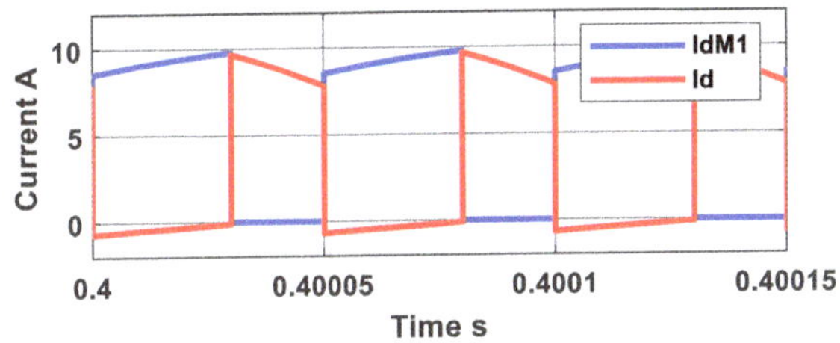

(**b**)

Figure 9. (**a**) The switch M_1 voltage (blue), and the diode D (red) voltage (**b**). The switch M_1 current (blue), and the diode D (red) current.

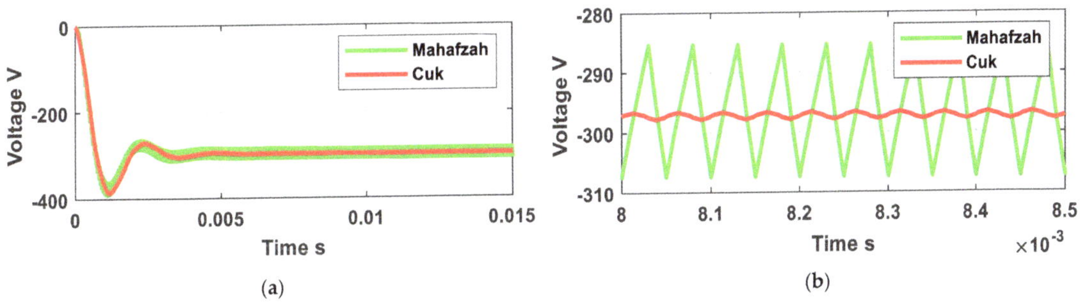

Figure 10. (**a**) The output voltage of both converters, (**b**) the zoomed in.

Figure 11a compares the two capacitor voltages in the proposed and Cuk converter. As noticed from Figure 11b, the coupling capacitor Cuk converter has a much higher applied voltage than its counterpart in the proposed converter. Similar to the Cuk converter, the proposed converter has the boundary characteristics shown in Figure 12. The coupling capacitor is selected to endure the applied voltage across its terminal in the Cuk converter. The critical value that separates the two modes is plotted in the cyan curve. The voltage gain as a function of the duty cycle and K value is given by (25). Then, according to (25), the critical value between CCM and DCM is given as described in (26). Accordingly, $K_{critical}$ is equal to 0.16.

$$V_G(D,K) = \begin{cases} \frac{-D}{(1-D)} & K > K_{critical} \\ \frac{-D}{\sqrt{K}} & K > K_{critical} \end{cases} \quad (25)$$

$$K_{critical} = (1-D)^2 \quad (26)$$

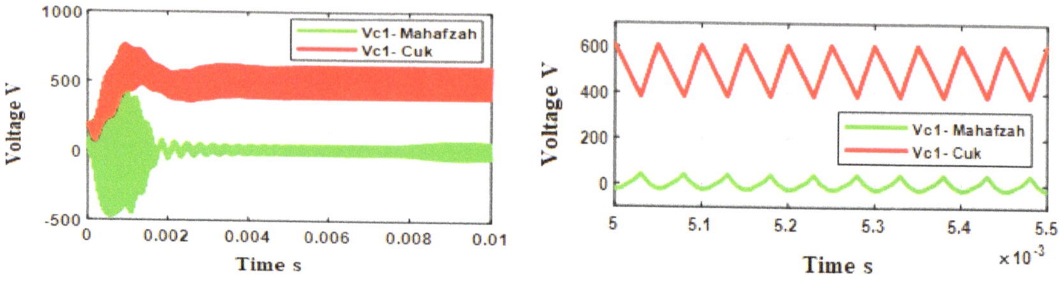

(**a**) (**b**)

Figure 11. (**a**) The coupling capacitor voltage of both converters, (**b**) zoomed in during steady state.

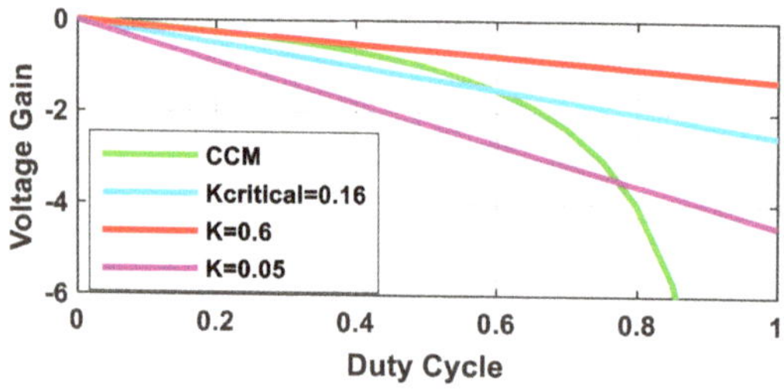

Figure 12. Characteristics of the proposed converter.

As illustrated in Figure 11 and in Equation (19), the coupling capacitor value of the proposed converter is noticeably reduced by five times compared to the coupling capacitor value for the cuck converter under the same operating conditions. The loss components of the proposed converter can be divided into conduction losses, switching losses, and control losses [23,34]. It should be noted that these losses are associated with semiconductor devices. Table 3 illustrates all loss components and provides the related equation.

Table 3. Loss calculation of the proposed Mahafzah converter.

Loss Component	Equation	Note
Conduction Loss	$P_{M1} = i_d^2 R_{ds-on} D$ $P_D = (V_f i_d + i_d^2 R_f)(1-D)$	R_{on}: MOSFET on-state resistance
Switching Loss	$P_{M1} = 05 f_s C_{oss}(0.5V_{in} + V_o)^2$ $P_D = 05 f_s C_d(0.5V_{in} + V_o)^2$	C_{oss} is M_1 output capacitance
Control Loss	$P_{gates} = Q_g V_{gs} f_s$	Q_g is the gate charge of M_1
Passive Devices	$R_{L1} = R_{1dc}\left(\frac{DT_s V_{in}}{L_1}\right)^2$ $R_{L2} = R_{2dc}\left(\frac{DT_s V_o}{L_2}\right)^2$	Losses in each L_1 and L_2 based on using their DC resistance. The losses in the coupling capacitor are ignored due to its small ERS

Using the presented equations in Table 2, the efficiency of the proposed converter is calculated when changing the load simultaneously. The efficiency of the proposed converter is compared with the Cuk converter, as illustrated in Figure 13. The efficiency of both converters is calculated based on the equations presented in [22–24,34]. The efficiency calculation considers all the loss components, including the conduction, switching, and control losses. The efficiency is also calculated when the load current is changed from 10% up to 100% of the rated current. As seen in Figure 13, the proposed converter has better efficiency than the Cuk converter when the load is increased. The efficiency of the proposed converter reaches around 88% at full load conditions, while the Cuk converter efficiency reaches 87% at the same rated conditions.

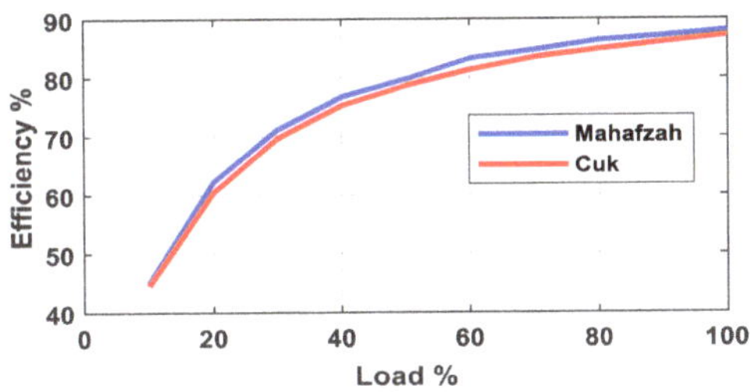

Figure 13. The efficiency of both converters.

5. Experimental Results

5.1. Experiment Setup

The experimental setup for testing the proposed configuration is illustrated in the circuit diagram (Figure 14). To validate the proposed converter, a low voltage–low power prototype (Figure 15) was constructed using the available equipment in the laboratory. The input voltage was set to 12 V by a standalone battery, and low-power inductors with an approximate value of 1.2 mH were selected. The load resistance was set to 100 Ω, resulting in an output voltage of −18 V and a converter power of 3.24 W. The available IRF540N MOSFET was utilized as the switch, controlled by an N-type transistor 2N3904 through an Arduino-based chip, allowing for adjustment of the duty cycle and switching frequency. Prior to conducting the measurements, the proposed converter was re-simulated using MATLAB/ Simulink with the parameters provided in Table 4.

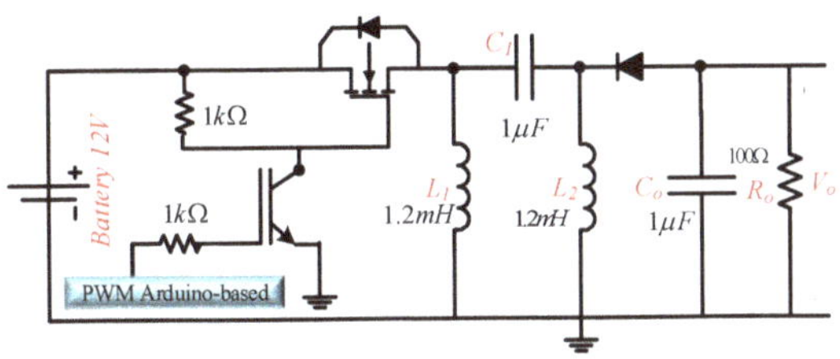

Figure 14. Circuit diagram of the experimental setup (new configuration proposed).

Figure 15. The 3.24 W/−18 V converter prototype.

Table 4. Selected parameters of the prototype.

The Selected Parameters for Testing and Validation	
Parameter	**Value**
P_{in}/P_o	3.24 W
V_{in}	12 V
V_o	−18 V
MOSFET	IRF540N
Driving Transistor	2N3904
Diode	1N4007
$L_1 = L_2$	1.2 mH
C_1	1 µF
Duty Cycle	70%
f_s	20 kHz

5.2. Experimental Results and Discussion

The output voltage is shown in Figure 16 (Ch2). The load voltage is recorded at −18 V with a low voltage ripple. The driving voltage of IRF540N is shown in Figure 16 (Ch1). The applied voltage reaches a peak of 12 V, with a duty cycle is about 70% (35 µs). Accordingly, the inductor L_1 voltage is plotted in Figure 17.

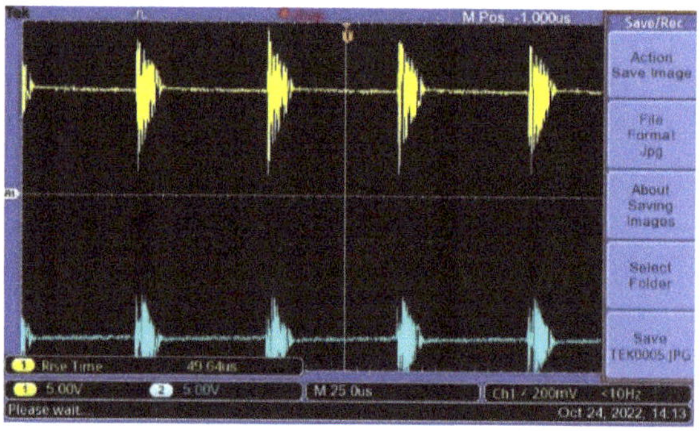

Figure 16. Ch1: Input Voltage, Ch2: Output Voltages.

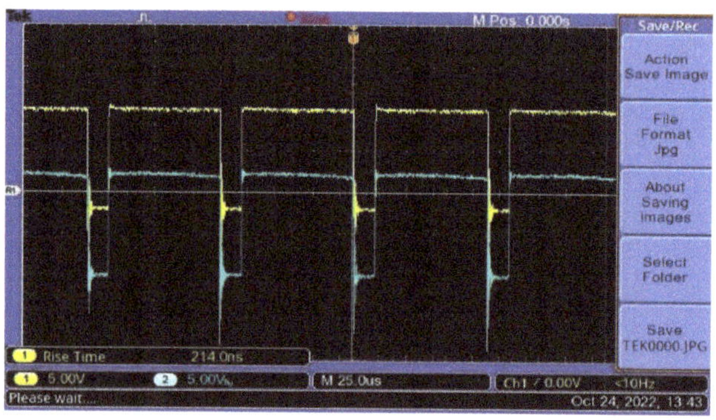

Figure 17. Ch$_1$: V_{L1}, Ch$_2$: V_{L2}.

As observed in Figure 17 (Ch1), the inductor voltage reaches 12 V (the input voltage) during the switch M_1's turn-on time. However, when the switch is turned off (with an off time of approximately 15 µs), the inductor voltage decreases to −6 V. Similarly, during the switch M_1's turn-on time, the inductor L_2 exhibits a voltage of a 3 V across its terminals. Conversely, when the switch M_1 is turned off, the inductor L_2 displays −10 V, see Figure 17 (Ch2). In addition, the coupling capacitor has 7 V across its terminal, which corresponds to the difference between the input and output voltage. Consequently, the rated voltage of the selected voltage of C_1 should be around 15 V. This confirms that the selected coupling capacitor has a lower rated voltage than the same one in the Cuk converter (in the Cuk converter case, the rating voltage of the coupling capacitor must be selected around 45 V. This reduces the selected rated voltage of C_1 in the proposed converter by 66.67% compared to the same capacitor of Cuk converter, as seen in Figure 18.

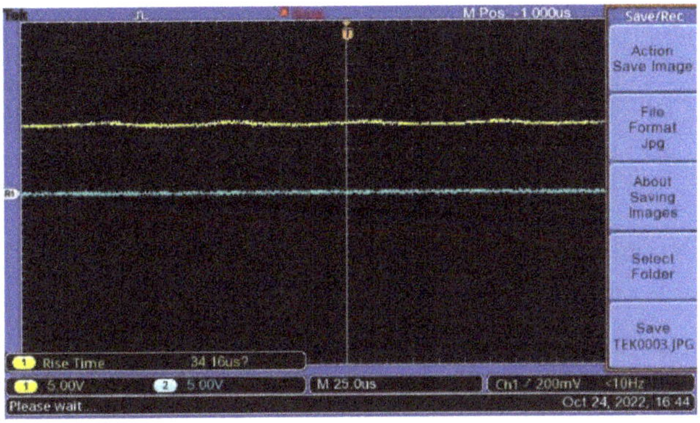

Figure 18. C_1 Voltage.

The drain-source voltage of the IRF540N is depicted in Figure 19, while the diode voltage is plotted in Figure 20. The MOSFET is operated with a duty cycle of 70%. It can be seen from these Figures that during the switch turn-off period, there is some ringing present in the voltage waveform. This ringing is related to some reasons, such as: one reason is the resonance between L_1 and the MOSFET's parasitic capacitance during the energy transfer period. It is not possible to resonate L_2 and C_1 with L_1, because the resonant frequency of this combination is about 28 kHz. Similarly, there is no possible resonant between L_2 and C_1 with L_1, as their resonance frequency is about 40 kHz, significantly lower than the frequency depicted in Figures 19 and 20.

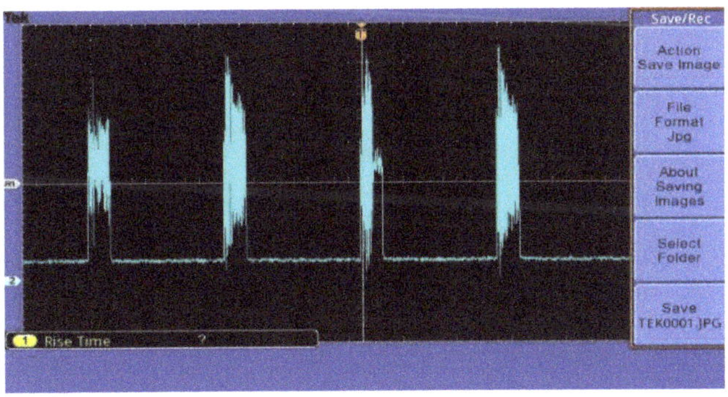

Figure 19. Drain-source voltage.

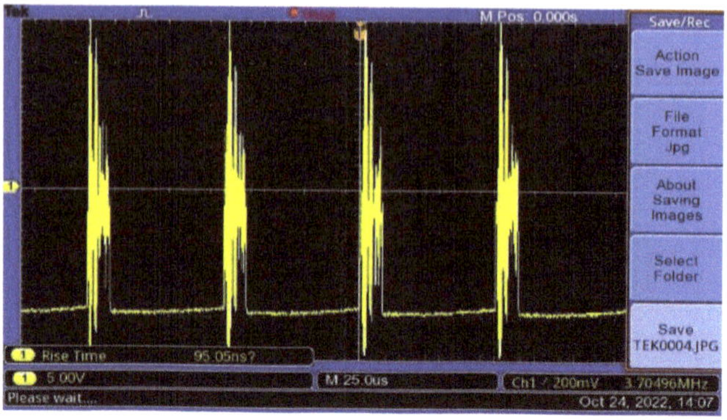

Figure 20. Diode voltage.

The second reason is the possible resonance between the inductance of L_1 and the capacitance of parasitic capacitance of the used passive prob. The third possible reason may be related to the poor copper board used which causes some EMI issues. However, these reasons can be easily overcome with very good PCB design and using advanced measuring devices. In sum, Table 5 provides a comprehensive comparison between the Cuk converter and the Mahafzah converter, considering their main features under the same operating conditions.

Table 5. A comprehensive compression between Cuk and Mahafzah converters.

Parameters	Converter Topology	
	Cuk Converter	**Mahafzah Converter**
Component Count	Same	Same
Coupling Capacitor Voltage	High	Reduced (√)
Efficiency	Low	Improved (√)
Ripple in Vo	Low (√)	High
Transient Period	Long	Short (√)

"√" indicates which converter is better for each parameter.

Overall, based on the information provided in the table, the Mahafzah converter demonstrates certain advantages over the Cuk converter in terms of the reduced coupling capacitor voltage, improved efficiency, and shorter transient period. However, it is important to note that the table does not provide specific quantitative values or detailed explanations for each feature, making it difficult to conduct a thorough analysis without further information.

6. Conclusions

This paper proposed a DC-DC converter based on the Cuk converter, namely the Mahafzah converter. The proposed converter maintains the same component counts, duty cycle, voltage gain, and inverted output voltage as the Cuk converter, but with a different arrangement and new design. The new configuration of the proposed converter offers the advantage of reducing the rating voltage of the coupling capacitor, resulting in a smaller size and lower cost. Additionally, it operates at a lower rated voltage of the coupling capacitor. Moreover, the currents of the semiconductor devices are reduced compared to those in the devices in the Cuk converter. As a result, the losses in the proposed converter are reduced, leading to improved efficiency. The operating modes and mathematical equations governing the currents and voltages of the inductors, coupling capacitor, switch, and diode are thoroughly discussed in this paper. A design example is presented to verify the effectiveness of the proposed configuration. The design is validated through simulation to evaluate the operation, performance, and efficiency of the converter. Experimental tests are also conducted to validate the simulation results. For this purpose, a low voltage–low power prototype (12/−18 V, 3.24 W) is built to verify the operation and validate the proposed converter waveforms. The results demonstrate the excellent performance of the new converter, as evidenced by the matching between the simulation and design calculation results. Furthermore, the proposed converter exhibits higher efficiency than the Cuk converter under load variations, with an efficiency of 88% at rated load conditions, surpassing the Cuk converter by 1%. All the highlighted insights of this new design will hopefully lead to increased efforts toward the development of advanced energy conversion for electric vehicles and hybrid renewable energy systems.

Author Contributions: Conceptualization, K.A.M.; methodology, K.A.M. and A.Q.A.-S.; software, K.A.M. and N.N.; validation, M.A.H. and A.Q.A.-S.; formal analysis, K.A.M., A.Q.A.-S., T.S.B. and N.N.; experiment, K.A.M. and A.Q.A.-S.; investigation, K.A.M. and A.Q.A.-S.; writing—original draft preparation, K.A.M. and A.Q.A.-S.; writing—review and editing, K.A.M., T.S.B., N.N., A.Q.A.-S. and M.A.H.; project administration, M.A.H. All authors have read and agreed to the published version of the manuscript.

Funding: This research received no external funding.

Institutional Review Board Statement: Not applicable.

Informed Consent Statement: Not applicable.

Data Availability Statement: Not applicable.

Conflicts of Interest: The authors declare no conflict of interest.

References

1. Cao, L.; Loo, K.H.; Lai, Y.M. Output-Impedance Shaping of Bidirectional DAB DC–DC Converter Using Double-Proportional-Integral Feedback for Near-Ripple-Free DC Bus Voltage Regulation in Renewable Energy Systems. *IEEE Trans. Power Electron.* **2016**, *31*, 2187–2199. [CrossRef]
2. Zeng, J.; Du, X.; Yang, Z. A Multiport Bidirectional DC–DC Converter for Hybrid Renewable Energy System Integration. *IEEE Trans. Power Electron.* **2021**, *36*, 12281–12291. [CrossRef]
3. He, P.; Khaligh, A. Comprehensive Analyses and Comparison of 1 kW Isolated DC–DC Converters for Bidirectional EV Charging Systems. *IEEE Trans. Transp. Electrif.* **2017**, *3*, 147–156. [CrossRef]
4. Heydari-Doostabad, H.; O'Donnell, T. A Wide-Range High-Voltage-Gain Bidirectional DC–DC Converter for V2G and G2V Hybrid EV Charger. *IEEE Trans. Ind. Electron.* **2022**, *69*, 4718–4729. [CrossRef]
5. Prabhakaran, P.; Agarwal, V. Novel Boost-SEPIC Type Interleaved DC–DC Converter for Mitigation of Voltage Imbalance in a Low-Voltage Bipolar DC Microgrid. *IEEE Trans. Ind. Electron.* **2020**, *67*, 6494–6504. [CrossRef]
6. Gui, Y.; Han, R.; Guerrero, J.M.; Vasquez, J.C.; Wei, B.; Kim, W. Large-Signal Stability Improvement of DC–DC Converters in DC Microgrid. *IEEE Trans. Energy Convers.* **2021**, *36*, 2534–2544. [CrossRef]
7. Hazucha, P.; Moon, T.S.; Schrom, G.; Paillet, F.; Gardner, D.; Rajapandian, S.; Karnik, T. High Voltage Tolerant Linear Regulator with Fast Digital Control for Biasing of Integrated DC–DC Converters. *IEEE J. Solid-State Circuits* **2007**, *42*, 66–73. [CrossRef]
8. Subasinghage, K.; Gunawardane, K.; Kularatna, N.; Lie, T.T. Extending the Supercapacitor-Assisted Low-Dropout Regulator (SCALDO) Technique to a Split-Rail DC–DC Converter Application. *IEEE Access* **2019**, *7*, 124034–124047. [CrossRef]
9. Liu, M.; Zhang, D.; Zhou, Z. Linear Regulator Design Considerations of the Serial Linear-Assisted Switching Converter Used as Envelope Amplifier. *IEEE Trans. Power Electron.* **2016**, *31*, 3673–3689. [CrossRef]
10. Czarkowski, D.; Kazimierczuk, M.K. Linear circuit models of PWM flyback and buck/boost converters. *IEEE Trans. Circuits Syst. I Fundam. Theory Appl.* **1992**, *39*, 688–693. [CrossRef]
11. Bhatia, D.; Xue, L.; Li, P.; Wu, Q.; Bashirullah, R. High-voltage tolerant digitally aided DCM/PWM multiphase DC-DC boost converter with integrated Schottky diodes in 0.13 µm 1.2 V digital CMOS process. *IEEE J. Solid-State Circuits* **2013**, *48*, 774–789. [CrossRef]
12. Do, H. Zero-Voltage-Switching Synchronous Buck Converter with a Coupled Inductor. *IEEE Trans. Ind. Electron.* **2011**, *58*, 3440–3447. [CrossRef]
13. Azer, P.; Emadi, A. Generalized State Space Average Model for Multi-Phase Interleaved Buck, Boost and Buck-Boost DC–DC Converters: Transient, Steady-State and Switching Dynamics. *IEEE Access* **2020**, *8*, 77735–77745. [CrossRef]
14. Memon, A.H.; Baloach, M.H.; Sahito, A.A.; Soomro, A.M.; Memon, Z.A. Achieving High Input PF for CRM Buck-Buck/Boost PFC Converter. *IEEE Access* **2018**, *6*, 79082–79093. [CrossRef]

15. Siwakoti, Y.P.; Mostaan, A.; Abdelhakim, A.; Davari, P.; Soltani, M.N.; Khan, N.H.; Li, L.; Blaabjerg, F. High-Voltage Gain Quasi-SEPIC DC–DC Converter. *IEEE J. Emerg. Sel. Top. Power Electron.* **2019**, *7*, 1243–1257. [CrossRef]
16. Anand, A.; Singh, B. Modified Dual Output Cuk Converter-Fed Switched Reluctance Motor Drive with Power Factor Correction. *IEEE Trans. Power Electron.* **2019**, *34*, 624–635. [CrossRef]
17. Singh, B.; Kushwaha, R. Power Factor Preregulation in Interleaved Luo Converter-Fed Electric Vehicle Battery Charger. *IEEE Trans. Ind. Appl.* **2021**, *57*, 2870–2882. [CrossRef]
18. Sarani, S.; Zarchi, H.A.; Delavaripour, H. Ripple-Free Input Current Flyback Converter Using a Simple Passive Circuit. *IEEE Trans. Ind. Electron.* **2022**, *69*, 2557–2564. [CrossRef]
19. Li, G.; Xia, J.; Wang, K.; Deng, Y.; He, X.; Wang, Y. Hybrid Modulation of Parallel-Series LLC Resonant Converter and Phase Shift Full-Bridge Converter for a Dual-Output DC–DC Converter. *IEEE J. Emerg. Sel. Top. Power Electron.* **2019**, *7*, 833–842. [CrossRef]
20. Ma, X.; Wang, P.; Bi, H.; Wang, Z. A Bidirectional LLCL Resonant DC–DC Converter with Reduced Resonant Tank Currents and Reduced Voltage Stress of the Resonant Capacitor. *IEEE Access* **2020**, *8*, 125549–125564. [CrossRef]
21. Wang, C. A Novel ZCS-PWM Flyback Converter With a Simple ZCS-PWM Commutation Cell. *IEEE Trans. Ind. Electron.* **2008**, *55*, 749–757. [CrossRef]
22. Zhang, F.; Yan, Y. Novel Forward–Flyback Hybrid Bidirectional DC–DC Converter. *IEEE Trans. Ind. Electron.* **2009**, *56*, 1578–1584. [CrossRef]
23. Mahafzah, K.A.; Krischan, K.; Muetze, A. Efficiency enhancement of a three phase Hard Switching Inverter under light load conditions. In Proceedings of the IECON 2016—42nd Annual Conference of the IEEE Industrial Electronics Society, Florence, Italy, 23–26 October 2016; pp. 3372–3377. [CrossRef]
24. Shen, Y.; Wang, H.; Al-Durra, A.; Qin, Z.; Blaabjerg, F. A Structure-Reconfigurable Series Resonant DC–DC Converter with Wide-Input and Configurable-Output Voltages. *IEEE Trans. Ind. Appl.* **2019**, *55*, 1752–1764. [CrossRef]
25. Mahafzah, K.A.; Obeidat, M.A.; Al-Shetwi, A.Q.; Ustun, T.S. A Novel Synchronized Multiple Output DC-DC Converter Based on Hybrid Flyback-Cuk Topologies. *Batteries* **2022**, *8*, 93. [CrossRef]
26. Basić, M.; Dujić, D. Hybrid modular multilevel converter for variable DC link voltage operation. *CPSS Trans. Power Electron. Appl.* **2021**, *6*, 178–190. [CrossRef]
27. Safayatullah, M.; Elrais, M.T.; Ghosh, S.; Rezaii, R.; Batarseh, I. A Comprehensive Review of Power Converter Topologies and Control Methods for Electric Vehicle Fast Charging Applications. *IEEE Access* **2022**, *10*, 40753–40793. [CrossRef]
28. Alatai, S.; Salem, M.; Ishak, D.; Das, H.S.; Alhuyi Nazari, M.; Bughneda, A.; Kamarol, M. A Review on State-of-the-Art Power Converters: Bidirectional, Resonant, Multilevel Converters and Their Derivatives. *Appl. Sci.* **2021**, *11*, 10172. [CrossRef]
29. Richelli, A.; Salem, M.; Colalongo, L. A Review of Fully Integrated and Embedded Power Converters for IoT. *Energies* **2021**, *14*, 5419. [CrossRef]
30. Zhang, N.; Sutanto, D.; Muttaqi, K.M. A review of topologies of three-port DC–DC converters for the integration of renewable energy and energy storage system. *Renew. Sustain. Energy Rev.* **2016**, *56*, 388–401. [CrossRef]
31. Salem, M.; Jusoh, A.; Idris, N.R.N.; Das, H.S.; Alhamrouni, I. Resonant power converters with respect to passive storage (LC) elements and control techniques–An overview. *Renew. Sustain. Energy Rev.* **2018**, *91*, 504–520. [CrossRef]
32. Pop-Calimanu, I.M.; Popescu, S.; Lascu, D. A New SEPIC-Based DC–DC Converter with Coupled Inductors Suitable for High Step-Up Applications. *Appl. Sci.* **2021**, *12*, 178. [CrossRef]
33. Kushwaha, R.; Singh, B. A Modified Bridgeless Cuk Converter based EV Charger with Improved Power Quality. In Proceedings of the 2019 IEEE Transportation Electrification Conference and Expo (ITEC), Detroit, MI, USA, 19–21 June 2019; pp. 1–6. [CrossRef]
34. Mahafzah, K.A.; Krischan, K.; Muetze, A. Efficiency enhancement of a three phase Soft Switching Inverter under light load conditions. In Proceedings of the IECON 2016—42nd Annual Conference of the IEEE Industrial Electronics Society, Florence, Italy, 23–26 October 2016; pp. 3378–3383. [CrossRef]

where ω_c is the cut-off frequency of the filter. In time domain, this equation can be expressed as follows

$$\frac{di_{Lf}}{dt} = -\omega_c i_{Lf} + \omega_c i_L. \tag{2}$$

The filtered signal i_{Lf} is used both for current-mode control and for obtaining the average PV power after multiplying it by the PV voltage v_{pv}. For current-mode control, the signal $R_s i_{Lf}$ is compared with the reference signal $R_s I_{\text{ref}} - v_{\text{ramp}}$, where $v_{\text{ramp}} = m_a t \bmod T$ is the ramp compensating signal, $m_a = V_M/T$ is its slope, V_M is its amplitude and T is its period. The control logic compares the signal $R_s i_{Lf}$ with the signal $R_s I_{\text{ref}} - v_{\text{ramp}}$ in such a way that at the beginning of each switching cycle with the period dictated by the clock signal CLK, the switch S is turned ON and OFF whenever the signal $R_s i_{Lf}$ reaches the signal $R_s I_{\text{ref}} - v_{\text{ramp}}$. The binary signal u is a result of this switching decision and it takes the value 1 when the switch is turned ON and 0 when it is turned OFF. The current reference $R_s I_{\text{ref}}$ is provided by the MPPT.

A flowchart for current-mode control proposed in this work is shown in Figure 3. The program loop starts with the initialization of the voltage at the PV terminal, the current through the inductor and the voltage at the load (Block 2). These initial values are used to set the initial value of the reference voltage. In practice, the initial reference voltage is imposed by $V_{\text{ref,init}} = 0.85 \times V_{\text{MPPT,max}}$ [18]. Therefore, the choice of the initial values in the system must be made taking into account this condition on the reference voltage. When the irradiation is uniform, the P&O MPPT algorithm and CMC keep operating at the MPP (Block 3, Block 4, and Block 5). The P&O algorithm verifies fluctuations of power and voltage of PV array and determines the set-point voltage V_{ref} constantly. To maximize the output power from the PV array, its output voltage needs to be maintained at the level determined by the P&O algorithm. The resulting reference current I_{ref} from the reference voltage will produce the control signal that will be used by the pulse width modulation (PWM) for the boost converter switch. During the switching of the boost converter, the fourth-order Runge–Kutta algorithm is used to determine the dynamics taken by the converter (the values of the converter states and the values of the duty cycle) over time (Block 7, Block 8, Block 9). It should be noted that all numerical simulations in this paper are performed using MATLAB/SIMULINK. Other software such as C++, Fortran or Python could also be used because it is the simulation of ordinary differential equations [25–29].

2.2. Mathematical Modeling

Considering the system shown in Figure 2, applying Kirchhoff's laws, we obtain the following system of equations:

$$\frac{dv_{pv}}{dt} = \frac{i_{pv}}{C_1} - \frac{i_L}{C_1}, \tag{3}$$

$$\frac{di_L}{dt} = \frac{v_{pv}}{L} - \frac{r}{L} i_L - \frac{v_2}{L}(1-u), \tag{4}$$

$$\frac{dv_2}{dt} = \frac{E - v_2}{RC_2} + \frac{i_L}{C_2}(1-u), \tag{5}$$

where v_{pv} and i_{pv} represent the voltage and current of the PV source, respectively. These two variables are related with a highly nonlinear and implicit Equation (3). i_L is the current of the inductor, and u is the state of the switch. C_1 is the capacitance directly connected to the PV and its role is to smooth the voltage at the output of the PV. L represents the inductance, V_{CC} the output voltage of the battery. r is the equivalent resistance in series with the coil and the equivalent resistance that can be felt at the output of the converter is R.

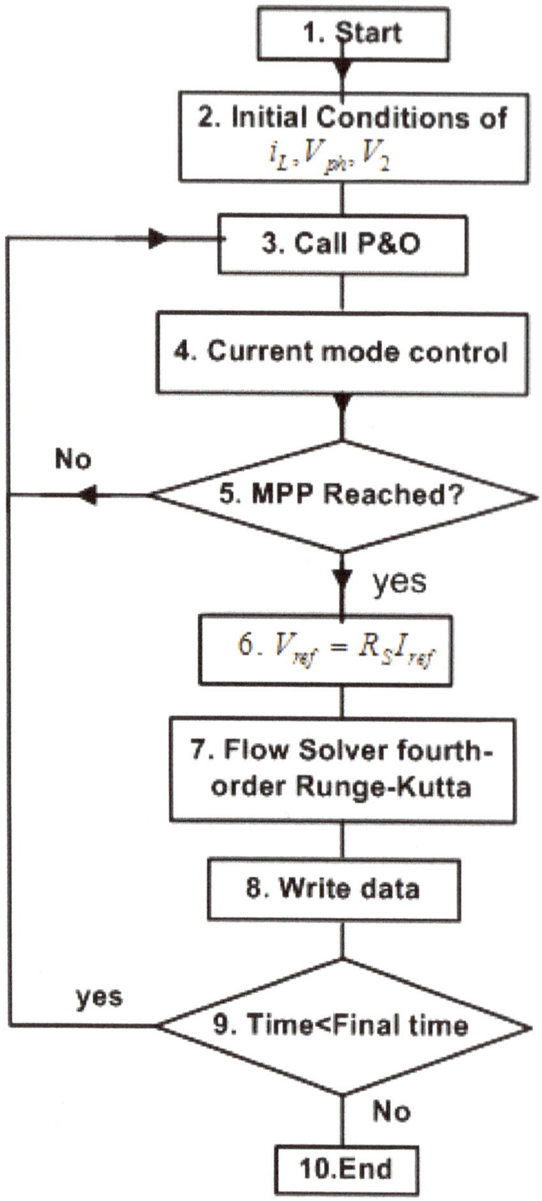

Figure 3. Flowchart of the proposed technique.

A PV generator represents a fundamental power source of a PV system. The output current/voltage characteristics depend on the solar irradiance and temperature. The PV generator has nonlinear electrical model with a single maximum power point (MPP). The performance of a PV generator is evaluated under standard test conditions, where the irradiance is normalized at 1 kW/m^2 and the temperature is defined at 25 °C. A PV generator is characterized by its current–voltage characteristic (I–V) which can be subdivided into three operating zones, a linear zone with a practically constant current, a concave zone with almost constant voltage and an MPP which is the desired point for operation. Since this is the optimal point, the nonlinear characteristic (I–V) is linearized close to the MPP using Taylor's series expansion and ignoring high-order terms. The (I–V) equation of the PV model can be approximated by the following linear Norton equivalent model [11]:

$$\hat{i}_{pv} \approx 2i_{mpp} - \frac{I_{\mathrm{mpp}}}{V_{\mathrm{mpp}}}\hat{v}_{pv} + 0(\hat{v}_{pv}^2). \tag{6}$$

Although numerical simulations could be performed using the nonlinear PV model, the linear model is useful to perform steady-state analysis, for controller design, stability analysis and for prediction of bifurcations. The state-space model of the power stage (3)–(5) together with (2) describing the current sensor and the switching logic determining the value of the binary control signal u appearing in (4) and (5) represent the closed-loop model of the PV system. While the nonlinear model of the PV generator can only be used for performing numerical simulations, the switched model with linearized PV generator model can be used for mathematically predicting the onset of period-doubling bifurcation in the system.

2.3. Steady-State Analysis

Obtaining the steady-state duty cycle D requires performing steady-state analysis. Under MPPT conditions and in steady-state operation, the following equalities hold

$$V_2(1-D) = V_{\mathrm{mpp}} - rI_{\mathrm{mpp}}, \qquad V_2 = E + RI_{\mathrm{mpp}}(1-D). \tag{7}$$

Solving both equations for the steady-state duty cycle D, one obtains

$$D = 1 - \frac{1}{2RI_{\mathrm{mpp}}}\left(\sqrt{4RI_{\mathrm{mpp}}(V_{\mathrm{mpp}} - rI_{\mathrm{mpp}}) + E^2} - E\right). \tag{8}$$

Note that when the parasitic resistance r and R are negligible, one has which is the well-known expression of the duty cycle of an ideal boost converter with input V_{mpp} and output E. Since, under MPPT control and steady-state operation, the PV current will be equal to I_{mpp} and since the average capacitor voltage at the input port of the converter is zero, the inductor current average value in steady state will be also equal to I_{mpp}. For this reason, the reference current under steady-state operation must be given by the following expression

$$I_{\mathrm{ref}} = I_{\mathrm{mpp}} + (\frac{m_a}{R_s} + \frac{m_1}{2})DT - \frac{1}{R_s}\frac{m_1(1-e^{-DT\omega_c})}{\omega_c}, \tag{9}$$

where

$$m_1 = \frac{V_{\mathrm{mpp}} - rI_{\mathrm{mpp}}}{L}, \quad m_a = \frac{V_M}{T}. \tag{10}$$

Note that if $m_a = 0$ (no ramp compensation) and $\omega_c \to \infty$ (ideal current sensor), the previous expression becomes

$$i_{\mathrm{ref}} = I_{\mathrm{mpp}} + \frac{m_1}{2}DT = I_{\mathrm{mpp}} + \Delta i_L, \tag{11}$$

where $\Delta i_L = m_1 DT/2$ is the ripple amplitude of the inductor current. Since i_{ref} is the peak value under the previous conditions, this guarantees that the average value of the inductor current will coincide with I_{mpp} in steady state and will also ensure the PV current in the sense of the average steady-state input capacitor current is zero.

Figure 4 shows the evolution of the reference current i_{ref} in terms of the ramp amplitude V_M and for two values of the output voltage E.

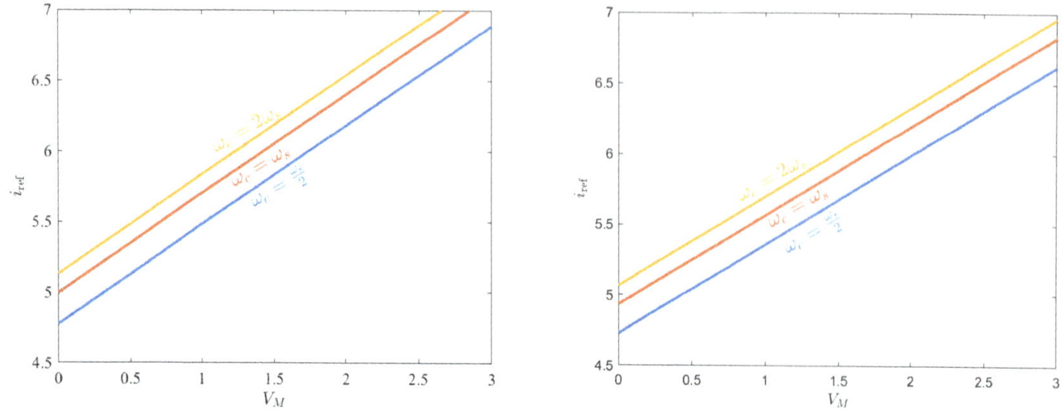

(**a**) (**b**)

Figure 4. The evolution of the current reference i_{ref} in terms of the ramp amplitude V_M according to (9). (**a**) $E = 60$ V. (**b**) $E = 48$ V.

We notice from the above tables, that, whatever the value of ω_c, the reference current increases with the amplitude of the ramp VM. Moreover, as ω_c increases, the curves i_{ref} in terms of V_M tend to be confused.

It is good to note that to maintain the same average inductor current and to make it equal to the MPP current, the current reference in the peak of the current loop control has been adapted according to (9).

2.4. Floquet Theory

2.4.1. The Piecewise Linear State-Space Switched Model Close to the Maximum Power Point

The periodic equation to be considered for our purposes will be expressed in the state-space representation (or matrix representation). This form lends itself well to the calculation of the solution of a differential equation system. The dynamics will be expressed as a function of a vector of variables, which we will call the state vector. Assuming that our subsystem is linear and time-invariant, the evolution of each subsystem is defined by:

$$\begin{aligned} \dot{x} &= A_1 x + B_1 W \text{ if } u = 1, \\ \dot{x} &= A_0 x + B_0 W \text{ if } u = 0. \end{aligned} \quad (12)$$

where $x = (x_1 x_2 x_3 x_4)^T$, A_1, B_1, A_0, B_0 are the state and input matrices corresponding to the different switch states and the external input parameters vector are given by:

$$A_1 = \begin{pmatrix} -\frac{1}{R_{pN}C_1} & -\frac{1}{C_1} & 0 & 0 \\ \frac{1}{L} & -\frac{1}{L}r & 0 & 0 \\ 0 & 0 & -\frac{1}{RC_2} & 0 \\ 0 & w_c k_p & 0 & -w_c \end{pmatrix}, \quad (13)$$

$$A_0 = \begin{pmatrix} -\frac{1}{R_{pN}C_1} & -\frac{1}{C_1} & 0 & 0 \\ \frac{1}{L} & -\frac{1}{L}r & -\frac{1}{L} & 0 \\ 0 & \frac{1}{C_2} & -\frac{1}{RC_2} & 0 \\ 0 & w_c k_p & 0 & -w_c \end{pmatrix}, \quad (14)$$

$$B_1 = \begin{pmatrix} \frac{2}{C_1} & 0 \\ 0 & 0 \\ 0 & \frac{1}{RC_2} \\ 0 & 0 \end{pmatrix}, \tag{15}$$

$$B_0 = \begin{pmatrix} \frac{2}{C_1} & 0 \\ 0 & 0 \\ 0 & \frac{1}{RC_2} \\ 0 & 0 \end{pmatrix}, \tag{16}$$

$$W = \begin{pmatrix} i_{mpp} \\ E \end{pmatrix}. \tag{17}$$

The switching condition in Figure 2 is given by:

$$h(x,t) = R_s\left(i_{ref} - Kx\right) - \frac{V_M}{T}t, \tag{18}$$

with $\frac{V_M}{T}t = V_{ramp}$, the vector K is given by:

$$K = \begin{pmatrix} 0 \\ 0 \\ 0 \\ -1 \end{pmatrix}^T. \tag{19}$$

2.4.2. Stability Analysis Using Floquet Theory

The differential equations describing the dynamics of switching converters are time-periodic with the switching period T determining the periodicity of solutions at the fast switching scale. Floquet theory has been widely used in the analysis of the stability of dynamical systems [30] in general and in switching converters in particular [11,31,32]. For DC-DC converters, the stability dynamics at the fast switching cycle can be accurately predicted by analyzing the stability of the fixed points of the Poincare map of the system using its Jacobian matrix or using Floquet theory combined with the Filippov method which leads to the same results as the Poincare map [33]. The main tool for studying the stability of periodic orbits using Floquet theory is the principal fundamental matrix or the monodromy matrix M. This matrix plays a key role in the accurate stability analysis of switching systems [34–36]. The dynamics in the vicinity of a quasi-static periodic orbit can be expressed in the monodromy matrix as follows:

$$\hat{x}(t+T) = M\hat{x}(t) \quad \forall t, \tag{20}$$

where the overhat stands for small signal variations. Its eigenvalues are called the characteristic multipliers or Floquet multipliers and it can be seen that they determine the amount of contraction or expansion near a periodic orbit and hence they determine the stability of these periodic orbits. It can be obtained by computing the state transition matrices before and after each switch and the saltation matrix that describes the behaviors of the solution 7 switching [24] which are described in the following.

Let $X(DT) = (I - \Phi)^{-1}\Psi$ be the steady-state value of X at time instant DT, where $\Phi = \Phi_1\Phi_0,\ \Phi_1 = e^{A_1DT},\ \Phi_0 = e^{A_0(1-D)T}$,

$$\Psi = e^{A_1DT}A_0^{-1}\left(e^{A_0(1-D)T} - I\right)B_0W + A_1^{-1}\left(e^{A_1DT} - I\right)B_1W. \tag{21}$$

The monodromy matrix use [11] can be expressed as follows $P = \Phi_0\,S\,\Phi_1$

$$P = \Phi_0\,S\,\Phi_1, \tag{22}$$

where S is saltation matrix and is given by:

$$S = I + \frac{(A_0X(DT) - A_1X(DT))\ K^T}{K^T\ (A_1X(DT) + B_1W) + m_a}. \tag{23}$$

Equations (22) and (23) will allow us to determine the Floquet multipliers. According to Floquet's theory, if only one multiplier has a value equal to −1, then the dynamics of the system presents period doubling bifurcation. In general, the occurrence of the critical point of this bifurcation has two regions: a region of subharmonic oscillation and a region of periodic oscillation of period-1. The second region described above is the stability region in power systems [24].

3. Results and Discussions

3.1. *Floquet Theory on the Stability*

3.1.1. Simulations Results in MATLAB/SIMULINK Software

The monodromy matrix is explained in Equation (23). Using a code developed in the MATLAB software, we find the eigenvalues noted as λ of this monodromy matrix. These eigenvalues are solutions of equation $\det(P - \lambda I) = 0$. Figure 5 is the representation in the complex plane of the Floquet multipliers when the amplitude V_M varies from 0.0 to 3, for three values of the cut frequency $f_c = \frac{f_s}{2}, f_c = f_s, f_c = 2f_s$ with $f_s = 50kHz$ and for two values of the output voltage $(E = 60\ V \text{and} E = 48\ V)$. It should be noted that the period-1 orbit will lose stability and bifurcate into period-2 (sub-harmonic oscillation) if the system has the phenomenon of smooth period-doubling when a control parameter varies. Moreover, at this value of the control parameter, period-1 is destroyed and period-2 is created. Indeed, one of the Floquet multipliers is approximately equal to 1, which allows us to say that at this value of the amplitude V_M the period-1 orbit destabilizes and gives way to a sub-harmonic oscillation of period 2. In Figure 5, the followings remarks can be noticed:

Remark 1. *We find that for several values of the gain V_M, the moduli of the Floquet multipliers are located in the circle of unit radius synonymous with the stability of the period-1 orbit. In addition, for other ranks of the parameter V_M, the Floquet multipliers are located outside the circle of unit radius, which is synonymous with destabilization of the periodic orbit of period-1.*

Remark 2. *We observe that when for E $= 60\,V$, E $= 48\,V$, f_c is increasing, i.e., $\left(f_c = \frac{f_s}{2}\right) < (f_c = f_s) < (f_c = 2f_s)$, the critical values V_M (allowing us to set the boundary between period-1 oscillations and subharmonic oscillations) evolve in a decreasing manner, i.e., $V_{M_{\frac{f_s}{2}}} > V_{M_{f_s}} > V_{M_{2f_s}}$. We can conclude that the stability zone increases with the cut-off frequency, this conclusion justifies once again the bifurcation diagrams in Figure 6.*

3.1.2. Simulations Results in PSIM Software

Figure 7 shows the results of the stability study of our system in the PSIM software. The PSIM software is a power electronics simulation software. We notice a similarity between the results obtained with PSIM and those obtained with MATLAB/SIMULINK. In the following, we will use the bifurcation diagrams for a more detailed study.

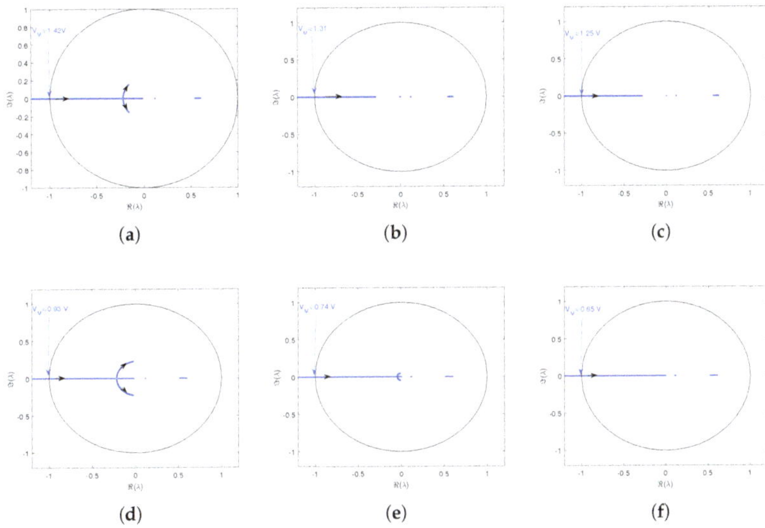

Figure 5. MATLAB simulation of evolution of the Floquet multipliers of the PV system by taking the amplitude of the carrier signal amplitude V_M as a bifurcation parameter for different values of the current sensor bandwidth ω_c and DC output voltage E. The critical values of V_M at which period doubling bifurcation takes place are indicated. (**a**) $f_c = \frac{f_s}{2}$, $E = 60$ V; (**b**) $f_c = f_s$, $E = 60$ V; (**c**) $f_c = 2f_s$, $E = 60$ V; (**d**) $f_c = \frac{f_s}{2}$, $E = 48$ V; (**e**) $f_c = f_s$, $E = 48$ V; (**f**) $f_c = 2f_s$, $E = 48$ V.

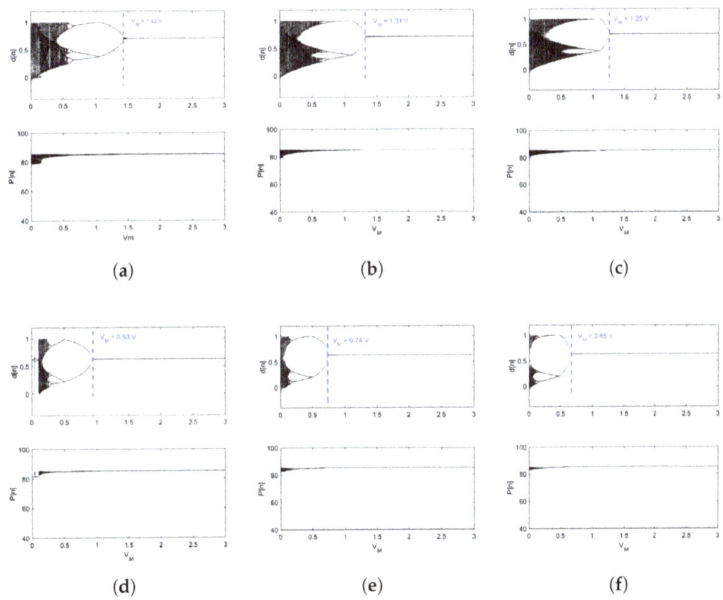

Figure 6. Bifurcation diagrams in MATLAB software by taking the amplitude of the carrier signal amplitude V_M as a bifurcation parameter for different values of the current sensor bandwidth ω_c and DC output voltage E. (**a**) $f_c = \frac{f_s}{2}$, $E = 60$ V. (**b**) $f_c = f_s$, $E = 60$ V. (**c**) $f_c = 2f_s$, $E = 60$ V. (**d**) $f_c = \frac{f_s}{2}$, $E = 48$ V. (**e**) $f_c = f_s$, $E = 48$ V. (**f**) $f_c = 2f_s$, $E = 48$ V.

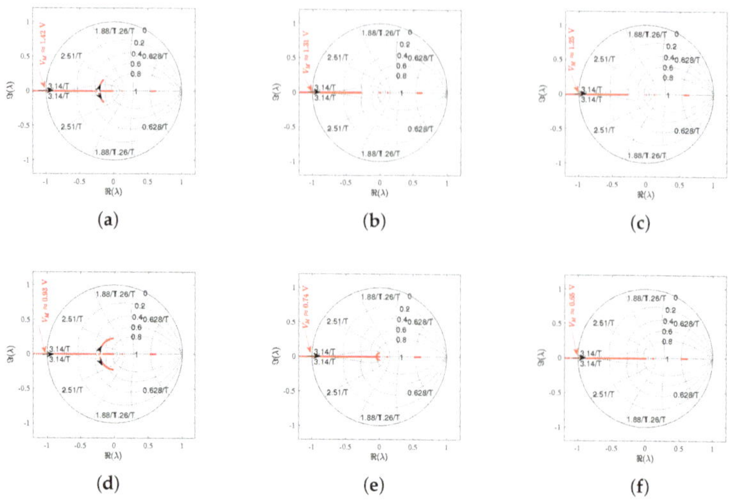

Figure 7. PSIM software simulation of Evolution of the Floquet multipliers of the PV system by taking the amplitude of the carrier signal amplitude V_M as a bifurcation parameter for different values of the current sensor bandwidth ω_c and DC output voltage E. The critical values of V_M at which period doubling bifurcation takes place are indicated. (**a**) $f_c = \frac{f_s}{2}$, $E = 60$ V. (**b**) $f_c = f_s$, $E = 60$ V. (**c**) $f_c = 2f_s$, $E = 60$ V. (**d**) $f_c = \frac{f_s}{2}$, $E = 48$ V. (**e**) $f_c = f_s$, $E = 48$ V. (**f**) $f_c = 2f_s$, $E = 48$ V.

3.2. Bifurcation Behavior from the Nonlinear Circuit-Level Switched Model with the Linear Model of the PV Generator from MATLAB/SIMULINK Software

Bifurcation diagrams constitute appropriate means of recapitulating different transitions to chaos in the system in terms of different parameter values. The maximum Lyapunov exponent is complementary with bifurcation diagram, it is the tool that allows us to conclude if the system is chaotic or not and there are computed following the well-known Wolf algorithm [37]. In this section, the numerically dynamics behavior of our system is performed in MATLAB/SIMULINK software by integrating system (3) using the most frequent fourth-order Runge–Kutta scheme [38] adapted to DC-DC converters which offers a better accuracy for solving single-step differential equations unlike the discretization method developed in the literature. First, the system has been carefully studied through simulations using the linear model of the PV generator in MATLAB software. We used a fixed time step equal to $h = 5 \times 10^{-8}$, a total number of iterations $N = 10^6$ and a transient phase-cut to $N' = 8 \times 10^5$ The bifurcation diagram and graphs of maximal Lyapunov exponents presented below are obtained for three values of the cut-off frequency, i.e., $f_c = \frac{f_s}{2}, f_c = f_s, f_c = 2f_s \text{with} f_s = 50$ kHz , and for two values of DC output voltage ($E = 60\,V$ and $E = 48\,V$), we use ramp amplitude V_M as a bifurcation parameter. The local maximas of the duty cycle and the local maximas of the output power of the generator made it possible to plot bifurcation diagrams. The dynamics of the MPPT algorithm is neglected because it is usually much slower than the converter dynamics. The values of the parameters used for this study are shown in Tables 1 and 2.

Table 1. Boost converter parameters.

Parameters	Values
C_1	10 µF
L	200 µH
r	100 mΩ
E	48 V and 60 V
R_2	200 mΩ
R_s	1 Ω
C_2	47 µF
i_{ref}	updated according to (9)
f_s	50 kHz
f_c	variable
V_M	Variable

Table 2. Parameter of the PV generator (BP 585 module).

Parameters	Values
Maximum power P_{mpp}	85.17 W
Voltage at maximum power V_{mpp}	18.28 V
Current at maximum power V_{mpp}	4.66 A
Maximum power P_{max}	85.17 W
Short-circuit current I_{sc}	5 A
Open-circuit voltage V_{oc}	21.1 V

Figure 6 shows the bifurcation diagrams when the V_M parameter varies for different values of cut-off frequency and output voltage. For $E = 60$ V and $\left(f_c = \frac{f_s}{2}, f_c = f_s, f_c = 2f_s\right)$, we can observe in Figure 6a–c inverse period-doubling phenomena from chaos to period-1 chaos $\rightarrow$ *period* $- 8 \rightarrow$ *period* $- 4 \rightarrow$ *period* $- 2 \rightarrow$ *period* $- 1$ and when the ordinate axis is a representation of the local maxima of the duty cycle. For a representation of the local maxima of the power output of pv on the ordinate axis of the same figure Figure 3 (a_1, b_1, c_1) show that we start directly from chaos in period-1, we also remark that the system stabilizes at the maximum power point, which shows that our MPPT controller used is good. In the same figure, we can see that when the cut-off frequency increases, i.e., $\left(f_c = \frac{f_s}{2}\right) < (f_c = f_s) < (f_c = 2f_s)$ the critical value V_M represents the boundary between the periodic oscillations of period-1 and the subharmonic oscillations decrease. In other words, when the cut-off frequency increases, the system tends to lose its subharmonic behaviour. Thus, the cut-off frequency increases with the stability range, which is interesting for this study since the desired behavior for a pv energy conversion chain is the periodic behavior.

For $E = 48$ V and $f_c \in \left(\frac{f_s}{2}, f_s, 2f_s\right)$, Figure 3 (d_1, e_1, f_1), we can also observe inverse period-doubling phenomena from chaos to period 1 and a faster loss of chaos. We can, therefore, conclude in this part that it is desirable to choose $E = 48$ V as the output voltage of the battery instead of $E = 60\ V$ because it is at this value and for different values of the cut-off frequency that the system tends to quickly lose subharmonic behaviors since the desire behavior for this type of application is the periodic behavior.

Figure 8 shows the Lyapunov exponent graphs complementary with the bifurcation diagrams in Figure 6. These graphs are obtained for three values of the cut-off frequency and for two values of the output voltage and in the V_M zone between $0 \leq V_M \leq 1$, in this area, we observe in Figure 8 negative values of the maximum Lyapunov exponent which correspond to regular oscillations in the system and positive values corresponding to subharmonic oscillations. We can also remark that the cut-off frequency increases with stability, which justifies the bifurcation diagrams of Figure 8.

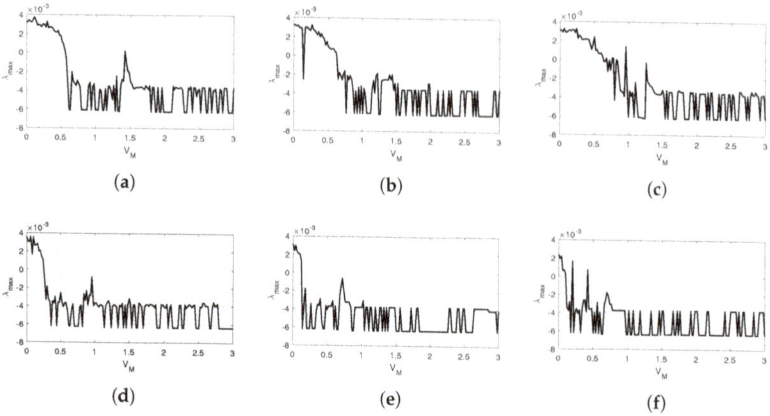

Figure 8. Lyapunov exponent diagrams in MATLAB/SIMULINK software by taking the amplitude of the carrier signal amplitude V_M as a bifurcation parameter for different values of the current sensor bandwidth ω_c and DC output voltage E. (**a**) $f_c = \frac{f_s}{2}$, $E = 60$ V. (**b**) $f_c = f_s$, $E = 60$ V. (**c**) $f_c = 2f_s$, $E = 60$ V. (**d**) $f_c = \frac{f_s}{2}$, $E = 48$ V. (**e**) $f_c = f_s$, $E = 48$ V. (**f**) $f_c = 2f_s$, $E = 48$ V.

The bifurcation diagrams Figure 6 and Lyapunov exponent in Figure 8 show an overlap between regular and irregular behaviors in the subharmonic oscillation zone. In addition, the bifurcation diagrams in Figure 6 show border collision [39–41] respectively to the critical values $V_M = 1.42\,\mathrm{V}$, $V_M = 1.31\,\mathrm{V}$, $V_M = 1.25\,\mathrm{V}$ (see Figure 6a–c) and $V_M = 0.93\,\mathrm{V}$, $V_M = 0.74\,\mathrm{V}$, $V_M = 0.65\,\mathrm{V}$ (see Figure 6d,e,f).

Remark 3. *We note that for E = 60 V, the critical values V_M, Figure 6a–c, are superior to those of E = 48 V Figure 6d–f; this means that the range of stability is greater when E = 48 V compared with E = 60 V. Therefore, it is desirable to choose E = 48 V as the battery output voltage instead of E = 60 V because it is at this value and for different values of the cut-off frequency that the system tends to quickly lose the subharmonic behaviours and increases the stability range.*

3.3. Bifurcation Behavior from the Nonlinear Circuit-Level Switched Model with the Nonlinear Model of the PV Generator from PSIM Software

The circuit diagram for the current-mode-controlled boost converter is constructed using the PSIM simulation software. The selection of component parameters is consistent with the numerical simulation in Tables 1 and 2. In this part, nonlinear phenomena such as bifurcations are exhibited, using the nonlinear and linear models of the PV generator, the V_M bifurcation parameter, the different values of the output voltage ($E = 60\,\mathrm{V}$ and $E = 48\,\mathrm{V}$) and the different values of the cut-off frequency $f_c = \frac{f_s}{2}, f_c = f_s, f_c = 2f_s$ with$f_s = 50$ kHz being equal to those used in numerical simulations in MATLAB. Comparing the simulation bifurcation of the nonlinear model of the pv generator in Figure 9 (PSIM simulation) with that in Figure 6 (MATLAB simulation), it can be seen that the results are in almost perfect agreement.

Note here that the results presented in these two figures are identical, for example, for E = 60 and $\left(f_c = \frac{f_s}{2}\right)$, the critical value obtained is the same in both cases and is equal to $V_M = 1.42\,\mathrm{V}$. We want to show through these two graphs that linearizing appropriately the PV generator model does not affect the accuracy of the model in predicting the period-doubling bifurcation of the system. The aim is to study the linear model of PV generator, which is going to help us to study the stability of our system, as it is very difficult to study stability with the nonlinear model of PV.

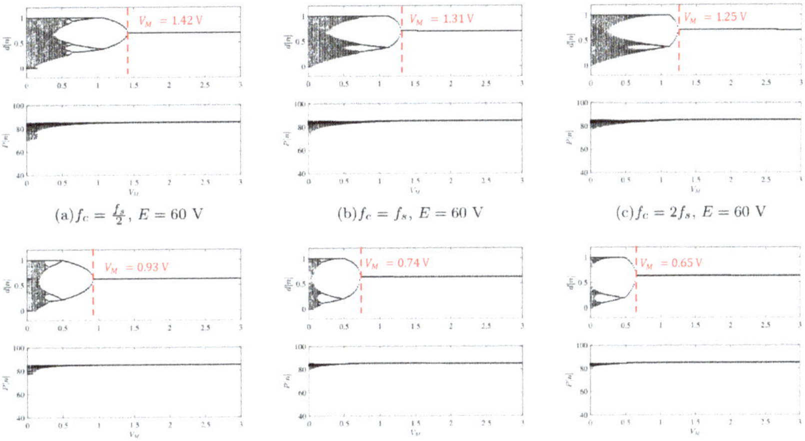

(a) $f_c = \frac{f_s}{2}$, $E = 60$ V (b) $f_c = f_s$, $E = 60$ V (c) $f_c = 2f_s$, $E = 60$ V

(d) $f_c = \frac{f_s}{2}$, $E = 48$ V (e) $f_c = f_s$, $E = 48$ V (f) $f_c = 2f_s$, $E = 48$ V

Figure 9. Bifurcation diagrams obtained by taking the amplitude of the carrier signal amplitude V_M as a bifurcation parameter for different values of the current sensor bandwidth ω_c and DC output voltage E. The linearized model of the PV generator close to the MPP was used.

Remark 4. *Figure 9 shows the bifurcation with the nonlinear model of the PV generator substituted by its linearized model constituting a linear resistance* $V_{\text{mpp}}/I_{\text{mpp}}$ *and a current source* $2I_{\text{mpp}}$. *It can be observed that the matching between these bifurcation diagrams and those obtained in Figure 6 is excellent. It is widely believed among the power electronics community that with values of steady-state duty cycles of less than* 0.5, *no external ramp is needed to achieve a stable system. However, from the previous bifurcation diagrams, it is clear that the system is still prone to period-doubling bifurcation even for* $D < 0.5$ *if the cut-off frequency of the current sensor is relatively low. Therefore, a ramp slope compensation is also necessary for duty cycle values less than* 0.5.

3.4. Stability Boundaries in the Parameter Space

Now that the nonlinear behaviour of the PV system under peak current-mode control is understood and the mechanisms of losing stability are known, the stability boundary for useful practical design will be determined.

The focus of this paper is on the period-doubling bifurcation boundary. The results helps in setting the design parameters as they show the stable region in the parametric space.

One of the ways to locate the subharmonic oscillation boundary is by using the expression of the characteristic equation $\det(\mathbf{P} - \lambda\mathbf{I}) = 0$, imposing the period-doubling condition in the eigenvalue λ and solving the resulting equation together with the steady-state condition. Therefore, at the boundary of this bifurcation, the following conditions hold

$$\det(\mathbf{P} + \mathbf{I}) = 0, \tag{24a}$$

$$\mathbf{x}(DT) - \mathbf{x}((D+1)T) = \mathbf{0}, \tag{24b}$$

where $\mathbf{0} \in \mathbb{R}^4$ is a null vector. Note that the integral state variable can be determined by (24b).

Figure 10 presents the stability study in two dimensions of the photovoltaic system obtained in MATLAB/SIMULINK. It clearly shows the regions of stability (period-1 oscillations) and instability (subharmonic oscillations) of the system in the E-plane for three values of the cut-off frequency: blue $f_c = f_s/2$, red $f_c = f_s$ and yellow $f_c = 2\,f_s$. The other system parameters are taken as in Tables 1 and 2. We see on this map that when $20\text{ V} \leq E \leq 30\text{ V}$, the system does not show period-doubling phenomenon except for $f_c = f_s/2$. In this case, this map is not sufficient to conclude on the system stability. Moreover, for $30\ \text{V} \leq E \leq 70\ \text{V}$, the stability domain of the photovoltaic system increases with the filter cut-off frequency. Finally, for $70\ \text{V} \leq E \leq 90\ \text{V}$, the filter cut-off frequency has no influence on the stability of the PV system. These results are more complete than the results obtained in the previous sections, so this map will be a crucial tool in the decision-making process for engineers in industries.

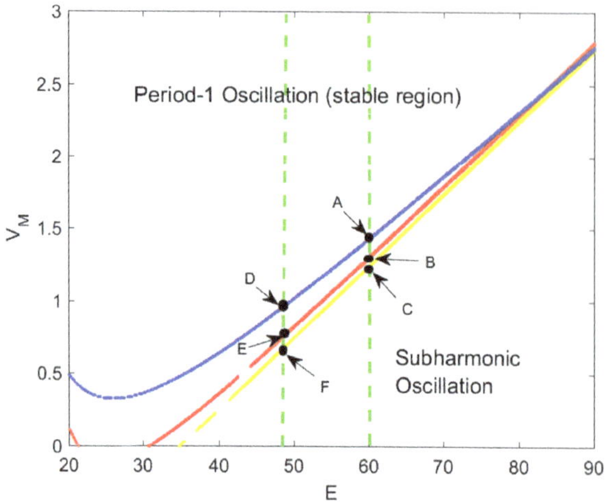

Figure 10. Stability boundaries in the plane $E - V_M$ for different values of cut-off frequency in MATLAB/SIMULINK software. f_s is fixed at $f_s = 50$ kHz. Where: A(60v,1.42v), B(60v,1.42v), C(60v,1.25v), D(48v,0.93v), E(48v,0.74v), and F(48v,0.65v) are the period splitting type bifurcation occur in the system. A and D are for fc = fs/2 ; B and E are for fc = fs ; and C and E are for fc = 2fs

Figure 11a shows the critical curves in the plane $D - V_M$ for different values of f_c. The conventional boundary with $\omega_c \to \infty$ is also shown. Figure 11b shows the critical curves in the plane $E - V_M$ for different values of f_c. The critical points for $E = 48$ and $E = 60$ are indicated in the figure. The critical values of the ramp amplitude for the different values of E and f_c remarkably coincide with the critical values obtained by the bifurcation diagrams in the previous section. For instance, for point A, one has $E = 60$ V and $f_c = f_s/2$ and the critical point of V_M is 1.42 V agreeing with the bifurcation diagram of Figure 6a. Above the curves, the system is stable and below it, it exhibits period-doubling bifurcation. The critical curve in the plane $D - V_M$ passes very close to the point $(0, \frac{1}{2})$ as long as f_c is relatively small. This curve becomes concave in this plane when V_M is increased. By increasing the ramp amplitude V_M, the region of stability is widened for $D > \frac{1}{2}$ and is reduced for $D < \frac{1}{2}$. Furthermore, the stable region becomes wider when increasing the cut-off frequency. These last results obtained with PSIM software are in perfect agreement with those of Figure 10 obtained with MATLAB/SIMULINK software.

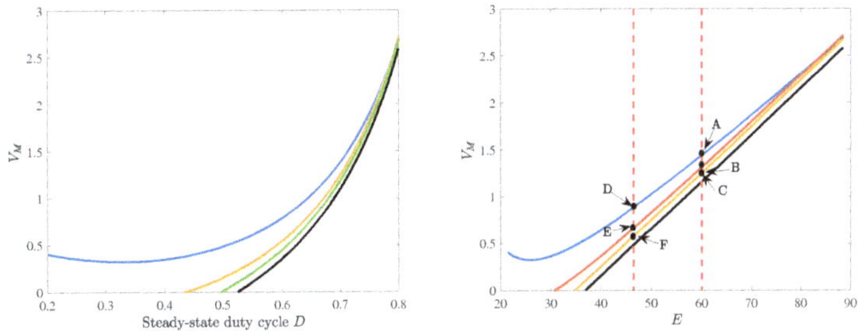

(**a**) (**b**)

Figure 11. Stability boundaries for different values of cut-off frequency in PSIM software: (**a**) in the plane $D - V_M$; (**b**) in the plane $E - V_M$. Where: A(60v,1.42v), B(60v,1.42v), C(60v,1.25v), D(48v,0.93v), E(48v,0.74v), and F(48v,0.65v) are the period splitting type bifurcation occur in the system. A and D are for fc=fs/2 ; B and E are for fc = fs ; and C and E are for fc = 2fs.

4. Conclusions

This paper focuses on the nonlinear behaviour and stability of the current-mode-controlled boost converter with the battery load. First, numerical analysis of its state equations, bifurcation diagrams and Lyapunov exponent was conducted in MATLAB/SIMULINK Software using a linear model of PV. Secondly, analogy simulations using PSIM were performed using a nonlinear models of the PV generator. The stability of a boost converter supplied by a PV panel was studied. To make an analytical study possible, the nonlinear PV generator has been linearizing around its MPPT. The simulation results considered three values of the cut-off frequency and two values of the output voltage. They showed through the bifurcation diagrams and the Lyapunov exponent that the system presents nonlinear phenomena such as chaos and periodic motion, which are influenced by system parameters and topological structure. The numerical results obtained in MATLAB/SIMULINK software are remarkable in their agreement with the analogy simulations in PSIM. We can also mention that linearizing appropriately the PV generator model does not affect the accuracy of the model in predicting the period-doubling bifurcation of the system. In general, we have seen that the stability increases with the frequency of the cut which is interesting about this study since the type of behaviour desired for a PV is the period-1.

Author Contributions: Conceptualization, E.R.M.K., A.S.T.K.; Data curation, M.S.S., T.T.T., A.R.M., M.T., Z.I.K.; Formal analysis, E.R.M.K., A.S.T.K., M.S.S., T.T.T., A.T.A., A.R.M., M.T., Z.I.K.; Investigation, T.T.T., A.T.A., A.R.M., M.T. Methodology, E.R.M.K., A.S.T.K., M.S.S., T.T.T., A.T.A., A.R.M., M.T., Z.I.K.; Resources, E.R.M.K., A.S.T.K., M.S.S., T.T.T., A.T.A., A.R.M., M.T., Z.I.K.; Software, A.S.T.K., M.S.S., T.T.T., A.R.M., M.T., Z.I.K.; Supervision, E.R.M.K.; Validation, M.S.S., A.T.A., A.R.M.; Visualization, A.T.A. and Z.I.K.; Writing – original draft, E.R.M.K., A.S.T.K.; Writing – review & editing, E.R.M.K., A.S.T.K., M.S.S., T.T.T., A.T.A., A.R.M., M.T., Z.I.K. All authors have read and agreed to the published version of the manuscript.

Funding: This research was funded by Prince Sultan University, Riyadh, Saudi Arabia.

Institutional Review Board Statement: Not applicable.

Informed Consent Statement: Not applicable

Data Availability Statement: Not applicable.

Acknowledgments: The authors would like to thank Prince Sultan University, Riyadh, Saudi Arabia for funding the Article Processing Charges (APCs) of this publication. Special acknowledgments are given to Automated Systems & Soft Computing Lab (ASSCL), Prince Sultan University, Riyadh, Saudi Arabia.

Conflicts of Interest: The authors declare no conflict of interest.

References

1. Srinivasarao, P.; Peddakapu, K.; Mohamed, M.; Deepika, K.K.; Sudhakar, K. Simulation and experimental design of adaptive-based maximum power point tracking methods for photovoltaic systems. *Comput. Electr. Eng.* **2021**, *89*, 106910. [CrossRef]
2. Kishore, D.K.; Mohamed, M.; Sudhakar, K.; Peddakapu, K. Swarm intelligence-based MPPT design for PV systems under diverse partial shading conditions. *Energy* **2023**, *265*, 126366. [CrossRef]
3. Kishore, D.K.; Mohamed, M.; Sudhakar, K.; Peddakapu, K. An improved grey wolf optimization based MPPT algorithm for photovoltaic systems under diverse partial shading conditions. *Proc. J. Phys. Conf. Ser.* **2022**, *2312*, 012063. [CrossRef]
4. Syafiq, A.; Pandey, A.; Adzman, N.; Abd Rahim, N. Advances in approaches and methods for self-cleaning of solar photovoltaic panels. *Sol. Energy* **2018**, *162*, 597–619. [CrossRef]
5. Bourourou, F.; Tadjer, S.A.; Habi, I. Wind Power Conversion Chain Harmonic Compensation using APF Based on FLC. *Alger. J. Renew. Energy Sustain. Dev.* **2020**, *2*, 75–83. [CrossRef]
6. Rana, K.; Kumar, V.; Sehgal, N.; George, S.; Azar, A.T. Efficient maximum power point tracking in fuel cell using the fractional-order PID controller. In *Renewable Energy Systems*; Azar, A.T., Kamal, N.A., Eds.; Advances in Nonlinear Dynamics and Chaos (ANDC); Academic Press: Cambridge, MA, USA, 2021; pp. 111–132.
7. Fekik, A.; Azar, A.T.; Kamal, N.A.; Serrano, F.E.; Hamida, M.L.; Denoun, H.; Yassa, N. Maximum Power Extraction from a Photovoltaic Panel Connected to a Multi-cell Converter. In Proceedings of the International Conference on Advanced Intelligent Systems and Informatics, Cairo, Egypt, 19–21 October 2020; Hassanien, A.E.; Slowik, A., Snášel, V., El-Deeb, H., Tolba, F.M., Eds.; Springer International Publishing: Cham, Switzerland, 2021; pp. 873–882.
8. Ammar, H.H.; Azar, A.T.; Shalaby, R.; Mahmoud, M.I. Metaheuristic Optimization of Fractional Order Incremental Conductance (FO-INC) Maximum Power Point Tracking (MPPT). *Complexity* **2019**, *2019*, 1–13. [CrossRef]
9. Amara, K.; Malek, A.; Bakir, T.; Fekik, A.; Azar, A.T.; Almustafa, K.M.; Bourennane, E.B.; Hocine, D. Adaptive neuro-fuzzy inference system based maximum power point tracking for stand-alone photovoltaic system. *Int. J. Model. Identif. Control* **2019**, *33*, 311–321. [CrossRef]
10. Ben Smida, M.; Sakly, A.; Vaidyanathan, S.; Azar, A.T. Control-Based Maximum Power Point Tracking for a Grid-Connected Hybrid Renewable Energy System Optimized by Particle Swarm Optimization. In *Advances in System Dynamics and Control*; Azar, A.T., Vaidyanathan, S., Eds.; Advances in Systems Analysis, Software Engineering, and High Performance Computing (ASASEHPC); IGI Global: Hershey, PA, USA, 2018; pp. 58–89.
11. El Aroudi, A. A new approach for accurate prediction of subharmonic oscillation in switching regulators—Part I: Mathematical derivations. *IEEE Trans. Power Electron.* **2016**, *32*, 5651–5665. [CrossRef]
12. Femia, N.; Petrone, G.; Spagnuolo, G.; Vitelli, M. A technique for improving P&O MPPT performances of double-stage grid-connected photovoltaic systems. *IEEE Trans. Ind. Electron.* **2009**, *56*, 4473–4482.
13. Pu, Q.; Zhu, X.; Liu, J.; Cai, D.; Fu, G.; Wei, D.; Sun, J.; Zhang, R. Integrated optimal design of speed profile and fuzzy PID controller for train with multifactor consideration. *IEEE Access* **2020**, *8*, 152146–152160. [CrossRef]
14. Hosseini, S.; Taheri, S.; Pouresmaeil, E.; Espinoza-Trejo, D.R. Enhancement of A Photovoltaic Inverter Efficiency Using A Shade-Tolerant MPPT. In Proceedings of the IECON 2021—47th Annual Conference of the IEEE Industrial Electronics Society, IEEE, Toronto, ON, Canada, 13–16 October 2021; pp. 1–5.
15. Garcerá, G.; Figueres, E.; Mocholí, A. Novel three-controller average current mode control of DC-DC PWM converters with improved robustness and dynamic response. *IEEE Trans. Power Electron.* **2000**, *15*, 516–528. [CrossRef]
16. Estcourt, C.; Stirrup, O.; Mapp, F.; Copas, A.; Howarth, A.; Owusus, M.W.; Low, N.; Saunders, J.; Mercer, C.; Flowers, P.; et al. O18. 3 Characteristics and outcomes of people who used Accelerated Partner Therapy for chlamydia in the LUSTRUM cluster cross-over randomised control trial. *Sex. Transm. Infect.* **2021**, *97*, A58.
17. Bianconi, E.; Calvente, J.; Giral, R.; Mamarelis, E.; Petrone, G.; Ramos-Paja, C.A.; Spagnuolo, G.; Vitelli, M. A fast current-based MPPT technique employing sliding mode control. *IEEE Trans. Ind. Electron.* **2012**, *60*, 1168–1178. [CrossRef]
18. Hosseini, S.; Taheri, S.; Farzaneh, M.; Taheri, H. A high-performance shade-tolerant MPPT based on current-mode control. *IEEE Trans. Power Electron.* **2019**, *34*, 10327–10340. [CrossRef]
19. Weidong, J.; Wang, L.; Wang, J.; Zhang, X.; Wang, P. A carrier-based virtual space vector modulation with active neutral-point voltage control for a neutral-point-clamped three-level inverter. *IEEE Trans. Ind. Electron.* **2018**, *65*, 8687–8696. [CrossRef]
20. Garcia-Teruel, A.; Forehand, D. A review of geometry optimisation of wave energy converters. *Renew. Sustain. Energy Rev.* **2021**, *139*, 110593. [CrossRef]
21. Guo-Hua, Z.; Bo-Cheng, B.; Jian-Ping, X.; Yan-Yan, J. Dynamical analysis and experimental verification of valley current controlled buck converter. *Chin. Phys. B* **2010**, *19*, 050509. [CrossRef]
22. El Aroudi, A.; Al-Numay, M.; Garcia, G.; Al Hossani, K.; Al Sayari, N.; Cid-Pastor, A. Analysis of nonlinear dynamics of a quadratic boost converter used for maximum power point tracking in a grid-interlinked PV system. *Energies* **2018**, *12*, 61. [CrossRef]
23. Erickson, R.W.; Maksimović, D. Principles of steady-state converter analysis. In *Fundamentals of Power Electronics*; Springer: Berlin/Heidelberg, Germany, 2020; pp. 15–41.
24. El Aroudi, A.; Zhioua, M.; Al-Numay, M.; Garraoui, R.; Al-Hosani, K. Stability Analysis of a Boost Converter with an MPPT Controller for Photovoltaic Applications. In Proceedings of the Mediterranean Conference on Information & Communication Technologies, Saidia, Morocco, 7–9 May 2015; Springer: Berlin/Heidelberg, Germany, 2016; pp. 483–491.

25. Dongmo Wamba, M.; Montagner, J.P.; Romanowicz, B. Imaging deep-mantle plumbing beneath La Réunion and Comores hot spots: Vertical plume conduits and horizontal ponding zones. *Sci. Adv.* **2023**, *9*, eade3723. [CrossRef]
26. Wamba, M.; Montagner, J.P.; Romanowicz, B.; Barruol, G. Multi-Mode Waveform Tomography of the Indian Ocean Upper and Mid-Mantle Around the Réunion Hotspot. *J. Geophys. Res. Solid Earth* **2021**, *126*, e2020JB021490. [CrossRef]
27. Kagho, L.; Dongmo, M.; Pelap, F. Dynamics of an Earthquake under Magma Thrust Strength. *J. Earthq.* **2015**, *2015*, 1–9. [CrossRef]
28. Pelap, F.; Fomethe, A.; Dongmo, M.; Kagho, L.; Tanekou, G.; Makenne, Y. Direction effects of the pulling force on the first order phase transition in a one block model for earthquakes. *J. Geophys. Eng.* **2014**, *11*, 045007. [CrossRef]
29. Dongmo, M.; Kagho, L.; Pelap, F.; Tanekou, G.; Makenne, Y.; Fomethe, A. Water effects on the first-order transition in a model of earthquakes. *Int. Sch. Res. Not.* **2014**, *2014*. [CrossRef]
30. Giaouris, D.; Elbkosh, A.; Banerjee, S.; Zahawi, B.; Pickert, V. Stability of switching circuits using complete-cycle solution matrices. In Proceedings of the 2006 IEEE International Conference on Industrial Technology, Mumbai, India, 16–18 August 2006; pp. 1954–1959.
31. Giaouris, D.; Banerjee, S.; Zahawi, B.; Pickert, V. Stability analysis of the continuous-conduction-mode buck converter via Filippov's method. *IEEE Trans. Circuits Syst. I Regul. Pap.* **2008**, *55*, 1084–1096. [CrossRef]
32. Giaouris, D.; Banerjee, S.; Zahawi, B.; Pickert, V. Control of fast scale bifurcations in power-factor correction converters. *IEEE Trans. Circuits Syst. II Express Briefs* **2007**, *54*, 805–809. [CrossRef]
33. Tucker, W. Computing accurate Poincaré maps. *Phys. D Nonlinear Phenom.* **2002**, *171*, 127–137. [CrossRef]
34. Giaouris, D.; Banerjee, S.; Stergiopoulos, F.; Papadopoulou, S.; Voutetakis, S.; Zahawi, B.; Pickert, V.; Abusorrah, A.; Al Hindawi, M.; Al-Turki, Y. Foldings and grazings of tori in current controlled interleaved boost converters. *Int. J. Circuit Theory Appl.* **2014**, *42*, 1080–1091. [CrossRef]
35. Sahan, B.; Vergara, A.N.; Henze, N.; Engler, A.; Zacharias, P. A single-stage PV module integrated converter based on a low-power current-source inverter. *IEEE Trans. Ind. Electron.* **2008**, *55*, 2602–2609. [CrossRef]
36. Deane, J.H.; Hamill, D.C. Instability, subharmonics and chaos in power electronic systems. In Proceedings of the 20th Annual IEEE Power Electronics Specialists Conference, Milwaukee, WI, USA, 26–29 June 1989; pp. 34–42.
37. Wolf, A.; Swift, J.B.; Swinney, H.L.; Vastano, J.A. Determining Lyapunov exponents from a time series. *Phys. D Nonlinear Phenom.* **1985**, *16*, 285–317. [CrossRef]
38. Zhang, S.; Deng, Z.; Li, W. A precise Runge–Kutta integration and its application for solving nonlinear dynamical systems. *Appl. Math. Comput.* **2007**, *184*, 496–502. [CrossRef]
39. Banerjee, S.; Ranjan, P.; Grebogi, C. Bifurcations in two-dimensional piecewise smooth maps-theory and applications in switching circuits. *IEEE Trans. Circuits Syst. I Fundam. Theory Appl.* **2000**, *47*, 633–643. [CrossRef]
40. Robert, B.; Robert, C. Border collision bifurcations in a one-dimensional piecewise smooth map for a PWM current-programmed H-bridge inverter. *Int. J. Control* **2002**, *75*, 1356–1367. [CrossRef]
41. Wolf, D.M.; Varghese, M.; Sanders, S.R. Bifurcation of power electronic circuits. *J. Frankl. Inst.* **1994**, *331*, 957–999. [CrossRef]

sustainability

MDPI

Article

P2P Coordinated Control between SPV and STATCOM in a Microgrid for Power Quality Compensation Using LSTM–Genetic Algorithm

Durgamadhab Swain [1,2], Meera Viswavandya [3], Ritesh Dash [4], Kalvakurthi Jyotheeswara Reddy [4], Dhanamjayulu Chittathuru [5,*], Arunkumar Gopal [5], Baseem Khan [6,*] and Manam Ravindra [7]

1 Research Scholar, Department of Electrical Engineering, Biju Patnaik University of Technology, Rourkela 769015, India; durgamadhab17@gmail.com
2 Department of Electrical Engineering, Ajay Binay Institute of Technology, Cuttack 753014, India
3 Department of Electrical Engineering, Odisha University of Technology and Research, Bhubaneswar 751029, India; mviswavandya@outr.ac.in
4 School of Electrical and Electronics Engineering, REVA University, Bengaluru 560064, India; rdasheee@gmail.com (R.D.); jyothireddy.kalvakurthi@gmail.com (K.J.R.)
5 School of Electrical Engineering, Vellore Institute of Technology, Vellore 632014, India; g.arunkumar@vit.ac.in
6 Department of Electrical and Computer Engineering, Hawassa University, Hawassa P.O. Box 05, Ethiopia
7 Department of Electrical and Electronics Engineering, Aditya College of Engineering, Surampalem 533437, India; ravieeejntu@gmail.com
* Correspondence: dhanamjayulu.c@vit.ac.in (D.C.); baseemkh@hu.edu.et (B.K.)

Abstract: The deployment of a static synchronous compensator within a microgrid can facilitate voltage and reactive power regulation, leading to enhanced stability and reliability. Within a microgrid setting, the effectiveness of a STATCOM in balancing the power supply is influenced by several factors, including the system configuration, the operating conditions, and the specific requirements of the power grid. The capacity, response time, and magnitude of system disturbances also play a role in determining the STATCOM's ability to balance the power supply. To ensure the successful integration of a STATCOM within a microgrid, coordinating the control system with other distributed energy resources (DER), especially when multiple control strategies are employed, can be a challenging task. Therefore, a meticulously designed control system is indispensable to guarantee the microgrid's efficient and effective operation. The use of GA in LSTM tuning can accelerate the process of identifying the optimal hyperparameters for a specific task, obviating the need for time-consuming and computationally expensive grid searches or manual tuning. This method can be particularly advantageous when handling large data sets and complex models. In this paper, an attempt has been made to model the STATCOM to communicate with the microgrid, tuned using LSTM–GA, for the effective calculation of real and reactive power support during grid disturbances.

Keywords: algorithm; GA; PSO; PSO–LSTM; search space

Citation: Swain, D.; Viswavandya, M.; Dash, R.; Reddy, K.J.; Chittathuru, D.; Gopal, A.; Khan, R.; Ravindra, M. P2P Coordinated Control between SPV and STATCOM in a Microgrid for Power Quality Compensation Using LSTM–Genetic Algorithm. *Sustainability* **2023**, *15*, 10913. https://doi.org/10.3390/su151410913

Academic Editor: Noradin Ghadimi

Received: 13 May 2023
Revised: 3 July 2023
Accepted: 8 July 2023
Published: 12 July 2023

1. Introduction

A microgrid is a type of electrical system that can operate independently or in coordination with the main grid. It consists of one or more distributed energy resources (DERs), such as solar panels, wind turbines, batteries, or generators, which are used to generate or store electricity [1]. Microgrids are designed to provide reliable, efficient, and eco-friendly power to local communities, businesses, and institutions, particularly in remote or off-grid areas where access to the main grid is limited or unreliable. Additionally, they can function as a backup source of power during emergencies, such as grid outages or natural disasters.

Voltage fluctuations are a common power quality issue in microgrids, especially those that incorporate renewable energy sources such as solar and wind. These sources have variable outputs, causing voltage fluctuations that can negatively impact the stability and performance of the microgrid [2]. Voltage fluctuation can be addressed based on voltage regulation. Voltage regulation is the process of keeping the voltage level in an electrical system stable and constant. This process is particularly important in microgrids, where voltage regulation is critical to ensuring dependable and efficient system operation, especially when intermittent renewable energy sources are present. To ensure voltage regulation in a microgrid, voltage regulators such as automatic voltage regulators (AVRs) or static VAR compensators (SVCs) can be used. These devices can regulate voltage levels in real time by increasing or decreasing the reactive power output of the system.

A static synchronous compensator (STATCOM) is a power electronics device that is frequently used for voltage regulation and reactive power compensation in electrical power systems. It is a type of flexible AC transmission system (FACTS) device that can introduce reactive power into the system to enhance the power quality and voltage stability. With the capacity to supply both capacitive and inductive reactive power, the STATCOM can react quickly to changes in system conditions, making it a versatile device that is appropriate for various applications, including microgrids [3].

Ping He et al. [4], in their paper, present a coordinated control strategy for a PSS and STATCOM, two critical power system devices. The goal of this approach is to enhance power system stability and damping, particularly in the presence of disturbances such as faults or sudden load changes. The study employs a multi-machine power system model and simulation techniques to evaluate the effectiveness of the proposed coordinated control strategy in various scenarios. According to the simulation outcomes, the strategy significantly improves the power system stability and damping and outperforms other control methods that disregard the coordination between the PSS and STATCOM.

Kaliaperumal Rukmani et al. [5] introduce a new approach to optimize the allocation of distribution static compensators (D-STATCOMs) in distribution systems where there is uncertainty. D-STATCOMs are crucial in enhancing the power quality and stability of distribution systems by injecting reactive power. The proposed method involves a combination of fuzzy logic and particle swarm optimization (PSO) algorithms to determine the optimal locations and sizes of D-STATCOMs. Fuzzy logic is utilized to manage uncertainties in the system parameters, while the PSO algorithm is utilized to locate the optimal solution.

Tariq, M. et al. [6], in their article, describe a new approach to voltage regulation and power quality improvement using static synchronous compensators. The proposed method involves adjusting the phase angle between the current and voltage using a simple PI controller to control the output voltage of the STATCOM. The effectiveness of the proposed method is evaluated through simulation studies, which show that it can successfully regulate the voltage and improve the power quality.

Anil Bharadwaj et al. [7] propose a novel approach to tuning PI and PID controllers in power systems equipped with various types of flexible AC transmission system devices, including a STATCOM, SSSC, and UPFC. The proposed method aims to minimize the damping of oscillations in the power system by adjusting the parameters of the controllers. To evaluate the performance of the proposed tuning method, the authors use a multi-machine power system model and conduct simulation studies. The results indicate that the proposed method can effectively improve the damping of oscillations in the power system and outperform other tuning methods that do not consider the presence of FACTS devices.

Sarath Perera et al. [8] present a framework for the reduction of power network oscillations with the use of static synchronous compensators and the synthesis of $H2/H\infty$ controllers. The framework employs an $H2/H\infty$ synthesis technique to design the controller and improve the stability of the power system. The paper evaluates the effectiveness of the proposed framework through simulations using a power system model with a STATCOM, and the results indicate that the framework can effectively reduce oscillations in the power system and enhance its stability.

Claudia Battistelli et al. [9] suggest using the whale optimization algorithm (WOA) to develop power system stabilizers for multi-machine power systems. The goal of this method is to enhance the stability of the system by designing power system stabilizers that minimize oscillations. The authors evaluate the proposed method by simulating it using a multi-machine power system model, and the findings show that the WOA-based stabilizers can effectively damp oscillations in the power system and enhance its stability [7].

Liangce He et al. [10] propose a method to optimize the economic and environmental performance of an integrated regional energy system by incorporating an integrated demand response into the environmental economic dispatch process. The proposed method optimizes the dispatch of different energy sources to minimize the total cost and emissions of the system while considering the impact of DR on the load demand. The effectiveness of the proposed method is evaluated through simulations, which show that it can effectively reduce the total cost and emissions of the system while considering the impact of DR.

D. Ranamuka et al. [11,12] propose a strategy for the control of the power flow in distribution systems using the coordinated control of distributed solar–PV and battery energy storage units. The objective is to enhance the stability and efficiency of the distribution systems through real-time power flow control via distributed energy resources. The proposed method is evaluated using a distribution system model with solar–PV and battery energy storage units, and the simulation results demonstrate its effectiveness in improving the stability and efficiency of the distribution system [13,14].

The paper [15,16] proposes a power system stabilizer (PSS) design for the damping of low-frequency oscillations in a multi-machine power system with the integration of renewable power generation. The proposed PSS design is based on eigenvalue analysis and aims to optimize the damping of low-frequency oscillations in the system. To evaluate the effectiveness of the proposed PSS design, simulations are conducted using a multi-machine power system model with renewable power generation. The simulation results indicate that the proposed PSS design can effectively dampen low-frequency oscillations and enhance the stability of the power system [17,18]. As observed in the literature review, GA searches for a solution in the space based on probabilistic principles. Therefore, there is no sufficient guarantee that the system will always be optimized to the global optimal solution. Depending on the complexity of the STATCOM control problem and the specific fitness function used, there is a risk that GA may become trapped in local optima and fail to discover the best possible controller settings.

The performance of particle swarm optimization (PSO) in controlling STATCOMs in power systems may be limited by multiple factors [19,20]. One of these factors is the high sensitivity of the PSO algorithm to the initial conditions, which may lead to suboptimal solutions or becoming stuck in local optima. A solution to this problem can be achieved by starting with a good initial population and modifying the parameters of the PSO algorithm [21,22]. Another limitation is the inability of the PSO STATCOM to handle uncertainties in the power system, such as variations in renewable energy sources or changes in load demand. This may result in suboptimal STATCOM operation and reduced performance in controlling the power system [23,24].

Hyperparameter selection in the case of optimization is a critical challenge as it is the hyperparameters that will determine the area of optimization. Overfitting is a common problem in the case of data fitting, which is primarily due to the involvement of long-term dependencies. To avoid such conditions, it is necessary to store past data in the system's memory for the easy analysis and prediction of situations under abnormal conditions.

PSO, inspired by swarm behavior, is a population-based optimization algorithm that seeks optimal solutions in a problem space. In contrast, long short-term memory (LSTM) is a type of recurrent neural network (RNN) renowned for capturing and learning temporal dependencies in sequential data. The integration of PSO with LSTM involves utilizing PSO as a training algorithm to optimize LSTM's weights and biases. PSO achieves this by iteratively updating the particle positions and velocities based on personal and global best solutions, enabling LSTM to discover optimal parameter values for enhanced predictive accuracy. However, empirical studies consistently demonstrate LSTM's superiority over PSO-based LSTM models. LSTM's inherent capability to capture long-term dependencies and handle sequential data empowers it in time series prediction tasks. LSTM exhibits stronger learning capabilities and generalization power compared to PSO, which primarily focuses on optimization.

The combination of LSTM and genetic algorithm (GA) can have diverse applications, including time-series prediction, anomaly detection, and optimization problems [25,26]. For time-series prediction, LSTM can be utilized to forecast future values based on historical data, while GA can be used to optimize the hyperparameters of the LSTM model, such as the learning rate and the number of LSTM cells. In anomaly detection, LSTM can learn the regular patterns in data and identify any deviations from them, and GA can optimize the threshold for the detection of anomalies [27,28]. In optimization problems, LSTM can serve as a surrogate model for the evaluation of the objective function, and GA can search for the optimal solution.

In this paper, an attempt has been made to design a STATCOM, particularly in a microgrid, to provide voltage and reactive power support under variations in environmental parameters. We have analyzed various grid disturbances related to the injected current, voltage, and harmonics when utilizing a STATCOM. These disturbances include voltage sags and swells, which are temporary voltage decreases or increases caused by faults or abrupt changes in load demand. The current injected by the STATCOM is carefully examined to address and stabilize the voltage levels during such disturbances. Furthermore, we investigate the impact of harmonics on the grid, which refer to additional frequencies that can distort the sinusoidal waveform of the power supply. Through an evaluation of the harmonic content within the injected voltage, the effectiveness of the STATCOM in mitigating harmonic distortion and enhancing the power quality is assessed. By considering these grid disturbances, the study provides valuable insights into the STATCOM's performance and capabilities in effectively managing voltage fluctuations and harmonics, thus ensuring a reliable and efficient power supply. LSTM has been used to store the previous memory and historical data of the power quality issues and the amount of reactive power support, whereas genetic algorithm provides support for hyperparameter optimization. The objectives of the research can be summarized as follows.

- Design of LSTM–GA mathematical modeling for STATCOM microgrid analysis. The limitation of GA modeling in handling convex optimization can be best addressed using this LSTM–GA model. LSTM will provide memory-level access to the past historical data along with the lookup table.
- Modeling of the proposed system using MATLAB Simulink model and its validation under dynamic non-linear loading conditions and PV output variation with respect to environmental parameters.
- A detailed comparative analysis with another established benchmarking model in terms of stability analysis; validation has been carried out with a step function.

2. STATCOM Architecture and Analysis

STATCOMs play a vital role in power systems by regulating the voltage and providing reactive power support. These devices are built on a voltage source converter (VSC) architecture and are connected to the grid through phase reactors and a step-up transformer. The VSC incorporates insulated gate bipolar transistors (IGBTs) in a modular multi-level converter (MMC) configuration.

The primary objective of a STATCOM is to control the voltage at its point of connection to the grid. It achieves this by generating or absorbing reactive power, effectively regulating the system voltage and enhancing power system stability. Unlike conventional reactive power compensation devices, such as capacitor banks or synchronous condensers, STATCOMs offer the advantages of a rapid response and continuous control across a wide range of operating conditions.

At the core of a STATCOM lies the VSC, comprising power semiconductor switches, particularly IGBTs, which can swiftly switch on and off to produce the desired voltage waveform. The MMC configuration within the VSC allows for high-voltage operation and facilitates the generation of a high-quality voltage with minimal harmonic distortion. By converting DC power into AC power, the VSC allows the STATCOM to inject or absorb reactive power into the grid as required.

Through the dynamic adjustment of the voltage waveform produced by the VSC, the STATCOM actively regulates the grid voltage, compensates for reactive power imbalances, mitigates voltage flicker, improves the power factor, and bolsters the overall power system stability and reliability. The control system of the STATCOM continuously monitors the system conditions and employs feedback control algorithms to adapt the voltage output of the VSC accordingly. The presence of a step-up transformer enables the STATCOM to connect to the high-voltage transmission grid, ensuring appropriate voltage levels for effective power transmission and distribution. Additionally, the phase reactors provide impedance and protection to the STATCOM system while limiting the flow of short-circuit currents.

The diagram in Figure 1 illustrates the principle behind the controlled switching of the IGBT valves in a single sub-module, which generates the fundamental three-level waveform utilized by the STATCOM to control the contribution of reactive power to the grid. However, the performance, rating, and controllability of the STATCOM can be significantly enhanced by the connection of multiple sub-modules in series.

When these sub-modules are interconnected, it enables the construction of a voltage waveform with an improved resolution. This means that the voltage can be precisely controlled with finer granularity, resulting in the more accurate regulation of the reactive power. The increased resolution enhances the overall performance of STATCOM, enabling it to provide more efficient and precise control over reactive power compensation in the grid. Figure 2 presents the STATCOM vector analysis under the stand-by, inductive, and capacitive modes, respectively.

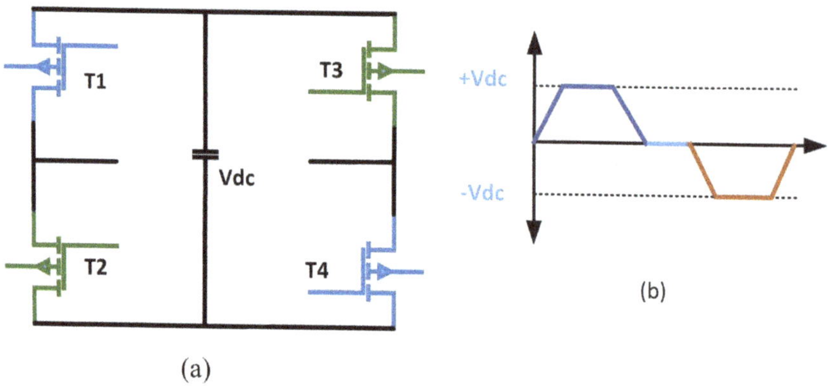

Figure 1. STATCOM. (**a**) Typical full-bridge sub-module, (**b**) output waveform.

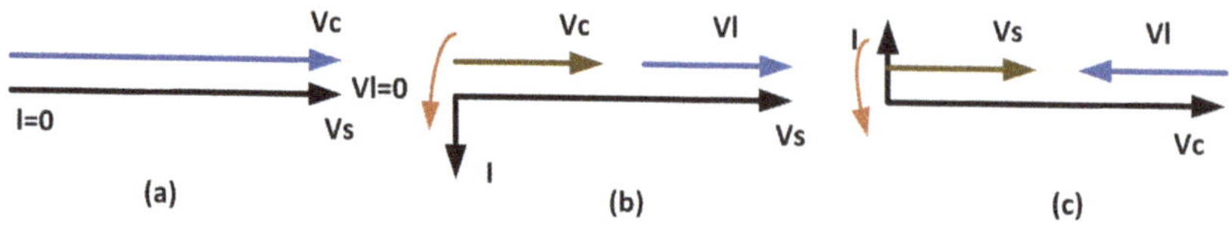

Figure 2. STATCOM. (**a**) Standby mode, (**b**) inductive mode, (**c**) capacitive mode.

Furthermore, the series connection of multiple sub-modules allows for the higher rating capability of the STATCOM. By combining the voltage outputs of each sub-module, the overall voltage level and capacity of the STATCOM can be increased. This expanded rating capability enables the STATCOM to handle larger reactive power demands and provide enhanced support to the power system (Figure 3).

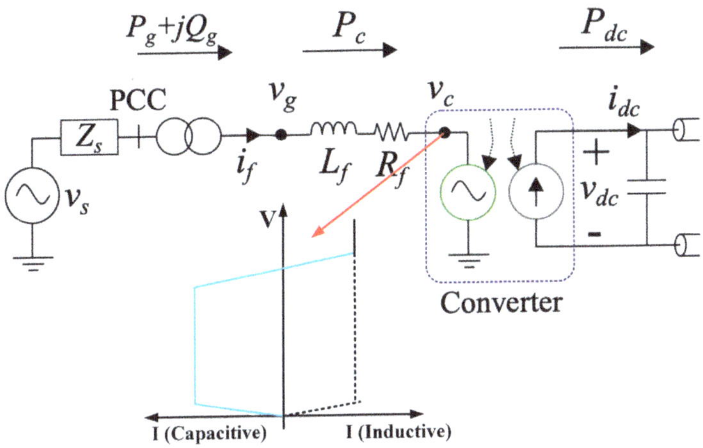

Figure 3. STATCOM along with IV characteristics [8].

Two control loops are generally involved in a VSC-based STATCOM architecture. The outer control loop will control the V_{ac} or V_{dc} depending on the control requirements and the number of variables involved in the system. The inner control loop controls the actual response with the set values. Therefore, the current control loop becomes

$$V_c^d = \omega_g L_f i_f^g - (K_p - \frac{K_i}{s})(i_f^{d*} - i_f^d) + V_g^d \quad (1)$$

$$V_c^q = -\omega_g L_f i_f^g - (K_p - \frac{K_i}{s})(i_f^{q*} - i_f^q) + V_g^q \quad (2)$$

Equations (1) and (2) represent the direct and quadrature axis control of the voltage waveform in the inner current control loop.

3. Problem Formulation and Solution Methodology

The single-line multi-machine bus (SIMB) microgrid implementing a STATCOM is shown in Figure 4.

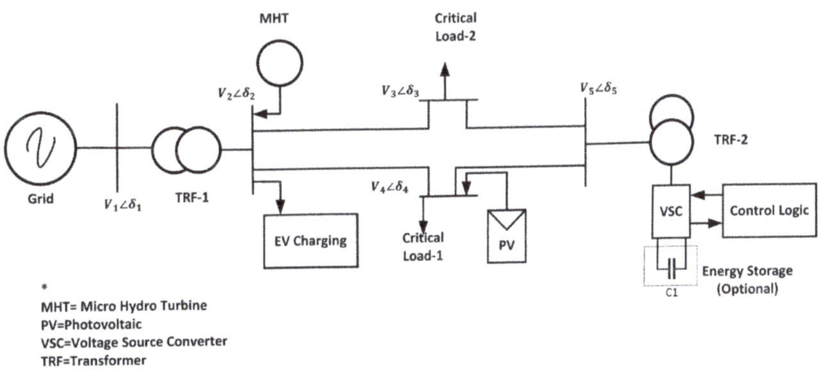

Figure 4. SIMB architecture for microgrid.

This architecture is widely used for STATCOM performance evaluation under the microgrid model. The model mainly consists of an infinite bus connected to a synchronous generator at one side through a two-winding transformer; at the other side, it is connected to a microgrid architecture supported by a STATCOM and PV source. The STATCOM under investigation is a GTO-based voltage source converter (VSC). Here, the VSC will generate a controllable voltage in accordance with leakage reactance. Then, the voltage difference between the STATCOM terminal and bus will determine the type of power exchange, such as active power or reactive power, between the two devices. The amount of reactive power exchange can be controlled using the voltage and phase angle δ.

Therefore, the nonlinear equation between the STATCOM's voltage and current becomes

$$I_{LO}^{*} = I_{Load} + jI_{Loq} \tag{3}$$

and

$$V_0 = CV_{dc}(cos\theta + jSin\theta) \tag{4}$$

In Equation (3), I_{LO}, I_{Load} and I_{Loq} represent the STATCOM's load current and the load current with respect to the d-axis and q-axis, respectively. Similarly, V_{dc} represents DC-ref voltage at the input of the STATCOM. Further, Equation (4) can be modified as follows:

$$V_0 = \frac{C^2}{C_{dc}} V_{dc} \angle\theta [I_{Load} Cos\theta + I_{loq} Sin\theta] \tag{5}$$

Equation (5) represents the output voltage equation of the STATCOM, designed based on the d- and q-axis current levels. Here, "c" represents the ratio between the AC and DC voltages. In order to calculate the virtual electrical torque, the speed deviation has been taken into consideration, $\delta\omega$. Therefore, the new damping controller can be designed as a lead–lag compensating controller.

Based on Equation (5), it is understood that the proper tuning of the STATCOM parameters, such as θ and C, is required to minimize the damping at the injected voltage level. Therefore, the optimization objective function can be formulated as

$$J = \Sigma_{i=1}^{N_p} \int_0^t |\delta\omega_i| t dt \tag{6}$$

In Equation (6), t represents the simulation time for the model and N_p represents the size of the population in genetic algorithm. The objective is to minimize the cost function and thereby improve the settling time and overshoot.

During cost function optimization using GA, there is a state called the fitness function or value evaluation, which requires a probability evaluation of each chromosome pair in the objective function. This requires an iteration to be run in order to evaluate the fitness value. Therefore, in order to reduce the optimization time in evaluating the constraints at each step, long short-term memory (LSTM) has been introduced. LSTM will hold the best solution for subsequent levels of iteration and thereby reduce the optimization time by t_{n-best}, where t_n is the total duration of the iteration, which has been reduced to t_{n-best} based on the best solution.

Figure 5 shows the LSTM–GA architecture for STATCOM PI controller optimization. As observed, GA produces two sets of optimized data related to the electrical torque reference value and C value, which is simply the ratio of the DC injected voltage to the AC injected voltage. The LSTM encoder will hold the best-optimized value during GA iteration, along with the time stamp. Therefore, during the decoding process, the same time stamp can be utilized for reactive power generation and subsequent voltage support, along with SSR damping.

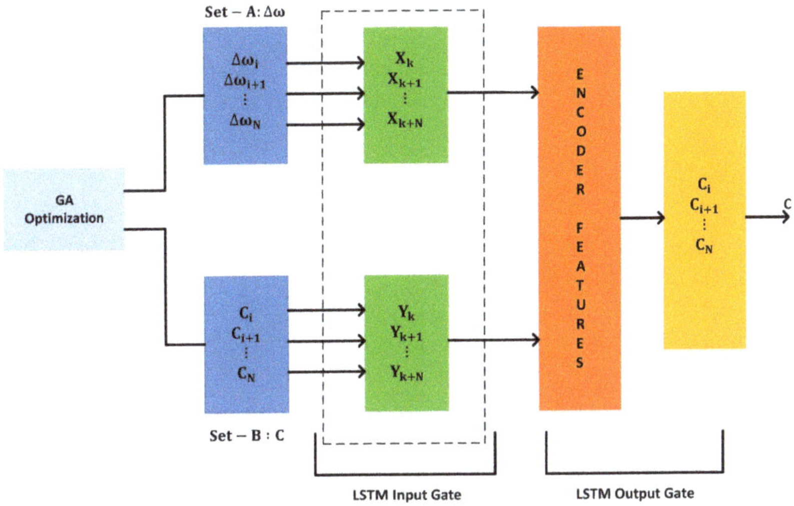

Figure 5. LSTM–GA architecture for STATCOM PI controller optimization.

In order to properly train the model, it is highly recommended to activate each layer in the LSTM architecture using the appropriate activation function. The inputs to the activation layer are represented by f and V, while the output of the LSTM network provides the reactive power compensation factor in $V\angle\delta$. Generally, each layer in the activation module evaluates the weighted sum of all its input connections and maps it to an output, as shown in Equation (7), where λ represents the layer of the input module.

$$V_{\lambda_j} = \sum_i V\angle\delta_{\lambda_{ji}}\, y_i \quad for\ \lambda \in \{f, V, Q, P\} \tag{7}$$

The recurrent connection between each layer λ can be varied by changing the unit of i. By applying the squashing function $f_{\lambda i}$ on y_i, the output for each layer λ can be modeled as

$$y_{\lambda_j} = f_{\lambda i}(V_{\lambda j}) \quad for\ \lambda \in \{f, V, Q\} \tag{8}$$

In Equation (6), a cascaded inner current control loop has been implemented to interconnect LSTM and GA with the PI control features. Each memory cell unit in LSTM holds the previous state in the same proportion as the activation in the forget layer gate. Therefore, the current state vector S_{cj} updates itself based on the modulus of the activation function at the input gate. Hence,

$$S_{cj} = Y_{\psi j}\hat{S_{cj}} + y_{ij}x_{cj} \tag{9}$$

Based on Equation (9), the learning rate can be made more effective by designing each layer to track the activity flow over time. To achieve this, an eligibility trace module has been provided to trace the most recent activity value, as presented in Equation (8).

$$\sum \lambda_{ji} = Y_i \ldots\ldots \forall \lambda \in [v\angle\delta, \theta] \tag{10}$$

Similarly, the memory cell and forget gate can be modeled as

$$\sum P_{ji} = Y_{\epsilon i}\hat{\epsilon}P_{ji} + Y_{lj}Yi \tag{11}$$

To develop a robust algorithm to track the required output against the predicted output, a time range of [0, 1] has been implemented. The cross-entropy function has been used to quantify the error between the predicted and actual outputs of the LSTM module. The pseudo-code for the implementable algorithm is presented in Algorithm 1.

GA requires the appropriate encoding and representation of solutions as chromosomes. In convex optimization (Algorithm 1), the solutions are typically represented as vectors or matrices, and the optimization algorithms exploit the convex structure. The mapping of these representations to the chromosome structure is used in GA.

Algorithm 1 Voltage and angle error evaluation—training pseudo-code

Require: V_{er},I_{der}, θ_s
Ensure: 3 Variable state transition pattern
  for $V_t = 1$ **do**
    for `f=-1` **do**
      for $V_{er} = 0$ **do**
        `Evaluate` V_{er} `and` θ_s
      end for
    end for
  end for
  Weight w_i=min(δT_{er}, δW_{er},θ_s)
  $\theta_x V_x = \max(\theta_{x-1} V_{x-1}, \theta_x V_x)$
  if $(M^{t_0}_{L-1} - M^{t_1}_{L}) \leq 0.02$ **then**
    break
  else
    $(M^{t_0}_{L-1} - M^{t_1}_{L}) \leq 0.02$
  end if

4. Result Analysis

The proposed LSTM–GA model has been designed using MATLAB Simulink as per the architecture shown in Figure 6.

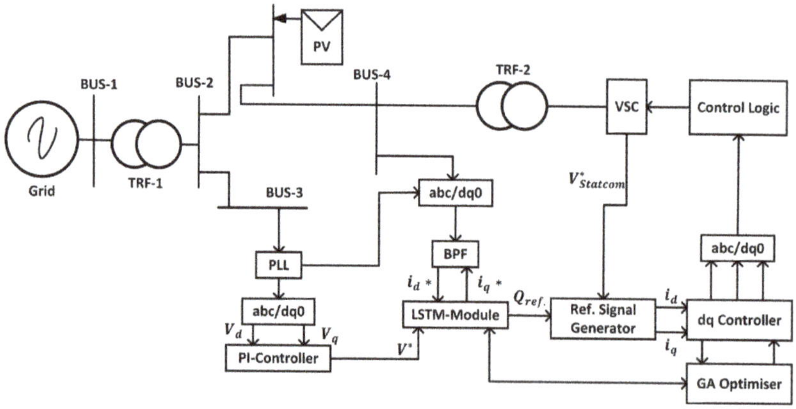

Figure 6. Simulink model for STATCOM microgrid coordinated control action using LSTM and GA.

Here, the LSTM module will process the i_d, i_q, and v^* values to determine the required amount of reactive power for the grid, which the STATCOM needs to produce. LSTM will share the hyperplane with the GA-optimized module to evaluate the agent position and initialize the chromosome parameter with four different variables.

Table 1 shows the STATCOM model's parameters, which have been used in the Simulink model to perform the analysis. All the set points, especially the AC voltage reference magnitude and DC voltage set point, are assumed to be per unit. All the AC and DC regulator gains have been evaluated based on a genetic algorithm ensemble with the classical Ziegler–Nichols method.

Table 1. STATCOM module parameters for Simulink model used in microgrid architecture.

Sr. No.	Name of Parameter	Rating	Remarks
1	Coupling capacitor	420 micro F	Storage
2	PWM frequency	2.33 kHz	Under Modulation
3	Coupling TFR	210:800	Centre Tap
4	AC ref. voltage	1 pu	707 V
5	DC ref. voltage	750 V	-
6	AC voltage regulator gain	[0.52 0.39]	ZNM
7	DC voltage regulator gain	[0.03 0.27]	ZNM-GA
8	Current regulator gain	[0.11 0.17]	GA-LSTM

As observed, the regulator gain for the AC voltage is [0.52 0.39], which was evaluated through the Ziegler–Nicholas method (ZNM), and the corresponding DC voltage gain was evaluated through ZNM and GA. Moreover, the current regulator gain, which is a function of both the AC voltage and DC voltage regulation gains, has been evaluated through the ZNM–LSTM ensemble with GA. Thus, a single hyperplane can be maintained throughout the analysis.

Figures 7 and 8 present the LSTM–GA performance analysis for two different values of μ, i.e., 0.11 and 0.18, respectively. As observed, the objective of LSTM–GA is to forecast the required reactive power support for the microgrid, which is 15.58% and 12.10%, respectively, in this case. With the increase in the chromosome size, the system is able to accurately predict the amount of required reactive power support for the grid. In the subsequent discussion, the performance analysis has been carried out with μ of 0.18.

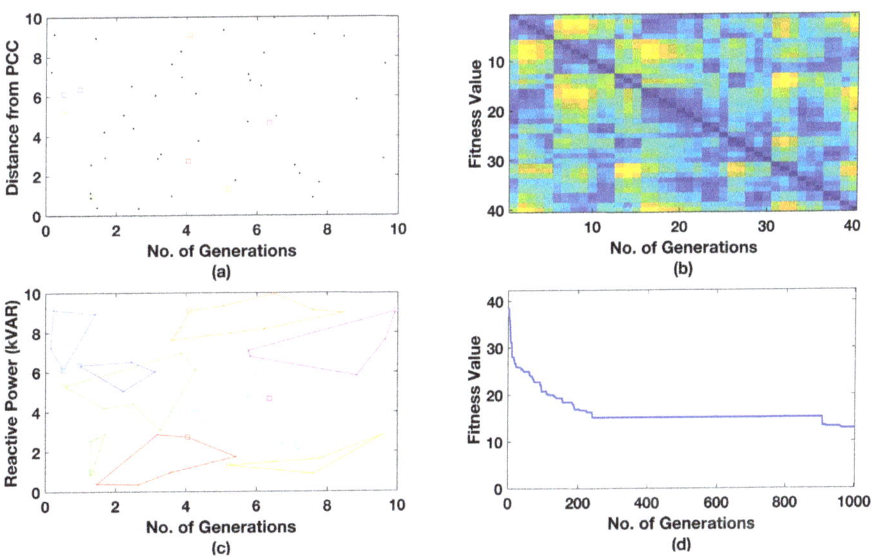

Figure 7. GA performance with $\mu = 0.11$. (**a**) FACTS location initialization, (**b**) reactive power support range, (**c**) percentage of reactive power support, (**d**) best solution history.

Figure 9 presents the STATCOM DC link voltage for three different models (a) the Fuzzy–PI STATCOM, (b) PSO–PI STATCOM, and (c) proposed LSTM–GA–PI STATCOM (LGPS). As observed, the "LGPS" model produces a standard optimized DC link voltage of 700.24 V (Figure 9c), which is 0.32% less compared to the fuzzy model (Figure 9a) and 0.23% less compared to the PSO STATCOM (Figure 9b) model. This reduction in the voltage percentage will also reduce the voltage stress on the switch.

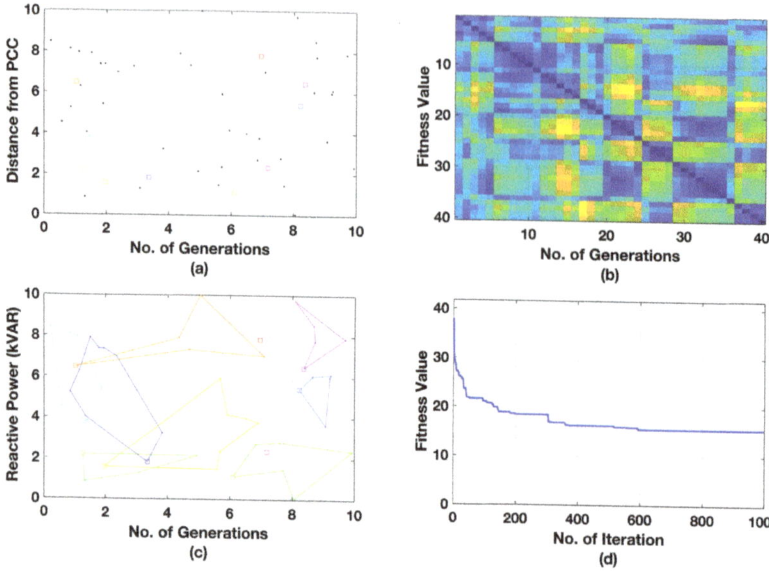

Figure 8. GA performance with $\mu = 0.18$. (**a**) FACTS location initialization, (**b**) reactive power support range, (**c**) percentage of reactive power support, (**d**) best solution history.

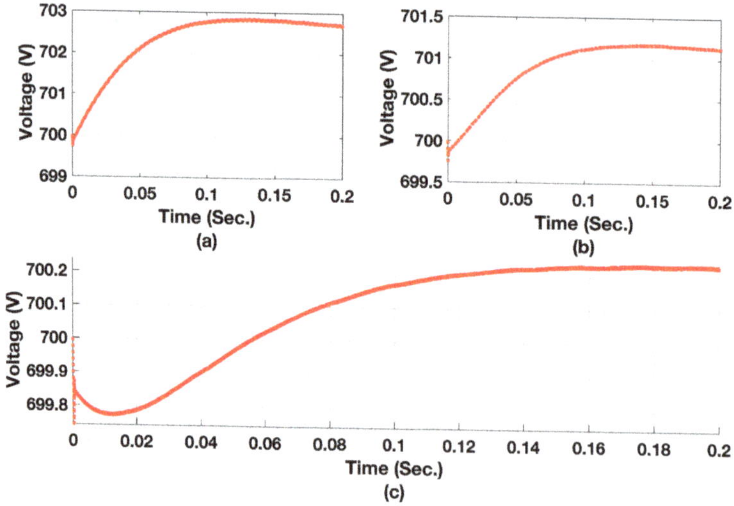

Figure 9. STATCOM DC link voltage obtained from microgrid side. (**a**) Fuzzy PI Controller (**b**) PSO-PI Controller (**c**) LSTM-GA-PI controller.

Figure 10 presents the STATCOM DC link current for all the models. Here, it is observed that with the Fuzzy STATCOM model, the system exhibits sub-synchronous resonance (SSR) between 0.002 and 0.005 s, and a similar SSR was also noticed with the PSO STATCOM, from 0.006 to 0.008 s. However, an SSR limit of 2.8% was noticed with the hyperplane concept using LSTM and GA. As compared to the Fuzzy and PSO STATCOM models, the SSR has been reduced by 7.2% and 9.43% with the proposed "LGPS" model. The SSR also reduces the voltage swell at the point of common coupling and thereby indirectly supplies the reactive power compensation in the line.

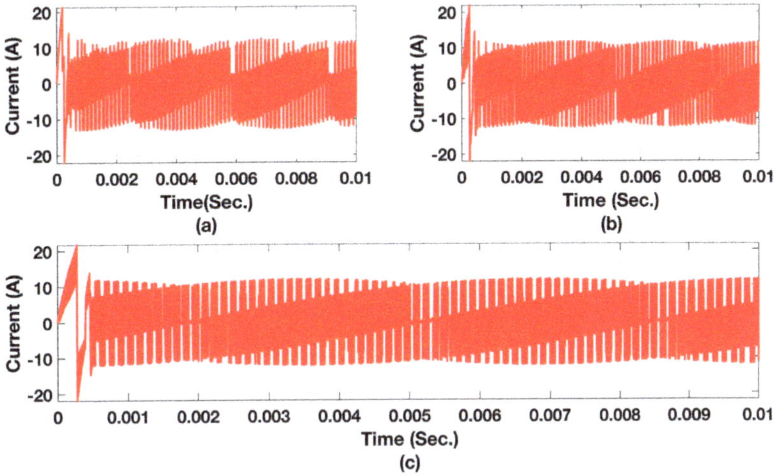

Figure 10. STATCOM DC link current obtained from microgrid side. (**a**) Fuzzy–PI STATCOM, (**b**) PSO–PI STATCOM, and (**c**) LGPS.

Figure 11 presents the STATCOM injected current and Figure 12 presents the STATCOM injected voltage at the point of common coupling. As observed, the injected current using the proposed model is 11.23 Amp. Similarly, a voltage level of 188 V has been maintained at the PCC, against 200 V and 197 V in the case of the fuzzy and PSO-enabled PI controllers. The THD levels of all three models for the injected current are shown in Figure 13. With the proposed model, the THD has been reduced to 11.44%, against 15.04% in Fuzzy–PI STATCOM and 12.39% in PSO–PI STATCOM.

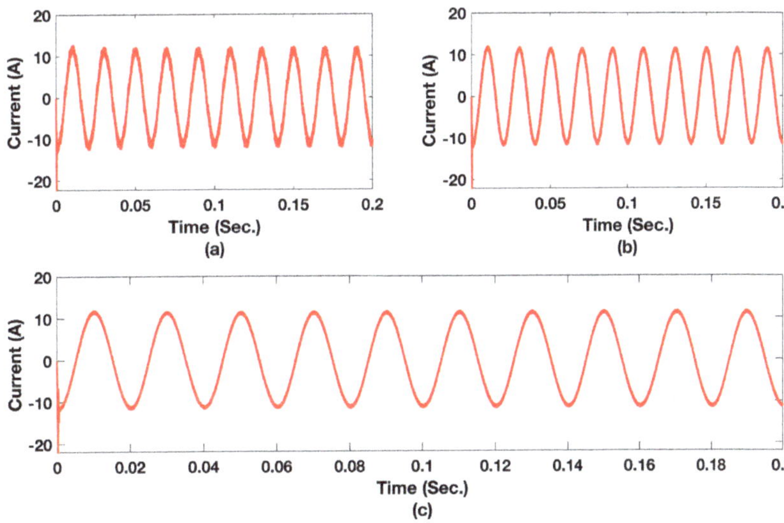

Figure 11. STATCOM injected current at PCC into microgrid. (**a**) Fuzzy–PI STATCOM, (**b**) PSO–PI STATCOM, and (**c**) LGPS.

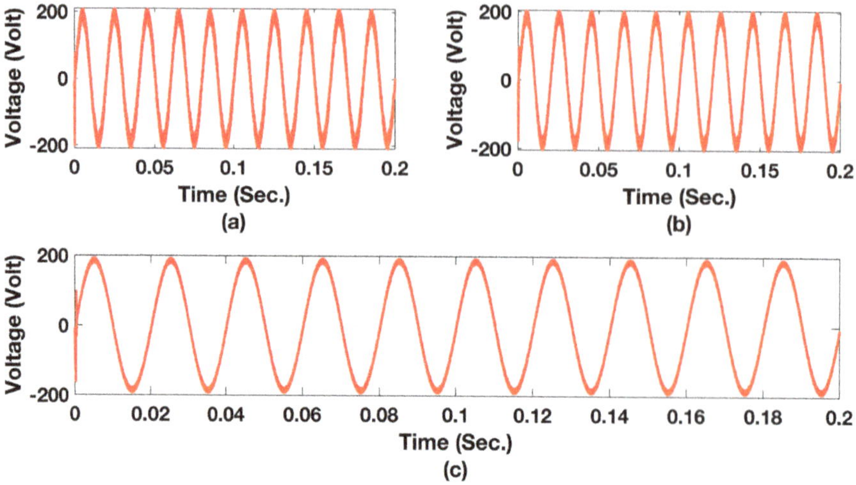

Figure 12. STATCOM injected voltage at PCC into microgrid. (**a**) Fuzzy–PI STATCOM, (**b**) PSO–PI STATCOM, and (**c**) LGPS.

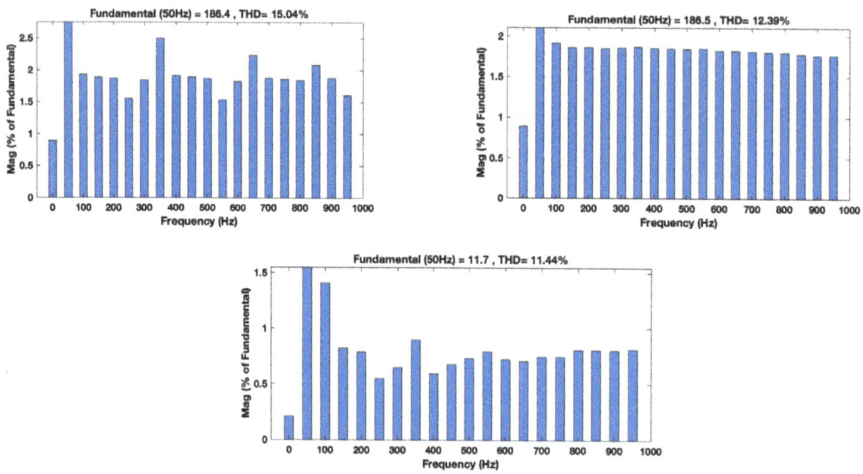

Figure 13. Total harmonic distortion of current waveform at the terminal of PCC.

Figure 14 presents the voltage waveform of the r-phase of the microgrid. In Figure 14a, it is observed that the voltage is 252 V with harmonic content of 12.3% and that of for Figure 14b, PSO-PI controller, the voltage becomes 238 V with harmonic content of 10.78%. However, with proposed controller Figure 14c the voltage is maintained at 231.7 V. The percentage of harmonics injected by the STATCOM becomes 18% and that in the proposed model becomes 12.03%.

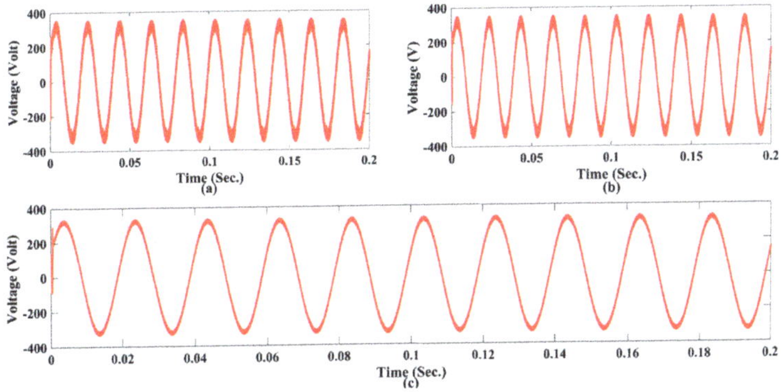

Figure 14. Voltage waveform of r-phase of microgrid. (**a**) Fuzzy PI Controller (**b**) PSO-PI Controller (**c**) LSTM-GA-PI controller.

5. Discussion

The P2P coordinated control between the SPV and STATCOM in a microgrid for power quality compensation using LSTM–genetic algorithm has been analyzed experimentally (MATLAB simulation) with two benchmarking models, the Fuzzy–PI and PSO–PI models. On analyzing the model, the observations are as follows.

Table 2 shows the power quality analysis of the STATCOM microgrid. It is observed that maximum harmonics have been produced with the Fuzzy–PI STATCOM of the order of 15.43%, and the least harmonics produced amount to 11.22%, with the proposed model. In all three cases, the broad band has been maintained for the notch. As compared to all the other algorithms, with the LSTEM–GA–PI STATCOM, the lowest DC offset was observed. Similarly, Table 3 presents the time-domain analysis of the STATCOM–PI controller. By testing these different control algorithms against a step function input, the time-domain analysis allowed for a comparison of their performance in terms of how well they respond to sudden changes and achieve the desired system behavior. The proposed algorithm produces 8.84% of overshoot, which is also the lowest among all the benchmarking models.

Table 2. Power quality analysis of STATCOM microgrid.

Sr. No.	Technique	Power Quality Attribute	Magnitude
01	Fuzzy–PI STATCOM	DC Offset	0.21%
		Harmonic Current	15.34%
		Inter Harmonics	1.87%
		Notching	Broad Band
		Noise	0.82%
02	PSO–PI STATCOM	DC Offset	0.14%
		Harmonic Current	12.72%
		Inter Harmonics	1.25%
		Notching	Broad Band
		Noise	0.57%
03	LSTM–GA–PI STATCOM	DC Offset	0.07%
		Harmonic Current	11.22%
		Inter Harmonics	0.87%
		Notching	Broad Band
		Noise	0.44%

Table 3. Time-domain analysis of STATCOM–PI controller.

Sr. No.	Technique	Parameters	Magnitude	Remarks
01	Fuzzy–PI STATCOM	Delay Time Rise Time Peak Time Settling Time Max. Overshoot	0.58 0.62 0.77 2.23 14.44%	Marginally Stable Critically Damped
02	PSO–PI STATCOM	Delay Time Rise Time Peak Time Settling Time Max. Overshoot	0.49 0.53 0.65 1.90 12.27%	Asymptotically Stable Critically Damped
03	LSTM–GA–PI STATCOM	Delay Time Rise Time Peak Time Settling Time Max. Overshoot	0.35 0.38 0.47 1.36 8.84%	Stable

Figure 15 presents the voltage and current performance of the STATCOM at the PCC. Figure 15a presents the voltage and current waveform for the Fuzzy–PI STATCOM. As observed, the current has undergone oscillations from 0.3 s to 0.4 s. This is due to the unavailability of internal memory and also the inability of the controller to dynamically assign the reference voltage for the grid-side converter of the DFIG. Similarly, for the PSO–PI controller, the oscillations are less as compared to the GA–PI controller, as presented in Figure 15b. However, with the proposed LSTM–GA–PI STATCOM (Figure 15c), the oscillations have been damped out completely. This is because of the presence of a memory unit in the feedback loop. The maximum peak overshoot in the current waveform is 0.58 pu, as compared to 0.98 pu and 0.77pu in the Fuzzy–PI and PSO–PI controllers, respectively.

Figure 16 shows the DC offset voltage analysis for the Fuzzy–PI, PSO–PI, and LSTM–GA–PI STATCOM controllers. As observed, the initial oscillation presents negative slope characteristics for LSTM–GA–PI as compared to the other controllers. Similarly, the second transition event from 0.2 s to 0.3 s shows fewer oscillations for the DC offset.

Figure 17 presents the power quality analysis of the STATCOM's injected real and reactive power for the Fuzzy–PI, PSO–PI, and LSTM–GA–PI STATCOMs. As observed in Figure 17a, the reactive power has been absorbed by the STATCOM for three cycles, whereas, for the PSO–PI controller in Figure 17b, it shows oscillations with a time-varying negative slope. However, with the LSTM–GA–PI controller, the reactive power has been injected at 25% so as to reduce the burden on the DFIG stator.

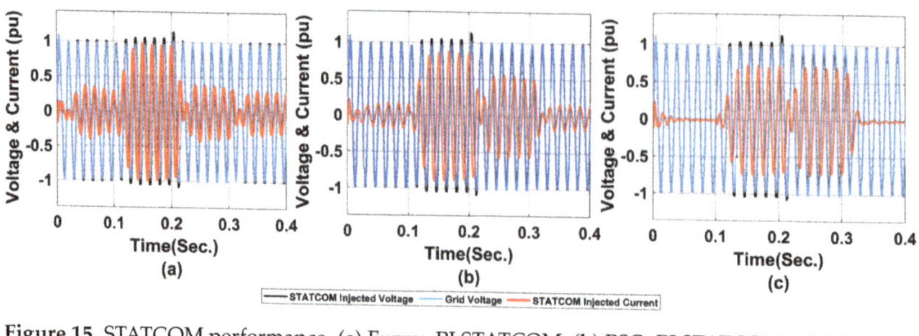

Figure 15. STATCOM performance. (**a**) Fuzzy–PI STATCOM, (**b**) PSO–PI STATCOM, (**c**) LSTM–GA–PI STATCOM.

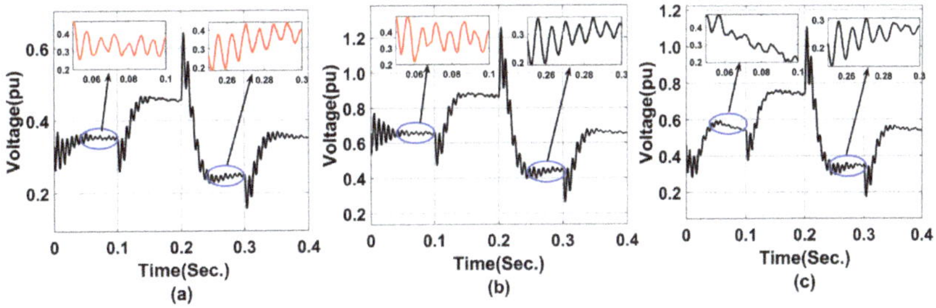

Figure 16. DC offset voltage analysis for DFIG controller. (**a**) Fuzzy–PI, (**b**) PSO–PI, (**c**) LSTM–GA-PI.

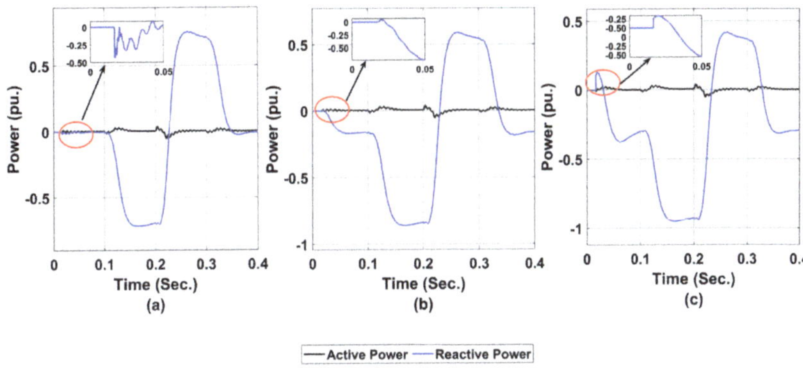

Figure 17. Power quality analysis of STATCOM injected real and reactive power. (**a**) Fuzzy–PI, (**b**) PSO–PI, (**c**) LSTM–GA–PI.

The integration of solar photovoltaic systems in a microgrid represents the utilization of clean and renewable energy sources. This reduces the reliance on fossil fuels and conventional power generation, resulting in lower greenhouse gas emissions and promoting environmental sustainability. A microgrid is a localized and decentralized energy system that can operate independently or in conjunction with the main power grid. Coordinated control between the SPV and STATCOM enhances the microgrid's power quality by ensuring stable voltage and frequency levels. This improved power quality enables the efficient operation of connected devices, minimizes electrical disturbances, optimizes energy consumption, and reduces waste, contributing to sustainability.

The utilization of advanced control techniques, such as LSTM and genetic algorithm, underscores the intelligence of the control system. By leveraging machine learning and optimization algorithms, such as LSTM and genetic algorithm, respectively, the microgrid can adapt to changing conditions, optimize the energy flow, and minimize losses. This intelligent control approach enhances the overall performance and energy efficiency of the microgrid, maximizing the utilization of the available renewable energy resources and contributing to sustainability.

Microgrids are designed to operate autonomously during grid disruptions and enhance the resilience against natural disasters and other disturbances. By incorporating coordinated control between the SPV and STATCOM, the microgrid effectively compensates for power quality issues and maintains a stable energy supply. This increased energy independence improves the microgrid's resilience and reduces the reliance on the main power grid. Ultimately, it contributes to the overall sustainability of the energy system by ensuring a reliable and uninterrupted power supply, particularly during critical situations.

6. Conclusions

The optimal and coordinated performance of a STATCOM and microgrid using LSTM–genetic algorithm has been evaluated in this article using simulation methods under normal and abnormal operating conditions. Both the AC and DC voltage gains of the STATCOM were optimized using LSTM–GA. It was observed that when carefully tuning the parameters, the DC offset for the LSTM–GA STATCOM was reduced significantly to 0.07%, compared to 0.21%, and it also avoids SSR to an extent of 17%.

The harmonic and inter-harmonic components using the LSTM–GA methodology reduce the burden on the transmission line and thereby reduce the overheating of the conductor in a microgrid system under load variation conditions. In order to maintain a proper system balance with respect to the IEC and IEEE standards, the notching level of the broad band range has been maintained. During time-domain analysis, the proposed LSTM–PI–GA model shows a shorter settling time as compared to the other two benchmarking models under the step-changing mode of operation.

The optimized STATCOM device has been presented as a dependable solution to improve the stability of microgrid systems, regardless of whether they are functioning normally or abnormally. This device is capable of suppressing transient oscillations in power and frequency while managing voltage fluctuations caused by external disturbances or changes in load demand. Overall, the STATCOM is a highly effective resource in ensuring the consistent and stable performance of microgrid systems.

Author Contributions: Conceptualization, D.S.; Methodology, D.S.; Software, D.S. and K.J.R.; Validation, M.V., K.J.R., A.G. and M.R.; Formal analysis, D.S., M.V., R.D., K.J.R., A.G., B.K. and M.R.; Investigation, D.S., M.V., R.D., K.J.R. and D.C.; Resources, K.J.R., D.C., A.G., B.K. and M.R.; Data curation, D.S., M.V., R.D., K.J.R., A.G. and B.K.; Writing—original draft, D.S., R.D., D.C. and B.K.; Writing—review & editing, D.S., M.V., R.D., K.J.R., D.C. and M.R.; Visualization, D.C., A.G. and B.K.; Supervision, D.C.; Project administration, D.C.; Funding acquisition, B.K. All authors have read and agreed to the published version of the manuscript.

Funding: This research received no external funding.

Institutional Review Board Statement: Not applicable.

Informed Consent Statement: Not applicable.

Data Availability Statement: Data will be available on request.

Conflicts of Interest: The authors declare no conflict of interest.

References

1. Pullins, S. Why microgrids are becoming an important part of the energy infrastructure. *Electr. J.* **2019**, *32*, 17–21. [CrossRef]
2. Eid, B.M.; Rahim, N.A.; Selvaraj, J.; El Khateb, A.H. Control Methods and Objectives for Electronically Coupled Distributed Energy Resources in Microgrids: A Review. *IEEE Syst. J.* **2016**, *10*, 446–458. [CrossRef]
3. Cagnano, A.; De Tuglie, E.; Mancarella, P. Microgrids: Overview and guidelines for practical implementations and operation. *Appl. Energy* **2019**, *258*, 114039. [CrossRef]
4. He, P.; Fang, Q.; Jin, H.; Ji, Y.; Gong, Z.; Dong, J. Coordinated design of PSS and STATCOM-POD based on the GA-PSO algorithm to improve the stability of wind-PV-thermal-bundled power system, *Int. J. Electr. Power Energy Syst.* **2022**, *141*, 108208. [CrossRef]
5. Kaliaperumal Rukmani, D.; Thangaraj, Y.; Subramaniam, U.; Ramachandran, S.; Madurai Elavarasan, R.; Das, N.; Baringo, L.; Imran Abdul Rasheed, M. A New Approach to Optimal Location and Sizing of DSTATCOM in Radial Distribution Networks Using Bio-Inspired Cuckoo Search Algorithm. *Energies* **2020**, *13*, 4615. [CrossRef]
6. Tariq, M.; Zaheer, H.; Mahmood, T. Modeling and Analysis of STATCOM for Renewable Energy Farm to Improve Power Quality and Reactive Power Compensation. *Eng. Proc.* **2021**, *12*, 44. [CrossRef]
7. Bharadwaj, A.; Maiti, S.; Dhal, N.; Chakraborty, S.; Pillai, S.K. Chapter 10—E-STATCOM (energy storage+STATCOM): A solution to integrate large-scale wind farms into the grid at medium and high power levels. In *Power Quality in Modern Power Systems*; Sanjeevikumar, P., Sharmeela, C., Holm-Nielsen, J.B., Sivaraman, P., Eds.; Academic Press: Cambridge, MA, USA, 2021; pp. 283–310, ISBN 9780128233467. [CrossRef]
8. Perera, S.; Elphick, S. Chapter 2—Steady-state voltage in low voltage networks. In *Applied Power Quality*; Perera, S., Elphick, S., Eds.; Elsevier: Amsterdam, The Netherlands, 2023; pp. 19–48. ISBN 9780323854672. [CrossRef]

9. Battistelli, C.; Monti, A. Chapter 5—Dynamics of modern power systems. In *Converter-Based Dynamics and Control of Modern Power Systems*; Monti, A., Milano, F., Bompard, E., Guillaud, X., Eds.; Academic Press: Cambridge, MA, USA, 2021; pp. 91–124. ISBN 9780128184912. [CrossRef]
10. He, L.; Lu, Z.; Geng, L.; Zhang, J.; Li, X.; Guo, X. Environmental economic dispatch of integrated regional energy system considering integrated demand response. *Int. J. Electr. Power Energy Syst.* **2020**, *116*, 105525. [CrossRef]
11. Ranamuka, D.; Muttaqi, K.M.; Sutanto, D. Flexible AC Power Flow Control in Distribution Systems by Coordinated Control of Distributed Solar-PV and Battery Energy Storage Units. *IEEE Trans. Sustain. Energy* **2020**, *11*, 2054–2062. [CrossRef]
12. Patnaik, B.; Swain, S.C.; Dash, R.K. An Experimental Analysis of Solar PV on Higher Concentration of Methane. In Proceedings of the 2022 IEEE India Council International Subsections Conference (INDISCON), Bhubaneswar, India, 15–17 July 2022; pp. 1–5. [CrossRef]
13. Sera, D.; Kerekes, T.; Teodorescu, R.; Blaabjerg, F. Improved MPPT algorithms for rapidly changing environmental conditions. In Proceedings of the 2006 12th International Power Electronics and Motion Control Conference, Portoroz, Slovenia, 30 August–1 September 2006; pp. 1614–1619.
14. Narsingoju, K.; Busa, V.; Kumar, G.V. Simulation analysis of maximum power control of photovoltaic power system. *Int. J. Adv. Electr. Electron. Eng.* **2012**, *1*, 9–14.
15. Patnaik, B.; Swain, S.C.; Dash, R. A Study on Effect of GHG on the Performance of Grid Connected Solar PV System. In Proceedings of the 2022 IEEE India Council International Subsections Conference (INDISCON), Bhubaneswar, India, 15–17 July 2022; pp. 1–4. [CrossRef]
16. Sakib, N.; Kabir, M.W.; Subbir, M.; Alam, S. A comparative study of flower pollination algorithm and bat algorithm on continuous optimization problems. *Int. J. Appl. Inf. Syst.* **2014**, *7*, 13–19. [CrossRef]
17. Yang, B.; Zhong, L.; Zhang, X.; Shu, H.; Yu, T.; Li, H.; Jiang, L.; Sun, L. Novel bio-inspired memetic salp swarm algorithm and application to MPPT for PV systems considering partial shading condition. *J. Clean. Prod.* **2019**, *215*, 1203–1222. [CrossRef]
18. Kumar, C.; Rao, R. A novel global MPP tracking of a photovoltaic system based on whale optimization algorithm. *Int. J. Renew. Energy Dev.* **2016**, *5*, 225–232. [CrossRef]
19. Mohanty, S.; Subudhi, B.; Ray, P.K. A new MPPT design using grey wolf optimization technique for photovoltaic system under partial shading conditions. *IEEE Trans. Sustain. Energy* **2016**, *7*, 181–188. [CrossRef]
20. Mansoor, M.; Feroz, Mirza, A.; Ling, Q. Harris hawk optimization-based MPPT control for PV systems under partial shading conditions. *J. Clean. Prod.* **2020**, *274*, 122857. [CrossRef]
21. El-Helw, H.M.; Magdy, A.; Marei, M.I. A hybrid maximum power point tracking technique for partially shaded photovoltaic arrays. *IEEE Access* **2017**, *5*, 11900–11908. [CrossRef]
22. Shen, H.; Xu, L.; Xu, Y.; Cao, J. A novel hybrid PSO-LSTM approach for improving MPPT performance of PV system under partially shaded conditions. *Sol. Energy* **2019**.
23. Kong, W.C.; Dong, Z.Y.; Jia, Y.W.; Hill, D.J.; Xu, Y.; Zhang, Y. Short-Term Residential Load Forecasting Based on LSTM Recurrent Neural Network. *IEEE Trans. Smart Grid* **2019**, *10*, 841–851. [CrossRef]
24. Ali, A.N.; Saied, M.H.; Mostafa, M.Z.; Abdel-Moneim, T.M. A survey of maximum PPT techniques of PV systems. In Proceedings of the 2012 IEEE Energytech, Cleveland, OH, USA, 29–31 May 2012; pp. 1–10.
25. Abd-Elazim, S.M.; Ali, E.S. Imperialist competitive algorithm for optimal STATCOM design in a multimachine power system. *Int. J. Electr. Power Energy Syst.* **2016**, *76*, 136–146. [CrossRef]
26. Ballaji, A.; Dash, R.; Subburaj, V.; Kalvakurthi, J.R.; Swain, D.; Swain, S.C. Design & Development of MPPT Using PSO with Predefined Search Space Based on Fuzzy Fokker Planck Solution. *IEEE Access* **2022**, *10*, 80764–80783. [CrossRef]
27. Ramadan, A.; Kamel, S.; Hassan, M.H.; V´eliz, T.M.; Eltamaly, A.M. Parameter Estimation of Static/Dynamic Photovoltaic Models Using a Developed Version of Eagle Strategy Gradient-Based Optimizer. *Sustainability* **2021**, *13*, 13053. [CrossRef]
28. Rivera, D.; Guillen, D.; Mayo-Maldonado, J.C.; Valdez-Resendiz, J.E.; Escobar, G. Power grid dynamic performance enhancement via STATCOM data-driven control. *Mathematics* **2021**, *9*, 2361. [CrossRef]

sustainability

MDPI

Article

Long-Term Field Observation of the Power Generation and System Temperature of a Roof-Integrated Photovoltaic System in South Korea

Muhammad Hanif Ainun Azhar [1,†], Salh Alhammadi [1,†], Seokjin Jang [1,†], Jitaek Kim [2], Jungtaek Kim [2] and Woo Kyoung Kim [1,*]

[1] School of Chemical Engineering, Yeungnam University, 280 Daehak-ro, Gyeongsan 38541, Gyeongbuk, Republic of Korea; mhanifainun@yu.ac.kr (M.H.A.A.); saleh@ynu.ac.kr (S.A.); 21811313@yu.ac.kr (S.J.)

[2] Roser Roofing System, 43, Nae-Ri-19, Ammyang-myeon, Gyeongsan 38539, Gyeongbuk, Republic of Korea; jkim77@roser.com (J.K.); jtkim@roser.com (J.K.)

* Correspondence: wkim@ynu.ac.kr

† These authors contributed equally to this work.

Abstract: A miniature house roof-integrated photovoltaic (PV) system in South Korea was monitored for 2.5 years. System performance was evaluated through power generation, solar irradiance, and system temperature. The comparison of each month's power generation and solar irradiance revealed a parallel correlation over the entire observation period. The internal module temperature was almost always higher than the roof rear and module rear temperatures by 1–2 and 1–5 °C, respectively, while the temperature behind the PV modules was the lowest among the three temperatures, showing that the installation of PV modules as a roofing system does not affect the temperature of the roofing system. The system temperatures affected the power conversion efficiency; a maximum of 11.42% was achieved when the system temperatures were the lowest, and a minimum of 5.24% was achieved when the system temperatures were the highest. Hence, half of the anticipated generated power was lost due to the temperature fluctuation. Overall, installing PV modules as an entire roofing system is possible with this configuration due to the minimum effect on the roof temperature. However, PV system temperature control is essential for maintaining the power generation performance of the PV modules.

Keywords: building-integrated photovoltaics; roof-integrated photovoltaics; photovoltaic module temperature; building-integrated photovoltaic system temperature

check for updates

Citation: Azhar, M.H.A.; Alhammadi, S.; Jang, S.; Kim, J.; Kim, J.; Kim, W.K. Long-Term Field Observation of the Power Generation and System Temperature of a Roof-Integrated Photovoltaic System in South Korea. *Sustainability* **2023**, *15*, 9493. https://doi.org/10.3390/su15129493

Academic Editors: Prince Winston David and Praveen Kumar B

Received: 26 April 2023
Revised: 8 June 2023
Accepted: 9 June 2023
Published: 13 June 2023

1. Introduction

The energy issue is among the top contemporary global crises [1]. Renewable and clean sources of energy need to be capitalized given the aggravated global security issues as of February 2022 and the electricity demand rising by 5.9% to 1400 terawatt hours (TWh), complemented by 14.6 gigatons of CO_2 emissions from burning fossil fuels for electricity and heat production [1,2]. Photovoltaic (PV) systems, which utilize a renewable energy source in the form of sunlight, experienced increased capacity and market size. The PV module market increased to 310 gigawatts (GW) in 2022 [3], with the PV module production capacity by the end of 2021 being over 470 gigawatt-peak (GWp) [4]. PV systems covered over 5% of the global electricity generation in 2021, reducing annual CO_2 emissions by as much as 1100 megatons during the year [5].

Continuous improvements in the PV system application method are required to further boost the use of PV systems in the future. A few strategies have been implemented to diversify the PV system applications, such as vertically installed PV [6], agriphotovoltaics [7], floating photovoltaics [8], building-applied photovoltaics (BAPVs) [9], and building-integrated photovoltaics (BIPVs) [10]. Using a PV system as an intrinsic part of a building, a BIPV system takes part in the building structure while generating power [10,11]. In BIPVs, PV systems mainly serve as façades and/or roofing systems for buildings [10–12]. In a façade system, PV modules can serve as curtains, glazing, or spandrel panels; in a BIPV roofing system, PV modules can serve as roofing tiles, shingles, standing seams, or skylights [11]. The long-term cost offset is beneficial by replacing one functional component of a building with a working PV system [13]. The BIPV roofing system is viable for use over residential houses, as there are many areas with passively functioning roofs that PV systems can replace to create hundreds of GWp [14]. Due to these advantages, BIPV implementation has been widespread in recent years [10,14]. The policies of several countries have played a major role in promoting BIPV implementation. A few examples include tax-free self-consumption of BIPV power in France [15] and Japan [16], incentives for BIPV installation in China [17], and BIPV installation subsidies in Korea [18]. Government aid has directly contributed to the rapid growth of the number of BIPV installations, with Grand View Research forecasting that the global BIPV market will increase from USD 19.82 billion in 2022 to USD 88.38 billion in 2030 [19]. More studies are required to complement this growing industry and understand the factors contributing to system performance.

Several factors affect the performance of PV modules, namely PV module manufacturing design, cell technology type, solar irradiance, wind direction and speed, mounting configuration, and temperature [20–22]. Appropriate mounting configuration can limit PV module degradation, as shown by Jordan et al., who studied the performance of aluminum back surface field (Al-BSF)–based PV modules installed in Las Vegas, USA, with different mounting configurations [23]. The authors reported reduced heat transfer into the modules with a lower amount of metal roof and increased heat transfer with a rack-mounting configuration, leading to a lower PV module degradation rate. Manufacturing design effects due to insulations behind the PV modules were studied by Gok et al., who evaluated the performance of glass/back-sheet- and glass/glass-based crystalline silicon (c-Si) PV modules installed in two different mounting configurations in Canobbio, Switzerland [24]. The authors reported that higher operating temperatures significantly impacted the glass/back-sheet module, with the performance loss rate (PLR) varying from 0.01%/year for ventilation to −0.42%/year for insulation. Conversely, the glass/glass module displayed an unexpected opposite trend, with the PLR varying from −0.10%/year for ventilation to 0.26%/year for insulation. IV measurements revealed that the reduced performance of the glass/back-sheet module originated from the deterioration of the fill factor through increased resistance, whereas the rise in the short-circuit current (I_{sc}) was the primary driver of the insulated glass/glass module performance improvement. Kumar et al. studied different cell technology types, focusing on three different types of PV technologies: crystalline silicon (c-Si), copper indium selenide (CIS), and cadmium telluride (CdTe) modules, as BIPVs and BAPVs in Malaysia [21]. The energy-generating performance was different among the three technologies, with the c-Si, CIS, and CdTe modules generating peaks of 4240, 4280, and 4490 kWh, respectively. Singh et al. reported on the performance of high-efficiency heterojunctions with intrinsic thin-layer (HIT)-technology-based PV modules depending on different climatic conditions [25]. The authors showed that HIT technology-based PV modules were more efficient at cold partition temperatures. The impacts of urban heat islands (UHIs) and urban air pollution on BIPV energy efficiency have been extensively studied. For instance, Wang et al. reported that the UHI and solar radiation absorption caused by smog could reduce the overall PV energy generation in urban locations by more than 10% compared to that in rural locations [26]. The effects of the tilt angle and wind speed were studied by Dabaghzadeh et al., who studied the convective cooling of a PV system by modeling different tilt angles and wind speeds [27]. The authors observed the lowest temperature for a tilt angle of 45°, aided by optimal convective cooling, irrespective of the wind speed.

BIPV research has mainly focused on temperature, power generation, and the correlation between the two parameters. Kumar et al. reported on different temperature and performance losses for different PV module types, with the c-Si BIPV exhibiting 13.6% reduced performance, the CIS system exhibiting 12.8% reduced performance, and the CdTe system exhibiting 8.8% reduced power generation efficiency [21]. Poulek et al. compared the temperature of BIPV modules to that of PV modules with conventional configurations and its relation to energy production [28]. The authors used a modeling approach to compare BIPV modules with free-standing PV modules in four different climates while conducting a field study in Prague, Czech Republic. The field study indicated that the temperature of the BIPV modules was higher by more than 5 °C compared to that of the conventionally installed PV modules. At the same time, a difference of 3–5% in energy production due to the increased module temperature was evident. Using their model, the authors observed that climate differences from cold to hot temperatures had a negligible effect on the BIPV performance degradation; however, the PV module degradation in area with very hot temperatures was rapid. Kim et al. studied BIPV module temperatures with different insulations by conducting a field study on a miniature house with a tilted BIPV roofing system in Daejeon, South Korea [29]. The locations of the insulation were different: One system had insulation behind the PV modules (i.e., a warm roofing system), while another system had insulation behind the ceiling (i.e., a cold roofing system). The comparison of the two systems showed that the BIPV power generation of the cold roofing system was 7% higher than that of the cold roofing system. D'Orazio et al. analyzed different BIPV configurations in Ancona, Italy [30]. The authors compared rack-mounted high-ventilated, moderate-ventilated, and non-ventilated BIPV systems to see the effect of natural ventilation with the addition of air gaps. Rack-mounted PV modules exhibited the lowest PV module and air-back temperatures. In contrast, non-ventilated BIPVs exhibited the highest temperatures. The authors found that an air gap of 0.04 m between the PV modules and the roof was sufficient to create a difference of less than 3% in annual power generation prowess. Kaplanis et al. conducted a modeling study of the aging effect of BIPV and BAPV systems [31]. PV module aging was predicted to increase the PV module temperature relative to the case of the reference modules. Given the expected increase in PV module temperature when installed in a BIPV system compared to the case of a conventional installation, the aging effect might become more prominent.

Several studies have focused on decreasing the PV module temperature in BIPV systems. Mittelman et al. modeled a cooling channel in an attempt to increase the performance of PV modules in BIPVs and observed that adding 0.02–0.20 m air space between the PV modules and the roof can decrease their temperature by 10–20 °C and thus increase their energy-generating performance by 1–2% [32]. Other attempts to decrease the PV module temperature incorporated phase change materials (PCMs). Karthikeyan et al. studied the effect of a non-contact composite PCM on the optimal PCM thickness and observed that the optimum PCM thickness was 2.5 cm, yielding an average of 6.7 °C PV module temperature reduction [32]. Hasan et al. studied five different PCMs to decrease the BIPV module temperature [33]. The authors used $CaCl_2$ as PCM and achieved an 18 °C lower BIPV module temperature for 30 min. In a more extended observation of 5 h, a 10 °C temperature reduction was maintained.

Considering the findings of previous studies on BIPV and PV modules in general, the issue of system temperature and power in BIPVs will persist with the growth of the BIPV industry, especially concerning how temperature affects the PV performance or the building [10]. Several short-term studies on the effect of BIPV module temperature on module performance—involving either modeling or field investigations—have been conducted [21,28–33]; however, only a few studies have reported on the effect of PV modules installed on top of roof tile systems and how the temperature of the roof tiles with attached PV modules compares to that under roof tiles without attached PV modules. The specific weather and climate of the regions where BIPVs are installed are also important for comparison. While temperature-reducing efforts have been generally fruitful, there are several associated drawbacks. Natural circulation might allow dust to collect in the air gap, reducing heat transfer. At the same time, several PCMs are costly and hazardous to the environment [34]. Additionally, PCMs can achieve limited BIPV module cooling due to their short cooling duration. Because of all these issues and how long-term studies (i.e., studies longer than a year) are not available, a miniature house BIPV system was manufactured to better understand the long-term temperature conditions of the roofing system with metal tiles and attached PV modules without dedicated ventilations and PCMs for 2.5 years in South Korea. The comparison of the temperature of the roof tiles with that under the PV modules was used to assess the possibility of installing PV modules on the entire roofing system in this roofing configuration with metal tiles. The contributions of this study are as follows:

- Insight into how a BIPV roofing system in South Korea performed for 2.5 years in terms of its power generation and system temperature.
- More detailed and specific information on the BIPV system temperature based on the PV module internal temperature and the comparison between the temperature of the roof tiles and that of the roof tiles behind the PV modules.
- Observations regarding the changes in PV module power conversion prowess during long-term use in the BIPV roofing system.

The findings of this study address questions regarding the effect of PV module installation on the roof temperature and the PV module's performance over time.

2. Materials and Methods

This study observed and analyzed a miniature house roof-integrated PV system. The utilized solar roofing system was manufactured by Roser (Gyeongsan, South Korea) (Figure 1). The system was a sloped, unventilated roofing system and was 187.5 cm in length, 154 cm in width, and 133 cm in height. Four PV modules with 536×536 mm^2 dimensions were attached to the top of the roof. The modules were surrounded by more roof tiles to form the roof of the miniature house, covering a total area of 187.5×149.0 cm^2. The system was installed in 2019 at the Yeungnam University Photovoltaic Power Systems R&BD Demonstration Complex, Gyeongsan, South Korea (location: 35.82° N, 128.76° E). The system was south-facing, had no obstructions in front of it resulting in shading, and was tilted by 35° to the ground. The PV modules were tested under standard conditions (STC), having a peak power of 40 W. Other parameters of the PV modules are outlined in Table 1.

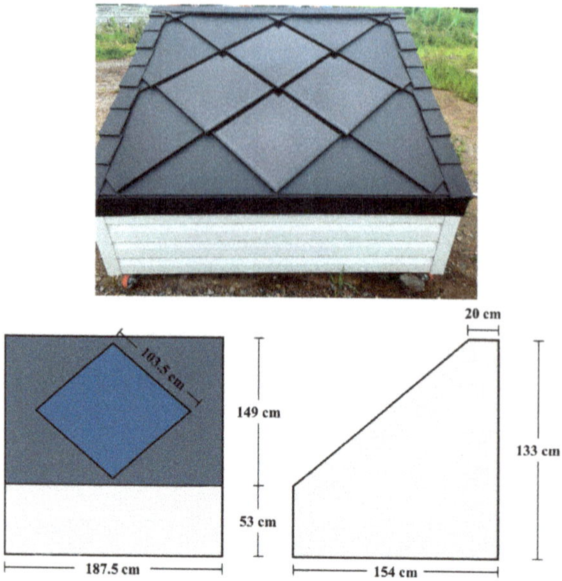

Figure 1. Installed solar roofing system (PV module depicted as the blue square) and its dimensions.

Table 1. Studied solar module label specifications.

PV Parameters		Value
Peak Power (P_m)	(W)	40
Open-Circuit Voltage (V_{OC})	(V)	5.75
Short-Circuit Current (I_{SC})	(A)	8.96
Maximum Voltage (V_m)	(V)	4.76
Maximum Current (I_m)	(A)	8.42

Three parameters were monitored to assess the roof-integrated PV system's performance: solar irradiance, power generation, and system temperature. Solar irradiance data, the primary source of electrical generation from PV modules, were obtained using a CMP6 first-class pyranometer (<5% daily uncertainty) installed near the BIPV system. The PV modules were connected to a programmable DC electronic load LODA LF 600C (accuracy ±0.05%) to monitor the power generation of the PV modules for the duration of the study. Three thermocouples were installed inside the roofing system at different locations to monitor changes in the system's temperatures. The three types of temperature data collected were the module internal temperature (MI), module rear temperature (MR), and roof rear temperature (RR). The thermocouple measuring MI was placed between the PV modules and the thin roof layer beneath. MR and RR were measured to assess the conditions of the roof tiles with and without attached PV modules, respectively. The thermocouple measuring MR was placed behind the thin roof under the PV modules while the thermocouple measuring RR was fixed behind a roof tile with no attached PV modules. The locations of the thermocouples are shown in Figure 2. Solar irradiance, power generation, and system temperature data were subsequently averaged at a 30-s interval and stored in separate daily files by the monitoring system software Data Gather.

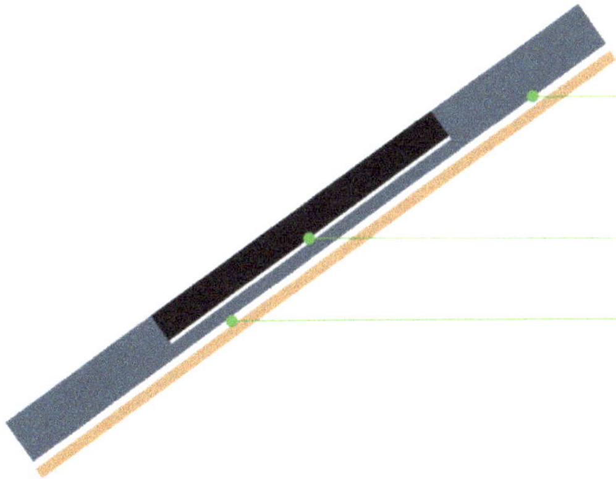

Figure 2. Schematic diagram of the thermocouples (green dots) locations inside the roofing system.

From the obtained solar irradiance data, monthly totals were created to elucidate the total irradiance during each month over the observation period. Monthly totals were created for the power generation data to elucidate the total power generation of the BIPV system during each month over the observation period. The power generation data were subsequently used to obtain the power yield as follows:

$$\text{Monthly Power Yield} = \frac{\text{Total power production in one month}}{\text{Total power capacity of the PV system}} \tag{1}$$

The monthly power yield was subsequently calculated using the total monthly solar irradiance to determine the system efficiency as follows:

$$\text{System Efficiency} = \frac{\text{Monthly Power Yield}}{\text{Total Monthly Solar Irradiance}} \times 100\% \tag{2}$$

System temperature data were analyzed based on monthly averaging throughout the study.

3. Results and Discussion

3.1. Power Generation Interdependence on Solar Irradiance

The power generation performance of the BIPV system and solar irradiance were examined for 2.5 years. To elucidate the diurnal cycles of the two parameters, Figure 3 represents two selected days, namely 14 February and 2 June 2021, with the graphs depicting the data on a minute basis. Power generation generally followed the solar irradiance trend, as PV modules use solar irradiance as their energy source to generate electricity. The different solar irradiance values between the two days are mainly due to seasonal differences. In South Korea, February is winter, and June is summer, characterized by different sun paths and, thus, different irradiance intensities [8]. The total generated power and solar irradiance for each month are shown in Figure 4, with more detailed data from 2020 until the middle of 2022 shown in Table 2. Power generation and solar irradiance exhibited similar trends, with fluctuations in solar irradiance being accompanied in most cases by proportional fluctuations in generated power. The detailed power generation and solar irradiance values in Table 2 reveal that the highest power generation occurred in the earlier part of each year, with the highest power generation being evident in April or May. Total solar irradiance exhibited a slightly shifted trend, with the highest solar irradiance observed in the months of May to July of each year except in 2020.

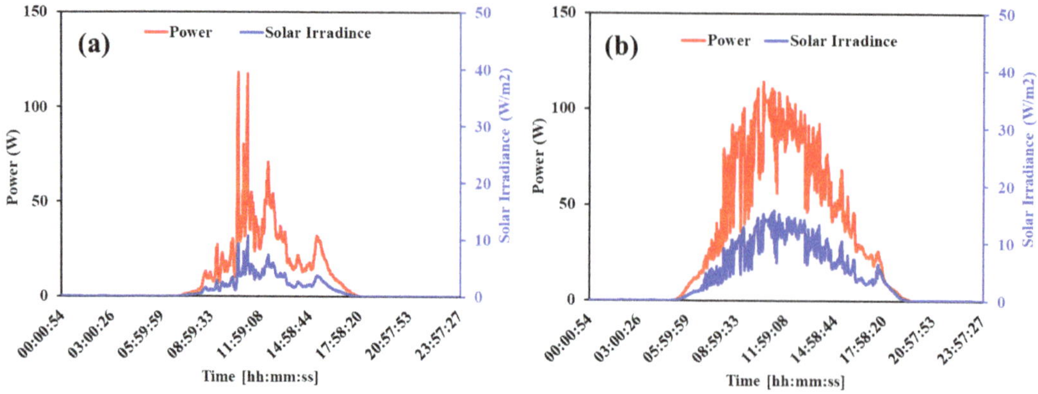

Figure 3. Daily power and solar irradiance on (**a**) 14 February 2021 and (**b**) 2 June 2021.

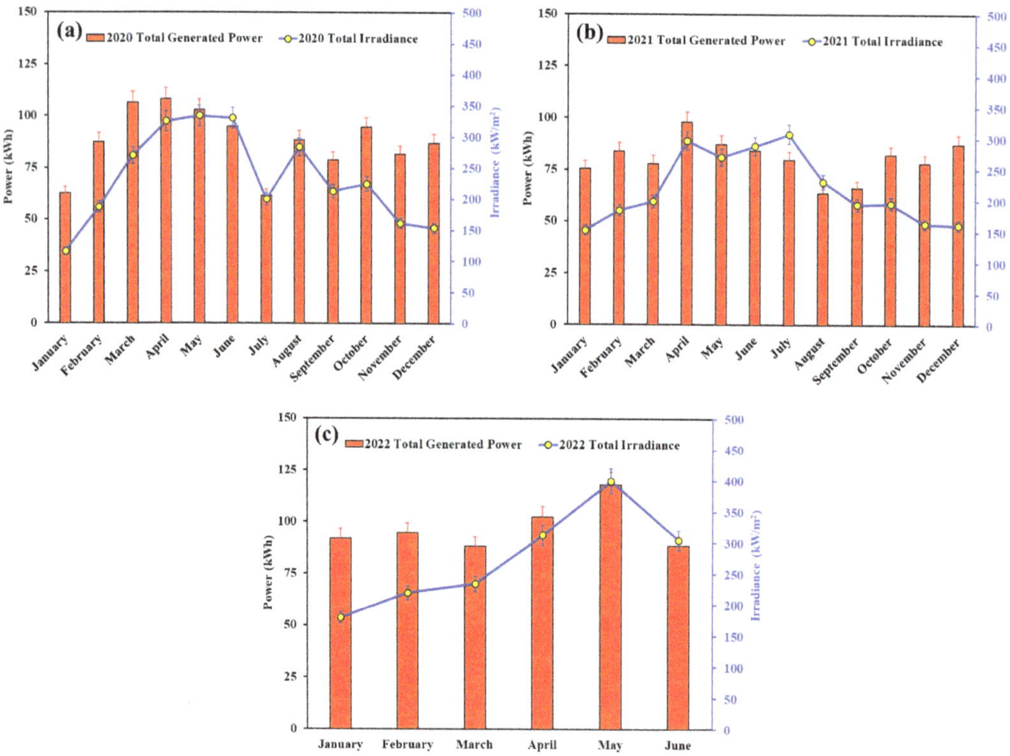

Figure 4. Total generated power and solar irradiance in (**a**) 2020, (**b**) 2021, and (**c**) 2022.

Table 2. Total average generated power and solar irradiance from January 2020 to June 2022.

Month	Total Power Generated (kWh)			Total Solar Irradiance (kW/m^2)		
	2020	2021	2022	2020	2021	2022
January	59.60	75.47	92.07	115.36	152.49	179.26
February	87.43	83.80	94.55	187.13	183.87	219.12
March	106.27	77.94	88.33	270.03	198.21	234.04
April	108.11	98.02	102.35	324.94	296.54	312.20
May	102.79	87.19	118.06	333.76	270.32	399.13
June	94.94	84.00	88.61	330.04	288.26	304.50
July	61.53	79.84	-	199.68	307.39	-
August	88.53	63.71	-	283.95	230.40	-
September	78.77	66.17	-	212.90	193.92	-
October	94.59	82.26	-	224.66	195.57	-
November	81.65	78.22	-	160.98	163.03	-
December	86.99	87.39	-	153.60	160.73	-

Solar irradiance variations throughout the year were attributed to changes in the solar position. The path of the sun is longer in spring and summer, resulting in longer days and higher total solar irradiance values [8]. The uncertain nature of the weather resulted in several cloudy and rainy days affecting the amount of solar irradiance throughout the observation period, with the relatively lower values in July 2020 and June 2022 being attributed to increased occurrences of such uncertain weather conditions. As solar irradiance is among the main influences on a PV system's power output, the power generation data followed the fluctuations in solar irradiance input throughout the observation period, and this agrees with Tina et al., whose floating PV system produced power generation values proportional to the solar irradiance values during optimal PV system operation [8]. However, the highest generated power value was not produced during the solar irradiance peak, as observed in May 2020 and July 2021, and this was due to another factor, namely system temperature, influencing the power generation of the PV modules [7,8,14]. The effect of temperature on the BIPV system's performance is discussed in the following section.

3.2. System Temperature

Three system temperatures were examined and processed to obtain the average system temperature data. Figure 5 shows the diurnal variations of the system temperatures on 14 February and 2 June 2021 based on a minute interval. All temperatures increase in the middle of the day and decrease towards the end of the day; however, the trend of MI is distinctly different from those of the two roof temperatures, MR and RR, with the latter trends being quite similar. The trend similarity was also reported by Saleh et al. [33] for the PV module temperature and Chung and Park [34] for the roof temperature. Similar to the case in Figure 3, the difference in temperature between the two days was due to seasonal differences, in South Korea, February is winter and June is summer. The analyzed data from 2020 until the middle of 2022 are represented in Figure 6 and the temperature data from 2020 to 2022 are summarized in Table 3. The trend shown in Figure 6 displays an increase from the start of the year until the middle of the year and a subsequent decline until the end of the year. The variation shown throughout the year is caused by seasonal changes, as South Korea experiences four seasons throughout the year. The increases in system temperatures were correlated with—and caused by—the seasonal changes from a cold winter to a warm spring to an even warmer summer. Subsequently, the system temperature dropped in autumn, and the lowest temperature occurred in winter, thereby completing the annual cycle.

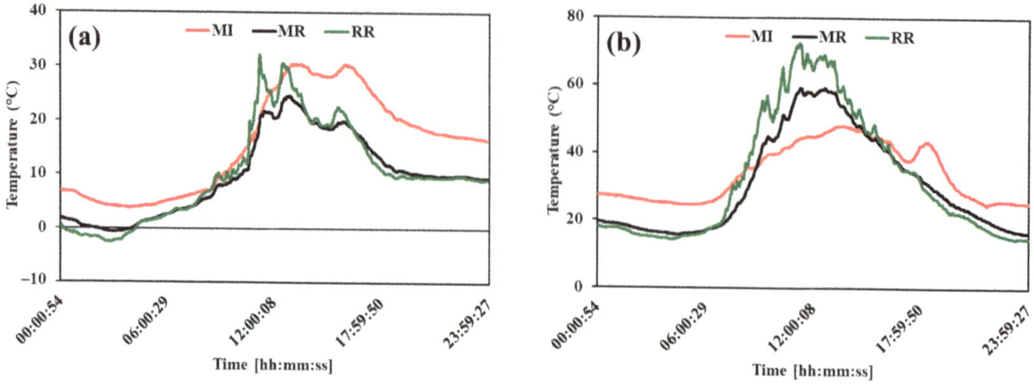

Figure 5. Diurnal variations of the system temperatures on (**a**) 14 February and (**b**) 2 June 2021.

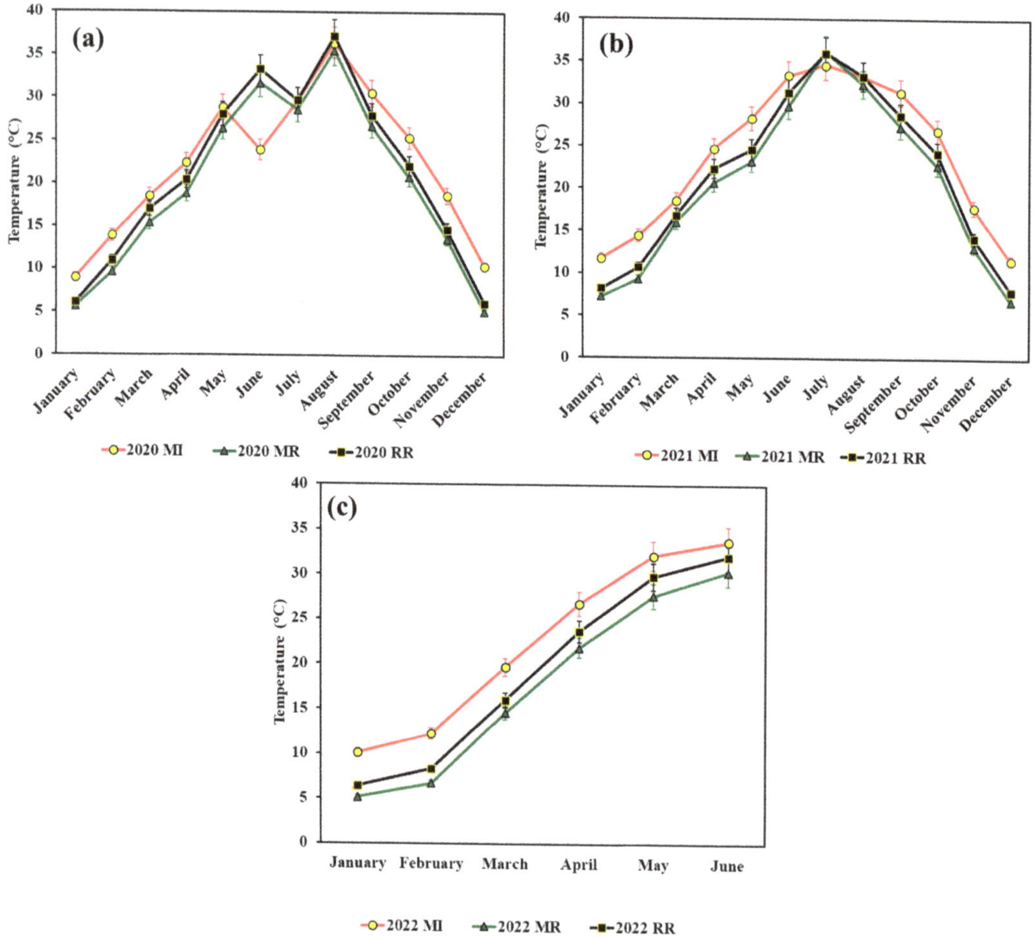

Figure 6. Average system temperature in (**a**) 2021, (**b**) 2021, and (**c**) 2022.

Table 3. Average system temperature from January 2020 to June 2022.

Month	2020			2021			2022		
	MI (°C)	MR (°C)	RR (°C)	MI (°C)	MR (°C)	RR (°C)	MI (°C)	MR (°C)	RR (°C)
January	8.96	5.61	6.09	11.73	7.17	8.19	10.09	5.11	6.39
February	13.88	9.65	10.96	14.39	9.31	10.67	12.23	6.68	8.31
March	18.47	15.32	16.99	18.52	15.96	16.77	19.66	14.57	15.99
April	22.38	18.83	20.43	24.62	20.61	22.27	26.77	21.87	23.69
May	28.86	26.43	28.06	28.27	23.11	24.53	32.18	27.67	29.85
June	32.92	31.67	33.32	33.35	29.74	31.38	33.74	30.32	32.04
July	29.66	28.61	29.72	34.65	36.21	36.08	-	-	-
August	36.49	35.59	37.29	33.29	32.35	33.31	-	-	-
September	30.59	26.71	28.01	31.39	27.34	28.65	-	-	-
October	25.43	20.78	22.17	26.89	22.79	24.28	-	-	-
November	18.71	13.65	14.73	17.89	13.16	14.26	-	-	-
December	10.46	5.18	6.16	11.65	6.85	8.01	-	-	-

The average value of MI was higher than MR and RR throughout the year, except in the middle of the year, from June to August, when those of RR matched its values. The higher MI compared to the other system temperatures was caused by the activity of the PV modules. The absorption of solar energy in the form of light and its conversion to electricity was conducted at the PV modules' capacity, with the other form of solar energy, i.e., heat energy, being absorbed. Heat absorption was subsequently aggravated by the dark color of the PV modules, thereby increasing the heat absorption capacity of the PV modules [11,12]. A high MI value is characteristic of most BIPV systems, as roofing systems encapsulate them without enough air ventilation compared to the cases of conventionally installed PV modules and more traditional BAPV systems [11,12].

The MR values were always the lowest among the values of the three measured temperatures and were always approximately 1–2 °C lower than the RR values and 1–5 °C lower than the MI values. The annual differences between RR and MR are shown in Figure 7, where MR is almost always lower than RR throughout the 2.5-year observation period, except in July 2021. The small differences between MR and RR were likely due to the different roof tile configurations used in this system when the PV modules were installed on top of the tiles. The installation of PV modules on the roof tiles left a little gap which facilitated the heat transfer caused by more wind. In contrast, the original roof tiles retained more heat because the gap was nonexistent [35–37]. The finding implies that the PV system could be installed as the entire roofing system with this configuration without sacrificing the roof temperature.

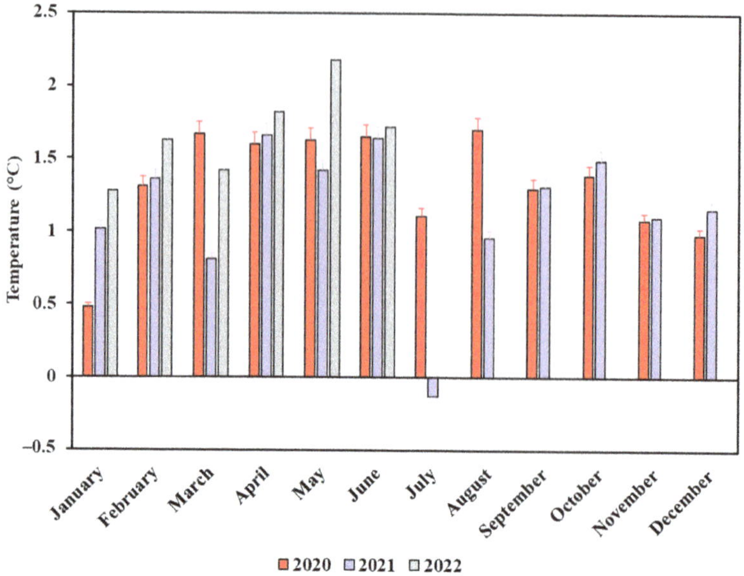

Figure 7. Monthly differences between the roof rear temperature (RR) and module rear temperature (MR) throughout the observation period.

3.3. *Effect of System Temperature on Power Conversion*

To determine how the system temperature affects the power-generation efficiency of this BIPV system, monthly power conversion and system temperature were compared and analyzed, and tabulated in Table 4. Figure 8 shows that the trends of system temperature and power conversion are opposite to each other. The system temperature trends followed the seasonal changes in South Korea, i.e., the system temperature increased in spring and summer and decreased in autumn and winter. Conversely, power conversion decreased in spring and summer and increased in autumn and winter.

Table 4. Power conversion from January 2020 to June 2022.

Month	Power Conversion (%)		
	2020	**2021**	**2022**
January	10.42	9.98	10.35
February	10.07	10.17	9.63
March	7.93	7.93	7.61
April	6.93	6.89	6.83
May	6.21	6.50	5.96
June	5.99	6.07	6.06
July	6.21	5.24	
August	6.29	5.57	
September	7.71	7.11	
October	8.49	8.48	
November	10.57	9.99	
December	11.42	10.96	

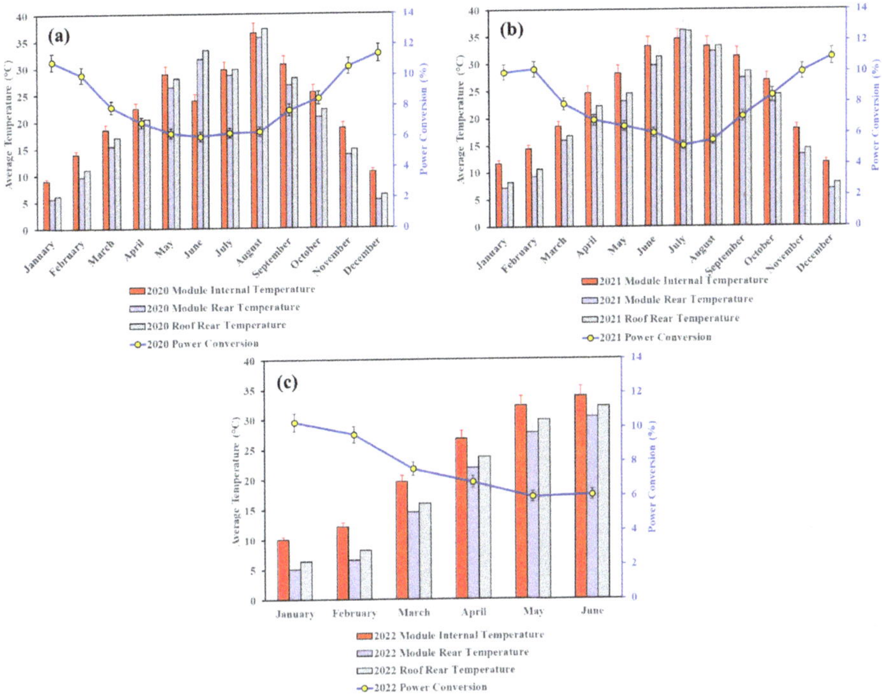

Figure 8. Power conversion and system temperature in (**a**) 2020, (**b**) 2021, and (**c**) 2022.

As mentioned previously, the system temperature is one of the factors affecting the power generation performance of a BIPV system [21,26–33,35–37]. In this BIPV roofing system, it was apparent that the system temperature affected the performance of the PV modules. The highest efficiency of 11.42% was achieved in December 2020, when the system temperature was the lowest. In contrast, the lowest efficiency of 5.24% occurred in July 2021, when the system temperature was 10 °C above the ambient temperature. This correlation shows how a system temperature increase negatively impacts the PV module efficiency, dropping to half its maximum value at the peak system temperature. A PV temperature increase leads to an increase in carrier concentration, thereby enhancing the rate of carrier recombination and leading to decreased open-current voltage (VOC), fill factor (FF), and thus performance [38].

Figure 9 shows that during the 2.5-year observation period, there was no significant decline in system performance regarding power conversion. The largest drop in power conversion was in July 2021, which decreased by ~1% compared to the previous year. Such a large decline in efficiency was not evident in the rest of the months when efficiency decreased on average only by 0.25% compared to the values in the previous years. These findings confirm that this particular BIPV system can maintain its performance with a negligible effect on efficiency over 2.5 years.

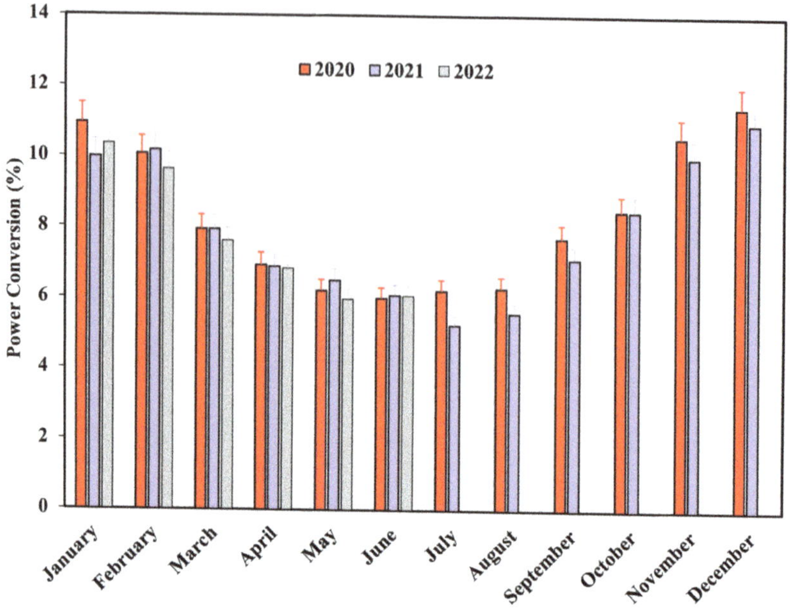

Figure 9. Comparison of monthly power conversion from January 2020 to June 2022.

4. Conclusions

The long-term performance of a BIPV system of 160 Wp was evaluated using outdoor monitoring in South Korea for 2.5 years. This study revealed the intercorrelations between power generation, solar irradiance, and system temperature. While solar irradiance mainly affected power generation, system temperatures contributed to power generation fluctuations, following the seasonal changes from cold to warm and again to cold temperatures throughout the year. Results showed that half of the BIPV system's performance was lost due to temperature fluctuations. Therefore, addressing this issue is crucial, particularly in hot regions. Factors such as wind direction, ventilation, and the type of PV technology could play important roles in achieving better economic value for the installed PV system. The temperature behind the PV modules was lower than that behind the roof tiles, implying that installing such a BIPV system will not increase the temperature of the roofing system. This finding suggests that the comfort of the room underneath the PV modules may be uncompromised by using this configuration in the climate of South Korea. Therefore, expanding the use of PV modules as an entire roofing system by using this configuration has great potential. This study proves that one of the generally associated temperature concerns—the roofing temperature—is not necessarily concerned with this BIPV system configuration.

Author Contributions: Conceptualization, W.K.K., M.H.A.A., S.A. and S.J.; Methodology, W.K.K., M.H.A.A., S.A. and S.J.; Data curation, M.H.A.A., S.A. and S.J.; Formal analysis, M.H.A.A., S.A. and S.J.; Resources, W.K.K., J.K. (Jitaek Kim) and J.K. (Jungtaek Kim); Funding acquisition, W.K.K.; Supervision, W.K.K.; Writing—original draft, M.H.A.A., S.A. and S.J.; Writing—review and editing, W.K.K. All authors have read and agreed to the published version of the manuscript.

Funding: This research was supported by the 2021 Yeungnam University Research Grant.

Institutional Review Board Statement: Not applicable.

Informed Consent Statement: Not applicable.

Data Availability Statement: Not applicable.

Conflicts of Interest: The authors declare no conflict of interest.

References

1. International Energy Agency. World Energy Outlook 2022. Available online: https://www.iea.org/reports/world-energy-outlook-2022 (accessed on 7 February 2023).
2. International Energy Agency. Global Energy Review: CO2 Emissions in 2021. Available online: https://www.iea.org/reports/global-energy-review-co2-emissions-in-2021-2 (accessed on 7 February 2023).
3. Scully, J. PV Industry Production Hits 310GW of modules in 2022; What about 2023? Available online: https://www.pv-tech.org/pv-industry-production-hits-310gw-of-modules-in-2022-what-about-2023/ (accessed on 10 March 2023).
4. Verband Deutscher Mashinen- und Anlagenbau. International Technology Roadmap for Photovoltaic (ITRPV). Available online: https://www.vdma.org/international-technology-roadmap-photovoltaic (accessed on 7 February 2023).
5. Masson, G.; Bosch, E.; Kaizuka, I.; Jäger-Waldau, A.; Donoso, J. Snapshot of Global PV Markets 2022 Task 1 Strategic PV Analysis and Outreach PVPS. Available online: https://www.researchgate.net/publication/360562245_Snapshot_of_Global_PV_Markets_2022_Task_1_Strategic_PV_Analysis_and_Outreach_PVPS (accessed on 7 February 2023).
6. Chudinzow, D.; Nagel, S.; Güsewell, J.; Eltrop, L. Vertical bifacial photovoltaics—A complementary technology for the European Electricity Supply? *Appl. Energy* **2020**, *264*, 114782. [CrossRef]
7. Pascaris, A.S.; Schelly, C.; Pearce, J.M. A first investigation of agriculture sector perspectives on the opportunities and barriers for agrivoltaics. *Agronomy* **2020**, *10*, 1885. [CrossRef]
8. Tina, G.M.; Bontempo-Scavo, F.; Merlo, L.; Bizzarri, F. Comparative analysis of monofacial and bifacial photovoltaic modules for Floating power plants. *Appl. Energy* **2021**, *281*, 116084. [CrossRef]
9. Baumann, T.; Nussbaumer, H.; Klenk, M.; Dreisiebner, A.; Carigiet, F.; Baumgartner, F. Photovoltaic systems with vertically mounted bifacial PV modules in combination with green roofs. *Sol. Energy* **2019**, *190*, 139–146. [CrossRef]
10. Corti, P.; Bonomo, P.; Frontini, F.; Mace, P.; Bosch, E. Building Integrated Photovoltaics: A Practical Handbook for Solar Buildings' Stakeholders. Available online: https://solarchitecture.ch/wp-content/uploads/2020/11/201022_BIPV_web_V01.pdf (accessed on 7 February 2023).
11. Eiffert, P.; Kiss, G.J. Building-Integrated Photovoltaics for Commercial and Institutional Structures: A Sourcebook for Architects and Engineers. Available online: https://www.nrel.gov/docs/fy00osti/25272.pdf (accessed on 7 February 2023).
12. Roberts, S.; Guariento, N. *Building Integrated Photovoltaics: A Handbook*; Birkhauser: Basel, Switzerland, 2009; ISBN 9783034604864.
13. James, T.; Goodrich, A.; Woodhouse, M.; Margolis, R.; Ong, S. Building-Integrated Photovoltaics (BIPV) in the Residential Sector: An Analysis of Installed Rooftop System Prices. Available online: https://www.nrel.gov/docs/fy12osti/53103.pdf (accessed on 7 February 2023).
14. Kuhn, T.E.; Erban, C.; Heinrich, M.; Eisenlohr, J.; Ensslen, F.; Neuhaus, D.H. Review of Technological Design Options for Building Integrated Photovoltaics (BIPV). *Energy Build.* **2021**, *231*, 110381. [CrossRef]
15. De L'epine, M.; Salel, D. National Survey Report of PV Power Applications in France 2021. Available online: https://iea-pvps.org/wp-content/uploads/2022/07/IEA-PVPS-NSR-France-EN-2021-v5.pdf (accessed on 9 March 2023).
16. Yamazaki, M.; Ikki, O. National Survey Report of PV Power Applications in JAPAN 2021. Available online: https://iea-pvps.org/wp-content/uploads/2022/11/NSR_Japan_2021.pdf (accessed on 9 March 2023).
17. Fang, L.; Honghua, X.; Yang, Y. National Survey Report of PV Power Applications in CHINA 2021. Available online: https://iea-pvps.org/wp-content/uploads/2022/11/PVPS-National-Survey-Report-China-2021.pdf (accessed on 9 March 2023).
18. Park, C.; Yang, H.; Won, C. National Survey Report of PV Power Applications in Korea 201. Available online: https://iea-pvps.org/wp-content/uploads/2021/01/NSR_Korea_2019.pdf (accessed on 9 March 2023).
19. Building-Integrated Photovoltaics Market Size Report, 2030. Available online: https://www.grandviewresearch.com/industry-analysis/building-integrated-photovoltaics-bipv-market (accessed on 10 March 2023).
20. Huld, T.; Amillo, A. Estimating PV module performance over large geographical regions: The role of irradiance, air temperature, wind speed and solar spectrum. *Energies* **2015**, *8*, 5159–5181. [CrossRef]
21. Kumar, N.M.; Sudhakar, K.; Samykano, M. Performance comparison of BAPV and BIPV systems with C-si, CIS and CdTe photovoltaic technologies under tropical weather conditions. *Case Stud. Therm. Eng.* **2019**, *13*, 100374. [CrossRef]
22. Liu, Z.; Zhang, Y.; Yuan, X.; Liu, Y.; Xu, J.; Zhang, S.; He, B. A comprehensive study of feasibility and applicability of building integrated photovoltaic (BIPV) systems in regions with high solar irradiance. *J. Clean. Prod.* **2021**, *307*, 127240. [CrossRef]
23. Jordan, D.C.; Deline, C.; Deceglie, M.; Silverman, T.J.; Luo, W. PV Degradation—Mounting & Temperature. In Proceedings of the IEEE 46th Photovoltaic Specialists Conference (PVSC46), Chicago, IL, USA, 16–21 June 2019. [CrossRef]
24. Gok, A.; Ozkalay, E.; Friesen, G.; Frontini, F. The Influence of Operating Temperature on the Performance of BIPV Modules. *IEEE J. Photovolt.* **2020**, *10*, 1371–1378. [CrossRef]
25. Kumar Singh, Y.; Dubey, S.; Pandey, K.; Sastry, O.S. Performance of HIT Module in Different Climatic Condition in India. *Mater. Today Proc.* **2019**, *17*, 321–328. [CrossRef]
26. Skandalos, N.; Kapsalis, V.; Karamanis, D. The Effect of Local Climatic Conditions on the Building Integration of Photovoltaics. *IOP Conf. Ser. Earth Environ. Sci.* **2022**, *1123*, 012020. [CrossRef]
27. Dabaghzadeh, N.; Eslami, M. Temperature distribution in a photovoltaic module at various mounting and wind conditions: A complete CFD modeling. *J. Renew. Sustain. Energy* **2019**, *11*, 053503. [CrossRef]

28. Poulek, V.; Matuška, T.; Libra, M.; Kachalouski, E.; Sedláček, J. Influence of Increased Temperature on Energy Production of Roof Integrated PV Panels. *Energy Build.* **2018**, *166*, 418–425. [CrossRef]
29. Kim, H.; Boafo, F.E.; Kim, J.; Kim, J. Investigating the effect of roof configurations on the performance of BIPV System. *Energy Procedia* **2015**, *78*, 1974–1979. [CrossRef]
30. D'Orazio, M.; Di Perna, C.; Di Giuseppe, E. Performance Assessment of Different Roof Integrated Photovoltaic Modules under Mediterranean Climate. *Energy Procedia* **2013**, *42*, 183–192. [CrossRef]
31. Kaplanis, S.; Kaplani, E.; Kaldellis, J.K. PV Temperature and Performance Prediction in Free-Standing, BIPV and BAPV Incorporating the Effect of Temperature and Inclination on the Heat Transfer Coefficients and the Impact of Wind, Efficiency and Ageing. *Renew. Energy* **2022**, *181*, 235–249. [CrossRef]
32. Mittelman, G.; Alshare, A.; Davidson, J.H. A Model and Heat Transfer Correlation for Rooftop Integrated Photovoltaics with a Passive Air Cooling Channel. *Solar Energy* **2009**, *83*, 1150–1160. [CrossRef]
33. Saleh, I.M.; Abufares, H.M.; Snousi, H.M. Three-Year Performance Evaluation of Single Junction Amorphous Solar Cells Grid-Connected Power Station in Libya. *Conf. Pap. Eng.* **2013**, *2013*, 950195. [CrossRef]
34. Chung, M.H.; Park, J.C. An Experimental Study on the Thermal Performance of Phase-Change Material and Wood-Plastic Composites for Building Roofs. *Energies* **2017**, *10*, 195. [CrossRef]
35. Karthikeyan, V.; Sirisamphanwong, C.; Sukchai, S.; Sahoo, S.K.; Wongwuttanasatian, T. Reducing PV module temperature with radiation based PV module incorporating composite phase change material. *J. Energy Storage* **2020**, *29*, 101346. [CrossRef]
36. Hasan, A.; McCormack, S.; Huang, M.; Norton, B. Evaluation of Phase Change Materials for Thermal Regulation Enhancement of Building Integrated Photovoltaics. *Sol. Energy* **2010**, *84*, 1601–1612. [CrossRef]
37. Wai, L.Z.; Yusoff, M.; Irwanto, M.; Razak, A.; Ibrahim, S.; Zhubir, N. Investigation of Solar Panel Performance Based on Different Wind Velocity Using ANSYS Software. *Indones. J. Electr. Eng. Comput. Sci.* **2016**, *1*, 456–463. [CrossRef]
38. Amin, A.A.; Al-Maghrabi, M.A. The analysis of temperature effect for MC-si photovoltaic cells performance. *Silicon* **2017**, *10*, 1551–1555. [CrossRef]

sustainability

MDPI

Article

Energy Performance Analysis of Photovoltaic Integrated with Microgrid Data Analysis Using Deep Learning Feature Selection and Classification Techniques

Sana Qaiyum [1,*], Martin Margala [1], Pravin R. Kshirsagar [2], Prasun Chakrabarti [3] and Kashif Irshad [4]

1 School of Computing and Informatics, University of Louisiana at Lafayette, Lafayette, LA 70504, USA; martin.margala@louisiana.edu
2 Department of Data Science, Tulsiramji Gaikwad Patil College of Engineering and Technology, Nagpur 441108, India; pravinrk88@yahoo.com
3 Department of Computer Science and Engineering, ITM SLS Baroda University, Vadodara 391510, India; drprasun.cse@gmail.com
4 Interdisciplinary Research Centre for Renewable Energy and Power System, King Fahad University of Petroleum and Minerals, Dhahran 31261, Saudi Arabia; kashif.irshad@kfupm.edu.sa
* Correspondence: sqaiyum.cs@gmail.com; Tel.: +1-966571179728

Abstract: Microgrids are an essential element of smart grids, which contain distributed renewable energy sources (RESs), energy storage devices, and load control strategies. Models built based on machine learning (ML) and deep learning (DL) offer hope for anticipating consumer demands and energy production from RESs. This study suggests an innovative approach for energy analysis based on the feature extraction and classification of microgrid photovoltaic cell data using deep learning algorithms. The energy optimization of a microgrid was carried out using a photovoltaic energy system with distributed power generation. The data analysis has been carried out for feature analysis and classification using a Gaussian radial Boltzmann with Markov encoder model. Based on microgrid energy optimization and data analysis, an experimental analysis of power analysis, energy efficiency, quality of service (QoS), accuracy, precision, and recall has been conducted. The proposed technique attained power analysis of 88%, energy efficiency of 95%, QoS of 77%, accuracy of 93%, precision of 85%, and recall of 77%.

Keywords: energy analysis; microgrid; photovoltaic cell; deep learning; distributed power generation

Citation: Qaiyum, S.; Margala, M.; Kshirsagar, P.R.; Chakrabarti, P.; Irshad, K. Energy Performance Analysis of Photovoltaic Integrated with Microgrid Data Analysis Using Deep Learning Feature Selection and Classification Techniques. *Sustainability* **2023**, *15*, 11081. https://doi.org/10.3390/su151411081

Academic Editors: Prince Winston David and Praveen Kumar B

Received: 30 May 2023
Revised: 6 July 2023
Accepted: 10 July 2023
Published: 15 July 2023

1. Introduction

Current demand for energy consumption is predicated on burning fossil fuels to provide dependable and resilient energy networks. Unquestionably, one of the most significant issues for scientists and engineers is the requirement for energy. Energy production techniques from the preceding century are now acknowledged as being inappropriate because of rising atmospheric carbon dioxide (CO_2) emissions [1]. The significant contribution of non-renewable energy sources raises environmental issues since they release greenhouse gases into the atmosphere, which have detrimental effects on human health and the ecosystem. As they are generally accessible, clean, and free of pollution, renewable energy sources (RESs) have gained popularity and attention across the globe. By 2030, it is anticipated that RESs like wind, solar, hydro, geothermal, and biomass would dominate world's electricity production and surpass all other energy sources [2]. Among the renewables, photovoltaics have experienced the fastest increase over the past few decades. Currently, wind energy production exceeds that of photovoltaic (PV), although PV is more widely accessible and wind turbines require extremely specialized site conditions. As an alternative, photovoltaics have become a viable option in the fight against climate change. The process of turning light into electricity using semiconducting materials that display the photovoltaic effect is known as photovoltaic (PV). The three main PV cell technologies—monocrystalline silicon, polycrystalline silicon, and thin film—control the global market. Among these, thin film solar cells offer several advantages over traditional silicon-based PV cells. They can be manufactured using low-cost techniques, such as roll-to-roll deposition or printing methods, which can potentially reduce production costs and enable large-scale manufacturing. Thin film solar cells are also lightweight and flexible, making them suitable for various applications, including building-integrated photovoltaics, portable electronics, and off-grid installations. However, the efficiency of thin film solar cells is generally lower than that of silicon-based cells [3].

The amazing rise in human living standards and the resulting rise in electricity demand have led to a number of super-large-scale power system flaws that are now more obvious than ever [4]. The traditional fossil fuel-based power plants are unable to provide enough energy to keep up with the rising demand for electricity. Early in the twenty-first century, the idea of a microgrid (MG) for integrating clean renewable energy sources (RESs) was put forth. A small-scale local power system called a microgrid is made up of electric loads, control systems, and distributed energy resources (DERs). Energy storage systems (ESSs) and RERs are both used in the microgrid's power generation or DERs. A new sort of contemporary active power distribution system for the use and advancement of renewable energy is the microgrid [5]. However, microgrids make it more challenging to maintain a balance between energy production and consumption, and the incorporation of RESs complicates power grid operations.

PV plant power generation frequently experiences significant fluctuations, including voltage irregularities, reserve power flow issues, and power distribution problems. Additionally, energy users show unpredictable usage of power due to a variety of factors, including alterations in the environment and user activity. Therefore, it is important to analyze the performance of the system to provide better consumer services and to maintain a reliable and sustainable system.

Accurate forecasting of PV panels is a challenging task since it depends only on the weather conditions such as temperature, humidity, etc. [6]. Prediction can be carried out using many techniques such as physical mode, machine learning (ML), and deep learning (DL) [2]. Each prediction method has its own advantages and disadvantages. Physical methods, for instance, can anticipate the shifting patterns of the environment with high precision, but they need a lot of processing capacity since they require a massive amount of data. Physical techniques encounter unforeseen estimating errors and are inappropriate for short-term forecasting horizons, which raises further problems. Similar to this, the majority of the statistical models used to anticipate renewable energy are linear in design, which restricts their application to forecasting issues with wider time horizons. ML-based prediction provides findings that are more accurate than those produced by statistical and physical methods owing to its data mining and feature extraction capabilities. However, ML-based prediction models require shallow models as the foundation of their learning strategies. Trees, regressors, and neural networks with zero or one hidden layer are examples of common shallow patterns. Often, extensive knowledge and expertise are needed while training such shallow models. Therefore, it is frequently difficult to investigate shallow structures theoretically. Hence, shallow models have certain disadvantages in real-world applications. Lately, it has been determined that DL-based approaches to energy generation and power load forecasting performed better than ML-based approaches. Unlike ML, DL-based methods do not have problems with manually chosen feature selection, complex samples, or ineffective generalization competence [7]. Consequently, it is impossible to ignore the dynamics of renewable resource energy generation behavior. The limited use of MGs in literature work prevents real-world data from being taken into account while controlling energy distribution. For better energy management of real-world data, a complete framework is needed. Furthermore, power trading between different market players is completely ignored, and prior statistical information of uncertain renewable resource energy production was assumed to be perfectly understood. This is the driving force behind the suggested method, which combines deep learning approaches with distributed energy management to enhance the effectiveness and dependability of the proposed system.

Objectives of this work are as follows:

1. To propose novel method for energy analysis based on a microgrid photovoltaic system by feature extraction and classification using deep learning techniques.
2. The energy optimization of a microgrid using a photovoltaic energy system with distributed power generation.
3. The data analysis for feature analysis and classification using a Gaussian radial Boltzmann with Markov encoder model.

2. Related Works

The following is a list of the most significant research topics on energy management that have been studied over the course of the last few years. The research conducted in [8] examined the most effective way to manage energy in renewable microgrids by taking into consideration batteries, solar panels, and wind turbines. The work in [9] analyzes the impact that wind and solar hybrid power generation technologies have on the performance of a sustainable microgrid. It is illustrated that a larger penetration of hybrid renewable energy sources might bring possible answers, provided they are controlled in the appropriate manner. The effects of uncertainty that are connected with RESs may be simulated by utilizing a stochastic technique that is based on Monte Carlo as described in [10]. The administration and maintenance of renewable microgrids is discussed in [11] from the perspective of a data-driven paradigm. When it comes to microgrids, ref. [12] makes use of the same research, but this time with a concentration on WSNs. It has been discovered that a microgrid that runs on renewable energy sources and fitted with a number of sensors is susceptible to being hacked. The creation of hydrogen by a renewable microgrid and the recovery of thermal energy by a fuel cell are the topics of discussion in [13]. By adopting multi-objective switching, the power losses and related costs with renewable sources may be minimized and optimized, as shown in [14]. Research of a similar kind is shown in [15] with regard to the influence that switching has on the most appropriate management strategy and course of action for microgrids. In [16], the authors conducted an analysis to see how the functioning of microgrids will be impacted by electric cars. According to [17], it is important to estimate charging needs for electric cars based on a detailed inspection and analysis of their erratic conduct in a complex environment. In order to do this, it is necessary to examine and analyze the erratic behavior of electric vehicles. Ref. [18] goes into more detail on the advantages and disadvantages of these kinds of systems, as well research that is relevant to both. In most cases, the energy management system (EMS) is not just concerned with preserving the energy balance inside the microgrid but it also has additional objectives in mind. The reduction of operating expenses, pollutants, and losses might be among these aims, in addition to a great number of other objectives, the attainment of which is contingent on the reason for the creation of such systems. A good number of these management systems additionally integrate various objectives in a manner that is multi-objective [19]. The probabilistic solar radiation forecasting that was produced in [20] made use of an analogue ensemble approach. In [21], it was recommended that ensemble forecasting based on empirical biasing be used to anticipate the geographical as well as temporal day ahead total daily radiation. This would be possible via the use of empirical biasing. In [22], Las-so was used to provide a radiation prediction for the next 5 min.

3. System Model

This section proposes a novel method in energy analysis based on microgrid photovoltaic cell data analysis by feature extraction and classification using deep learning techniques (Figure 1a). The energy optimization of the microgrid is carried out using a photovoltaic-based energy system with distributed power generation. The data analysis has been carried out for feature analysis and classification using a Gaussian radial Boltzmann with Markov encoder model. It is an algorithm for learning that assists individuals in finding intriguing characteristics hidden inside datasets that are made up of binary vectors. In networks with multiple layers of feature detectors, the learning process is often quite slow. However, the pace of the algorithm may be increased by adding a learning layer to the feature detectors in the network. The plug-and-play electric grid is one of the outstanding features of the microgrid since it can operate both independently and cooperatively with the power grids. The structural organization and linkages of the microgrid are shown in Figure 1b. With the capacity to choose the quantity and type of renewable energy sources that may be incorporated into the system, these tiny grids offer energy with improved stability, security, and resilience. It follows that microgrids have the capacity to effectively integrate a variety of diverse sources of distributed power generation, particularly renewable power sources. The microgrid is a small-scale local power system which integrates the clean renewable energy sources (RESs) made up of electric loads, control systems, and distributed energy resources (DERs). Energy storage systems (ESSs) and RERs are both used in a microgrid's power generation or DERs. A new sort of contemporary active power distribution system for the use and advancement of renewable energy is the microgrid.

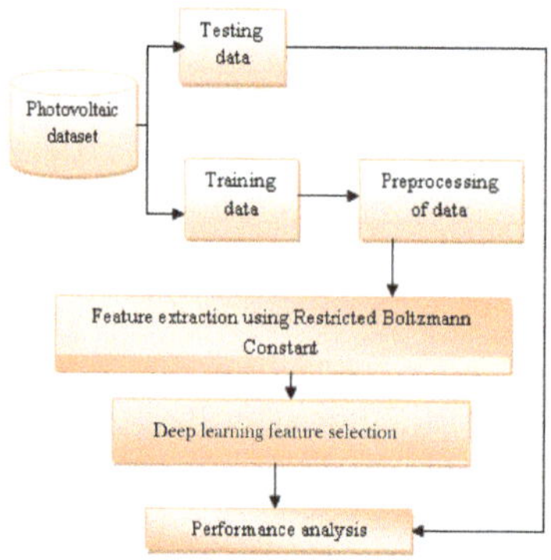

(a)

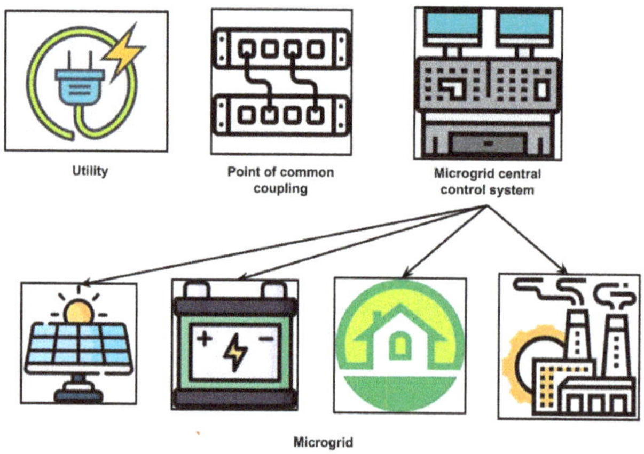

(b)

Figure 1. (**a**) The proposed energy analysis technique. (**b**) Structure of the microgrid.

Photovoltaic-based energy system with distributed power generation:

The method is based on a series of newly developed 3D matrix-based equations that calculate the system's overall load at the time of observation, the amount of power supplied to home DC, SLs from the utility's main DC grid, and the efficiency of the entire system. Let "NT" be number of time samples within the observational period of time "T." The following sets of indices, p q, and r, are provided in Equation (1) throughout the modeling.

$$
\begin{gathered}
\{\text{superset or subset}\} = \\
\left\{\begin{array}{c} p \supseteq P_1 = \{1,2,\ldots,x\}\text{or}p \supset P_1' = \{1,2,\ldots,x\} \\ q \supseteq Q = \{1,2,\ldots,y\}\text{or}q \supset Q' = \{1,2,\ldots,y\} \\ r \supseteq R = \{1,2,\ldots,n\}\text{or}r \supset R' = \{1,2,\ldots,n\} \end{array}\right\} \\
L_i = \sum_{j=1}^{N} |V_i||V_j| \angle (G_{ij}\cos(\varphi_i - \varphi_j) + B_{ij}\sin(\varphi_i - \varphi_j)) \\
M_i = \sum_{j=1}^{N} |V_i||V_j| \angle (G_{ij}\sin(\varphi_i - \varphi_j) - B_{ij}\cos(\varphi_i - \varphi_j))
\end{gathered} \tag{1}
$$

where "x" is the largest number of possible loads in "y" and SL at the time "t_n" has maximum samples ("n") in the system. The number "x" of DC loads that exist in the "y" MGs at time "t_n" is shown as (2). It is a huge 3D matrix of order [x, y] because each component of the matrix is a column matrix of NT of order 1.

$$
\begin{aligned}
\text{DC}_{\text{LyMGs}}(t_n) &= \left[[p_{\text{D}ij}(t_n)]_{\text{NT}\times 1}\right]_{x\times y} \\
&= \begin{bmatrix} [p_{\text{D11}}(t_n)]_{\text{NT}\times 1} & \cdots & [p_{\text{D1}y}(t_n)]_{\text{NT}\times 1} \\ \vdots & \ddots & \vdots \\ [p_{\text{D}x1}(t_n)]_{\text{NT}\times 1} & \cdots & [p_{\text{D}xy}(t_n)]_{\text{NT}\times 1} \end{bmatrix}
\end{aligned} \tag{2}
$$

This is [pD11 (t_n)]. The power of the first DC load present in the first MG at time "t_n" is represented by the column matrix NT × 1, which is further enlarged as Equation (3).

$$
\begin{gathered}
[p_{\text{D}ij}(t_n)]_{\text{NT}\times 1} = \begin{bmatrix} p_{\text{D}ij}(t_1) \\ p_{\text{Dij}}(t_2) \\ \vdots \\ p_{\text{D}ij}(t_n) \end{bmatrix}_{\text{NT}\times 1} \\
\text{AC}_{\text{L}_{\text{yMGs}}}(t_n) = \left[[p_{\text{Aij}}(t_n)]_{\text{NT}\times 1}\right]_{x\times y} \\
\text{I}_{\text{LyMGs}}(t_n) = \left[[p_{\text{Lij}}(t_n)]_{\text{NT}\times 1}\right]_{x\times y} \\
\text{VSD}_{\text{L}_{\text{yMGs}}}(t_n) = \left[[p_{\text{VSD}ij}(t_n)]_{\text{NT}\times 1}\right]_{x\times y}
\end{gathered} \tag{3}
$$

Equation (4) provides a matrix that includes rated power of converters connected to appropriate loads.

$$
(\text{DC}/\text{DC})_{\text{rating}}(\text{t}_\text{n}) = \left[[p_{\text{rDij}}(\text{t}_\text{n})]_{\text{NT}\times 1}\right]_{x\times y} \tag{4}
$$

Efficiency of the converter is evaluated as (5). The coefficients matrices' generalized form is provided by (4). An efficiency-based 3D matrix can finally be displayed as (4). Similar matrices would be created for converters connected to A and VSD loads.

$$
\begin{gathered}
\eta_{\text{dd}_y}(t_n) = \alpha_{mij}(t_n)\cdot\left(\frac{p_{\text{D}ij}(t_n)}{p_{\text{rD}ij}(t_n)}\right)^n + \alpha_{(n-1)jj}(t_n)\cdot(\cdots)^{n-1} + \cdots + \\
\alpha_{0ij}(t_n)\cdot(\cdots)^0, \\
\alpha_{nij}(t_n) = \begin{bmatrix} \alpha_{n11}(\text{t}_\text{n}) & \cdots & \alpha_{n1y}(t_n) \\ \vdots & \ddots & \vdots \\ \alpha_{nx1}(t_n) & \cdots & \alpha_{nxy}(t_n) \end{bmatrix} = \begin{bmatrix} \alpha_{nij}(t_1) \\ \alpha_{nij}(t_2) \\ \vdots \\ \alpha_{nij}(t_n) \end{bmatrix}. \\
\eta_{(\text{DC}/\text{DC})}(t_n) = \left[[\eta_{\text{dd}_{ij}}(t_n)]_{\text{NT}\times 1}\right]_{x\times y}
\end{gathered} \tag{5}
$$

Since each SL has its independent solar energy structure, therefore a PEC converter with an MPPT base connects it to the storage system. Solar energy can be expressed as Equation (6).

$$\begin{array}{c} p_{\text{solar}}(t_s) = \left[[p_{sj}(t_s)]_{\text{NT}\times 1} \right]^*_{y\times 1} \\ p_{\text{s,MGs}}(t_s) = [\lambda_{\text{SD}j}]_{1\times y} [\beta_j(t_s)]_{y\times \text{NT}} \left[[t_{sm} \cdots t_{s0}]_{\text{NT}\times 1} \right]^t \end{array} \quad (6)$$

In Equations (7) and (8), the first matrix is a matrix of conversion factors, the next matrix is a matrix of coefficients obtained from the fitting of curves, and the final matrix is a transposed time matrix with some power.

$$\begin{array}{c} [\lambda_{\text{SD}-j}]_{1\times y} = [\lambda_{\text{SD}-1} \quad \lambda_{\text{SD}-2} \quad \cdots \quad \lambda_{\text{SD}-y}]_{1\times y} \\ [\beta_j(t_n)]_{y\times \text{NT}} = \begin{bmatrix} \beta_{m1}(t_n) & & \cdots \\ \vdots & & \ddots \\ & & \beta_{01}(t_n) \\ & & \vdots \\ & & \beta_{my}(t_n) \cdots \beta_{0y}(t_n) \end{bmatrix}_{y\times \text{NT}} \\ \beta_{nj(t_n)} = \begin{bmatrix} \beta_{nj(t_1)} \\ \beta_{nj}(t_2) \\ \vdots \\ \beta_{nj(t_n)} \end{bmatrix}_{\text{NT}\times 1} \\ [t_{sm} \cdots t_{s0}]^t = \left[[t^m_{1m} \cdots t^m_{nm}] \cdots [t^0_{10} \cdots t^0_{n0}] \right]^t \end{array} \quad (7)$$

$$\begin{array}{c} P_{i,R_x=0} \approx \frac{V_i V_j}{X_i} [\sin\varphi_{ij}] \\ Q_{i,R_x=0} \approx \frac{V_i^2 - V_i V_j \cos\varphi_{ij}}{X_i} \\ \begin{bmatrix} f_D \\ f_Q \end{bmatrix} = \begin{bmatrix} \cos(\varphi i) & -\sin(\varphi i) \\ \sin(\varphi i) & \cos(\varphi i) \end{bmatrix} \begin{bmatrix} f_d \\ f_q \end{bmatrix} \end{array} \quad (8)$$

Distributed secondary control layers are used as shown in Equation (9) to adjust frequency and voltage anomalies.

$$\begin{array}{c} \overline{\omega_{avg}} = \frac{\sum_{i=1}^{N} \omega_{DGi}}{\omega_i = \left(\omega_{ref} - \overline{\omega_{avg}}\right)} \\ \omega_i = \left(\omega_{ref} - \overline{\omega_{avg}}\right) \\ \omega_i = k_{pf}\omega_i + k_{if} \int \omega_i dt \end{array} \quad (9)$$

Equation (10) can be used to express load voltage regulation methods.

$$sV_i = k_{pf}V_i + k_{if} \int V_i dt \quad (10)$$

From measured output current and voltage, instantaneous power is written as $p = v_{\text{odii}} \cdot i_{odi} + v_{\text{oqi}} i \cdot i_{\text{oqi}} i$ and $q = v_{\text{odi}} . i_{\text{oqii}} - v_{\text{oqi}} i \cdot i_{\text{odi}}$. By linearization, a tiny signal that represents active power is generated as shown in (11).

$$\begin{array}{c} \Delta P_i = -\omega_{ci} \Delta P_i \\ +\omega_{ci} \left(I_{odi} \Delta v_{odi} + I_{oqi} \Delta v_{oqi} + V_{odi} \Delta i_{odi} + V_{oqi} \Delta i_{oqi} \right) \\ v_{\text{odqi}} = [\; v_{\text{odi}} \quad v_{\text{oqi}} \;]^T, i_{\text{odqi}} = [\; i_{\text{odi}} \quad i_{\text{oqi}} \;]^T \end{array} \quad (11)$$

Equation (12) can be used to define the algebraic modeling for the voltage controller and current controller,

$$\begin{array}{c} i^*{}_{ldi} = F_i \cdot v_{odi} - \omega_b \cdot C_{fi} \cdot \Delta v_{oqi} + K_{PVi}\left(v^*_{odi} - v^*_{odi}\right) + K_{IVi}\varphi_{di} \\ i^*{}_{lqi} = F_i \cdot v_{oqi} + \omega_b \cdot C_{fi} \cdot \Delta v_{oqi} + K_{PVi}\left(v^*{}_{oqi} - v^*{}_{oqi}\right) + K_{IVi}\varphi_{qi} \\ v^*{}_{idi} = -\omega_b \cdot L_{fi} \cdot i_{lqi} + K_{PCi}(i^* ldi - i_{ldi}) + K_{IC} \cdot \gamma_{di} \\ v^*{}_{iqi} = \omega_b \cdot L_{fi} + K_{PCi}\left(i^*{}_{lqi} - i_{lqi}\right) + K_{IC} \cdot \Delta\gamma_{qi} \end{array} \tag{12}$$

$$\begin{array}{c} \frac{di_{ldqi}}{dt} = -\frac{R_{fi}}{L_{fi}} \cdot i_{ldqi} + \omega_i \cdot i_{ldqi} + \frac{1}{L_{fi}} \cdot v_{idqi} - \frac{1}{L_{fi}} \cdot v_{odqi} \\ \frac{dv_{odqi}}{dt} = \omega_i \cdot v_{oqqi} + \frac{1}{C_{fi}} \cdot i_{ldqi} - \frac{1}{C_{fi}} \cdot i_{odqi} \\ \frac{di_{\mathrm{odqi}}}{dt} = -\frac{R_{ci}}{L_{ci}} \cdot i_{\mathrm{odqi}} + \omega_i \cdot i_{\mathrm{odqi}} + \frac{1}{L_{ci}} \cdot v_{\mathrm{odiq}} - \frac{1}{L_{ci}} \cdot v_{\mathrm{obdqi}} \end{array} \tag{13}$$

By using reverse transformation, Equation (14) transforms bus voltage back into an *i*th specific inverter reference frame.

$$\begin{array}{c} [\Delta u_{\mathrm{Muq}}] = [T\gamma^{-1}] \cdot [\Delta u_{b\mathrm{DQ}}] + [T_\sigma^{-1}][\Delta\delta], \text{ where, } T_\sigma^{-1} = \\ \begin{bmatrix} -U_{bD}\sin(\delta) + U_{bQ}\cos(\delta) \\ -U_{bD}\cos(\delta) - U_{bQ}\sin(\delta) \end{bmatrix} \end{array} \tag{14}$$

P-N junction diodes are utilized in the structure of the PV module. As they are semiconductor devices, they can convert the energy that is taken in into usable electrical power. These diodes can convert incident light into electrical energy when it reaches their surface. Figure 2 depicts the basic construction, connections, and functionality of a PV module:

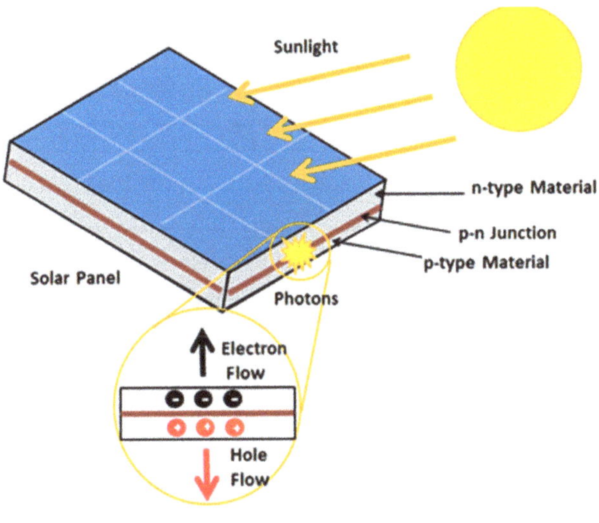

Figure 2. Working of Photovoltaic.

As seen in the image below, there exist two distinct layers of silicon: a negative N layer and a boron-doped positive P layer. The PV module clad with tempered glass captures solar energy when subjected to sunlight. The energy collected eventually rises above the band gap energy level, causing electrons to pass across that band on their way from the conduction band to the valence band. The conduction band's electrons can therefore move freely and create electron–hole pairs. The electricity generated during the process is used to power the load because the motion of electrons is what causes the passage of electric current. An array configuration is not sufficient to produce enough electricity because it suffers from multiple losses. The maximum power point tracking (MPPT) technique is the best way to maximize each string's efficiency. By employing such control strategies, PV modules can produce the maximum amount of electricity achievable.

4. Methodology

Feature analysis and classification using Gaussian radial Boltzmann with Markov encoder model:

The weighted sum of the densities of the M component parts is known as the mixture density. The *i*th component density is denoted by the expression $s(x;\theta i)$, where θ_i stands for the component parameters. With the restrictions that $\pi_i \geq 0$ and $\sum_{i=1}^{M} \pi_i = 1$, we use $\pi_{\rm i}$ to signify the weighting factor or "mixing proportions" of the ith component in combination. The chance that a data sample belongs to the *i*th mixture component is represented by s(i), and M _i = 1. Equations (15) and (16) are then used to define an M component mixture density (16),

$$s(x) = \sum_{i=1}^{M} \pi_i p(x;\theta_i), i = 1, \ldots, M \quad (15)$$

$$s(\mathrm{x}) = \sum_{c=1}^{C} \pi_c f_c(\mathrm{x} \mid \theta) \quad (16)$$

The mixture model has a vector of parameters, $\underline{\theta} = \{\theta_1, \ldots, \theta_M, \pi_1, \ldots \pi_M\}$

Hidden variables are treated as a latent variable, or z, in mixture models. It accepts values 1 through M as a discrete set that satisfies the conditions $z_M \epsilon \{0,1\}$ and $\sum_M z_M = 1$. A conditional distribution *p(x | z)* and a marginal distribution *p(z)* are how we define the joint distribution *p(x, z)*, i.e., from Equation (17),

$$p(z,x) = p(z)p(x \mid z) \quad (17)$$

Mixing coefficients k are used to specify the marginal distribution across z, as illustrated in Equation (18),

$$p(z_k = 1) = \pi_k \quad (18)$$

Equations (19) and (20) define probability density function of X.

$$p(x \mid \mu_k, \Sigma_k) = \frac{1}{\sqrt{2\pi\left|\Sigma^{-1}\right|}} \exp\left(-\frac{1}{2}(x-\mu_x)\Sigma_x^{-1}(x-\mu_x)^T\right) \quad (19)$$

$$f_c(\mathrm{x} \mid \mu_c, \Sigma_c) = \frac{1}{(2\pi)^{\frac{1}{t}}|\Sigma_c|^{\frac{1}{2}}} \exp\left(-\frac{1}{2}(\mathrm{x}-\mu_c)^t \Sigma_c^{-1}(\mathrm{x}-\mu_c)\right) \quad (20)$$

where μ_x is an $(\mu_{x1}, \ldots, \mu_{xN})$ and Σ_x covariance matrix and Σ_x is a vector of means (_x1,..., _xN). Equations (21)–(23), which are linear superpositions of Gaussians, can be used to represent Gaussian mixture distribution.

$$p(x) = \sum_{k=1}^{K} \pi_k p(x \mid \mu_k, \Sigma_k) \quad (21)$$

$$\begin{gathered} \hat{\pi}_c = \tfrac{n_c}{n}, \\ \dot{\mu}_c = \tfrac{1}{n_c} \textstyle\sum_{(i,y_i=c_j)} \mathrm{x}_i, p(x \mid z_k = 1) = p(x \mid \mu_k, \Sigma_k), \\ \hat{\Sigma}_c = \tfrac{1}{(n_c-1)} \textstyle\sum_{(i|y_i=c)} (\mathbf{x}_i - \mu_c)(\mathbf{x}_i - \mu_c)^t \end{gathered} \quad (22)$$

Conditional distribution of x for a specific value of z is a Gaussian, according to Equation (23):

$$p(x \mid z) = \prod_{k=1}^{K} p(x \mid \mu_k, \Sigma_k)^{z_k} \quad (23)$$

By adding joint distribution of all possible states of z to obtain Equation (24), one may determine the marginal distribution of x.

$$p(x) = \sum_{z} p(z)p(x \mid z) = \sum_{k=1}^{K} \pi_k p(x \mid \mu_k, \Sigma_k) \tag{24}$$

The "posterior probability" on a mixture component for a specific data vector is a significant derived quantity and is indicated in Equation (25):

$$\gamma(z_{nk}) = \frac{\pi_k N(x_n \mid \mu_k, \Sigma_k)}{\sum_{j=1}^{K} \pi_j N(x_n \mid \mu_j, \Sigma_j)} = \frac{p(z_k = 1)p(x \mid z_k = 1)}{\sum_{j=1}^{K} p(z_j = 1)p(x \mid z_j = 1)} \tag{25}$$

Maximal likelihood is the learning objective of RBMs, which are energy-based approaches. Equation (26), in its combined structure, defines the energy of its hidden parameters (*e*) and visible parameters (*f*).

$$En(\mathrm{e}, \mathrm{f}; \theta) = -\sum_{ij} \mathrm{W}_{ij}\mathrm{f}_i\mathrm{e}_j - \sum_{i} \mathrm{b}_i\mathrm{f}_i - \sum_{j} \mathrm{a}_j\mathrm{e}_j. \tag{26}$$

θ represents the element *W(a, b)*. Using Equation (27), one can determine the combined probability of v and h.

$$P_\theta(\mathrm{f}, \mathrm{e}) = \frac{1}{Z(\theta)} \exp(-En(\mathrm{f}, \mathrm{e}; \theta)). \tag{27}$$

In this context, the partition function is denoted by *Z(θ)*. The previous equation can be rewritten as Equation (28).

$$P_\theta(\mathrm{f}, \mathrm{e}) = \frac{1}{Z(\theta)} \exp\left(\sum_{i=1}^{D} \sum_{j=1}^{F} \mathrm{W}_{ij}\mathrm{f}_i\mathrm{e}_j + \sum_{i=1}^{D} \mathrm{f}_i\mathrm{b}_i + \sum_{j=1}^{F} \mathrm{e}_j\mathrm{a}_j\right) \tag{28}$$

Maximizing the probability function *P(f)* is the goal. The edge distribution of *P(f, e)* makes it easy to calculate *P(f)* by Equation (29):

$$P_\theta(\mathrm{f}) = \frac{1}{Z(\theta)} \sum_{\mathrm{h}} \exp\left[\mathrm{f}^T\mathrm{Wh} + \mathrm{a}^T\mathrm{h} + \mathrm{b}^T\mathrm{f}\right] \tag{29}$$

The RBM parameters are derived (*f*) by optimizing *P*. By optimizing log(*P(f)*) = L(), we can obtain maximum *P(f)* using Equation (30):

$$\begin{aligned} L(\theta) &= \tfrac{1}{N} \textstyle\sum_{n=1}^{N} \log P_\theta\left(\mathrm{f}^{(n)}\right) \\ \tfrac{\partial L(\theta)}{\partial \mathrm{W}_{ij}} &= \tfrac{1}{N} \textstyle\sum_{n=1}^{N} \tfrac{\partial}{\partial \mathrm{W}_{ij}} \log\left(\sum_{\mathrm{h}} \exp\left[\mathrm{f}^{(n)T}\mathrm{Wh} + \mathrm{a}^T\mathrm{h} + \mathrm{b}^T\mathrm{f}^{(n)}\right]\right) \\ &\quad - \tfrac{\partial}{\partial \mathrm{W}_{ij}} \log Z(\theta) = E_{P_{\mathrm{dut}}}\left[\mathrm{f}_i\mathrm{e}_j\right] - E_{P_\theta}\left[\mathrm{f}_i\mathrm{e}_j\right] \end{aligned} \tag{30}$$

The original purpose of stochastic gradient descent was to maximize *L(θ)*. Next, Equation (30) is used to calculate the *L(θ)* derivative for W.

The formula's first part is easy to evaluate. Across all datasets, the values of fi and ej are averaged. It is computationally challenging to solve the remaining part of the equation, which comprises all 2 | *f* | + | *e* | possible values of f and e. The formula's second part is Equation (31).

$$\sum_{f,e} \mathrm{f}_i\mathrm{e}_j P_\theta(\mathrm{f}, \mathrm{e}) \tag{31}$$

Monte Carlo simulations are used to estimate gradient as shown in the following equation:

$$\begin{aligned} \Delta \mathrm{a}_i &= f_i^{(0)} - f_i^{(k)} \\ \Delta \mathrm{b}_i &= P\left(\mathrm{e}_j = 1 \mid \mathrm{f}^{(0)}\right) - P\left(\mathrm{e}_j = 1 \mid \mathrm{f}^{(k)}\right) \\ \Delta \mathrm{W}_{ij} &= P\left(\mathrm{e}_j = 1 \mid \mathrm{f}^{(0)}\right)\mathrm{f}_i^{(0)} - P\left(\mathrm{e}_j = 1 \mid \mathrm{f}^{(k)}\right)\mathrm{f}_i^{(k)} \end{aligned} \tag{32}$$

where *f(0)* I is sample value and *f(k)* I is a sample that satisfies distribution *P(f)* identified by sampling. Lastly, Equation (33) provides the parameter update equation.

$$\begin{aligned} \mathrm{a}_i &= \mathrm{a}_i + \Delta \mathrm{a}_i \\ \mathrm{b}_j &= \mathrm{b}_j + \Delta \mathrm{b}_j \\ \mathrm{W}_{ij} &= \mathrm{W}_{ij} + \Delta \mathrm{W}_{ij} \end{aligned} \tag{33}$$

The probability distribution is shown below by Equation (34).

$$\begin{aligned} P\left(V, e^{(1)}, e^{(2)}\right) &= \tfrac{1}{Z(\theta)} \exp - E\left(V, e^{(0)}, e^{(\Omega)}; \theta\right) \\ \mid P\left(V, e^{(1)}, e^{(2)}\right) &= -V^T W^{(1)} e^{(1)} - V^T W^{(2)} e^{(2)} + b \end{aligned} \tag{34}$$

Encoders and decoders are essential elements of its design. Encoder and decoder both implement standard matrix multiplication. As a normalizing function, an encoder gradient function is utilized. After adjusting the weight and biases of the autoencoder, Equation (35) operates network training.

$$e^{(0)} = a\left(b^{(n)} + V^T W^{(n)}\right) \tag{35}$$

$$e^{(n)} = \sigma\left(b_i^{(n)} + e^{(n-1)T} W^{(n)}\right)$$

where $n = 1, 2, 3, \ldots, m$.

Consider training an HSI datacube with two hidden layers using Equations (36) and (37).

$$P\left(V_i = 1; e^{(2)}, e^{(2)}\right) = \alpha V^T W_1^{(1)} + \alpha V^T W_1^{(2)} \tag{36}$$

When n = 1:

$$\begin{aligned} P\left(V_i = 1; e^{(1)}, e^{(2)}\right) &= \alpha V^T W_1^{(1)} + \alpha V^T W_i^{(2)} \\ P\left(V_i = 2; e^{(1)}, e^{(2)}\right) &= a V^T W_2^{(1)} + a V^T W_2^{(2)} \end{aligned} \tag{37}$$

Mean-field value is represented by Equation (38):

$$P(x) = \sum_{h=1,2} Q\left(e^{(1)}, e^{(2)}\right) \log\left(\frac{e^{(i)}, e^{(2)}}{P\left(e^{(1)}, e^{(2)}\right)}\right) \tag{38}$$

where Gibbs energy is represented by Equation (39):

$$E(x) = \frac{1}{Z(D)} \exp(-P(x)) \tag{39}$$

By indicating a weight change, (40) and (41) present a new weight value. Each stratum is assigned the b = 0 bias.

$$\Delta e_i^{(1)} = \alpha \sum_i V_i W^{(1)} \tag{40}$$

$$\Delta e_i^{(2)} = \alpha \sum_i V_i W^{(2)} \tag{41}$$

Convolution filters and the weights of fully connected layers are two model parameters that are optimized using the gradient descent approach. It is essential to classify the image into the correct category since the final layer has a significant impact on classification outcomes. This is carried out by properly linking the weights from the prior layers. Here, in order to improve classification accuracy, the training of the final weight vector is optimized utilizing a newly created modified whale optimization method. The number of search agents is limited to 50, the utmost number of iterations is limited to 100, and the final parameter (vector a) is linearly modified between [0,2].

5. Performance Analysis

To computationally evaluate the cloud DNSE algorithm, the simulation set contains sensitivity based on error in measurement, grid variables, DNSE efficiency, and pseudo measurements. Reading the voltages at the node from a smart meter is the initial stage in this endeavor since smart meters are not set up to determine the voltages in the LV grid under examination. If the powers and voltages at all nodes are being monitored concurrently, this phase can be avoided. Table 1 displays dispersed networks for the low-voltage grid's resistance, reactance, and admittance by both series and shunt in accordance with their network types.

Table 1. Distributed Networks of LV Grid.

Distributed Networks	Resistance Ω/km	Series Reactance Ω/km	Shunt Admittance μS/km
Type 1	0.207	0.072	204.2
Type 2	0.320	0.075	175.9
Type 3	0.727	0.087	125.6

Table 2 gives analysis based on various circuit models. The circuit models analyzed are resistance, reactance and admittance network type in terms of power analysis, energy efficiency, QoS, accuracy, precision, and recall. The energy management system (EMS) and long short-term memory (LSTM) networks are compared.

Table 2. Analysis based on various circuit models.

Circuit Model	Techniques	Power Analysis	Energy Efficiency	QoS	Accuracy	Precision	Recall
Resistance	EMS	79	88	66	81	77	66
	LSTM	81	89	68	83	79	68
	Proposed method	83	92	71	85	81	72
Reactance	EMS	82	89	69	82	79	69
	LSTM	85	93	72	84	82	73
	Proposed method	87	95	75	86	83	75
Admittance network type	EMS	84	91	75	85	81	72
	LSTM	86	93	76	91	83	75
	Proposed method	88	95	77	93	85	77

Figure 3a–f give analysis based on resistance type circuit model. The proposed technique attained power analysis of 83%, energy efficiency of 92%, QoS of 71%, accuracy of 85%, precision of 81%, and recall of 72%, EMS achieved a 79% power analysis, an 88% energy efficiency, a 66% quality of service, an 81% accuracy, a 77% precision, and a 66% recall, and power analysis of 83%, energy efficiency of 92%, QoS of 71%, accuracy of 85%, and precision of 81% were all achieved with LSTM.

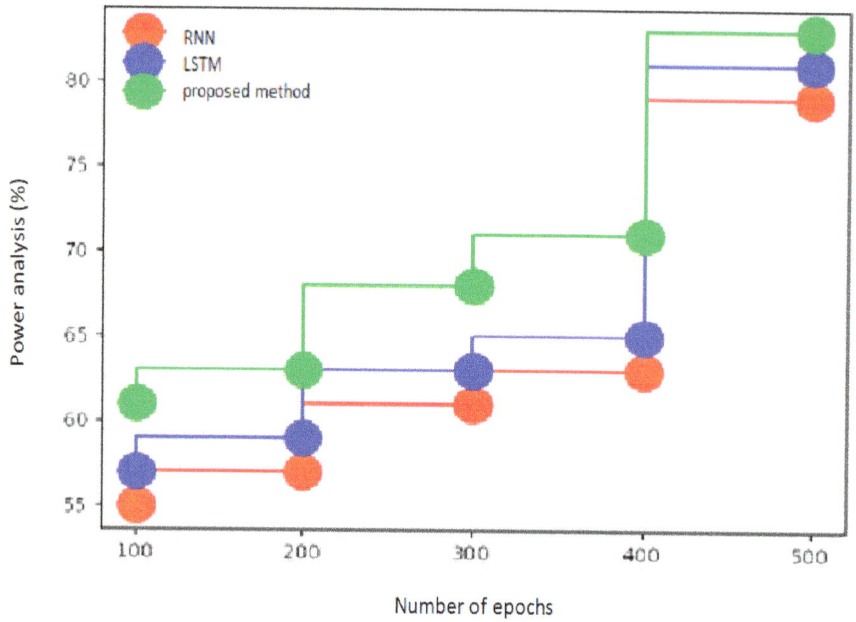

(a)

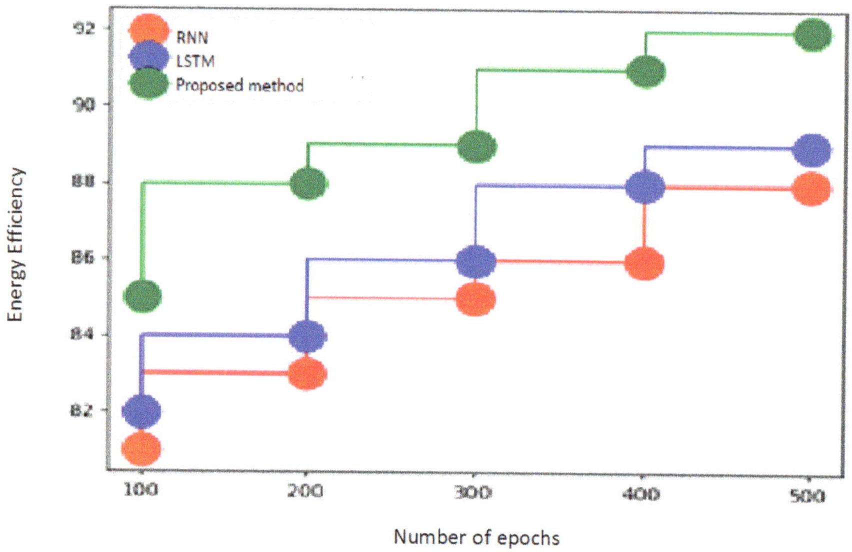

(b)

Figure 3. *Cont.*

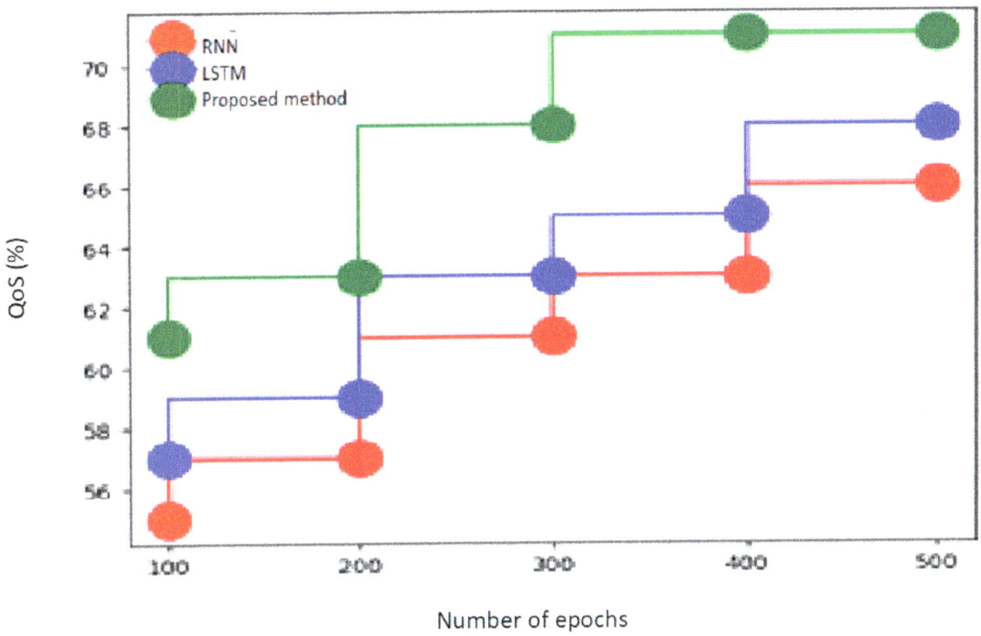

(c)

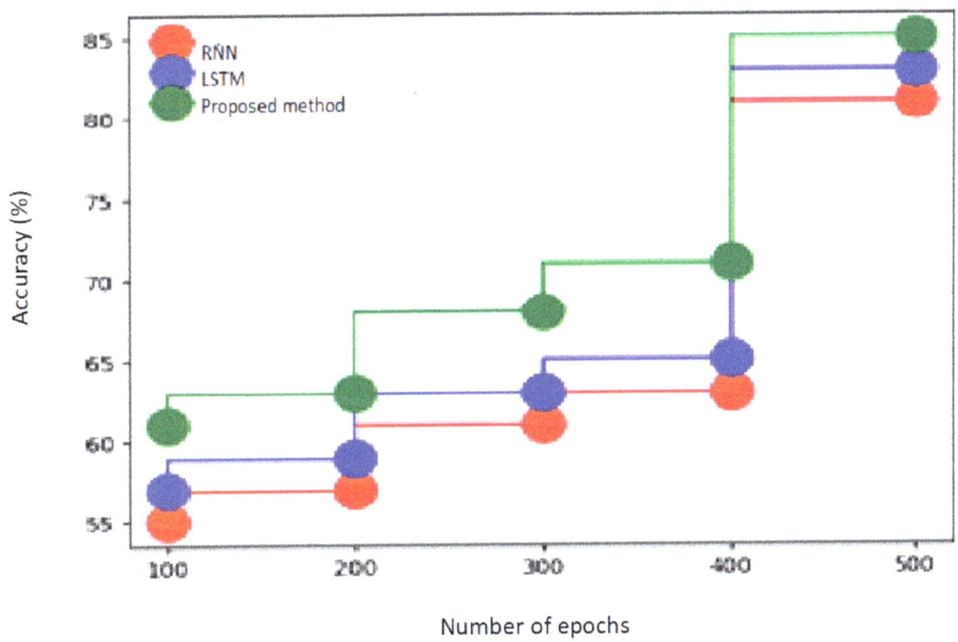

(d)

Figure 3. *Cont.*

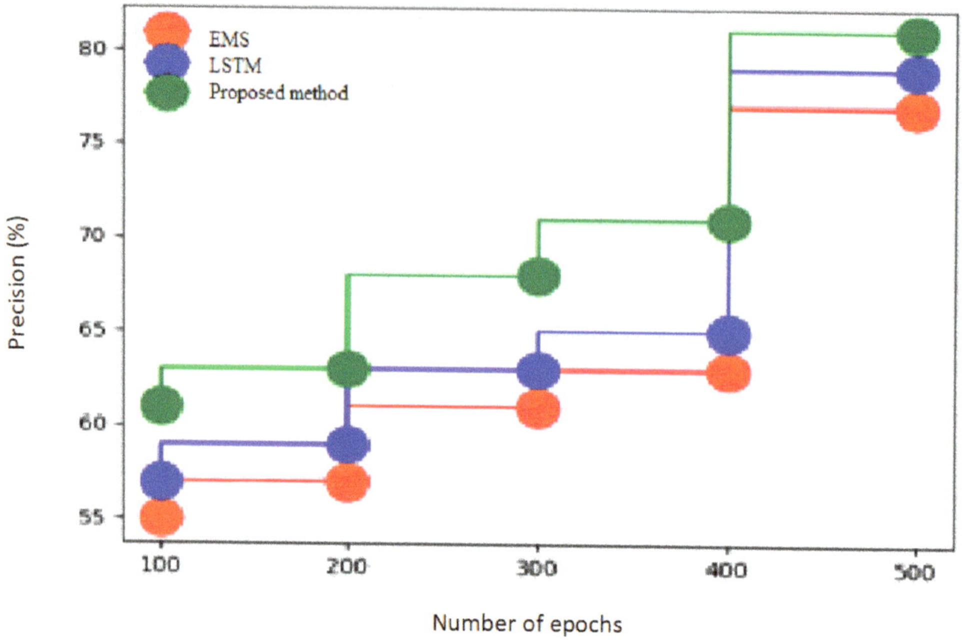

(e)

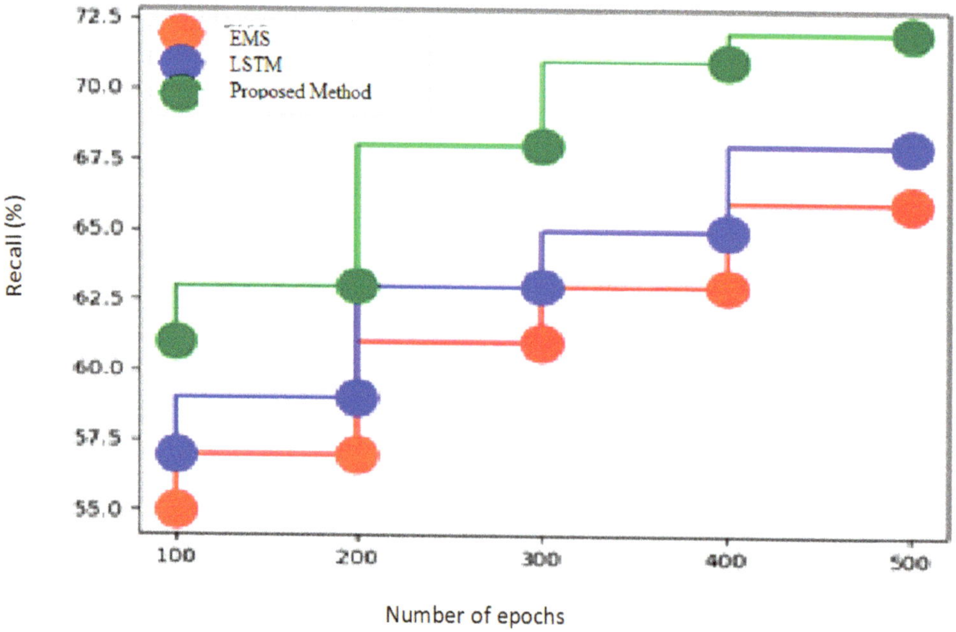

(f)

Figure 3. (**a–f**) Analysis for resistance type circuit model in terms of power analysis, energy efficiency, QoS, accuracy, precision, recall.

Figure 4a–f show analysis for reactance circuit model. The proposed technique attained power analysis of 87%, energy efficiency of 95%, QoS of 75%, accuracy of 86%, precision of 83%, and recall of 75%, EMS achieved a power analysis of 82%, an energy efficiency of 89%, a QoS of 69%, an accuracy of 82%, a precision of 79%, and a recall of 69%, and 85% power analysis, 93% energy efficiency, 72% QoS, 82% accuracy, 81% precision, and 72% recall were achieved by LSTM.

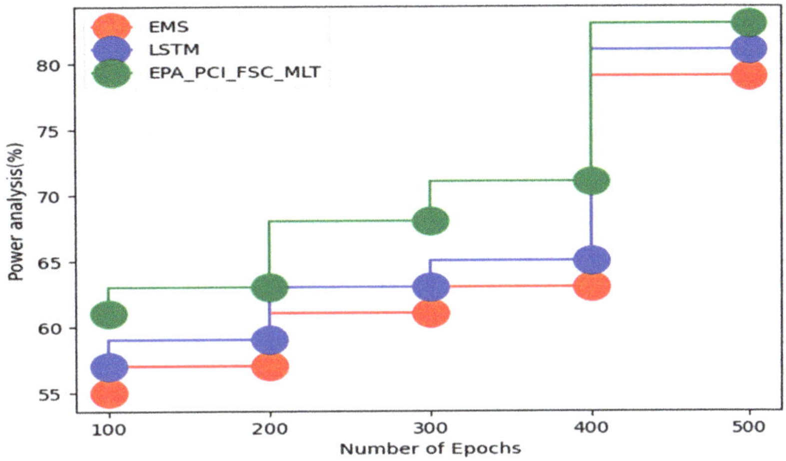

(a)

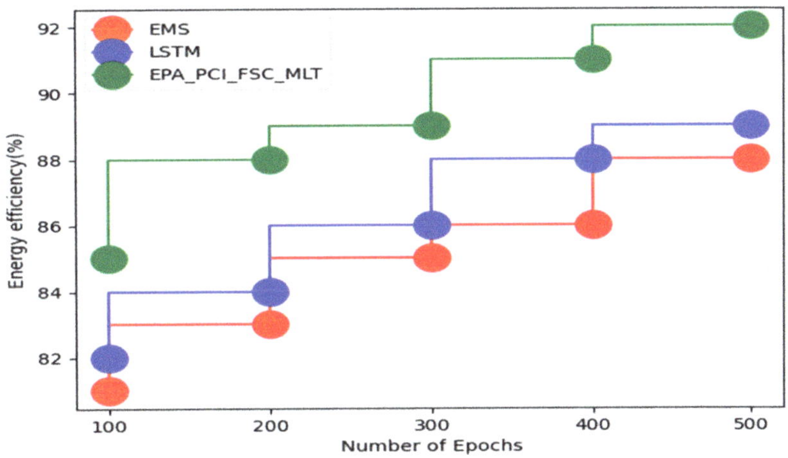

(b)

Figure 4. *Cont.*

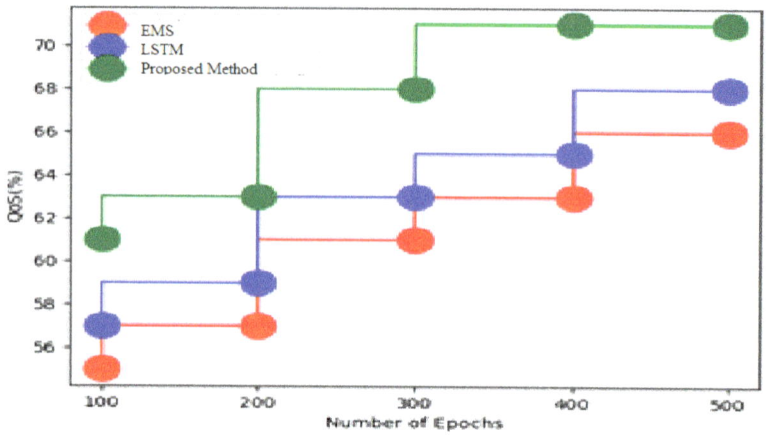

(c)

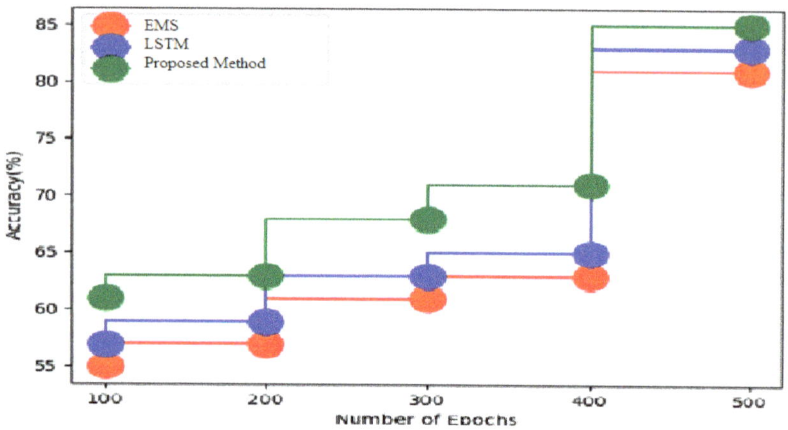

(d)

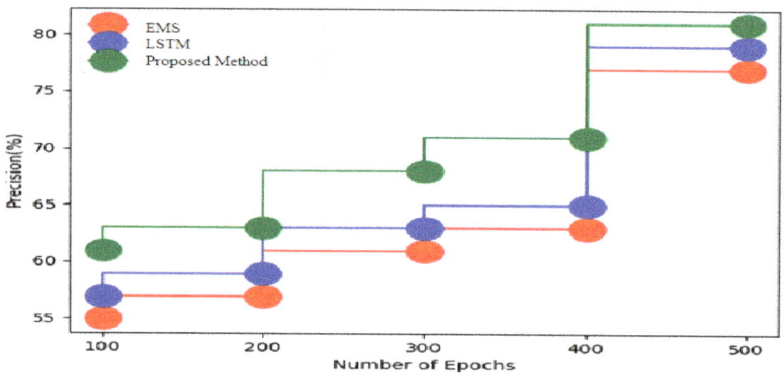

(e)

Figure 4. *Cont.*

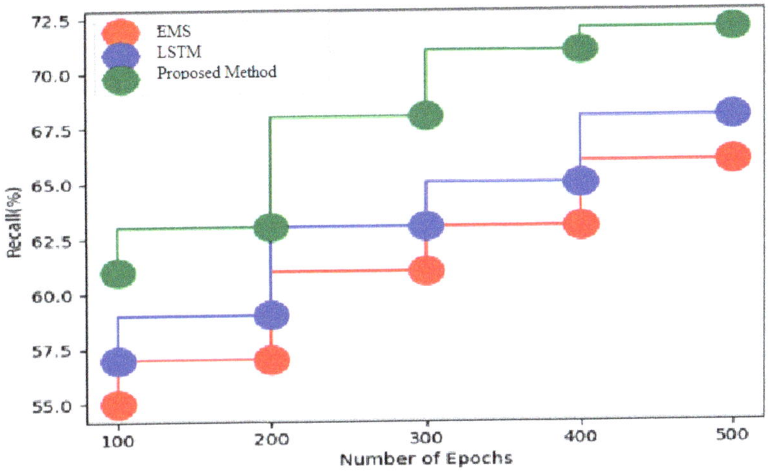

(f)

Figure 4. (**a**–**f**) Analysis for reactance circuit model in terms of power analysis, energy efficiency, QoS, accuracy, precision, recall.

Figure 5a–f give analysis based on admittance network type circuit model. Power analysis of 88%, energy efficiency of 95%, QoS of 77%, accuracy of 93%, precision of 85%, recall of 77%, and QoS of 93% were achieved with the proposed technique. In comparison to LSTM, EMS achieved power analysis of 86%, energy efficiency of 93%, QoS of 76%, accuracy of 85%, precision of 81%, and recall of 72%. EMS also achieved energy efficiency of 91%, QoS of 75%, accuracy of 85%, and recall of 72%.

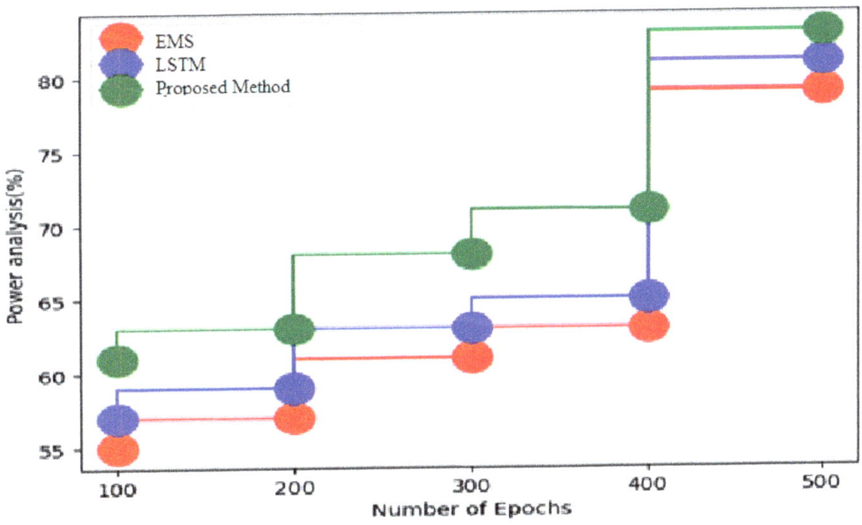

(a)

Figure 5. *Cont.*

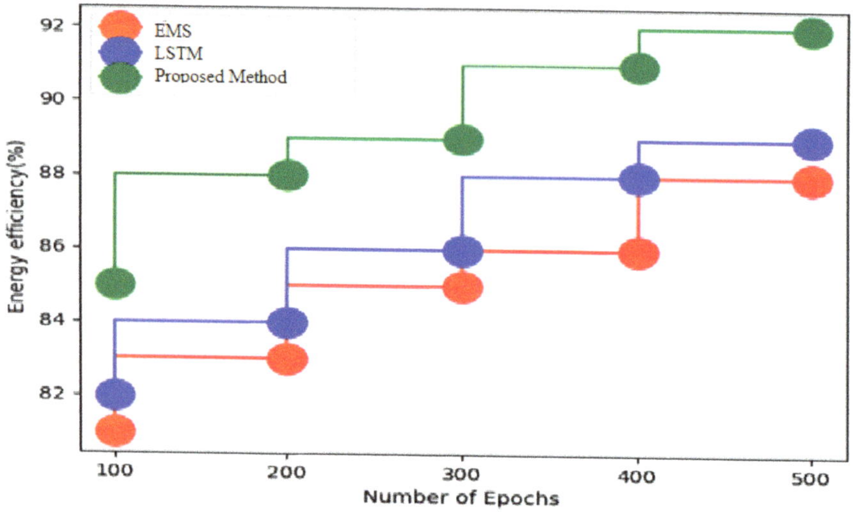

(b)

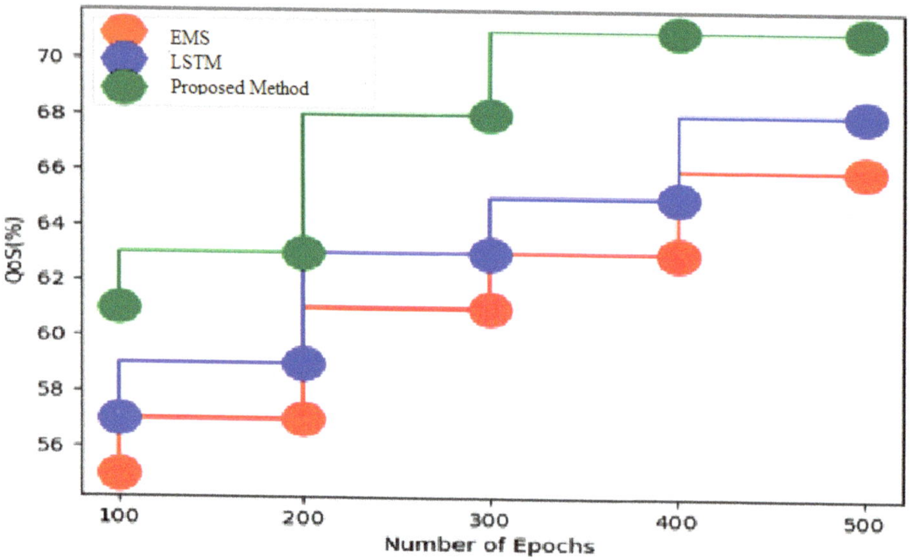

(c)

Figure 5. *Cont.*

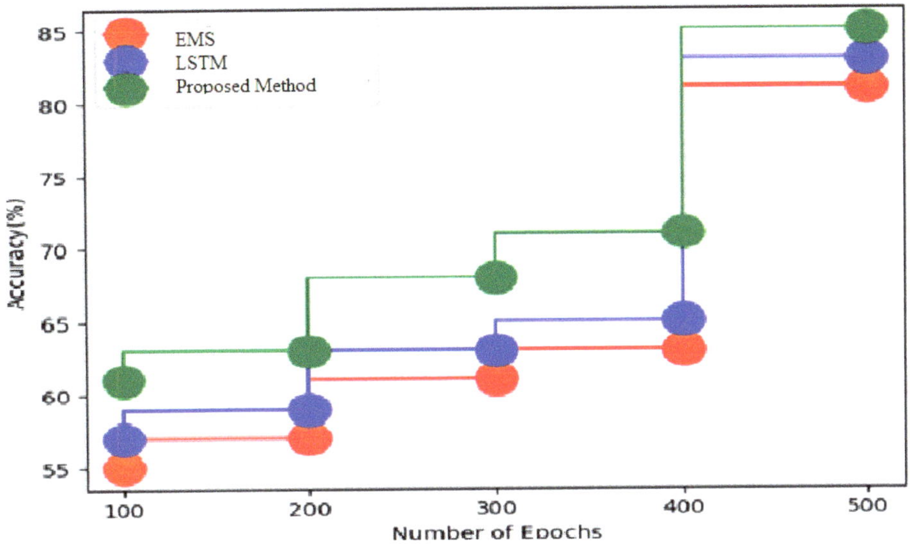

(d)

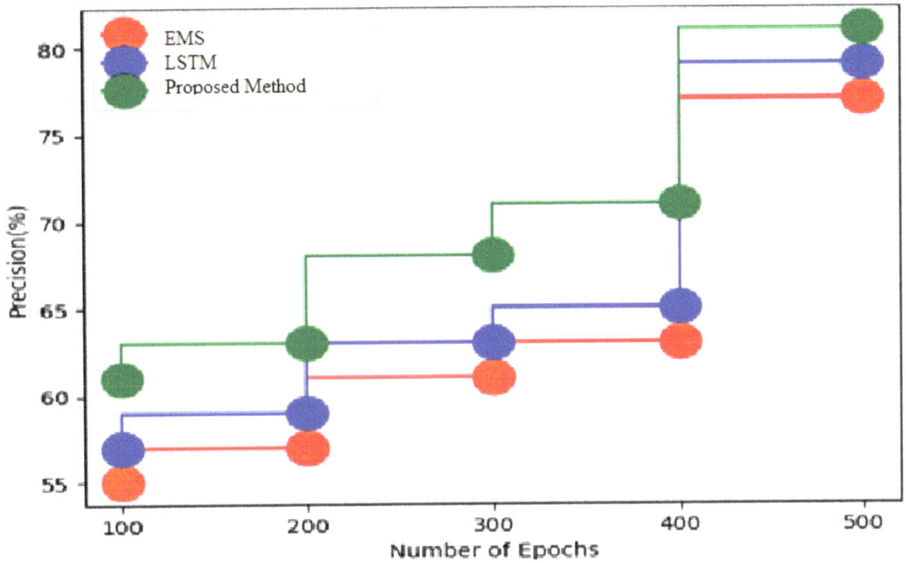

(e)

Figure 5. *Cont.*

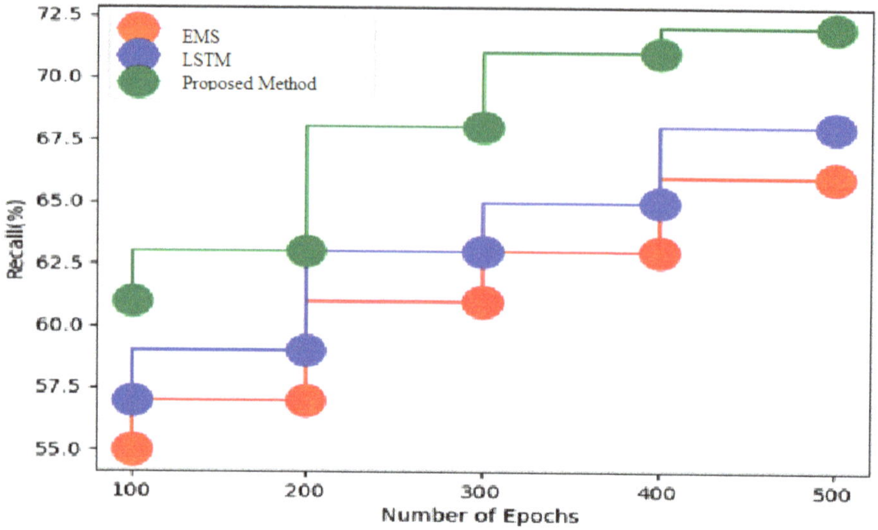

(f)

Figure 5. (**a–f**) Analysis for admittance network type model in terms of power analysis, energy efficiency, QoS, accuracy, precision, recall.

6. Conclusions

This study proposes an idea for energy analysis based on a microgrid photovoltaic system using a deep learning method. The energy optimization of the microgrid was carried out using a photovoltaic-based energy system with distributed power generation. The data analysis has been carried out for feature analysis and classification using a Gaussian radial Boltzmann with Markov encoder model. When taking into account an MG with photovoltaics (PVs), solar radiations abruptly increase in intensity during the day. That will boost MG production at a specific moment. Table 2 presents the comparative analysis of the proposed method with EMS and LSTM and it was shown to achieve higher power analysis, energy efficiency, QoS, accuracy, precision, and recall for different circuit models. In power analysis, the proposed method achieves 5% and 2.4% increases compared to EMS and LSTM models, respectively, for resistance models. Similarly, it has shown an increase of 6% and 2.3% and 4.7% and 2.3% for reactance and admittance circuit models.

Author Contributions: Methodology, S.Q., M.M., P.R.K. and K.I.; Validation, P.C.; Formal analysis, S.Q.; Investigation, S.Q. and P.R.K.; Data curation, P.C.; Writing—original draft, M.M.; Writing—review & editing, K.I.; Visualization, M.M.; Supervision, M.M. and P.C.; Project administration, P.R.K. All authors have read and agreed to the published version of the manuscript.

Funding: This research received no external funding.

Institutional Review Board Statement: Not Applicable.

Informed Consent Statement: Not Applicable.

Data Availability Statement: The data presented in this study are available on request from the corresponding author.

Conflicts of Interest: The authors declare no conflict of interest.

References

1. Rafati, A.; Joorabian, M.; Mashhour, E.; Shaker, H.R. Machine learning-based very short-term load forecasting in microgrid environment: Evaluating the impact of high penetration of PV systems. *Electr. Eng.* **2022**, *104*, 2667–2677. [CrossRef]
2. Aslam, S.; Herodotou, H.; Mohsin, S.M.; Javaid, N.; Ashraf, N.; Aslam, S. A survey on deep learning methods for power load and renewable energy forecasting in smart microgrids. *Renew. Sustain. Energy Rev.* **2021**, *144*, 110992. [CrossRef]
3. Scarpulla, M.A.; McCandless, B.; Phillips, A.B.; Yan, Y.; Heben, M.J.; Wolden, C.; Xiong, G.; Metzger, W.K.; Mao, D.; Krasikov, D.; et al. CdTe-based thin film photovoltaics: Recent advances, current challenges and future prospects. *Sol. Energy Mater. Sol. Cells* **2023**, *255*, 112289. [CrossRef]
4. Rosero, D.G.; Díaz, N.L.; Trujillo, C.L. Cloud and machine learning experiments applied to the energy management in a microgrid cluster. *Appl. Energy* **2021**, *304*, 117770. [CrossRef]
5. Deshpande, K.; Möhl, P.; Hämmerle, A.; Weichhart, G.; Zörrer, H.; Pichler, A. Energy Management Simulation with Multi-Agent Reinforcement Learning: An Approach to Achieve Reliability and Resilience. *Energies* **2022**, *15*, 7381. [CrossRef]
6. Yamashita, D.Y.; Vechiu, L.; Gaubert, J.-P. A review of hierarchical control for building microgrids. *Renew. Sustain. Energy Rev.* **2020**, *118*, 109523. [CrossRef]
7. Wang, H.; Lei, Z.; Zhang, X.; Zhou, B.; Peng, J. A review of deep learning for renewable energy forecasting. *Energy Conversion and Management* **2019**, *198*, 111799. [CrossRef]
8. Bidgoli, M.A.; Ahmadian, A. Multi-stage optimal scheduling of multi-microgrids using deep-learning artificial neural network and cooperative game approach. *Energy* **2022**, *239*, 122036. [CrossRef]
9. Jalli, R.K.; Mishra, S.P.; Dash, P.K.; Naik, J. Fault analysis of photovoltaic based DC microgrid using deep learning randomized neural network. *Appl. Soft Comput.* **2022**, *126*, 109314. [CrossRef]
10. Zhang, J.; Wu, H.; Bassoli, R.; Bonetto, R.; Fitzek, F.H. Deep learning-based energy optimization for electric vehicles integrated smart micro grid. In Proceedings of the ICC 2022-IEEE International Conference on Communications, Seoul, Republic of Korea, 16–20 May 2022; IEEE: Piscataway, NJ, USA, 2022; pp. 2187–2193.
11. Shojaeighadikolaei, A.; Ghasemi, A.; Bardas, A.G.; Ahmadi, R.; Hashemi, M. Weather-Aware Data-Driven Microgrid Energy Management Using Deep Reinforcement Learning. In Proceedings of the 2021 North American Power Symposium (NAPS), College Station, TX, USA, 14–16 November 2021; IEEE: Piscataway, NJ, USA, 2021; pp. 1–6.
12. Alavi, S.A.; Mehran, K.; Vahidinasab, V.; Catalão, J.P. Forecast-based consensus control for DC microgrids using distributed long short-term memory deep learning models. *IEEE Trans. Smart Grid* **2021**, *12*, 3718–3730. [CrossRef]
13. Lu, R.; Bai, R.; Ding, Y.; Wei, M.; Jiang, J.; Sun, M.; Xiao, F.; Zhang, H.T. A hybrid deep learning-based online energy management scheme for industrial microgrid. *Appl. Energy* **2021**, *304*, 117857. [CrossRef]
14. Emara, D.; Ezzat, M.; Abdelaziz, A.Y.; Mahmoud, K.; Lehtonen, M.; Darwish, M.M. Novel control strategy for enhancing microgrid operation connected to photovoltaic generation and energy storage systems. *Electronics* **2021**, *10*, 1261. [CrossRef]
15. Widodo, D.A.; Iksan, N. Renewable Energy Generation Forecasting on Smart Home Micro Grid using Deep Neural Network. In Proceedings of the 2021 International Conference on Artificial Intelligence and Mechatronics Systems (AIMS), Bandung, Indonesia, 28–30 April 2021; IEEE: Piscataway, NJ, USA, 2021; pp. 1–4.
16. Krishnan, V.A.; Balamurugan, P. An efficient DLN2-CRSO approach based dynamic stability enhancement in micro-grid system. *Appl. Energy* **2022**, *322*, 119432. [CrossRef]
17. Fang, X.; Zhao, Q.; Wang, J.; Han, Y.; Li, Y. Multi-agent deep reinforcement learning for distributed energy management and strategy optimization of microgrid market. *Sustain. Cities Soc.* **2021**, *74*, 103163. [CrossRef]
18. Shirzadi, N.; Nasiri, F.; El-Bayeh, C.; Eicker, U. Optimal dispatching of renewable energy-based urban microgrids using a deep learning approach for electrical load and wind power forecasting. *Int. J. Energy Res.* **2022**, *46*, 3173–3188. [CrossRef]
19. Kanwal, S.; Khan, B.; Ali, S.M. Machine learning based weighted scheduling scheme for active power control of hybrid microgrid. *Int. J. Electr. Power Energy Syst.* **2021**, *125*, 106461. [CrossRef]
20. Zjavka, L. Power quality 24-hour prediction using differential, deep and statistics machine learning based on weather data in an off-grid. *J. Frankl. Inst.* **2022**. [CrossRef]
21. Bagheri, F.; Dagdougui, H.; Gendreau, M. Stochastic optimization and scenario generation for peak load shaving in Smart District microgrid: Sizing and operation. *Energy Build.* **2022**, *275*, 112426. [CrossRef]
22. Lan, T.; Jermsittiparsert, K.; TAlrashood, S.; Rezaei, M.; Al-Ghussain, L.; AMohamed, M. An advanced machine learning based energy management of renewable microgrids considering hybrid electric vehicles' charging demand. *Energies* **2021**, *14*, 569. [CrossRef]

sustainability

Review

Parameter Extraction of Solar Photovoltaic Cell and Module Models with Metaheuristic Algorithms: A Review

Zaiyu Gu [1], Guojiang Xiong [1,2,*] and Xiaofan Fu [1]

[1] College of Electrical Engineering, Guizhou University, Guiyang 550025, China
[2] Guizhou University Institute of Engineering Investigation and Design Co., Ltd., Guiyang 550025, China
* Correspondence: gjxiongee@foxmail.com

Abstract: As the photovoltaic (PV) market share continues to increase, accurate PV modeling will have a massive impact on the future energy landscape. Therefore, it is imperative to convert difficult-to-understand PV systems into understandable mathematical models through equivalent PV models. However, the multi-peaked, non-linear, and strongly coupled characteristics of PV models make it challenging to extract accurate parameters of PV models. Metaheuristics can address these challenges effectively regardless of gradients and function forms, and have gained increasing attention in solving this issue. This review surveys different metaheuristics to the PV model parameter extraction and explains multiple algorithms' behavior. Some frequently used performance indicators to measure the effectiveness, robustness, accuracy, competitiveness, and resources consumed are tabulated and compared, and then the merits and demerits of different algorithms are outlined. The patterns of variation in the results extracted from different external environments were analyzed, and the corresponding literature was summarized. Then, challenges for both metaheuristics and application scenarios are analyzed. Finally, corresponding perspectives on future research are summarized as a valid reference for technological advances in PV model parameter extraction.

Keywords: PV model; parameter extraction; metaheuristic

check for updates

Citation: Gu, Z.; Xiong, G.; Fu, X. Parameter Extraction of Solar Photovoltaic Cell and Module Models with Metaheuristic Algorithms: A Review. *Sustainability* **2023**, *15*, 3312. https://doi.org/10.3390/su15043312

Academic Editors: Luis Hernández-Callejo, Prince Winston David and Praveen Kumar B

Received: 6 January 2023
Revised: 2 February 2023
Accepted: 7 February 2023
Published: 10 February 2023

1. Introduction

Fossil fuels' total reserves are limited, and their overuse has threatened human health and the ecological environment. Thus, developing renewable energy sources is an extremely urgent concern [1–5]. Renewable energy, including the energy sources of solar, hydro, wind, geothermal, and biomass energy [6–8], is inexhaustible or short-term renewable. Solar energy is a form of energy that contains a tremendous amount of energy and has the potential to meet all the energy requirements of current human activities [9]. As a result, solar energy has been employed in varied applications such as desalination, heating plants, and photovoltaic (PV) power generation [10,11]. Due to the clean and widespread availability of electrical energy in various fields, PV power generation is an important project for developing renewable energy sources [12].

Accurate modeling is essential for the assessment, efficiency improvement, fault analysis, and simulation of PV systems [13–15]. A PV system consists of an aggregation of PV cells, and they are typically modeled with equivalent circuits, mainly including single diode (SDM), double diode (DDM), and triple diode (TDM) models [16–18]. These equivalent circuits can simulate PV cells' electrical characteristics. They have five, seven, and nine parameters to be extracted, respectively. As the number of diodes increases, more parameters to be extracted are involved, which results in more computational difficulty. The challenges faced by the issue include not only the multiplication of solution complexity due to multiple unknown parameters but also the coupling between electrical quantities, leading to a highly implicit function [4,19–21]. Moreover, the non-linear characteristics are challenging to solve due to the exponential functions in the characteristic equations. These challenges render determining accurate PV models a puzzle.

Extracting proper parameters of PV models is a thorny issue, and it is primarily solved by three types of methods: point-specific-based methods, traditional numerical optimization methods, and metaheuristic methods [19]. The first category, also referred to as analytical methods, relies heavily on the analytical treatment of the models to reduce the parameters and on specific points to deduce the model parameters [13,22,23]. They generally have low accuracy, especially when there is noise on specific data points. The second category is also known as the deterministic methods, which extensively use the idea of gradients. They are highly exploitable and computationally fast but are sensitive to initialization settings, and the accuracy of the solutions can be insufficient [24–26]. That dilemma is because the PV model's mathematical formulation is implicit, has exponential functions, and requires extraction of multiple parameters. As a result, the mentioned issue has multi-peaked, non-linear, and strongly coupled characteristics, which pose a significant challenge to solving the issue using deterministic methods. Unlike the above two categories, natural phenomena inspire the third class of methods: metaheuristics. They do not rely on gradients and detailed data, are conceptually simple and computationally convenient, and can solve complex optimization issues with high accuracy [27–31]. Therefore, scholars have identified the merits of metaheuristics and applied them to many problems.

Nowadays, the metaheuristics for this paper's problems have evolved considerably, and it is necessary to review the current developments in parameter extraction techniques. Recently, several reviews have partially covered the application of metaheuristics in this area. Abbassi et al. [19] comprehensively described and summarized different indicators and cases and briefly assessed the results. However, the authors were biased towards a broad overview of different methods and ignored details about the metaheuristics' application mechanisms. They merely measured the indicators' presence, without specific results to give the methods' effectiveness. Oliva et al. [32] undertook a dedicated review, tabulated each indicator's results, and described the details of some metaheuristics. Nevertheless, the work mainly focused on PV cells, with insufficient attention to PV modules, and ignored a review of the TDM and algorithmic settings. Venkateswari et al. [33] summarized the indicators and case names, described improved concepts, and compared some metaheuristics. However, they just summarized the minimum root mean square error (RMSE) results and lacked data on other indicators. Li et al. [20] overviewed the environmental factors' presence and surveyed the results of various approaches. However, the review mainly focused on the SDM and DDM and lacked the algorithmic settings of metaheuristics. Overall, the available reviews mainly highlighted the statistics of the RMSE values for SDM and DDM. Specific data on other indicators, i.e., the total number of fitness evaluations (TNFES), the sum of individual absolute errors (SIAE), and the mean, maximum, and standard deviation (STD) of RMSE, were unavailable for judging different methods' performance in computational resources, accuracy, reliability, and robustness. We also note the following shortcomings in past reviews: (a) a lack of holistic evaluation of metaheuristics in recent years for cells and modules, (b) no discussion or literature screening of the situation when the temperature changes, and (c) omission of a presentation of data changes when partial shade is applied.

A holistic view of this type of research takes time to establish for researchers unfamiliar with this area. Meanwhile, the available reviews should include the results of the last several years of study. However, although some reviews comprehensively summarize all solutions to the problem, they mention too few metaheuristics and need more numerical details. Others focus on PV cells and modules, but omit the analysis of metaheuristics. These shortcomings make their conclusions rather one-sided and make it difficult for the reader to understand the research results from multiple dimensions. Therefore, a persuasive article that considers the model's various aspects, the parameter settings, and the evaluation metrics and integrates the results of a large number of applications of metaheuristics to the problem is needed to present the recent research results. This paper provides a comprehensive and detailed summary and analysis of the application of metaheuristics to model PV accurately in recent years. Specifically, the metaheuristics are categorized and their rationale is outlined. The algorithmic settings are summarized, and the results are compared and ranked in various indicators. The variation of the parameters in different environments is studied, and a brief description of the relevant literature in recent years is given. Some cell models that are temporarily not in widespread use today but are of high research value are analyzed. Then, their advantages and disadvantages are analyzed, and the remaining challenges are analyzed. Eventually, future directions for research are summarized in solution approaches and application scenarios.

This work's main contributions are as follows:

- The mathematical models of current commonly used SDM, DDM, TDM, and PV modules are explained;
- The characteristics of each metaheuristic method and their enhancements and applications are outlined;
- The statistical results of RMSE, TNFES, SIAE and algorithmic settings of selected metaheuristics are summarized and compared;
- The output characteristics of the PV system are discussed for the dynamic temperature, irradiance, and partial shading, and the variation in parameters and RMSE are analyzed;
- Existing challenges and possible future work focuses are analyzed and provided.

The remainder is briefly sketched as follows. The PV cell's mathematical model and the evaluation indicators are explained in Section 2. Section 3 illustrates different metaheuristics. Section 4 provides an overall analysis of different methods, existing challenges, and possible research directions. Finally, Section 5 gives the conclusion.

2. PV Models and Problem Formulations

Several PV models and their corresponding equivalent circuits are revealed in the first part of this section, to quantify the electrical characteristics of PV systems. Directly comparing PV models' parameters extracted by different methods is not easy. To objectively appraise the extracted results of different methods, the second part of this section gives several indicators commonly used to evaluate the experimental results.

2.1. PV Models

SDM, DDM, and TDM models have been widely used by researchers in recent years [20]. In general, more diodes in a circuit represent a more accurate model, but also increase the model complexity [33].

2.1.1. SDM

Figure 1a mentions the equivalent schematic of the SDM. The output voltage and current are V and I, respectively, and the electrical expression of I is shown below [34,35].

$$I = I_{ph} - I_{sh} - I_{sd} = I_{ph} - \frac{V + IR_s}{R_{sh}} - I_{ssd}\left[\exp\left(\frac{q(V + IR_s)}{nkT}\right) - 1\right] \tag{1}$$

where I_{ph}, I_{sh}, I_{sd}, and I_{ssd} represent the photogenerated line current, shunt resistor line current, diode line current, and diode saturation current, respectively. R_s and R_{sh} represent the series resistance and branch resistance, respectively. n represents the ideal factor. T, k, and q represent the Boltzmann constant (1.3806×10^{-23} J/K), absolute temperature, and unit charge (1.6022×10^{-19} C).

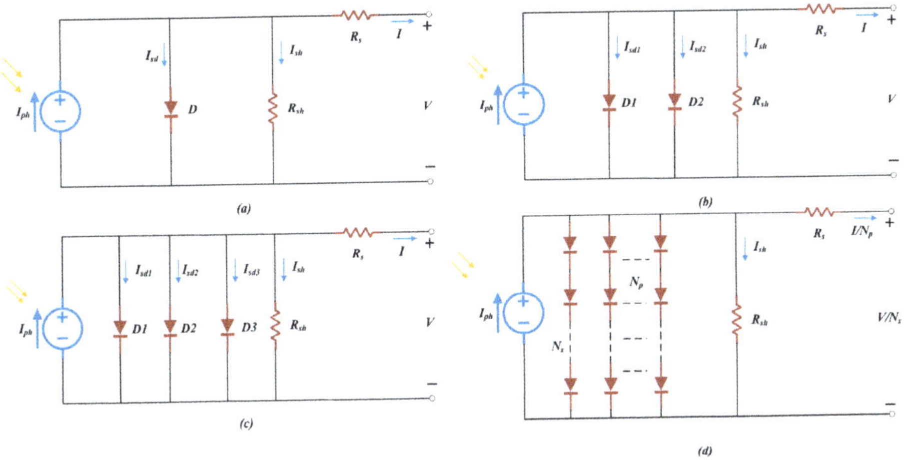

Figure 1. PV models' circuits: (**a**) SDM; (**b**) DDM; (**c**) TDM; (**d**) PV module.

The above demonstrates that accurate modeling requires estimating the values of I_{ph}, I_{ssd}, n, R_s, and R_{sh}.

2.1.2. DDM

Figure 1b mentions the equivalent schematic of the DDM. After adding a diode, below is the electrical expression of I [36,37].

$$I = I_{ph} - I_{sh} - I_{sd1} - I_{sd2} = I_{ph} - \frac{V + IR_s}{R_{sh}} - I_{ssd1}\left[\exp\left(\frac{q(V + IR_s)}{n_1 kT}\right) - 1\right] - I_{ssd2}\left[\exp\left(\frac{q(V + IR_s)}{n_2 kT}\right) - 1\right] \quad (2)$$

where I_{sd1} and I_{sd2} represent the first and second diode line currents, respectively, I_{ssd1} and I_{ssd2} represent the corresponding diode saturation currents, and n_1 and n_2 represent the corresponding ideal factors.

This model needs to estimate the values of I_{ph}, I_{ssd1}, I_{ssd2}, n_1, n_2, R_s, and R_{sh}.

2.1.3. TDM

Figure 1c mentions the equivalent schematic of the TDM. Below is the electrical expression of I [38–40].

$$I = I_{ph} - I_{sh} - \sum_{j=1\to 3} I_{sdj} = I_{ph} - \frac{V + IR_s}{R_{sh}} - \sum_{j=1\to 3} I_{ssdj}\left[\exp\left(\frac{q(V + IR_s)}{n_j kT}\right) - 1\right] \quad (3)$$

where I_{sdj}, I_{ssdj}, and n_j represent the jth diode line current, the saturation current, and the ideal factor, respectively.

The TDM requires estimating the values of I_{ph}, I_{ssd1}, I_{ssd2}, I_{ssd3}, n_1, n_2, n_3, R_s, and R_{sh}.

2.1.4. PV Module

Figure 1d mentions the equivalent schematic of the PV module based on the SDM. A PV module composed of $N_s \times N_p$ cells inherently has a high complexity. Therefore, using the SDM to construct PV modules is the first choice for most researchers. Equation (4) is the electrical expression of the PV module's current [4,41].

$$I = I_{ph}N_p - \frac{V + IR_sN_s/N_p}{R_{sh}N_s/N_p} - I_{ssd}N_p\left[\exp\left(\frac{q(V + IR_sN_s/N_p)}{nN_skT}\right) - 1\right] \quad (4)$$

The PV module has the same parameters as the SDM (I_{ph}, I_{ssd}, n, R_s, and R_{sh}).

2.1.5. PV Model Review

Although the SDM, with its simple structure and fair accuracy, is presented at the very beginning of this section, it is not the earliest cell model. It is a development of the ideal PV cell model (IPCM). Compared to the IPCM, which has a straightforward structure consisting of only a current source and diode, the SDM simulates the flow resistance, electrode resistance, and surface contact resistance, explains the physical behavior, and is widely used in this problem [42]. To further improve the accuracy of the model's simulated conduct at low irradiance, a diode is added to the DDM to represent the loss of current in the depletion region. However, the added unknown parameters increase the difficulty of the solution. TDM has the potential to achieve higher accuracy than DDM after calculating the leakage current and grain boundaries with the addition of a diode. Again, the solution difficulty increases as the dimensionality of the problem increase.

In addition, there are many less commonly used improved diode models, such as the modified 3-diode model [43], the SDM with capacitance [44], the Generalized Multi-Dimension Diode Model [45], the Modified SDM (MSDM) [46], the Four Diode Model (FDM) [47], the Modified DDM (MDDM) [48] and the Modified TDM (MTDM) [49]. We note that metaheuristics have recently been used to solve the FDM and the modified SDM, DDM, and TDM models. Thus, it would be a trend for future research to consider these four models to find a cell model that matches the proposed method to achieve a balance between solution difficulty and accuracy.

For the modules, in addition to the SDM presented in Section 2.1.4, the use of DDM and TDM formations are also options considered by the researchers. Their accuracy and solution difficulty performance are similar to their performance in the cell model. The appropriate model-building module must be selected to fit the specific needs. In this paper, considering that counting all the above models would cause duplication of content, excessive length, and difficulty reading, only the computational results of the modules composed of SDM components are summarized. The increased accuracy, increased difficulty in solving, and increased computational resources due to the increase in diodes will be reflected in the computational results of the cell model.

In addition, several specific PV models exist to achieve accurate modeling of PV systems in specific situations. They are not commonly used for the time being, but are of great interest. The dynamic PV model is one of them. It considers underdamped currents, switching frequency harmonics, varying loads, and resonance of cables, and is more suitable for grid-connected operation [50,51]. Its equivalent circuit diagram is shown in Figure 2 [52].

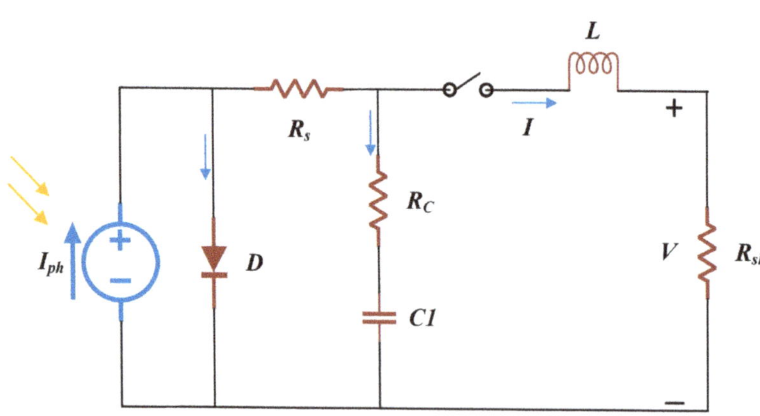

Figure 2. Dynamic model's circuits.

The model's output current is shown as follows [53]:

$$\begin{cases} I(s) = \frac{a_{21}(s+b_1)+b_2(s-a_{11})}{(s-a_{11})(s-a_{22})-a_{21}a_{12}} \cdot \frac{V_{OC}}{s} \\ \begin{pmatrix} a_{11} & a_{12} \\ a_{21} & a_{22} \end{pmatrix} = \begin{pmatrix} \frac{-1}{C(R_s+R_C)} & \frac{-R_s}{C(R_s+R_C)} \\ \frac{R_s}{L(R_s+R_C)} & \frac{-(R_C R_L+R_C R_s+R_L R_s)}{L(R_s+R_C)} \end{pmatrix}, \begin{pmatrix} b_1 \\ b_2 \end{pmatrix} = \begin{pmatrix} \frac{1}{C(R_s+R_C)} \\ \frac{R_C}{L(R_s+R_C)} \end{pmatrix} \end{cases} \tag{5}$$

where s is the time, R_s and the open circuit voltage V_{oc} are usually known, the inductor L, the resistor R_C, and the capacitor C are unknown. Therefore, C, R_C, and L are the parameters to be extracted.

2.2. Problem Formulations

RMSE between the measured data and the calculated data usually serves as the objective function [54–56]:

$$\text{RMSE} = \sqrt{\frac{1}{N}\sum_{k=1}^{N} f^2(V, I, x)} \tag{6}$$

where x represents the solution vector and N represents the actual data's amount, and $f(V, I, x)$ calculates the current error in the following way.

For SDM:

$$\begin{cases} f(V, I, x) = I_{ph} - \frac{V+IR_s}{R_{sh}} - I_{ssd}\left[exp\left(\frac{q(V+IR_s)}{nkT}\right) - 1\right] - I \\ x = \left(I_{ph}, I_{ssd}, R_s, R_{sh}, n\right) \end{cases} \tag{7}$$

For DDM:

$$\begin{cases} f(V, I, x) = I_{ph} - \frac{V+IR_s}{R_{sh}} - I_{ssd1}\left[\exp\left(\frac{q(V+IR_s)}{n_1 kT}\right) - 1\right] - I_{ssd2}\left[\exp\left(\frac{q(V+IR_s)}{n_2 kT}\right) - 1\right] - I \\ x = (I_{PV}, I_{ssd1}, I_{ssd2}, R_s, I_{sh}, n_1, n_2) \end{cases} \tag{8}$$

For TDM:

$$\begin{cases} f(V, I, x) = I_{ph} - \frac{V+IR_s}{R_{sh}} - \sum_{j\to 3} I_{ssdj}\left[\exp\left(\frac{q(V+IR_s)}{n_j kT}\right) - 1\right] - I \\ x = (I_{PV}, I_{ssd1}, I_{ssd2}, I_{ssd3}, R_s, I_{sh}, n_1, n_2, n_3) \end{cases} \tag{9}$$

For PV module:

$$\begin{cases} f(V, I, x) = I_{ph}N_p - \frac{V+IR_s N_s/N_p}{R_{sh}N_s/N_p} - I_{ssd}N_p\left[\exp\left(\frac{q\left(V+IR_s N_s/N_p\right)}{nN_s kT}\right) - 1\right] - I \\ x = \left(I_{ph}, I_{ssd}, R_s, R_{sh}, n\right) \end{cases} \tag{10}$$

For the objective function RMSE, its computation requires solving methods with the ability to solve implicit functions. Commonly used are deterministic and metaheuristic methods. Several deterministic methods, including Newton Raphson [24], Lambert W function [25], Levenberg Marquardt [57], and Berndt–Hall–Hall–Hausman [58], have successfully solved the non-linear problem. However, it does not mean that deterministic methods can tackle the challenge of initial value sensitivity well. Due to challenges such as non-linearity and non-convexity, metaheuristics are considered to be the best solution for solving this issue.

2.3. Indicators Summary

Varied algorithmic settings substantially affect the results of metaheuristic methods and various indicators can evaluate the results from diverse aspects. Hence, we summarize the approach and case settings and the performance evaluation indicators. Usually, the literature has drawn characteristic curves to visualize the accuracy of the extracted parameters. Nevertheless, when the parameters' difference is not very large, some general and objective indicators are used as the basis for evaluating the advantages and disadvantages of different methods. Here, we highlight the commonly used indicators to compare them:

- Individual absolute error (IAE): it represents the difference between the actual and simulated current values [28,30];
- Sum of IAEs (SIAE) and mean IAEs (MIAE): they are more holistic in evaluating the accuracy of the simulated data [29,59];
- RMSE: it focuses on overall assessment of the data's dispersion [31,60];
- Friedman test (FT), Wilcoxon rank sum test (WRT), and Wilcoxon signed rank test (WST): they broaden evaluation scales from statistical perspectives;

$$\mathrm{IAE} = |f(V, I, x)| \tag{11}$$

$$\mathrm{SIAE} = \sum_{k=1}^{N} IAE \tag{12}$$

$$\mathrm{MIAE} = \frac{1}{N}\sum_{k=1}^{N} IAE \tag{13}$$

- In addition, a few works in the literature also use evaluation indicators such as the sum of squares of power, current, and voltage errors (ERR) [61].

3. Methods and Results

Metaheuristics have no special data or environment requirements and have high robustness and accuracy in this studied issue, which is also the reason that they have been frequently used. Different metaheuristics were inspired by various things when they were developed. Figure 3 categorizes the metaheuristics into four genres by the type each one simulates, i.e., evolution-based methods (GA, DE, JAYA), human social activity-based methods (GSK, SDO, TLBO), animal activity-based methods (PSO, ABC, GWO, WOA, HHO), and natural phenomenon-based methods (TGA, SOS, FPOA). In this section, the widely used metaheuristics for solving this issue, namely GA, DE, PSO, ABC, GWO, JAYA, TLBO, and WOA, are selected and briefly described. They share a high degree of similarity in the optimization process. For brevity, Figure 4 gives the general flowchart of metaheuristics.

3.1. GAs

The survival of the fittest phenomenon inspires the evolutionary algorithm, i.e., genetic algorithm. A solution is encoded as binary chromosomes, and all chromosomes are updated through iteration and fitness assessment. Selection, crossover, and mutation are the iteration's three primary operations. The first operation is related to the fitness value and usually uses roulette, random traversal sampling, and ranked selection. The second operation improves exploitation by changing the subsequence of random loci between chromosomes, and the third operation improves exploration by changing genes on individual chromosomes [62].

In [63], the authors used GA in 30XLS and 34XLS PV modules. Characteristic curves were plotted to visualize the accuracy. However, the method of validating the results was relatively simple. In [64], an adaptive genetic algorithm (AGA) was designed, employing the Pearson residual reduction and minimum mean square error reduction techniques. Relevant manufacturer data at different temperatures verified the AGA's accuracy. However, it lacked the comparison under different light intensities, and the validation was too homogeneous. For intelligent algorithms, more data-based optimization often means more accurate results. Therefore, Harrag et al. [65] combined genetic algorithms with neural networks and proposed a metaheuristic based on genetic neural networks (GNN). GNN's effectiveness was verified on the SDM and DDM with the *RMSE*.

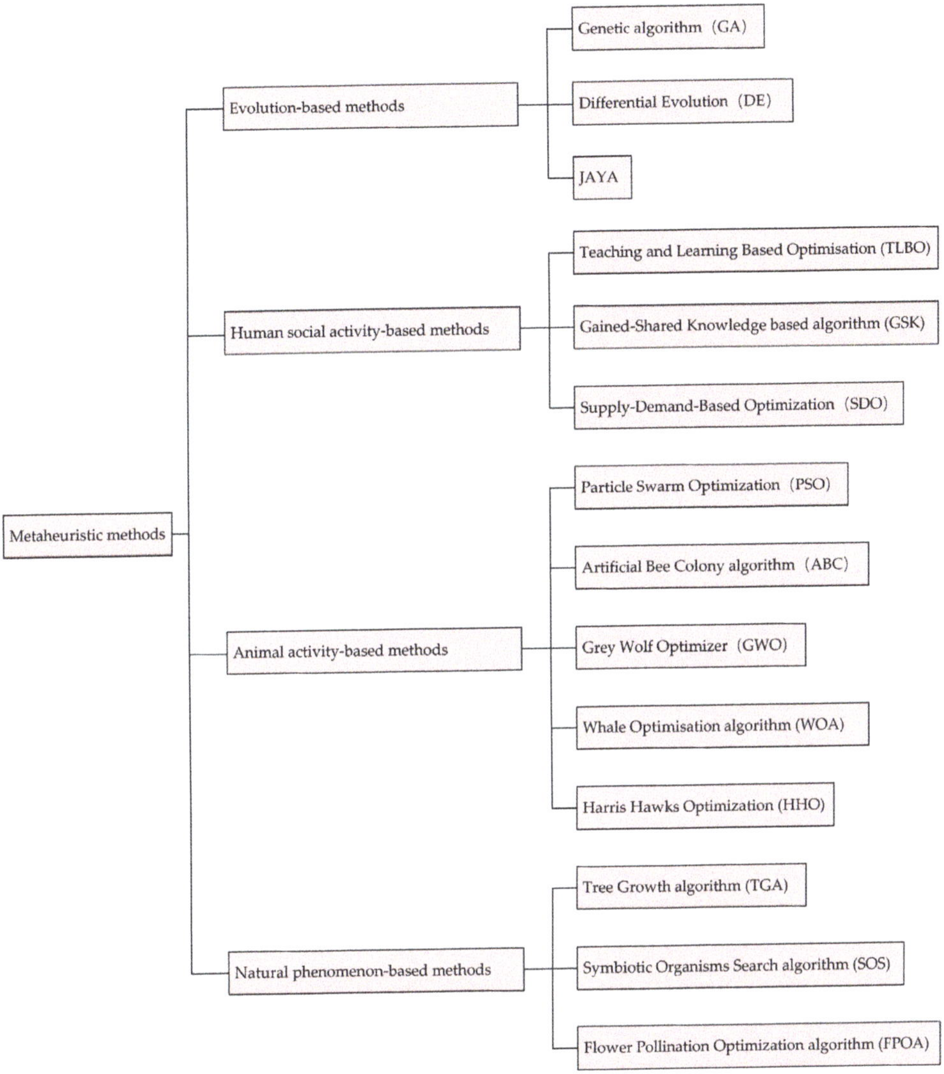

Figure 3. Metaheuristic methods' genres.

Table 1 lists essential information on GA variants. Among them, the squared error for GA was 5.8297×10^{-8} and 3.0751×10^{-7}, which is highly accurate, but there is a lack of comparison algorithms to judge the competitiveness of this result. AGA did not give any numerical RMSE values. The minimum *RMSE* for GNN reached the order of 1×10^{-3}, yet almost all recent state-of-the-art algorithms reached the order of 1×10^{-4}. The GA variants' performance is not ranked in this section, as the current GA variants did not use the same metric function.

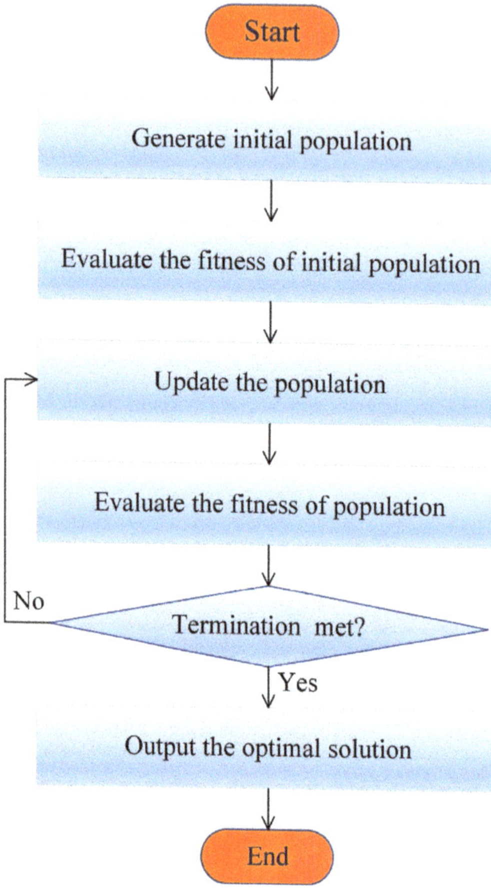

Figure 4. Metaheuristics' general flowchart.

Table 1. GAs' essential information and metrics.

Method	Main Contributors	Case	Algorithmic Parameter	Indicator	TNFES	Run
GA [63]	Harrag et al., CCNS Laboratory, Department of Electronics, Faculty of Technology, Ferhat Abbas University	30XLS 34XLS	NP = 100, CP = 0.5, MP = 0.02	SE	10,000	-
AGA [64]	Kumari et al., School of Electrical Engineering, VIT University	-	C1 = 0.01, C2 = 0.001	-	-	-
GNN [65]	Wang et al., Zhengzhou University of Aeronautics	SDM DDM	NP = 30 NP = 50	RMSE RMSE	9000 15,000	80 80

3.2. DEs

DE is fast in converging, simple in structure, and easy to implement [66,67]. As a population-based metaheuristic, DE has the same three operations with GA. DE individuals achieve mutation by adding different weight coefficients to the product of the difference between two individuals. The crossover is used to produce a trial vector from the target individual and the mutant vector. The selection usually chooses a greedy selection scheme to retain fitter individuals.

In [68], an improved adaptive DE (IADE) with exponential scaling factor (F) and crossover rate (CR) based on automatic performance updates was presented. The results' accuracy was verified using PV data with different temperatures and light intensities in terms of mean RMSE and fitted plots. Biswas et al. [61] designed a novel successful history-based DE (L-SHADE) with a linear reduced population size (NP) technique. Its parameter estimation was implemented using three particular points. The results showed that the error was almost zero. In [23], Chin et al. designed a differential evolution based on three points to improve the speed and accuracy of L-SHADE. In [69], an enhanced adaptive differential evolution (EJADE) was implemented by cross-ranking and dynamic population reduction techniques, and the algorithm's reliability was verified well. Xiong et al. [70] designed a new method (QILDE) for developing optimal value fields by adding quadratic interpolation to the crossover step. Applications of QILDE to six different PV models showed its strong competitiveness in different cases. In [71], a new method (EBLSHADE) based on SHADE with the linear population size reduction technique and greedy variation technique was designed. Its practical application in PV models demonstrated its importance in optimizing PV model parameters. In [72], dynamic control factors, including mutation and crossover, were designed and introduced into DE to form the new method called DEDCF. In [73], the authors designed a directed permutation differential evolution (DPDE) using the information on the direction of movement of populations and individuals, and applied it to a solar cell model. Hu et al. [41] designed a novel DE (RLDE) with reinforcement learning that adjusts the value of F by the Q-learning to achieve automatic parameter tuning, and compared RLDE with other methods, showing its superior robustness and accuracy. A heterogeneous differential evolution (HDE) was built in [74] with two improved mutation methods, a heterogeneous technique and an information exchange technique. It was demonstrated that the performance of HDE was representative in multiple dimensions through its application to the problems covered in this study. Kharchouf et al. [75] introduced Lambert's W function and metaheuristic techniques to DE for preferential F and CR, and named the method MSDE. It demonstrated high success through application. In [76], a novel DE (FADE) capable of optimizing F and CR was designed by employing fuzzy selection techniques and adaptive parameter tuning techniques. SIAE and RMSE demonstrated its excellent accuracy and robustness.

Tables 2 and 3 show the essential information and numerical metrics for each DE's variant, respectively. It is noticeable that there are many recent studies on DE, and most of them have obtained excellent performance. Regarding resource consumption, DE3P has the least, at 2500, followed by EBLSHADE, DEDCF, MSDE, EJADE, QILDE, RLDE, L-SHADE, DPDE, HDE, FADE, and IADE, respectively. Since ERRs were rarely used, data for WRT, WST, FT, and IAE were unavailable for statistics, and SIAE and MIAE are similar, we tabulate specific data for SIAE and various types of RMSE in Table 3 for comparison. To achieve a comprehensive accuracy comparison across multiple cases, the SDM, DDM, and Photowatt-PWP201 with the minimum RMSE values are used for the combined ranking. According to the FT results, MSDE (1.333) ranks first, followed by DEDCF (1.667), EJADE (4.333), QILDE (4.333), RLDE (4.333), HDE (4.667), DPDE (5.333), and EBLSHADE (5.833). However, EBLSHADE achieves excellent accuracy even though it is in last place, so future research in DE could further focus on reducing resource consumption and achieving improved performance in multiple accuracy evaluation metrics.

3.3. PSOs

PSO is a hot topic in artificial intelligence. The particle's new position is a combination of the current position and the updated velocity. The updating of the velocity is composed of three parts, and the first part is the current velocity scaled by the weight factor (w). The second part is the individual best position to steer the current position under the weight of the learning factor (c_1) and a random variable ($r1$). The third part is the global best position to steer the current position under the weight of the learning factor (c_2) and a random variable ($r2$). The $r1$ and $r2$ are unrelated, as are the c_1 and c_2 [77,78].

Table 2. DEs' essential information and metrics.

Method	Main Contributors	Case	Algorithmic Parameter	Indicator	TNFES	Run
IADE [68]	Jiang et al., School of Computer Engineering, Nanyang Technological University	SDM Photowatt-PWP201 SL80CE-36M	Iteration = 8000, $a = \ln 2$, $b = 0.5$	RMSE	-	30 30 -
L-SHADE [61]	Biswas et al., School of Electrical and Electronic Engineering, Nanyang Technological University	Kyocera KC200GT Shell SQ85 Shell ST40	$NP = 50$, $F = rand\ (0.1, 0.5)$, $CR = rand\ (0.1, 0.5)$	ERR	50,000	30
DE3P [23]	Chin et al., Centre of Electrical Energy Systems, School of Electrical Engineering, Universiti Teknologi Malaysia	SDM Photowatt-PWP201 STM6-40/36 STP6-120/36	$NP = 50$, $F = 0.7$, $CR = 0.8$	RMSE SIAE MIAE	2500	35
EJADE [69]	Li et al., School of Computer Engineering, Hubei University of Arts and Science	SDM DDM Photowatt-PWP201 STM6-40/36 STP6-120/36	$NP_{max} = 50$, $NP_{min} = 4$	RMSE	10,000 20,000 10,000 15,000 15,000	30
QILDE [70]	Xiong et al., Guizhou Key Laboratory of Intelligent Technology in Power System, College of Electrical Engineering, Guizhou University	SDM DDM Photowatt-PWP201 STM6-40/36 STP6-120/36 Sharp ND-R250A5	$F = rand\ (0.1, 1)$, $CR = rand\ (0, 1)$	RMSE FT	10,000 20,000 10,000 30,000 30,000 30,000	50 50 50 50 50 50
EBLSHADE [71]	Song et al., School of Computer Science and Technology, Shandong Technology and Business University	SDM DDM Photowatt-PWP201 STM6-40/36 STP6-120/36	$NP = 50$, $H = 100$, $w1 = 0.2$, $w2 = 0.6$, $pmin = 0.05$, $pmax = 0.2$	RMSE IAE	4000 10,000 5000 10,000 15,000	30 30 30 30 30
DEDCF [72]	Parida et al., Department of Electrical Engineering, ITER, Siksha O Anusandhan	SDM DDM Photowatt-PWP201	$NP = 10D$, $F = rand\ (0.1, 0.9)$, $CR = rand\ (0, 1)$	RMSE MIAE	10,000 14,000 10,000	50 50 50
DPDE [73]	Gao et al., Faculty of Engineering, University of Toyama	SDM DDM TDM Photowatt-PWP201 STM6-40/36 STP6-120/36	$NP = 18D$, $H = 5$, $p = 0.11$	RMSE SIAE WRT FT	50,000	30
RLDE [41]	Hu et al., School of Computer Science, China University of Geosciences	SDM DDM Photowatt-PWP201 STM6-40/36 STP6-120/36	$NP = 30$, $f = -0.1$ or 0 or 0.1, $CR = 0.9$	RMSE	30,000	30
HDE [74]	Wang et al., School of Software, Yunnan University	SDM DDM TDM Photowatt-PWP201 STM6-40/36 STP6-120/36	$NP = 30$, $p = 0.1$	RMSE WRT FT	50,000	30
MSDE [75]	Kharchouf et al., University Abdelmalek Essadi, FSTT	SDM DDM Photowatt-PWP201 STM6-40/36	$NP = 10D$, $F = 0.7$, $CR = 0.8$	RMSE	10,000 14,000 10,000 10,000	30
FADE [76]	Dang et al., Institute for Electrical Power and Integrated Energy of Shaanxi Province, Xi'an University of Technology	Photowatt-PWP201 STM6-40/36 STP6-120/36	$NP = 25$, $uF^{init} = 0.7$, $CR^{init} = 0.5$	RMSE SIAE	75,000	30

Ben et al. [79] applied PSO to the SDM and compared it with other methods, concluding that PSO outperformed other methods with data supporting. In [80], Ni et al. presented an adaptive elite mutation technique for PSO (PSO-AEM) for a domain search of the optimal global position of PSO, and found that PSO-AEM had a faster speed and higher accuracy. Merchaoui et al. [81] found that PSO was prone to premature convergence, so an adaptive mutation technique was proposed and introduced into PSO to form an improved MPSO. MPSO achieved good IAE and RMSE values and fitted the characteristic curves well at different temperatures and light intensities. In [82], Guaranteed Convergent Particle Swarm Optimization (GCPSO) was presented to avoid premature convergence. In [83], an enhanced leader PSO (ELPSO) using five mutation operators to enhance the leader was designed, following the idea that a high-quality leader could pull the solution towards the excellent region. The identification results showed that ELPSO effectively improved the quality of PSO solutions. In [84], the authors presented an improved PSO (SAIW-PSO) which used the simulated annealing technique to control w and introduced a deterministic method for optimizing the current values. The fitting results supported the view that SAIW-PSO was accurate, fast, and effective. Kiani et al. [85] designed a dynamic inertia weight PSO (DEDIWPSO) with a double exponential function to mitigate the premature convergence. This method demonstrated excellent validity, reliability, and accuracy in the issue covered in this work. The authors in [86] implemented PSO in parallel (PPSO) on a modern graphics processing unit (GPU). They demonstrated the very high accuracy and short elapsed time of PPSO by estimating multiple PV models' parameters. In [87], an enhanced PSO (PSO-ST) was developed using sinusoidal chaos and tangential chaos techniques to adjust the weight and learning factors. Inspired by cuckoo search random reselect parasitic nests, Fan et al. [88] developed a new method (PSOCS) by combining the random reselection strategy with PSO. The application results showed PSOCS's stability and effectiveness.

Table 3. DEs' experiment results.

Method	Case	SIAE	MIN RMSE	Mean RMSE	MAX RMSE	STD of RMSE	Rank
IADE [68]	SDM	-	9.8900×10^{-4}	-	-	-	N/A
	Photowatt-PWP201	-	2.4000×10^{-3}	-	-	-	
	SL80CE-36M	-	1.15×10^{-2}	-	-	-	
DE3P [23]	SDM	0.0172	8.1291×10^{-4}	-	-	-	N/A
	Photowatt-PWP201	0.0505	2.422747×10^{-3}	-	-	-	
	STM6-40/36	0.0210	1.774×10^{-3}	-	-	-	
	STP6-120/36	0.2091	1.4091×10^{-2}	-	-	-	
EJADE [69]	SDM	-	9.8602×10^{-4}	9.8602×10^{-4}	9.8602×10^{-4}	5.13×10^{-17}	4.333
	DDM	-	9.8248×10^{-4}	9.8363×10^{-4}	9.8602×10^{-4}	1.36×10^{-6}	
	Photowatt-PWP201	-	2.4251×10^{-3}	2.4251×10^{-3}	2.4251×10^{-3}	3.27×10^{-17}	
	STM6-40/36	-	1.7298×10^{-3}	1.7298×10^{-3}	1.7298×10^{-3}	5.94×10^{-18}	
	STP6-120/36	-	1.6601×10^{-2}	1.6601×10^{-2}	1.6601×10^{-2}	2.33×10^{-17}	
QILDE [70]	SDM	0.01770381	9.8602×10^{-4}	9.8603×10^{-4}	9.8616×10^{-4}	2.7839×10^{-8}	4.333
	DDM	0.01731807	9.8248×10^{-4}	9.8480×10^{-4}	9.8968×10^{-4}	1.5868×10^{-6}	
	Photowatt-PWP201	0.04178701	2.4251×10^{-3}	2.4257×10^{-3}	2.4370×10^{-3}	2.2436×10^{-6}	
	STM6-40/36	0.02177419	1.7298×10^{-3}	1.7298×10^{-3}	1.7298×10^{-3}	1.1295×10^{-17}	
	STP6-120/36	0.27797426	1.6601×10^{-2}	1.6601×10^{-2}	1.6601×10^{-2}	2.8518×10^{-14}	
	Sharp ND-R250A5	0.21759981	1.1183×10^{-2}	1.1183×10^{-2}	1.1183×10^{-2}	5.1647×10^{-10}	
EBLSHADE [71]	SDM	-	9.8602×10^{-4}	9.8602×10^{-4}	-	1.9169×10^{-15}	5.833
	DDM	-	9.8295×10^{-4}	9.8574×10^{-4}	-	1.2825×10^{-6}	
	Photowatt-PWP201	-	2.4251×10^{-3}	2.4251×10^{-3}	-	2.8821×10^{-17}	
	STM6-40/36	-	1.7298×10^{-3}	1.7298×10^{-3}	-	6.40591×10^{-14}	
	STP6-120/36	-	1.6601×10^{-2}	1.6601×10^{-2}	-	8.0544×10^{-16}	
DEDCF [72]	SDM	-	7.730062×10^{-4}	-	-	-	2
	DDM	-	7.419648×10^{-4}	-	-	-	
	Photowatt-PWP201	-	2.05296×10^{-3}	-	-	-	
DPDE [73]	SDM	0.02153	$9.86021877891470 \times 10^{-4}$	$9.86021877891542 \times 10^{-4}$	$9.86021877891588 \times 10^{-4}$	$2.57114481592195 \times 10^{-17}$	5.333
	DDM	0.021276	$9.82484827161920 \times 10^{-4}$	$9.82549779378988 \times 10^{-4}$	$9.83081420487992 \times 10^{-4}$	$1.51333797156833 \times 10^{-7}$	
	TDM	0.021275	$9.82484851785319 \times 10^{-4}$	$9.83096769943567 \times 10^{-4}$	$9.86188097663681 \times 10^{-4}$	$1.02284590208062 \times 10^{-6}$	
	Photowatt-PWP201	0.048924	$2.42507486809506 \times 10^{-3}$	$2.42507486809511 \times 10^{-3}$	$2.42507486809514 \times 10^{-3}$	$1.82238517018742 \times 10^{-17}$	
	STM6-40/36	0.021903	$1.72981370994065 \times 10^{-3}$	$1.72981370994068 \times 10^{-3}$	$1.72981370994070 \times 10^{-3}$	$1.09732017119964 \times 10^{-17}$	
	STP6-120/36	0.317128	$1.66006031250851 \times 10^{-2}$	$1.66006031250854 \times 10^{-2}$	$1.66006031250855 \times 10^{-2}$	$7.66886076234863 \times 10^{-17}$	
RLDE [41]	SDM	-	9.8602×10^{-4}	9.8602×10^{-4}	9.8602×10^{-4}	3.4834×10^{-17}	4.333
	DDM	-	9.8248×10^{-4}	9.8695×10^{-4}	9.8457×10^{-4}	1.7498×10^{-6}	
	Photowatt-PWP201	-	2.4251×10^{-3}	2.4251×10^{-3}	2.4251×10^{-3}	6.3084×10^{-17}	
	STM6-40/36	-	1.7298×10^{-3}	1.7298×10^{-3}	1.7298×10^{-3}	1.5784×10^{-17}	
	STP6-120/36	-	1.6601×10^{-2}	1.6601×10^{-2}	1.6601×10^{-2}	1.9764×10^{-16}	
HDE [74]	SDM	0.021527	$9.86021877891313 \times 10^{-4}$	$9.86021877891456 \times 10^{-4}$	$9.86021877891534 \times 10^{-4}$	$4.56994495305984 \times 10^{-17}$	4.667
	DDM	0.021275	$9.82484851785123 \times 10^{-4}$	$9.84154478759700 \times 10^{-4}$	$9.86021877891565 \times 10^{-4}$	$1.67264373173134 \times 10^{-6}$	
	TDM	0.021275	$9.82484851785213 \times 10^{-4}$	$9.82852008467139 \times 10^{-4}$	$9.88358683960422 \times 10^{-4}$	$1.08111146060101 \times 10^{-6}$	
	Photowatt-PWP201	0.048924	$2.42507486809496 \times 10^{-4}$	$2.42507486809504 \times 10^{-4}$	$2.42507486809510 \times 10^{-3}$	$3.15406568173825 \times 10^{-17}$	
	STM6-40/36	0.021903	$1.72981370994065 \times 10^{-3}$	$1.72981370994068 \times 10^{-3}$	$1.72981370994070 \times 10^{-3}$	$7.89430228096153 \times 10^{-18}$	
	STP6-120/36	0.31713	$1.66006031250847 \times 10^{-2}$	$1.66006031250851 \times 10^{-2}$	$1.66006031250855 \times 10^{-2}$	$1.86128634500124 \times 10^{-16}$	
MSDE [75]	SDM	-	7.7692×10^{-4}	-	-	-	1.333
	DDM	-	7.63×10^{-4}	-	-	-	
	Photowatt-PWP201	-	1.7298×10^{-3}	-	-	-	
	STM6-40/36	-	2.0529×10^{-3}	-	-	-	
FADE [76]	Photowatt-PWP201	0.0489237	2.42507×10^{-3}	2.42507×10^{-3}	2.42507×10^{-3}	-	N/A
	STM6-40/36	0.0219033	1.72981×10^{-3}	1.72981×10^{-3}	1.72981×10^{-3}	-	
	STP6-120/36	0.3171278	1.66006×10^{-2}	1.66006×10^{-2}	1.66006×10^{-2}	-	

The "N/A" means that there is insufficient data to support an average algorithm ranking using the Friedman Test on the three cases: SDM, DDM, and Photowatt-PWP201.

Tables 4 and 5 combine the essential information and numerical metrics of the PSO's variants. In the past five years, there have been numerous studies on PSO. Regarding resource consumption, PSO-AEM has the lowest TNFES of 10,000, followed by PSOCS, PSO, ELPSO, MPSO, SAIW-PSO, DEDIWPSO, PSO-ST, GCPSO, and PPSO. Regarding the ranking of MIN RMSE metrics, DEDIWPSO is first, followed by PSO-ST, GCPSO, MPSO, PPSO, and PSOCS. Although DEDIWPSO has the highest accuracy, it consumes massive computational resources. Hence, a considerable reduction in computational resource consumption while keeping accuracy constant is worthy of further research.

Table 4. PSOs' essential information and metrics.

Method	Main Contributors	Case	Algorithmic Parameter	Indicator	TNFES	Run
PSO [79]	Ben et al., Laboratory of Electronics, Signal Processing and Physical Modeling, Faculty of Sciences of Agadir Ibn Zohr University	SDM	NP = 50, Iteration = 1000, w = 0.4, $c_1 = c_2 = 2$	RMSE IAE	-	-
PSO-AEM [80]	Ni et al., Institute of Equipment Supervision and Inspection; Suzhou Nuclear Power Research Institute	-	NP = 50	-	10,000	-
MPSO [81]	Merchaoui et al., Electrical Department, National Engineering School of Monastir, University of Monastir	SDM DDM Photowatt-PWP201 IFRI250-60	NP = 60, Iteration = 2000, w = 0.4, $c_1 = c_2 = 2$	RMSE IAE	-	-
GCPSO [82]	Nunes et al., Department of Electromechanical Engineering, University of Beira Interior	SDM DDM Photowatt-PWP201 Sharp ND-R250A5	NP = 20D, Iteration = 10,000, w = 0.55, $c_1 = 1$, $c_2 = 2$	RMSE SIAE	-	100
ELPSO [83]	Rezaee et al., Department of Electrical Engineering, Lashtenesha-Zibakenar Branch, Islamic Azad University	SDM DDM STM6-40/36	NP = 991, $c_1 = 1$, $c_2 = 2$ NP = 1489, $c_1 = 1$, $c_2 = 2$ NP = 991, $c_1 = 1$, $c_2 = 2$	RMSE IAE	101,000 151,500 101,000	30
SAIW-PSO [84]	Kiani et al., Department of Electrical Engineering, University of Engineering and Technology, Taxila	SDM DDM	NP = 100, Iteration = 10,000,	RMSE	-	100
DEDIWPSO [85]	Kiani et al., Department of Electrical Engineering, University of Engineering and Technology, Taxila	SDM DDM Photowatt-PWP201 JKM330P-72	NP = 100, Iteration = 10,000, w^{init} = 0.8	RMSE IAE	-	30
PPSO [86]	Gao et al., Department of Electrical and Computer Engineering, National University of Singapore	SDM DDM Photowatt-PWP201	DDM: NP = 6400, Others: NP = 3200, w = 0.5, c_1 = 2.5, c_2 = 1.6	RMSE	640,000 2,560,000 640,000	30
PSO-ST [87]	Kiani et al., Department of Electrical Engineering, University of Engineering and Technology, Taxila	SDM DDM Photowatt-PWP201 JKM330P-72	NP = 100, Iteration = 10,000,	RMSE SIAE	-	30
PSOCS [88]	Fan et al., College of Electrical and Electronic Engineering, Wenzhou University	SDM DDM Photowatt-PWP201 SM55 KC200GT ST40	NP = 30	RMSE	20,000	30 -

Table 5. PSOs' experiment results.

Method	Case	SIAE	MIN RMSE	Mean RMSE	MAX RMSE	STD of RMSE	Rank
MPSO [81]	SDM	-	7.7301×10^{-4}	-	-	-	4
	DDM	-	7.4444×10^{-4}	-	-	-	
	Photowatt-PWP201	-	2.0530×10^{-3}	-	-	-	
	IFRI250-60	-	7.5589×10^{-3}	-	-	-	
GCPSO [82]	SDM	0.01763274	7.730063×10^{-4}	7.730063×10^{-4}	7.730065×10^{-4}	4.055839W-11	2.667
	DDM	0.01637239	7.182745×10^{-4}	7.301380×10^{-4}	7.417141×10^{-4}	5.371802×10^{-6}	
	Photowatt-PWP201	0.04400032	2.046535×10^{-3}	2.046535×10^{-3}	2.046536×10^{-3}	1.105194×10^{-10}	
	Sharp ND-R250A5	0.21867809	7.697717×10^{-3}	7.697717×10^{-3}	7.697719×10^{-3}	2.395516×10^{-10}	
ELPSO [83]	SDM	-	7.7301×10^{-4}	7.7314×10^{-4}	7.7455×10^{-4}	3.4508×10^{-7}	N/A
	DDM	-	7.4240×10^{-4}	7.5904×10^{-4}	7.9208×10^{-4}	9.4291×10^{-6}	
	STM6-40/36	-	2.1803×10^{-3}	2.2503×10^{-3}	3.7160×10^{-3}	2.9211×10^{-4}	
SAIW-PSO [84]	SDM	-	7.73006×10^{-4}	7.73006×10^{-4}	7.73006×10^{-4}	5.49562×10^{-15}	N/A
	DDM	-	7.41937×10^{-4}	7.42261×10^{-4}	7.54275×10^{-4}	1.41853×10^{-6}	
DEDIWPSO [85]	SDM	-	7.730062×10^{-4}	7.730062×10^{-4}	7.730062×10^{-4}	5.18668×10^{-15}	1.5
	DDM	-	7.182306×10^{-4}	7.187462×10^{-4}	7.318100×10^{-4}	2.486129×10^{-6}	
	Photowatt-PWP201	-	2.03992×10^{-3}	2.03992×10^{-3}	2.03992×10^{-3}	2.995389×10^{-15}	
	JKM330P-72	-	4.3113×10^{-2}	4.3113×10^{-2}	4.3113×10^{-2}	-	
PPSO [86]	SDM	-	9.8602×10^{-4}	9.8602×10^{-4}	9.8602×10^{-4}	7.0798×10^{-13}	5.167
	DDM	-	9.8248×10^{-4}	9.8323×10^{-4}	9.8602×10^{-4}	1.3436×10^{-6}	
	Photowatt-PWP201	-	2.4250×10^{-3}	2.4250×10^{-3}	2.4250×10^{-3}	2.8947×10^{-13}	
PSO-ST [87]	SDM	0.0214710	7.73006×10^{-4}	7.73006×10^{-4}	7.73006×10^{-4}	5.18622×10^{-15}	1.833
	DDM	0.0212734	7.183701×10^{-4}	7.187382×10^{-4}	7.218291×10^{-4}	1.318531×10^{-6}	
	Photowatt-PWP201	0.055499	2.03992×10^{-3}	2.03992×10^{-3}	2.03992×10^{-3}	2.91529×10^{-15}	
	JKM330P-72	-	4.3114×10^{-2}	4.3114×10^{-2}	4.3114×10^{-2}	6.2983×10^{-17}	
PSOCS [88]	SDM	-	9.8602×10^{-4}	9.8602×10^{-4}	9.8603×10^{-4}	1.7459×10^{-9}	5.833
	DDM	-	9.8297×10^{-4}	1.0286×10^{-3}	1.4133×10^{-4}	9.9217×10^{-5}	
	Photowatt-PWP201	-	2.4251×10^{-3}	2.4252×10^{-3}	2.4282×10^{-3}	5.9113×10^{-7}	
	SM55	-	3.8067×10^{-3}	-	-	-	
	KC200GT	-	2.5402×10^{-2}	-	-	-	
	ST40	-	7.3431×10^{-4}	-	-	-	

The "N/A" means that there is insufficient data to support an average algorithm ranking using the Friedman Test on the three cases: SDM, DDM, and Photowatt-PWP201.

3.4. ABCs

ABC was designed with several key members: a nectar source, nectar, and three types of bees [89]. The nectar amount from the flower represents the function value, and the food location means the solution. The nectar source and employed and onlooker bees are in quantity the same and the nectar source corresponds to the employed bees. Onlooker bees rely on nectar and employed bees to find flowers, and scout bees randomly fly to seek flowers near the hive [90].

In [91], the authors combined TLBO and ABC to design a method (TLABC) that included three search phases. The employed bee stage combined a teaching mechanism, the onlooker bee stage combined a learning mechanism, and the reconnaissance bee combined a generalized reversal mechanism. In [92], Wu et al. designed a new ABC (ABCTRR) by combining ABCs' exploiting capability with the trust-region reflective technique's exploiting capability. In [93], a new algorithm (IABC) was designed to solve ABC's early convergence issue by dividing the employed bee into two parts, one unchanged and the other searching the domain of the optimal global position. The identified parameters illustrated the high accuracy of IABC. For the integration of exploitation and exploration well, Tefek [94] combined ABC with a local search method to develop a new approach (ABC-Ls). Comparison revealed that ABC-Ls were more accurate, faster, and more stable. In [95], the authors compared ABC with PSO, showing that ABC outperformed PSO in all aspects of the results. In [96], a fitness distance balance mechanism was applied to TLABC to reconstruct a new method (FDB-TLABC). Experimental results confirmed the excellent performance of FDB-TLABC.

In Table 6, ABC-TRR has the least TNFES, followed by ABC, TLABC, IABC, ABC-Ls, and FDB-TLABC. There is an order-of-magnitude difference in resource consumption between ABC-TRR and the other variants of ABC. Table 7 compiles the experimental results. FDB-TLABC ranks first in combined MIN RMSE, followed by ABC-Ls, ABC-TRR, and TLABC. Therefore, achieving another increase in accuracy with less resource consumption for ABC is a priority for future research.

3.5. GWOs

GWO is a population-based metaheuristic with only two parameters [97]. Chase, encirclement, harassment, and attack are the hunt's four phases. Based on wolf rank, four types of wolves are included in GWO, with alpha being the strongest, followed by beta, delta, and omega. Wolves' mean solutions are in the solution space and are allowed to reposition. GWO only keeps the three optimal solutions, with other wolves responsible for position updating.

Table 6. ABCs' essential information and metrics.

Method	Main Contributors	Case	Algorithmic Parameter	Indicator	TNFES	Run
TLABC [91]	Chen et al., School of Electrical and Information Engineering, Jiangsu University	SDM DDM Photowatt-PWP201	*NP* = 50, *limit* = 200, scale factor *F* = *rand* (0, 1)	RMSE SIAE	50,000	30
ABC-TRR [92]	Wu et al., College of Physics and Information Engineering, Fuzhou University	SDM DDM Photowatt-PWP201	*NP* = 10, *limit* = 10 *NP* = 10, *limit* = 20 *NP* = 10, *limit* = 10	RMSE SIAE	1000 5000 1000	1000
IABC [93]	Xu et al., College of Mathematics and Physics, Inner Mongolia University for Nationalities	SDM DDM	*NP* = 50, *limit* = 50	RMSE IAE	50,000	-
ABC-Ls [94]	Tefek et al., Department of Computer Engineering, Osmaniye Korkut Ata University	SDM DDM Photowatt-PWP201	*NP* = 100, *limit* = 250 *NP* = 100, *limit* = 500 *NP* = 100, *limit* = 250	RMSE IAE	50,000	30
Best-so-far ABC [95]	Garoudja et al., Centre de Développement des Technologies Avancées, CDTA	SDM LG395N2W	*NP* = 150, *limit* = 750	RMSE	35,000	-
FDB-TLABC [96]	Duman et al., Electrical Engineering, Engineering and Natural Sciences Faculty, Bandirma Onyedi Eylul University	SDM DDM Photowatt-PWP201 STM6-40/36 STP6-120/36	*NP* = 50, *limit* = 200, scale factor *F* = *rand* (0, 1)	RMSE SIAE MIAE	50,000 70,000 50,000 50,000 50,000	51 51 51 - -

Table 7. ABCs' experiment results.

Method	Case	SIAE	MIN RMSE	Mean RMSE	MAX RMSE	STD of RMSE	Rank
TLABC [92]	SDM	0.02152738	9.86022×10^{-4}	9.98523×10^{-4}	1.03970×10^{-3}	1.86022×10^{-5}	3.667
	DDM	0.00135397	9.84145×10^{-4}	1.05553×10^{-3}	1.05553×10^{-3}	1.55034×10^{-4}	
	Photowatt-PWP201	0.04880919	2.42507×10^{-3}	2.42647×10^{-3}	2.44584×10^{-3}	3.99568×10^{-6}	
ABC-TRR [92]	SDM	0.02152687	9.860219×10^{-4}	9.860219×10^{-4}	9.860219×10^{-4}	6.15×10^{-17}	3
	DDM	0.02127522	9.824849×10^{-4}	9.825556×10^{-4}	9.860219×10^{-4}	4.95×10^{-7}	
	Photowatt-PWP201	0.04892367	2.425075×10^{-3}	2.425075×10^{-3}	2.425075×10^{-3}	9.68×10^{-17}	
IABC [93]	SDM	-	9.8602×10^{-4}	-	-	-	N/A
	DDM	-	9.8248×10^{-4}	-	-	-	
ABC-Ls [94]	SDM	-	9.8602×10^{-4}	-	-	-	2
	DDM	-	9.8257×10^{-4}	-	-	-	
	Photowatt-PWP201	-	2.4251×10^{-4}	-	-	-	
Best-so-far ABC [95]	SDM	-	0.027	-	-	-	N/A
	LG395N2W	-	0.013	-	-	-	
FDB-TLABC [96]	SDM	0.017633	7.7301×10^{-4}	-	-	-	1.333
	DDM	0.017001	7.4194×10^{-4}	-	-	-	
	Photowatt-PWP201	-	2.054×10^{-3}	-	-	-	
	STM6-40/36	-	1.7319×10^{-3}	-	-	-	
	STP6-120/36	-	1.4251×10^{-2}	-	-	-	

The "NA" means that there is insufficient data to support an average algorithm ranking using the Friedman Test on the three cases: SDM, DDM, and Photowatt-PWP201.

Vinod et al. [98] pioneered the use of GWO for the SDM, and the results showed that GWO had a high degree of accuracy. The study [99] found that more populations performed better, so a multi-group grey wolf optimizer (MGGWO) was developed. The results showed that MGGWO was excellent in speed and accuracy. A new GWO (OLBGWO) was designed in [100], which combined an orthogonal learning mechanism to improve the local exploration capability of GWO. OLBGWO's performance was evaluated in different PV models, and the results showed its excellent speed and accuracy. In [101], an improved GWO (I-GWO) was developed by introducing a hunting search mechanism based on dimensional learning. Ramadan et al. [102] introduced a domain search strategy to implement an improved GWO (IGWO) and demonstrated the algorithm's accuracy in two PV cases.

The relevant information and experimental results of the variants of GWO are summarized in Tables 8 and 9. I-GWO has the lowest resource consumption, followed by OLBGWO, GWO, MGGWO, and IGWO. Regarding overall accuracy ranking, OLBGWO is first and I-GWO is second. It is worth noting that MGGWO achieves a MIN RMSE of 4×10^{-4} on the SDM, a value not performed by any of the other algorithms counted. Variants of GWO use more computational resources, so there is much room for improvement in reducing the consumption of computational resources for GWO.

3.6. *JAYAs*

JAYA, which means victory in Sanskrit, combines survival of the fittest with the leader leading the population [103]. A key feature of JAYA is that there are no control parameters and no initial derivation information. When updating iteratively, the superior solution is approached quickly, and the inferior solution is moved away quickly.

Table 8. GWOs' essential information and metrics.

Method	Main Contributors	Case	Algorithmic Parameter	Indicator	TNFES	Run
GWO [98]	Vinod et al., Department of Electrical Engineering, Speciality of Optmization in Engineering, National Institute of Technology, Silchar, India	SDM	$NP = 50$	RMSE, IAE	50,000	-
MGGWO [99]	AlShabi et al., Mechanical and Nuclear Engineering Department, University of Sharjah, Sharjah, UAE	SDM	$NP = 20$	RMSE, MIAE	1,000,000	-
OLBGWO [100]	Xavier et al., Bule Hora University	SDM DDM Photowatt-PWP201 ST40 KC200GT	$NP = 30$, Orthogonal experiment levels: 3, Orthogonal experiment factors: 4	RMSE SIAE WRT	30,000	30
I-GWO [101]	Yesilbudak, Department of Electrical and Electronics Engineering, Faculty of Engineering and Architecture, Nevsehir Haci Bektas V eli University	SDM DDM TDM Photowatt-PWP201	$NP = 15$	RMSE IAE	25,000	50
IGWO [102]	Ramadan et al., Department of Electrical Engineering, Faculty of Engineering, Aswan University	TDM Photowatt-PWP201	$NP = 1000$, Iteration = 5000, $r1 = rand$, $r2 = rand$	RMSE	-	30

Table 9. GWOs' experiment results.

Method	Case	SIAE	MIN RMSE	Mean RMSE	MAX RMSE	STD of RMSE	Rank
GWO [98]	SDM	-	9.94378×10^{-4}	-	-	-	N/A
MGGWO [99]	SDM	-	4×10^{-4}	-	-	-	N/A
OLBGWO [100]	SDM DDM Photowatt-PWP201 ST40 KC200GT	- - - - -	9.86×10^{-4} 9.83×10^{-4} 2.4×10^{-3} 9.5666×10^{-4} 2.48×10^{-2}	9.86×10^{-4} 9.85×10^{-4} 2.4×10^{-3} - -	9.86×10^{-4} 9.86×10^{-4} 2.4×10^{-3} - -	1.4×10^{-8} 1.78×10^{-6} 2.4284×10^{-9} - -	1.333
I-GWO [101]	SDM DDM TDM Photowatt-PWP201	0.02152728 0.02127500 0.02128348 0.04892353	9.8602×10^{-4} 9.824852×10^{-4} 9.8251×10^{-4} 2.425075×10^{-3}	- - - -	- - - -	- - - -	1.667
IGWO [102]	TDM Photowatt-PWP201	- -	9.8331×10^{-4} 2.4276291×10^{-3}	9.84×10^{-4} 2.432×10^{-3}	9.85×10^{-4} 2.438×10^{-3}	6.60404×10^{-7} 5.26003×10^{-6}	N/A

The "N/A" means that there is insufficient data to support an average algorithm ranking using the Friedman Test on the three cases: SDM, DDM, and Photowatt-PWP201.

In [104], the authors designed an improved JAYA (IJAYA) that adaptively adjusted weights and optimized the algorithm performance using chaotic elite learning methods. IJAYA showed highly competitive performance in several PV models with excellent accuracy and reliability. An improved JAYA (EOJAYA) was developed in [105] by introducing an elite opposition mechanism to modify the update scheme. In [106], the Nelder-Mead algorithm was introduced to boost JAYA and this method's effectiveness was verified well in the SDM. In [107], a PGJAYA was designed to digitize the performance of individuals in a probabilistic manner as a guide to improve the search method. Adaptive chaotic perturbation techniques were employed to elevate the solution's overall quality. The PV model parameters estimated by PGJAYA proved its accuracy and robustness. Luu and Nguyen [108] introduced an adaptive population size mechanism to form a modified JAYA (MJA), and verified its performance and feasibility in the SDM and DDM. Jian et al. [109] developed a modified JAYA (LCJAYA) by introducing a logical chaotic mapping mechanism and a chaotic mutation mechanism in the update phase and search strategy of JAYA, respectively. LCJAYA's reliability and accuracy was verified in different PV cases. In [110], a simple improved JAYA (CLJAYA) was designed by integrating learning techniques, and its efficiency and accuracy was demonstrated in benchmark functions and PV models.

In [111], the authors improved a new JAYA (EJAYA) using an adaptive operator mechanism, a population size adjustment mechanism, and an opposition learning technique. The extraction of PV parameters demonstrated the effectiveness of EJAYA under different conditions. An enhanced chaotic JAYA (CJAYA) was developed in [112] by introducing an adaptive weighting strategy and three chaotic mechanisms including sine, tent, and logistic mappings. Saadaoui et al. [113] improved JAYA (MLJAYA) through three techniques: adaptive weighting, multiple learning, and chaotic perturbation. Jian and Cao [114] developed a chaotic second-order oscillation JAYA (CSOOJAYA) by using second-order oscillation factors, chaotic logistic mapping, and a mutation mechanism. The behavior of CSOOJAYA in solving the studied issue was demonstrated with good reliability and accuracy.

The essential information and experimental results of the variants of JAYA are summarized in Tables 10 and 11. Among them, the TNFES of EJAYA ranks first with 30,000, followed by CLJAYA, IJAYA, PGJAYA, LCJAYA, CJAYA, CSOOJAYA, EO-Jaya, and Jaya-NM. Regarding overall accuracy ranking, CLJAYA ranks first, followed by LCJAYA, EJAYA, MLJAYA, PGJAYA, CSOOJAYA, and IJAYA in order. In terms of computational resources, the JAYA variants consume more. Regarding specific values of FT, the difference between most variants is small, so further research on JAYA could go towards reducing the consumption of computational resources.

Table 10. JAYAs' essential information and metrics.

Method	Main Contributors	Case	Algorithmic Parameter	Indicator	TNFES	Run
IJAYA [104]	Yu et al., School of Electrical Engineering, Zhengzhou University	SDM DDM Photowatt-PWP201	$NP = 20$	RMSE IAE	50,000	30
EO-Jaya [105]	Wang et al., Department of Systems Engineering and Engineering Management, City University of Hong Kong	SDM DDM	$NP = 150$	RMSE	1,500,000	50
Jaya-NM [106]	Luo et al., School of Computer and Communication Engineering, University of Science and Technology Beijing (USTB)	SDM	$NP = 150$	RMSE	1,500,000	-
PGJAYA [107]	Yu et al., School of Electrical Engineering, Zhengzhou University	SDM DDM Photowatt-PWP201	$NP = 20$	RMSE	50,000	30
MJA [108]	Luu et al., Faculty of Electronics Technology, Industrial University of Ho Chi Minh City	SDM DDM	$NP^{init} = 10D$, $NP^{min} = D$, $r = rand\ (-0.5, 0.5)$,	RMSE	-	30
LCJAYA [109]	Jian et al., School of Optical Electrical and Computer Engineering, University of Shanghai for Science and Technology	SDM DDM Photowatt-PWP201	$NP = 20$	RMSE	50,000	30
CLJAYA [110]	Zhang et al., School of Electrical and Information Engineering, Tianjin University	SDM DDM Photowatt-PWP201	$NP = 20$	RMSE MIAE	20,000 50,000 30,000	-
EJAYA [111]	Yang et al., School of Computer Science, China University of Geosciences	SDM DDM Photowatt-PWP201 STM6-40/36 STP6-120/36	$NP = 30$, rate $Ra = 0.3$	RMSE WST	30,000	30
CJAYA [112]	Premkumar et al., Department of Electrical and Electronics Engineering, GMR Institute of Technology	SDM DDM STM6-40/36 STP6-120/36	$NP = 30$ $NP = 50$ $NP = 80$ $NP = 80$	RMSE IAE WST	50,000	30
MLJAYA [113]	Saadaoui et al., Laboratory of Materials and Renewable Energies, Faculty of Science, Ibn Zohr University	SDM DDM Photowatt-PWP201	$NP = 30$, $F = 3randn$	RMSE SIAE	-	30
CSOOJAYA [114]	Jian et al., School of Optical Electrical and Computer Engineering, University of Shanghai for Science and Technology	SDM DDM Photowatt-PWP201	$NP = 20$	RMSE IAE	50,000	30

Table 11. JAYAs' experiment results.

Method	Case	SIAE	MIN RMSE	Mean RMSE	MAX RMSE	STD of RMSE	Rank
IJAYA [104]	SDM	-	9.8603×10^{-4}	9.9204×10^{-4}	1.0622×10^{-3}	1.4033×10^{-5}	6.5
	DDM	-	9.8293×10^{-4}	1.0269×10^{-3}	1.4055×10^{-3}	9.8325×10^{-5}	
	Photowatt-PWP201	-	2.4251×10^{-3}	2.4289×10^{-3}	2.4393×10^{-3}	3.7755×10^{-6}	
EO-Jaya [105]	SDM	-	9.8603×10^{-4}	-	-	-	N/A
	DDM	-	9.8262×10^{-4}	-	-	-	
Jaya-NM [106]	SDM	-	9.8602×10^{-4}	-	-	-	N/A
PGJAYA [107]	SDM	-	9.8602×10^{-4}	9.8602×10^{-4}	9.8602×10^{-4}	1.4485×10^{-9}	3.833
	DDM	-	9.8263×10^{-4}	9.8582×10^{-4}	9.9499×10^{-4}	2.5375×10^{-6}	
	Photowatt-PWP201	-	2.425075×10^{-3}	2.425144×10^{-3}	2.426764×10^{-3}	3.071420×10^{-7}	
MJA [108]	SDM	-	9.860218×10^{-4}	9.860218×10^{-4}	9.860218×10^{-4}	1.99×10^{-17}	N/A
	DDM	-	9.824848×10^{-4}	9.8260×10^{-4}	9.860218×10^{-4}	6.46×10^{-7}	
LCJAYA [109]	SDM	-	9.8602×10^{-4}	9.8602×10^{-4}	9.8602×10^{-4}	5.6997×10^{-16}	3.5
	DDM	-	9.8250×10^{-4}	9.8308×10^{-4}	9.8602×10^{-4}	1.3118×10^{-6}	
	Photowatt-PWP201	-	2.425075×10^{-3}	2.425075×10^{-3}	2.425075×10^{-3}	2.415229×10^{-16}	
CLJAYA [110]	SDM	-	9.8602×10^{-4}	-	-	-	3.167
	DDM	-	9.8249×10^{-4}	-	-	-	
	Photowatt-PWP201	-	2.425075×10^{-3}	-	-	-	
EJAYA [111]	SDM	-	9.8602×10^{-4}	9.8602×10^{-4}	9.8602×10^{-4}	6.80×10^{-17}	3.5
	DDM	-	9.8248×10^{-4}	9.8448×10^{-4}	9.8602×10^{-4}	1.51×10^{-6}	
	Photowatt-PWP201	-	2.4251×10^{-3}	2.4251×10^{-3}	2.4251×10^{-4}	6.39×10^{-17}	
	STM6-40/36	-	1.7298×10^{-3}	1.7298×10^{-3}	1.7298×10^{-3}	1.47×10^{-17}	
	STP6-120/36	-	1.6601×10^{-2}	1.6601×10^{-2}	1.6601×10^{-2}	2.68×10^{-16}	
CJAYA [112]	SDM	-	9.8625×10^{-4}	9.8878×10^{-4}	9.8991×10^{-4}	4.5584×10^{-8}	N/A
	DDM	-	1.0145×10^{-3}	1.01458×10^{-3}	1.0365×10^{-3}	7.5514×10^{-5}	
	STM6-40/36	-	1.7242×10^{-3}	1.7289×10^{-3}	1.7845×10^{-3}	1.4751×10^{-7}	
	STP6-120/36	-	1.6285×10^{-2}	1.6299×10^{-2}	1.6302×10^{-2}	3.2565×10^{-7}	
MLJAYA [113]	SDM	0.01781248	9.8602×10^{-4}	9.8602×10^{-4}	9.8602×10^{-4}	-	3.667
	DDM	0.0176	9.8294×10^{-4}	1.0618×10^{-3}	1.42102×10^{-3}	-	
	Photowatt-PWP201	0.04686375	2.4250748×10^{-3}	2.44395×10^{-3}	2.49419×10^{-3}	-	
CSOOJAYA [114]	SDM	-	9.860219×10^{-4}	9.860219×10^{-4}	9.860219×10^{-4}	4.717305×10^{-17}	3.833
	DDM	-	9.824849×10^{-4}	9.824849×10^{-4}	9.824849×10^{-4}	5.576332×10^{-17}	
	Photowatt-PWP201	-	2.425075×10^{-3}	2.425075×10^{-3}	2.425075×10^{-3}	2.699858×10^{-17}	

The "N/A" means that there is insufficient data to support an average algorithm ranking using the Friedman Test on the three cases: SDM, DDM, and Photowatt-PWP201.

3.7. TLBOs

TLBO is a group metaheuristic developed based on the influence of teachers on students [115]. TLBO assumes that student outcomes are related to teacher competence. As the best in the group, the teacher teaches the students and raises the group's average achievement by a random factor. Students learn from each other at random coefficients during the learning phase and are led by the better of the two at random.

Chen et al. [116] suggested a generalized opposition-based learning mechanism for TLBO (GOTLBO). GOTLBO was demonstrated with excellent performance in benchmark functions and parameter extraction cases. To target different stages' effectiveness, Yu et al. [117] developed a self-adaptive TLBO (SATLBO) concerning elite learning mechanisms in the teacher stage and diverse learning mechanisms in the learner stage. SATLBO achieved competitive RMSE values in several PV models. Ramadan et al. [118] developed an enhanced TLBO (ETLBO) with controlled parameters replacing random parameter values and highlighted its effectiveness and competitiveness by extracting PV model parameters. Xiong et al. [21] developed an either/or TLBO (EOTLBO). To improve the generalizability of the method, EOTLBO replaced the mean with the learner median at the teacher stage. A random learner was added to the EOTLBO at the learner stage to improve the exploration capacity. The authors argued that it was inefficient for individuals to go through both teacher and learner stages, so EOTLBO implemented an either/or mechanism to choose one stage based on a chaotic map. EOTLBO showed excellent competitiveness, accuracy, and reliability. Abdel-Basset et al. [119] designed a modified TLBO (MTLBO). Individuals in both stages were divided into three strata of ground performance. Individual updates within each stratum did not interfere with each other. MTLBO was demonstrated with high accuracy in five PV models. Li et al. [120] developed an optimized TLBO (DMTLBO). The authors introduced the idea of dynamic self-adaption to the teacher stage and the idea of inter-comparison to the learner stage to further explore the capabilities of each stage. DMTLBO's accuracy, speed, and competitiveness were confirmed in different cases.

The essential information and experimental results of the TLBO variants are summarized in Tables 12 and 13. In the crucial information, GOTLBO has the least computational resources, followed by EOTLBO, SATLBO, MTLBO, DMTLBO, and ETLBO. In the accuracy ranking, EOTLBO comes first, followed by DMTLBO, MTLBO, and SATLBO. GOTLBO and ETLBO are not included because of missing values for some of the selected cases in the ranking. A direct comparison of the values in Table 13 reveals that the MIN RMSE of GOTLBO and ETLBO, which are early variants, struggle to outperform the other TLBO variants of recent years. An upward trend in the improvement of TLBO can be observed. However, the consumption of computational resources, unlike the development of accuracy, does not decrease significantly with the approaching number of years. Therefore, a reduction in the use of computational resources needs to be considered in future studies of TLBO.

Table 12. TLBOs' essential information and metrics.

Method	Main Contributors	Case	Algorithmic Parameter	Indicator	TNFES	Run
GOTLBO [116]	Chen et al., School of Electrical and Information Engineering, Jiangsu University	SDM DDM	*NP* = 20, SDM: *Jr* = 0.1, DDM: *Jr* = 0	RMSE	10,000 20,000	30
SATLBO [117]	Yu et al., Key Laboratory of Advanced Control and Optimization for Chemical Processes, Ministry of Education, East China University of Science and Technology	SDM DDM Photowatt-PWP201	*NP* = 40	RMSE	50,000	30
ETLBO [118]	Ramadan et al., Department of Electrical Engineering, Faculty of Engineering, Aswan University	SDM DDM STM6-40/36 STP6-120/36	*NP* = 200, Iteration = 5000,	RMSE IAE	-	-
EOTLBO [21]	Xiong et al., Guizhou Key Laboratory of Intelligent Technology in Power System, College of Electrical Engineering, Guizhou University	SDM DDM Photowatt-PWP201 Sharp ND-R250A5	*NP* = 50	RMSE WRT FT	20,000	50
MTLBO [119]	Abdel-Basset et al., Faculty of Computers and Informatics, Zagazig University	SDM DDM Photowatt-PWP201 STM6-40/36 STP6-120/36	*NP* = 50	RMSE	50,000	30
DMTLBO [120]	Li et al., Guizhou Key Laboratory of Intelligent Technology in Power System, College of Electrical Engineering, Guizhou University	SDM DDM Photowatt-PWP201 STM6-40/36 STP6-120/36	*NP* = 50	RMSE SIAE	50,000	30

Table 13. TLBOs' experiment results.

Method	Case	SIAE	MIN RMSE	Mean RMSE	MAX RMSE	STD of RMSE	Rank
GOTLBO [116]	SDM	-	9.87442×10^{-4}	1.33488×10^{-3}	1.98244×10^{-3}	2.99407×10^{-4}	N/A
	DDM	-	9.83177×10^{-4}	1.24360×10^{-3}	1.78774×10^{-3}	2.09115×10^{-4}	
SATLBO [117]	SDM	-	9.86022×10^{-4}	9.87795×10^{-4}	9.94939×10^{-6}	2.30015×10^{-6}	3.667
	DDM	-	9.828037×10^{-4}	9.981111×10^{-4}	1.047045×10^{-3}	1.951533×10^{-5}	
	Photowatt-PWP201	-	2.425075×10^{-3}	2.425428×10^{-3}	2.429130×10^{-3}	7.410517×10^{-7}	
ETLBO [118]	SDM	-	9.86022×10^{-4}	-	-	-	N/A
	DDM	-	9.8241×10^{-4}	-	-	-	
	STM6-40/36	-	1.7759×10^{-3}	-	-	-	
	STP6-120/36	-	1.6172×10^{-2}	-	-	-	
EOTLBO [21]	SDM	-	$9.86021878 \times 10^{-4}$	$9.86021878 \times 10^{-4}$	$9.86021878 \times 10^{-4}$	$4.12665088 \times 10^{-17}$	1.667
	DDM	-	$9.82484852 \times 10^{-4}$	$9.84733697 \times 10^{-4}$	$9.89424104 \times 10^{-4}$	$1.69176118 \times 10^{-6}$	
	Photowatt-PWP201	-	$2.42507487 \times 10^{-3}$	$2.42507487 \times 10^{-3}$	$2.42507487 \times 10^{-3}$	$3.61995116 \times 10^{-17}$	
	Sharp ND-R250A5	-	$1.11833356 \times 10^{-2}$	$1.11839904 \times 10^{-2}$	$1.12154997 \times 10^{-2}$	$4.54767027 \times 10^{-6}$	
MTLBO [119]	SDM	-	9.860219×10^{-4}	9.860219×10^{-4}	9.860219×10^{-4}	$1.9292748 \times 10^{-17}$	2.667
	DDM	-	9.824849×10^{-4}	9.824855×10^{-4}	9.825026×10^{-4}	3.3000000×10^{-9}	
	Photowatt-PWP201	-	2.4250749×10^{-3}	2.4250749×10^{-3}	2.4250749×10^{-3}	$1.3070107 \times 10^{-17}$	
	STM6-40/36	-	1.7298137×10^{-3}	1.7298137×10^{-3}	1.7298137×10^{-3}	$5.9363718 \times 10^{-18}$	
	STP6-120/36	-	$1.66006031 \times 10^{-2}$	$1.66006031 \times 10^{-2}$	$1.66006031 \times 10^{-2}$	$8.0041380 \times 10^{-17}$	
DMTLBO [120]	SDM	0.0178	9.8602×10^{-4}	9.8602×10^{-4}	9.8602×10^{-4}	2.07×10^{-17}	2
	DDM	0.0176	9.8248×10^{-4}	9.8406×10^{-4}	9.8638×10^{-4}	1.53×10^{-6}	
	Photowatt-PWP201	0.0411	2.4251×10^{-3}	2.4251×10^{-3}	2.4251×10^{-3}	2.15×10^{-17}	
	STM6-40/36	0.0215	1.7298×10^{-3}	1.7298×10^{-3}	1.7298×10^{-3}	5.74×10^{-14}	
	STP6-120/36	0.2741	1.6601×10^{-2}	1.6601×10^{-2}	1.6601×10^{-2}	4.55×10^{-10}	

The "N/A" means that there is insufficient data to support an average algorithm ranking using the Friedman Test on the three cases: SDM, DDM, and Photowatt-PWP201.

3.8. WOAs

WOA consists of an attack prey phase responsible for exploitation and a search prey phase responsible for exploration [121,122]. The bubble net attack consists of two mechanisms, i.e., encircling prey and spiral update position, both of which have the same probability of being selected. The encircling prey mechanism can determine any position between the present and best individuals within a specific range related to the parameter a, which decreases from 2 to 1 as the optimization proceeds. In the spiral position update, the individual's position is determined by the spiral equation between the whale and the prey. In the search phase, individuals are updated similarly to the encircling prey mechanism, except that a random individual replaces the optimal individual.

An improved WOA (IWOA) was developed in [123] to address the premature convergence of WOA. IWOA adjusted the encircling prey mechanism and modified the updating search phase to enhance the exploration, diversity, and robustness. Experiments in different PV models showed that IWOA extracted parameters with fast convergence, high quality, good robustness, and competitiveness. In [124], Elazab et al. pioneered the application of WOA to this studied problem. Comparisons with other algorithms demonstrated that WOA can fit PV data more accurately. To further enhance the ability of WOA to cope with the studied problem, Xiong et al. [18] developed a variant of WOA (MCSWOA) by modifying the search strategy of WOA using DE's mutation equation. A crossover operator was designed to improve the algorithm's applicability in different dimensions. A selection operator was designed to ensure that the optimization process would not worsen at any time. The perfect convergence curves, RMSE values, SIAE values, and ranking indicated that MCSWOA was characterized by high accuracy, competitiveness, and fast convergence. Pourmousa et al. [125] designed a Springy WOA (SWOA) by adding a deletion stage to the WOA. Peng et al. [126] developed a new approach (ISNMWOA) by combining the Nelder-Mead simplex technique with WOA. The results demonstrated that ISNMWOA's performance was significantly higher than WOA and it ran faster than other high-performance methods.

The essential information and experimental results of the variants of GWO are summarized in Tables 14 and 15. WOA has the least computational resources, followed by ISNMWOA, MCSWOA, IWOA, and SWOA, in order. In Table 15, SWOA has the highest overall MIN RMSE ranking, followed by ISNMWOA, IWOA, and MCSWOA. SWOA has high accuracy but consumes a lot of computational resources, with 5000 iterations at a population size of 30. The accuracy of ISNMWOA is close to that of SWOA, and TNFES at 20,000 is much lower than SWOA but still needs further improvement.

Table 14. WOAs' essential information and metrics.

Method	Main Contributors	Case	Algorithmic Parameter	Indicator	TNFES	Run
WOA [124]	Elazab et al., Electrical Power and Machines Department, Faculty of Engineering, Ain Shams University	KC200GT	NP = 30, Iteration = 500,	-	15,000	-
IWOA [123]	Xiong et al., Guizhou Key Laboratory of Intelligent Technology in Power System, College of Electrical Engineering, Guizhou University	SDM DDM Photowatt-PWP201	NP = 50, Iteration = 2000,	RMSE SIAE WRT, FT	-	50
MCSWOA [18]	Xiong et al., Guizhou Key Laboratory of Intelligent Technology in Power System, College of Electrical Engineering, Guizhou University	SDM DDM Photowatt-PWP201 STM6-40/36 STP6-120/36 Sharp ND-R250A5	NP = 50	RMSE SIAE FT	50,000	50
SWOA [125]	Pourmousa et al., Department of Electrical Engineering, Iran University of Science and Technology	SDM DDM TDM Photowatt-PWP201	NP = 30, Iteration = 5000,	RMSE IAE	-	30
ISNMWOA [126]	Peng et al., Department of Computer Science and Artificial Intelligence, Wenzhou University	SDM DDM TDM Photowatt-PWP201	NP = 30	RMSE SIAE	20,000	-

Table 15. WOAs' experiment results.

Method	Case	SIAE	MIN RMSE	Mean RMSE	MAX RMSE	STD of RMSE	Rank
IWOA [123]	SDM	0.01770338	9.860219×10^{-4}	9.860219×10^{-4}	9.860219×10^{-4}	5.12×10^{-16}	2.667
	DDM	0.01735511	9.824849×10^{-4}	9.826140×10^{-4}	9.860219×10^{-4}	9.86×10^{-5}	
	Photowatt-PWP201	0.04176116	2.425075×10^{-3}	2.425075×10^{-3}	2.425075×10^{-3}	2.90×10^{-17}	
MCSWOA [18]	SDM	0.01770381	9.8602×10^{-4}	9.8602×10^{-4}	9.8602×10^{-4}	4.8373×10^{-10}	3.167
	DDM	0.01730633	9.8250×10^{-4}	1.0078×10^{-3}	1.1903×10^{-3}	3.7264×10^{-5}	
	Photowatt-PWP201	0.04178694	2.4251×10^{-3}	2.4252×10^{-3}	2.4270×10^{-3}	3.2927×10^{-7}	
	STM6-40/36	0.02177346	1.7298×10^{-3}	1.7311×10^{-3}	1.7364×10^{-3}	1.0774×10^{-6}	
	STP6-120/36	0.27780418	1.6601×10^{-2}	1.6632×10^{-2}	1.6741×10^{-2}	2.6486×10^{-5}	
	Sharp ND-R250A5	0.21759970	1.1183×10^{-2}	1.1187×10^{-2}	1.1244×10^{-2}	9.1358×10^{-6}	
SWOA [125]	SDM	-	9.8602×10^{-4}	9.8602×10^{-4}	9.8602×10^{-4}	-	2
	DDM	-	9.8249×10^{-4}	9.8250×10^{-4}	9.8251×10^{-4}	-	
	TDM	-	9.8033×10^{-4}	9.8051×10^{-4}	9.8154×10^{-4}	-	
	Photowatt-PWP201	-	2.4250×10^{-3}	2.4250×10^{-3}	2.4250×10^{-3}	-	
ISNMWOA [126]	SDM	0.021527008	9.8602×10^{-4}	-	-	-	2.167
	DDM	0.021275213	9.8248×10^{-4}	-	-	-	
	TDM	0.021275347	9.8248×10^{-4}	-	-	-	
	Photowatt-PWP201	0.048923833	2.4251×10^{-3}	-	-	-	

3.9. Hybrids

The above methods used for the studied problem are partially dominated by a single metaheuristic algorithm. In addition to them, hybrid approaches that combine two and more metaheuristics are also popular for solving this problem. The motivation behind the hybrid approaches is integrating diverse features of different algorithms to equilibrate the global and local search abilities.

In [127], Xiong et al. devised an approach (DE/WOA) that took full advantages of DE and WOA to balance diversity and convergence. Long et al. [128] developed an approach (GWOCS) introducing the opposing learning mechanism of cuckoo search (CS) for the three optimal individuals preserved by GWO to achieve improved performance. The results of benchmark functions and PV models supported the authors' expectations of performance improvement. Rizk et al. [129] developed a new method (PSOGWO) by mixing GWO and PSO to make full use of their exploration and exploitation advantages. Different PV models demonstrated the excellent performance of PSOGWO. Li et al. [130] designed a DE-based adaptive TLBO (ATLDE) by mixing DE with TLBO and adjusting the teaching and learning stages using a ranking probability mechanism. Experimental results supported ATLDE's competitiveness. In [131], the authors effectively combined DE with Harris Hawks Optimization (HHO) to form a new method (HHODE), and demonstrated the effectiveness of the improvement using RMSE values for the extracted PV parameters. Yu et al. [132] devised a new method (HAJAYADE) by replacing the two parameters of JAYA adaptively. Then, the method combined DE and introduced a mutational operator and an adaptive chaos mechanism to ensure its performance. Devarapalli et al. [133] improved the updated approach of a hybrid of GWO and sine cosine algorithm (HGWOSCA) to gain an enhanced method (EHGWOSCA). Singh et al. [47] hybridized the Dingo Optimizer and PSO to form a new hybrid algorithm (HPSODOX) and developed a four-diode PV model to reveal HPSODOX's performance. The results supported the validity of the algorithm improvement. Weng et al. [134] integrated a Backtracking Search Algorithm with TLBO to form a new method (TLBOABC) and verified the method's effectiveness well.

The essential information and experimental results of the hybrid methods are summarized in Tables 16 and 17. TLBOBSA has the lowest computational resource consumption, followed by ATLDE, DE/WOA, GWOCS, and HAJAYADE. TLBOBSA has the highest overall ranking for MIN RMSE, followed by DE/WOA, HAJAYADE, and GWOCS. TLBOBSA ranks the highest in resource consumption and accuracy, indicating that a suitable hybrid scheme can achieve significant performance. It should be noted that the MIN RMSE of HPSODOX, although very small, needs more basic information, and there are no repeated runs for the experiment, so it is impossible to evaluate the performance of this method for the time being.

3.10. Others

New methods usually lead to breakthroughs in specific problems, since they bring different search mechanisms. Therefore, researchers favor novel approaches and their variants in exploring the PV model parameter extraction, and have provided some new approaches.

Naeijian et al. [135] developed a Whippy Harris Hawk Optimization (WHHO) that handled the worst individual by adding elimination cycles to improve all-around performance. The simulation results demonstrated the fast convergence of WHHO and the high robustness and accuracy for the extracted parameters. Xiong et al. [4] used a Gaining-Sharing Knowledge-based algorithm (GSK) for the issue addressed in this work for the first time. They demonstrated the high accuracy, robustness, and competitiveness of GSK in different PV models. Sallam et al. [136] developed an improved GSK (IGSK) using a boundary constraint processing mechanism, a linear population size reduction technique, and knowledge rate adaptive technology. Xiong et al. [137] applied Supply and Demand Based Optimization (SDO) and pioneered a comparison between SDO and several advanced methods in extracting PV model parameters, which powerfully demonstrated the feasibility and competitiveness of SDO. Diad et al. [138] used a Tree Growth Algorithm (TGA) to tackle the issue, and the RMSE values showed the TGA's good accuracy. Abbassi et al. [139] provided PV model parameters extracted by a Salp Swarm Algorithm (SSA) and demonstrated its accuracy and competitiveness with multiple metrics. Sharma et al. [140] solved this problem using Tunicate Swarm Algorithm (TSA) and verified TSA's accuracy, feasibility, and competitiveness with simulations. Gupta et al. [141] designed a chaotic TSA (CTSA) to tackle the issue, and the results supported its accuracy and competitiveness. Ramadan et al. [142] developed Chaotic Game Optimization (CGO) for the issue and confirmed its good performance. Long et al. [143] designed a Hybrid Seagull Optimization (HSOA) with three mechanisms, differential mutation, memory-guided and non-linear control, and tested it in different PV models. Shaban et al. [144] employed Rungakuta Optimizer (RUN) to tackle the issue. The simulation results demonstrated RUN's excellent competitiveness, convergence, and robustness. In [145], the authors used a Flower Pollination Optimization Algorithm (FPOA) for the TDM's parameters with industrial samples. The results supported the high-performance of FPOA in the TDM. In [146], the authors used the Symbiotic Organisms Search (SOS) method to tackle the issue. The results powerfully demonstrated the superiority of SOS.

Table 16. Hybrids' essential information and metrics.

Method	Main Contributors	Case	Algorithmic Parameter	Indicator	TNFES	Run
DE/WOA [127]	Xiong et al., Guizhou Key Laboratory of Intelligent Technology in Power System, College of Electrical Engineering, Guizhou University	SDM DDM Photowatt-PWP201	$NP = 40$, $F = rand\ (0.1, 1)$, $CR = rand\ (0, 1)$	RMSE MIAE	50,000	50
GWOCS [128]	Long et al., Key Laboratory of Economics System Simulation, Guizhou University of Finance and Economics	SDM DDM Photowatt-PWP201 STM6-40/36	$NP = 30$	RMSE IAE FT	50,000	30
PSOGWO [129]	Rezk et al., College of Engineering at Wadi Addawaser, Prince Sattam Bin Abdulaziz University	Photowatt-PWP201 STE4/100 FSM	Iteration = 1200 Iteration = 6000 Iteration = 2000	RMSE MIAE	-	-
ATLDE [130]	Li et al., School of Computer Science, China University of Geosciences	SDM DDM STM6-40/36 STP6-120/36	$NP = 50$, $F = rand$, $CR = 0.9$	RMSE SIAE WRT	30,000	30
HHODE [131]	Ndi et al., Technology and Applied Sciences Laboratory, University of Douala	SDM DDM	Iteration = 3000	RMSE	-	20
HAJAYADE [132]	Yu et al., School of Management Science and Engineering, Nanjing University of Information Science and Technology	SDM DDM Photowatt-PWP201 STM6-40/36 STP6-120/36	$NP = 20$, $CR = 0.5$	RMSE WST	50,000	30
EHGWOSCA [133]	Devarapalli et al., Department of EEE, Lendi Institute of Engineering and Technology	SDM DDM Shell S75 Shell CS6K280M Shell ST40	Iteration = 500	ERR	-	30
HPSODOX [47]	Singh et al., Electrical and Instrumentation Engineering Department, Thapar Institute of Engineering and Technology	SDM DDM TDM FDM	-	RMSE FT	-	-
TLBOBSA [134]	Weng et al., Department of Computer Science and Artificial Intelligence, Wenzhou University	SDM DDM TDM Photowatt-PWP201	$NP = 30$	RMSE SIAE	20,000	30

Table 17. Hybrids' experiment results.

Method	Case	SIAE	MIN RMSE	Mean RMSE	MAX RMSE	STD of RMSE	Rank
DE/WOA [127]	SDM	0.01770392	9.860219×10^{-4}	9.860219×10^{-4}	9.860219×10^{-4}	3.545178×10^{-17}	2.333
	DDM	0.01731808	9.824849×10^{-4}	9.829703×10^{-4}	9.860377×10^{-4}	9.152178×10^{-7}	
	Photowatt-PWP201	0.04178725	2.425075×10^{-3}	2.425092×10^{-3}	2.425442×10^{-3}	6.270718×10^{-8}	
GWOCS [128]	SDM	-	9.8607×10^{-4}	9.8874×10^{-4}	9.9095×10^{-4}	2.4696×10^{-6}	3.5
	DDM	-	9.8334×10^{-4}	9.9411×10^{-4}	1.0017×10^{-3}	9.5937×10^{-6}	
	Photowatt-PWP201	-	2.4251×10^{-3}	2.4261×10^{-3}	2.4275×10^{-3}	1.1967×10^{-6}	
	STM6-40/36	-	1.7337×10^{-3}	1.7457×10^{-3}	1.7528×10^{-3}	1.0447×10^{-5}	
PSOGWO [129]	Photowatt-PWP201	0.06292	3.06×10^{-3}	-	-	-	N/A
	STE4/100	0.00384	3.0574×10^{-4}	-	-	-	
	FSM	0.16023	9.14×10^{-3}	-	-	-	
ATLDE [130]	SDM	0.0177	9.8602×10^{-4}	9.8602×10^{-4}	9.8602×10^{-4}	2.44×10^{-17}	N/A
	DDM	0.0173	9.8218×10^{-4}	9.8372×10^{-4}	9.8603×10^{-4}	1.37×10^{-6}	
	STM6-40/36	0.0218	1.7298×10^{-3}	1.7298×10^{-3}	1.7298×10^{-3}	8.22×10^{-18}	
	STP6-120/36	0.2780	1.6601×10^{-2}	1.6601×10^{-2}	1.6601×10^{-2}	1.02×10^{-16}	
HHODE [131]	SDM	-	1.4664×10^{-3}	-	-	-	N/A
	DDM	-	1.5978×10^{-3}	-	-	-	
HAJAYADE [132]	SDM	-	9.8602×10^{-4}	9.8602×10^{-4}	9.8602×10^{-4}	0	2.833
	DDM	-	9.8294×10^{-4}	9.8641×10^{-4}	9.96×10^{-4}	2.8534×10^{-6}	
	Photowatt-PWP201	-	2.4251×10^{-3}	2.4251×10^{-3}	2.4251×10^{-3}	3.2215×10^{-15}	
	STM6-40/36	-	1.7298×10^{-3}	1.7298×10^{-3}	1.7298×10^{-3}	3.6569×10^{-16}	
	STP6-120/36	-	1.6601×10^{-2}	1.6601×10^{-2}	1.6606×10^{-2}	9.2421×10^{-7}	
HPSODOX [47]	SDM	-	6.4923×10^{-9}	-	-	-	N/A
	DDM	-	6.5120×10^{-9}	-	-	-	
	TDM	-	6.5424×10^{-9}	-	-	-	
	FDM	-	6.5656×10^{-9}	-	-	-	
TLBOBSA [134]	SDM	0.021526887	9.86902×10^{-4}	9.8602×10^{-4}	9.8603×10^{-4}	5.64965×10^{-10}	1.667
	DDM	0.021312577	9.8155×10^{-4}	1.1334×10^{-3}	2.2181×10^{-3}	3.0012×10^{-4}	
	TDM	0.021263898	9.82553×10^{-4}	1.2081×10^{-3}	3.0608×10^{-3}	4.9433×10^{-4}	
	Photowatt-PWP201	0.048923676	2.42507×10^{-3}	2.42535×10^{-3}	2.43167×10^{-3}	1.21238×10^{-6}	

The "N/A" means that there is insufficient data to support an average algorithm ranking using the Friedman Test on the three cases: SDM, DDM, and Photowatt-PWP201.

Most of the above methods are applications of newly proposed metaheuristics in recent years, and their essential information and experimental results are summarized in Tables 18 and 19. SSA has the smallest TNFES, followed by IGSK, RUN, GSK, SDO, TSA, HSOA, CTSA, SOS, WHHO, and TGA. WHHO and TGA achieve the same combined MIN RMSE ranking, followed by GSK, IGSK, HSOA, and SOS, in that order. It is worth noting that RUN, as the original algorithm, obtained more accurate parameter values with not many computational resources. TGA achieved the most efficient MIN RMSE values for DDM and TDM, and GSK received enough accuracy to compare with many advanced algorithms with not many computational resources. This suggests that exploring the application of new methods may make it easier to achieve a solution to the issue.

Table 18. Other methods' essential information and metrics.

Method	Main Contributors	Case	Algorithmic Parameter	Indicator	TNFES	Run
WHHO [135]	Naeijian et al., Department of Electrical Engineering, Babol Noshirvani University of Technology	SDM DDM TDM Photowatt-PWP201	$NP = 30$, Iteration = 5000,	RMSE IAE	-	30
GSK [4]	Xiong et al., Guizhou Key Laboratory of Intelligent Technology in Power System, College of Electrical Engineering, Guizhou University	SDM DDM Photowatt-PWP201 STM6-40/36 STP6-120/36	$NP = 30$, $k_r = 0.9$, $k_f = 0.5$, $K = 10$, $p = 0.1$	RMSE SIAE FT	30,000 50,000 30,000 30,000 30,000	30
IGSK [136]	Sallam et al., The Faculty of Computers and Information, Zagazig University	SDM DDM Photowatt-PWP201 STM6-40/36 STP6-120/36	$NP^{init} = 25$, $k_r = 0.9$, $k_f = 0.5$, $K = 10$, $p = 0.1$	RMSE WST	10,000 20,000 10,000 15,000 15,000	30
SDO [137]	Xiong et al., Guizhou Key Laboratory of Intelligent Technology in Power System, College of Electrical Engineering, Guizhou University	SDM DDM PVM 752 GaAs STM6-40/36 STP6-120/36	$NP = 20$	RMSE SIAE WRT FT	50,000	50
TGA [138]	Diab et al., Electrical Engineering Department, Faculty of Engineering, Minia University	SDM DDM TDM PVM 752 GaAs Photowatt-PWP201 STE 20/100	$NP = 500$, Iteration = 500,	RMSE	-	-
SSA [139]	Abbassi et al., University of Kairouan, Institute of Applied Sciences and Technology of Kasserine (ISSATKas)	TITAN-12-50	$NP = 30$, Iteration = 100,	RMSE IAE	-	30
TSA [140]	Sharma et al., Research and Development Department, University of Petroleum and Energy Studies	Photowatt-PWP201	$NP = 30$	RMSE, SIAE, FT	50,000	30
CGO [142]	Ramadan et al., Department of Electrical Engineering, Faculty of Engineering, Aswan University	TDM Photowatt-PWP201	Iteration = 1000	RMSE IAE	-	15
HSOA [143]	Long et al., Key Laboratory of Economics System Simulation, Guizhou University of Finance and Economics	SDM DDM Photowatt-PWP201	$NP = 30$, $f_{cmax} = 2$, $f_{cmin} = 0$, $F = 0.5$	RMSE SIAE FT	50,000	20
RUN [144]	Shaban et al., Faculty of Computers and Information, Minia University	SDM DDM TDM	$NP = 30$, Iteration = 1000, $a = 20$, $b = 12$	RMSE IAE FT	-	30
FPOA [145]	Chellaswamy et al., Department of ECE, Lords Institute of Engineering and Technology	Sample2, Sample5	$\beta = 1.45$, $S_p = 0.85$	MIAE	-	-
CTSA [141]	Gupta et al., Electrical and Instrumentation Engineering Department, Thapar Institute of Engineering and Technology	DDM TDM	$NP = 50$, Iteration = 1000	RMSE SIAE	-	-
SOS [146]	Xiong et al., Guizhou Key Laboratory of Intelligent Technology in Power System, College of Electrical Engineering, Guizhou University	SDM DDM Photowatt-PWP201	$NP = 50$	RMSE SIAE WRT	50,000	50

Table 19. Other methods' experiment results.

Method	Case	SIAE	MIN RMSE	Mean RMSE	MAX RMSE	STD of RMSE	Rank
WHHO [135]	SDM	-	9.8602×10^{-4}	9.8602×10^{-4}	9.8602×10^{-4}	-	2.667
	DDM	-	9.82487×10^{-4}	9.8249×10^{-4}	9.8250×10^{-4}	-	
	TDM	-	9.80751×10^{-4}	9.8085×10^{-4}	9.8149×10^{-4}	-	
	Photowatt-PWP201	-	2.4250×10^{-3}	2.4250×10^{-3}	2.4250×10^{-3}	-	
GSK [4]	SDM	0.0174	9.8602×10^{-4}	9.8602×10^{-4}	9.8602×10^{-4}	2.18×10^{-17}	3
	DDM	0.0175	9.8248×10^{-4}	9.8280×10^{-4}	9.8602×10^{-4}	8.72×10^{-7}	
	Photowatt-PWP201	0.0411	2.4251×10^{-3}	2.4251×10^{-3}	2.4251×10^{-3}	1.04×10^{-9}	
	STM6-40/36	0.0218	1.7298×10^{-3}	1.7298×10^{-3}	1.7298×10^{-3}	6.25×10^{-18}	
	STP6-120/36	0.2829	1.6601×10^{-2}	1.6601×10^{-2}	1.6601×10^{-2}	1.44×10^{-16}	
IGSK [136]	SDM	-	9.8602188×10^{-4}	9.8602188×10^{-4}	9.8602188×10^{-4}	$3.5821018 \times 10^{-17}$	3.33
	DDM	-	9.8248485×10^{-4}	9.8272774×10^{-4}	9.8602188×10^{-4}	8.9578942×10^{-7}	
	Photowatt-PWP201	-	2.4250749×10^{-3}	2.4250749×10^{-3}	2.4250749×10^{-3}	$2.9226647 \times 10^{-17}$	
	STM6-40/36	-	1.7298137×10^{-3}	1.7298137×10^{-3}	1.7298137×10^{-3}	$7.0155794 \times 10^{-18}$	
	STP6-120/36	-	1.6600603×10^{-2}	1.6600603×10^{-2}	1.6600603×10^{-2}	$1.7069489 \times 10^{-16}$	
SDO [137]	SDM	0.01770381	9.8602×10^{-4}	9.8603×10^{-4}	9.8616×10^{-4}	2.5141×10^{-8}	N/A
	DDM	0.01730620	9.8250×10^{-4}	9.8822×10^{-4}	1.0271×10^{-3}	8.8518×10^{-6}	
	PVM 752 GaAs	0.00593491	2.3487×10^{-4}	3.1727×10^{-4}	3.7700×10^{-4}	2.7687×10^{-5}	
	STM6-40/36	0.02177419	1.7298×10^{-3}	1.7703×10^{-3}	1.9500×10^{-3}	4.5108×10^{-5}	
	STP6-120/36	0.27797428	1.6601×10^{-2}	1.6683×10^{-2}	1.6866×10^{-2}	7.1751×10^{-5}	
TGA [138]	SDM	-	$9.750530454421328 \times 10^{-4}$	-	-	-	2.667
	DDM	-	$8.488244232381 \times 10^{-4}$	-	-	-	
	TDM	-	$8.251052783901371 \times 10^{-4}$	-	-	-	
	PVM 752 GaAs	-	$9.037521972258222 \times 10^{-4}$	-	-	-	
	Photowatt-PWP201	-	$3.819491771269 \times 10^{-3}$	-	-	-	
	STE 20/100	-	$9.28071173 \times 10^{-4}$	-	-	-	
SSA [139]	TITAN-12-50(366)	-	2.9681 × 10−04	-	-	-	N/A
	TITAN-12-50(810.2)	-	1.5777 × 10−06	-	-	-	
TSA [140]	Photowatt-PWP201	0.0594	5.06×10^{-4}	1.45×10^{-3}	2.34×10^{-2}	1.25×10^{-3}	N/A
CGO [142]	TDM	-	9.82×10^{-4}	9.82×10^{-4}	9.82×10^{-4}	1.24841×10^{-9}	N/A
	Photowatt-PWP201	-	2.425075×10^{-3}	2.425092×10^{-3}	2.4251×10^{-3}	1.44688×10^{-8}	
HSOA [143]	SDM	0.0177065	9.8602×10^{-4}	1.0479×10^{-3}	1.1683×10^{-3}	5.3832×10^{-5}	4
	DDM	0.017402	9.8376×10^{-4}	1.1175×10^{-3}	1.7642×10^{-3}	1.9107×10^{-4}	
	Photowatt-PWP201	0.041788	2.4251×10^{-3}	2.4251×10^{-3}	2.4253×10^{-3}	4.1556×10^{-8}	
RUN [144]	SDM	-	9.86242×10^{-4}	1.479894×10^{-3}	2.444572×10^{-3}	4.30699×10^{-4}	N/A
	DDM	-	9.87168×10^{-4}	1.481762×10^{-3}	2.947571×10^{-3}	5.14117×10^{-4}	
	TDM	-	9.89133×10^{-4}	1.581238×10^{-3}	6.239595×10^{-3}	1.078762×10^{-3}	
CTSA [141]	DDM	0.2621	1.0239×10^{-8}	2.1185×10^{-8}	9.6017×10^{-8}	3.9865×10^{-8}	N/A
	TDM	0.0075	1.0036×10^{-6}	3.4906×10^{-6}	9.4766×10^{-6}	2.7057×10^{-6}	
SOS [146]	SDM	0.0181	9.8609×10^{-4}	1.0245×10^{-3}	1.1982×10^{-3}	5.2184×10^{-5}	5.333
	DDM	0.0182	9.8518×10^{-4}	1.0627×10^{-3}	1.3498×10^{-3}	9.6141×10^{-5}	
	Photowatt-PWP201	0.0421	2.4251×10^{-3}	2.4361×10^{-3}	2.5103×10^{-3}	1.7503×10^{-5}	

The "N/A" means that there is insufficient data to support an average algorithm ranking using the Friedman Test on the three cases: SDM, DDM, and Photowatt-PWP201.

4. Whole Analysis and Research Prospects

This section presents metaheuristic methods in solving the studied problem. We collect their data for an overall analysis and give some research prospects.

4.1. Data Analysis

In the third part, the final results of many methods are relatively convergent. For SDM, the RMSE is mainly distributed around 9.8206×10^{-4} and the rest is concentrated around 7.7301×10^{-4}. The DDM's primary distribution is around 9.8248×10^{-4}, with a secondary allocation of 7.42×10^{-4} to 7.1823×10^{-4}. For the TDM, the main distribution is between 9.8331×10^{-4} and 9.8033×10^{-4}, with higher precision than the main distribution interval, being 8.2511×10^{-4} for TGA and 6.5424×10^{-9} for HPSODOX. For Photowatt-PWP201, the main distribution is around 2.4251×10^{-3}, the secondary distribution is around 2.0399×10^{-3}, and the best-performing TSA reaches 5.06×10^{-4}. STM6-40/36 is mainly distributed at 1.7298×10^{-3}. STP6-120/36 is primarily distributed at 1.6601×10^{-2} nearby; the best-performing FDB-TLABC achieved 1.4251×10^{-2}. However, the different approaches rarely use the same cases and evaluation indicators, and the results may differ between models. Therefore, some well-performed variants of metaheuristics that used the RMSE indicators are selected for further comparison in Table 20, i.e., ABC-TRR, RLDE, OLBWOA, CSOOJAYA, DEDIWPSO, EOTLBO, IWOA, TLBOBSA, IGSK, HSOA, and SOS.

Table 20. Various methods' RMSE results.

Method	Case	MIN RMSE	Mean RMSE	MAX RMSE	STD of RMSE	Rank
ABC-TRR [92]	SDM	9.860219×10^{-4}	9.860219×10^{-4}	9.860219×10^{-4}	6.15×10^{-17}	5.958
	DDM	9.824849×10^{-4}	9.825556×10^{-4}	9.860219×10^{-4}	4.95×10^{-7}	
	Photowatt-PWP201	2.425075×10^{-3}	2.425075×10^{-3}	2.425075×10^{-3}	9.68×10^{-17}	
RLDE [41]	SDM	9.8602×10^{-4}	9.8602×10^{-4}	9.8602×10^{-4}	3.4834×10^{-17}	5.125
	DDM	9.8248×10^{-4}	9.8695×10^{-4}	9.8457×10^{-4}	1.7498×10^{-6}	
	Photowatt-PWP201	2.4251×10^{-3}	2.4251×10^{-3}	2.4251×10^{-3}	6.3084×10^{-17}	
OLBGWO [100]	SDM	9.86×10^{-4}	9.86×10^{-4}	9.86×10^{-4}	1.4×10^{-8}	4.583
	DDM	9.83×10^{-4}	9.85×10^{-4}	9.86×10^{-4}	1.78×10^{-6}	
	Photowatt-PWP201	2.4×10^{-3}	2.4×10^{-3}	2.4×10^{-3}	2.4284×10^{-9}	
CSOOJAYA [114]	SDM	9.860219×10^{-4}	9.860219×10^{-4}	9.860219×10^{-4}	4.717305×10^{-17}	4.917
	DDM	9.824849×10^{-4}	9.824849×10^{-4}	9.824849×10^{-4}	5.576332×10^{-17}	
	Photowatt-PWP201	2.425075×10^{-3}	2.425075×10^{-3}	2.425075×10^{-3}	2.699858×10^{-17}	
DEDIWPSO [85]	SDM	7.730062×10^{-4}	7.730062×10^{-4}	7.730062×10^{-4}	5.18668×10^{-15}	2.5
	DDM	7.182306×10^{-4}	7.187462×10^{-4}	7.318100×10^{-4}	2.486129×10^{-6}	
	Photowatt-PWP201	2.03992×10^{-3}	2.03992×10^{-3}	2.03992×10^{-3}	2.995389×10^{-15}	
EOTLBO [21]	SDM	$9.86021878 \times 10^{-4}$	$9.86021878 \times 10^{-4}$	$9.86021878 \times 10^{-4}$	$4.12665088 \times 10^{-17}$	4.5
	DDM	$9.82484852 \times 10^{-4}$	$9.84733697 \times 10^{-4}$	$9.89424104 \times 10^{-4}$	$1.69176118 \times 10^{-6}$	
	Photowatt-PWP201	$2.42507487 \times 10^{-3}$	$2.42507487 \times 10^{-3}$	$2.42507487 \times 10^{-3}$	$3.61995116 \times 10^{-17}$	
IWOA [123]	SDM	9.860219×10^{-4}	9.860219×10^{-4}	9.860219×10^{-4}	5.12×10^{-16}	6.375
	DDM	9.824849×10^{-4}	9.826140×10^{-4}	9.860219×10^{-4}	9.86×10^{-5}	
	Photowatt-PWP201	2.425075×10^{-3}	2.425075×10^{-3}	2.425075×10^{-3}	2.90×10^{-17}	
TLBOBSA [134]	SDM	9.86902×10^{-4}	9.8602×10^{-4}	9.8603×10^{-4}	5.64965×10^{-10}	8.292
	DDM	9.8155×10^{-4}	1.1334×10^{-3}	2.2181×10^{-3}	3.0012×10^{-4}	
	Photowatt-PWP201	2.42507×10^{-3}	2.42535×10^{-3}	2.43167×10^{-3}	1.21238×10^{-6}	
IGSK [136]	SDM	9.8602188×10^{-4}	9.8602188×10^{-4}	9.8602188×10^{-4}	$3.5821018 \times 10^{-17}$	4.333
	DDM	9.8248485×10^{-4}	9.8272774×10^{-4}	9.8602188×10^{-4}	8.9578942×10^{-7}	
	Photowatt-PWP201	2.4250749×10^{-3}	2.4250749×10^{-3}	2.4250749×10^{-3}	$2.9226647 \times 10^{-17}$	
HSOA [143]	SDM	9.8602×10^{-4}	1.0479×10^{-3}	1.1683×10^{-3}	5.3832×10^{-5}	9.333
	DDM	9.8376×10^{-4}	1.1175×10^{-3}	1.7642×10^{-3}	1.9107×10^{-4}	
	Photowatt-PWP201	2.4251×10^{-3}	2.4251×10^{-3}	2.4253×10^{-3}	4.1556×10^{-8}	
SOS [146]	SDM	9.8609×10^{-4}	1.0245×10^{-3}	1.1982×10^{-3}	5.2184×10^{-5}	10.083
	DDM	9.8518×10^{-4}	1.0627×10^{-3}	1.3498×10^{-3}	9.6141×10^{-5}	
	Photowatt-PWP201	2.4251×10^{-3}	2.4361×10^{-3}	2.5103×10^{-3}	1.7503×10^{-5}	

The variants of metaheuristics that used the SIAE indicators are selected for further comparison in Figure 5, i.e., SOS, HSOA, GSK, TLBOBSA, DE/WOA, ISNMWOA, MCSWOA, IWOA, DMTLBO, PSO-ST, GCPSO, MLJAYA, I-GWO, HDE, DPDE, QILDE, ABC-TRR, and TLABC. Moreover, these methods were generally tested in the SDM, DDM, and Photowatt-PWP201 module. Here, the module only means the Photowatt-PWP201.

- The STD of RMSE reflects the results' robustness, MIN RMSE means the results' accuracy, and other RMSEs denote the range and sharpness of the fluctuations in the results. The SDM, DDM and Photowatt-PWP201 models of DEDIWPSO had the MIN RMSE (7.730062×10^{-4}, 7.182306×10^{-4}, and 2.03992×10^{-3}), mean RMSE (7.730062×10^{-4}, 7.187462×10^{-4}, and 2.03992×10^{-3}), MAX RMSE (7.730062×10^{-4}, 7.3181×10^{-4}, and 2.03992×10^{-3}) and STD (5.18668×10^{-15}, 2.486129×10^{-6}, and 2.995389×10^{-15}). It is followed by IGSK with MIN RMSE (9.8602188×10^{-4}, 9.8248485×10^{-4}, and 2.4250749×10^{-3}), mean RMSE (9.8602188×10^{-4}, 9.8272774×10^{-4}, and 2.4250749×10^{-3}), MAX MRSE (9.8602188×10^{-4}, 9.8602188×10^{-4}, and 2.4250749×10^{-3}) and STD ($3.5821018 \times 10^{-17}$, 8.9578942×10^{-7}, and $2.9226647 \times 10^{-17}$). Then, EOTLBO, OLBGWO, CSOOJAYA, RLDE, ABC-TRR, IWOA, TLBOBSA, HSOA, and SOS followed.
- Figure 4 shows the combined FT ranking for the SDM, DDM, and Photowatt-PWP201. It combines the absolute accuracy of the methods in a wide range of cases. GSK ranks first, followed by MCSWOA, IWOA, GCPSO, QILDE, DE/WOA, DMTLBO, HSOA, MLJAYA, SOS, TLABC, PSO-ST, ABC-TRR, I-GWO, HDE, TLBOBSA, ISNMWOA, and DPDE. GSK, as a new method achieving the highest accuracy, demonstrates the need to explore the performance of new schemes in this issue. It is worth noting that the rankings of the same methods in different PV models may differ, which indicates that different PV models have varied preferences for algorithms.
- TNFES is related to the computational resources consumed, with a lower TNFES representing a lower computational burden. For the SDM and module, ABC-TRR had the fewest TNFES (1000) while other methods basically used a TNFES of 50,000. For the DDM, ABC-TRR had the fewest TNFES (5000), while most of the rest consumed a TNFES of 50,000.

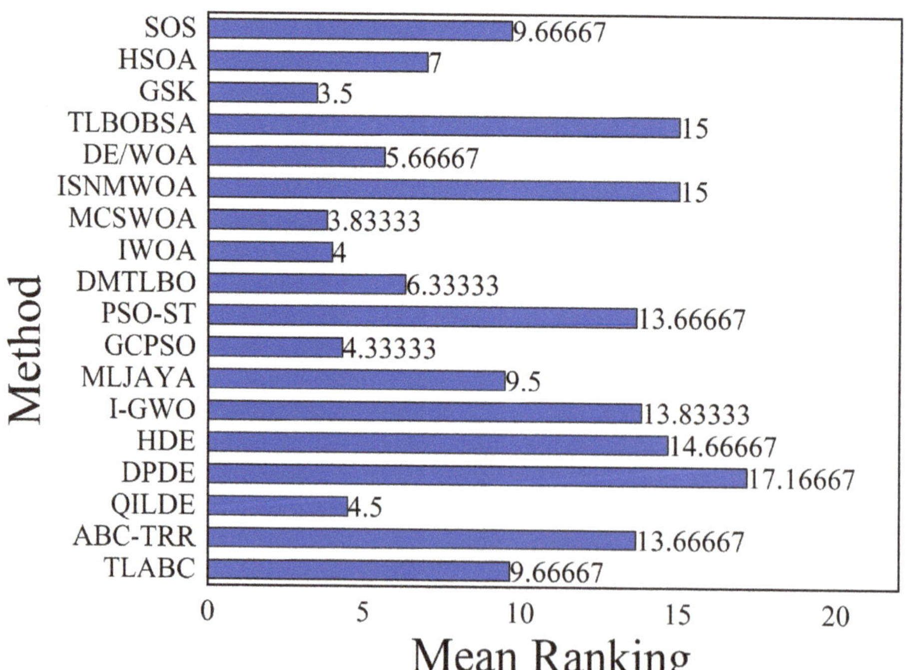

Figure 5. Various methods' Friedman Test.

4.2. Analysis of Temperature and Irradiance Influences

When the irradiance or temperature changes, the current output of the PV cell will also change, and therefore several unknown parameters representing the output characteristics of the PV cell will also change. The GSK algorithm with high accuracy is used in this section to identify the sampled data at different temperatures or irradiances in order to explore their patterns. The data are taken from the KC200GT module in Simulink.

4.2.1. Uniform Irradiance and Temperature

Eight cases under uniform conditions were set up to explore the effects of irradiance and temperature separately. The cases can be divided into five irradiances at 25 °C: 1000, 800, 600, 400, and 200 W/m^2 and four temperatures at 1000 W/m^2: 25, 40, 55, and 70 °C. Their I-V and P-V output characteristics are shown in Figures 6 and 7. In the figures, the output current increases with increasing irradiance, and the maximum power point voltage decreases with increasing temperature.

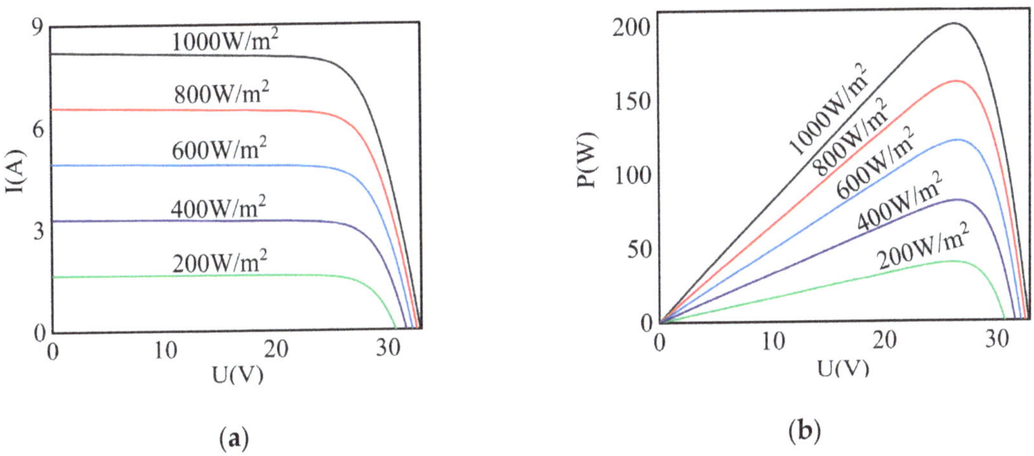

Figure 6. Characteristic curves in various irradiance: (**a**) I-V (**b**) P-V.

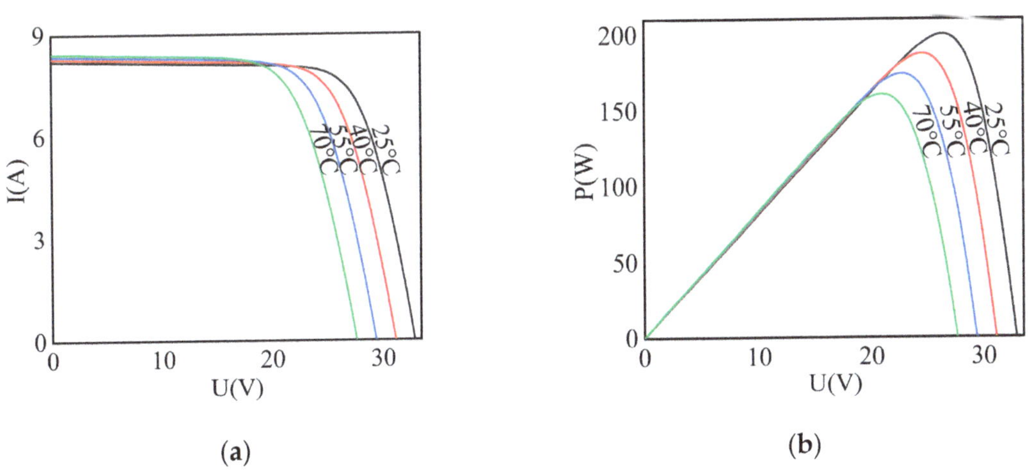

Figure 7. Characteristic curves in various temperatures: (**a**) I-V (**b**) P-V.

From the above characteristic plots, it is evident that when environmental factors change, corresponding parameters change accordingly to achieve a high degree of fit to the output curve. The unknown parameters extracted using GSK are illustrated in Table 21. When the irradiance is the variable, I_{ph} increases linearly with increasing irradiance, and R_s decreases in a non-linear fashion with increasing irradiance. When the temperature is the variable, I_{ph} increases weakly with increasing temperature, and I_{ssd} increases in a non-linear manner. Meanwhile, the RMSE increases with decreasing temperature, indicating that the lower the temperature, the lower the identification result's accuracy.

Table 21. Parameters of the KC200GT at different irradiances and temperatures.

Radiation /W/m^2	Temperature /°C	I_{ph}/A	I_{ssd}/μA	n	R_s/Ω	R_{sh}/Ω	RMSE
Variable	Fixed						
1000	25	8.22920506	$2.19226333 \times 10^{-10}$	0.34555194	149.79495733	52.64769156	$2.87908987 \times 10^{-3}$
800	25	6.58249378	$2.57655463 \times 10^{-10}$	0.34314866	190.38069917	52.99842155	$2.40659465 \times 10^{-3}$
600	25	4.93738274	$2.27177693 \times 10^{-10}$	0.34433472	250.19011038	52.72470592	$3.70428705 \times 10^{-3}$
400	25	3.29180014	$1.99109819 \times 10^{-10}$	0.34972198	372.27107651	52.42424407	$1.44743443 \times 10^{-3}$
200	25	1.64555637	$2.50815014 \times 10^{-10}$	0.34381397	769.17560620	52.94945965	$1.23547582 \times 10^{-3}$
Fixed	Variable						
1000	25	8.22811095	$2.49012735 \times 10^{-10}$	0.34410634	152.34953496	52.92528529	$5.10117026 \times 10^{-3}$
1000	40	8.30308470	$2.50259970 \times 10^{-9}$	0.34496529	149.56870789	52.70878667	$4.12556209 \times 10^{-3}$
1000	55	8.37565108	$2.31628311 \times 10^{-8}$	0.34480573	153.53100022	52.79148663	$8.96621362 \times 10^{-3}$
1000	70	8.45187588	$1.62391869 \times 10^{-7}$	0.34518787	146.10502751	52.62434432	$1.10992599 \times 10^{-2}$

Some methods counted in Section 3 simulated PV modules at different irradiances and temperatures. The methods are gathered together, as illustrated in Table 22. The methods' quantity is 22, indicating that the proportion of methods discussing these cases is low and that more consideration needs to be placed on these cases in future research work. Most of the 22 methods discussed irradiance and temperature together, and the cases they used most frequently are SM55, ST40, and KC200GT. Thus, other cases could be added to these three implementations in the future so that further generalizability can be demonstrated.

Table 22. Various methods with different irradiance and temperature experiments.

Method	Case	Radiation	Temperature	Describe
FDB-TLABC [96]	SM55, ST40, KC200GT	√	√	Experiments were designed for five sets of irradiances at 25 °C and three sets of temperature at 1000 W/m^2, with RMSEs consistently lying in the order of 1×10^{-5} in the three modules, much better than L-SHADE, LSHADE-EPSIN, and LSHADE-SPACMA.
IADE [68]	SL80CE-36M	√	√	Four sets of discriminative parameters and minimum RMSEs (0.0115, 0.006, 0.0071, 0.0154) were obtained from experiments fitting PV data for four different sets of environmental parameters at two temperatures and two irradiances in random combinations.

Table 22. *Cont.*

Method	Case	Radiation	Temperature	Describe
DE3P [23]	SM55, RSM50, ST40	√	√	Experiments were carried out with five sets of irradiances at constant temperature and three sets of temperature at constant irradiance, with a maximum RMSE of 0.0148 in the results, which is still an acceptable error.
EJADE [69]	SM55, KC200GT	√	√	The optimal average RMSE was obtained consistently with eight competing algorithms for experiments at different irradiances and temperatures. The RMSEs were of order 1×10^{-4} at 25 °C for 200~800 W/m^2 and 1×10^{-3} for the other experiments.
AGA [64]	-	-	√	A PV cell fitting experiment at different temperatures was designed, and the initial and post-simulation parameter values for the standard case were given.
GWO [98]	-	-	√	Ten sets of experiments at different temperatures (−5 °C~45 °C) were designed and showed an enormous advantage in comparison experiments with MMA, with RMSEs almost of order 1×10^{-3} overall.
OLBGWO [100]	ST40, KC200GT	√	√	The experimental design was the same as that of FDB-TLABC. The ST40 module's RMSEs were at or near the 1×10^{-4} order of magnitude. In the KC200GT module, the RMSEs were at or near the 1×10^{-3} order of magnitude.
EJAYA [111]	SM55, KC200GT	√	√	The experimental design was the same as EJADE. The SM55 experiments' RMSEs were in order 1×10^{-4}, and the other experiments' RMSEs were in order 1×10^{-3}.
MPSO [81]	SM55, ST40, KC200GT	√	√	The experimental design was the same as FDB-TLABC. In the KC200GT, the RMSEs were of order 1×10^{-3}; in the other experiments, the RMSEs were of order 1×10^{-4}.
GCPSO [82]	Sharpe ND-R250A5	√	√	Five experiments with different temperatures and irradiances were designed to obtain high fitting accuracy, with an RMSE of order 1×10^{-3}.
DEDIWPSO [85]	JKM330P	√	√	Experiments were designed for five different irradiances and temperatures, RMSE values were obtained consistently, and all RMSEs were of order 1×10^{-3}.
PSO-ST [87,88]	JKM330P	√	√	The same experimental design as DEDIWPSO, with RMSEs of order 1×10^{-3} and standard deviations of RMSEs on order 1×10^{-17}.
PSOCS [88]	SM55, ST40, KC200GT	√	√	The experimental design was the same as FDB-TLABC, with RMSE concentrated at the order of magnitude 1×10^{-2} and 1×10^{-3}.
EOTLBO [21]	Sharpe ND-R250A5	√	√	The experimental design was the same as GCPSO, with RMSEs concentrated at orders 1×10^{-2} and 1×10^{-3}, and significantly better than the ten comparative algorithms in the text.
MTLBO [119]	SM55, ST40	√	√	The experimental design was the same as FDB-TLABC, whose RMSEs were concentrated on orders 1×10^{-3} and 1×10^{-4} and converged slightly faster than ITLBO.

Table 22. *Cont.*

Method	Case	Radiation	Temperature	Describe
WOA [124]	KC200GT	√	√	The fitting experiments were implemented with SDM, DDM, and TDM. The SDM error was 1.6%, the DDM error was 0.3%, and the TDM error was 0.08%. It indicates that, with sufficient computational resources, TDM > DDM > SDM in terms of accuracy.
ISNMWOA [126]	SM55, ST40, KC200GT	√	√	The experimental design was the same as FDB-TLABC, with the RMSEs concentrated on orders 1×10^{-3} and 1×10^{-4}. It showed that ISNMWOA still has high accuracy at low temperatures and irradiance.
SWOA [125]	SM55, SW255, KC200GT	√	√	Experiments were designed for five irradiances and seven temperatures. The RMSEs were concentrated around 1×10^{-2} for the irradiance experiments and around 1×10^{-3} for the temperature experiments.
DE/WOA [127]	JAM6-60-295W-4BB	√	√	Experiments with five irradiances and four temperatures were implemented. Significantly better RMSEs were consistently achieved compared to seven competing algorithms, and all results were concentrated around 1×10^{-5}.
HPSODOX [47]	-	\	√	Seven sets of experiments from −5 to 25 °C were designed. Of these, the RMSEs were located in order 1×10^{-9} at 25 °C and in order 1×10^{-8} at different temperatures.
TLBOBSA [134]	SM55, KC200GT	√	√	The experimental design was the same as EJADE. The experimental results were similar to EJAYA and slightly worse overall.
IGSK [136]	SM55, ST40	√	√	The experimental design was the same as MTLBO, with 11 RMSEs at the 1×10^{-4} order of magnitude and 6 RMSEs at the 1×10^{-3} order of magnitude in 17 experiments.

The "√" means that there are temperature or irradiance experiments in the literature.

4.2.2. Partial Shade Conditions

Four groups of KC200GTs were connected in series to obtain the multi-peak curve exhibited by the output of the PV power system when partially shaded (PSC). Four sets of comparison tests were designed: standard case (STC: 4 × 1000 W/m^2), type I partial shading (PSC-1: 1000, 800, 400, 400 W/m^2), type II partial shading (PSC-2: 800, 600, 400, 200 W/m^2), and type III partial shading (PSC-3: 800, 600, 400, 400 W/m^2). The output characteristics are shown in Figure 8. In the figure, STC has a single peak, PSC-1 and PSC-3 have three peaks, and PSC-2 has four. Additionally, STC has only one irradiance, PSC-1 and PSC-3 have three irradiances, and PSC-2 has four irradiances. Therefore, the PV's peaks are related to the irradiance types on the series-connected PV modules.

The mathematical models developed in Section 2 cannot generate multiple inflection points. Thus, the characteristic curve of the PSC fitted using these mathematical models will still have only one inflection point, and the accuracy of the fit will be very low. It is reflected in a large minimum RMSE. The extracted parameters are shown in Table 23, and it is clear that the RMSE at STC is much lower than that at PSC. Although the corresponding mathematical model was developed by Chellaswamy et al. [147], it requires human judgment and input of the number of modules to be shaded, which is difficult to achieve in reality. Therefore, more mathematical models need to be developed in future work to improve the accuracy of the parameters of the extracted PSCs. It is important to note that, due to the presence of parallel diodes in the system, the PV modules are in an idle state when the output current of the system is more significant than its photogenerated current.

The mathematical models developed to simulate the output characteristics of the PSC must take this critical point into account.

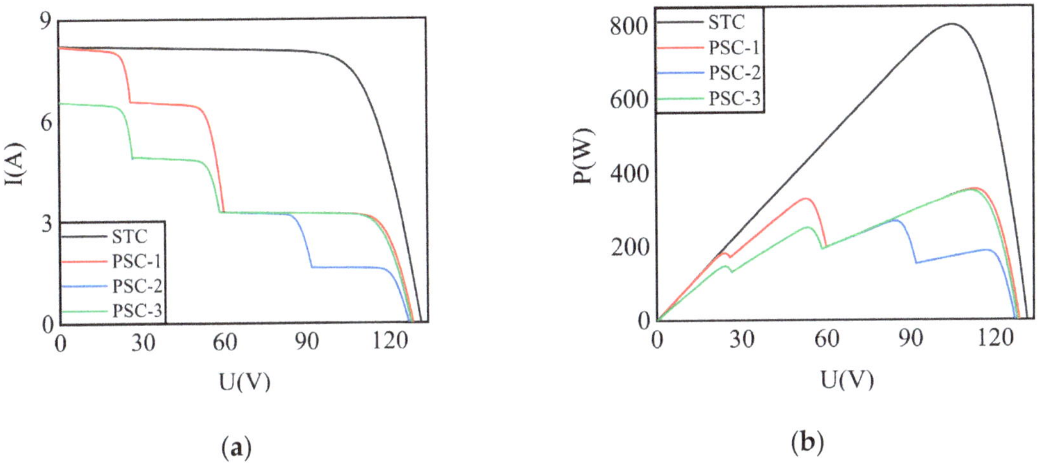

Figure 8. Characteristic curves in partial shade conditions: (**a**) I-V (**b**) P-V.

Table 23. Parameters of the KC200GT at partial shade conditions.

Case	I_{ph}/A	I_{ssd}/μA	n	R_s/Ω	R_{sh}/Ω	RMSE
STC	8.22879884	$2.32498946 \times 10^{-10}$	1.37930864×10^{0}	602.77198763	211.10041272	$1.31085496 \times 10^{-6}$
PSC-1	8.40661915	$3.20394383 \times 10^{-15}$	$1.62587931 \times 10^{-16}$	18.94997935	149.17780560	$6.96889061 \times 10^{-1}$
PSC-2	6.93947342	$1.16187272 \times 10^{-14}$	$2.40441463 \times 10^{-16}$	20.81282985	155.76285151	$3.71532656 \times 10^{-1}$
PSC-3	6.52880635	$5.19579219 \times 10^{-12}$	$1.10570546 \times 10^{-14}$	28.77275463	188.22179994	$4.55796025 \times 10^{-1}$

4.3. Analysis of Modified Diode Models' Works

The MSDM, MDDM, and MTDM all consider the quasi-neutral zone's losses. It is reflected in the circuit diagram by selecting a diode branch and adding a series resistor R_{sm}. The improved model adds an unknown parameter compared to the pre-improved model. Their circuit diagram is shown in Figure 9.

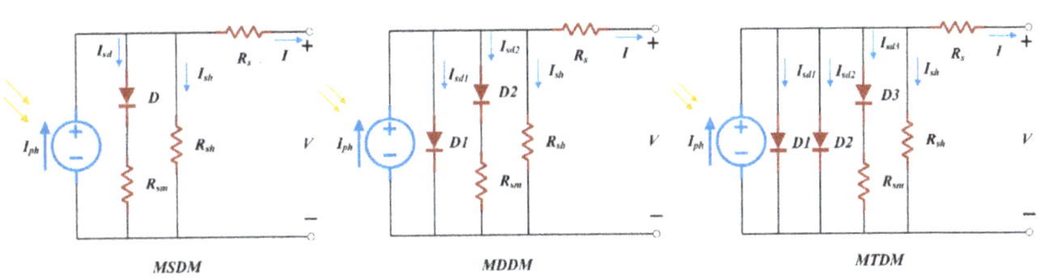

Figure 9. Modified diode models' circuits.

Their output current changes to [48,49]:

$$I = I_{ph} - \frac{V + IR_s}{R_{sh}} - \sum_{j=1\rightarrow(nD-1)} I_{ssdj}\left[\exp\left(\frac{q(V+IR_s)}{n_j kT}\right) - 1\right] - I_{ssdnD}\left[\exp\left(\frac{q(V+IR_s - I_{sdnD}R_{sm})}{n_{nD}kT}\right) - 1\right] \quad (14)$$

where nD represents the number of diodes in the cell model.

In this subsection, two papers from the last three years are chosen to present the results of metaheuristic approaches to solving the above models. Ramadan et al. [48] improved the Bald Eagle Search algorithm (IBES), employing decay equations to achieve adaptive learning factors. Abdelminaam et al. [49] pioneered the use of Turbulent Flow Optimization of Water (TFWO) for the parameter extraction of PV cells with a new objective function (PE5DSSE). Their extraction results are illustrated in Table 24.

Table 24. Results of the modified diode models.

Parameter	IBES MSDM	TFWO MSDM	IBES MDDM	TFWO MDDM	IBES MTDM	TFWO MTDM
I_{ph}/A	0.760713	0.760774525	0.760494	0.760783023	0.760473235	0.760780283
R_s/Ω	0.032091	0.037372671	0.015196	0.036835645	0.013865736	0.036749141
R_{sh}/Ω	54.30519	53.7186078	54.05261	55.8909553	55.47156858	55.52672891
R_{sm}/Ω	0.00352	0.5	0.02792	0.01025276	0.027870684	0.5
$I_{ssd1}/\mu\mathrm{A}$	3.71×10^{-7}	3.23×10^{-7}	1.00×10^{-10}	9.17×10^{-7}	1.00×10^{-10}	7.63×10^{-7}
$I_{ssd2}/\mu\mathrm{A}$	-	-	6.69×10^{-7}	2.07×10^{-7}	7.52×10^{-7}	2.47×10^{-9}
$I_{ssd3}/\mu\mathrm{A}$	-	-	-	-	1.00×10^{-10}	2.24×10^{-7}
n_1	1.4835	1.48118376	1.00	1.999992291	1.133059042	2
n_2	-	-	1.525277	1.443600817	1.537322148	2
n_3	-	-	-	-	1.004574508	1.450312839
PE5DSSE	-	2.5278×10^{-5}	-	2.51×10^{-5}	-	2.509×10^{-5}
MIN RMSE	9.61×10^{-4}	-	7.49×10^{-4}	-	7.39055×10^{-4}	-
Mean RMSE	1.507×10^{-3}	-	1.201×10^{-3}	-	7.64×10^{-4}	-
MAX RMSE	2.847×10^{-3}	-	3.378×10^{-3}	-	7.81×10^{-4}	-
STD of RMSE	7.61×10^{-4}	-	8.95×10^{-4}	-	2.21×10^{-5}	-

In Table 24, for MSDM, the parameter that differs most between IBES and TFWO is R_{sm}. For MDDM, IBES and TFWO are similar in I_{ph} and R_{sh}, and the other parameters differ more. For MTDM, IBES and TFWO are almost identical in I_{ph} and R_{sh}, and the other parameters differ more. As they use different objective functions, it is impossible to compare the accuracy of the two.

In IBES, the MIN RMSE is 9.88×10^{-4} for TDM and 9.86×10^{-4} for SDM and DDM. In TFWO, the PE5DSSE is 2.5278×10^{-5} for SDM, 2.51×10^{-5} for DDM and 2.51×10^{-5} for TDM. It indicates that the addition of Rsh did improve the accuracy by a small margin. Therefore, applying MSDM, MDDM, MTDM, and the PV module models constructed from them to future studies will be an effective way to improve the accuracy further.

4.4. Analysis of Dynamic Models' Works

The above results are for static models. This subsection starts with several representative metaheuristics for solving dynamic models to analyze their parameter extraction results.

Yousri et al. [52] developed CHCLPSO by combining heterogeneous integrated learning PSO with chaotic optimization techniques. HROA was developed along similar lines to CHCLPSO, a hybrid of the chaotic mapping mechanism with the Rao_1 algorithm by Wang et al. [53]. Elaziz et al. [51] developed EMPA by an effective combination of DE and the Marine Predator algorithm.

For the results of the dynamic model, CHCLPSO provides parameters of $R_C = 7.3149\ \Omega$, $C = 3.81307 \times 10^{-7}$ F, and $L = 7.3251 \times 10^{-6}$ H. EMPA provides parameters of $R_C = 7.315\ \Omega$, $C = 3.1831 \times 10^{-7}$ F, and $L = 7.3251 \times 10^{-6}$ H. Their difference is insignificant, indicating that both methods have similar solving power. The MIN and Mean RMSEs for CHCLPSO are 8.45045×10^{-3}, and the STD is 1.13566×10^{-12}. The MIN, Mean, and MAX RMSEs for HROA are 6.709393×10^{-3}, and the STD is 5.209153×10^{-18}. The Mean RMSE for EMPA makes it clear that HROA has the best accuracy and robustness, followed by EMPA and CHCLPSO. However, CHCLPSO is at the same level of accuracy as EMPA, and both have a minor STD. This indicates that EMPA and CHCLPSO have converged early, and their further improvement needs to start from exploration. For HROA, it achieves the optimal

RMSE value, but 6.709393 × 10^{-3} is still a significant error and there is room for further optimization of the accuracy of the solution.

It is worth mentioning that the model of dynamics is suitable for grid-connected operation. However, there has been little research related to it since its introduction, and especially little research on metaheuristic methods to optimize the dynamic model. Therefore, it has broad application and research prospects and is a crucial research direction for the future.

4.5. Whole Analysis

Pursuant to scholarly opinion and statistical results, Table 25 analyses the positive and negative properties of various metaheuristics. They can help beginners to understand cutting-edge research.

Table 25. Various methods' positive and negative properties.

Type	Positive	Negative
GAs	• Using probabilistic mutation techniques • Fast handling of non-linear problems [63] • Easily contribute to the convergence and accuracy of other methods [65]	• Reliance on the initialized populations' quality • Lower accuracy of solution than advanced methods
DEs	• Simple and precise implementation • Steady and fast • Extensible, with many variants • Employing adequate parameter tweaking mechanisms ensures an overall improvement in the algorithm's capabilities in specific problems [41,68,76]	• The parameters' decision shapes the results • Computing resources are underutilized
PSOs	• Straightforward code • Fast merit search • Low fluctuant solution and efficient • Supports parallel operation for faster and greater accuracy [86] • For the problems in this paper, PSO secured quality solutions [82,83]	• Excessive parameters and empirical reliance • Converge prematurely • Prone to converge to local optimum in multi-peaked issues
ABCs	• Superb exploration [92] • Rapid convergence [89] • Simplicity implementation [93] • Fits PV characteristic curves more accurately than PSO [95] • Premium performance in combination with alternative methods [91,94]	• Weak exploitation • Parameters and performance are strongly correlated
GWOs	• A few parameters • Flexibility and simplicity • Well-aligned exploration and exploitation [97] • Tackling PV parameter estimation issues with small errors [98]	• Poor handling issues with numerous variables • Exploitation requires reinforcement [100]

Table 25. *Cont.*

Type	Positive	Negative
JAYAs	• No parameters • Efficient and succinct • Adaptive control factor optimizes accuracy and stability [104,107] • Mixing different methods of consideration facilitates performance improvement [106,111]	• Weak exploration [108] • Pseudo-random operators restricted pervasiveness • Performance degradation in multi-dimensional issues [103]
TLBOs	• No parameters • Universal in optimization issues • Competitive in large scale issues [115] • Diverse variants enhance behavior when employed for specific problems [118,120]	• Slow convergence [116] • Mandatory structures squandering resources [21] • Inadequate balance of exploration and exploitation [117,118]
WOAs	• A few parameters • Simple structure • Intense exploitation competency [123] • Variant with outstanding solutions quality [125,126]	• Premature convergence [18] • Poor in convergence and precision • Performance degradation in complex issues
GSKs	• Intense exploration competency [4] • Competitive in multidimensional issues • Fits PV characteristic curve accurately [136]	• Excessive parameters • Weak exploitation
SDOs	• A few parameters • Simple structure • Well-balanced exploration and exploitation [137]	• Poor in convergence • Needs improvement in solution quality [39]
HHOs	• Fewer mechanisms, simpler calculations [148] • Fast convergence [135] • Suitable for multimodal scenarios [55]	• Excessive parameters • Premature convergence
TGAs	• Simple structure • High accuracy of identification results • Highly competitive [138]	• Excessive parameters • Slow convergence • Excessive consumption of computational resources
SOSs	• No parameters • Simple structure • Superb exploration [146]	• Weak exploitation • Excessive resources consumption
FPOAs	• Fewer parameters • Easy to implement • Simple structure • More accurate than PSO and DE [145]	• Premature convergence • Slow convergence

For the different applied metaheuristics, we find the following challenges.

1. The promotion of GA has been rare in recent years, and accuracy is supposed to be enhanced.
2. DE's convergence rate and PSO's accuracy could improve.
3. ABC is weakly exploited and significant in parameter settings.
4. GWO and WOA have few parameters and struggle with multi-dimensional issues.
5. JAYA and TLBO's promotions are flawed in accuracy.
6. Hybrid approaches may complicate the implementation and introduce additional parameters.

7. New approaches are not sufficiently balanced for specific issues. For example, GSK, SDO, TGA, and SOS are under-exploited, and HHO and FPOA are under-explored.

The challenges above are all tailored to specific metaheuristics. Moreover, several additional challenges remain for the parameter extraction problem.

1. TNFES is a sign of computational resources, yet its value is almost pitched at 50,000. Reducing TNFES without compromising accuracy is imperative.
2. More diodes in the cell model may increase the extraction accuracy. Recently, a four diode model was proposed [47] and the results showed good fitting effect. However, more diodes also indicate more parameters that need to be extracted and solutions are also more intractable. Hence, selecting a suitable PV model for an algorithm is challenging.
3. Some of the literature used too few PV cases to demonstrate metaheuristics' generalizability.
4. Metaheuristics' effectiveness is devoid of practical engineering applications.
5. More and exact measured data means more accurate extraction results, but obtaining sufficient high-precision measurements is challenging and costly.
6. In engineering, running time is pivotal. Hence, cutting running times is a challenge.
7. Multiple matrices are imperative to signal the competitiveness of metaheuristic results, yet some of the literature adopted few matrices for comparison.

4.6. Research Prospects

The previous section summarizes the challenges in studies, and this section suggests some research directions. They are an essential reference for researchers in developing their plans.

For specific metaheuristics:

1. Exploration techniques such as chaotic mapping and second-order oscillation mechanisms can be considered to incorporate into GA. They are envisaged to augment accuracy and robustness.
2. DE might be combined with exploitation-based metaheuristics, such as the Search Backtracking Algorithm, or with search mechanisms that accelerate the convergence. PSO demands more diversity-raising search mechanisms such as trust region reflection, taboo search, and fitness distance balance. Additionally, studies on adapting their parameters are well-tried.
3. ABC considers introducing neighborhood search and adaptive mechanisms to speed up the convergence.
4. For GWO and WOA, adaptive operators could be inserted to improve applicability in the face of high-dimensional issues.
5. JAYA and TLBO could borrow the exploration-type mechanisms in CSOOJAYA, MTLBO, and EBLSHADE to improve the overall performance.
6. Hybrid methods can identify contributing components through component analysis and remove unimportant components to alleviate implementation redundancy.
7. New methods can adopt adaptive learning, neighborhood search, chaotic mapping, and algorithmic blending techniques to enhance their behavior.

Regardless of the specific techniques, any approach to raise the metaheuristics is to employ complementary improvements to balance exploration and exploitation and, ultimately, fit to the studied issue.

In addition to research directions for metaheuristics, some potential directions for application scenarios include the following areas:

1. For the parameter extraction, diminishing computational resources' consumption is at stake. Reducing TNFES while maintaining the same accuracy by introducing different techniques, i.e., local search and reinforcement learning, is a direction worthy of further research.

2. Some methods are feasible for low-dimensional issues, and some deliver better performance for high-dimensional issues. Meanwhile, the selection of MSDM, MDDM, MTDM, and FDM with 6, 8, 10, and 11 unknown parameters to be included in the cell model is a future research direction for further performance improvement. Hence, it would be interesting to pick the right algorithm improvement to render PV models with desirable accuracy.
3. For the issue of too few employed cases, more cases are considered in future research to reveal the methods' generalizability. Examples include cases at different temperatures and irradiances and cases in partial shade.
4. The real-time extraction of PV models' parameters at different operating conditions is highly suggested. It is excellent work to accurately model the dynamics of photovoltaics for practical engineering problems.
5. Faced with the problem of little measured data, inserting deep learning techniques such as neural networks to eliminate erroneous data and expand the amount of data for metaheuristic methods is an effective way to facilitate the extraction accuracy.
6. The graphical processing unit (GPU) allows different solutions to be updated simultaneously to raise the efficiency. Thus, metaheuristic methods' speed improvements can be geared toward direct runtime reductions through GPU-like devices.
7. More performance evaluation indicators can demonstrate metaheuristic methods' overall effectiveness more comprehensively. Therefore, introducing more multifaceted indicators is necessary to enhance persuasiveness.

5. Conclusions

PV generation is playing a more significant role in the future energy landscape. Meanwhile, accurate PV models can support the PV systems' accurate assessment, efficiency improvement, fault analysis, and simulation. Thus, this paper reviewed different metaheuristics employed in the PV model parameters extraction. In our work, (a) the PV models and problem formulations were explained; (b) different metaheuristics and their developments and applications were summarized; (c) the algorithmic parameter settings, various evaluation indicators, independent running numbers, and computational resources (TNFES) were assembled; (d) the final results of various algorithms were compared, and especially RMSE and SIAE were ranked; (e) the unknown parameters and RMSE variation patterns in different environments were analyzed; and (f) a comprehensive analysis of the challenges encountered by metaheuristics in solving the studied issue was presented, and some ideas for future research were outlined.

This study can assist beginners in gaining an introductory and systematic perspective on the issue. It may also provide a reference direction for further research when unfamiliar researchers understand the application of metaheuristics to this engineering problem.

Author Contributions: Conceptualization, Z.G. and G.X.; methodology, G.X.; formal analysis, G.X. and X.F.; writing—original draft preparation, Z.G.; writing—review and editing, G.X. and X.F.; supervision, G.X. All authors have read and agreed to the published version of the manuscript.

Funding: This research was funded by the National Natural Science Foundation of China, grant number 52167007, the Natural Science Foundation of Guizhou Province, grant number QiankeheBasic-ZK [2022] General121, the Innovation Foundation of Guizhou University Institute of Engineering Investigation and Design Co., Ltd., grant number GuiDaKanCha [2022]03, and the Modern Power System and Its Digital Technology Engineering Research Center, grant number QianJiaoJi [2022]043.

Institutional Review Board Statement: Not applicable.

Informed Consent Statement: Not applicable.

Data Availability Statement: Not applicable.

Conflicts of Interest: The authors declare no conflict of interest.

References

1. Li, Y.; Chiu, Y.H.; Lin, T.Y. Research on New and Traditional Energy Sources in OECD Countries. *Int. J. Environ. Res. Public Health* **2019**, *16*, 1122. [CrossRef]
2. Ridha, H.M.; Heidari, A.A.; Wang, M.; Chen, H. Boosted mutation-based Harris hawks optimizer for parameters identification of single-diode solar cell models. *Energy Convers. Manag.* **2020**, *209*, 112660. [CrossRef]
3. You, V.; Kakinaka, M. Modern and traditional renewable energy sources and CO(2) emissions in emerging countries. *Environ. Sci. Pollut. Res.* **2022**, *29*, 17695–17708. [CrossRef] [PubMed]
4. Xiong, G.; Li, L.; Mohamed, A.W.; Yuan, X.; Zhang, J. A new method for parameter extraction of solar photovoltaic models using gaining–sharing knowledge based algorithm. *Energy Rep.* **2021**, *7*, 3286–3301. [CrossRef]
5. Lin, S.; Zhang, C.; Ding, L.; Zhang, J.; Liu, X.; Chen, G.; Wang, S.; Chai, J. Accurate Recognition of Building Rooftops and Assessment of Long-Term Carbon Emission Reduction from Rooftop Solar Photovoltaic Systems Fusing GF-2 and Multi-Source Data. *Remote Sens.* **2022**, *14*, 3144. [CrossRef]
6. Dixit, S. Solar technologies and their implementations: A review. *Mater. Today Proc.* **2020**, *28*, 2137–2148. [CrossRef]
7. Kant Bhatia, S.; Palai, A.K.; Kumar, A.; Kant Bhatia, R.; Kumar Patel, A.; Kumar Thakur, V.; Yang, Y.H. Trends in renewable energy production employing biomass-based biochar. *Bioresour. Technol.* **2021**, *340*, 125644. [CrossRef]
8. Qazi, A.; Hussain, F.; Rahim, N.A.B.D.; Hardaker, G.; Alghazzawi, D.; Shaban, K.; Haruna, K. Towards Sustainable Energy: A Systematic Review of Renewable Energy Sources, Technologies, and Public Opinions. *IEEE Access* **2019**, *7*, 63837–63851. [CrossRef]
9. Rojas, D.; Rivera, M.; Wheeler, P. Basic Principles of Solar Energy. In Proceedings of the 2021 IEEE CHILEAN Conference on Electrical, Electronics Engineering, Information and Communication Technologies (CHILECON), Valparaíso, Chile, 6–9 December 2021; pp. 1–6.
10. Saidur, R.; BoroumandJazi, G.; Mekhlif, S.; Jameel, M. Exergy analysis of solar energy applications. *Renew. Sustain. Energy Rev.* **2012**, *16*, 350–356. [CrossRef]
11. Zheng, Y.; Weng, Q. Modeling the Effect of Green Roof Systems and Photovoltaic Panels for Building Energy Savings to Mitigate Climate Change. *Remote Sens.* **2020**, *12*, 2402. [CrossRef]
12. Chen, D.-Y.; Peng, L.; Zhang, W.-Y.; Wang, Y.-D.; Yang, L.-N. Research on Self-Supervised Building Information Extraction with High-Resolution Remote Sensing Images for Photovoltaic Potential Evaluation. *Remote Sens.* **2022**, *14*, 5350. [CrossRef]
13. Humada, A.M.; Darweesh, S.Y.; Mohammed, K.G.; Kamil, M.; Mohammed, S.F.; Kasim, N.K.; Tahseen, T.A.; Awad, O.I.; Mekhilef, S. Modeling of PV system and parameter extraction based on experimental data: Review and investigation. *Sol. Energy* **2020**, *199*, 742–760. [CrossRef]
14. Kumari, P.A.; Geethanjali, P. Parameter estimation for photovoltaic system under normal and partial shading conditions: A survey. *Renew. Sustain. Energy Rev.* **2018**, *84*, 1–11. [CrossRef]
15. Pillai, D.S.; Rajasekar, N. Metaheuristic algorithms for PV parameter identification: A comprehensive review with an application to threshold setting for fault detection in PV systems. *Renew. Sustain. Energy Rev.* **2018**, *82*, 3503–3525. [CrossRef]
16. Humada, A.M.; Hojabri, M.; Mekhilef, S.; Hamada, H.M. Solar cell parameters extraction based on single and double-diode models: A review. *Renew. Sustain. Energy Rev.* **2016**, *56*, 494–509. [CrossRef]
17. Selem, S.I.; El-Fergany, A.A.; Hasanien, H.M. Artificial electric field algorithm to extract nine parameters of triple-diode photovoltaic model. *Int. J. Energy Res.* **2020**, *45*, 590–604. [CrossRef]
18. Xiong, G.; Zhang, J.; Shi, D.; Zhu, L.; Yuan, X.; Yao, G. Modified Search Strategies Assisted Crossover Whale Optimization Algorithm with Selection Operator for Parameter Extraction of Solar Photovoltaic Models. *Remote Sens.* **2019**, *11*, 2795. [CrossRef]
19. Abbassi, R.; Abbassi, A.; Jemli, M.; Chebbi, S. Identification of unknown parameters of solar cell models: A comprehensive overview of available approaches. *Renew. Sustain. Energy Rev.* **2018**, *90*, 453–474. [CrossRef]
20. Li, S.; Gong, W.; Gu, Q. A comprehensive survey on meta-heuristic algorithms for parameter extraction of photovoltaic models. *Renew. Sustain. Energy Rev.* **2021**, *141*, 110828. [CrossRef]
21. Xiong, G.; Zhang, J.; Shi, D.; Zhu, L.; Yuan, X. Parameter extraction of solar photovoltaic models with an either-or teaching learning based algorithm. *Energy Convers. Manag.* **2020**, *224*, 113395. [CrossRef]
22. Ibrahim, H.; Anani, N. Evaluation of Analytical Methods for Parameter Extraction of PV modules. *Energy Procedia* **2017**, *134*, 69–78. [CrossRef]
23. Chin, V.J.; Salam, Z. A New Three-point-based Approach for the Parameter Extraction of Photovoltaic Cells. *Appl. Energy* **2019**, *237*, 519–533.
24. Abbassi, A.; Gammoudi, R.; Ali Dami, M.; Hasnaoui, O.; Jemli, M. An improved single-diode model parameters extraction at different operating conditions with a view to modeling a photovoltaic generator: A comparative study. *Sol. Energy* **2017**, *155*, 478–489. [CrossRef]
25. Nguyen, H.; Nguyen, D.; Ngo, A.P.; Thomas, C. Solar PV Modeling with Lambert W Function: An Exponential Cone Programming Approach. In Proceedings of the 2022 IEEE Kansas Power and Energy Conference (KPEC), Manhattan, KS, USA, 25–26 April 2022; pp. 1–5.
26. Sharadga, H.; Hajimirza, S.; Cari, E.P.T. A Fast and Accurate Single-Diode Model for Photovoltaic Design. *IEEE J. Emerg. Sel. Top. Power Electron.* **2021**, *9*, 3030–3043. [CrossRef]

27. Tina, G.M.; Ventura, C.; Ferlito, S.; De Vito, S. A State-of-Art-Review on Machine-Learning Based Methods for PV. *Appl. Sci.* **2021**, *11*, 7550. [CrossRef]
28. Eslami, M.; Akbari, E.; Seyed Sadr, S.T.; Ibrahim, B.F. A novel hybrid algorithm based on rat swarm optimization and pattern search for parameter extraction of solar photovoltaic models. *Energy Sci. Eng.* **2022**, *10*, 2689–2713. [CrossRef]
29. Huang, T.; Zhang, C.; Ouyang, H.; Luo, G.; Li, S.; Zou, D. Parameter identification for photovoltaic models using an improved learning search algorithm. *IEEE Access* **2020**, *8*, 116292–116309. [CrossRef]
30. Rizk-Allah, R.M.; Hassanien, A.E. Locomotion-based Hybrid Salp Swarm Algorithm for Estimation of Fuzzy Representation-based Photovoltaic Modules. *J. Mod. Power Syst. Clean Energy* **2021**, *9*, 384–394. [CrossRef]
31. Yu, K.; Liang, J.J.; Qu, B.Y.; Cheng, Z.; Wang, H. Multiple learning backtracking search algorithm for estimating parameters of photovoltaic models. *Appl. Energy* **2018**, *226*, 408–422. [CrossRef]
32. Oliva, D.; Elaziz, M.A.; Elsheikh, A.H.; Ewees, A.A. A review on meta-heuristics methods for estimating parameters of solar cells. *J. Power Sources* **2019**, *435*, 126683. [CrossRef]
33. Venkateswari, R.; Rajasekar, N. Review on parameter estimation techniques of solar photovoltaic systems. *Int. Trans. Electr. Energy Syst.* **2021**, *31*, e13113. [CrossRef]
34. Chen, Z.; Lin, W.; Wu, L.; Long, C.; Peijie, L.; Cheng, S. A capacitor based fast I-V characteristics tester for photovoltaic arrays. *Energy Procedia* **2018**, *145*, 381–387. [CrossRef]
35. Toledo, F.J.; Blanes, J.M. Geometric properties of the single-diode photovoltaic model and a new very simple method for parameters extraction. *Renew. Energy* **2014**, *72*, 125–133. [CrossRef]
36. Li, D.; Yang, B.; Li, L.; Li, Q.; Deng, J.; Guo, C. Recent Photovoltaic Cell Parameter Identification Approaches: A Critical Note. *Front. Energy Res.* **2022**, *10*, 902749. [CrossRef]
37. Younis, A.; Bakhit, A.; Onsa, M.; Hashim, M. A comprehensive and critical review of bio-inspired metaheuristic frameworks for extracting parameters of solar cell single and double diode models. *Energy Rep.* **2022**, *8*, 7085–7106. [CrossRef]
38. Sun, L.; Wang, J.; Tang, L. A Powerful Bio-Inspired Optimization Algorithm Based PV Cells Diode Models Parameter Estimation. *Front. Energy Res.* **2021**, *9*, 675925. [CrossRef]
39. Shaheen, A.M.; El-Seheimy, R.A.; Xiong, G.; Elattar, E.; Ginidi, A.R. Parameter identification of solar photovoltaic cell and module models via supply demand optimizer. *Ain Shams Eng. J.* **2022**, *13*, 101705. [CrossRef]
40. Qais, M.H.; Hasanien, H.M.; Alghuwainem, S. Parameters extraction of three-diode photovoltaic model using computation and Harris Hawks optimization. *Energy* **2020**, *195*, 117040. [CrossRef]
41. Hu, Z.; Gong, W.; Li, S. Reinforcement learning-based differential evolution for parameters extraction of photovoltaic models. *Energy Rep.* **2021**, *7*, 916–928. [CrossRef]
42. Jordehi, A.R. Parameter estimation of solar photovoltaic (PV) cells: A review. *Renew. Sustain. Energy Rev.* **2016**, *61*, 354–371. [CrossRef]
43. Nishioka, K.; Sakitani, N.; Uraoka, Y.; Fuyuki, T. Analysis of multicrystalline silicon solar cells by modified 3-diode equivalent circuit model taking leakage current through periphery into consideration. *Sol. Energy Mater. Sol. Cells* **2007**, *91*, 1222–1227. [CrossRef]
44. Suskis, P.; Galkin, I. Enhanced Photovoltaic Panel Model for MATLAB-Simulink Environment Considering Solar Cell Junction Capacitance. In Proceedings of the 2013 39th Annual Conference of the IEEE Industrial Electronics Society (IECON), Vienna, Austria, 10–13 November 2013; pp. 1613–1618.
45. Soon, J.J.; Low, K.-S. Optimizing Photovoltaic Model for Different Cell Technologies Using a Generalized Multidimension Diode Model. *IEEE Trans. Ind. Electron.* **2015**, *62*, 6371–6380. [CrossRef]
46. Rawa, M.; Calasan, M.; Abusorrah, A.; Alhussainy, A.A.; Al-Turki, Y.; Ali, Z.M.; Sindi, H.; Mekhilef, S.; Aleem, S.H.E.A.; Bassi, H. Single Diode Solar Cells—Improved Model and Exact Current–Voltage Analytical Solution Based on Lambert's W Function. *Sensors* **2022**, *22*, 4173. [CrossRef] [PubMed]
47. Singh, B.; Singla, M.K.; Nijhawan, P. Parameter estimation of four diode solar photovoltaic cell using hybrid algorithm. *Energy Sources Part A Recovery Util. Environ. Eff.* **2022**, *44*, 4597–4613. [CrossRef]
48. Ramadan, A.; Kamel, S.; Hassan, M.H.; Khurshaid, T.; Rahmann, C. An Improved Bald Eagle Search Algorithm for Parameter Estimation of Different Photovoltaic Models. *Processes* **2021**, *9*, 1127. [CrossRef]
49. Abdelminaam, D.S.; Said, M.; Houssein, E.H. Turbulent Flow of Water-Based Optimization Using New Objective Function for Parameter Extraction of Six Photovoltaic Models. *IEEE Access* **2021**, *9*, 35382–35398. [CrossRef]
50. Di Piazza, M.C.; Luna, M.; Vitale, G. Dynamic PV Model Parameter Identification by Least-Squares Regression. *IEEE J. Photovolt.* **2013**, *3*, 799–806. [CrossRef]
51. Abd Elaziz, M.; Thanikanti, S.B.; Ibrahim, I.A.; Lu, S.; Nastasi, B.; Alotaibi, M.A.; Hossain, M.A.; Yousri, D. Enhanced Marine Predators Algorithm for identifying static and dynamic Photovoltaic models parameters. *Energy Convers. Manag.* **2021**, *236*, 113971. [CrossRef]
52. Yousri, D.; Allam, D.; Eteiba, M.B.; Suganthan, P.N. Static and dynamic photovoltaic models' parameters identification using Chaotic Heterogeneous Comprehensive Learning Particle Swarm Optimizer variants. *Energy Convers. Manag.* **2019**, *182*, 546–563. [CrossRef]
53. Wang, S.; Yu, Y.; Hu, W. Static and dynamic solar photovoltaic models' parameters estimation using hybrid Rao optimization algorithm. *J. Clean. Prod.* **2021**, *315*, 128080. [CrossRef]

54. Ginidi, A.R.; Shaheen, A.M.; El-Sehiemy, R.A.; Elattar, E. Supply demand optimization algorithm for parameter extraction of various solar cell models. *Energy Rep.* **2021**, *7*, 5772–5794. [CrossRef]
55. Chen, H.; Jiao, S.; Wang, M.; Heidari, A.A.; Zhao, X. Parameters identification of photovoltaic cells and modules using diversification-enriched Harris hawks optimization with chaotic drifts. *J. Clean. Prod.* **2020**, *244*, 118778. [CrossRef]
56. Song, S.; Wang, P.; Heidari, A.A.; Zhao, X.; Chen, H. Adaptive Harris hawks optimization with persistent trigonometric differences for photovoltaic model parameter extraction. *Eng. Appl. Artif. Intell.* **2022**, *109*, 104608. [CrossRef]
57. Ridha, H.M.; Hizam, H.; Mirjalili, S.; Othman, M.L.; Ya'acob, M.E.; Ahmadipour, M.; Ismaeel, N.Q. On the problem formulation for parameter extraction of the photovoltaic model: Novel integration of hybrid evolutionary algorithm and Levenberg Marquardt based on adaptive damping parameter formula. *Energy Convers. Manag.* **2022**, *256*, 115403. [CrossRef]
58. Mohammed Ridha, H.; Hizam, H.; Mirjalili, S.; Lutfi Othman, M.; Effendy Ya'acob, M.; Ahmadipour, M. Novel parameter extraction for Single, Double, and three diodes photovoltaic models based on robust adaptive arithmetic optimization algorithm and adaptive damping method of Berndt-Hall-Hall-Hausman. *Sol. Energy* **2022**, *243*, 35–61. [CrossRef]
59. Xiong, G.; Zhang, J.; Shi, D.; Zhu, L.; Yuan, X.; Tan, Z. Winner-leading competitive swarm optimizer with dynamic Gaussian mutation for parameter extraction of solar photovoltaic models. *Energy Convers. Manag.* **2020**, *206*, 112450. [CrossRef]
60. Ortiz-Conde, A.; Trejo, O.; Garcia-Sanchez, F.J. Direct Extraction of Solar Cell Model Parameters Using Optimization Methods. In Proceedings of the 2021 IEEE Latin America Electron Devices Conference (LAEDC), Mexico, Mexico, 19–21 April 2021; pp. 1–6.
61. Biswas, P.P.; Suganthan, P.N.; Wu, G.; Amaratunga, G.A.J. Parameter estimation of solar cells using datasheet information with the application of an adaptive differential evolution algorithm. *Renew. Energy* **2019**, *132*, 425–438. [CrossRef]
62. Katoch, S.; Chauhan, S.S.; Kumar, V. A review on genetic algorithm: Past, present, and future. *Multimed Tools Appl.* **2021**, *80*, 8091–8126. [CrossRef] [PubMed]
63. Harrag, A.; Messalti, S. Extraction of Solar Cell Parameters Using Genetic Algorithm. In Proceedings of the 2015 4th International Conference on Electrical Engineering (ICEE), Boumerdes, Algeria, 13–15 December 2015; pp. 13–15.
64. Kumari, P.A.; Geethanjali, P. Adaptive Genetic Algorithm Based Multi-Objective Optimization for Photovoltaic Cell Design Parameter Extraction. *Energy Procedia* **2017**, *117*, 432–441. [CrossRef]
65. Wang, L.; Chen, Z.; Guo, Y.; Hu, W.; Chang, X.; Wu, P.; Han, C.; Li, J. Accurate Solar Cell Modeling via Genetic Neural Network-Based Meta-Heuristic Algorithms. *Front. Energy Res.* **2021**, *9*, 696204. [CrossRef]
66. Storn, R.; Price, K. Differential Evolution-A Simple and Efficient Heuristic for Global Optimization over Continuous Spaces. *J. Glob. Optim.* **1997**, *11*, 341–359. [CrossRef]
67. Luo, Q.; Peng, W.; Wu, G.; Xiao, Y. Orbital Maneuver Optimization of Earth Observation Satellites Using an Adaptive Differential Evolution Algorithm. *Remote Sens.* **2022**, *14*, 1966. [CrossRef]
68. Jiang, L.L.; Maskell, D.L.; Patra, J.C. Parameter estimation of solar cells and modules using an improved adaptive differential evolution algorithm. *Appl. Energy* **2013**, *112*, 185–193. [CrossRef]
69. Li, S.; Gu, Q.; Gong, W.; Ning, B. An enhanced adaptive differential evolution algorithm for parameter extraction of photovoltaic models. *Energy Convers. Manag.* **2020**, *205*, 112443. [CrossRef]
70. Xiong, G.; Zhang, J.; Shi, D.; Zhu, L.; Yuan, X. Parameter extraction of solar photovoltaic models via quadratic interpolation learning differential evolution. *Sustain. Energy Fuels* **2020**, *4*, 5595–5608. [CrossRef]
71. Song, Y.; Wu, D.; Wagdy Mohamed, A.; Zhou, X.; Zhang, B.; Deng, W.; Khalil, A.M. Enhanced Success History Adaptive DE for Parameter Optimization of Photovoltaic Models. *Complexity* **2021**, *2021*, 6660115. [CrossRef]
72. Parida, S.M.; Rout, P.K. Differential evolution with dynamic control factors for parameter estimation of photovoltaic models. *J. Comput. Electron.* **2021**, *20*, 330–343. [CrossRef]
73. Gao, S.; Wang, K.; Tao, S.; Jin, T.; Dai, H.; Cheng, J. A state-of-the-art differential evolution algorithm for parameter estimation of solar photovoltaic models. *Energy Convers. Manag.* **2021**, *230*, 113784. [CrossRef]
74. Wang, D.; Sun, X.; Kang, H.; Shen, Y.; Chen, Q. Heterogeneous differential evolution algorithm for parameter estimation of solar photovoltaic models. *Energy Rep.* **2022**, *8*, 4724–4746. [CrossRef]
75. Kharchouf, Y.; Herbazi, R.; Chahboun, A. Parameter's extraction of solar photovoltaic models using an improved differential evolution algorithm. *Energy Convers. Manag.* **2022**, *251*, 114972. [CrossRef]
76. Dang, J.; Wang, G.; Xia, C.; Jia, R.; Li, P. Research on the parameter identification of PV module based on fuzzy adaptive differential evolution algorithm. *Energy Rep.* **2022**, *8*, 12081–12091. [CrossRef]
77. Zhang, Y.; Wang, S.; Ji, G. A Comprehensive Survey on Particle Swarm Optimization Algorithm and Its Applications. *Math. Probl. Eng.* **2015**, *2015*, 931256. [CrossRef]
78. Dong, C.; Meng, X.; Guo, L.; Hu, J. 3D Sea Surface Electromagnetic Scattering Prediction Model Based on IPSO-SVR. *Remote Sens.* **2022**, *14*, 4657. [CrossRef]
79. Ben Hmamou, D.; Elyaqouti, M.; Arjdal, E.; Chaoufi, J.; Saadaoui, D.; Lidaighbi, S.; Aqel, R. Particle swarm optimization approach to determine all parameters of the photovoltaic cell. *Mater. Today Proc.* **2022**, *52*, 7–12. [CrossRef]
80. Ni, B.; Zou, P.; Chen, Y.; Zhang, Z. Identification of Solar Cell Model Parameters based on PSO with Adaptive Elite Mutation. In Proceedings of the 2018 Chinese Automation Congress (CAC), Xi'an, China, 30 November–2 December 2018; pp. 1340–1344.
81. Merchaoui, M.; Sakly, A.; Mimouni, M.F. Particle swarm optimisation with adaptive mutation strategy for photovoltaic solar cell/module parameter extraction. *Energy Convers. Manag.* **2018**, *175*, 151–163. [CrossRef]

82. Nunes, H.G.G.; Pombo, J.A.N.; Mariano, S.J.P.S.; Calado, M.R.A.; Felippe de Souza, J.A.M. A new high performance method for determining the parameters of PV cells and modules based on guaranteed convergence particle swarm optimization. *Appl. Energy* **2018**, *211*, 774–791. [CrossRef]
83. Rezaee Jordehi, A. Enhanced leader particle swarm optimisation (ELPSO): An efficient algorithm for parameter estimation of photovoltaic (PV) cells and modules. *Sol. Energy* **2018**, *159*, 78–87. [CrossRef]
84. Kiani, A.T.; Nadeem, M.F.; Ahmed, A.; Sajjad, I.A.; Haris, M.S.; Martirano, L. Optimal Parameter Estimation of Solar Cell using Simulated Annealing Inertia Weight Particle Swarm Optimization (SAIW-PSO). In Proceedings of the 2020 IEEE International Conference on Environment and Electrical Engineering and 2020 IEEE Industrial and Commercial Power Systems Europe (EEEIC/I&CPS Europe), Madrid, Spain, 9–12 June 2020; pp. 1–6.
85. Kiani, A.T.; Nadeem, M.F.; Ahmed, A.; Khan, I.; Elavarasan, R.M.; Das, N. Optimal PV Parameter Estimation via Double Exponential Function-Based Dynamic Inertia Weight Particle Swarm Optimization. *Energies* **2020**, *13*, 4037. [CrossRef]
86. Gao, S.; Xiang, C.; Lee, T.H. Highly Efficient Photovoltaic Parameter Estimation Using Parallel Particle Swarm Optimization on a GPU. In Proceedings of the 2021 IEEE 30th International Symposium on Industrial Electronics (ISIE), Kyoto, Japan, 20–23 June 2021; pp. 1–7.
87. Kiani, A.T.; Nadeem, M.F.; Ahmed, A.; Khan, I.A.; Alkhammash, H.I.; Sajjad, I.A.; Hussain, B. An Improved Particle Swarm Optimization with Chaotic Inertia Weight and Acceleration Coefficients for Optimal Extraction of PV Models Parameters. *Energies* **2021**, *14*, 2980. [CrossRef]
88. Fan, Y.; Wang, P.; Heidari, A.A.; Chen, H.; HamzaTurabieh; Mafarja, M. Random reselection particle swarm optimization for optimal design of solar photovoltaic modules. *Energy* **2022**, *239*, 121865. [CrossRef]
89. Karaboga, D.; Basturk, B. A powerful and efficient algorithm for numerical function optimization: Artificial bee colony (ABC) algorithm. *J. Glob. Optim.* **2007**, *39*, 459–471. [CrossRef]
90. Yan, J.; Chen, Y.; Zheng, J.; Guo, L.; Zheng, S.; Zhang, R. Multi-Source Time Series Remote Sensing Feature Selection and Urban Forest Extraction Based on Improved Artificial Bee Colony. *Remote Sens.* **2022**, *14*, 4859. [CrossRef]
91. Chen, X.; Xu, B.; Mei, C.; Ding, Y.; Li, K. Teaching–learning–based artificial bee colony for solar photovoltaic parameter estimation. *Appl. Energy* **2018**, *212*, 1578–1588. [CrossRef]
92. Wu, L.; Chen, Z.; Long, C.; Cheng, S.; Lin, P.; Chen, Y.; Chen, H. Parameter extraction of photovoltaic models from measured I-V characteristics curves using a hybrid trust-region reflective algorithm. *Appl. Energy* **2018**, *232*, 36–53. [CrossRef]
93. Xu, L.; Bai, L.; Bao, H.; Jiang, J. Parameter Identification of Solar Cell Model Based on Improved Artificial Bee Colony Algorithm. In Proceedings of the 2021 13th International Conference on Advanced Computational Intelligence (ICACI), Wanzhou, China, 14–16 May 2021; pp. 239–244.
94. Tefek, M.f.F. Artificial bee colony algorithm based on a new local search approach for parameter estimation of photovoltaic systems. *J. Comput. Electron.* **2021**, *20*, 2530–2562. [CrossRef]
95. Garoudja, E.; Filali, W. Photovoltaic Module Parameters Extraction Using Best-so-Far ABC Algorithm. In Proceedings of the 2019 International Conference on Advanced Electrical Engineering (ICAEE), Algiers, Algeria, 19–21 November 2019; pp. 1–5.
96. Duman, S.; Kahraman, H.T.; Sonmez, Y.; Guvenc, U.; Kati, M.; Aras, S. A powerful meta-heuristic search algorithm for solving global optimization and real-world solar photovoltaic parameter estimation problems. *Eng. Appl. Artif. Intell.* **2022**, *111*, 104763. [CrossRef]
97. Faris, H.; Aljarah, I.; Al-Betar, M.A.; Mirjalili, S. Grey wolf optimizer: A review of recent variants and applications. *Neural Comput. Appl.* **2017**, *30*, 413–435. [CrossRef]
98. Vinod, A.; Sinha, A.K. Estimation of parameters for one diode solar PV cell using grey wolf optimizer to obtain exact V-I characteristics. *J. Eng. Res.* **2019**, *7*, 1–19.
99. AlShabi, M.; Ghenai, C.; Bettayeb, M.; Ahmad, F.F.; El Haj Assad, M. Multi-group grey wolf optimizer (MG-GWO) for estimating photovoltaic solar cell model. *J. Therm. Anal. Calorim.* **2020**, *144*, 1655–1670. [CrossRef]
100. Xavier, F.J.; Pradeep, A.; Premkumar, M.; Kumar, C. Orthogonal Learning-Based Gray Wolf Optimizer for Identifying the Uncertain Parameters of Various Photovoltaic Models. *Optik* **2021**, *247*, 167973. [CrossRef]
101. Yesilbudak, M. Parameter Extraction of Photovoltaic Cells and Modules Using Grey Wolf Optimizer with Dimension Learning-Based Hunting Search Strategy. *Energies* **2021**, *14*, 5735. [CrossRef]
102. Ramadan, A.-E.; Kamel, S.; Khurshaid, T.; Oh, S.-R.; Rhee, S.-B. Parameter Extraction of Three Diode Solar Photovoltaic Model Using Improved Grey Wolf Optimizer. *Sustainability* **2021**, *13*, 6963. [CrossRef]
103. Zitar, R.A.; Al-Betar, M.A.; Awadallah, M.A.; Doush, I.A.; Assaleh, K. An Intensive and Comprehensive Overview of JAYA Algorithm, its Versions and Applications. *Arch. Comput. Methods Eng.* **2022**, *29*, 763–792. [CrossRef] [PubMed]
104. Yu, K.; Liang, J.J.; Qu, B.Y.; Chen, X.; Wang, H. Parameters identification of photovoltaic models using an improved JAYA optimization algorithm. *Energy Convers. Manag.* **2017**, *150*, 742–753. [CrossRef]
105. Wang, L.; Huang, C. A novel Elite Opposition-based Jaya algorithm for parameter estimation of photovoltaic cell models. *Optik* **2018**, *155*, 351–356. [CrossRef]
106. Luo, X.; Cao, L.; Wang, L.; Zhao, Z.; Huang, C. Parameter identification of the photovoltaic cell model with a hybrid Jaya-NM algorithm. *Optik* **2018**, *171*, 200–203. [CrossRef]
107. Yu, K.; Qu, B.; Yue, C.; Ge, S.; Chen, X.; Liang, J. A performance-guided JAYA algorithm for parameters identification of photovoltaic cell and module. *Appl. Energy* **2019**, *237*, 241–257. [CrossRef]

108. Luu, T.V.; Nguyen, N.S. Parameters extraction of solar cells using modified JAYA algorithm. *Optik* **2020**, *203*, 164034. [CrossRef]
109. Jian, X.; Weng, Z. A logistic chaotic JAYA algorithm for parameters identification of photovoltaic cell and module models. *Optik* **2020**, *203*, 164041. [CrossRef]
110. Zhang, Y.; Ma, M.; Jin, Z. Comprehensive learning Jaya algorithm for parameter extraction of photovoltaic models. *Energy* **2020**, *211*, 118644. [CrossRef]
111. Yang, X.; Gong, W. Opposition-based JAYA with population reduction for parameter estimation of photovoltaic solar cells and modules. *Appl. Soft Comput.* **2021**, *104*, 107218. [CrossRef]
112. Premkumar, M.; Jangir, P.; Sowmya, R.; Elavarasan, R.M.; Kumar, B.S. Enhanced chaotic JAYA algorithm for parameter estimation of photovoltaic cell/modules. *ISA Trans.* **2021**, *116*, 139–166. [CrossRef] [PubMed]
113. Saadaoui, D.; Elyaqouti, M.; Assalaou, K.; hmamou, D.B.; Lidaighbi, S. Multiple learning JAYA algorithm for parameters identifying of photovoltaic models. *Mater. Today Proc.* **2022**, *52*, 108–123. [CrossRef]
114. Jian, X.; Cao, Y. A Chaotic Second Order Oscillation JAYA Algorithm for Parameter Extraction of Photovoltaic Models. *Photonics* **2022**, *9*, 131. [CrossRef]
115. Rao, R.V.; Savsani, V.J.; Vakharia, D.P. Teaching–learning-based optimization: A novel method for constrained mechanical design optimization problems. *Comput.-Aided Des.* **2011**, *43*, 303–315. [CrossRef]
116. Chen, X.; Yu, K.; Du, W.; Zhao, W.; Liu, G. Parameters identification of solar cell models using generalized oppositional teaching learning based optimization. *Energy* **2016**, *99*, 170–180. [CrossRef]
117. Yu, K.; Chen, X.; Wang, X.; Wang, Z. Parameters identification of photovoltaic models using self-adaptive teaching-learning-based optimization. *Energy Convers. Manag.* **2017**, *145*, 233–246. [CrossRef]
118. Ramadan, A.; Kamel, S.; Korashy, A.; Yu, J. Photovoltaic Cells Parameter Estimation Using an Enhanced Teaching–Learning-Based Optimization Algorithm. *Iran. J. Sci. Technol. Trans. Electr. Eng.* **2019**, *44*, 767–779. [CrossRef]
119. Abdel-Basset, M.; Mohamed, R.; Chakrabortty, R.K.; Sallam, K.; Ryan, M.J. An efficient teaching-learning-based optimization algorithm for parameters identification of photovoltaic models: Analysis and validations. *Energy Convers. Manag.* **2021**, *227*, 113614. [CrossRef]
120. Li, L.; Xiong, G.; Yuan, X.; Zhang, J.; Chen, J. Parameter Extraction of Photovoltaic Models Using a Dynamic Self-Adaptive and Mutual- Comparison Teaching-Learning-Based Optimization. *IEEE Access* **2021**, *9*, 52425–52441. [CrossRef]
121. Mohammed, H.M.; Umar, S.U.; Rashid, T.A. A Systematic and Meta-Analysis Survey of Whale Optimization Algorithm. *Comput. Intell. Neurosci.* **2019**, *2019*, 8718571. [CrossRef]
122. Li, H.; Chang, J.; Xu, F.; Liu, Z.; Yang, Z.; Zhang, L.; Zhang, S.; Mao, R.; Dou, X.; Liu, B. Efficient Lidar Signal Denoising Algorithm Using Variational Mode Decomposition Combined with a Whale Optimization Algorithm. *Remote Sens.* **2019**, *11*, 126. [CrossRef]
123. Xiong, G.; Zhang, J.; Shi, D.; He, Y. Parameter extraction of solar photovoltaic models using an improved whale optimization algorithm. *Energy Convers. Manag.* **2018**, *174*, 388–405. [CrossRef]
124. Elazab, O.S.; Hasanien, H.M.; Elgendy, M.A.; Abdeen, A.M. Parameters estimation of single- and multiple-diode photovoltaic model using whale optimisation algorithm. *IET Renew. Power Gener.* **2018**, *12*, 1755–1761. [CrossRef]
125. Pourmousa, N.; Ebrahimi, S.M.; Malekzadeh, M.; Gordillo, F. Using a novel optimization algorithm for parameter extraction of photovoltaic cells and modules. *Eur. Phys. J. Plus* **2021**, *136*, 470. [CrossRef]
126. Peng, L.; He, C.; Heidari, A.A.; Zhang, Q.; Chen, H.; Liang, G.; Aljehane, N.O.; Mansour, R.F. Information sharing search boosted whale optimizer with Nelder Mead simplex for parameter estimation of photovoltaic models. *Energy Convers. Manag.* **2022**, *270*, 116246. [CrossRef]
127. Xiong, G.; Zhang, J.; Yuan, X.; Shi, D.; He, Y.; Yao, G. Parameter extraction of solar photovoltaic models by means of a hybrid differential evolution with whale optimization algorithm. *Sol. Energy* **2018**, *176*, 742–761. [CrossRef]
128. Long, W.; Cai, S.; Jiao, J.; Xu, M.; Wu, T. A new hybrid algorithm based on grey wolf optimizer and cuckoo search for parameter extraction of solar photovoltaic models. *Energy Convers. Manag.* **2020**, *203*, 112243. [CrossRef]
129. Rezk, H.; Arfaoui, J.; Gomaa, M.R. Optimal Parameter Estimation of Solar PV Panel Based on Hybrid Particle Swarm and Grey Wolf Optimization Algorithms. *Int. J. Interact. Multimed. Artif. Intell.* **2021**, *6*, 145. [CrossRef]
130. Li, S.; Gong, W.; Wang, L.; Yan, X.; Hu, C. A hybrid adaptive teaching–learning-based optimization and differential evolution for parameter identification of photovoltaic models. *Energy Convers. Manag.* **2020**, *225*, 113474. [CrossRef]
131. Ndi, F.E.; Perabi, S.N.; Ndjakomo, S.E.; Abessolo, G.O. Harris Hawk Optimization Combined with Differential Evolution for the Estimation of Solar Cell Parameter. *Int. J. Photoenergy* **2022**, *2022*, 7021658. [CrossRef]
132. Yu, X.; Wu, X.; Luo, W. Parameter Identification of Photovoltaic Models by Hybrid Adaptive JAYA Algorithm. *Mathematics* **2022**, *10*, 183. [CrossRef]
133. Devarapalli, R.; Rao, B.V.; Al-Durra, A. Optimal parameter assessment of Solar Photovoltaic module equivalent circuit using a novel enhanced hybrid GWO-SCA algorithm. *Energy Rep.* **2022**, *8*, 12282–12301. [CrossRef]
134. Weng, X.; Liu, Y.; Heidari, A.A.; Cai, Z.; Lin, H.; Chen, H.; Liang, G.; Alsufyani, A.; Bourouis, S. Boosted backtracking search optimization with information exchange for photovoltaic system evaluation. *Energy Sci. Eng.* **2022**, *11*, 267–298. [CrossRef]
135. Naeijian, M.; Rahimnejad, A.; Ebrahimi, S.M.; Pourmousa, N.; Gadsden, S.A. Parameter estimation of PV solar cells and modules using Whippy Harris Hawks Optimization Algorithm. *Energy Rep.* **2021**, *7*, 4047–4063. [CrossRef]
136. Sallam, K.M.; Hossain, M.A.; Chakrabortty, R.K.; Ryan, M.J. An improved gaining-sharing knowledge algorithm for parameter extraction of photovoltaic models. *Energy Convers. Manag.* **2021**, *237*, 114030. [CrossRef]

137. Xiong, G.; Zhang, J.; Shi, D.; Yuan, X. Application of Supply-Demand-Based Optimization for Parameter Extraction of Solar Photovoltaic Models. *Complexity* **2019**, *2019*, 3923691. [CrossRef]
138. Diab, A.A.Z.; Sultan, H.M.; Aljendy, R.; Al-Sumaiti, A.S.; Shoyama, M.; Ali, Z.M. Tree Growth Based Optimization Algorithm for Parameter Extraction of Different Models of Photovoltaic Cells and Modules. *IEEE Access* **2020**, *8*, 119668–119687. [CrossRef]
139. Abbassi, R.; Abbassi, A.; Heidari, A.A.; Mirjalili, S. An efficient salp swarm-inspired algorithm for parameters identification of photovoltaic cell models. *Energy Convers. Manag.* **2019**, *179*, 362–372. [CrossRef]
140. Sharma, A.; Dasgotra, A.; Tiwari, S.K.; Sharma, A.; Jately, V.; Azzopardi, B. Parameter Extraction of Photovoltaic Module Using Tunicate Swarm Algorithm. *Electronics* **2021**, *10*, 878. [CrossRef]
141. Gupta, J.; Nijhawan, P.; Ganguli, S. Parameter extraction of solar PV cell models using novel metaheuristic chaotic tunicate swarm algorithm. *Int. Trans. Electr. Energy Syst.* **2021**, *31*, 13244. [CrossRef]
142. Ramadan, A.; Kamel, S.; Hussein, M.M.; Hassan, M.H. A New Application of Chaos Game Optimization Algorithm for Parameters Extraction of Three Diode Photovoltaic Model. *IEEE Access* **2021**, *9*, 51582–51594. [CrossRef]
143. Long, W.; Jiao, J.; Liang, X.; Xu, M.; Tang, M.; Cai, S. Parameters estimation of photovoltaic models using a novel hybrid seagull optimization algorithm. *Energy* **2022**, *249*, 123760. [CrossRef]
144. Shaban, H.; Houssein, E.H.; Pérez-Cisneros, M.; Oliva, D.; Hassan, A.Y.; Ismaeel, A.A.K.; AbdElminaam, D.S.; Deb, S.; Said, M. Identification of Parameters in Photovoltaic Models through a Runge Kutta Optimizer. *Mathematics* **2021**, *9*, 2313. [CrossRef]
145. Chellaswamy, C.; Taha; Mohammed, S.; Rajasree, R.Y.; Mohammad, J.; Gulshan, S. A Novel Optimization Method for Parameter Extraction of Industrial Solar Cells. In Proceedings of the 2019 Innovations in Power and Advanced Computing Technologies (i-PACT), Vellore, India, 22–23 March 2019; pp. 1–6.
146. Xiong, G.; Zhang, J.; Yuan, X.; Shi, D.; He, Y. Application of Symbiotic Organisms Search Algorithm for Parameter Extraction of Solar Cell Models. *Appl. Sci.* **2018**, *8*, 2155. [CrossRef]
147. Chellaswamy, C.; Ramesh, R. Parameter extraction of solar cell models based on adaptive differential evolution algorithm. *Renew. Energy* **2016**, *97*, 823–837. [CrossRef]
148. Heidari, A.A.; Mirjalili, S.; Faris, H.; Aljarah, I.; Mafarja, M.; Chen, H. Harris hawks optimization: Algorithm and applications. *Future Gener. Comput. Syst.* **2019**, *97*, 849–872. [CrossRef]

sustainability MDPI

Article

A New Hybrid Multi-Population GTO-BWO Approach for Parameter Estimation of Photovoltaic Cells and Modules

Hossam Hassan Ali [1], Mohamed Ebeed [2], Ahmed Fathy [3,4,*], Francisco Jurado [5], Thanikanti Sudhakar Babu [6] and Alaa A. Mahmoud [1]

1 Electrical Department, Faculty of Technology and Education, Sohag University, Sohag 82524, Egypt; hossam_hassan123@techedu.sohag.edu.eg (H.H.A.); alaa-abd-el_samee@techedu.sohag.edu.eg (A.A.M.)
2 Department of Electrical Engineering, Faculty of Engineering, Sohag University, Sohag 82524, Egypt; mebeed@eng.sohag.edu.eg
3 Electrical Engineering Department, Faculty of Engineering, Jouf University, Sakaka 72388, Saudi Arabia
4 Electrical Power & Machine Department, Faculty of Engineering, Zagazig University, Zagazig 44519, Egypt
5 Department of Electrical Engineering, University of Jaén, 23700 Linares, Spain; fjurado@ujaen.es
6 Department of Electrical and Electronics Engineering, Chaitanya Bharathi Institute of Technology, Hyderabad 500075, India; sudhakarbabu@ieee.org
* Correspondence: afali@ju.edu.sa

Abstract: Modeling the photovoltaic (PV) generating unit is one of the most important and crucial tasks when assessing the accurate performance of the PV system in power systems. The modeling of the PV system refers to the assigning of the optimal parameters of the PV's equivalent circuit. Identifying these parameters is considered to be a complex optimization problem, especially with the deviation of the solar irradiance and the ambient temperature. In this regard, this paper proposes a novel hybrid multi-population gorilla troops optimizer and beluga whale optimization (HMGTO-BWO) model to evaluate the optimal parameters of the PV cell/panel; it is based on a multi-population strategy to improve its diversity and to avoid the stagnation of the conventional GTO. The BWO explorative and exploitative powers, which are based on synchronized motion and Lévy flight, are used. The suggested HGTO-BWO is implemented to minimize the root mean square error (RMSE) between the simulated and measured data for each cell/panel represented by a double diode model (DDM) and triple diode model (TDM). The proposed HGTO-BWO is investigated according to the standard and CEC-2019 benchmark functions, and the obtained results are compared with seven other optimization techniques in terms of statistical analysis, convergence characteristics, boxplots, and the Wilcoxon rank sum test. The minimum obtained RMSE values of the PVW 752 cell were 2.0886×10^{-4} and 1.527×10^{-1} for the DDM and TDM, respectively. Furthermore, the minimum fetched fitness value for the STM6-40/36 modules was 1.8032×10^{-3}. The obtained results proved the effectiveness and preference of the suggested HGTO-BWO in estimating the parameters of the PV modules.

Keywords: multi-population; HGTO-BWO; parameters estimation; PV cell/panel

check for updates

Citation: Ali, H.H.; Ebeed, M.; Fathy, A.; Jurado, F.; Babu, T.S.; A. Mahmoud, A. A New Hybrid Multi-Population GTO-BWO Approach for Parameter Estimation of Photovoltaic Cells and Modules. *Sustainability* **2023**, *15*, 11089. https://doi.org/10.3390/su151411089

Academic Editor: Idiano D'Adamo

Received: 29 June 2023
Revised: 13 July 2023
Accepted: 14 July 2023
Published: 16 July 2023

1. Introduction

Renewable energy sources (RESs) like wind and solar should be considered in order to mitigate the effects of climate change and rising temperatures, as well as to protect the planet from the pollution and destruction produced by traditional fossil energy [1]. The process of ecological transition involves identifying consumption and sustainable community models to reduce harmful emissions and to create reliance on power generation from renewable sources [2]. One of the aims of the sustainable development goals (SDGs), especially the seventh goal, is to obtain modern energy which is sustainable and highly reliable at the lowest cost [3]. There is a great deal of interest in RESs due to the enormous financial and environmental problems associated with traditional energy sources like fossil fuels. It is essential to transform the solar energy into different forms that may be utilized in daily life with the assistance of an appropriate device to exploit it [4,5]. Even though solar energy is abundant, its expansion is hampered by problems like fractional shadow, high construction cost, weather variation, and the need for costly storage. As a result, photovoltaic (PV) modeling is necessary to estimate the performance of a PV system before installation. Furthermore, the prediction of PV panel operating attributes is critical in solar PV system design, evaluation, simulation analysis, and control. Also, modeling aids in comprehending the functioning precept and attributes of the solar PV system under variable meteorological situations. The PV solar system is useful for capturing the solar energy and converting it into electrical power [5–7]; it has penetrated into many applications [5]. Moreover, the economic implications of the decreased lifetime and its causes are presented in [8]. One of the scientists' priorities is to improve the efficiency and dependability of these technologies. Understanding the mechanisms of power absorption and conversion in solar cells, as well as correct modelling, can help in forecasting and designing them properly. One of the most critical challenges that researchers are facing is how to build a reliable model of the solar panel [9–11].

Changes in temperature and sun irradiance have significant impacts on the performance of PV systems [12]. Therefore, to maximize the performance of these systems, adequate mathematical models are required that precisely replicate the PV system behavior under several operational scenarios. Three of the most common PV system models, the single, double, and triple diode models (SDM, DDM, and TDM), are used [13,14].

The parameters of the SDM are simple to estimate as it only has five parameters, but its performance suffers from minimal irradiance scales and as a consequence of temperature changes. The DDM includes seven unknown parameters; it employs a second diode to achieve current reunification and to deal with other non-idealities [15]. However, the DDM suffers from some defects in recombining the current and other non-idealities. The final model is TDM, with nine ungiven parameters; it was introduced in [16]. Unfortunately, the nine parameters should be calculated as the manufacturers do not directly give them. To decrease the difference between the measured assessed power–voltage (*P-V*) and current–voltage (*I-V*) curves, the issue is converted into an optimization problem with a nonlinear objective function and a significant number of local minima.

Researchers are interested in employing metaheuristic algorithms to estimate the PV model parameters due to their notable success in handling various real-world optimization problems [5–7]. A hybrid seagull optimization algorithm architecture (HSOA) has been described for assessing the PV model parameters and developing a nonlinear control factor, which is dependent on the cosine function, to stabilize exploitation and exploration capabilities [1]. A springy whale optimization algorithm is described as an enhanced optimization technique to determine the parameters of PV cell/panel models [9]. Changes have been made to the way that the whales move in order to improve the algorithm performance. This helped the algorithm avoid the local solution, and the algorithm convergence speed was enhanced. In [13], an improved cuckoo search optimizer (ICSO) and a modified cuckoo search optimizer (MCSO) are implemented to solve the parameter evaluation issue of a PV system. Solar cell parameters have been evaluated through a genetic neural network (GNN) strategy [14]. The PV module characteristics have been identified with the aid of the tabu search optimizer (TSO) [15]; moreover, the lightning search algorithm, pattern search (PS), gravity search algorithm (GSA), genetic algorithm (GA), and PSO have been applied and compared to the presented approach [16].

In order to define the values of the ungiven parameters, the sooty tern optimization (STO) approach was developed for parameter evaluation of the PV cells/modules [17]. The hybrid particle swarm optimization (PSO) and rat search algorithm have been presented and combined as a hybrid approach for extracting the parameters of hybrid systems, including those of fuel cells and solar PVs [18]. The presented approach in that work reduced the likelihood of a local minimum and increased the algorithm accuracy. In [19], the animals migration optimizer (AMO) was introduced to construct the SDM of a PV system. The approach capacity for producing prompt, dependable, and consistent outcomes has been considered. In [20], a chaotic WOA for estimating the solar cell parameters was introduced; the key benefit of this method is that its parameters are automatically computed and adjusted using chaotic maps. In [21], a mathematical model for PV solar cells was created using the equilibrium optimizer (EO). The results using the EO have been compared with Harries hawk optimization (HHO), the teaching learn-based optimizer (TLBO), and PSO. In [22], the many approaches employed in constructing the SDM, DDM, and TDM of PV systems were reviewed and compared in terms of pros and cons.

The fractional-order Darwinian PSO methodology was used in [23] to enhance the conventional PSO method in evaluating the electrical parameters of PV cells/modules. To assign the solar cell parameters, the authors in [24] presented a hybrid honey badger algorithm and GTO [25]. These algorithms reduced the root mean square error (RMSE) between the simulated and measured results. In [26–28], a marine predatory animal (MPA) algorithm is described for computing the parameters of PV cells/panels in constant and varying weather situations. An improved stochastic fractal search algorithm has been used to solve the parameter appreciation of SDM solar cells and PV panels [29]. The authors in [30] presented the computational optimization method for extracting the parameters of solar cells/panels using an enhanced arithmetic optimization algorithm. In order to study the DDM-based circuit of a PV panel, practical tests to obtain the measured *I-V* and *P-V* characteristics have been conducted while considering various statistical analyses to determine the average, maximum, minimum, and standard deviations. A quick and efficient method for collecting the solar cell/panel parameters from the datasheet is provided in [31]. A niche PSO using a parallel computing technique was presented in [32] to identify the PV panel parameters. A multi-agent system (MAS) has been combined with CSO to estimate the parameters of various PV cells [33]. The circuits of SDM, DDM, and TDM for PV cells have been analyzed using the atomic orbital search to determine the ungiven parameters [34]. The tree seed algorithm has been used to calculate the parameters of the STM6-40/36 PV panel with different maximum fitness evaluations [35]. Moreover, a heterogeneous mechanism for the differential evolution algorithm (DE) [36], population diversity controlled DE [37], the artificial parameter-less optimization algorithm [38], random reselection PSO [39], the arithmetic operation algorithm based on the Newton–Raphson and Lambert W approaches [40], and adaptive slime mold [41] have been utilized to construct different equivalent circuits of PV cells/panels. A mayfly algorithm [42], northern goshawk optimization [43], and Newton–Raphson (NR) with an enhancement of a tuna swarm optimizer by a chaotic tent map [44] have been presented to evaluate the parameters of a TDM circuit. The parameters of a PV equivalent circuit were resolved by a chimp optimization algorithm with a robust niching approach [45], hybrid PSO with a gravitational search algorithm [46], chaos game optimization [47], an improved gradient-based optimizer based on sine cosine [48], DE enhanced by a chaotic map [49], and the predict output-based backpropagation neural network with EO [50]. Furthermore, the forensic-based investigation algorithm [51], the supply–demand optimizer [52], the enhanced hunger games search via the Laplacian Nelder–Mead approach [53], the Rao-1 optimization-based chaotic sequence [54], the arithmetic optimization algorithm-based guaranteed convergence and modified third-order NR [55], and the hybridized wind-driven optimization with fruit fly optimization [56] have been used to compute the parameters of various types of PV models.

Most of the reported studies have limitations, such as the falling into local optima, the requirement for numerous controlling parameters, and the complexity in implementation, in addition to the use of absolute algorithms without fundamental changes or modifications. The motivation of this study is to introduce a novel hybrid multi-population gorilla troops optimizer and beluga whale optimization (HGTO-BWO) to determine the PV cell/panel parameters such that all the gaps in the previous works are covered.

GTO is characterized by its ability to solve real-world problems with limited and unknown search space. On the other hand, the BWO has better stability, good convergence accuracy, stronger search ability, and a faster convergence rate. Therefore, hybridization between GTO and BWO results in a strong optimizer which is able to solve the handled problem with good efficiency. Table 1 provides a comparison of the recent work published in 2023 with regard to parameter estimations of PVs. The multi-population technique is applied to enhance the algorithm performance and avoid early convergence through dividing the entire population into many subgroups to preserve population variety. Different subgroups can be discovered throughout the whole search area and can reach the optimal solution efficiently by searching in different locations inside the search area at one time. Moreover, the optimization techniques can be easily and efficiently incorporated into multi-population methods [57,58]. The following are the major contributions of this article:

- A novel approach of hybrid multi-population GTO-BWO is proposed in this work.
- The classical and CEC-C06 2019 benchmark functions are utilized to test and assess the proposed technique's performance.
- The proposed HGTO-BWO is implemented to determine the ungiven parameters of TDM and DDM equivalent circuits of PV cells/panels.
- A comparison is made with TSA, the grey wolf optimizer (GWO), the whale optimization algorithm (WOA), the sine cosine algorithm (SCA), harmony search (HS), beluga whale optimization (BWO), and the artificial gorilla troops optimizer (GTO).
- The fetched results assure the effectiveness and validity of the suggested HGTO-BWO.

Table 1. A comparison of recent work published in 2023.

Ref.	Obj. Function	Model Type	Algorithm	Remark
[59]	RMSE	SDM, DDM and TDM	Hybrid chaotic NSO-PS	One type of PV is R.T.C used in all case studies; complexity and improved performance
[60]	RMSE	SDM and DDM	Growth optimizer	Ability to determine ungiven PV model parameters; low convergence
[61]	RMSE	SDM, DDM, and TDM	Chaos game optimization with least squares	Speed convergence and the RMSE values are similar to those of some other methods
[62]	Non-linear square with RMSE	SDM	GWO	Complexity in obj. function
[63]	RMSE	SDM, DDM, and TDM	Improved moth–flame algorithms	Low obj. function
Proposed	RMSE	SDM, DDM and TDM	HGTO-BWO	High performance and efficiency; avoids local optimum; fast convergence

The rest of this article is as follows. Section 2 describes the mathematical model of solar PVs. Section 3 illustrates the problem expression, while the proposed hybrid multi-population GTO and BWO algorithm is presented in Section 4. The testing of the benchmark functions is presented in Section 5, and the application of the PV parameter estimation is given in Section 6. The conclusions are clarified in Section 7.

2. Modeling of Solar Photovoltaic (PV)

A solar PV cell is typically described through an electrical analogous circuit that includes current source, resistors, and a diode. Numerous PV cell modeling systems have evolved due to nonlinearity. The models of a PV cell are divided into three categories: single, double, and triple diode models. The prediction accuracy of the *I-V* curve is defined by the number of diodes in the model. Also, adding another diode, from one to three, enhances the model performance and precision at minimal irradiance levels. Similarly, the growth of modeling results in the development of the TDM. The model of the analogous circuit, its equations, and the specifications of the ungiven parameters are shown in Figures 1 and 2. As the number of diodes grows, the number of model parameters to be evaluated grows and then the complexity of the problem is increased [64,65].

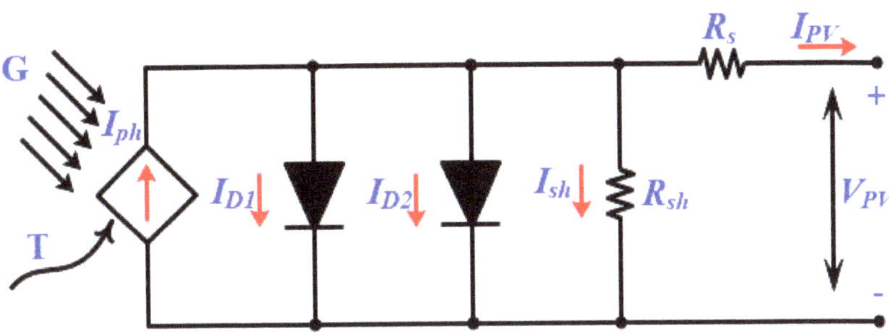

Figure 1. DDM equivalent circuit.

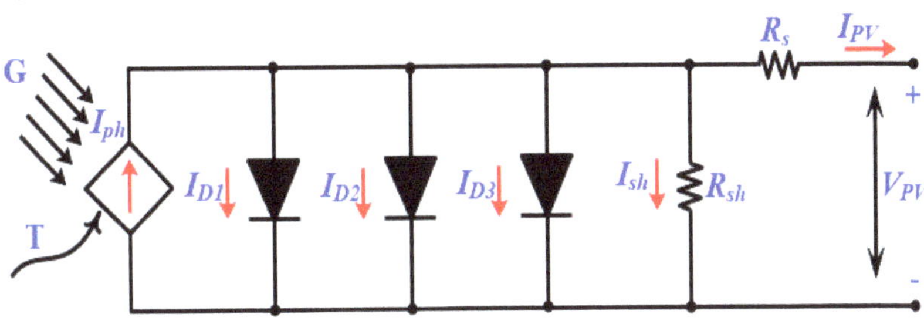

Figure 2. TDM equivalent circuit.

2.1. Double Diode Model (DDM)

The DDM uses dual diodes and dual resistors coupled in a series and shunted to the diode; this configuration is designed to compensate for the losses. The DDM of a solar cell is shown in Figure 1; with this concept, a second diode is added to reduce the transmission losses caused by the depletion layer carrier recombination and surface recombination, as specified by I_{d2} [29,66]. The component of the current is represented by the current of the first diode I_{d1}.

The DDM can be formulated as follows:

$$I_{PV} = I_{ph} - I_{d1} - I_{d2} - I_{sh} \tag{1}$$

$$I_{PV} = I_{ph} - I_{d1}\left[exp\left[\frac{q[V_{PV} + R_s I_{PV}]}{A_1 KT}\right] - 1\right] - I_{d2}\left[exp\left[\frac{q[V_{PV} + R_s I_{PV}]}{A_2 KT}\right] - 1\right] - \left[\frac{V_{PV} + R_s I_{PV}}{R_{sh}}\right] \tag{2}$$

This model has seven parameters to be computed; they are provided as a vector, as given in Equation (3).

$$x = \left[A_1 A_2 R_s R_{sh} I_{d1} I_{d2} I_{ph}\right] \tag{3}$$

where I_{d1}, I_{d2}, and I_{ph} are the reversal saturation currents of the diodes and photon current; q is the electronic charge; A_1 and A_2 are the diodes' ideality factors; T denotes the temperature in Kelvin; K refers to the Boltzmann constant; and R_{sh} and R_s are the shunt and series resistances.

2.2. Triple Diode Model (TDM)

Another model described in this work is the TDM; this model includes a current source, two resistors, and triple diodes, as shown in Figure 2. Dual diodes are considered in the model and are similar to those of the DDM, due to the reassembly and connection losses, while the third diode is due to the losses of the reassembly flow zones and boundaries [65,67].

The TDM can be expressed by following equations:

$$I_{PV} = I_{ph} - I_{d1} - I_{d2} - I_{d3} - I_{sh} \tag{4}$$

$$\begin{aligned} I_{PV} = I_{ph} - I_{d1}\left[exp\left[\frac{q[V_{PV}+R_sI_{PV}]}{A_1KT}\right]-1\right] - I_{d2}\left[exp\left[\frac{q[V_{PV}+R_sI_{PV}]}{A_2KT}\right]-1\right] \\ -I_{d3}\left[exp\left[\frac{q[V_{PV}+R_sI_{PV}]}{A_3KT}\right]-1\right] - \left[\frac{V_{PV}+R_sI_{PV}}{R_{sh}}\right] \end{aligned} \tag{5}$$

There are nine parameters to be evaluated in this model. The following vector can be used to represent them:

$$x = \left[A_1A_2A_3R_sR_{sh}I_{d1}I_{d2}I_{d3}I_{ph}\right] \tag{6}$$

2.3. PV Panel Model

The PV panel comprises numerous cells coupled in a series or parallel to produce greater voltage and current. Figure 3 illustrates the PV panel equivalent circuit.

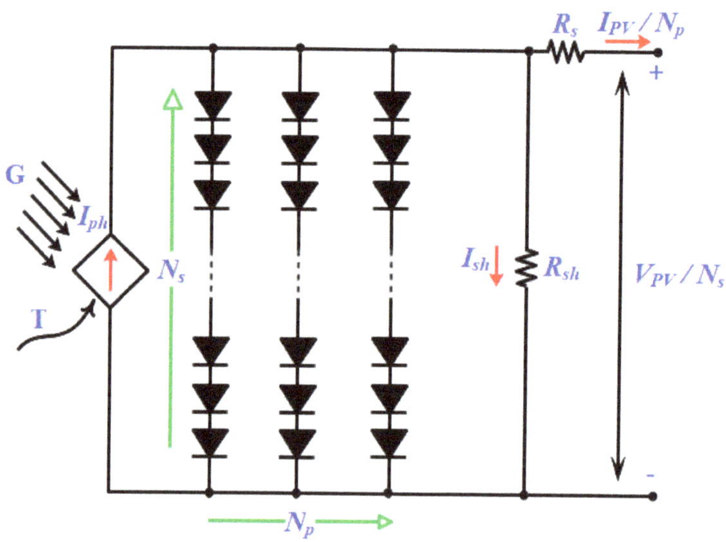

Figure 3. Solar PV module equivalent circuit.

The cells coupled in a series produce the same current. As a result, the panel output current can be written as given in Equation (7).

$$\begin{aligned} I_{PV} = I_{ph}\times N_p - I_{d1}\times N_p\left[exp\left[\frac{q[V_{PV}/N_s+R_s/N_p\times I_{PV}]}{A_1KT}\right]-1\right] \\ -I_{d2}\times N_P\left[exp\left[\frac{q[V_{PV}/N_s+R_s/N_p\ I_{PV}]}{A_2KT}\right]-1\right] - [V_{PV}/N_s + R_s/N_p\times I_{PV}R_{sh}] \end{aligned} \tag{7}$$

where N_s denotes the number of solar cells combined in a series, while N_p denotes the number of cells connected in parallel [67,68]. Seven parameters should be calculated in the PV panel circuit; these are A_1, A_2, R_s, R_{sh}, I_{d1}, I_{d2}, and I_{ph}.

3. Problem Expression

In order to find the PV cell/panel equivalent circuit parameters, an optimization problem was formulated and solved to mitigate the RMSE between the measured current (I_M) and the simulated one (I_S). In the optimization formula, the unknown parameters are defined as design variables; the fitness value can be formulated as follows:

$$RMSE = \sqrt{\frac{1}{N_m}\sum_{i=1}^{N_m} f_{PV}(V_{PV}, I_{PV}, x)^2} = \sqrt{\frac{1}{N_m}\sum_{i=1}^{N_m} (I_M - I_S)^2} \quad (8)$$

where N_m indicates the number of measured patterns, f_{PV} refers to the PV model function, and I_M and I_S are the measured and simulated currents, respectively. The DDM objective function contains seven unknown parameters; it can be written as,

$$\begin{aligned} f_{DDM}(V_{PV}, I_{PV}, x) &= I_{PV} - x_7 + x_5\left(exp\left(\tfrac{q(V_{PV}+I_{PV}x_3)}{x_1KT}\right) - 1\right) + x_6\left(exp\left(\tfrac{q(V_{PV}+I_{PV}x_3)}{x_2KT}\right) - 1\right) \\ &+ \tfrac{(V_{PV}+I_{PV}x_3)}{x_4} \end{aligned} \quad (9)$$

where $x = \left[A_1 A_2 R_s R_{sh} I_{d1} I_{d2} I_{ph}\right]$.

On the other hand, the TDM objective function comprises nine unknown parameters, which can be expressed as,

$$\begin{aligned} f_{TDM}(V_{PV}, I_{PV}, x) &= I_{PV} - x_9 + x_6\left(exp\left(\tfrac{q(V_{PV}+I_{PV}x_4)}{x_1KT}\right) - 1\right) + x_7\left(exp\left(\tfrac{q(V_{PV}+I_{PV}x_4)}{x_2KT}\right) - 1\right) \\ &+ x_8\left(exp\left(\tfrac{q(V_{PV}+I_{PV}x_4)}{x_3KT}\right) - 1\right) + \tfrac{(V_{PV}+I_{PV}x_4)}{x_5} \end{aligned} \quad (10)$$

where $x = \left[A_1 A_2 A_3 R_s R_{sh} I_{d1} I_{d2} I_{d3} I_{ph}\right]$.

Finally, the TDM of the PV panel objective function can be written as,

$$\begin{aligned} f_{PV\_p}(V_{PV}, I_{PV}, x) &= I_{PV} - x_7 * N_p + x_5 * N_p\left[exp\left[\tfrac{q\left[V_{PV}/N_s + x_3/N_p * I_{PV}\right]}{x_1KT}\right] - 1\right] + x_6 \\ &* N_P\left[exp\left[\tfrac{q\left[V_{PV}/N_s + x_3/N_p\ I_{PV}\right]}{x_2KT}\right] - 1\right] + \left[\tfrac{V_{PV}/N_s + x_3/N_p * I_{PV}}{x_5}\right] \end{aligned} \quad (11)$$

4. The Proposed Solution Methodology

This section describes and explains the main aspects of GTO, BWO, and the proposed HMGTO-BWO.

4.1. Gorilla Troops Optimizer (GTO)

The GTO is an efficient optimization algorithm that was inspired by the social life of gorillas, including their movements and lifestyles [69]. The leader in a gorilla group is known as a silverback and all the males and females follow it. The young male gorillas are known as blackbacks; they help the silverback and act as backup protection for the group. Two phases of exploitation and exploration form the GTO. Three operators are used in the exploration phases; the first operator is the migration to new locations, while the second operator is based on the movement of other gorillas; the third operator is dependent on the motion of the groups to known areas. In the GTO, the parameter X refers to the gorilla position, and the GX denotes the candidate gorilla locations, while the best solution position is represented as the silverback position. The exploitation phase is based on three motions of the gorillas, including their motion to a new unknown area, their motion to each other, and their movement to unknown locations. Mathematically, the exploration phase of the GTO can be described as follows:

$$GX(t+1) = \begin{cases} (UB - LB) \times r_1 + LB, & rand < p \\ (r_2 - C) \times X_r(t) + L \times H, & rand \geq 0.5 \\ X(i) - L \times (L \times (X(t) - GX_r(t)) + r_3 \times (X(t) - GX_r(t))) & rand < 0.5 \end{cases} \tag{12}$$

where r_1, r_2, and r_3 are random values in the range [0, 1]; UB and LB are the upper and lower limits of the variables; the P operator is a generated random value; and C, L, and H are operators that can be computed as follows:

$$C = F \times \left(1 - \frac{t}{t_{\max}}\right) \tag{13}$$

$$F = \cos(2 \times r_4) + 1 \tag{14}$$

$$L = C \times l \tag{15}$$

$$H = Z \times X(t) \tag{16}$$

$$Z = [-C, C] \tag{17}$$

where $t_{\max}$ *and* t are the maximum and current iterations, and r_4 is a random value in the range [0, 1]. The exploitation phase in this algorithm is based on the motion of the followers to the silverback gorilla. However, when the silverback dies or becomes ill, the male blackback gorillas become leaders; these gorillas fight to obtain the female gorillas. The exploitation phase mimics the motion of the males and females to the silverback. In addition to that, when the silverback dies or becomes old, the blackback gorilla males become leaders. Thus, the group may follow the silverback or the blackback gorilla males. The transition between the two movements can be adjusted using two operators, C and W. In the case that $C \geq W$, the gorillas update their locations with respect to the silverback as follows:

$$GX(t+1) = M \times L \times (X(t) - X_{\text{best}}) + X(t) \tag{18}$$

$$M = \left(\left|\frac{1}{N}\sum_{i=1}^{N} GX_i(t)\right|^g\right)^{\frac{1}{g}} \tag{19}$$

$$g = 2^L \tag{20}$$

where X_{best} represents the silverback's location. If $C < W$, the other gorillas follow the adult males; this may be described as follows:

$$GX(i) = X_{\text{silverback}} - (X_{\text{best}} \times Q - X(t) \times Q) \times A \tag{21}$$

$$Q = 2 \times r_5 - 1 \tag{22}$$

$$A = \beta \times E \tag{23}$$

where r_5 represents a random value in the range [0, 1], β denotes a predefined operator, and E is a random value obtained from the normal distribution. The GTO's pseudocode is depicted in Algorithm 1.

Algorithm 1 The Pseudocode of GTO

Start GTO
  Input : Set the parameters of the GTO (N, $t_{\max}$, UB, LB
  Output : The best position of the population (X_{best}) the corresponding fitness function.
  Initialize the populations and calculate the objective functions and assign the best result.
  While t < $t_{\max}$
    Update the values of the C, L using Equations (13) and (15).
    Update the positions of the gorillas according to (12).
    Compute the fitness function and assign the best solution.
    If $C \geq W$
      Update the positions of the gorillas using Equation (18).
    Otherwise
      Update the positions of the gorillas using Equation (21).
    end
    Calculate the objective functions for the new locations and include them,
    if their values are better than the previous solutions
  End while
End GTO

4.2. Beluga Whale Optimization (BWO)

BWO is a new optimizer that was conceptualized from the motion, preying, and behavior of beluga whales (BWs) in the seas and oceans [70]. BWs are social creatures that share information and communicate together to search for food locations. Initially, the fitness function is expressed as follows:

$$F_X = \begin{bmatrix} f(x_{1,1}, x_{1,2}, \ldots, x_{1,d}) \\ f(x_{2,1}, x_{2,2}, \ldots, x_{2,d}) \\ \vdots \\ f(x_{n,1}, x_{n,2}, \ldots, x_{n,d}) \end{bmatrix} \tag{24}$$

The swimming motion of the two BW pairs represents the exploration phase, which may be mathematically described as follows:

$$X_{i,j}^{t+1} = \begin{cases} X_{i,p_j}^t + \left(X_{r,p_1}^t - X_{i,p_j}^t\right)(1+r_1)\sin(2\pi r_2), & j = \text{even} \\ X_{i,p_j}^t + \left(X_{r,p_1}^t - X_{i,p_j}^t\right)(1+r_1)\cos(2\pi r_2), & j = \text{odd} \end{cases} \tag{25}$$

where X_{r,p_1}^t is a whale selected randomly from the generated BWs. The BWO exploitation phase is conceptualized from the hunting and preying process of BWs. They update their locations based on the best solution using the Levy flight strategy, as follows:

$$X_i^{t+1} = r_3 X_{\text{best}}^t - r_4 X_i^t + C_1 \cdot L_F \cdot \left(X_r^t - X_i^t\right) \tag{26}$$

$$C_1 = 2r_4(1 - t/t_{\max}) \tag{27}$$

where X_{best}^t represents the best location, X_r^t refers to a randomly selected BW, and L_F is a Lévy flight function, which can be determined as follows:

$$L_F = 0.05 \times \frac{u \times \sigma}{|v|^{1/\beta}} \tag{28}$$

$$\sigma = \left(\frac{\sin(\pi\beta/2) \times \Gamma(1+\beta)}{\beta \times \Gamma((1+\beta)/2) \times 2^{(\beta-1)/2}}\right)^{1/\beta} \tag{29}$$

where u and v are random variables, and β is an adaptive variable used to enable the transition between the exploitation and the exploration phases; it can be calculated as,

$$B_f = B_0(1 - t/2t_{\max}) \tag{30}$$

where B_0 is a random value in the range [0, 1]. If $B_f > 0.5$, the BWs update their locations in the exploration phase; otherwise, they update their locations in the exploitation manner. The final stage of the BWO is based on the whale fall of BWs when they have been attacked by the killer whales. The dead BWs are deposited on a deep seabed. This stage is represented as follows:

$$X_i^{t+1} = r_5 X_i^t - r_6 X_r^t + r_7 X_{\text{step}} \tag{31}$$

$$X_{\text{step}} = (U_b - L_b)\exp(-C_2 t/t_{\max}) \tag{32}$$

$$C_2 = 2W_f \times n \tag{33}$$

$$W_f = 0.1 - 0.05t/t_{\max} \tag{34}$$

where r_5, r_6, and r_7 denote random variables in the range [0, 1]. The pseudocode of BWO is depicted in Algorithm 2.

Algorithm 2 The Pseudocode of the BWO

Start BWO
  Input : Set the parameters of the BWO (N, $t_{\max}$, UB, LB).
  Output : The best position (X_{best}) of the populations and the corresponding fitness function.
  While t < $t_{\max}$
    Update the values of the using C_1, B_f, and W_f using Equations (27), (30) and (34).
    If $B_f > 0.5$
      // Exploration phase
      Update the locations of the BWs using Equation (25).
    Otherwise
      // Exploitation phase
      Update the locations of the gorillas using Equation (26).
    end
    Compute the fitness functions for the new positions and select the best result.
    If $B_f \geq W_f$
      // whale fall
      Update the locations of the BWs using Equation (31).
    End
    Compute the fitness functions for the new positions and select the best result.
  End while
End BWO

4.3. The Proposed Hybrid Multi-Population GTO and BWO

The proposed HMGTO-BWO is introduced to solve complex and nonlinear optimization issues. The following steps describe the procedure of the proposed HMGTO-BWO:

Step 1: Define the parameters of the proposed HMGTO-BWO as well as the constraints of the problem.

Step 2: Generate a set of populations randomly.

Step 3: Divide the populations into three subpopulations (N_1, N_2, N_3), where $N_1 = N_2 = N/3$ and $N_3 = N - (N1 + N2)$, where N, N_1, N_2, and N_3 are numbers of the population, the first subpopulation, the second subpopulation, and the third subpopulation, respectively.

Step 4: Update the populations in each subpopulation group based on the GTO, as illustrated in Section 4.1.

Step 5: Accept the new updated subpopulations if their values are better than those of the old populations.

Step 6: Combine the three subpopulations as one vector; it represents the initial populations of the BWO technique.

Step 7: Update the populations based on BWO, including the swimming motion, the Levy flight motion, and the fall of BWs.

Step 8: Repeat Step 3 to Step 7 until the stopping criterion is satisfied.

The step procedures of the suggested algorithm are depicted in Figure 4.

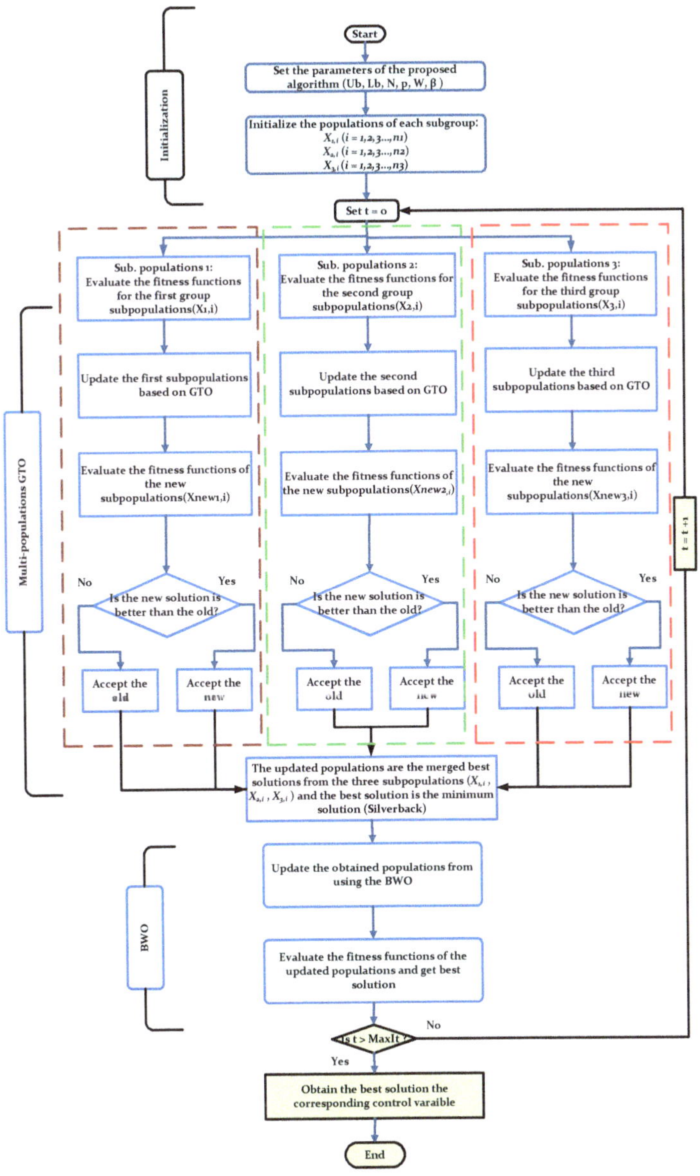

Figure 4. Flowchart of the proposed HMGTO-BWO.

The HGTO-BWO computational complexity is based on the initialization, fitness assessment, and updating of the silverbacks and BW, and it can be described as follows:

$$O(HGTO - BWO) = (Sub.Population1 + Sub.Population2 + Sub.Population3)_{GTO} + BWO \tag{35}$$

$$\begin{aligned} O(HGTO - BWO) &= \left[O\left(t_{max} \times \tfrac{1}{3}N_1\right) + O\left(t_{max} \times \tfrac{1}{3}N_1 \times D\right) \times 2\right] + \left[O\left(t_{max} \times \tfrac{1}{3}N_2\right) + O\left(t_{max} \times \tfrac{1}{3}N_2 \times D\right) \times 2\right] \\ &+ \left[O\left(t_{max} \times \tfrac{1}{3}N_3\right) + O\left(t_{max} \times \tfrac{1}{3}N_3 \times D\right) \times 2\right] + O(N \times (1 + 1.1 \times t_{max})) \\ &= O(\mathrm{N} \times (1 + t_{max} + TD) \times 2 + (1 + 1.1 \times t_{max})) \end{aligned} \tag{36}$$

where D is the dimension of the problem.

5. Testing of Benchmark Function

For a fair comparison between the suggested HGTO-BWO and the other algorithmic approaches, the maximum iterations number was set to 500; the population size was assigned to 30; and 30 runs were conducted for each considered optimizer. The proposed HGTO-BWO was investigated via the traditional benchmark functions and CEC 2019 functions. The fetched results were compared to TSA, GWO, WOA, SCA, HS, BWO, and GTO. The algorithms' parameters are presented in Table A1 in Appendix A.

5.1. Traditional Benchmark Functions

The proposed HGTO-BWO was investigated via the solving of various traditional benchmark functions [71]; F1 to F13 have constant dimensions of 30, while the functions F14 to F23 have different dimensions. These functions are divided into the functions from F1 to F7, which are unimodal; F8 to F13, which are multi-modal; and F14 to F23, which are composites (See Supplementary Materials). Table 2 shows the statistical results of all the traditional benchmark functions; it includes the worst, average, best, standard deviation (std), and *p*-value. The values given in bold indicate the best solutions obtained by the proposed HGTO-BWO approach.

Table 2. Statistical analysis of traditional benchmark functions solved via the HGTO-BWO approach and other techniques.

Function No	Algorithm	Worst	Mean	Best	std	*p*-Value
F1	TSA	6.062×10^{-21}	8.593×10^{-22}	1.829×10^{-24}	1.367×10^{-21}	1.21×10^{-12}
	GWO	3.300×10^{-26}	2.105×10^{-27}	4.759×10^{-29}	6.022×10^{-27}	1.21×10^{-12}
	WOA	6.553×10^{-70}	2.364×10^{-71}	2.778×10^{-83}	1.197×10^{-70}	1.21×10^{-12}
	SCA	4.516×10^{2}	3.588×10^{1}	2.339×10^{-2}	1.003×10^{2}	1.21×10^{-12}
	HS	3.079×10^{3}	2.476×10^{3}	1.562×10^{3}	3.939×10^{2}	1.21×10^{-12}
	BWO	1.925×10^{-257}	1.272×10^{-258}	1.274×10^{-272}	$\mathbf{0.00 \times 10^{0}}$	1.21×10^{-12}
	GTO	$\mathbf{0.00 \times 10^{0}}$	$\mathbf{0.00 \times 10^{0}}$	$\mathbf{0.00 \times 10^{0}}$	$\mathbf{0.00 \times 10^{0}}$	1.21×10^{-12}
	HGTO-BWO	$\mathbf{0.00 \times 10^{0}}$	$\mathbf{0.00 \times 10^{0}}$	$\mathbf{0.00 \times 10^{0}}$	$\mathbf{0.00 \times 10^{0}}$	NAN
F2	TSA	4.979×10^{-13}	1.128×10^{-13}	1.047×10^{-14}	1.206×10^{-13}	3.02×10^{-11}
	GWO	4.515×10^{-16}	9.566×10^{-17}	1.246×10^{-17}	8.300×10^{-17}	3.02×10^{-11}
	WOA	1.156×10^{-48}	7.100×10^{-50}	1.076×10^{-56}	2.454×10^{-49}	3.02×10^{-11}
	SCA	1.346×10^{-1}	2.449×10^{-2}	8.107×10^{-5}	3.647×10^{-2}	3.02×10^{-11}
	HS	1.370×10^{1}	1.057×10^{1}	7.436×10^{0}	1.673×10^{0}	3.02×10^{-11}
	BWO	1.849×10^{-129}	6.467×10^{-131}	1.310×10^{-137}	3.37×10^{-130}	3.02×10^{-11}
	GTO	1.835×10^{-190}	6.261×10^{-192}	3.521×10^{-206}	$\mathbf{0.00 \times 10^{0}}$	3.02×10^{-11}
	HGTO-BWO	$\mathbf{6.916 \times 10^{-247}}$	$\mathbf{2.305 \times 10^{-248}}$	$\mathbf{5.079 \times 10^{-268}}$	$\mathbf{0.00 \times 10^{0}}$	NAN

Table 2. *Cont.*

Function No	Algorithm	Worst	Mean	Best	std	*p*-Value
F3	TSA	2.826×10^{-3}	3.329×10^{-4}	1.572×10^{-8}	7.279×10^{-4}	1.21×10^{-12}
	GWO	3.698×10^{-4}	2.062×10^{-5}	7.248×10^{-9}	6.741×10^{-5}	1.21×10^{-12}
	WOA	7.244×10^{4}	3.723×10^{4}	2.987×10^{3}	1.716×10^{4}	1.21×10^{-12}
	SCA	2.274×10^{4}	9.511×10^{3}	1.459×10^{3}	5.359×10^{3}	1.21×10^{-12}
	HS	3.362×10^{4}	2.686×10^{4}	1.965×10^{4}	3.983×10^{3}	1.21×10^{-12}
	BWO	1.555×10^{-242}	1.060×10^{-243}	2.864×10^{-257}	$\mathbf{0.00 \times 10^{0}}$	1.21×10^{-12}
	GTO	$\mathbf{0.00 \times 10^{0}}$	$\mathbf{0.00 \times 10^{0}}$	$\mathbf{0.00 \times 10^{0}}$	$\mathbf{0.00 \times 10^{0}}$	1.21×10^{-12}
	HGTO-BWO	$\mathbf{0.00 \times 10^{0}}$	$\mathbf{0.00 \times 10^{0}}$	$\mathbf{0.00 \times 10^{0}}$	$\mathbf{0.00 \times 10^{0}}$	NAN
F4	TSA	7.663×10^{-1}	2.588×10^{-1}	1.258×10^{-2}	2.234×10^{-1}	3.02×10^{-11}
	GWO	7.987×10^{-6}	8.996×10^{-7}	4.183×10^{-8}	1.485×10^{-6}	3.02×10^{-11}
	WOA	9.425×10^{1}	5.178×10^{1}	7.053×10^{-2}	3.113×10^{1}	3.02×10^{-11}
	SCA	5.758×10^{1}	3.099×10^{1}	1.005×10^{1}	1.173×10^{1}	3.02×10^{-11}
	HS	4.145×10^{1}	3.622×10^{1}	3.042×10^{1}	2.199×10^{0}	3.02×10^{-11}
	BWO	3.880×10^{-126}	3.095×10^{-127}	2.571×10^{-133}	7.804×10^{-127}	3.02×10^{-11}
	GTO	1.694×10^{-192}	8.353×10^{-194}	9.838×10^{-208}	$\mathbf{0.00 \times 10^{0}}$	3.02×10^{-11}
	HGTO-BWO	$\mathbf{6.349 \times 10^{-238}}$	$\mathbf{2.187 \times 10^{-239}}$	$\mathbf{5.440 \times 10^{-257}}$	$\mathbf{0.00 \times 10^{0}}$	NAN
F5	TSA	2.889×10^{1}	2.838×10^{1}	2.609×10^{1}	7.737×10^{-1}	2.37×10^{-12}
	GWO	2.852×10^{1}	2.682×10^{1}	2.566×10^{1}	7.761×10^{-1}	2.37×10^{-12}
	WOA	2.877×10^{1}	2.794×10^{1}	2.728×10^{1}	5.032×10^{-1}	2.37×10^{-12}
	SCA	4.983×10^{5}	4.696×10^{4}	1.048×10^{2}	9.858×10^{4}	2.37×10^{-12}
	HS	1.768×10^{6}	1.070×10^{6}	6.292×10^{5}	2.773×10^{5}	2.37×10^{-12}
	BWO	8.153×10^{-6}	1.346×10^{-6}	1.618×10^{-9}	2.176×10^{-6}	2.37×10^{-12}
	GTO	2.477×10^{1}	2.445×10^{0}	6.987×10^{-8}	7.461×10^{0}	2.37×10^{-12}
	HGTO-BWO	$\mathbf{1.395 \times 10^{-28}}$	$\mathbf{5.853 \times 10^{-30}}$	$\mathbf{0.00 \times 10^{0}}$	$\mathbf{2.608 \times 10^{-29}}$	NAN
F6	TSA	4.820×10^{0}	3.817×10^{0}	2.590×10^{0}	6.004×10^{-1}	1.212×10^{-12}
	GWO	1.754×10^{0}	8.393×10^{-1}	6.737×10^{-5}	4.021×10^{-1}	1.212×10^{-12}
	WOA	9.864×10^{-1}	3.725×10^{-1}	1.355×10^{-1}	1.979×10^{-1}	1.212×10^{-12}
	SCA	2.624×10^{2}	2.762×10^{1}	4.337×10^{0}	4.997×10^{1}	1.212×10^{-12}
	HS	3.225×10^{3}	2.582×10^{3}	1.422×10^{3}	4.447×10^{2}	1.212×10^{-12}
	BWO	2.490×10^{-13}	1.879×10^{-14}	1.153×10^{-17}	4.703×10^{-14}	1.212×10^{-12}
	GTO	7.369×10^{-7}	1.292×10^{-7}	7.310×10^{-11}	1.675×10^{-7}	1.212×10^{-12}
	HGTO-BWO	$\mathbf{0.00 \times 10^{0}}$	$\mathbf{0.00 \times 10^{0}}$	$\mathbf{0.00 \times 10^{0}}$	$\mathbf{0.00 \times 10^{0}}$	NAN
F7	TSA	2.022×10^{-2}	9.471×10^{-3}	1.816×10^{-3}	4.734×10^{-3}	3.02×10^{-11}
	GWO	6.588×10^{-3}	1.963×10^{-3}	6.607×10^{-4}	1.299×10^{-3}	3.02×10^{-11}
	WOA	1.522×10^{-2}	3.099×10^{-3}	5.474×10^{-5}	3.995×10^{-3}	3.16×10^{-10}
	SCA	4.586×10^{-1}	8.768×10^{-2}	9.960×10^{-3}	9.341×10^{-2}	3.02×10^{-11}
	HS	1.093×10^{0}	7.266×10^{-1}	3.535×10^{-1}	1.790×10^{-1}	3.02×10^{-11}
	BWO	4.232×10^{-4}	1.563×10^{-4}	$\mathbf{3.893 \times 10^{-7}}$	1.160×10^{-4}	1.70×10^{-2}
	GTO	3.524×10^{-4}	1.005×10^{-4}	1.298×10^{-5}	8.193×10^{-5}	4.12×10^{-1}
	HGTO-BWO	$\mathbf{3.263 \times 10^{-4}}$	$\mathbf{8.538 \times 10^{-5}}$	2.321×10^{-6}	$\mathbf{7.039 \times 10^{-5}}$	NAN
F8	TSA	-4.628×10^{3}	-5.706×10^{3}	-6.921×10^{3}	5.559×10^{2}	1.720×10^{-12}
	GWO	-3.023×10^{3}	-6.026×10^{3}	-7.397×10^{3}	9.273×10^{2}	1.720×10^{-12}
	WOA	-6.738×10^{3}	-1.051×10^{4}	$\mathbf{-1.257 \times 10^{4}}$	1.872×10^{3}	1.720×10^{-12}
	SCA	-3.240×10^{3}	-3.726×10^{3}	-4.747×10^{3}	3.438×10^{2}	1.720×10^{-12}
	HS	-1.130×10^{4}	-1.160×10^{4}	-1.196×10^{4}	1.851×10^{2}	1.720×10^{-12}
	BWO	$\mathbf{-1.257 \times 10^{4}}$	$\mathbf{-1.257 \times 10^{4}}$	$\mathbf{-1.257 \times 10^{4}}$	$\mathbf{1.548 \times 10^{-8}}$	4.562×10^{-11}
	GTO	$\mathbf{-1.257 \times 10^{4}}$	$\mathbf{-1.257 \times 10^{4}}$	$\mathbf{-1.257 \times 10^{4}}$	1.868×10^{-5}	4.562×10^{-11}
	HGTO-BWO	$\mathbf{-1.257 \times 10^{4}}$	$\mathbf{-1.257 \times 10^{4}}$	$\mathbf{-1.257 \times 10^{4}}$	3.317×10^{-3}	NAN

Table 2. *Cont.*

Function No	Algorithm	Worst	Mean	Best	std	*p*-Value
F9	TSA	2.745×10^{2}	1.924×10^{2}	1.364×10^{2}	3.806×10^{1}	1.212×10^{-12}
	GWO	1.195×10^{1}	2.920×10^{0}	5.684×10^{-14}	3.451×10^{0}	1.188×10^{-12}
	WOA	5.684×10^{-14}	1.895×10^{-15}	$\mathbf{0.00 \times 10^{0}}$	1.038×10^{-14}	3.337×10^{-1}
	SCA	1.144×10^{2}	3.067×10^{1}	1.303×10^{-2}	3.118×10^{1}	1.212×10^{-12}
	HS	6.129×10^{1}	5.208×10^{1}	3.510×10^{1}	6.576×10^{0}	1.212×10^{-12}
	BWO	$\mathbf{0.00 \times 10^{0}}$	$\mathbf{0.00 \times 10^{0}}$	$\mathbf{0.00 \times 10^{0}}$	$\mathbf{0.00 \times 10^{0}}$	NAN
	GTO	$\mathbf{0.00 \times 10^{0}}$	$\mathbf{0.00 \times 10^{0}}$	$\mathbf{0.00 \times 10^{0}}$	$\mathbf{0.00 \times 10^{0}}$	NAN
	HGTO-BWO	$\mathbf{0.00 \times 10^{0}}$	$\mathbf{0.00 \times 10^{0}}$	$\mathbf{0.00 \times 10^{0}}$	$\mathbf{0.00 \times 10^{0}}$	NAN
F10	TSA	1.926×10^{1}	1.676×10^{0}	1.077×10^{-12}	3.633×10^{0}	1.212×10^{-12}
	GWO	1.714×10^{-13}	1.039×10^{-13}	6.839×10^{-14}	2.216×10^{-14}	1.112×10^{-12}
	WOA	7.994×10^{-15}	4.796×10^{-15}	$\mathbf{8.882 \times 10^{-16}}$	2.529×10^{-15}	1.233×10^{-9}
	SCA	2.035×10^{1}	1.628×10^{1}	1.762×10^{-02}	7.960×10^{0}	1.212×10^{-12}
	HS	1.198×10^{1}	1.022×10^{1}	9.181×10^{0}	7.230×10^{-1}	1.212×10^{-12}
	BWO	$\mathbf{8.882 \times 10^{-16}}$	$\mathbf{8.882 \times 10^{-16}}$	$\mathbf{8.882 \times 10^{-16}}$	$\mathbf{0.00 \times 10^{0}}$	NAN
	GTO	$\mathbf{8.882 \times 10^{-16}}$	$\mathbf{8.882 \times 10^{-16}}$	$\mathbf{8.882 \times 10^{-16}}$	$\mathbf{0.00 \times 10^{0}}$	NAN
	HGTO-BWO	$\mathbf{8.882 \times 10^{-16}}$	$\mathbf{8.882 \times 10^{-16}}$	$\mathbf{8.882 \times 10^{-16}}$	$\mathbf{0.00 \times 10^{0}}$	NAN
F11	TSA	7.086×10^{-2}	1.139×10^{-2}	$\mathbf{0.00 \times 10^{0}}$	1.437×10^{-2}	3.453×10^{-7}
	GWO	1.466×10^{-2}	2.227×10^{-3}	$\mathbf{0.00 \times 10^{0}}$	4.603×10^{-3}	1.104×10^{-2}
	WOA	1.334×10^{-1}	8.450×10^{-3}	$\mathbf{0.00 \times 10^{0}}$	3.220×10^{-2}	1.608×10^{-1}
	SCA	1.395×10^{0}	9.139×10^{-1}	3.508×10^{-1}	3.097×10^{-1}	1.212×10^{-12}
	HS	3.395×10^{1}	2.408×10^{1}	1.598×10^{1}	4.187×10^{0}	1.212×10^{-12}
	BWO	$\mathbf{0.00 \times 10^{0}}$	$\mathbf{0.00 \times 10^{0}}$	$\mathbf{0.00 \times 10^{0}}$	$\mathbf{0.00 \times 10^{0}}$	NAN
	GTO	$\mathbf{0.00 \times 10^{0}}$	$\mathbf{0.00 \times 10^{0}}$	$\mathbf{0.00 \times 10^{0}}$	$\mathbf{0.00 \times 10^{0}}$	NAN
	HGTO-BWO	$\mathbf{0.00 \times 10^{0}}$	$\mathbf{0.00 \times 10^{0}}$	$\mathbf{0.00 \times 10^{0}}$	$\mathbf{0.00 \times 10^{0}}$	NAN
F12	TSA	1.763×10^{01}	8.358×10^{0}	1.229×10^{0}	4.909×10^{0}	1.2×10^{-12}
	GWO	8.906×10^{-2}	4.194×10^{-2}	1.302×10^{-2}	1.904×10^{-2}	1.2×10^{-12}
	WOA	4.461×10^{0}	1.803×10^{-1}	6.551×10^{-3}	8.092×10^{-1}	1.2×10^{-12}
	SCA	7.264×10^{6}	2.860×10^{5}	6.013×10^{-1}	1.328×10^{6}	1.2×10^{-12}
	HS	2.270×10^{5}	5.098×10^{4}	1.847×10^{3}	5.045×10^{4}	1.2×10^{-12}
	BWO	6.621×10^{-13}	7.690×10^{-14}	1.117×10^{-16}	1.543×10^{-13}	1.2×10^{-12}
	GTO	1.505×10^{-7}	3.881×10^{-8}	3.088×10^{-10}	4.298×10^{-8}	1.2×10^{-12}
	HGTO-BWO	$\mathbf{1.571 \times 10^{-32}}$	$\mathbf{1.571 \times 10^{-32}}$	$\mathbf{1.571 \times 10^{-32}}$	$\mathbf{5.567 \times 10^{-48}}$	NAN
F13	TSA	4.576×10^{0}	2.978×10^{0}	1.628×10^{0}	6.748×10^{-1}	1.2×10^{-12}
	GWO	1.141×10^{0}	6.990×10^{-1}	1.132×10^{-1}	2.570×10^{-1}	1.2×10^{-12}
	WOA	1.704×10^{0}	6.341×10^{-1}	6.034×10^{-2}	3.425×10^{-1}	1.2×10^{-12}
	SCA	2.870×10^{6}	2.229×10^{5}	2.964×10^{0}	6.961×10^{5}	1.2×10^{-12}
	HS	2.840×10^{6}	1.215×10^{6}	3.453×10^{5}	5.884×10^{5}	1.2×10^{-12}
	BWO	8.110×10^{-12}	4.565×10^{-13}	2.230×10^{-15}	1.484×10^{-12}	1.2×10^{-12}
	GTO	6.478×10^{-2}	4.357×10^{-3}	8.216×10^{-12}	1.414×10^{-2}	1.2×10^{-12}
	HGTO-BWO	$\mathbf{1.35 \times 10^{-32}}$	$\mathbf{1.35 \times 10^{-32}}$	$\mathbf{1.35 \times 10^{-32}}$	$\mathbf{5.567 \times 10^{-48}}$	NAN
F14	TSA	1.830×10^{1}	9.339×10^{0}	9.98×10^{-1}	4.957×10^{0}	1.21×10^{-12}
	GWO	1.267×10^{1}	4.948×10^{0}	9.98×10^{-1}	4.243×10^{0}	1.21×10^{-12}
	WOA	1.076×10^{1}	3.154×10^{0}	9.98×10^{-1}	3.622×10^{0}	1.21×10^{-12}
	SCA	1.076×10^{1}	2.114×10^{0}	9.98×10^{-1}	2.498×10^{0}	1.21×10^{-12}
	HS	$\mathbf{9.980 \times 10^{-1}}$	$\mathbf{9.980 \times 10^{-1}}$	9.98×10^{-1}	1.873×10^{-7}	1.21×10^{-12}
	BWO	1.992×10^{0}	1.064×10^{0}	9.98×10^{-1}	2.522×10^{-1}	1.21×10^{-12}
	GTO	$\mathbf{9.980 \times 10^{-1}}$	$\mathbf{9.980 \times 10^{-1}}$	9.98×10^{-1}	5.831×10^{-17}	1.61×10^{-01}
	HGTO-BWO	$\mathbf{9.980 \times 10^{-1}}$	$\mathbf{9.980 \times 10^{-1}}$	9.98×10^{-1}	$\mathbf{0.00 \times 10^{0}}$	NAN

Table 2. *Cont.*

Function No	Algorithm	Worst	Mean	Best	std	*p*-Value
F15	TSA	1.103×10^{-1}	1.267×10^{-2}	3.077×10^{-4}	2.256×10^{-2}	6.542×10^{-10}
	GWO	2.036×10^{-2}	6.373×10^{-3}	**3.075×10^{-4}**	9.316×10^{-3}	1.651×10^{-9}
	WOA	1.590×10^{-3}	7.336×10^{-4}	3.223×10^{-4}	3.649×10^{-4}	3.411×10^{-9}
	SCA	1.624×10^{-3}	1.029×10^{-3}	5.172×10^{-4}	3.986×10^{-4}	9.499×10^{-10}
	HS	1.562×10^{-2}	2.036×10^{-3}	6.734×10^{-4}	2.721×10^{-3}	4.490×10^{-10}
	BWO	7.762×10^{-4}	**3.639×10^{-4}**	3.091×10^{-4}	9.203×10^{-5}	8.286×10^{-9}
	GTO	**1.223×10^{-3}**	3.991×10^{-4}	**3.075×10^{-4}**	2.794×10^{-4}	7.665×10^{-1}
	HGTO-BWO	**1.223×10^{-3}**	3.685×10^{-4}	**3.075×10^{-4}**	2.323×10^{-4}	NAN
F16	TSA	-1.000×10^{0}	-1.028×10^{0}	-1.032×10^{0}	9.652×10^{-3}	6.319×10^{-12}
	GWO	**-1.032×10^{0}**	-1.032×10^{0}	-1.032×10^{0}	3.067×10^{-8}	6.319×10^{-12}
	WOA	**-1.032×10^{0}**	-1.032×10^{0}	-1.032×10^{0}	2.284×10^{-9}	6.319×10^{-12}
	SCA	-1.031×10^{0}	-1.032×10^{0}	-1.032×10^{0}	3.080×10^{-5}	6.319×10^{-12}
	HS	-1.031×10^{0}	-1.032×10^{0}	-1.032×10^{0}	1.986×10^{-4}	6.319×10^{-12}
	BWO	-1.03×10^{0}	-1.031×10^{0}	-1.032×10^{0}	3.914×10^{-4}	6.319×10^{-12}
	GTO	**-1.032×10^{0}**	-1.032×10^{0}	-1.032×10^{0}	**6.321×10^{-16}**	7.639×10^{-1}
	HGTO-BWO	**-1.032×10^{0}**	-1.032×10^{0}	-1.032×10^{0}	6.388×10^{-16}	NAN
F17	TSA	3.985×10^{-1}	3.980×10^{-1}	3.979×10^{-1}	1.141×10^{-4}	1.21×10^{-12}
	GWO	3.988×10^{-1}	**3.979×10^{-1}**	3.979×10^{-1}	1.706×10^{-4}	1.21×10^{-12}
	WOA	3.980×10^{-1}	**3.979×10^{-1}**	3.979×10^{-1}	2.229×10^{-5}	1.21×10^{-12}
	SCA	4.262×10^{-1}	4.010×10^{-1}	3.979×10^{-1}	6.015×10^{-3}	1.21×10^{-12}
	HS	3.997×10^{-1}	3.981×10^{-1}	3.979×10^{-1}	4.448×10^{-4}	1.21×10^{-12}
	BWO	4.087×10^{-1}	4.011×10^{-1}	3.979×10^{-1}	2.804×10^{-3}	1.21×10^{-12}
	GTO	**3.979×10^{-1}**	**3.979×10^{-1}**	3.979×10^{-1}	**0.00×10^{0}**	NAN
	HGTO-BWO	**3.979×10^{-1}**	**3.979×10^{-1}**	3.979×10^{-1}	**0.00×10^{0}**	NAN
F18	TSA	3.0×1^{01}	6.600×10^{0}	3.0×10^{0}	9.335×10^{0}	5.21×10^{-12}
	GWO	3.0×10^{0}	3.0×10^{0}	3.0×10^{0}	3.081×10^{-5}	5.21×10^{-12}
	WOA	3.0×10^{0}	3.0×10^{0}	3.0×10^{0}	6.025×10^{-5}	5.21×10^{-12}
	SCA	3.0×10^{0}	3.0×10^{0}	3.0×10^{0}	1.173×10^{-4}	5.21×10^{-12}
	HS	3.006×10^{0}	3.001×10^{0}	3.0×10^{0}	1.271×10^{-3}	5.21×10^{-12}
	BWO	6.659×10^{0}	3.914×10^{0}	3.009×10^{0}	9.835×10^{-1}	5.21×10^{-12}
	GTO	3.0×10^{0}	3.0×10^{0}	3.0×10^{0}	**9.257×10^{-16}**	3.146×10^{-02}
	HGTO-BWO	3.0×10^{0}	3.0×10^{0}	3.0×10^{0}	1.414×10^{-15}	NAN
F19	TSA	-3.862×10^{0}	**-3.863×10^{0}**	**-3.86×10^{0}**	1.091×10^{-4}	7.57×10^{-12}
	GWO	-3.855×10^{0}	-3.862×10^{0}	**-3.86×10^{0}**	2.427×10^{-3}	7.57×10^{-12}
	WOA	-3.834×10^{0}	-3.855×10^{0}	**-3.86×10^{0}**	7.989×10^{-3}	7.57×10^{-12}
	SCA	-3.843×10^{0}	-3.854×10^{0}	-3.862×10^{0}	3.439×10^{-3}	7.57×10^{-12}
	HS	-3.863×10^{0}	**-3.863×10^{0}**	**-3.86×10^{0}**	2.946×10^{-5}	7.57×10^{-12}
	BWO	-3.852×10^{0}	-3.858×10^{0}	-3.862×10^{0}	3.045×10^{-3}	7.57×10^{-12}
	GTO	**-3.863×10^{0}**	**-3.863×10^{0}**	**-3.86×10^{0}**	**2.612×10^{-15}**	1.000
	HGTO-BWO	**-3.863×10^{0}**	**-3.863×10^{0}**	**-3.86×10^{0}**	**2.612×10^{-15}**	NAN
F20	TSA	-2.840×10^{0}	-3.248×10^{0}	-3.321×10^{0}	1.106×10^{-1}	2.646×10^{-7}
	GWO	-3.103×10^{0}	-3.262×10^{0}	**-3.322×10^{0}**	7.766×10^{-2}	2.646×10^{-7}
	WOA	-3.055×10^{0}	-3.248×10^{0}	**-3.322×10^{0}**	9.266×10^{-2}	6.828×10^{-7}
	SCA	-1.454×10^{0}	-2.918×10^{0}	-3.220×10^{0}	3.302×10^{-1}	2.319×10^{-11}
	HS	**-3.203×10^{0}**	-3.302×10^{0}	**-3.322×10^{0}**	**4.509×10^{-2}**	6.360×10^{-6}
	BWO	-3.169×10^{0}	-3.267×10^{0}	-3.317×10^{0}	4.963×10^{-2}	2.667×10^{-6}
	GTO	**-3.203×10^{0}**	**-3.298×10^{0}**	**-3.322×10^{0}**	4.837×10^{-2}	8.819×10^{-1}
	HGTO-BWO	**-3.203×10^{0}**	**-3.298×10^{0}**	**-3.322×10^{0}**	4.837×10^{-2}	NAN

Table 2. *Cont.*

Function No	Algorithm	Worst	Mean	Best	std	p-Value
F21	TSA	-2.603×10^{0}	-7.082×10^{0}	-1.012×10^{1}	3.387×10^{0}	1.406×10^{-11}
	GWO	-2.682×10^{0}	-9.564×10^{0}	-1.015×10^{1}	1.828×10^{0}	1.406×10^{-11}
	WOA	-2.585×10^{0}	-8.285×10^{0}	-1.015×10^{1}	2.950×10^{0}	1.406×10^{-11}
	SCA	-4.973×10^{-1}	-2.264×10^{0}	-5.364×10^{0}	1.829×10^{0}	1.406×10^{-11}
	HS	-2.629×10^{0}	-5.921×10^{0}	-1.015×10^{1}	3.755×10^{0}	1.406×10^{-11}
	BWO	-1.013×10^{1}	$\mathbf{-1.015 \times 10^{1}}$	-1.015×10^{1}	$4.298 \times 10-3$	1.406×10^{-11}
	GTO	$\mathbf{-1.015 \times 10^{1}}$	$\mathbf{-1.015 \times 10^{1}}$	-1.015×10^{1}	$\mathbf{5.827 \times 10^{-15}}$	3.882×10^{-3}
	HGTO-BWO	$\mathbf{-1.015 \times 10^{1}}$	$\mathbf{-1.015 \times 10^{1}}$	-1.015×10^{1}	6.506×10^{-15}	NAN
F22	TSA	-1.826×10^{0}	-7.195×10^{0}	-1.034×1^{01}	3.555×10^{0}	6.387×10^{-12}
	GWO	-5.088×10^{0}	-1.022×10^{1}	-1.04×10^{1}	9.702×10^{-1}	6.387×10^{-12}
	WOA	-1.836×10^{0}	-8.237×10^{0}	-1.04×10^{1}	3.179×10^{0}	6.387×10^{-12}
	SCA	$-9.062 \times 10-1$	-4.124×10^{0}	-7.417×10^{0}	1.629×10^{0}	6.387×10^{-12}
	HS	-2.75×10^{0}	-5.982×10^{0}	-1.04×10^{1}	3.471×10^{0}	6.387×10^{-12}
	BWO	-1.039×10^{1}	$\mathbf{-1.04 \times 10^{1}}$	-1.04×10^{1}	4.324×10^{-3}	6.387×10^{-12}
	GTO	$\mathbf{-1.04 \times 10^{1}}$	$\mathbf{-1.04 \times 10^{1}}$	-1.04×10^{1}	$\mathbf{8.080 \times 10^{-16}}$	1.000
	HGTO-BWO	$\mathbf{-1.04 \times 10^{1}}$	$\mathbf{-1.04 \times 10^{1}}$	-1.04×10^{1}	$\mathbf{8.080 \times 10^{-16}}$	NAN
F23	TSA	-1.671×10^{0}	-5.757×10^{0}	-1.041×10^{1}	3.745×10^{0}	7.574×10^{-12}
	GWO	-2.422×10^{0}	-1.026×10^{1}	$\mathbf{-1.054 \times 10^{1}}$	1.481×10^{0}	7.574×10^{-12}
	WOA	-1.672×10^{0}	-8.079×10^{0}	$\mathbf{-1.054 \times 10^{1}}$	3.345×10^{0}	7.574×10^{-12}
	SCA	-9.436×10^{-1}	-3.886×10^{0}	-5.962×10^{0}	1.374×10^{0}	7.574×10^{-12}
	HS	-2.518×10^{1}	-6.977×10^{0}	$\mathbf{-1.054 \times 10^{1}}$	3.631×10^{0}	7.574×10^{-12}
	BWO	-1.052×10^{1}	-1.053×10^{1}	$\mathbf{-1.054 \times 10^{1}}$	3.777×10^{-3}	7.574×10^{-12}
	GTO	$\mathbf{-1.054 \times 10^{1}}$	$\mathbf{-1.054 \times 10^{1}}$	$\mathbf{-1.054 \times 10^{1}}$	$\mathbf{1.189 \times 10^{-15}}$	3.648×10^{-1}
	HGTO-BWO	$\mathbf{-1.054 \times 10^{1}}$	$\mathbf{-1.054 \times 10^{1}}$	$\mathbf{-1.054 \times 10^{1}}$	2.356×10^{-15}	NAN

5.2. CEC-C06 2019 Benchmark Functions

To verify the efficiency and validation of the suggested HGTO-BWO, it was tested and assessed with the modern functions of CEC-C06 2019 [72,73]. The CEC-C06 2019 benchmark function comprises 10 functions, where CEC01 to CEC03 have variable dimensions, while CEC04 to CEC10 have constant dimensions. The statistical analysis during the solving of CEC01 to CEC10 via the proposed HGTO-BWO and others was conducted, and the fetched results are tabulated in Table 3. The bold values given in Table 3 refer to the best results obtained through the proposed approach.

Table 3. Statistical analysis of CEC-C06 2019 benchmark functions solved via the proposed hybrid approach and others.

Function No	Algorithm	Worst	Mean	Best	Std	p-Value
CEC01	TSA	3.841×10^{9}	2.210×10^{8}	7.038×10^{4}	7.170×10^{8}	3.020×10^{-11}
	GWO	4.821×10^{9}	3.646×10^{8}	4.506×10^{4}	9.087×10^{8}	3.020×10^{-11}
	WOA	1.447×10^{11}	2.554×10^{10}	3.498×10^{6}	4.112×10^{10}	3.020×10^{-11}
	SCA	3.788×10^{10}	8.327×10^{9}	2.425×10^{7}	8.725×10^{9}	3.020×10^{-11}
	HS	9.231×10^{10}	2.119×10^{10}	3.137×10^{9}	1.969×10^{10}	3.020×10^{-11}
	BWO	8.286×10^{4}	6.219×10^{4}	4.906×10^{4}	8.071×10^{3}	3.020×10^{-11}
	GTO	3.981×10^{4}	3.769×10^{4}	3.549×10^{4}	$\mathbf{9.246 \times 10^{2}}$	3.387×10^{-2}
	HGTO-BWO	$\mathbf{3.889 \times 10^{4}}$	$\mathbf{3.704 \times 10^{4}}$	$\mathbf{3.224 \times 10^{4}}$	1.215×10^{3}	NAN

Table 3. *Cont.*

Function No	Algorithm	Worst	Mean	Best	Std	p-Value
CEC02	TSA	1.956×10^{1}	1.840×10^{1}	1.735×10^{1}	7.225×10^{-1}	1.061×10^{-11}
	GWO	1.734×10^{1}	1.734×10^{1}	$\mathbf{1.734 \times 10^{1}}$	2.575×10^{-4}	1.061×10^{-11}
	WOA	1.747×10^{1}	1.735×10^{1}	$\mathbf{1.734 \times 10^{1}}$	2.329×10^{-2}	1.061×10^{-11}
	SCA	1.772×10^{1}	1.749×10^{1}	1.737×10^{1}	8.718×10^{-2}	1.061×10^{-11}
	HS	1.147×10^{2}	5.717×10^{1}	2.004×10^{1}	2.245×10^{1}	1.061×10^{-11}
	BWO	1.780×10^{1}	1.755×10^{1}	1.744×10^{1}	8.285×10^{-2}	1.061×10^{-11}
	GTO	$\mathbf{1.734 \times 10^{1}}$	$\mathbf{1.734 \times 10^{1}}$	$\mathbf{1.734 \times 10^{1}}$	$6.032 \times 10-^{13}$	1.506×10^{-1}
	HGTO-BWO	$\mathbf{1.734 \times 10^{1}}$	$\mathbf{1.734 \times 10^{1}}$	$\mathbf{1.734 \times 10^{1}}$	$\mathbf{1.108 \times 10^{-14}}$	NAN
CEC03	TSA	1.271×10^{1}	1.27×10^{1}	1.27×10^{1}	1.298×10^{-3}	1.720×10^{-12}
	GWO	1.27×10^{1}	1.27×10^{1}	1.27×10^{1}	1.158×10^{-5}	1.720×10^{-12}
	WOA	1.27×10^{1}	1.27×10^{1}	1.27×10^{1}	2.173×10^{-6}	1.720×10^{-12}
	SCA	1.27×10^{1}	1.27×10^{1}	1.27×10^{1}	1.158×10^{-4}	1.720×10^{-12}
	HS	1.27×10^{1}	1.27×10^{1}	1.27×10^{1}	6.272×10^{-7}	1.720×10^{-12}
	BWO	1.27×10^{1}	1.27×10^{1}	1.27×10^{1}	1.160×10^{-4}	1.720×10^{-12}
	GTO	1.27×10^{1}	1.27×10^{1}	1.27×10^{1}	$\mathbf{3.688 \times 10^{-15}}$	1.000
	HGTO-BWO	1.27×10^{1}	1.27×10^{1}	1.27×10^{1}	3.917×10^{-15}	NAN
CEC04	TSA	9.665×10^{3}	4.501×10^{3}	1.712×10^{2}	2.214×10^{3}	3.020×10^{-11}
	GWO	4.212×10^{3}	2.409×10^{2}	2.141×10^{1}	7.933×10^{2}	4.204×10^{-1}
	WOA	6.893×10^{2}	3.777×10^{2}	1.490×10^{2}	1.541×10^{2}	3.020×10^{-11}
	SCA	4.111×10^{3}	1.666×10^{3}	7.541×10^{2}	8.432×10^{2}	3.020×10^{-11}
	HS	1.427×10^{2}	7.831×10^{1}	4.261×10^{1}	2.816×10^{1}	4.218×10^{-4}
	BWO	1.211×10^{4}	7.614×10^{3}	2.383×10^{3}	2.226×10^{3}	3.020×10^{-11}
	GTO	4.388×10^{2}	1.015×10^{2}	3.383×10^{1}	7.859×10^{1}	5.264×10^{-4}
	HGTO-BWO	$\mathbf{1.264 \times 10^{2}}$	$\mathbf{5.574 \times 10^{1}}$	$\mathbf{2.189 \times 10^{1}}$	$\mathbf{2.474 \times 10^{1}}$	NAN
CEC05	TSA	4.893×10^{0}	3.033×10^{0}	1.560×10^{0}	8.286×10^{-1}	4.504×10^{-11}
	GWO	1.891×10^{0}	1.455×10^{0}	1.055×10^{0}	2.840×10^{-1}	1.221×10^{-2}
	WOA	2.786×10^{0}	1.862×10^{0}	1.287×10^{0}	3.340×10^{-1}	4.616×10^{-10}
	SCA	2.442×10^{0}	2.188×10^{0}	2.004×10^{0}	$\mathbf{9.594 \times 10^{-2}}$	3.020×10^{-11}
	HS	1.495×10^{0}	$\mathbf{1.278 \times 10^{0}}$	1.099×10^{0}	1.014×10^{-1}	2.170×10^{-1}
	BWO	4.149×10^{0}	3.617×10^{0}	2.867×10^{0}	3.177×10^{-1}	3.020×10^{-11}
	GTO	1.871×10^{0}	1.249×10^{0}	$\mathbf{1.032 \times 10^{0}}$	2.017×10^{-1}	4.643×10^{-1}
	HGTO-BWO	$\mathbf{1.780 \times 10^{0}}$	1.261×10^{0}	1.047×10^{0}	1.684×10^{-1}	NAN
CEC06	TSA	1.184×10^{1}	1.106×10^{1}	9.047×10^{0}	6.621×10^{-1}	3.020×10^{-11}
	GWO	1.210×10^{1}	1.095×10^{1}	9.122×10^{0}	7.040×10^{-1}	3.020×10^{-11}
	WOA	1.141×10^{1}	9.831×10^{0}	7.546×10^{0}	9.329×10^{-1}	4.504×10^{-11}
	SCA	1.218×10^{1}	1.101×10^{1}	9.583×10^{0}	7.444×10^{-1}	3.020×10^{-11}
	HS	1.118×10^{1}	9.057×10^{0}	6.929×10^{0}	1.404×10^{0}	2.922×10^{-9}
	BWO	1.200×10^{1}	1.090×10^{1}	9.780×10^{0}	$\mathbf{5.720 \times 10^{-1}}$	3.020×10^{-11}
	GTO	1.207×10^{1}	8.029×10^{0}	5.137×10^{0}	1.503×10^{0}	3.835×10^{-6}
	HGTO-BWO	$\mathbf{7.922 \times 10^{0}}$	$\mathbf{6.194 \times 10^{0}}$	$\mathbf{3.711 \times 10^{0}}$	1.058×10^{0}	NAN
CEC07	TSA	1.035×10^{3}	5.909×10^{2}	1.111×10^{2}	2.431×10^{2}	1.010×10^{-8}
	GWO	1.091×10^{3}	5.154×10^{2}	-1.534×10^{1}	3.448×10^{2}	4.639×10^{-5}
	WOA	1.429×10^{3}	6.162×10^{2}	1.365×10^{2}	3.478×10^{2}	2.602×10^{-8}
	SCA	1.140×10^{3}	7.929×10^{2}	4.640×10^{2}	$\mathbf{1.563 \times 10^{2}}$	3.020×10^{-11}
	HS	7.648×10^{2}	2.673×10^{2}	$\mathbf{-2.514 \times 10^{2}}$	2.779×10^{2}	5.943×10^{-2}
	BWO	1.183×10^{3}	8.537×10^{2}	5.020×10^{2}	1.761×10^{2}	3.020×10^{-11}
	GTO	1.098×10^{3}	3.736×10^{2}	-1.016×10^{2}	2.990×10^{2}	1.680×10^{-3}
	HGTO-BWO	$\mathbf{4.571 \times 10^{2}}$	$\mathbf{1.388 \times 10^{2}}$	-1.506×10^{2}	1.838×10^{2}	NAN

Table 3. Cont.

Function No	Algorithm	Worst	Mean	Best	Std	p-Value
CEC08	TSA	7.051×10^{0}	6.333×10^{0}	5.427×10^{0}	4.589×10^{-1}	9.063×10^{-8}
	GWO	6.584×10^{0}	4.984×10^{0}	3.182×10^{0}	8.750×10^{-1}	3.711×10^{-1}
	WOA	7.060×10^{0}	5.965×10^{0}	4.795×10^{0}	5.548×10^{-1}	1.493×10^{-4}
	SCA	6.828×10^{0}	6.055×10^{0}	4.697×10^{0}	4.921×10^{-1}	1.337×10^{-5}
	HS	$\mathbf{6.117 \times 10^{0}}$	$\mathbf{4.841 \times 10^{0}}$	$\mathbf{2.763 \times 10^{0}}$	9.617×10^{-1}	2.062×10^{-1}
	BWO	6.843×10^{0}	6.311×10^{0}	5.657×10^{0}	$\mathbf{2.931 \times 10^{-1}}$	2.922×10^{-9}
	GTO	6.668×10^{0}	5.235×10^{0}	3.694×10^{0}	6.907×10^{-1}	9.000×10^{-1}
	HGTO-BWO	6.183×10^{0}	5.159×10^{0}	3.630×10^{0}	7.707×10^{-1}	NAN
CEC09	TSA	1.579×10^{3}	2.795×10^{2}	3.468×10^{0}	4.309×10^{2}	3.020×10^{-11}
	GWO	6.206×10^{0}	4.419×10^{0}	3.038×10^{0}	8.261×10^{-1}	3.020×10^{-11}
	WOA	2.169×10^{1}	5.703×10^{0}	3.495×10^{0}	3.128×10^{0}	3.020×10^{-11}
	SCA	3.871×10^{2}	1.150×10^{2}	9.989×10^{0}	9.173×10^{1}	3.020×10^{-11}
	HS	5.387×10^{0}	3.612×10^{0}	2.680×10^{0}	6.163×10^{-1}	4.077×10^{-11}
	BWO	1.964×10^{3}	1.230×10^{3}	6.707×10^{2}	3.098×10^{2}	$3.020 \times 10-^{11}$
	GTO	4.680×10^{0}	2.803×10^{0}	2.413×10^{0}	4.563×10^{-1}	1.091×10^{-5}
	HGTO-BWO	$\mathbf{2.804 \times 10^{0}}$	$\mathbf{2.476 \times 10^{0}}$	$\mathbf{2.369 \times 10^{0}}$	$\mathbf{1.148 \times 10^{-1}}$	NAN
CEC10	TSA	2.065×10^{1}	2.047×10^{1}	2.022×10^{1}	1.108×10^{-1}	3.338×10^{-11}
	GWO	2.065×10^{1}	2.050×10^{1}	2.031×10^{1}	$\mathbf{8.815 \times 10^{-2}}$	3.020×10^{-11}
	WOA	2.052×10^{1}	2.028×10^{1}	2.007×10^{1}	1.117×10^{-1}	2.371×10^{-10}
	SCA	2.064×10^{1}	2.049×10^{1}	2.028×10^{1}	9.279×10^{-2}	3.020×10^{-11}
	HS	2.054×10^{1}	2.033×10^{1}	2.008×10^{1}	1.281×10^{-1}	1.464×10^{-10}
	BWO	2.061×10^{1}	2.044×10^{1}	2.020×10^{1}	1.031×10^{-1}	3.338×10^{-11}
	GTO	2.031×10^{1}	1.954×10^{1}	3.734×10^{0}	2.988×10^{0}	2.921×10^{-2}
	HGTO-BWO	$\mathbf{2.023 \times 10^{1}}$	$\mathbf{1.937 \times 10^{1}}$	$\mathbf{7.754 \times 10^{-13}}$	3.658×10^{0}	NAN

6. Application of HGTO-BWO: Parameter Estimation of PV Cell/Module

The proposed HGTO-BWO was applied to solve a vital problem regarding the identification of the optimal parameters of the PV cell/panel equivalent circuit with the aid of experimental data. The topic is very important as it is necessary to establish a reliable model of a PV system that simulates reality. This helps many researchers to conduct their work in the constructed circuit via the proposed methodology.

The parameters were computed in standard operating conditions for the PVM752 cell, STM6-40/36 panel, and PWP-201 module. Also, the double diode models for the KC200GT and MSX60 were constructed under various solar irradiances and temperatures. Table 4 shows the considered upper and lower limits of the design variables.

Table 4. The upper and lower limits of design variables for various PV cell/models.

Parameters	PVW 752		STM6-40/36		PWP-201		MSX60		KC200GT	
	Lb	Ub	Lb	Ub	Lb	Ub	Lb	Ub	Lb	Ub
$A_{1,2,3}$	1	2	1	60	1	50	1	2	1	2
R_s	0	0.8	0	0.36	0	2	0	2	0	2
R_{sh}	0	1000	0	1000	0	2000	0	500	0	500
I_{d1}, I_{d2}, I_{d3}	0	1×10^{-6}	0	50×10^{-6}	0	50×10^{-6}	0	10×10^{-6}	0	10×10^{-6}
I_{ph}	0	0.5	0	2	0	2	0	8	0	16.4

6.1. Case 1: Constant Weather Conditions

6.1.1. PVW 752 Cell

The proposed HGTO-BWO was employed to determine the parameters of the DDM and TDM for a PVM752 GaAs thin film cell at 25 °C and 1000 W/m^2; the electric characteristics of the PVM752 cell and the measured *I-V* data are presented in [74]. The convergence curves obtained by the considered optimizers for both models are shown in Figure 5. The

RMSE value, statistical analysis, and optimal parameters of the DDM obtained through the proposed HGTO-BWO in comparison to the others are illustrated in Table 5. The least obtained RMSEs for the DDM and TDM were achieved by HGTO-BWO with values of 2.0886×10^{-4} and 1.527×10^{-4}, respectively. On the other hand, the GTO approach ranked second, achieving fitness values of 4.6815×10^{-4} for the DDM and 2.278×10^{-4} for the TDM. The HS algorithm was the worst approach; it provided fitness values of 6.6870×10^{-1} for the DDM and 3.738 for the TDM. Figure 6 shows the estimated and measured *P-V* and *I-V* curves of the DDM and TDM. It is notable that the estimated curves converge closely with the measured data; this means the PV cell/panel performed well and converged with the real one. The statistical parameters including best, worst, mean, and std are shown in Figure 7. The results shown in Table 5 verify the effectiveness of the suggested approach in extracting the parameters with the least fitness values compared to the other algorithms. The values given in bold indicate the best solutions obtained by the proposed HGTO-BWO approach.

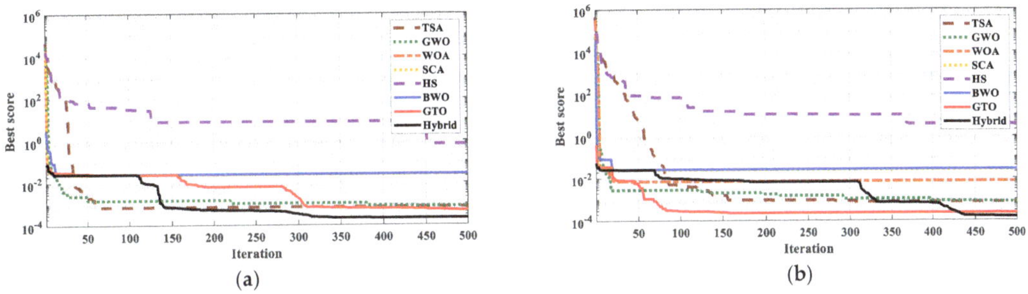

Figure 5. Convergence curves for PVW 752: (**a**) DDM and (**b**) TDM.

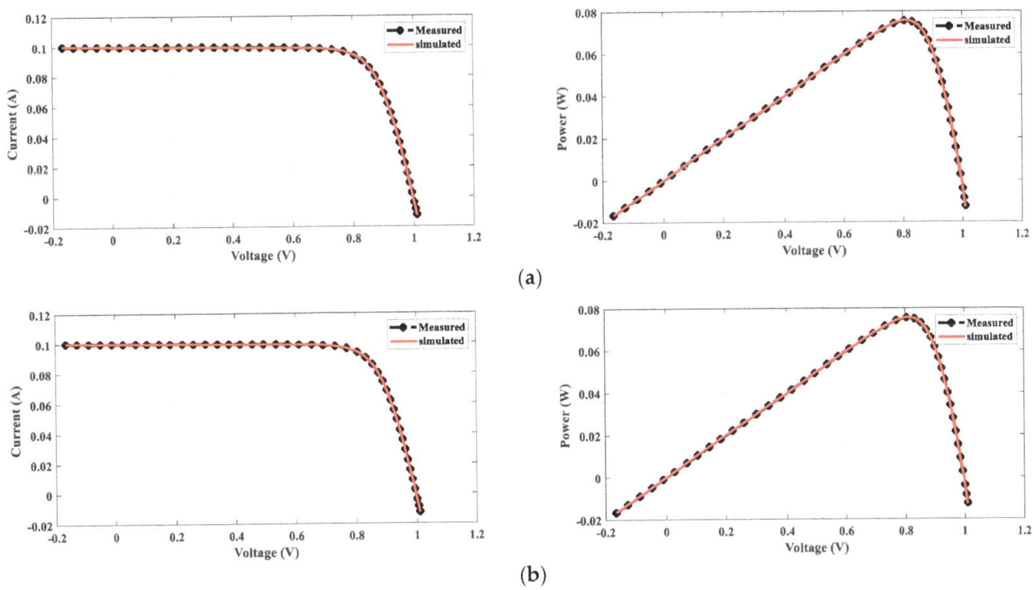

Figure 6. Measured and determined *I-V* and *P-V* curves for the PVM752 cell: (**a**) DDM and (**b**) TDM.

Table 5. Optimal parameters of PVW752 PV cell DDM and TDM models.

Alg.	TSA	GWO [75]	WOA [76]	SCA	HS	BWO	GTO	HGTO-BWO
DDM								
A_1	1.7120	1.3103	1.7975	2.0000	1.9999	1.0000	1.9910	1.3954
A_2	2.0000	2.0000	1.7975	2.0000	1.9998	1.0000	1.8486	1.8296
Rs	0.5153	0.5188	0.0000	0.0000	0.1255	0.0000	0.5764	0.6759
R_{sh}	1000.000	602.772	14.589	14.559	995.383	15.763	996.861	616.390
I_{d1}	2.173×10^{-16}	0.0×10^{0}	0.0×10^{0}	0.0×10^{0}	3.303×10^{-10}	0.0×10^{0}	0.0×10^{0}	4.698×10^{-14}
I_{d2}	3.824×10^{-10}	3.803×10^{-10}	0.0×10^{0}	0.0×10^{0}	7.259×10^{-9}	0.0×10^{0}	7.798×10^{-11}	2.822×10^{-11}
I_{ph}	0.1002	0.1004	0.1138	0.1139	0.5000	0.1096	0.1000	0.1001
RMSE	7.1892×10^{-4}	7.6797×10^{-4}	2.5400×10^{-2}	2.5400×10^{-2}	6.6870×10^{-1}	2.5504×10^{-2}	4.6815×10^{-4}	$\mathbf{2.0886 \times 10^{-4}}$
Worst	2.5405×10^{-2}	2.5416×10^{-2}	8.3219×10^{-2}	8.3219×10^{-2}	1.8163×10^{1}	8.1095×10^{-2}	2.5400×10^{-2}	$\mathbf{2.5400 \times 10^{-2}}$
Mean	3.3031×10^{-3}	2.3763×10^{-2}	3.8904×10^{-2}	2.7333×10^{-2}	6.1022×10^{0}	4.1477×10^{-2}	2.2662×10^{-2}	1.4662×10^{-2}
std	4.7330×10^{-3}	6.2492×10^{-3}	2.4866×10^{-2}	1.0555×10^{-2}	4.4800×10^{0}	1.4131×10^{-2}	6.5555×10^{-3}	1.1001×10^{-2}
p-value	7.9106×10^{-3}	1.6060×10^{-9}	2.1821×10^{-11}	2.262×10^{-11}	2.262×10^{-11}	2.262×10^{-11}	1.0147×10^{-3}	NAN
TDM								
A_1	1.9293	1.6268	1.9620	1.0000	1.9987	1.0000	1.6157	1.9993
A_2	2.0000	2.0000	1.9893	1.1864	1.9981	1.0000	1.0000	1.1530
A_3	2.0000	1.9859	1.9977	1.1541	1.9995	1.0000	1.0050	1.9992
R_s	0.5356	0.5161	0.0000	0.0000	0.0027	0.0131	0.6605	0.7163
R_{sh}	853.5804	628.6429	94.9593	14.6251	996.7450	19.8767	608.0249	720.9139
I_{d1}	2.471×10^{-11}	0.0×10^{0}	2.422×10^{-10}	0.0×10^{0}	6.784×10^{-9}	0.0×10^{0}	3.779×10^{-12}	1.708×10^{-10}
I_{d2}	3.088×10^{-10}	3.559×10^{-10}	0.0×10^{0}	0.0×10^{0}	1.092×10^{-9}	0.0×10^{0}	0.0×10^{0}	1.371×10^{-16}
I_{d3}	2.362×10^{-11}	2.097×10^{-11}	0.0×10^{0}	0.0×10^{0}	2.872×10^{-8}	0.0×10^{0}	0.0×10^{0}	2.894×10^{-20}
I_{ph}	0.1004	0.1003	0.0996	0.1137	0.4999	0.1015	0.1001	0.1000
RMSE	7.510×10^{-4}	7.603×10^{-4}	7.171×10^{-3}	2.540×10^{-2}	3.738×10^{0}	2.642×10^{-2}	2.278×10^{-4}	$\mathbf{1.527 \times 10^{-4}}$
Worst	1.4277×10^{-2}	1.7234×10^{-2}	3.1417×10^{-1}	3.1109×10^{-1}	9.7810×10^{-3}	1.4804×10^{0}	3.1880×10^{-3}	2.4999×10^{-3}
Mean	7.7371×10^{-3}	5.9997×10^{-3}	5.1015×10^{-2}	1.5508×10^{-1}	4.4487×10^{-3}	4.6697×10^{-1}	2.5146×10^{-3}	1.9473×10^{-3}
std	2.9796×10^{-3}	2.7148×10^{-3}	8.9905×10^{-2}	1.4850×10^{-1}	1.5576×10^{-3}	3.2987×10^{-1}	4.5441×10^{-4}	1.4158×10^{-4}
p-value	3.019×10^{-11}	3.019×10^{-11}	3.019×10^{-11}	3.019×10^{-11}	3.6897×10^{-11}	3.019×10^{-11}	3.2555×10^{-7}	NAN

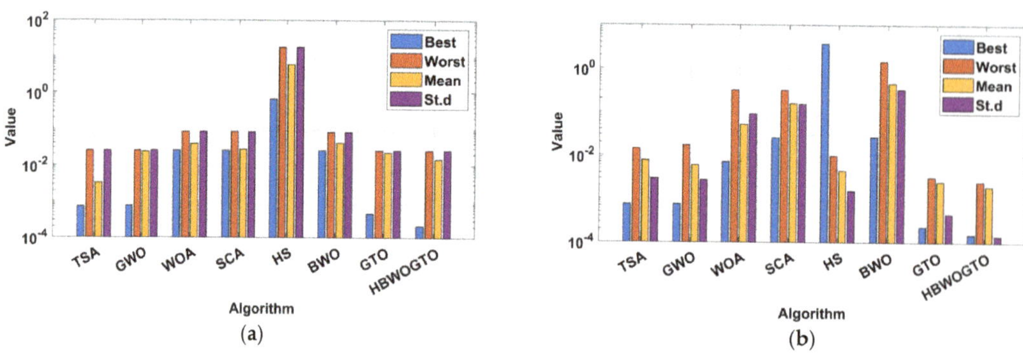

Figure 7. Bar chart of statistical analysis for the PVM752 cell: (**a**) DDM and (**b**) TDM.

6.1.2. PV Panel

The analyzed panels in this work were the Photowatt PWP-201 and STM6-40/36. The first one was investigated at 51 °C and 1000 W/m^2, while the PWP-201 module was investigated at 45 °C and 1000 W/m^2. The measured data for both considered panels are given in [54,77]. The fetched results for the DDMs of both panels are tabulated in Table 6. The proposed HGTO-BWO came in the first rank with the best RMSEs of 2.42508×10^{-3} for PWP-201 and 1.8032×10^{-3} for STM6-40/36, while the GTO approach was in the second rank with fitness values of 2.42511×10^{-3} and 1.88×10^{-3} for PWP-201 and STM6-40/36, respectively. The computed and measured *P-V* and *I-V* curves are shown in Figure 8. The obtained curves are completely consistent with the measured data, and this confirms the preference of the proposed method in obtaining an approved equivalent circuit that simulates reality. The mean ranking of the RMSE values is illustrated in

Figure 9. It is confirmed that the proposed HGTO-BWO has the greatest rank among the applied algorithms, while Figure 10 illustrates the bar chart of the statistical analysis of the DDM. Moreover, the TDM optimal parameters and statistical analysis achieved by the proposed HGTO-BWO and the others are presented in Table 7. It is observed that the best fitness values were attained by the proposed HGTO-BWO with values of 2.2068×10^{-3} for PWP-201 and 1.7435×10^{-3} for STM6-40/36, while the worst fitness values were obtained via BWO with 1.5132×10^{-1} and 2.4985×10^{-1} for PWP-201 and STM6-40/36, respectively. Figure 11 depicts the measured and computed *P-V* and *I-V* curves of the TDM; the computed data converge with the measured ones; this validates the competence of the proposed HGTO-BWO. The mean ranking of the RMSE values and the statistical analysis for the TDM are presented in Figures 12 and 13, respectively. The RMSE values in bold indicate the best solutions obtained by the proposed HGTO-BWO approach.

Table 6. The calculated parameters of DDM and statistical analysis for PWP-201 and STM6-40/36 panels.

Alg.	TSA	GWO [75]	WOA [76]	SCA	HS	BWO	GTO	HGTO-BWO
PWP201								
A_1	50.0000	50.0000	46.5652	50.0000	49.2825	1.0000	48.6368	48.6477
A_2	48.3674	46.5502	49.9999	42.6562	49.5258	50.0000	1.0000	48.6330
Rs	1.1616	1.1824	1.1658	0.9700	1.1852	1.8645	1.2013	1.2012
R_{sh}	880.7826	1220.8951	1999.9965	904.1055	1420.8958	109.5641	977.9933	982.7504
I_{d1}	4.343×10^{-6}	4.180×10^{-6}	0.0×10^{0}	4.661×10^{-6}	3.556×10^{-6}	0.0×10^{0}	3.477×10^{-6}	2.886×10^{-6}
I_{d2}	3.688×10^{-7}	2.946×10^{-7}	4.923×10^{-6}	0.0×10^{0}	5.916×10^{-7}	3.483×10^{-6}	0.0×10^{0}	5.988×10^{-7}
I_{ph}	1.0345	1.0303	1.0275	1.0158	1.0285	0.8628	1.0305	1.0305
RMSE	3.20292×10^{-3}	2.56748×10^{-3}	2.63181×10^{-3}	2.42124×10^{-2}	2.49283×10^{-3}	2.34014×10^{-1}	2.42511×10^{-3}	$\mathbf{2.42508} \times 10^{-3}$
Worst	1.4277×10^{-2}	1.7234×10^{-2}	3.1417×10^{-1}	3.1109×10^{-1}	9.7810×10^{-3}	1.4804×10^{0}	3.1880×10^{-3}	2.4999×10^{-3}
Mean	7.7371×10^{-3}	5.9997×10^{-3}	5.1015×10^{-2}	1.5508×10^{-1}	4.4487×10^{-3}	4.6697×10^{-1}	2.5146×10^{-3}	1.9473×10^{-3}
std	2.9796×10^{-3}	2.7148×10^{-3}	8.9905×10^{-2}	1.4850×10^{-1}	1.5576×10^{-3}	3.2987×10^{-1}	4.5441×10^{-4}	1.4158×10^{-4}
p-value	3.019×10^{-11}	3.019×10^{-11}	3.019×10^{-11}	3.019×10^{-11}	3.6897×10^{-11}	3.019×10^{-11}	3.2555×10^{-7}	NAN
STM6-40/36								
A_1	55.8706	54.3091	59.9590	60.0000	57.9886	54.6034	59.9962	60.0000
A_2	60.0000	55.3299	59.9590	26.9562	48.3514	54.8365	45.6821	43.5068
Rs	0.0902	0.1768	0.0190	0.0000	0.2181	0.0056	0.2450	0.2818
R_{sh}	929.6713	768.3818	999.3155	484.4850	649.1660	87.2447	611.4645	574.3640
I_{d1}	1.121×10^{-6}	1.564×10^{-6}	5.814×10^{-6}	5.764×10^{-6}	1.966×10^{-6}	7.834×10^{-7}	2.858×10^{-6}	2.855×10^{-6}
I_{d2}	3.006×10^{-6}	3.692×10^{-9}	0.000	0.000	1.356×10^{-7}	7.834×10^{-7}	5.815×10^{-8}	2.551×10^{-8}
I_{ph}	1.6596	1.6584	1.6613	1.6695	1.6616	1.8222	1.6635	1.6643
RMSE	3.8737×10^{-3}	3.2441×10^{-3}	4.3087×10^{-3}	7.5372×10^{-3}	2.1464×10^{-3}	7.7568×10^{-2}	1.8800×10^{-3}	$\mathbf{1.8032 \times 10^{-3}}$
Worst	1.4277×10^{-2}	1.7234×10^{-2}	3.1417×10^{-1}	3.1109×10^{-1}	9.7810×10^{-3}	1.4801×10^{0}	3.1880×10^{-3}	2.4999×10^{-3}
Mean	7.7371×10^{-3}	5.9997×10^{-3}	5.1015×10^{-2}	1.5508×10^{-1}	4.4487×10^{-3}	4.6697×10^{-1}	2.5146×10^{-3}	1.9473×10^{-3}
std	2.9796×10^{-3}	2.7148×10^{-3}	8.9905×10^{-2}	1.4850×10^{-1}	1.5576×10^{-3}	3.2987×10^{-1}	4.5441×10^{-4}	1.4158×10^{-4}
p-value	3.0199×10^{-11}	3.0199×10^{-11}	3.0199×10^{-11}	3.0199×10^{-11}	3.6897×10^{-11}	3.0199×10^{-11}	3.2555×10^{-7}	NAN

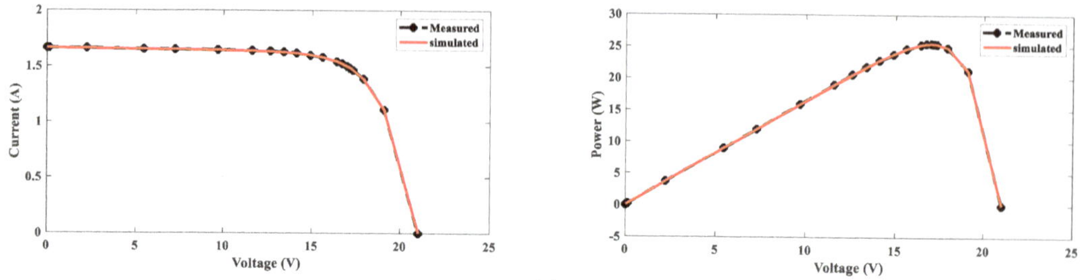

(a)

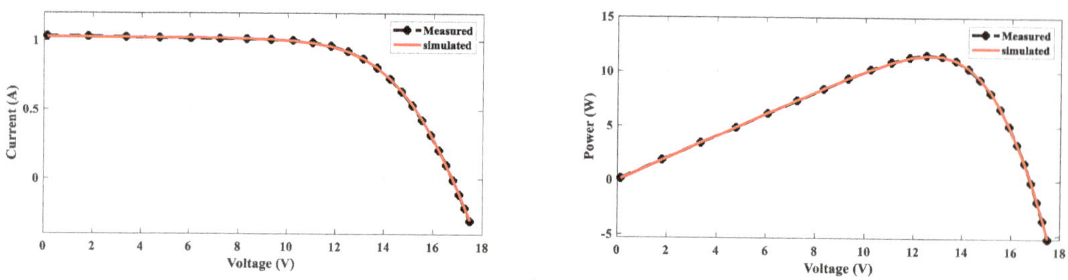

(b)

Figure 8. P-V and I-V curves for DDM: (a) STM6-40/36 and (b) PWP-201.

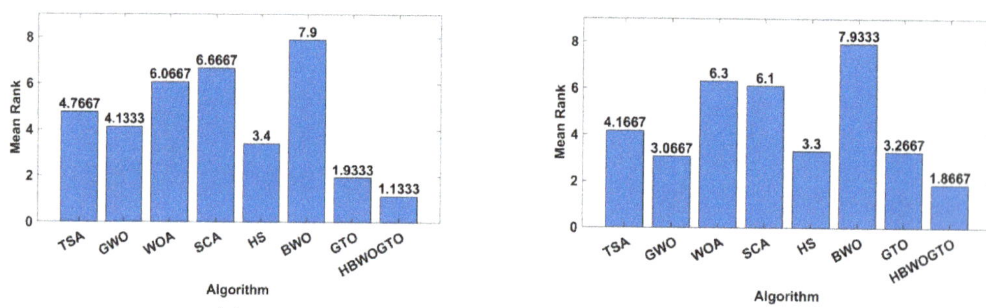

(a) (b)

Figure 9. The mean ranking RMSE DDM of the Friedman test: (a) STM6-40/36 panels and (b) PWP-201.

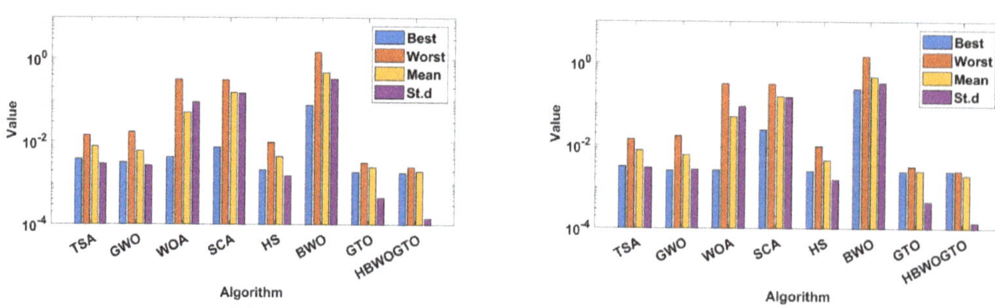

(a) (b)

Figure 10. The statistical analysis for the DDM: (a) STM6-40/36 panels and (b) PWP-201.

Table 7. TDM optimal parameters and statistical analysis for PWP-201 and STM6-40/36 panels.

Alg.	TSA	GWO [75]	WOA [76]	SCA	HS	BWO	GTO	HGTO-BWO
PWP201								
A_1	50.0000	47.4470	50.0000	50.0000	49.3165	42.8772	49.9856	44.4743
A_2	50.0000	49.7858	49.4766	6.9151	48.5919	1.0000	1.0000	14.9639
A_3	49.2038	47.8414	9.1965	4.9402	48.4106	42.0707	48.6492	47.8405
R_s	1.1754	1.1860	1.1625	1.1260	1.1985	1.7161	1.2011	1.2922
R_{sh}	1434.1924	1248.8693	2000.0000	116.3588	1117.9430	171.5803	984.7595	1030.1916
I_{d1}	4.635×10^{-8}	5.765×10^{-7}	4.924×10^{-6}	4.681×10^{-6}	8.715×10^{-7}	0.0×10^{0}	0.0×10^{0}	4.465×10^{-9}
I_{d2}	3.161×10^{-6}	3.603×10^{-6}	0.0×10^{0}	0.0×10^{0}	2.462×10^{-6}	0.0×10^{0}	0.0×10^{0}	4.699×10^{-20}
I_{d3}	1.405×10^{-6}	0.0×10^{0}	0.0×10^{0}	0.0×10^{0}	2.458×10^{-7}	3.997×10^{-7}	3.488×10^{-6}	2.730×10^{-6}
I_{ph}	1.0297	1.0291	1.0274	1.1189	1.0299	0.9035	1.0305	1.0297
RMSE	2.6051×10^{-3}	2.4976×10^{-3}	2.6738×10^{-3}	2.7293×10^{-2}	2.4956×10^{-3}	1.5132×10^{-1}	2.4251×10^{-3}	**2.2068** $\times 10^{-3}$
Worst	1.3165×10^{-1}	7.2474×10^{-3}	7.8391×10^{-1}	7.8391×10^{-1}	1.8873×10^{-2}	7.8391×10^{-1}	2.7425×10^{-1}	2.7425×10^{-1}
Mean	1.0840×10^{-2}	3.4417×10^{-3}	2.2133×10^{-1}	2.4793×10^{-1}	6.1577×10^{-3}	4.6665×10^{-1}	9.3345×10^{-2}	2.9811×10^{-2}
std	2.3150×10^{-2}	1.1862×10^{-3}	2.2042×10^{-1}	1.7304×10^{-1}	3.5373×10^{-3}	1.6864×10^{-1}	1.3011×10^{-1}	8.2878×10^{-2}
p-value	6.5238×10^{-7}	3.8338×10^{-6}	2.4324×10^{-9}	4.1950×10^{-10}	1.3848×10^{-6}	5.4773×10^{-11}	9.2129×10^{-3}	NAN
STM6-40/36								
A_1	60.0000	18.8859	27.4319	60.0000	58.0623	50.3727	59.8760	59.8001
A_2	60.0000	55.1060	59.9990	56.4263	57.8885	49.1217	45.3316	37.5844
A_3	60.0000	56.1273	59.9990	60.0000	55.5272	49.1043	59.9974	59.9783
R_s	0.0003	0.1634	0.0123	0.0000	0.0883	0.0003	0.2649	0.3455
R_{sh}	933.7652	1000.00	861.6239	577.2329	901.7309	54.1090	570.3259	595.6377
I_{d1}	1.215×10^{-6}	0.0×10^{0}	0.0×10^{0}	0.0×10^{0}	2.489×10^{-6}	3.932×10^{-8}	1.808×10^{-13}	3.100×10^{-6}
I_{d2}	2.828×10^{-6}	1.365×10^{-6}	5.857×10^{-6}	0.0×10^{0}	7.033×10^{-7}	3.932×10^{-8}	5.465×10^{-8}	1.456×10^{-9}
I_{d3}	1.804×10^{-6}	7.092×10^{-7}	0.0×10^{0}	5.828×10^{-6}	3.482×10^{-7}	3.932×10^{-8}	2.653×10^{-6}	8.272×10^{-9}
I_{ph}	1.6619	1.6569	1.6627	1.6656	1.6599	1.6349	1.6643	1.6643
RMSE	4.8420×10^{-3}	3.4043×10^{-3}	4.4379×10^{-3}	6.1808×10^{-3}	3.3162×10^{-3}	2.4985×10^{-1}	1.8235×10^{-3}	$\mathbf{1.7435 \times 10^{-3}}$
Worst	2.9048×10^{-2}	2.0280×10^{-2}	1.5378×10^{0}	1.5378×10^{0}	1.3253×10^{-2}	1.3756×10^{0}	4.5153×10^{-3}	3.2509×10^{-3}
Mean	9.9667×10^{-3}	6.9313×10^{-3}	3.2435×10^{-1}	2.3526×10^{-1}	5.9740×10^{-3}	4.4539×10^{-1}	2.5938×10^{-3}	2.0732×10^{-3}
std	5.2223×10^{-3}	3.9683×10^{-3}	5.6249×10^{-1}	2.8494×10^{-1}	2.9328×10^{-3}	2.5235×10^{-1}	6.5633×10^{-4}	3.0115×10^{-4}
p-value	3.0199×10^{-11}	3.019×10^{-11}	2.9822×10^{-11}	3.0199×10^{-11}	3.019×10^{-11}	3.0199×10^{-11}	1.8575×10^{-3}	NAN

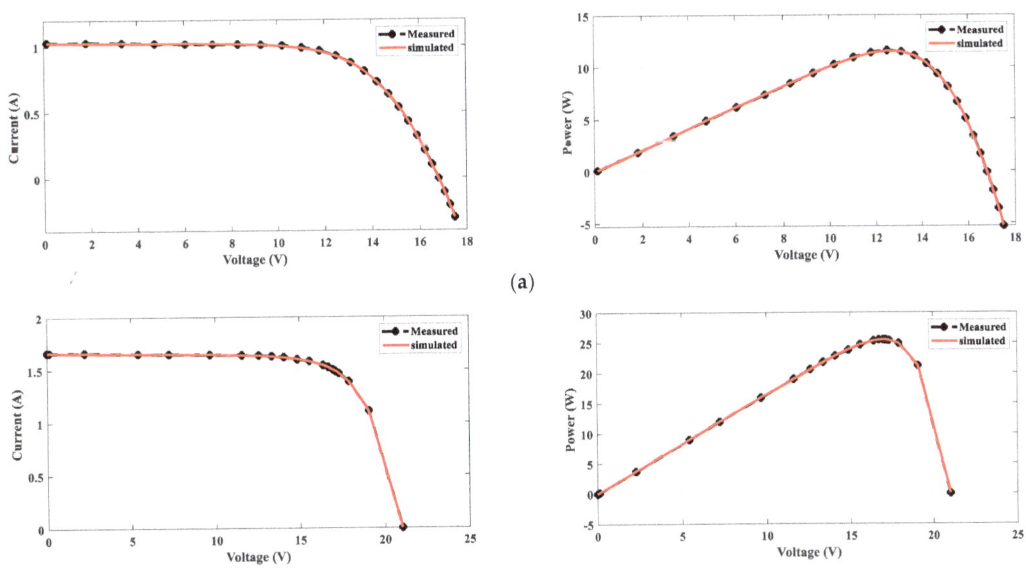

(b)

Figure 11. The *I-V* and *P-V* curves for TDM: (**a**) PWP-201 and (**b**) STM6-40/36.

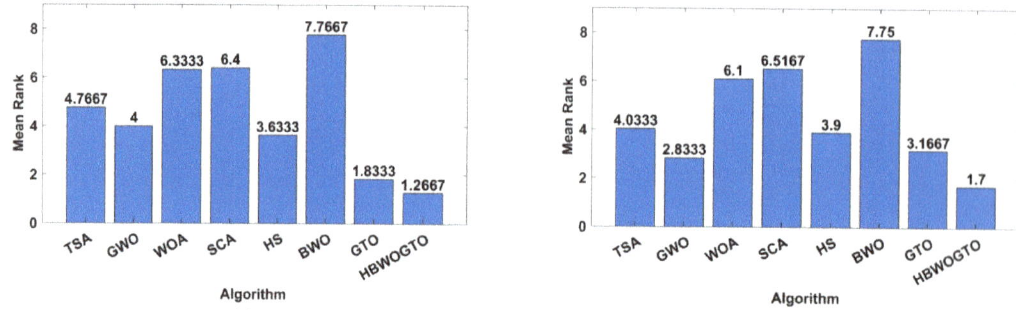

(a) (b)

Figure 12. The mean ranking RMSE TDM of the Friedman test: (**a**) STM6-40/36 panels and (**b**) PWP-201.

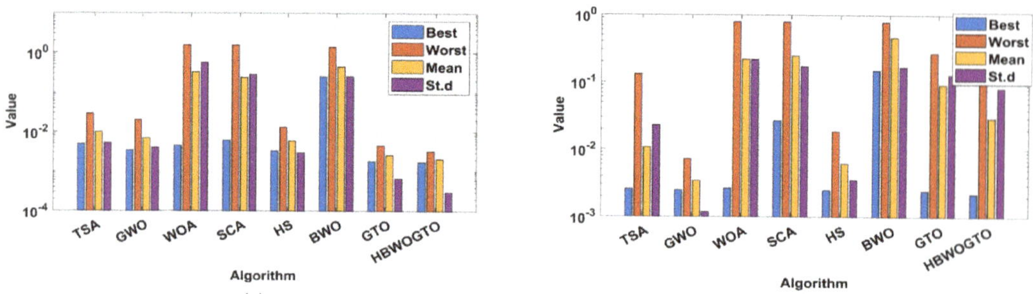

(a) (b)

Figure 13. The statistical analysis for the TDM: (**a**) STM6-40/36 panels and (**b**) PWP-201.

6.2. Case 2: Variable Weather Conditions

The changes in temperature and solar radiation should be considered during the design of the PV system as they have great influence on the system's efficiency [78]. The proposed HGTO-BWO constructed the DDM of MSX60 and KC200GT PV panels at various weather situations with the aid of the data given in [79]. Table 8 displays the statistical analysis of KC200GT in case A and B, where case A was conducted by operating the panel at 1000 W/m^2 and at the different temperatures of 25 °C, 50 °C, and 75 °C. In case B, the panel was operated at 25 °C and at the various irradiances of 1000 W/m^2, 800 W/m^2, 600 W/m^2, 400 W/m^2, and 200 W/m^2. The best RMSE values of 3.5092×10^{-3} and 1.6067×10^{-3} were obtained during operation at 25 °C and 50 °C via the proposed approach. Additionally, it achieved the minimum fitness values of 9.1596×10^{-4}, 6.3910×10^{-4}, 7.7891×10^{-4}, and 2.3850×10^{-4} during operation at 800 W/m^2, 600 W/m^2, 400 W/m^2, and 200 W/m^2, respectively. Figure 14 illustrates the simulated and measured data of *I-V* and *P-V* curves. The statistical parameters of the approach during the establishment of the circuit of KC200GT at irradiances of 1000 W/m^2 and a temperature of 25 °C are illustrated in Figure 15 while the statistical analyses of MSX60 during various temperature and irradiances are displayed in Table 9. The proposed approach achieved the best RMSE values for MSX60 of 1.0765×10^{-4}, 1.9324×10^{-4}, and 2.9790×10^{-5} at 1000 W/m^2 and temperatures of 25 °C, 50 °C, and 75 °C, respectively. Moreover, at 25 °C the fitness values were 1.1336×10^{-3} at 800 W/m^2; 6.7775×10^{-4} at 600 W/m^2; 2.4366×10^{-5} at 400 W/m^2; and 5.9828×10^{-5} at 200 W/m^2. The estimated and measured curves are shown in Figure 16. Moreover, the bar chart of the statistical analysis for MSX60 at 1000 W/m^2 and 25 °C is given in Figure 17. The curves confirmed the efficiency and reliability of the suggested HGTO-BWO technique in establishing the PV panel equivalent circuit at different operating conditions.

Table 8. Statistical analysis of KC200GT obtained via the proposed approach and others.

Alg.	TSA	GWO [75]	WOA [76]	SCA	HS	BWO	GTO	HGTO-BWO
25 °C −1000 W/m²								
Worst	2.9928×10^{-1}	3.0217×10^{-1}	4.8023×10^{-1}	1.9379	1.1325×10^{-1}	1.4025	4.8847×10^{-2}	4.4998×10^{-2}
Mean	1.2413×10^{-1}	1.4665×10^{-1}	2.0105×10^{01}	5.6411×10^{-1}	1.0267×10^{-1}	8.1383×10^{-1}	2.0235×10^{-2}	1.7427×10^{-2}
Best	7.4638×10^{-2}	7.4290×10^{-2}	5.6054×10^{-2}	2.7115×10^{-1}	8.4595×10^{-2}	3.3635×10^{-1}	4.3534×10^{-3}	$\mathbf{3.5092 \times 10^{-3}}$
std	5.0151×10^{-2}	6.5383×10^{-2}	1.0046×10^{-1}	5.5487×10^{-1}	5.7736×10^{-3}	3.1312×10^{-1}	1.2395×10^{-2}	1.0633×10^{-2}
p-value	3.019×10^{-11}	3.0199×10^{-11}	3.0199×10^{-11}	3.0199×10^{-11}	3.0199×10^{-11}	3.0199×10^{-11}	4.3764×10^{-1}	NAN
50 °C −1000 W/m²								
Worst	$3.6628 \times 10{-}^{1}$	4.1794×10^{-1}	5.4631×10^{-1}	1.8390	7.5862×10^{-2}	1.2964	9.7639×10^{-3}	6.5681×10^{-3}
Mean	8.4434×10^{-2}	1.1236×10^{-1}	1.7634×10^{-1}	6.2204×10^{-1}	5.7186×10^{-2}	7.9166×10^{-1}	5.5423×10^{-3}	3.5076×10^{-3}
Best	3.7404×10^{-2}	2.3983×10^{-2}	3.6157×10^{-2}	2.3277×10^{-1}	4.2992×10^{-2}	2.7384×10^{-1}	1.7663×10^{-3}	$\mathbf{1.6067 \times 10^{-3}}$
std	6.0767×10^{-2}	1.2209×10^{-1}	1.3871×10^{-1}	4.9262×10^{-1}	7.9771×10^{-3}	2.3545×10^{-1}	1.7398×10^{-3}	1.2797×10^{-3}
p-value	3.019×10^{-11}	3.0199×10^{-11}	3.0199×10^{-11}	3.0199×10^{-11}	3.0199×10^{-11}	3.0199×10^{-11}	7.7387×10^{-6}	NAN
75 °C −1000 W/m²								
Worst	6.2182×10^{-1}	1.0621×10^{-1}	7.1242×10^{-1}	1.7607	6.3285×10^{-2}	1.5777	2.2913×10^{-2}	1.6548×10^{-2}
Mean	9.0915×10^{-2}	4.9024×10^{-2}	1.7288×10^{-1}	5.6890×10^{-1}	2.5581×10^{-2}	7.3398×10^{-1}	8.9380×10^{-3}	7.5146×10^{-3}
Best	1.6338×10^{-2}	9.4445×10^{-3}	1.1059×10^{-2}	1.0406×10^{-1}	7.6656×10^{-3}	2.0249×10^{-1}	$\mathbf{6.6018 \times 10^{-3}}$	6.6031×10^{-3}
std	1.3247×10^{-1}	2.1144×10^{-2}	$2.2991 \times 10_{-1}$	3.8031×10^{-1}	1.5386×10^{-2}	3.1381×10^{-1}	3.3844×10^{-3}	1.8398×10^{-3}
p-value	3.3384×10^{-11}	3.3384×10^{-11}	3.3384×10^{-11}	3.0199×10^{-11}	1.4643×10^{-10}	3.0199×10^{-11}	1.0315×10^{-2}	NAN
25 °C −800 W/m²								
Worst	1.4449×10^{-1}	2.2284×10^{-1}	7.0500×10^{-1}	1.4772	1.5979×10^{-1}	1.1877	3.3746×10^{-2}	3.2627×10^{-2}
Mean	9.0134×10^{-2}	1.0398×10^{-1}	1.9967×10^{-1}	2.5799×10^{-1}	8.4035×10^{-2}	6.0816×10^{-1}	1.4743×10^{-2}	1.1191×10^{-2}
Best	3.5152×10^{-2}	4.3232×10^{-2}	6.5955×10^{-2}	1.4555×10^{-1}	6.0633×10^{-2}	2.8132×10^{-1}	1.2057×10^{-3}	$\mathbf{9.1596 \times 10^{-4}}$
std	2.6954×10^{-2}	3.9288×10^{-2}	1.5380×10^{-1}	2.3535×10^{-1}	1.8618×10^{-2}	2.6438×10^{-1}	9.0202×10^{-3}	7.1797×10^{-3}
p-value	3.019×10^{-11}	3.0199×10^{-11}	3.0199×10^{-11}	3.0199×10^{-11}	3.0199×10^{-11}	3.0199×10^{-11}	8.2357×10^{-2}	NAN
25 °C −600 W/m²								
Worst	1.0078×10^{-1}	1.0134×10^{-1}	1.0689	1.0695	9.9952×10^{-2}	1.1223	3.4077×10^{-2}	2.8480×10^{-2}
Mean	6.0087×10^{-2}	6.2787×10^{-2}	2.0351×10^{-1}	2.7402×10^{-1}	6.0285×10^{-2}	4.7936×10^{-1}	1.1981×10^{-2}	7.9571×10^{-3}
Best	2.5294×10^{-2}	2.9938×10^{-2}	6.4945×10^{-2}	6.0119×10^{-2}	4.9966×10^{-2}	2.5931×10^{-1}	1.2080×10^{-3}	$\mathbf{6.3910 \times 10^{-4}}$
std	1.6177×10^{-2}	1.6123×10^{-2}	1.8811×10^{-1}	3.6424×10^{-1}	1.1376×10^{-2}	1.7777×10^{-1}	7.7565×10^{-3}	5.7342×10^{-3}
p-value	3.3384×10^{-11}	3.0199×10^{-11}	3.0199×10^{-11}	3.0199×10^{-11}	3.0199×10^{-11}	3.0199×10^{-11}	1.0315×10^{-2}	NAN
25 °C −400 W/m²								
Worst	9.3298×10^{-2}	8.1703×10^{-2}	2.4730×10^{-1}	7.1507×10^{-1}	1.2053×10^{-1}	6.0211×10^{-1}	2.3817×10^{-2}	2.4774×10^{-2}
Mean	5.3040×10^{-2}	4.5807×10^{-2}	1.0975×10^{-1}	9.7278×10^{-2}	4.4209×10^{-2}	3.1765×10^{-1}	9.7556×10^{-3}	7.0046×10^{-3}
Best	2.3846×10^{-2}	1.2940×10^{-2}	2.9330×10^{-2}	3.0107×10^{-2}	2.1731×10^{-2}	1.1111×10^{-1}	1.4399×10^{-3}	$\mathbf{7.7891 \times 10^{-4}}$
std	1.7438×10^{-2}	1.8296×10^{-2}	5.9500×10^{-2}	1.1921×10^{-1}	1.8535×10^{-2}	1.2998×10^{-1}	6.7435×10^{-3}	5.4953×10^{-3}
p-value	3.6897×10^{-11}	6.0658×10^{-11}	3.0199×10^{-11}	3.0199×10^{-11}	3.6897×10^{-11}	3.0199×10^{-11}	$4.5146 \times 10{-}^{2}$	NAN
25 °C −200 W/m²								
Worst	4.7413×10^{-2}	5.4922×10^{-2}	3.4134×10^{-1}	3.3976×10^{-1}	6.1243×10^{-2}	2.5166×10^{01}	6.0751×10^{-3}	9.9241×10^{-3}
Mean	3.3437×10^{-2}	3.3552×10^{-2}	1.2126×10^{-1}	5.6296×10^{-2}	3.4866×10^{-2}	1.2785×10^{-1}	3.8762×10^{-3}	2.6626×10^{-3}
Best	8.0880×10^{-3}	7.1727×10^{-3}	4.4441×10^{-2}	1.3225×10^{-2}	1.3280×10^{-2}	5.2239×10^{-2}	3.8039×10^{-4}	$\mathbf{2.3850 \times 10^{-4}}$
std	1.1001×10^{-2}	1.3676×10^{-7}	9.2663×10^{-2}	5.4823×10^{-2}	1.0797×10^{-2}	5.6607×10^{-2}	1.6824×10^{-3}	2.2870×10^{-3}
p-value	4.5043×10^{-11}	3.6897×10^{-11}	3.0199×10^{-11}	3.0199×10^{-11}	3.0199×10^{-11}	3.0199×10^{-11}	1.9527×10^{-03}	NAN

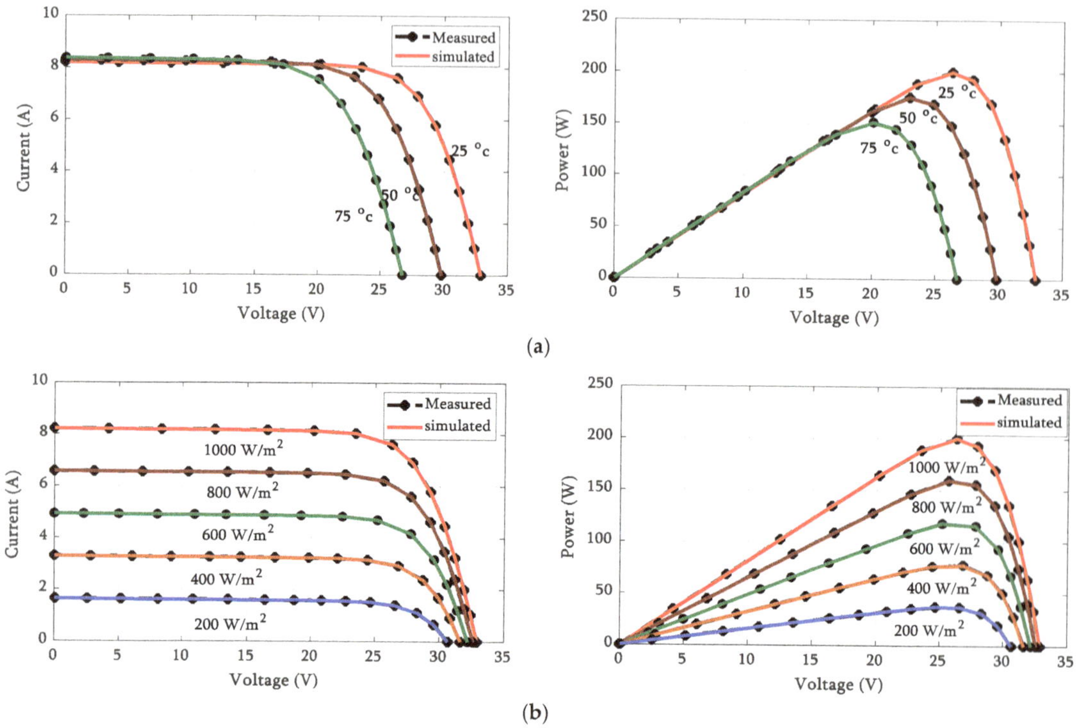

Figure 14. *P-V* and *I-V* curves of KC200GT: (**a**) constant irradiance of 1000 W/m^2 and (**b**) constant temperature of 25 °C.

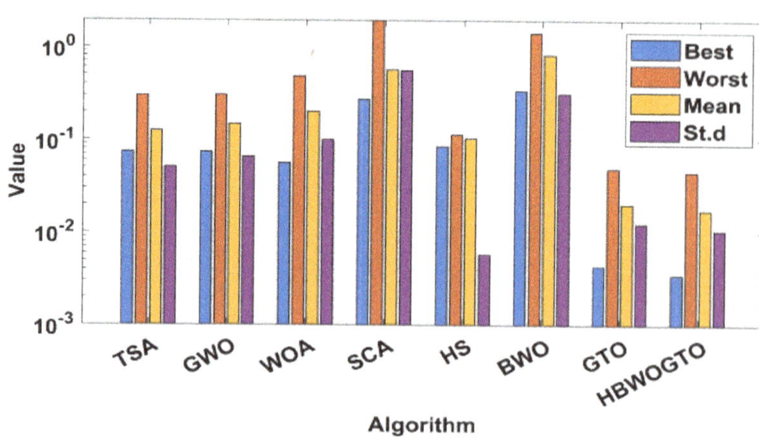

Figure 15. The statistical analysis for KC200Gt at 1000 W/m^2 and 25 °C.

Table 9. Statistical parameters of MSX60 panel obtained via the proposed approach and others.

Alg.	TSA	GWO [75]	WOA [76]	SCA	HS	BWO	GTO	HGTO-BWO
25 °C –1000 W/m²								
Worst	6.5103×10^{-2}	6.0081×10^{-2}	7.6505×10^{-1}	7.6471×10^{-1}	4.8401×10^{-2}	$6.2681 \times 10_{-1}$	2.1971×10^{-2}	1.7942×10^{-2}
Mean	4.4751×10^{-2}	4.3733×10^{-2}	1.3110×10^{-1}	1.2951×10^{-1}	3.8764×10^{-2}	3.2185×10^{-1}	1.0675×10^{-2}	7.8333×10^{-3}
Best	2.2794×10^{-2}	2.5944×10^{-2}	2.4565×10^{-2}	4.6448×10^{-2}	3.1247×10^{-2}	6.6376×10^{-2}	1.8986×10^{-3}	$\mathbf{1.0765 \times 10^{-4}}$
std	1.0134×10^{-2}	8.8505×10^{-3}	1.8165×10^{-1}	1.7389×10^{-1}	3.5080×10^{-3}	1.5338×10^{-1}	3.6012×10^{-3}	3.2102×10^{-3}
p-value	3.0199×10^{-11}	3.0199×10^{-11}	3.0199×10^{-11}	3.0199×10^{-11}	3.0199×10^{-11}	3.0199×10^{-11}	1.5846×10^{-4}	NAN
50 °C –1000 W/m²								
Worst	9.4706×10^{-2}	1.0820×10^{-1}	2.1100×10^{-1}	7.7056×10^{-1}	5.1888×10^{-2}	6.8428×10^{-1}	6.7625×10^{-3}	5.6571×10^{-3}
Mean	3.7155×10^{-2}	4.3804×10^{-2}	9.3860×10^{-2}	1.4572×10^{-1}	2.8151×10^{-2}	3.5886×10^{-1}	5.2558×10^{-3}	1.9660×10^{-3}
Best	1.8231×10^{-2}	7.3825×10^{-3}	1.8017×10^{-2}	6.7026×10^{-2}	2.3209×10^{-2}	1.3748×10^{-1}	2.3178×10^{-3}	$\mathbf{1.9324 \times 10^{-4}}$
std	2.0419×10^{-2}	3.0053×10^{-2}	5.3859×10^{-2}	1.2205×10^{-1}	5.3028×10^{-3}	1.5568×10^{-1}	1.4676×10^{-03}	1.4357×10^{-3}
p-value	3.0199×10^{-11}	3.0199×10^{-11}	3.0199×10^{-11}	3.0199×10^{-11}	3.0199×10^{-11}	3.0199×10^{-11}	8.4848×10^{-09}	NAN
75 °C –1000 W/m²								
Worst	1.6542×10^{-1}	1.3157×10^{-1}	2.5008×10^{-1}	2.3490×10^{-1}	3.6143×10^{-2}	4.9704×10^{-1}	5.2441×10^{-3}	1.4355×10^{-3}
Mean	3.0890×10^{-2}	2.6953×10^{-2}	8.4518×10^{-2}	1.6671×10^{-1}	1.2211×10^{-2}	2.5469×10^{-1}	1.0018×10^{-3}	2.7448×10^{-4}
Best	7.9526×10^{-3}	5.4140×10^{-3}	4.1026×10^{-3}	3.1053×10^{-2}	5.7215×10^{-3}	7.9537×10^{-2}	4.2011×10^{-5}	$\mathbf{2.9790 \times 10^{-5}}$
std	3.5408×10^{-2}	3.1096×10^{-2}	7.2347×10^{-2}	5.0708×10^{-2}	6.4175×10^{-3}	1.0582×10^{-1}	1.6193×10^{-3}	2.6447×10^{-4}
p-value	3.0199×10^{-11}	3.0199×10^{-11}	3.0199×10^{-11}	3.0199×10^{-11}	3.0199×10^{-11}	3.0199×10^{-11}	3.6322×10^{-1}	NAN
25 °C –800 W/m²								
Worst	5.2262×10^{-2}	5.8604×10^{-2}	6.1242×10^{-1}	6.1262×10^{-1}	4.0953×10^{-2}	4.4466×10^{-1}	1.2853×10^{-2}	1.2248×10^{-2}
Mean	3.5046×10^{-2}	3.1685×10^{-2}	1.0641×10^{-1}	7.3537×10^{-2}	2.9988×10^{-2}	2.2502×10^{-1}	8.6485×10^{-3}	6.6717×10^{-3}
Best	2.0629×10^{-2}	1.0734×10^{-2}	2.7649×10^{-2}	3.4265×10^{-2}	2.4251×10^{-2}	6.6000×10^{-2}	1.9350×10^{-3}	$\mathbf{1.1336 \times 10^{-3}}$
std	7.1751×10^{-3}	9.9180×10^{-3}	1.0964×10^{-1}	1.0260×10^{-1}	3.6823×10^{-3}	9.9871×10^{-2}	2.4842×10^{-3}	2.6766×10^{-3}
p-value	3.019×10^{-11}	3.6897×10^{-11}	3.019×10^{-11}	3.019×10^{-11}	3.019×10^{-11}	3.019×10^{-11}	2.2658×10^{-3}	NAN
25 °C –600 W/m²								
Worst	6.8702×10^{-2}	5.1465×10^{-2}	4.5790×10^{-1}	4.5562×10^{-1}	3.9280×10^{-2}	3.7030×10^{-1}	1.6311×10^{-2}	1.3913×10^{-2}
Mean	3.3254×10^{-2}	2.8439×10^{-2}	9.7793×10^{-2}	6.3663×10^{-2}	2.2989×10^{-2}	1.9982×10^{-1}	8.7540×10^{-3}	5.6637×10^{-3}
Best	9.8529×10^{-3}	1.0988×10^{-2}	1.5243×10^{-2}	2.4905×10^{-2}	1.3734×10^{-2}	6.0603×10^{-2}	1.0329×10^{-3}	$\mathbf{6.7775 \times 10^{-4}}$
std	1.3240×10^{-2}	1.0653×10^{-2}	1.0243×10^{-1}	7.5057×10^{-2}	7.5049×10^{-3}	8.0178×10^{-2}	3.4223×10^{-3}	3.1345×10^{-3}
p-value	6.0658×10^{-11}	4.0772×10^{-11}	3.0199×10^{-11}	3.0199×10^{-11}	3.6897×10^{-11}	3.0199×10^{-11}	1.4067×10^{-4}	NAN
25 °C –400 W/m²								
Worst	3.8949×10^{-2}	4.4247×10^{-2}	3.0055×10^{-1}	2.9852×10^{-1}	4.8025×10^{-2}	1.8624×10^{-1}	8.5444×10^{-3}	8.5110×10^{-3}
Mean	2.7298×10^{-2}	2.7571×10^{-2}	8.3476×10^{-2}	6.4845×10^{-2}	1.9225×10^{-2}	1.0319×10^{-1}	5.1743×10^{-3}	3.1605×10^{-3}
Best	1.1747×10^{-2}	5.9229×10^{-3}	2.0784×10^{-2}	1.2556×10^{-2}	6.7426×10^{-3}	4.3302×10^{-2}	8.5576×10^{-5}	$\mathbf{2.4366 \times 10^{-5}}$
std	7.4119×10^{-3}	1.0414×10^{-2}	5.8035×10^{-2}	7.9820×10^{-2}	8.3822×10^{-3}	3.8071×10^{-2}	2.2684×10^{-3}	1.9135×10^{-3}
p-value	3.019×10^{-11}	4.0772×10^{-11}	3.0199×10^{-11}	3.0199×10^{-11}	4.5043×10^{-11}	3.0199×10^{-11}	1.7836×10^{-4}	NAN
25 °C –200 W/m²								
Worst	2.9797×10^{-2}	2.6406×10^{-2}	7.1457×10^{-2}	1.4110×10^{-1}	3.2667×10^{-2}	1.6797×10^{-1}	7.5735×10^{-3}	3.2257×10^{-3}
Mean	1.6498×10^{-2}	1.8636×10^{-2}	3.4480×10^{-2}	2.7961×10^{-2}	1.7397×10^{-2}	5.8459×10^{-2}	1.7579×10^{-3}	1.0399×10^{-3}
Best	9.7357×10^{-3}	1.0787×10^{-3}	7.8442×10^{-3}	6.3339×10^{-3}	5.7370×10^{-3}	2.6227×10^{-2}	4.6560×10^{-4}	$\mathbf{5.9828 \times 10^{-5}}$
std	4.3918×10^{-3}	6.1085×10^{-3}	1.5446×10^{-2}	2.2766×10^{-2}	6.4824×10^{-3}	2.5259×10^{-2}	1.9112×10^{-3}	7.0129×10^{-4}
p-value	3.019×10^{-11}	1.3289×10^{-10}	3.0199×10^{-11}	3.0199×10^{-11}	3.0199×10^{-11}	3.0199×10^{-11}	2.8378×10^{-1}	NAN

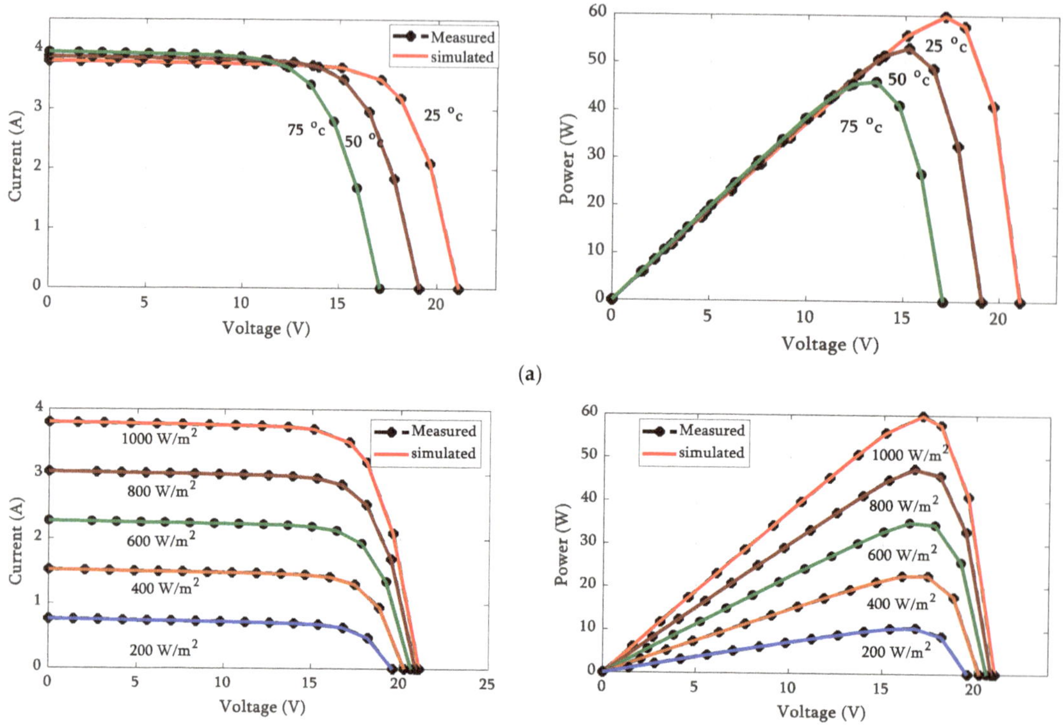

Figure 16. *I-V* and *P-V* curves of MSX60: (**a**) constant irradiance of 1000 W/m^2 and (**b**) constant temperature of 25 °C.

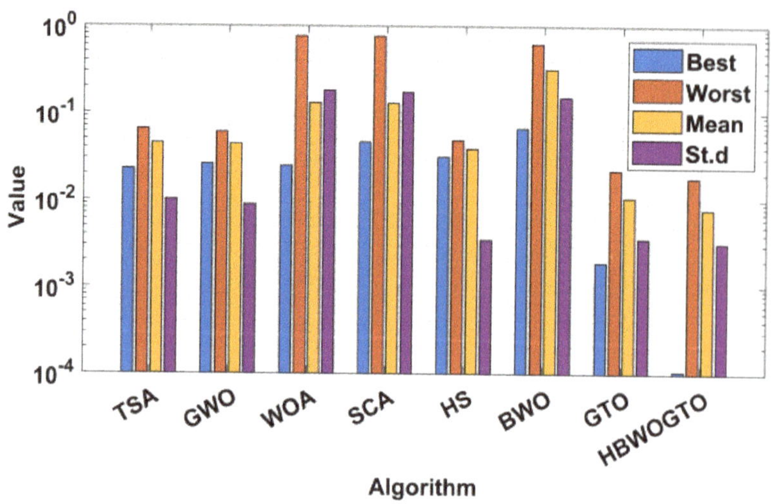

Figure 17. The statistical analysis for MSX60 at 1000 W/m^2 and 25 °C.

The fetched results demonstrated that the proposed HGTO-BWO is efficient in finding the optimal parameters of various models for the PV cell/panel as it outperformed the other regarded methodologies in all considered cases.

7. Conclusions

A new hybrid multi-population gorilla troops optimizer and beluga whale optimization (HGTO-BWO) was proposed to assign the PV cell/panel equivalent circuit by estimating its optimal parameters. In the proposed approach, a multi-population methodology was employed to improve the performance of the algorithm and to prevent it from falling into the local optima. The classical and CEC-C06 2019 benchmark functions were solved via the proposed approach to assess its performance. Two models, the double and triple diode models (DDM and TDM), were constructed for the PV cell/panel via minimizing the root mean square error (RMSE) between the simulated and measured currents. Various PV cells and panels operating in stable and variable weather situations were analyzed. Also, excessive comparison with TSA, SCA, GWO, WOA, HS, BWO, and GTO was conducted. The proposed approach findings can be summarized as follows:

- For the PVW 752 cell, the proposed HGTO-BWO achieved the best fitness values of 1.527×10^{-4} and 2.0886×10^{-4} for the TDM and DDM, respectively.
- The proposed approach achieved the lowest RMSEs of 2.42508×10^{-3} for the PWP-201 and 1.8032×10^{-3} for the STM6-40/36 DDM.
- The HGTO-BWO achieved the best fitness values of 2.2068×10^{-3} for the PWP-201 panel and 1.7435×10^{-3} for the STM6-40/36 TDM.
- For KC200GT, the minimum fitness values were 2.3850×10^{-4}, 7.7891×10^{-4}, 6.3910×10^{-4}, and 9.1596×10^{-4} during operation at 200 W/m^2, 400 W/m^2, 600 W/m^2, and 800 W/m^2, respectively.
- For MSX60, the proposed methodology realized the best RMSE values of 1.0765×10^{-4}, 1.9324×10^{-4}, and 2.9790×10^{-5} at 25 °C, 50 °C, and 75 °C, respectively, while at 25 °C the fitness values were 1.1336×10^{-3} at 800 W/m^2, 6.7775×10^{-4} at 600 W/m^2, 2.4366×10^{-5} at 400 W/m^2, and 5.9828×10^{-5} at 200 W/m^2.

The results revealed that the proposed approach can be recommended as an efficient optimizer when constructing the PV unit equivalent circuit via identifying its parameters. The proposed method requires a great effort to implement and program, which is considered a major obstacle during implementation; in addition, a lot of time is needed. Therefore, simplifying this method and reducing the time required will be of interest to the authors of the future works. Moreover, the validation of the proposed methodology in estimating the parameters of the PV array when operated under different conditions will be considered in the next work.

Supplementary Materials: The following supporting information can be downloaded at: https://www.mdpi.com/article/10.3390/su151411089/s1, Figure S1: The convergence curves of some traditional benchmark functions achieved by the proposed hybrid approach and others; Figure S2: The boxplots curves of some traditional benchmark functions obtained via the proposed hybrid approach and others; Figure S3: The convergence of some CEC-2019 functions achieved by the proposed hybrid method and others; Figure S4: The boxplots curves of some CEC-2019 functions obtained via the proposed hybrid approach and others; Figure S5: Performance of RMSE through iteration process for DDM (a) PWP-201 and (b) STM6-40/36; Figure S6: RMSE with numeral of iteration for TDM (a) STM6-40/36 panels and (b) PWP-201.

Author Contributions: Conceptualization, H.H.A. and A.F.; methodology, A.A.M. and H.H.A.; software, M.E. and F.J.; validation, A.F., T.S.B., and F.J.; formal analysis, H.H.A. and A.F.; investigation, M.E. and H.H.A.; data curation, M.E., F.J., and T.S.B.; writing—original draft preparation, A.A.M. and A.F.; writing—review and editing, A.F.; supervision, T.S.B. and F.J. All authors have read and agreed to the published version of the manuscript.

Funding: This research received no external funding.

Institutional Review Board Statement: Not applicable.

Informed Consent Statement: Not applicable.

Data Availability Statement: Not applicable.

Conflicts of Interest: The authors declare no conflict of interest.

Appendix A

Table A1. The studied algorithms' parameters.

Algorithms	Parameter	All Algorithms
TSA	$p_{min} = 1$, $p_{max} = 4$	
SCA	a = 2	
GWO	a = 2 to 0	Pop.size = 30
WOA	a = 2 to 0, $a_2 = -1$ to -2, b = 1	Max_Iter = 500
HS	HMCR = 0.8, PAR = 0.2, FW_d = 0.995	No. Run = 30
BWO	Wf = [0.1 0.05]	
GTO	β = 3, p = 0.03, w = 0.8	

References

1. Long, W.; Jiao, J.; Liang, X.; Xu, M.; Tang, M.; Cai, S. Parameters Estimation of Photovoltaic Models Using a Novel Hybrid Seagull Optimization Algorithm. *Energy* **2022**, *249*, 123760. [CrossRef]
2. D'Adamo, I.; Mammetti, M.; Ottaviani, D.; Ozturk, I. Photovoltaic Systems and Sustainable Communities: New Social Models for Ecological Transition. The Impact of Incentive Policies in Profitability Analyses. *Renew. Energy* **2023**, *202*, 1291–1304. [CrossRef]
3. D'Adamo, I.; Gastaldi, M.; Morone, P.; Ozturk, I. Economics and Policy Implications of Residential Photovoltaic Systems in Italy's Developed Market. *Util. Policy* **2022**, *79*, 101437. [CrossRef]
4. Ganesan, S.; David, P.W.; Balachandran, P.K.; Senjyu, T. Fault Identification Scheme for Solar Photovoltaic Array in Bridge and Honeycomb Configuration. *Electr. Eng.* **2023**. [CrossRef]
5. Ayyarao, T.S.L.V.; Kumar, P.P. Parameter Estimation of Solar PV Models with a New Proposed War Strategy Optimization Algorithm. *Int. J. Energy Res.* **2022**, *46*, 7215–7238. [CrossRef]
6. Shaheen, M.A.M.; Hasanien, H.M.; Alkuhayli, A. A Novel Hybrid GWO-PSO Optimization Technique for Optimal Reactive Power Dispatch Problem Solution. *Ain Shams Eng. J.* **2021**, *12*, 621–630. [CrossRef]
7. Vankadara, S.K.; Chatterjee, S.; Balachandran, P.K.; Mihet-Popa, L. Marine Predator Algorithm (MPA)-Based MPPT Technique for Solar PV Systems under Partial Shading Conditions. *Energies* **2022**, *15*, 6172. [CrossRef]
8. Libra, M.; Mrázek, D.; Tyukhov, I.; Severová, L.; Poulek, V.; Mach, J.; Šubrt, T.; Beránek, V.; Svoboda, R.; Sedláček, J. Reduced Real Lifetime of PV Panels—Economic Consequences. *Sol. Energy* **2023**, *259*, 229–234. [CrossRef]
9. Pourmousa, N.; Ebrahimi, S.M.; Malekzadeh, M.; Gordillo, F. Using a Novel Optimization Algorithm for Parameter Extraction of Photovoltaic Cells and Modules. *Eur. Phys. J. Plus* **2021**, *136*, 470. [CrossRef]
10. Nayagam, V.S.; Kumar, S.S.; Thiyagarajan, V.; Kamal, N.; Nisha, N.; Isaac, J.S.; Kassa, A. A Novel Optimization Algorithm for Modifying the Parameter Unit of Solar PV Cell. *Int. J. Photoenergy* **2022**, *2022*, 5240115. [CrossRef]
11. Prince Winston, D.; Kumaravel, S.; Praveen Kumar, B.; Devakirubakaran, S. Performance Improvement of Solar PV Array Topologies during Various Partial Shading Conditions. *Sol. Energy* **2020**, *196*, 228–242. [CrossRef]
12. Vankadara, S.K.; Chatterjee, S.; Balachandran, P.K. An Accurate Analytical Modeling of Solar Photovoltaic System Considering Rs and Rsh under Partial Shaded Condition. *Int. J. Syst. Assur. Eng. Manag.* **2022**, *13*, 2472–2481. [CrossRef]
13. Gude, S.; Jana, K.C. Parameter Extraction of Photovoltaic Cell Using an Improved Cuckoo Search Optimization. *Sol. Energy* **2020**, *204*, 280–293. [CrossRef]
14. Wang, L.; Chen, Z.; Guo, Y.; Hu, W.; Chang, X.; Wu, P.; Han, C.; Li, J. Accurate Solar Cell Modeling via Genetic Neural Network-Based Meta-Heuristic Algorithms. *Front. Energy Res.* **2021**, *9*, 1–14. [CrossRef]
15. Repalle, N.B.; Sarala, P.; Mihet-Popa, L.; Kotha, S.R.; Rajeswaran, N. Implementation of a Novel Tabu Search Optimization Algorithm to Extract Parasitic Parameters of Solar Panel. *Energies* **2022**, *15*, 4515. [CrossRef]
16. Sarjila, R.; Ravi, K.; Edward, J.B.; Kumar, K.S.; Prasad, A. Parameter Extraction of Solar Photovoltaic Modules Using Gravitational Search Algorithm. *J. Electr. Comput. Eng.* **2016**, *2016*, 2143572. [CrossRef]
17. Singh, A.; Sharma, A.; Rajput, S.; Mondal, A.K.; Bose, A.; Ram, M. Parameter Extraction of Solar Module Using the Sooty Tern Optimization Algorithm. *Electronics* **2022**, *11*, 564. [CrossRef]
18. Singla, M.K.; Nijhawan, P.; Oberoi, A.S. A Novel Hybrid Particle Swarm Optimization Rat Search Algorithm for Parameter Estimation of Solar PV and Fuel Cell Model. *COMPEL—Int. J. Comput. Math. Electr. Electron. Eng.* **2022**, *41*, 1505–1527. [CrossRef]
19. Ridhor, S.I.A.; Isa, Z.M.; Nayan, N.M. Parameter Extraction of PV Cell Single Diode Model Using Animal Migration Optimization. *Int. J. Electr. Eng. Appl. Sci.* **2020**, *3*, 1–6.

20. Oliva, D.; Abd El Aziz, M.; Ella Hassanien, A. Parameter Estimation of Photovoltaic Cells Using an Improved Chaotic Whale Optimization Algorithm. *Appl. Energy* **2017**, *200*, 141–154. [CrossRef]
21. Ramadan, T.; Kamel, S.; Neggaz, N.; Alghamdi, A.S. Developing Photovoltaic Cells Parameter Estimation Algorithm Based on Equilibrium Optimization Technique. *J. Eng. Res.* **2021**, *10*, 1–27. [CrossRef]
22. Saha, C.; Agbu, N.; Jinks, R. 2—Review Article of the Solar PV Parameters Estimation Using Evolutionary Algorithms. *MOJ Sol. Photoen Sys.* **2018**, *2*, 66–78. [CrossRef]
23. Ahmed, W.A.E.M.; Mageed, H.M.A.; Mohamed, S.A.E.; Saleh, A.A. Fractional Order Darwinian Particle Swarm Optimization for Parameters Identification of Solar PV Cells and Modules. *Alexandria Eng. J.* **2022**, *61*, 1249–1263. [CrossRef]
24. Rawa, M.; Abusorrah, A.; Al-Turki, Y.; Calasan, M.; Micev, M.; Ali, Z.M.; Mekhilef, S.; Bassi, H.; Sindi, H.; Abdel Aleem, S.H.E. Estimation of Parameters of Different Equivalent Circuit Models of Solar Cells and Various Photovoltaic Modules Using Hybrid Variants of Honey Badger Algorithm and Artificial Gorilla Troops Optimizer. *Mathematics* **2022**, *10*, 1057. [CrossRef]
25. Ginidi, A.; Ghoneim, S.M.; Elsayed, A.; El-Sehiemy, R.; Shaheen, A.; El-Fergany, A. Gorilla Troops Optimizer for Electrically Based Single and Double-Diode Models of Solar Photovoltaic Systems. *Sustainability* **2021**, *13*, 9459. [CrossRef]
26. Bayoumi, A.S.A.; El-Sehiemy, R.A.; Abaza, A. Effective PV Parameter Estimation Algorithm Based on Marine Predators Optimizer Considering Normal and Low Radiation Operating Conditions. *Arab. J. Sci. Eng.* **2022**, *47*, 3089–3104. [CrossRef]
27. El Sattar, M.A.; Al Sumaiti, A.; Ali, H.; Diab, A.A.Z. Marine Predators Algorithm for Parameters Estimation of Photovoltaic Modules Considering Various Weather Conditions. *Neural Comput. Appl.* **2021**, *33*, 11799–11819. [CrossRef]
28. Rezk, H.; Abdelkareem, M.A. Optimal Parameter Identification of Triple Diode Model for Solar Photovoltaic Panel and Cells. *Energy Rep.* **2022**, *8*, 1179–1188. [CrossRef]
29. Xu, S.; Qiu, H. A Modified Stochastic Fractal Search Algorithm for Parameter Estimation of Solar Cells and PV Modules. *Energy Rep.* **2022**, *8*, 1853–1866. [CrossRef]
30. Abbassi, A.; Ben Mehrez, R.; Touaiti, B.; Abualigah, L.; Touti, E. Parameterization of Photovoltaic Solar Cell Double-Diode Model Based on Improved Arithmetic Optimization Algorithm. *Optik (Stuttg).* **2022**, *253*, 168600. [CrossRef]
31. Muhammadsharif, F.F. A New Simplified Method for Efficient Extraction of Solar Cells and Modules Parameters from Datasheet Information. *Silicon* **2022**, *14*, 3059–3067. [CrossRef]
32. Lin, X.; Wu, Y. Parameters Identification of Photovoltaic Models Using Niche-Based Particle Swarm Optimization in Parallel Computing Architecture. *Energy* **2020**, *196*, 117054. [CrossRef]
33. Gude, S.; Jana, K.C. A Multiagent System Based Cuckoo Search Optimization for Parameter Identification of Photovoltaic Cell Using Lambert W-Function. *Appl. Soft Comput.* **2022**, *120*, 108678. [CrossRef]
34. Ali, F.; Sarwar, A.; Ilahi Bakhsh, F.; Ahmad, S.; Ali Shah, A.; Ahmed, H. Parameter Extraction of Photovoltaic Models Using Atomic Orbital Search Algorithm on a Decent Basis for Novel Accurate RMSE Calculation. *Energy Convers. Manag.* **2023**, *277*, 116613. [CrossRef]
35. Beşkirli, A.; Dağ, İ. Parameter Extraction for Photovoltaic Models with Tree Seed Algorithm. *Energy Rep.* **2023**, *9*, 174–185. [CrossRef]
36. Wang, D.; Sun, X.; Kang, H.; Shen, Y.; Chen, Q. Heterogeneous Differential Evolution Algorithm for Parameter Estimation of Solar Photovoltaic Models. *Energy Rep.* **2022**, *8*, 4724–4746. [CrossRef]
37. Yu, Y.; Wang, K.; Zhang, T.; Wang, Y.; Peng, C.; Gao, S. A Population Diversity-Controlled Differential Evolution for Parameter Estimation of Solar Photovoltaic Models. *Sustain. Energy Technol. Assess.* **2022**, *51*, 101938. [CrossRef]
38. Alanazi, M.; Alanazi, A.; Almadhor, A.; Rauf, H.T. Photovoltaic Models' Parameter Extraction Using New Artificial Parameterless Optimization Algorithm. *Mathematics* **2022**, *10*, 4617. [CrossRef]
39. Fan, Y.; Wang, P.; Heidari, A.A.; Chen, H.; HamzaTurabieh; Mafarja, M. Random Reselection Particle Swarm Optimization for Optimal Design of Solar Photovoltaic Modules. *Energy* **2022**, *239*, 121865. [CrossRef]
40. Ridha, H.M.; Hizam, H.; Mirjalili, S.; Othman, M.L.; Ya'acob, M.E.; Abualigah, L. A Novel Theoretical and Practical Methodology for Extracting the Parameters of the Single and Double Diode Photovoltaic Models (December 2021). *IEEE Access* **2022**, *10*, 11110–11137. [CrossRef]
41. Lin, H.; Ahmadianfar, I.; Amiri Golilarz, N.; Jamei, M.; Heidari, A.A.; Kuang, F.; Zhang, S.; Chen, H. Adaptive Slime Mould Algorithm for Optimal Design of Photovoltaic Models. *Energy Sci. Eng.* **2022**, *10*, 2035–2064. [CrossRef]
42. Chen, N.; Bi, W.; Xu, G.; Wu, Z.; Wu, M.; Luo, K. Mayfly Optimization Algorithm–Based PV Cell Triple-Diode Model Parameter Identification. *Front. Energy Res.* **2022**, *10*, 1–10. [CrossRef]
43. El-Dabah, M.A.; El-Sehiemy, R.A.; Hasanien, H.M.; Saad, B. Photovoltaic Model Parameters Identification Using Northern Goshawk Optimization Algorithm. *Energy* **2023**, *262*, 125522. [CrossRef]
44. Kumar, C.; Magdalin Mary, D. A Novel Chaotic-Driven Tuna Swarm Optimizer with Newton-Raphson Method for Parameter Identification of Three-Diode Equivalent Circuit Model of Solar Photovoltaic Cells/Modules. *Optik (Stuttg).* **2022**, *264*, 169379. [CrossRef]
45. Bo, Q.; Cheng, W.; Khishe, M.; Mohammadi, M.; Mohammed, A.H. Solar Photovoltaic Model Parameter Identification Using Robust Niching Chimp Optimization. *Sol. Energy* **2022**, *239*, 179–197. [CrossRef]
46. Gupta, J.; Hussain, A.; Singla, M.K.; Nijhawan, P.; Haider, W.; Kotb, H.; AboRas, K.M. Parameter Estimation of Different Photovoltaic Models Using Hybrid Particle Swarm Optimization and Gravitational Search Algorithm. *Appl. Sci.* **2023**, *13*, 249. [CrossRef]

47. Ramadan, A.; Kamel, S.; Hussein, M.M.; Hassan, M.H. A New Application of Chaos Game Optimization Algorithm for Parameters Extraction of Three Diode Photovoltaic Model. *IEEE Access* **2021**, *9*, 51582–51594. [CrossRef]
48. Yu, S.; Chen, Z.; Heidari, A.A.; Zhou, W.; Chen, H.; Xiao, L. Parameter Identification of Photovoltaic Models Using a Sine Cosine Differential Gradient Based Optimizer. *IET Renew. Power Gener.* **2022**, *16*, 1535–1561. [CrossRef]
49. Jiang, Y.; Luo, Q.; Zhou, Y. Improved Gradient-based Optimizer for Parameters Extraction of Photovoltaic Models. *IET Renew. Power Gener.* **2022**, *16*, 1602–1622. [CrossRef]
50. Wang, J.; Yang, B.; Li, D.; Zeng, C.; Chen, Y.; Guo, Z.; Zhang, X.; Tan, T.; Shu, H.; Yu, T. Photovoltaic Cell Parameter Estimation Based on Improved Equilibrium Optimizer Algorithm. *Energy Convers. Manag.* **2021**, *236*, 114051. [CrossRef]
51. Shaheen, A.M.; Ginidi, A.R.; El-Sehiemy, R.A.; Ghoneim, S.S.M. A Forensic-Based Investigation Algorithm for Parameter Extraction of Solar Cell Models. *IEEE Access* **2021**, *9*, 1–20. [CrossRef]
52. Shaheen, A.M.; El-Seheimy, R.A.; Xiong, G.; Elattar, E.; Ginidi, A.R. Parameter Identification of Solar Photovoltaic Cell and Module Models via Supply Demand Optimizer. *Ain Shams Eng. J.* **2022**, *13*, 101705. [CrossRef]
53. Yu, S.; Heidari, A.A.; He, C.; Cai, Z.; Althobaiti, M.M.; Mansour, R.F.; Liang, G.; Chen, H. Parameter Estimation of Static Solar Photovoltaic Models Using Laplacian Nelder-Mead Hunger Games Search. *Sol. Energy* **2022**, *242*, 79–104. [CrossRef]
54. Lekouaghet, B.; Boukabou, A.; Boubakir, C. Estimation of the Photovoltaic Cells/Modules Parameters Using an Improved Rao-Based Chaotic Optimization Technique. *Energy Convers. Manag.* **2021**, *229*, 113722. [CrossRef]
55. Ridha, H.M.; Hizam, H.; Mirjalili, S.; Othman, M.L.; Ya'acob, M.E.; Ahmadipour, M. Parameter Extraction of Single, Double, and Three Diodes Photovoltaic Model Based on Guaranteed Convergence Arithmetic Optimization Algorithm and Modified Third Order Newton Raphson Methods. *Renew. Sustain. Energy Rev.* **2022**, *162*, 112436. [CrossRef]
56. Ibrahim, I.A.; Hossain, M.J.; Duck, B.C. A Hybrid Wind Driven-Based Fruit Fly Optimization Algorithm for Identifying the Parameters of a Double-Diode Photovoltaic Cell Model Considering Degradation Effects. *Sustain. Energy Technol. Assessments* **2022**, *50*, 101685. [CrossRef]
57. Saha, A.K. Multi-Population-Based Adaptive Sine Cosine Algorithm with Modified Mutualism Strategy for Global Optimization. *Knowl. -Based Syst.* **2022**, *251*, 109326. [CrossRef]
58. Ma, H.; Shen, S.; Yu, M.; Yang, Z.; Fei, M.; Zhou, H. Multi-Population Techniques in Nature Inspired Optimization Algorithms: A Comprehensive Survey. *Swarm Evol. Comput.* **2019**, *44*, 365–387. [CrossRef]
59. Satria, H.; Syah, R.B.Y.; Nehdi, M.L.; Almustafa, M.K.; Adam, A.O.I. Parameters Identification of Solar PV Using Hybrid Chaotic Northern Goshawk and Pattern Search. *Sustainability* **2023**, *15*, 5027. [CrossRef]
60. Ben Aribia, H.; El-Rifaie, A.M.; Tolba, M.A.; Shaheen, A.; Moustafa, G.; Elsayed, F.; Elshahed, M. Growth Optimizer for Parameter Identification of Solar Photovoltaic Cells and Modules. *Sustainability* **2023**, *15*, 7896. [CrossRef]
61. Bogar, E. Chaos Game Optimization-Least Squares Algorithm for Photovoltaic Parameter Estimation. *Arab. J. Sci. Eng.* **2023**, *48*, 6321–6340. [CrossRef]
62. Rawat, N.; Thakur, P.; Singh, A.K.; Bhatt, A.; Sangwan, V.; Manivannan, A. A New Grey Wolf Optimization-Based Parameter Estimation Technique of Solar Photovoltaic. *Sustain. Energy Technol. Assess.* **2023**, *57*, 103240. [CrossRef]
63. Qaraad, M.; Amjad, S.; Hussein, N.K.; Badawy, M.; Mirjalili, S.; Elhosseini, M.A. Photovoltaic Parameter Estimation Using Improved Moth Flame Algorithms with Local Escape Operators. *Comput. Electr. Eng.* **2023**, *106*, 108603. [CrossRef]
64. Changmai, P.; Deka, S.; Kumar, S.; Babu, T.S.; Aljafari, B.; Nastasi, B. A Critical Review on the Estimation Techniques of the Solar PV Cell's Unknown Parameters. *Energies* **2022**, *15*, 7212. [CrossRef]
65. Yang, B.; Wang, J.; Zhang, X.; Yu, T.; Yao, W.; Shu, H.; Zeng, F.; Sun, L. Comprehensive Overview of Meta-Heuristic Algorithm Applications on PV Cell Parameter Identification. *Energy Convers. Manag.* **2020**, *208*, 112595. [CrossRef]
66. Ali, H.H.; Fathy, A.; Al-dhaifallah, M.; Abdelaziz, A.Y.; Ebeed, M. An Efficient Capuchin Search Algorithm for Extracting the Parameters of Different PV Cells / Modules. *Front. Energy Res.* **2022**, *10*, 1028816. [CrossRef]
67. Diab, A.A.Z.; Ezzat, A.; Rafaat, A.E.; Denis, K.A.; Abdelsalam, H.A.; Abdelhamid, A.M. Optimal Identification of Model Parameters for PVs Using Equilibrium, Coot Bird and Artificial Ecosystem Optimisation Algorithms. *IET Renew. Power Gener.* **2022**, *16*, 2172–2190. [CrossRef]
68. Fathy, A.; Rezk, H. Parameter Estimation of Photovoltaic System Using Imperialist Competitive Algorithm. *Renew. Energy* **2017**, *111*, 307–320. [CrossRef]
69. Abdollahzadeh, B.; Gharehchopogh, F.S.; Mirjalili, S. Artificial Gorilla Troops Optimizer: A New Nature-inspired Metaheuristic Algorithm for Global Optimization Problems. *Int. J. Intell. Syst.* **2021**, *36*, 5887–5958. [CrossRef]
70. Zhong, C.; Li, G.; Meng, Z. Beluga Whale Optimization: A Novel Nature-Inspired Metaheuristic Algorithm. *Knowledge-Based Syst.* **2022**, *251*, 109215. [CrossRef]
71. Mirjalili, S.; Mirjalili, S.M.; Lewis, A. Grey Wolf Optimizer. *Adv. Eng. Softw.* **2014**, *69*, 46–61. [CrossRef]
72. Szabo, R.; Gontean, A. Photovoltaic Cell and Module I-V Characteristic Approximation Using Bézier Curves. *Appl. Sci.* **2018**, *8*, 655. [CrossRef]
73. Abdullah, J.M.; Ahmed, T. Fitness Dependent Optimizer: Inspired by the Bee Swarming Reproductive Process. *IEEE Access* **2019**, *7*, 43473–43486. [CrossRef]
74. Premkumar, M.; Jangir, P.; Ramakrishnan, C.; Nalinipriya, G.; Alhelou, H.H.; Kumar, B.S. Identification of Solar Photovoltaic Model Parameters Using an Improved Gradient-Based Optimization Algorithm with Chaotic Drifts. *IEEE Access* **2021**, *9*, 62347–62379. [CrossRef]

75. Darmansyah; Robandi, I. Photovoltaic Parameter Estimation Using Grey Wolf Optimization. In Proceedings of the 2017 3rd International Conference on Control, Automation and Robotics (ICCAR), Nagoya, Japan, 24-26 April 2017; pp. 593–597. [CrossRef]
76. Elazab, O.S.; Hasanien, H.M.; Elgendy, M.A.; Abdeen, A.M. Whale Optimisation Algorithm for Photovoltaic Model Identification. *J. Eng.* **2017**, *2017*, 1906–1911. [CrossRef]
77. Naeijian, M.; Rahimnejad, A.; Ebrahimi, S.M.; Pourmousa, N.; Gadsden, S.A. Parameter Estimation of PV Solar Cells and Modules Using Whippy Harris Hawks Optimization Algorithm. *Energy Rep.* **2021**, *7*, 4047–4063. [CrossRef]
78. Libra, M.; Petrik, T.; Poulek, V.; Tyukhov, I.I.; Kourim, P. Changes in the Efficiency of Photovoltaic Energy Conversion in Temperature Range with Extreme Limits. *IEEE J. Photovolt.* **2021**, *11*, 1479–1484. [CrossRef]
79. Arias García, R.M.; Pérez Abril, I. Photovoltaic Module Model Determination by Using the Tellegen's Theorem. *Renew. Energy* **2020**, *152*, 409–420. [CrossRef]

sustainability

MDPI

Article

Solar Panel Cooling System Evaluation: Visual PROMETHEE Multi-Criteria Decision-Making Approach

Ateekh Ur Rehman

Department of Industrial Engineering, College of Engineering, King Saud University, Riyadh 11421, Saudi Arabia; arehman@ksu.edu.sa

Abstract: The energy sector is interested in sustainable solar power plants. It is obvious that the working temperature of solar panels, which is significantly higher than the specified working cell temperature in hot climes, has a significant impact on efficiency and longevity. The selection of solar panel cooling systems, on the other hand, is worrisome since the choice process incorporates ergonomic, technical, economic, and environmental issues. The goal of this research is to (1) present a multi-criteria decision-making approach that is both quantitative and qualitative in nature for selecting solar panel cooling systems; (2) outrank nine alternative solar panel cooling systems with eleven performance measures for each alternative to assist decision makers in determining the best viable choice; and (3) visualize the relationship between the different solar panel cooling systems and performance measures under consideration. The proposed approach is to compare and rank solar panel cooling systems, as well as their validation and evaluation through sensitivity analysis. When operating efficiency is prioritized, finned air cooling is shown to be the best solar panel cooling technique, whereas thermosiphon cooling is the best alternative when emission reduction criteria are prioritized. A comparison of the findings shows that phase change material cooling and forced convection cooling performed worst in almost all cases.

Keywords: solar energy; panel cooling systems; multi-criteria evaluation; Visual PROMETHEE

check for updates

Citation: Rehman, A.U. Solar Panel Cooling System Evaluation: Visual PROMETHEE Multi-Criteria Decision-Making Approach. *Sustainability* **2023**, *15*, 12947. https://doi.org/10.3390/su151712947

Academic Editor: Oz Sahin

Received: 31 July 2023
Revised: 22 August 2023
Accepted: 24 August 2023
Published: 28 August 2023

1. Introduction and Background Review

In pursuit of green technology innovations, the energy industry demonstrates a focus on long-term sustainability renewable energy generation. The goal is to generate and transfer power to major domestic and industrial customers by 2030. It is concerned with a number of factors, for example, energy generation, public awareness, demand, and the risk at which local and national transmission occurs. The manufacturers of solar panel modules are more interested in improving the efficiency of solar panels; for this, they need meticulous alternative selection and assessment. There are challenges and opportunities in solar modules [1,2] influenced by ambient temperature, solar radiation intensity, the solar panels' surface temperature, dust, and shading, among other factors that may be overcome by adopting a suitable cooling and cleaning system. Solar panel cooling approaches [3,4] use several physical ways employing various flow media to minimize the solar panels' surface temperature. These approaches are categorized as either passive (no external energy is needed) or active (additional energy is utilized to circulate the cooling fluid). Passive approaches [5,6] include the use of fins or expanded surfaces to facilitate heat transfer, the use of phase change material to absorb heat produced in the panel, the use of heat pipe cooling, and convection via natural circulation water or air cooling. Active approaches [7,8] involve the circulation of air or water over the panel surfaces, both with and without the assistance of fins. There are several factors to consider when comparing active and passive solar panel cooling systems. However, the comparative ease of operation depends on the specific cooling system being used, as well as factors such as the size and location of the system; the required level of maintenance; the effective performance, particularly in hot and humid environments; and the cost when evaluating different solar panel cooling systems. For example, studies [7,8] have shown that water spraying can reduce the temperature of solar panels by up to 23 °C and significantly increase their electrical efficiency. Temperature impact affects efficiency and panel life span despite greater energy generation [9,10]. Likewise, researchers [5,11,12] have documented the recovery of useful electrical power with considerable changes in the heat dissipation process of solar panels by using various passive or active panel cooling processes. Although cooling clearly increases renewable energy production [5], it necessitates an additional structure that can extract heat from the panel and distribute it elsewhere. Notably, the design and maintenance [13] of a cooling strategy can be expensive, and the cost of system maintenance may outweigh the benefits of increased power generation. When compared to silicon-based panels [14], thin-film solar panels [15] are less influenced by a rise in temperature. Also, the effectiveness of solar panel cooling systems may vary depending on various factors such as the climate, panel design, and the type of cooling approach used. Recently, studies [16] have shown that the use of nano-fluids improves the heat transfer coefficient, solar panel power, and system performance. Past studies [17,18] have also proposed the use of phase change material cooling and microchannel heat sink cooling. However, there is no conclusive evidence to suggest which solar panel cooling strategy is the most effective, as it depends on various factors [4]. It is obvious that selecting the best solar panel cooling system necessitates the use of mathematical tools to analyze the alternatives. As a result, multi-criteria decision analysis is the ideal tool and can be used well in certain scenarios. The PROMETHEE (Preference Ranking Organization METHod for Enrichment Evaluations) decision-making method compares options based on a set of criteria [19]. There are several studies that have applied the PROMETHEE in various ways, such as in the selection of the most appropriate variant of the solar water supply system [20], ranking sites for solar farms [21,22], evaluating the effectiveness of integrated shading devices for office buildings [23], and for decision making in solar plant locations [24], as it allows for the consideration of multiple criteria, and can help decision makers to assess energy technologies [25]. Similarly, it is evident that researchers justified the importance of reliability information to evaluate solar panel selection [26]. On the other hand, strategies are developed to select an efficient solar panel, and a comprehensive comparative analysis is presented [27].

Multiple qualitative and quantitative metrics were used to evaluate, assess, and rate the various solar cooling systems. For example, Mardani et al. [28] reviewed multi-criteria approaches in sustainable and renewable energy system problems. Wang et al. [29] used multi-criteria decision analysis as a key tool to evaluate renewable energy technologies in households. Seker and Kahraman [30] proposed a socio-economic evaluation model for sustainable solar panels by integrating the analytic hierarchy process and multiplicative multi-objective ratio analysis method. Similarly, Krysiak and Kluczek [31] assessed the sustainable development of photovoltaic modules using a multi-criteria decision-making method based on the analytic hierarchy process [32]. Similarly, a multi-attribute decision-making approach based on intuitionistic fuzzy logic is adopted to select and assess solar panels [33].

The motivation to evaluate solar panel cooling systems is to improve solar panel efficiency and output. When solar panels get too hot, their efficiency drops, resulting in less energy output. The temperature of the solar panels may be adjusted by integrating cooling devices, resulting in enhanced efficiency and output. Existing research has mostly focused on the installation of cooling systems in residential buildings and their influence on energy efficiency. However, research evaluating the effectiveness of various cooling technologies and their effects on the long-term durability of solar panels is lacking. Furthermore, research has been performed in certain geographic locations, and further studies that outrank cooling systems are needed.

The goal of this research is to present a multi-criteria decision-making approach that is both quantitative and qualitative in nature for selecting solar panel cooling systems, to assist decision makers in determining the best viable choice, and to visualize the relationship between the different solar panel cooling systems and performance measures under consideration. The proposed approach provides substantial support in comparing and ranking solar panel cooling systems, as well as their validation and evaluation through sensitivity analysis. Six different scenarios, depending on the subjective and objective importance given to each performance measure, are possible choices.

This paper is divided into eight sections. The Section 1 provides an introduction; the Section 2 presents various attributes, performance measures, and alternative solar panel cooling systems; and the Section 3 presents alternative solar panel cooling systems. Section 4 presents the adopted MCDM approach steps. Sections 5 and 6 then present the implementation of the proposed approach and sensitivity analysis using subjective and objective criterion weights, respectively. Section 7 discusses management benefits, while Section 8 concludes the paper with conclusions and future study directions.

2. Attributes and Performance Measures

There are several benefits of cooling solar panels, which include improving panel overall efficiency, reducing energy consumption, and extending the panel lifespan. By lowering the temperature of the panels, either through passive cooling techniques or more advanced cooling systems [4], the electrical conversion efficiency of the panels can be increased, resulting in higher energy production [34]. Additionally, reducing the temperature of the panels can also help to reduce the wear and tear on the system, leading to a longer lifespan for the panels and a higher return on investment [35]. Furthermore, by reducing energy consumption through more efficient cooling mechanisms, the environmental footprint of the solar system can be minimized, making it a more sustainable option for generating renewable energy [36]. It is decided to make use of a systematic multi-criteria analysis approach to identify the competitiveness of each alternative. The multiple attributes opted to evaluate alternative solar panel cooling systems are as follows: cooling effectiveness, energy efficiency, environmental impact, durability, noise, panel size and weight, and or cost as decision-making attributes (refer to Figure 1). Decision makers opt for multiple combinations of attributes to be evaluated in order to select the most suitable panel cooling system to improve overall efficiency.

As represented in Figure 1, the adopted performance measures to evaluate cooling techniques are briefly described below.

Energy Efficiency (PM01): Energy efficiency is crucial for solar panel cooling since it lowers the amount of energy used for cooling and raises the overall effectiveness of the solar panel system. The surface temperature of the solar panel can be managed by utilizing more energy-efficient cooling techniques, which helps to maximize the electrical conversion efficiency of the solar panel. This can result in considerable drops in energy usage while raising the solar system's output of energy [37]. In order to increase the effectiveness of solar panels, a number of researchers have used cutting-edge cooling techniques. The electrical efficiency is a function of the cell temperature [23], and it is mathematically estimated by researchers.

Cooling cost (PM02): The cost of cooling solar panels plays an important role in determining the overall cost-effectiveness and return on investment of the solar system [38]. While advanced cooling techniques may improve the electrical conversion efficiency and extend the lifespan of the solar panels, additional supplies and installation charges may also be required, raising the overall cost of the system. Therefore, it is important to consider the costs of cooling technologies and weigh them against the potential benefits in terms of energy production and system longevity [39].

Reliability factor (PM03): Reliability refers to the ability of components of the cooling system to operate without failure or malfunction over time [40]. If the cooling system fails, the temperature of the solar panels can rise, which can have detrimental effects on the energy output and lifespan of the system [41]. Thus, it is crucial for the cooling system to be reliable. Additionally, a reliable cooling system can help to minimize maintenance and repair costs, as well as reduce downtime, ensuring that the solar system operates over its expected lifetime with minimal interruptions [2,30].

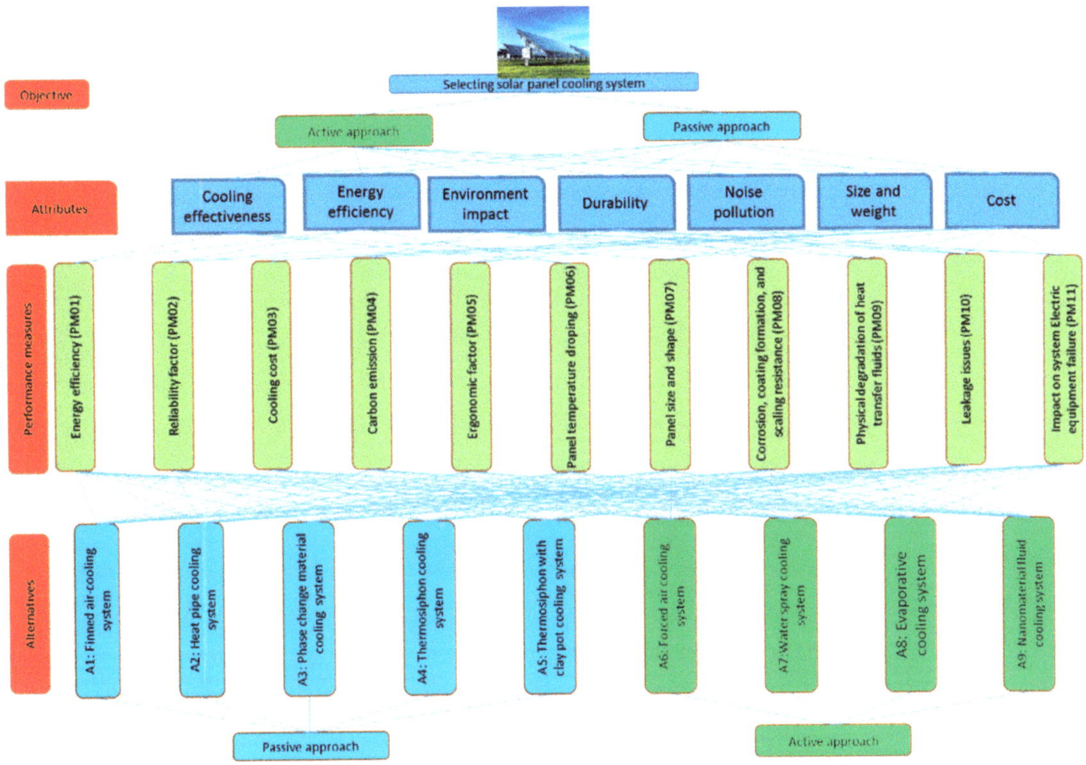

Figure 1. Attributes, their performance measures, and alternatives opted for in evaluation of solar panel cooling systems.

Carbon emission (PM04): The carbon footprint of the downstream processing and manufacturing of solar cooling systems is highlighted by researchers [17,42,43], indicating that any cooling system's environmental impact is an important consideration. Additionally, government policies promote solar energy modules due to their low-carbon emission profile. Therefore, employing more energy-efficient cooling systems in solar modules is an important task. Carbon emissions of the entire system can be reduced, making it a more sustainable option for generating renewable energy [44].

Ergonomic factor (PM05): In the development of solar cooling systems, the primary focus is on technology feasibility, sustainability, and energy efficiency. However, ergonomics could indirectly play a role in selecting solar cooling techniques by ensuring the safety and comfort of individuals involved in the installation, operation, and maintenance of the solar system [45]. By designing these cooling systems to be ergonomically efficient, risks associated with injury, fatigue, and discomfort could be reduced, ultimately improving the overall safety and performance of the system.

Panel temperature dropping (PM06): The operating temperature of a solar panel has a significant impact on its energy output and lifespan, and excessive temperatures can negatively affect overall efficiency and performance [35,36]. Therefore, reducing the operating temperature of the solar module is one of the primary approaches to increasing power generation [46]. Various cooling techniques are being explored to decrease the operating temperature of solar panels and increase their efficiency. The drop in the operating temperature achieved through these cooling approaches can lead to a significant increase in power output, ranging from 20% to 30%, depending on the cooling approach adopted.

Panel size and shape (PM07): The solar cooling system's effectiveness is often measured in terms of the decrease in operating temperature of the solar panels, and hence, the area of the panel indirectly impacts the cooling requirements [35]. Larger solar modules have more surface area that is exposed to the sun and absorb more energy, resulting in higher operating temperatures. Therefore, larger solar modules may potentially require more powerful cooling systems to maintain their operating temperatures within the desired range. However, the efficiency of solar panels also varies with their size, efficiency, and technology, which may be important factors in selecting a cooling technique [47,48].

Degradation resistance (PM08): It is evident that [49] there are multiple reasons why degradation, i.e., corrosion, coating formation, and scaling, can occur in solar cooling systems. For example, exposure to different environmental factors, including humidity, saltwater, and other pollutants, can lead to corrosion and scaling. Similarly, chemical reactions between the cooling medium and the solar panel materials can also cause corrosion and coating formation. For example, copper in the system can form copper salts that can affect the solar panel's performance. The materials with lower resistance to corrosion are more likely to corrode and form coatings. Incorrect system design or installation can also contribute to corrosion and coating formation. For example, using dissimilar metals in a cooling system can lead to galvanic corrosion.

Thermal decomposition (PM09): One of the potential challenges [16,50–52] that could arise in a solar cooling system is thermal decomposition, which means physical degradation of heat transfer fluids. Heat transfer fluids are used [53] in some cooling systems to improve heat transfer and reduce the operating temperature of solar panels. However, the fluids can suffer physical and thermal decomposition and degradation over time, which reduces their effectiveness. Some of the factors that can cause physical degradation in heat transfer fluids include the fluid's properties, such as viscosity and thermal stability, the operating temperature range, and the type of solar panels used. Additionally, exposure to ultraviolet radiation, oxygen, and other environmental factors can also contribute to fluid degradation. If the fluid is not replaced or maintained regularly, its ability to transfer heat effectively can decrease, which can negatively affect the solar panel's performance [39–41]. Therefore, it is crucial to ensure that the heat transfer fluid used in a solar cooling system is selected carefully and maintenance and replacement schedules are adhered to in order to maintain the system's optimal performance.

Leakage issues (PM10): Leakage issues are a significant challenge in solar cooling systems that use heat transfer fluids, as these fluids can leak through damaged piping or seals. The loss of fluid from the system not only reduces its effectiveness in cooling solar panels but can also potentially damage other components in the system. Moreover, the leakage of fluids can create safety concerns, as they may pose a risk to the environment or human health. Therefore, it is crucial to consider potential leakage issues when designing, installing, and maintaining solar cooling systems and take appropriate measures to prevent or mitigate them [1,3,8].

Impact of any equipment failure (PM11): Electric equipment failure/or any power supply failure can impact the overall performance and reliability of the system, which can indirectly affect the cooling system's ability to maintain the desired operating temperatures of the solar panels [54]. If the electric/non-electric equipment fails, it may lead to a complete shutdown of the system, which can have a domino effect on other components, including the cooling system. Additionally, if the cooling system is not designed or installed correctly [55,56], it may be more susceptible to failures caused by electric equipment failures, such as a power surge or overload, which may impact its ability to cool the solar panels effectively. Thus, the impact of any equipment failure on a solar cooling system depends on various factors, including the type and severity of the failure, the system's design and installation, and the overall maintenance and operation of the system [3,44].

Thus, various alternatives for solar panel cooling in solar power plant installations have been represented in Figure 1 and briefly described in the following section.

3. Solar Panel Cooling Systems: Alternatives

Solar panel cooling systems use several physical methods, such as sensible and latent heat storage and dissipation or heat convection techniques employing different media, and they are evaluated to examine the reduction in panel operating temperature. These systems are classified as having either a passive approach or an active approach (refer to Figure 1). Active cooling systems [3,4] often perform better in terms of temperature reduction than passive cooling systems. Passive cooling [3,4] does not require much energy to operate and instead relies on natural cooling, which takes longer to cool over time.

Finned air cooling (A1) employs a natural heat transfer process in which heat from the solar panel is transmitted to the fins and then travels through convection by natural wind movement [57]. Heat pipe solar cooling systems (A2) use heat pipes to dissipate heat from solar panels, reducing the temperature and increasing their efficiency. The heat pipes work by transferring excess heat from the panels to the cooler end of the system, where it is dissipated into the air or water [58]. This cooling system has been found to be an attractive option in hot climates [59]. Researchers [60] have conducted studies on the effectiveness of heat pipe solar cooling systems.

Phase change materials (A3) have been used to cool solar panels and increase their efficiency, particularly in hot climates. It works by absorbing the excess heat from the panels and storing it until the temperature drops, providing a buffer against temperature fluctuations [17]. When the panels start to heat up, the phase change material melts and absorbs the heat. As the temperature decreases, the phase change materials then solidify and release the stored heat. Thermosiphon cooling systems (A4) involve a closed-loop system containing a working fluid, such as acetone, that undergoes a phase change to cool a solar panel [61]. On the other hand, a thermosiphon with a clay pot cooling system (A5) uses the clay pot as the medium for evaporative cooling to cool a solar panel [61]. Both systems use the same basic principle of thermosiphon cooling, but they differ in the method of heat dissipation. The clay pot version of the system relies on the process of water evaporation using a simple clay pot placed on top of the solar panel, while the standard thermosiphon cooling system makes use of a working fluid to remove heat by undergoing a phase change in a closed-loop system. Both of these systems are effective in cooling solar panels and improving their performance, but the choice between them depends on factors such as the specific application, location, and cost considerations.

A forced air cooling system (A6) for solar panels is a type of cooling system that uses a fan to circulate air over the solar panels to reduce their temperature. The cool air can be produced in a number of ways, including compressor-cooled refrigerant or chilled water. This type of cooling system is often used in sunny areas where the heat from the sun can cause the solar panels to overheat, which can reduce their efficiency and lifespan [62–64]. By cooling the panels with forced air, their temperature can be regulated, which can help to maximize their energy output and improve their overall performance. This type of cooling system is efficient and cost-effective and is often used in residential and commercial solar power systems.

Evaporative cooling systems (A7) [65] and water spray cooling systems (A8) [66] both use water to cool an area, but the methods and efficiencies are different. Evaporative cooling systems rely on the process of evaporation to cool the air, whereas water spray cooling systems function by directing a spray of water into the air to lower air temperature. Evaporative cooling systems are more efficient in low-humidity areas, while water spray cooling systems are more efficient in high-humidity areas. Additionally, evaporative cooling systems are more energy-efficient than water spray systems, while water spray cooling systems are often used in outdoor settings and for cooling large areas. Both systems have unique advantages and limitations, and the choice between the two depends on the specific application. Nano-fluid cooling systems (A9) can be either active or passive. Some of the researchers studied [16,52,62] active cooling systems that use nano-fluids as a coolant in conjunction with traditional cooling methods such as air or water cooling, while other researchers [67] studied passive cooling systems that rely solely on a nano-fluid coolant to dissipate heat.

However, there is no conclusive evidence to suggest which solar panel cooling strategy is the most effective, as it depends on various factors. Different cooling techniques have been used successfully in a number of cases, making choosing difficult [4]. Thus, it is evident [40,44] that choosing the appropriate solar panel cooling system involves the use of a scientific instrument to evaluate the options. So, multi-criteria decision making (MCDM) is the best instrument and has been utilized well in these circumstances. The step-by-step details of the adopted methodologies are presented in the following section, followed by their application.

4. Adopted MCDM Methodology

The most commonly used multi-criteria decision-making approaches [68–70] include the analytic hierarchy process (AHP), elimination and choice-translating algorithm (ELECTRE), technique for order of preference by similarity to the ideal solution (TOPSIS), and Preference Ranking Organization METHod for Enrichment of Evaluations (PROMETHEE). Compared to AHP, TOPSIS, and ELECTRE, Visual PROMETHEE has the capability of incorporating decision making via positive and negative preference flow. Visual management of performance using the PROMETHEE technique is appealing in the assessment of alternatives due to concepts like preference flow, weights, geometrical analysis for interactive aid (GAIA) plane, and sensitivity analyses. Partially and completely ranking the options also aids in determining the preferred alternative. However, those who make decisions are frequently interested not just in rating options, but also in determining the superiority of one over another (if such a superiority exists). The adopted methodology for an effective selection of solar panel cooling systems is explained step-by-step below.

Step 1: Decision matrix [70]:

Supposing there are i alternative solar panel cooling systems and j evaluation measures in outranking these cooling panels. PM_{ij} is the j performance measure's value for solar panel cooling system i. The decision matrix's structure is shown in Table 1 below. There are j performance measures and i alternative solar panel cooling systems, and W_j is the amount of significance assigned to each assessment criterion j.

Table 1. Decision matrix: alternatives, performance measures, and their weights.

Alternative Solar Panel Cooling System i	Performance Measure j			
	1	2	...	j
1	PM_{11}	PM_{12}	...	PM_{1j}
2	PM_{21}	PM_{22}	...	PM_{2j}
...	...	...	...	...
i	PM_{i1}	PM_{i2}	...	PM_{ij}
Performance measure weightage W_j	W_1	W_2	...	W_j

Step 2: Performance measure weightage (W_j) [71–73]:

W_j estimates might be subjective or objective. Variations in assessment metrics are employed in the accepted strategy to evaluate the divergence in the ranking of various solar panel cooling systems. The adopted method takes into account both sorts of weights. The objective weights technique employs mathematical models like entropy calculation, such as in [74]; details of this are explained below. The decision matrix values of the jth performance measure for the ith solar panel cooling system are standardized using Equation (1) if the objective of the performance measure is maximization. In contrast, Equation (2) is opted for if the performance measure objective is minimization, wherein S_{ij} is the standardized value for the jth performance measure of the ith solar panel cooling system; PM_{ij} is the jth

evaluation measure's value for the ith solar panel cooling system (refer to Table 1). After standardization of all performance measures, the decision matrix is expressed in matrix form in Equation (3), as seen below.

$$S_{ij} = \left[\frac{PM_{ij} - \min\limits_{j} PM_{ij}}{\max\limits_{j} PM_{ij} - \min\limits_{j} PM_{ij}} \right] \quad (1)$$

$$S_{ij} = \left[\frac{\max\limits_{j} PM_{ij} - PM_{ij}}{\max\limits_{j} PM_{ij} - \min\limits_{j} PM_{ij}} \right] \quad (2)$$

$$S'_{ij} = \begin{bmatrix} S_{11} & S_{12} & \dots & S_{1j} \\ \vdots & \vdots & \vdots & \vdots \\ \vdots & \vdots & \vdots & \vdots \\ S_{i1} & S_{i2} & \dots & S_{ij} \end{bmatrix} \quad (3)$$

Entropy E_j, according to its definition, is determined using the following Equation (4), and W_j, the performance measure objective weight, is determined by using Equation (5).

$$E_j = -\frac{\sum_{i=1}^{m} [S_{ij} * \ln(S_{ij})]}{\ln(m)} \quad (4)$$

$$W_j = \frac{1 - E_j}{\left[1 - \sum_{j=1}^{n} E_j\right]} \quad (5)$$

In comparison, subjective weights [71] refer to the relative importance of performance measures in a multi-criteria decision making (MCDM) problem and are determined based on the judgment or opinion of the decision maker. In other words, the weights are not calculated mathematically but are assigned based on the subjective perception or expertise of the decision maker. The subjective weights are often obtained through surveys, interviews, expert opinions, or other qualitative methods.

Step 3: Outranking flow estimation [75]:

To start, we initially constructed a generalized preference function $PF^{j}_{i=a,i=b}$, where (a, b) is a pair of solar panel cooling systems and j is the performance measure. Each $PF^{j}_{i=a,i=b}$ lies between 0 and 1. For given performance measure j, if 'i = a' solar panel cooling system is evaluated over 'i = b', then any one of the following preferences occurs based on the $PF^{j}_{i=a,i=b}$ function value. If the $PF^{j}_{i=a,i=b}$ function value is exactly equal to zero, then there is no preference for option a over alternative b. If the $PF^{j}_{i=a,i=b}$ function value is close to zero, then the option 'a' solar panel cooling system has a weak preference over the 'b' solar panel cooling system. If the $PF^{j}_{i=a,i=b}$ function value is close to one, there is a substantial preference for the option 'a' solar panel cooling system over 'b'. Lastly, if the $PF^{j}_{i=a,i=b}$ function value is exactly equal to one, then there is a stringent preference for the option 'a' solar panel cooling system over 'b'.

Subsequently, using these preference function values, the preference index $PI_{i=a,i=b}$, which has a value range of 0 to 1, is calculated for each pair of choices using Equation (6), as below.

$$PI_{i=a,i=b} = \left[\sum_{j=1}^{j} \left(W^{j} \times PF^{j}_{i=a,i=b}\right)\right] \div \left[\sum_{j=1}^{j} W^{j}\right] \quad (6)$$

In Equation (6), W^j is the weight associated with each solar cooling system evaluation measure j, and the preference index $PI_{i=a,i=b}$ displays a preference for the option 'a' solar panel cooling system over option b, considering all j performance measures ($j \in 1$ to j). If $PI_{i=a,i=b}$ equals perfectly zero, then there is zero preference for alternative a over b; if $PI_{i=a,i=b}$ equals approximately zero, then there is a low preference for a over b; if $PI_{i=a,i=b}$

equals approximately to one, then there is a high preference for alternative a over b, and if $PI_{i=a,i=b}$ is exactly equal to zero, then there is a perfect preference for alternative a over b. Finally, using the preference index, the outranking flows F_a^+ and F_a^- are quantified using the following Equations (7) and (8), respectively, where i is the number of alternatives (i ϵ 1 to i), excluding alternative i = a.

$$F_a^+ = \frac{1}{i-1}\sum_{i=1}^{i} PI_{a,i} \quad (7)$$

$$F_a^- = \frac{1}{i-1}\sum_{i=1}^{i} PI_{i,a} \quad (8)$$

Step 4: Calculation of net outranking flow and final ranking [76]:

The aforementioned predicted outranking flows F_a^+ and F_a^- for each option are used to determine each alternative's dominance over the others. Positive ranking flow quantifies the 'a' solar panel cooling system's dominance over the other alternative solar panel cooling systems, whereas negative ranking flow quantifies alternative a's dominance over the other alternatives. F_a^+, F_b^+, F_a^-, and F_b^- estimated outranking flows for each option a and b and were utilized to determine which option is dominant over the others. By understanding the outranking flow for any two choices, outranking relations can be inferred. Thus, a partial ranking is determined based on the outranking relations between any two choices as follows: if {[($F_a^+ > F_b^+$) and ($F_a^- < F_b^-$)] or ($F_a^+ \geq F_b^+$) or ($F_a^- \leq F_b^-$)},then solar panel cooling system 'a' has preference over 'b'; if {($F_a^+ = F_b^+$)} and or {($F_a^- = F_b^-$)}, then alternative 'a' has preference over 'b'; and if the information is otherwise inconsistent, then alternative a is incompatible with b. Net outranking flow for alternative a is obtained using $F_a = F_a^+ - F_a^-$; while net outranking flow for alternative b is obtained using $F_b = F_b^+ - F_b^-$. A complete ranking is subsequently obtained as follows: if $F_a > F_b$, then alternative a has complete preference over alternative b; if $F_a = F_b$, then alternative a has complete indifference compared to alternative b. The complete ranking is obtained by ordering the alternatives in decreasing order of their net flow scores. The net flow score is a measure of the overall performance of an alternative, which takes into account the positive and negative outranking flows with respect to all other alternatives. The PROMETHEE I method is used to obtain the positive and negative outranking flows, whereas the PROMETHEE II method is used to obtain a complete ranking, and it considers all the alternatives and criteria involved in the decision-making problem. The final ranking obtained using the PROMETHEE provides a global view of the alternatives and facilitates the selection of the best option. The application of the above methodology for an efficient ranking of solar panel cooling systems is presented in the subsequent section.

5. Application of the MCDM Approach

The presented approach extends considerable support to comparing and ranking solar panel cooling systems along with their validation and sensitivity analyses. The step-wise application of the proposed multi-criteria decision-making approach for selecting solar panel cooling systems is presented as follows.

Step 1: This step is used to identify potential alternative solar panel cooling systems for evaluation. Each solar panel cooling system is evaluated using eleven performance measures; to start, equal weights are assigned to these measures. For the decision matrix on hand, refer to Table 2.

Table 2. Sample data for each alternative obtained via expert/published research.

		Performance Measures j										
Units		**$ PM01**	**PM02**	**PM03**	**PM04**	**PM05**	**PM06**	**PM07**	**PM08**	**PM09**	**PM10**	**PM11**
		%	**5-Point**	**USD/sq. m**	**Kg**	**5-Point**	**°C**	**Sq. m**	**Impact**	**Impact**	**5-Point**	**Impact**
Objective		**Maximize**	**Maximize**	**Minimize**	**Minimize**	**Minimize**	**Maximize**	**Minimize**	**Maximize**	**Minimize**	**Minimize**	**Minimize**
Weightages W_j		1	1	1	1	1	1	1	1	1	1	1
Alter-natives i	# A1: [77,78]	3.5	5	USD 58	High	1	12.50	25	Very high	Low	Low	Very low
	A2: [78–80]	7	2	USD 168	Low	5	14.20	670.84	Low	Average	High	Low
	A3: [81,82]	9	4	USD 1125	Low	3	23.00	16.73	Low	High	High	Low
	A4: [83–85]	7	2	USD 25	Low	4	23.00	670.84	High	High	High	Very high
	A5: [61]	9	1	USD 25	High	6	28.00	18.49	High	High	High	High
	A6: [86–88]	4	3	USD 68	High	2	22.00	77.44	Low	Low	Low	Very high
	A7: [88]	13.5	2	USD 75	High	7	40.00	56.25	Very low	Low	High	Very high
	A8: [89]	13.5	2	USD 75	High	8	40.00	56.25	Low	Low	High	Very high
	A9: [90,91]	17	1	USD 540	Very high	9	10.20	125.0	Low	Very high	High	High

Note: for # and $, refer to Figure 1.

In Table 2, the solar panel finned air-cooling system (A1) has an average of 3.5% energy efficiency at an estimated cooling cost per square meter of a solar panel of USD 58, and this panel cooling system is highly reliable. The effects of carbon emissions on the environment are highly alarming. In the opinion of researchers [92,93], the alternative A1 has an extremely low effect as well as very little usage of electrical equipment in the cooling system network and is very supportive in terms of corrosion resistance. Similarly, Franklin et al. [77] stated that the finned air-cooling system is good and practical for installation and maintenance due to reduced physical deterioration and leakage difficulties. In light of numerous assessment parameters, one can, therefore, obtain comparable readings for any other alternative solar panel cooling system networks. Thus, for a given performance measure, one can see differences in values corresponding to alternative cooling systems. Not a single alternative outperforms others in all eleven measures. For selected performance measures, the selective solar panel cooling system score is better than the others. In other words, no single solar panel cooling system yields the best overall performance measures. Considering the multiple-attribute measure decision situation presented above, the application of the approach adopted to evaluate and rank these solar panel cooling system is presented here below.

Step 2: At this step, six sets of performance measure weights (equal weights, objective weights using the entropy technique, and four subjective weights) are chosen. Set 1 signifies an equal weighting of all performance measures, Set 2 is objective weights using the entropy approach, and Set 3 to Set 6 are subjective weights; refer to Table 3 below.

Table 3. Performance measure weights to evaluate solar panel cooling systems.

Performance Measures j	# Set 1	Set 2	Set 3	Set 4	Set 5	Set 6
$ PM01	0.091	0.038	0.035	0.040	0.050	0.600
PM02	0.091	0.110	0.650	0.040	0.050	0.040
PM03	0.091	0.152	0.035	0.600	0.050	0.040
PM04	0.091	0.065	0.035	0.040	0.500	0.040

Table 3. *Cont.*

Performance Measures j	# Set 1	Set 2	Set 3	Set 4	Set 5	Set 6
PM05	0.091	0.043	0.035	0.040	0.050	0.040
PM06	0.091	0.051	0.035	0.040	0.050	0.040
PM07	0.091	0.141	0.035	0.040	0.050	0.040
PM08	0.091	0.076	0.035	0.040	0.050	0.040
PM09	0.091	0.056	0.035	0.040	0.050	0.040
PM10	0.091	0.204	0.035	0.040	0.050	0.040
PM11	0.091	0.064	0.035	0.040	0.050	0.040

Note: $ refer to Figure 1. # Set 1: equal weightage to all performance measures; Set 2: weightage to all performance measures using entropy approach; Set 3: 65% weightage to reliability, with others having 4% each; Set 4: 60% weight to cost, with others having 4% each; Set 5: 50% weightage to emission and others having 5% each; and Set 6: 60% weightage to efficiency and others having 4% each.

Step 3: In this step, as significantly large mathematical computations are needed, it is preferred to adopt the Visual PROMETHEE soft tool. So, using it, two preference flows (F_i^+ and F_i^-) and the net outranking flow (F_i) were obtained for each alternative solar panel cooling system i, as presented in Table 4 as follows.

Table 4. Partial and complete outranking flows for each alternative solar cooling system vs. the sets of weights.

			Alternatives i								
			# A1	A2	A3	A4	A5	A6	A7	A8	A9
Set of weightages	$ Set 1	F_i	0.4098	0.0597	−0.0854	−0.0283	−0.0668	0.2003	−0.2095	−0.1195	−0.1603
		F_i^+	0.5655	0.3053	0.2962	0.2435	0.2242	0.3986	0.1611	0.2170	0.2215
		F_i^-	0.1557	0.2456	0.3815	0.2718	0.2911	0.1983	0.3706	0.3365	0.3818
	Set 2	F_i	0.4382	0.077	−0.2293	0.0061	−0.1433	0.2852	−0.2007	−0.1375	−0.0958
		F_i^+	0.5948	0.2847	0.2295	0.2391	0.1707	0.466	0.1431	0.1807	0.2231
		F_i^-	0.1566	0.2077	0.4588	0.233	0.3141	0.1808	0.3438	0.3182	0.3189
	Set 3	F_i	0.6536	−0.0884	0.3437	−0.1223	−0.214	0.1807	−0.2689	−0.2343	−0.2500
		F_i^+	0.7135	0.1175	0.4906	0.0937	0.0863	0.2916	0.062	0.0835	0.0853
		F_i^-	0.0599	0.206	0.1469	0.2161	0.3003	0.1109	0.331	0.3178	0.3353
	Set 4	F_i	0.2610	0.0963	−0.562	0.0758	0.0588	0.1665	−0.0155	0.0241	−0.1050
		F_i^+	0.3295	0.2043	0.1303	0.1954	0.1869	0.2537	0.1476	0.1722	0.1319
		F_i^-	0.0685	0.1081	0.6923	0.1196	0.1281	0.0872	0.1631	0.1481	0.2369
	Set 5	F_i	0.3379	−0.1922	−0.4407	−0.4093	0.0757	0.2227	−0.0027	0.0468	0.3618
		F_i^+	0.4798	0.2804	0.1629	0.1339	0.2921	0.388	0.2574	0.2881	0.5718
		F_i^-	0.1419	0.4726	0.6036	0.5432	0.2163	0.1653	0.2601	0.2413	0.2100
	Set 6	F_i	0.1803	0.0263	−0.0376	−0.0125	−0.0294	0.0882	−0.0922	−0.0526	−0.0705
		F_i^+	0.2488	0.1343	0.1303	0.1071	0.0987	0.1754	0.0709	0.0955	0.0975
		F_i^-	0.0685	0.1081	0.1679	0.1196	0.1281	0.0872	0.1631	0.1481	0.168

Note: for $, refer to Table 3, and for #, refer to Figure 1. F_i: net preference flows; F_i^+: positive preference flows; and F_i^-: positive preference flows.

In Table 4, there are two preference flows (F_i^+ and F_i^-), and these values help to draw a partial ranking. It also shows incomparability between solar panel cooling system alternatives when both (F_i^+ and F_i^-) preference flows have conflicting rankings. Similarly, in Table 4, F_i is the net preference flow, and it is a complete ranking of solar panel cooling system alternatives. For example, in Table 4, for set 1, when the corresponding values for alternatives A1 and A2 are compared using positive outranking flow (F_i^+), it is evident that $(F_{A1}^+ = 0.5655) > (F_{A2}^+ = 0.3053)$; however, when A1 and A2 are compared using negative outranking flow (F_i^-), it is evident that $(F_{A1}^- = 0.1557) < (F_{A2}^- = 0.0.2456)$. This demonstrates that option A1 has a stronger preference than alternative A2. Similarly, when all alternatives are compared based on overall outranking flow (F_i), $F_{A1} = 0.4098$ and has the highest value when compared to the remaining eight alternatives, whereas alternative A7 has the lowest net outranking value, $F_{A7} = -0.2095$, implying that A1 is the first preference and A7 is the last preference.

Step 4: Using the previously calculated step 3, outranking flows, a ranking network diagram, and a geometrical analysis for the interactive plane are derived. Details are presented below.

Figure 2 shows the positive outranking flow F_i^+ in the left column and the negative outranking flow F_i^- in the right column for each alternative i. Outranking flows are arranged in such a way that the best are projected at the top of the column. The center column represents the net outranking flow F_i. For each alternative, a representative line is drawn from its F_i^+ to the corresponding F_i^- score. For any given two alternatives, if the representative lines are parallel, the alternative representing the top line is preferred. On the other hand, if the two lines intersect, the corresponding alternatives are incomparable. By correlating Table 4 and Figure 3, for outranking flow F_1^+, alternative A1 dominates all other alternatives; for outranking flow F_7^+, alternative A7, the water-sprayed solar panel cooling system, highly underperforms compared to all other alternatives; for outranking flow F_1^-, alternative A1, the finned air sink solar panel cooling system, dominates all other alternatives; and for outranking flow F_3^-, alternative A3 highly underperforms compared to others. Generally, these positive and negative outranking flows induce two different rankings. In order to circumvent this scenario, a complete ranking based on net flow F was obtained and is presented in Figure 3 as follows.

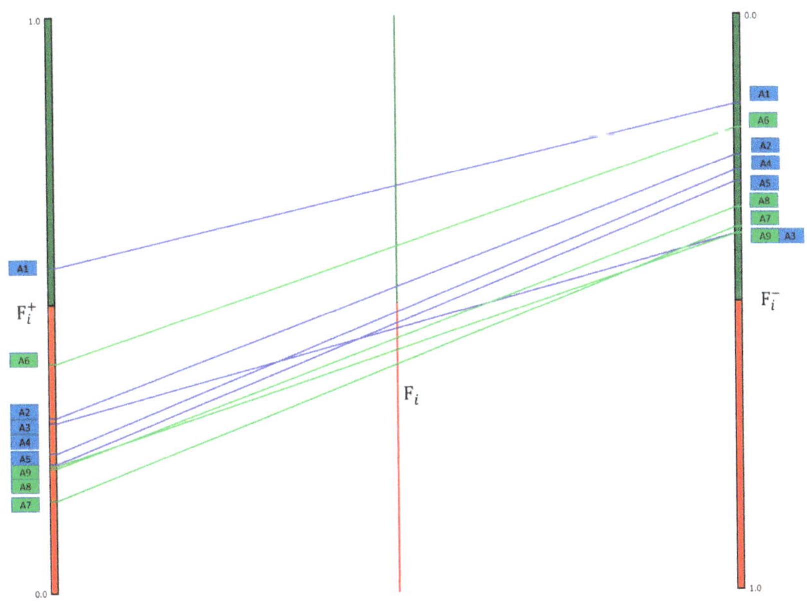

Figure 2. Partial ranking: as equal weightage to all performance measures (Set 1) (note: for notations, refer to Figure 1).

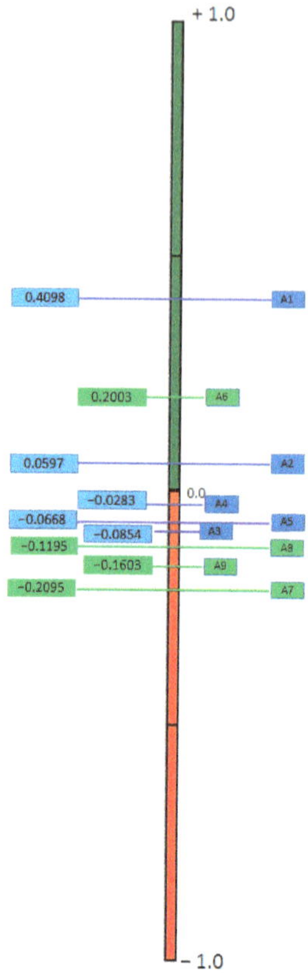

Figure 3. Complete ranking as equal weightage to all performance measures (Set 1) (note: for notations, refer to Figure 1).

From Figure 3, it is evident that alternative A1, the 'finned air-cooling system', surpasses all alternatives. Alternative A6, the forced air solar panel cooling system, performs as the next best option. Both of these options are best suited for dry, arid environments, while alternative A7, the water-sprayed solar panel cooling system, is least preferred over other alternatives. Subsequently, a network diagram was drawn (refer to Figure 4) in which each alternative is represented by a 'node' and its preference over other alternatives by an 'arrow'.

For example, in Table 4 for set 1, when the corresponding net outranking flow (F_i) values for all alternatives are compared, it is evident that (F_{A1} = 0.4098) > (F_{A6} = 0.2003) > (F_{A2} = 0.0597) > (F_{A4} = −0.0283) > (F_{A5} = −0.0668) > (F_{A3} = −0.0854) > (F_{A8} = −0.1195) > (F_{A9} = −0.1603) > (F_{A7} = −0.2095). Considering this net outranking relationship, the network diagram is drawn as presented in Figure 4. In Figure 4, it is evident that alternative A1, a passive cooling approach, is preferred over A6, which is an active cooling approach. When comparing all three active cooling approaches, A2, A4, and A3, approach A2 is preferred over both A4 and A3, whereas cooling approaches A4 and A3 are not comparable. Similarly, it is evident that passive cooling systems are outperforming active cooling systems, with the exception of forced air cooling system A6.

The Visual PROMETHEE represents the results in the GAIA (geometrical analysis for interactive assistant) plane (refer to Figure 5).

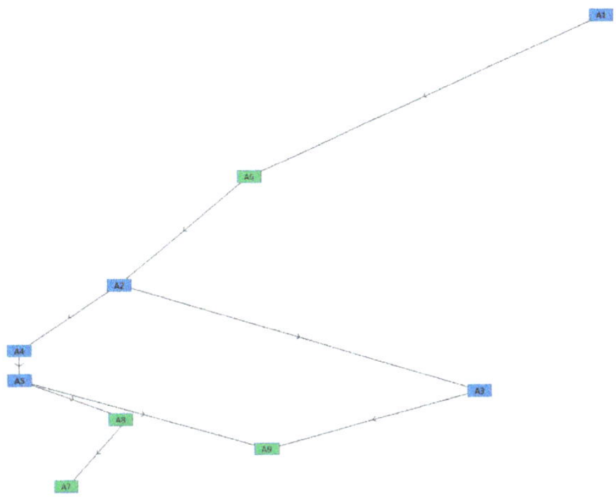

Figure 4. Network diagram as equal weightage to all performance measures (Set 1) (note: for notations, refer to Figure 1).

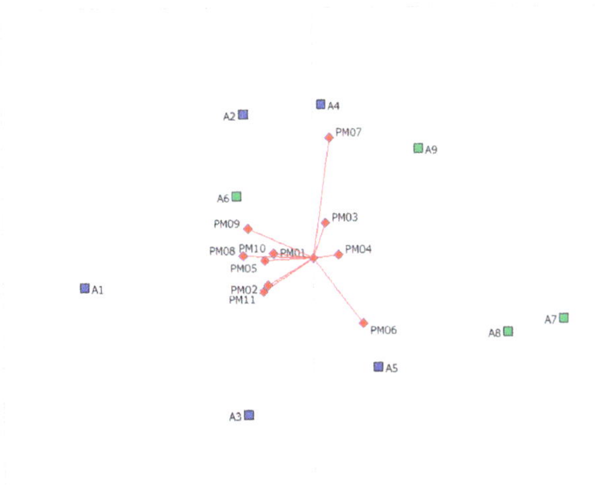

Figure 5. GAIA plane (note: for notations, refer to Figure 1).

In the GAIA plane (Figure 5), it is evident that solar panel cooling systems A1 and A3 are scoring opposite to A4 and A9; similarly, it is also apparent that the solar panel cooling systems A2 and A6 are scoring opposite to A5, A8 and A7. Solar panel cooling approach A1 scores better for measures PM02, PM05, and PM11 (refer to Figure 1). It is also observed that, as far as solar panel cooling systems A5, A8, and A7 are concerned, these cooling systems perform best for only one measure, i.e., panel temperature dropping. Similarly, when comparing active cooling approach A6 and passive cooling approach A2, both perform satisfactorily for the PM01, PM08, PM09, and PM10 performance measures. However, from Figure 5, it is evident that A6 scores better than A2. Thus, decision makers not typically having any pre-determined weights in mind warrants the need for sensitivity analysis. Hence, a feature of the Visual PROMETHEE software is adopted for this purpose. The details of the sensitivity analysis and results are presented here in the following section.

6. Sensitivity Analysis

Sensitivity analysis is performed to evaluate the deviation in the ranking of alternative solar panel cooling systems. Six sets of weights (equal weights, objective weights using the entropy approach, and four subjective scalings) of performance measure weights (refer to Table 3) are opted for in the sensitivity analysis. In subjective scaling, four scenarios are considered. Scenario 1 is based on technical experts who have knowledge about the workings of solar panel cooling systems; they prefer highly reliable solar cooling systems and so suggested having the highest importance of 65% to measure the reliability measure (PM02) compared to other measures. On the other hand, as in scenario 2, the management is keen on minimizing cooling system operating costs (PM03) as an economic concern and sets 60% weightages to cooling costs. In comparison, 60% weightages to environmental measures are seen in scenario 3, which is based on green and sustainable energy management; decision makers in this scenario are keen on minimizing carbon emissions by installing, operating, and maintaining environmentally friendly solar cooling systems. The last scenario, scenario 4, is that of an operational manager, whose target is to lower the amount of energy used for cooling and raise the overall effectiveness of the solar panel system, and suggested having 60% weightage to energy efficiency as crucial for the solar panel cooling system. On the contrary, objective weighting eliminates manmade disturbances and makes results accord more with the facts. The objective weights method makes use of mathematical models, such as entropy analysis [94]. Accordingly, the sensitivity analysis was performed relying on assigned entropy weights in the Visual PROMETHEE. The outcomes of the sensitivity analysis are presented below in Figure 6a–d.

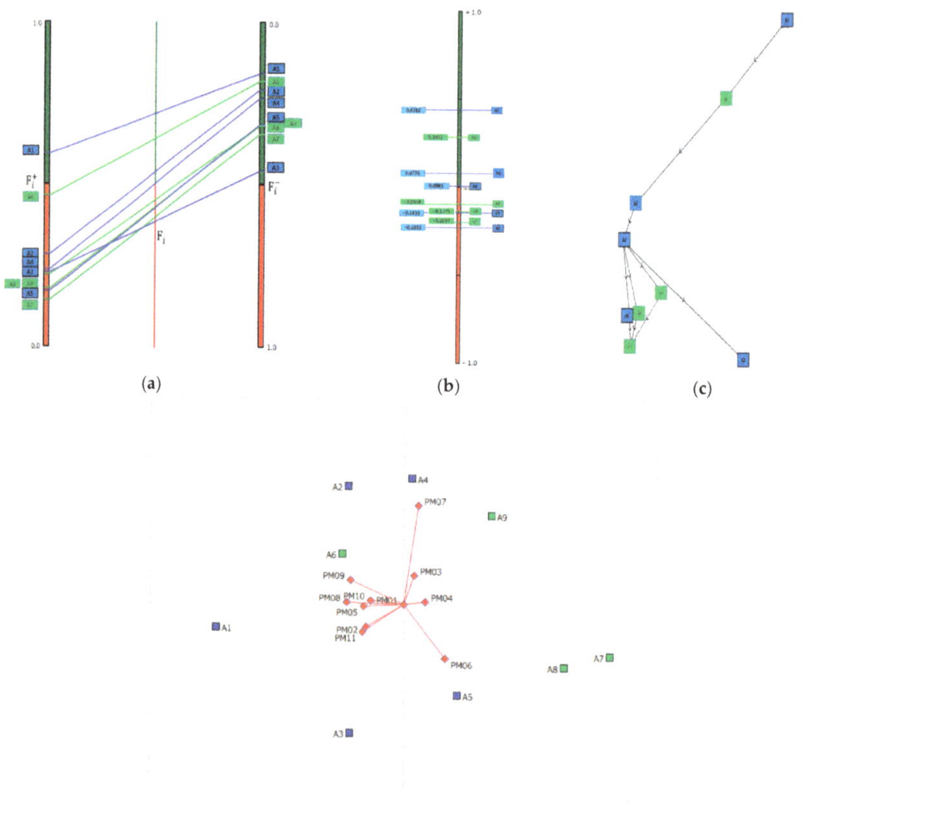

Figure 6. (**a**) Partial ranking, (**b**) complete ranking, (**c**) network, and (**d**) GAIA plane for performance weightages in Set 2 (refer to Table 3) (note: for notations, refer to Figure 1).

Similarly, sensitivity analysis was performed using subjective weightage Set 3. Below, Figure 7a–d represent corresponding outcomes representing partial ranking, complete ranking, the network diagram, and the GAIA plane, respectively.

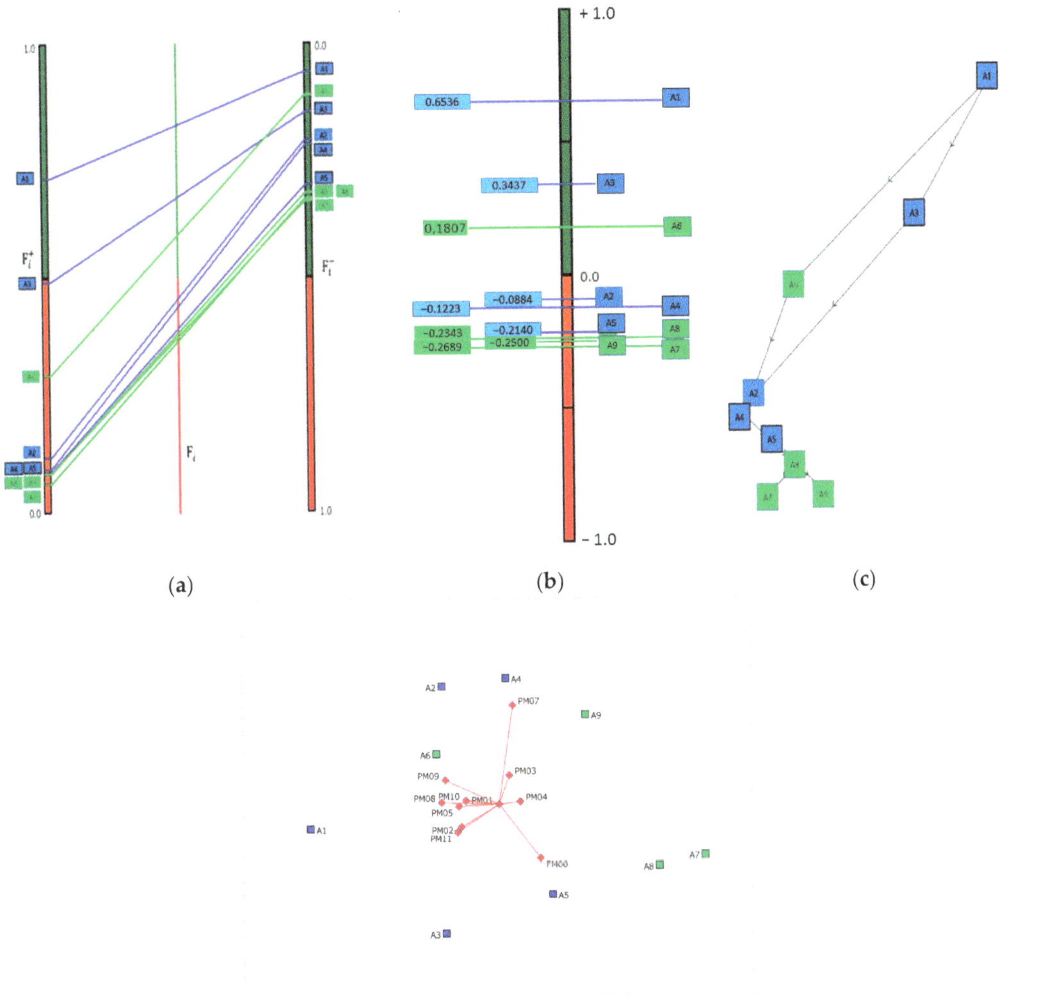

(**a**) (**b**) (**c**)

(**d**)

Figure 7. (**a**) Partial ranking, (**b**) complete ranking, (**c**) ranking network, and (**d**) GAIA plane for performance measure weightages in Set 3 (refer to Table 3) (note: for notations, refer to Figure 1).

In Figure 7, the decision maker's objective is to have a reliable solar cooling system to mitigate any level of operation risk. From Figure 7c, it is evident that A3, i.e., the phase change material cooling approach, is a more reliable solar panel cooling system compared to A2 (heat pipe cooling) and thermosiphon cooling with and without the material fluid pot (A2 and A4) in the category of passive cooling systems. However, A7, A8, and A9, active cooling systems, are the least reliable.

In the same way, Figure 8a–d below represent analysis corresponding to partial ranking, complete ranking, the network diagram, and the GAIA plane, respectively, for subjective weight applications. But, in this case, the highest weightage of 60% is set to cost measure (refer to Set 4 in Table 3). Here, when the management objective is to have a cost-effective solar cooling system, A3 is the least preferred cooling system compared to A2, A4, and A5 in the category of passive cooling systems. This is opposite to the finding from Figure 7. However, when there is management that has economic concerns, alternative A1 dominates all other alternatives on hand.

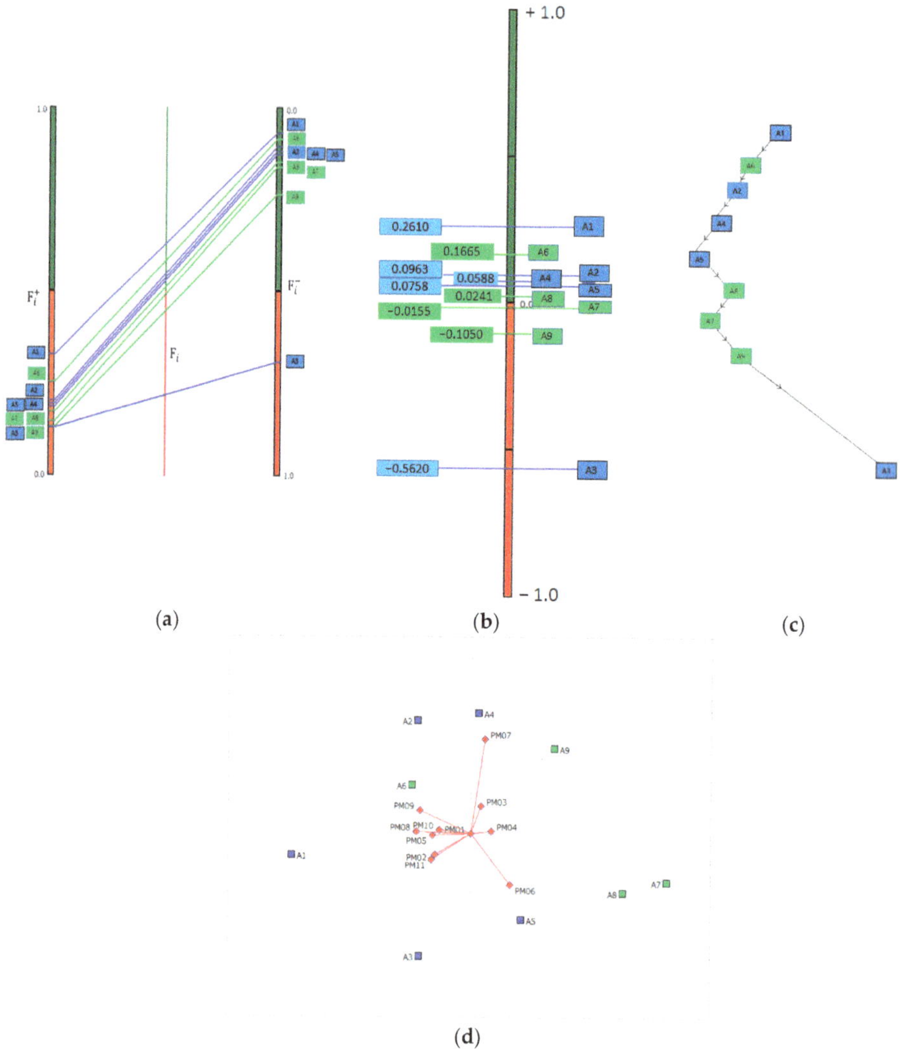

Figure 8. (**a**) Partial ranking, (**b**) complete ranking, (**c**) ranking network, and (**d**) GAIA plane for performance measure weightages in Set 4 (refer to Table 3) (note: for notations, refer to Figure 1).

Likewise, the following Figure 9a–d represent analysis corresponding to partial ranking, complete ranking, the network diagram, and the GAIA plane, respectively, for the highest subjective weightage of 60% to an evaluation measure PM04 (refer to Set 5 in Table 3). As the management objective is to have an environmentally friendly solar cooling system, the passive cooling approaches A2, A3, and A4 are the least preferred solar panel cooling systems compared to A1 and A5 in the same category. However, in this case, the nanomaterial fluid cooling system is still outranked by the finned air cooling system.

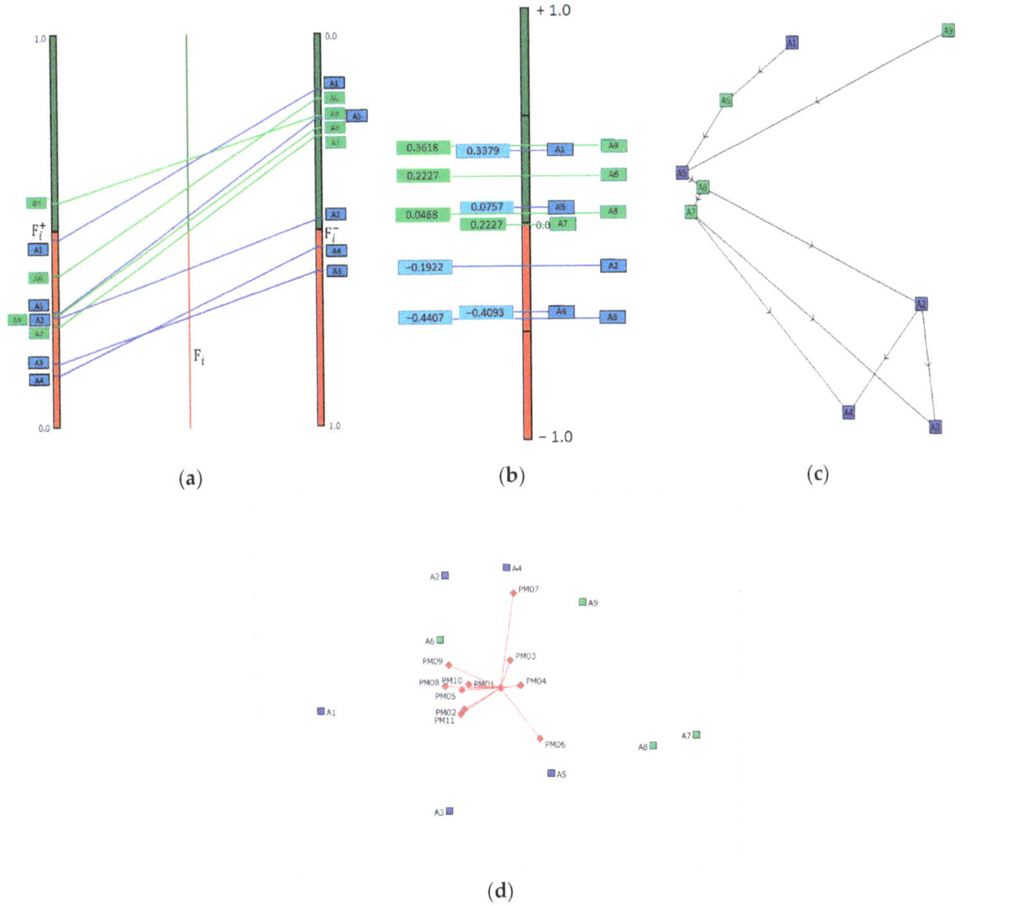

Figure 9. (**a**) Partial ranking, (**b**) complete ranking, (**c**) ranking network, and (**d**) GAIA plane for performance measure weightages in Set 5 (refer to Table 3) (note: for notations, refer to Figure 1).

Lastly, Figure 10a–d represent analyses corresponding to partial ranking, complete ranking, the network diagram, and the GAIA plane, respectively, for the highest subjective weightage of 60% set to an evaluation measure, PM01 (refer to Set 6 in Table 3). Here, the management objective is to have energy-efficient solar panel cooling systems where passive cooling systems outrank active cooling systems.

Thus, from the analysis, it is evident that for each scenario, the solar panel cooling system performance is sensitive to variation in performance measure weightages. The overall ranking for each alternative solar panel cooling system corresponding with multiple sets of weights assigned to performance measures is presented here in Table 5. Subsequently, the complete outranking flow (refer to Table 4) for each alternative is aggregated to drive overall ranking. Thus, after the sensitivity analysis, it is evident that solar panel cooling system alternatives A1 and A6 are found to be the best choice over other alternatives.

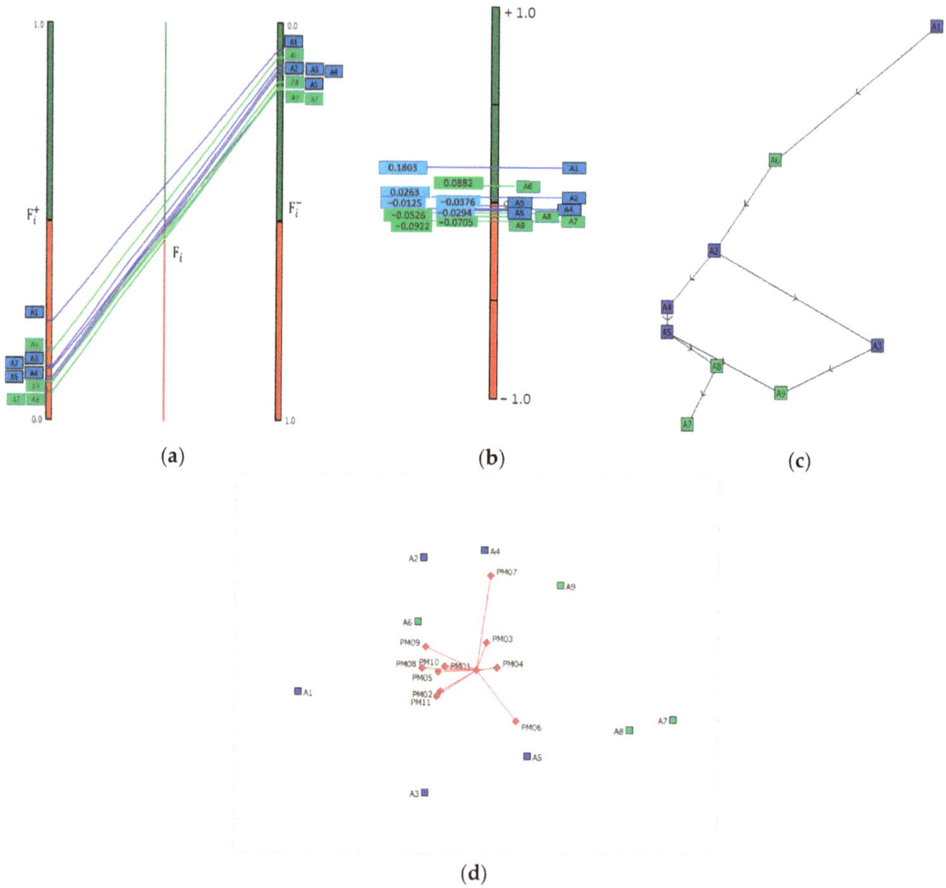

Figure 10. (**a**) Partial ranking, (**b**) complete ranking, (**c**) ranking network, and (**d**) GAIA plane for performance measure weightages in Set 6 (refer to Table 3) (note: for notations, refer to Figure 1).

Table 5. Ranking for each alternative solar panel cooling system corresponding with multiple sets of weights assigned to performance measures.

Set of Weightages		$ Set 1	Set 2	Set 3	Set 4	Set 5	Set 6	Overall Rank
Alternative solar panel cooling systems	# A1	1	1	1	1	2	1	1
	A2	3	3	4	3	7	3	3
	A3	6	9	2	9	9	6	9
	A4	4	4	5	4	8	4	7
	A5	5	7	6	5	4	5	4
	A6	2	2	3	2	3	2	2
	A7	9	8	9	7	6	9	8
	A8	7	6	7	6	5	7	6
	A9	8	5	8	8	1	8	5

Note: For $, refer to Table 3, and for #, refer to Figure 1.

7. The Managerial Perspectives of the Study

To stay competitive, solar panel manufacturing companies must focus on technological developments that are cost-effective and very responsive to market changes. These changes are changes in energy demand, changes in the cooling system parts and components as per customer demand, changes in the existing cooling system, large fluctuations in the energy demand and mix, changes in the government regulations that are related to safety and the environment, etc. It is well known that the performance of any solar panel module is susceptible to multiple factors. Traditional solar panel cooling systems are selected based on climate, type of panel, energy requirements, and cost. There is evidence that changing the cooling system from one configuration to another has an impact on the amount of energy generated by the solar panels; the amount of fluid required for cooling may change, which could impact the stability of the temperature of the solar panels and could, in turn, impact energy generation. However, the impact of changing a cooling system in solar energy modules will depend on the specific cooling system and the specific solar panel installed.

The study that is reported in the present paper clearly revealed that the performance of solar panel cooling systems is very likely to change whenever the environment is changed, either due to changes in the technical or non-technical operating scenarios. But, there is no guarantee that every change will lead to desirable conditions. It is also very risky to make a change without assessing the relative merit of the same from the viewpoint of system performance. The decision maker may find the methodology that is presented in this paper attractive for many reasons, as this method is capable of treating multi-criteria situations. It possesses an ability to incorporate decision making using threshold indifferences and preferences, as was explained in Sections 3 and 4. Concepts such as partial outranking, complete outranking, the graph and network diagram, and the feasibility of carrying out sensitivity analyses made this approach to the assessment of solar panel cooling system alternatives very attractive. The threshold weightage facilitates the comparison of alternatives using operational, environmental, and economic measures. One may compare alternatives using any sets of performance measures and their weights to decide the points of preference, indifference, and ignorance among choices. The concept of outranking relationships as a network diagram helps in the ordering of the nondominated alternative cooling systems. The presented approach aids in the synthesis of the preference relationship for each alternative solar panel cooling system in order to establish the required outranking relationship across solar panel cooling options in the context of all of the desired performance measures.

Thus, there are many decision situations in which the decision maker must choose among a finite number of alternatives, which are evaluated on a common set of multiple criteria. While evaluating the alternative solar panel cooling systems, it may be of interest to assess how similar and dissimilar the various solar panel cooling systems are and identify which performance measures play a significant role in establishing such relationships. This issue can be taken up with the help of the Visual PROMETHEE MCDM approach. The Visual PROMETHEE assesses the competitiveness of alternative solar panel cooling systems. Here, the adopted approach has the ability to incorporate positive and negative preferences. It synthesizes the preference relationships for each alternative to produce the desired outranking relationship between all of the alternatives. Concepts such as preference flow, weights, geometrical analysis for the interactive aid (GAIA) plane, as well as sensitivity analyses make this approach attractive in the assessment of solar panel cooling systems. Partial and complete ranking also helps identify the most preferred solar panel cooling system. However, decision makers are often interested not only in ranking solar panel cooling systems but also in establishing the superiority of one solar panel cooling system over another (if it exists). The PROMETHEE extends considerable support in this regard.

The practical significance of evaluating solar panel cooling systems within the context of solar power plant operation cannot be underestimated. Solar power plants may boost their energy production and, hence, their profitability by incorporating the proper cooling systems. There are numerous cooling solutions available, including both passive and active ones, and the method chosen will be determined by criteria such as climate, resource availability, and overall cost-effectiveness. This is especially relevant in areas with high ambient temperatures when solar panel efficiency can be severely reduced. Aside from the economic benefits, cooling measures can help to ensure the long-term viability of solar power facilities. The overall energy consumption of the power plant may be lowered by lowering the energy required to cool the panels, resulting in a smaller carbon footprint. This is consistent with global sustainability objectives.

8. Conclusions

This paper presents an approach for the assessment of solar panel cooling systems using the Visual PROMETHEE multi-criteria approach. The approach that was reported in the paper is found to be useful in ranking the choices among different alternative solar panel cooling systems using multiple performance measures. A total of nine possible alternative solar panel cooling systems were analyzed based on variations in cooling effectiveness, energy efficiency, environmental impact, durability, noise pollution, and system size, weight, and cost. Using this approach, outranking flows, partial ranking, complete ranking, the network, and the GAIA plane were obtained, with the objective being to explore the strong and weak points of each of the alternative solar panel cooling systems. The obtained results enabled the identification of groups of performance measures expressing similar preferences. For example, thermosiphon cooling and nanomaterial fluid cooling systems are preferred for their economic, ergonomic, and environmental measures. Meanwhile, thermosiphons with a clay pot cooling system, water spray cooling systems, and evaporative cooling systems are preferred only to drop solar panel temperature, which is also supported by Moharram et al. [94]. These systems are preferred for solar modules installed in very hot and dry climates. On the contrary, two passive cooling approaches, i.e., finned air cooling and phase change material cooling systems, have preference over reliability, ergonomics, and electric equipment failure risk measures. In short, the contributions of this study are as follows:

- It presented a multi-criteria decision-making approach that takes into account both quantitative and qualitative criteria;
- It outranked alternative solar panel cooling systems to determine the best viable choice using the PROMETHEE;
- It visualized the relationship between the different solar panel cooling systems and performance measures under consideration.

As can be seen from Table 5, the outranking of nine solar panel cooling systems is A1 > A6 > A2 > A4 > A5 > A3 > A8 > A9 > A7 when equal weights are assigned to each criterion. However, it is to be noted that equal weights to all the criteria are an exaggeration, and this is just used as a reference. On the other hand, when each criterion is weighted using the entropy approach, the preference of solar panel cooling system is A1 > A6 > A2 > A4 > A9 > A8 > A5 > A7 > A3. Sometimes, certain operating requirements need a maximization of the reliability of the cooling system, and in this case, the ranking is observed to be A1 > A3 > A6 > A2 > A4 > A5 > A8 > A9 > A7. On the other hand, when the need is to focus on minimizing cost, the preference is A1 > A6 > A2 > A4 > A5 > A8 > A7 > A9 > A3. In comparison, to maximize the efficiency of the solar plant, the outrank of nine alternatives is observed to be A1 > A6 > A2 > A4 > A5 > A3 > A8 > A9 > A7, and to minimize environmental impact by minimizing CO_2 emissions, the approach then suggests the preference order as A9 > A1 > A6 > A5 > A8 > A7 > A2 > A4 > A37, which shows that A9 is the best option. The overall outrank for nine alternatives is also derived as A1 > A6 > A2 > A5 > A9 > A8 > A4 > A7 > A3.

Thus, nine different solar panel cooling systems were identified, and the multi-criteria analysis tool Visual PROMETHEE was used. Six possible scenarios are examined based on the priority assigned to each performance metric. Under each situation, the best solar panel cooling strategy to the worst cooling approach has been ranked. When operational efficiency was given the most weight, finned cooling was revealed to be the best solar panel cooling method, whereas thermosiphon cooling was the best cooling solution when the emission reduction criteria were given the most weight. The second-best method was found to be forced air cooling under equal weights, entropy weights, and 60% weightage to economic and component failure risk. The third choice is a heat pipe passive cooling system. Future research directions are set to focus on the assessment of hybrid solar panel cooling systems that are better integrated with energy management systems for improved overall performance. Comparison experiments can be executed by comparing the proposed Visual PROMETHEE approach with other MCDM methods to show its effectiveness as a future scope. One can also include performance metrics such as maintainability, technical implementation challenges, and sustainable solutions. Similarly, one can involve the development of new testing protocols and standards that can be used to assess the reliability and performance of these systems over time.

Funding: This research was funded by the Deputyship for Research and Innovation, Ministry of Education in Saudi Arabia, through grant number (IFKSUOR3-079-1).

Data Availability Statement: Data are available in the manuscript.

Acknowledgments: The authors extend their appreciation to the Deputyship for Research and Innovation, "Ministry of Education" in Saudi Arabia, for funding this research (IFKSUOR3-079-1).

Conflicts of Interest: The author declares no conflict of interest.

References

1. Kapilan, N.; Nithin, K.C.; Chiranth, K.N. Challenges and Opportunities in Solar Photovoltaic System. *Mater. Today Proc.* **2022**, *62*, 3538–3543. [CrossRef]
2. Almeshaiei, E.; Al-Habaibeh, A.; Mina, N.; Akib, S. Rapid Evaluation of the Design and Manufacture of Cooling Systems of Photovoltaic Solar Panels. *Int. J. Interact. Des. Manuf.* **2023**, *17*, 197–214. [CrossRef]
3. Hernández-Callejo, L.; Gallardo-Saavedra, S.; Alonso-Gómez, V. A Review of Photovoltaic Systems: Design, Operation and Maintenance. *Solar Energy* **2019**, *188*, 426–440. [CrossRef]
4. Maleki, A.; Haghighi, A.; El Haj Assad, M.; Mahariq, I.; Alhuyi Nazari, M. A Review on the Approaches Employed for Cooling PV Cells. *Solar Energy* **2020**, *209*, 170–185. [CrossRef]
5. Hudișteanu, S.V.; Țurcanu, F.E.; Cherecheș, N.C.; Popovici, C.G.; Verdeș, M.; Huditeanu, I. Enhancement of PV Panel Power Production by Passive Cooling Using Heat Sinks with Perforated Fins. *Appl. Sci.* **2021**, *11*, 11323. [CrossRef]
6. Owhaib, W.; Qanadah, Y.; Al-Ajalin, H.; Tuffaha, A.; Al-Kouz, W. Photovoltaic Panel Efficiency Improvement by Using Direct Water Passive Cooling with Clay Pot. In Proceedings of the 2021 12th International Renewable Engineering Conference (IREC), Amman, Jordan, 14–15 April 2021; pp. 1–4.
7. El Hammoumi, A.; Chalh, A.; Allouhi, A.; Motahhir, S.; El Ghzizal, A.; Derouich, A. Design and Construction of a Test Bench to Investigate the Potential of Floating PV Systems. *J. Clean. Prod.* **2021**, *278*, 123917. [CrossRef]
8. Claus, R.; López, M. Key Issues in the Design of Floating Photovoltaic Structures for the Marine Environment. *Renew. Sustain. Energy Rev.* **2022**, *164*, 112502. [CrossRef]
9. Skoplaki, E.; Palyvos, J.A. On the Temperature Dependence of Photovoltaic Module Electrical Performance: A Review of Efficiency/Power Correlations. *Solar Energy* **2009**, *83*, 614–624. [CrossRef]
10. Al-Ghezi, M.K.; Ahmed, R.T.; Chaichan, M.T. The Influence of Temperature and Irradiance on Performance of the Photovoltaic Panel in the Middle of Iraq. *Int. J. Renew. Energy Dev.* **2022**, *11*, 501–553. [CrossRef]
11. Choudhary, K.; Jakhar, S.; Gakkhar, N.; Sangwan, K.S. Comparative Life Cycle Assessments of Photovoltaic Thermal Systems with Earth Water Heat Exchanger Cooling. *Procedia CIRP* **2022**, *105*, 255–260. [CrossRef]
12. Metwally, H.; Mahmoud, N.A.; Aboelsoud, W.; Ezzat, M. A Critical Review of Photovoltaic Panels Thermal Management: Criteria and Methods. *ASM* **2022**, *2*, 2. [CrossRef]
13. AlMallahi, M.N.; El Haj Assad, M.; AlShihabi, S.; Alayi, R. Multi-Criteria Decision-Making Approach for the Selection of Cleaning Method of Solar PV Panels in United Arab Emirates Based on Sustainability Perspective. *Int. J. Low-Carbon Technol.* **2022**, *17*, 380–393. [CrossRef]
14. Tembo, P.M.; Subramanian, V. Current Trends in Silicon-Based Photovoltaic Recycling: A Technology, Assessment, and Policy Review. *Solar Energy* **2023**, *259*, 137–150. [CrossRef]
15. Alami, A.H.; Alasad, S.; Aljaghoub, H.; Ayoub, M.; Alashkar, A.; Mdallal, A.; Hasan, R. Second-Generation Photovoltaics: Thin-Film Technologies. In *PV Technology and Manufacturing*; Alami, A.H., Ed.; Advances in Science, Technology & Innovation; Springer International Publishing: Cham, Switzerland, 2023; pp. 65–75. ISBN 978-3-031-31349-3.
16. Omri, M.; Selimefendigil, F.; Smaoui, H.T.; Kolsi, L. Cooling System Design for Photovoltaic Thermal Management by Using Multiple Porous Deflectors and Nanofluid. *Case Stud. Therm. Eng.* **2022**, *39*, 102405. [CrossRef]

17. Foteinis, S.; Savvakis, N.; Tsoutsos, T. Energy and Environmental Performance of Photovoltaic Cooling Using Phase Change Materials under the Mediterranean Climate. *Energy* **2023**, *265*, 126355. [CrossRef]
18. Fikri, M.A.; Samykano, M.; Pandey, A.K.; Kadirgama, K.; Kumar, R.R.; Selvaraj, J.; Rahim, N.A.; Tyagi, V.V.; Sharma, K.; Saidur, R. Recent Progresses and Challenges in Cooling Techniques of Concentrated Photovoltaic Thermal System: A Review with Special Treatment on Phase Change Materials (PCMs) Based Cooling. *Sol. Energy Mater. Sol. Cells* **2022**, *241*, 111739. [CrossRef]
19. De Smet, Y.; Lidouh, K. An Introduction to Multicriteria Decision Aid: The PROMETHEE and GAIA Methods. In *Business Intelligence: Second European Summer School, eBISS 2012, Brussels, Belgium, 15–21 July 2012, Tutorial Lectures*; Aufaure, M.-A., Zimányi, E., Eds.; Lecture Notes in Business Information Processing; Springer: Berlin/Heidelberg, Germany, 2013; pp. 150–176, ISBN 978-3-642-36318-4.
20. Đurin, B.; Nađ, L. Selection of the Technologically Most Appropriate Variant of the Solar Photovoltaic (PV) Water Supply System by Using Multi-Criteria Methods PROMETHEE and GAIA. *Appl. Water Sci.* **2018**, *8*, 122. [CrossRef]
21. Wu, Y.; Zhang, B.; Wu, C.; Zhang, T.; Liu, F. Optimal Site Selection for Parabolic Trough Concentrating Solar Power Plant Using Extended PROMETHEE Method: A Case in China. *Renew. Energy* **2019**, *143*, 1910–1927. [CrossRef]
22. Aldrin Wiguna, K.; Sarno, R.; Ariyani, N.F. Optimization Solar Farm Site Selection Using Multi-Criteria Decision Making Fuzzy AHP and PROMETHEE: Case Study in Bali. In Proceedings of the 2016 International Conference on Information & Communication Technology and Systems (ICTS), Surabaya, Indonesia, 12 October 2016; pp. 237–243.
23. A Multi-Dimensional Decision Support System for Choosing Solar Shading Devices in Office Buildings | SpringerLink. Available online: https://link.springer.com/chapter/10.1007/978-3-031-06825-6_168 (accessed on 21 August 2023).
24. Mirzaei, N.; Nowzarï, R. Applying a Combination of AHP, ANP, and PROMETHEE Methods to Find the Optimal Location for Solar Power Plant. *EJOSAT* **2021**, *32*, 1076–1085. [CrossRef]
25. Oberschmidt, J.; Geldermann, J.; Ludwig, J.; Schmehl, M. Modified PROMETHEE Approach for Assessing Energy Technologies. *Int. J. Energy Sect. Manag.* **2010**, *4*, 183–212. [CrossRef]
26. Tüysüz, N.; Kahraman, C. Picture Fuzzy Z-AHP: Application to Panel Selection of Solar Energy. In Proceedings of the Intelligent and Fuzzy Systems; Kahraman, C., Sari, I.U., Oztaysi, B., Cebi, S., Cevik Onar, S., Tolga, A.Ç., Eds.; Springer Nature Switzerlan: Cham, Switzerland, 2023; pp. 337–345.
27. Razzaque, A.; Masmali, I.; Hassan, R.; Shuaib, U.; Razaq, A.; Alhamzi, G.; Noor, S. A Modern Strategy for the Selection of Efficacious Solar Panels at Industrial Level under an Interval-Valued Picture Fuzzy Set Hybrid Environment. *Energies* **2023**, *16*, 5338. [CrossRef]
28. Mardani, A.; Jusoh, A.; Zavadskas, E.K.; Cavallaro, F.; Khalifah, Z. Sustainable and Renewable Energy: An Overview of the Application of Multiple Criteria Decision Making Techniques and Approaches. *Sustainability* **2015**, *7*, 13947–13984. [CrossRef]
29. Wang, C.-N.; Nguyen, V.T.; Thai, H.T.N.; Duong, D.H. Multi-Criteria Decision Making (MCDM) Approaches for Solar Power Plant Location Selection in Viet Nam. *Energies* **2018**, *11*, 1504. [CrossRef]
30. Seker, S.; Kahraman, C. Socio-Economic Evaluation Model for Sustainable Solar PV Panels Using a Novel Integrated MCDM Methodology: A Case in Turkey. *Socio Econ. Plan. Sci.* **2021**, *77*, 100998. [CrossRef]
31. Krysiak, M.; Kluczek, A. Application of Multi-Criteria Decision-Making Approach for Sustainability Assessment of Chosen Photovoltaic Modules. *Manag. Prod. Eng. Rev.* **2020**, *11*, 106–118. [CrossRef]
32. Saaty, R.W. The Analytic Hierarchy Process—What It Is and How It Is Used. *Math. Model.* **1987**, *9*, 161–176. [CrossRef]
33. Hussain, A.; Ullah, K.; Pamucar, D.; Haleemzai, I.; Tatić, D. Assessment of Solar Panel Using Multiattribute Decision-Making Approach Based on Intuitionistic Fuzzy Aczel Alsina Heronian Mean Operator. *Int. J. Intell. Syst.* **2023**, *2023*, e6268613. [CrossRef]
34. Peng, Z.; Herfatmanesh, M.R.; Liu, Y. Cooled Solar PV Panels for Output Energy Efficiency Optimisation. *Energy Convers. Manag.* **2017**, *150*, 949–955. [CrossRef]
35. Qi, L.; Jiang, M.; Lv, Y.; Zhang, Z.; Yan, J. Techno-Economic Assessment of Photovoltaic Power Generation Mounted on Cooling Towers. *Energy Convers. Manag.* **2021**, *235*, 113907. [CrossRef]
36. Lamnatou, C.h.r.; Chemisana, D. Photovoltaic-Green Roofs: A Life Cycle Assessment Approach with Emphasis on Warm Months of Mediterranean Climate. *J. Clean. Prod.* **2014**, *72*, 57–75. [CrossRef]
37. Abu-Rahmeh, T.M. Efficiency of Photovoltaic Modules Using Different Cooling Methods: A Comparative Study. *J. Power Energy Eng.* **2017**, *5*, 32–45. [CrossRef]
38. Sultan, S.M.; Tso, C.P.; Ervina, E.M.N. A New Production Cost Effectiveness Factor for Assessing Photovoltaic Module Cooling Techniques. *Int. J. Energy Res.* **2020**, *44*, 574–583. [CrossRef]
39. Semelane, S.; Nwulu, N.; Kambule, N.; Tazvinga, H. Economic Feasibility Assessment of Manufacturing Solar Panels in South Africa—A Case Study of Steve Tshwete Local Municipality. *Sustain. Energy Technol. Assess.* **2021**, *43*, 100945. [CrossRef]
40. Siah Chehreh Ghadikolaei, S. Solar Photovoltaic Cells Performance Improvement by Cooling Technology: An Overall Review. *Int. J. Hydrogen Energy* **2021**, *46*, 10939–10972. [CrossRef]
41. Rabelo, M.; Yousuf, H.; Kim, J.; Dao, V.-A.; Pham, D.P.; Yi, J. Progressive Cooling Techniques for Photovoltaic Module Efficiency and Reliability: Comparative Evaluation and Optimization. *Energy Rep.* **2022**, *8*, 8534–8545. [CrossRef]
42. Siah Chehreh Ghadikolaei, S. An Enviroeconomic Review of the Solar PV Cells Cooling Technology Effect on the CO_2 Emission Reduction. *Solar Energy* **2021**, *216*, 468–492. [CrossRef]
43. Şirin, C.; Goggins, J.; Hajdukiewicz, M. A Review on Building-Integrated Photovoltaic/Thermal Systems for Green Buildings. *Appl. Therm. Eng.* **2023**, *229*, 120607. [CrossRef]

44. Solar Energy and the Environment—U.S. Energy Information Administration (EIA). Available online: https://www.eia.gov/energyexplained/solar/solar-energy-and-the-environment.php (accessed on 6 July 2023).
45. Cuce, P.M.; Saxena, A.; Cuce, E.; Riffat, S. Applications of Solar PV Tree Systems with Different Design Aspects and Performance Assessment. *Int. J. Low Carbon Technol.* **2022**, *17*, 266–278. [CrossRef]
46. Sun, C.; Zou, Y.; Qin, C.; Zhang, B.; Wu, X. Temperature Effect of Photovoltaic Cells: A Review. *Adv. Compos. Hybrid Mater.* **2022**, *5*, 2675–2699. [CrossRef]
47. Hasan, K.; Yousuf, S.B.; Tushar, M.S.H.K.; Das, B.K.; Das, P.; Islam, M.S. Effects of Different Environmental and Operational Factors on the PV Performance: A Comprehensive Review. *Energy Sci. Eng.* **2022**, *10*, 656–675. [CrossRef]
48. Hamed, A.; Al-Ghussain, L.; Hassan, M.A.; Annuk, A. Techno-Economic Analysis for Optimal Configurations of PV Systems with Back Reflectors. *Energy Rep.* **2022**, *8*, 14979–14996. [CrossRef]
49. Aghaei, M.; Fairbrother, A.; Gok, A.; Ahmad, S.; Kazim, S.; Lobato, K.; Oreski, G.; Reinders, A.; Schmitz, J.; Theelen, M.; et al. Review of Degradation and Failure Phenomena in Photovoltaic Modules. *Renew. Sustain. Energy Rev.* **2022**, *159*, 112160. [CrossRef]
50. Muhamad Azim, M.K.; Arifutzzaman, A.; Saidur, R.; Khandaker, M.U.; Bradley, D.A. Recent Progress in Emerging Hybrid Nanomaterials towards the Energy Storage and Heat Transfer Applications: A Review. *J. Mol. Liq.* **2022**, *360*, 119443. [CrossRef]
51. Flamant, G.; Grange, B.; Wheeldon, J.; Siros, F.; Valentin, B.; Bataille, F.; Zhang, H.; Deng, Y.; Baeyens, J. Opportunities and Challenges in Using Particle Circulation Loops for Concentrated Solar Power Applications. *Prog. Energy Combust. Sci.* **2023**, *94*, 101056. [CrossRef]
52. Chavez Panduro, E.A.; Finotti, F.; Largiller, G.; Lervåg, K.Y. A Review of the Use of Nanofluids as Heat-Transfer Fluids in Parabolic-Trough Collectors. *Appl. Therm. Eng.* **2022**, *211*, 118346. [CrossRef]
53. Arias, I.; Cardemil, J.; Zarza, E.; Valenzuela, L.; Escobar, R. Latest Developments, Assessments and Research Trends for next Generation of Concentrated Solar Power Plants Using Liquid Heat Transfer Fluids. *Renew. Sustain. Energy Rev.* **2022**, *168*, 112844. [CrossRef]
54. Alami, A.H.; Rabaia, M.K.H.; Sayed, E.T.; Ramadan, M.; Abdelkareem, M.A.; Alasad, S.; Olabi, A.-G. Management of Potential Challenges of PV Technology Proliferation. *Sustain. Energy Technol. Assess.* **2022**, *51*, 101942. [CrossRef]
55. Aghaei, M.; Kumar, N.M.; Eskandari, A.; Ahmed, H.; de Oliveira, A.K.V.; Chopra, S.S. Chapter 5—Solar PV Systems Design and Monitoring. In *Photovoltaic Solar Energy Conversion*; Gorjian, S., Shukla, A., Eds.; Academic Press: Cambridge, MA, USA, 2020; pp. 117–145, ISBN 978-0-12-819610-6.
56. Shiravi, A.H.; Firoozzadeh, M. Performance Assessment of a Finned Photovoltaic Module Exposed to an Air Stream: An Experimental Study. *J. Braz. Soc. Mech. Sci. Eng.* **2022**, *44*, 535. [CrossRef]
57. Gao, Y.; Wu, D.; Dai, Z.; Wang, C.; Chen, B.; Zhang, X. A Comprehensive Review of the Current Status, Developments, and Outlooks of Heat Pipe Photovoltaic and Photovoltaic/Thermal Systems. *Renew. Energy* **2023**, *207*, 539–574. [CrossRef]
58. Elminshawy, N.A.S.; Mohamed, A.M.I.; Morad, K.; Elhenawy, Y.; Alrobaian, A.A. Performance of PV Panel Coupled with Geothermal Air Cooling System Subjected to Hot Climatic. *Appl. Therm. Eng.* **2019**, *148*, 1–9. [CrossRef]
59. Wu, S.-Y.; Zhang, Q.-L.; Xiao, L.; Guo, F.-H. A Heat Pipe Photovoltaic/Thermal (PV/T) Hybrid System and Its Performance Evaluation. *Energy Build.* **2011**, *43*, 3558–3567. [CrossRef]
60. Enhancing the Performance of Floating Photovoltaic System by Using Thermosiphon Cooling Method: Numerical and Experimental Analyses—ScienceDirect. Available online: https://www.sciencedirect.com/science/article/abs/pii/S1290072922002617 (accessed on 6 July 2023).
61. Gholami, Z.; Rahmati, M.H.; Arabhosseini, A.; Charzi, M. Combined Cooling of Photovoltaic Module Integrated with Thermoelectric Generators, by Using Earthenware Water Tank and Ultrasonic Humidifier: An Experimental Study. *Sustain. Energy Technol. Assess.* **2022**, *53*, 102601. [CrossRef]
62. Kozak-Jagieła, E.; Cisek, P.; Ocłoń, P. Cooling Techniques for PV Panels: A Review. *Sci. Radices* **2023**, *2*, 47–68. [CrossRef]
63. Amelia, A.R.; Irwan, Y.M.; Irwanto, M.; Leow, W.Z.; Gomesh, N.; Safwati, I.; Anuar, M.A.M. Cooling on Photovoltaic Panel Using Forced Air Convection Induced by DC Fan. *Int. J. Electr. Comput. Eng.* **2016**, *6*, 526. [CrossRef]
64. Hussien, A.; Eltayesh, A.; El-Batsh, H.M. Experimental and Numerical Investigation for PV Cooling by Forced Convection. *Alex. Eng. J.* **2023**, *64*, 427–440. [CrossRef]
65. Mahmood, D.M.N.; Aljubury, I.M.A. Experimental Investigation of a Hybrid Photovoltaic Evaporative Cooling (PV/EC) System Performance under Arid Conditions. *Results Eng.* **2022**, *15*, 100618. [CrossRef]
66. Zhou, N.; Chen, F.; Cao, Y.; Chen, M.; Wang, Y. Experimental Investigation on the Performance of a Water Spray Cooling System. *Appl. Therm. Eng.* **2017**, *112*, 1117–1128. [CrossRef]
67. Sohani, A.; Shahverdian, M.H.; Sayyaadi, H.; Samiezadeh, S.; Doranehgard, M.H.; Nizetic, S.; Karimi, N. Selecting the Best Nanofluid Type for A Photovoltaic Thermal (PV/T) System Based on Reliability, Efficiency, Energy, Economic, and Environmental Criteria. *J. Taiwan Inst. Chem. Eng.* **2021**, *124*, 351–358. [CrossRef]
68. Gnanasekaran, S.A.; Venkatachalam, N. A Review on Applications of Multi-Criteria Decision Making (MCDM) For Solar Panel Selection. *Int. J. Mech. Prod. Eng. Res. Dev.* **2019**, *9*, 11–20. [CrossRef]
69. Toloie-Eshlaghy, A.; Homayonfar, M. MCDM Methodologies and Applications: A Literature Review from 1999 to 2009. *Res. J. Int. Stud.* **2011**, *21*, 86–137.
70. Taherdoost, H.; Madanchian, M. Multi-Criteria Decision Making (MCDM) Methods and Concepts. *Encyclopedia* **2023**, *3*, 77–87. [CrossRef]

71. Zoraghi, N.; Amiri, M.; Talebi, G.; Zowghi, M. A Fuzzy MCDM Model with Objective and Subjective Weights for Evaluating Service Quality in Hotel Industries. *J. Ind. Eng. Int.* **2013**, *9*, 38. [CrossRef]
72. Wang, T.-C.; Lee, H.-D. Developing a Fuzzy TOPSIS Approach Based on Subjective Weights and Objective Weights. *Expert Syst. Appl.* **2009**, *36*, 8980–8985. [CrossRef]
73. Md Saad, R.; Ahmad, M.Z.; Abu, M.S.; Jusoh, M.S. Hamming Distance Method with Subjective and Objective Weights for Personnel Selection. *Sci. World J.* **2014**, *2014*, e865495. [CrossRef]
74. Lotfi, F.H.; Fallahnejad, R. Imprecise Shannon's Entropy and Multi Attribute Decision Making. *Entropy* **2010**, *12*, 53–62. [CrossRef]
75. Goswami, S.S. Outranking Methods: Promethee I and Promethee II. *Found. Manag.* **2020**, *12*, 93–110. [CrossRef]
76. Verma, P. Promethee: A Tool for Multi-Criteria Decision Analysis. In *Multi-Criteria Decision Analysis in Management*; IGI Global: Hershey, PA, USA, 2020; pp. 282–309. ISBN 978-1-79982-216-5.
77. Franklin, J.C.; Chandrasekar, M.; Sridharan, M. Performance of a Downstream Finned Solar Photovoltaic Thermal Air System. *J. Therm. Sci. Eng. Appl.* **2021**, *13*, 041028. [CrossRef]
78. Sharaf, M.; Yousef, M.S.; Huzayyin, A.S. Review of Cooling Techniques Used to Enhance the Efficiency of Photovoltaic Power Systems. *Environ. Sci. Pollut. Res.* **2022**, *29*, 26131–26159. [CrossRef]
79. Sabry, M.; Lashin, A. Performance of a Heat-Pipe Cooled Concentrated Photovoltaic/Thermoelectric Hybrid System. *Energies* **2023**, *16*, 1438. [CrossRef]
80. Jouhara, H.; Szulgowska-Zgrzywa, M.; Sayegh, M.A.; Milko, J.; Danielewicz, J.; Nannou, T.K.; Lester, S.P. The Performance of a Heat Pipe Based Solar PV/T Roof Collector and Its Potential Contribution in District Heating Applications. *Energy* **2017**, *136*, 117–125. [CrossRef]
81. Phase Change Material—An Overview | ScienceDirect Topics. Available online: https://www.sciencedirect.com/topics/materials-science/phase-change-material (accessed on 10 July 2023).
82. Sharaf, M.; Huzayyin, A.S.; Yousef, M.S. Performance Enhancement of Photovoltaic Cells Using Phase Change Material (PCM) in Winter. *Alex. Eng. J.* **2022**, *61*, 4229–4239. [CrossRef]
83. Mutombo, N.M.-A.; Inambao, F.; Bright, G. Performance Analysis of Thermosyphon Hybrid Photovoltaic Thermal Collector. *J. Energy S. Afr.* **2016**, *27*, 28–38. [CrossRef]
84. Zou, S.; Zhang, Q.; Ling, L.; Wang, J.; Ma, X.; Du, S.; Yue, C. Experimental Investigation on the Thermal Performance of a Water-Cooled Loop Thermosyphon System under Fan Failure Conditions. *Int. J. Refrig.* **2021**, *124*, 85–95. [CrossRef]
85. Shrivastava, A.; Prakash Arul Jose, J.; Dilip Borole, Y.; Saravanakumar, R.; Sharifpur, M.; Harasi, H.; Abdul Razak, R.K.; Afzal, A. A Study on the Effects of Forced Air-Cooling Enhancements on a 150 W Solar Photovoltaic Thermal Collector for Green Cities. *Sustain. Energy Technol. Assess.* **2022**, *49*, 101782. [CrossRef]
86. Valiente-Blanco, I.; Lopez-Pascual, D.; Diaz-Villar, P.; Mallol-Poyato, R.; Barragan, A.; Ocaña, M.; Granello, G.; Diez-Jimenez, E. Efficiency Improvement of Photovoltaic Solar Modules by Cooling Using an Underground Heat Exchanger. *J. Sol. Energy Eng.* **2022**, *144*, 061015. [CrossRef]
87. Dwivedi, P.; Sudhakar, K.; Soni, A.; Solomin, E.; Kirpichnikova, I. Advanced Cooling Techniques of P.V. Modules: A State of Art. *Case Stud. Therm. Eng.* **2020**, *21*, 100674. [CrossRef]
88. Haidar, Z.A.; Orfi, J.; Kaneesamkandi, Z. Experimental Investigation of Evaporative Cooling for Enhancing Photovoltaic Panels Efficiency. *Results Phys.* **2018**, *11*, 690–697. [CrossRef]
89. Murtadha, T.K.; dil Hussein, A.A.; Alalwany, A.A.H.; Alrwashdeh, S.S.; Al-Falahat, A.M. Improving the Cooling Performance of Photovoltaic Panels by Using Two Passes Circulation of Titanium Dioxide Nanofluid. *Case Stud. Therm. Eng.* **2022**, *36*, 102191. [CrossRef]
90. Ahmed, A.; Baig, H.; Sundaram, S.; Mallick, T.K. Use of Nanofluids in Solar PV/Thermal Systems. *Int. J. Photoenergy* **2019**, *2019*, e8039129. [CrossRef]
91. Rahman, N.M.A.; Haw, L.C.; Kamaluddin, K.A.; Abdullah, M.S.I. Investigating Photovoltaic Module Performance Using Aluminium Heat Sink and Forced Cold-Air Circulation Method in Tropical Climate Conditions. *Energy Rep.* **2023**, *9*, 2797–2809. [CrossRef]
92. Avoiding Corrosion in Liquid Cooling Systems. Available online: https://www.boydcorp.com/resources/resource-center/blog/avoiding-galvanic-corrosion.html (accessed on 11 July 2023).
93. Evidential Reasoning Approach with Multiple Kinds of Attributes and Entropy-Based Weight Assignment—ScienceDirect. Available online: https://www.sciencedirect.com/science/article/abs/pii/S0950705118304337 (accessed on 13 July 2023).
94. Moharram, K.A.; Abd-Elhady, M.S.; Kandil, H.A.; El-Sherif, H. Enhancing the Performance of Photovoltaic Panels by Water Cooling. *Ain Shams Eng. J.* **2013**, *4*, 869–877. [CrossRef]

www.ingramcontent.com/pod-product-compliance
Lightning Source LLC
Chambersburg PA
CBHW061547220926
51850CB00005B/274
9783036589558

MDPI
St. Alban-Anlage 66
4052 Basel
Switzerland
www.mdpi.com

Sustainability Editorial Office
E-mail: sustainability@mdpi.com
www.mdpi.com/journal/sustainability